D0927024

Technology

R. Thomas Wright

Professor Emeritus, Industry and Technology
Ball State University
Muncie, Indiana

Publisher
Goodheart-Willcox Company, Inc.
Tinley Park, Illinois
www.g-w.com

Manufactured in the United States of America.

Library of Congress Catalog Card Number 2006041213

ISBN-13: 978-1-59070-718-0
ISBN-10: 1-59070-718-4

2 3 4 5 6 7 8 9 – 08 – 11 10 09 08 07

The Goodheart-Willcox Company, Inc. Brand Disclaimer: Brand names, company names, and illustrations for products and services included in this text are provided for educational purposes only and do not represent or imply endorsement or recommendation by the author or the publisher.

The Goodheart-Willcox Company, Inc. Safety Notice: The reader is expressly advised to carefully read, understand, and apply all safety precautions and warnings described in this book or that might also be indicated in undertaking the activities and exercises described herein to minimize risk of personal injury or injury to others. Common sense and good judgment should also be exercised and applied to help avoid all potential hazards. The reader should always refer to the appropriate manufacturer's technical information, directions, and recommendations; then proceed with care to follow specific equipment operating instructions. The reader should understand these notices and cautions are not exhaustive.

The publisher makes no warranty or representation whatsoever, either expressed or implied, including but not limited to equipment, procedures, and applications described or referred to herein, their quality, performance, merchantability, or fitness for a particular purpose. The publisher assumes no responsibility for any changes, errors, or omissions in this book. The publisher specifically disclaims any liability whatsoever, including any direct, indirect, incidental, consequential, special, or exemplary damages resulting, in whole or in part, from the reader's use or reliance upon the information, instructions, procedures, warnings, cautions, applications, or other matter contained in this book. The publisher assumes no responsibility for the activities of the reader.

Library of Congress Cataloging-in-Publication Data

Wright, R. Thomas
Technology / by R. Thomas Wright.
p. cm.
ISBN-13: 978-1-59070-718-0
ISBN-10: 1-59070-718-4
1. Technology. I. Title.

T47.W74 2006
600—dc22

2006041213

Introduction

Technology will help you to understand:

- How people use technology to make our world work.
- Why technological systems work the way they do.
- In what ways technology affects both people and our planet.

It covers the seven areas of technological activity:

- Communication and information
- Transportation
- Construction
- Manufacturing
- Medical
- Agricultural and biotechnology
- Energy and power

In this book you will learn that technology is a reaction to problems and opportunities—a human adaptive system. You will learn that technological systems are made up of many parts that require tools. You will learn about the problem-solving and design process, especially the testing, evaluating, and communicating of design solutions.

Sections explore, in depth, the production of products and structures, communication and information, agricultural and biorelated technologies, transportation, and the use of energy. Because every system must have direction, the management of technological systems is covered. The examination of societal and personal views of technology rounds out the book.

Technology is illustrated with photographs, drawings, diagrams, and original artwork to help explain the concepts in the text. Most of these illustrations are in color. This material has been carefully selected to make technology easy to understand. Each chapter begins with objectives so you know what will be covered. Key words are in bold to help make you aware of them. Review questions and activities will improve your understanding. The activities between sections provide you with valuable hands-on experience.

Impacts, both positive and negative, accompany the use of technology. The only way that people in the modern world can choose and apply technology responsibly is to understand how technology develops and how the various technological systems interact.

A Student Activity Manual has activities and exercises that will give you important experience while fully enriching the concepts developed in the text.

A sound understanding of technology is vital for making wise choices. As you study, you will see the effects of your choices. These choices control how technology is used. Each person can make a difference to be sure that technology is used responsibly. With a solid understanding of technology, you can understand and take an active part in our human-built world.

About the Author

Dr. R. Thomas Wright is one of the leading figures in technology education curriculum development in the United States. He is the author or coauthor of many Goodheart-Willcox technology textbooks. Dr. Wright is the author of *Manufacturing* and *Automation Technology, Processes of Manufacturing, Exploring Manufacturing,* and *Technology.* He is the coauthor of *Exploring Production* with Richard M. Henak.

He has served the profession through many professional offices, including President of the International Technology Education Association (ITEA) and President of the Council on Technology Teacher Education (CTTE). His work has been recognized through the ITEA Academy of Fellows award and Award of Distinction, the CTTE Technology Teacher Educator of the Year, the Epsilon Pi Tau Laureate Citation and Distinguished Service Citation, the Sagamore of the Wabash Award from the Governor of Indiana, the Bellringer Award from the Indiana Superintendent of Public Instruction, the Ball State University Faculty of the Year Award and George and Frances Ball Distinguished Professorship, and the EEA-Ship Citation.

Dr. Wright's educational background includes a bachelor's degree from Stout State University, a master of science degree from Ball State University, and a doctoral degree from the University of Maryland. His teaching experience consists of 3 years as a junior high instructor in California and 37 years as a university instructor at Ball State University. In addition, he has also been a visiting professor at Colorado State University, Oregon State University, and Edith Cowan University in Perth, Australia.

Brief Contents

Contents

TSA Modular Activities

The Technology Student Association (TSA) is a non-profit national student organization devoted to teaching technology education to young people. TSA's mission is to inspire its student members to prepare for careers in a technology-driven economy and culture. The demand for technological expertise is escalating in American industry. Therefore, TSA's teachers strive to promote technological literacy, leadership, and problem solving to their student membership.

TSA Modular Activities are based on the Technology Student Association competitive events current at the time of writing. Please refer to the *Official TSA Competitive Events Guide,* which is periodically updated, for actual regulations for current TSA competitive events. TSA publishes two *Official TSA Competitive Events Guides:* one for middle school events and one for high school events.

To obtain additional information about starting a TSA chapter at your school, to order the *Official TSA Competitive Events Guide,* or to learn more about the Technology Student Association and technology education, contact TSA:

Technology Student Association
1914 Association Drive
Reston, VA 20191-1540
www.tsaweb.org

Features

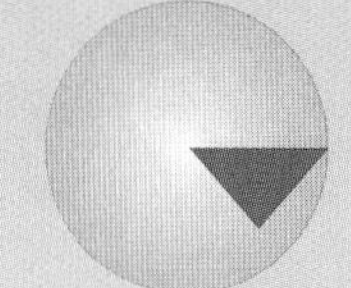

Section 1

Technology

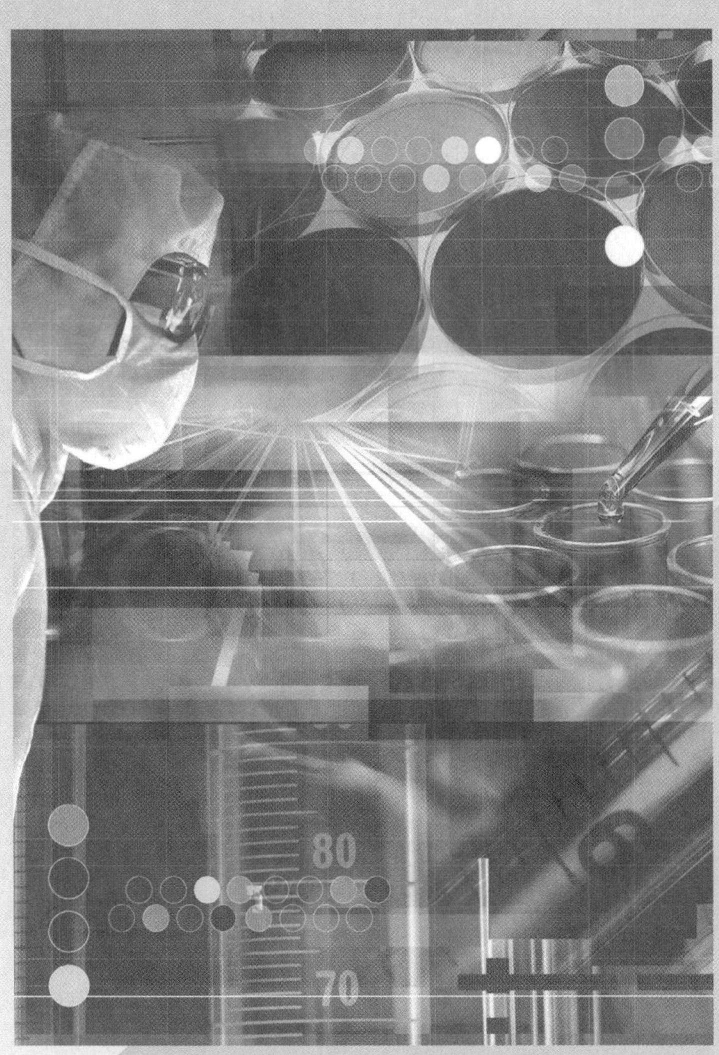

Tomorrow's Technology Today

Cloning

Identical twins often have stories of switching places with one another to avoid trouble—or to cause it. If you have ever wished for a twin to sit in your place on an exam day, your wish might not be altogether unrealistic. Modern science is now making it possible to create a human genetically identical to you, or a clone. Because every single gene is an exact copy of the original, a clone is, for all practical purposes, an identical twin of the original.

Recently, scientists have successfully cloned several types of plants and animals, including some endangered species. This is important because it opens up the possibility of repopulating species of plants and animals at risk of becoming extinct, and maybe even species that already are! The main purpose of cloning, though, is to mass-produce plants and animals with specific genetic qualities, such as the best tasting fruit or the ability to produce human insulin.

Cloning plants and animals has some undeniable benefits for humans and the quality of human life. Although there has been some debate about the ethical implications of plant and animal cloning, it is the idea of human cloning that has really started controversies among scientists, the government, and the public. Until recently, the idea of cloning humans was science fiction. Now, it is a real possibility.

Researchers are close to being able to successfully clone humans, but most cloning would not result in entirely new humans. Scientists are studying the possibility of using cloning to produce repair and replacement kits for people with severe medical problems. It may be possible to clone healthy cells and to fix diseased cells. This type of cloning is currently being used in stem cell research. A person's DNA is used to grow an embryonic clone, which is then used to grow stem cells. These cells can grow replacement organs and neurons to cure certain diseases.

Human cloning is a potential solution for couples with fertility problems. It also could theoretically be used to bring deceased relatives back to life. One couple has already paid $500,000 for a clone of their deceased baby daughter, which will be created using her preserved skin cells!

Cloning, especially human cloning, creates many ethical dilemmas. The success rate of animal cloning is currently less than 3 percent, and the process is very risky. Surviving clones often suffer from problematic genetic abnormalities. Some are born with defective lungs, hearts, blood vessels, or immune systems. It will be a more serious problem if human clones are created and born with these same problems. Cloning holds much promise, but also many risks. As interest in human cloning continues to grow, the ethical debate certainly will as well.

Chapter 1
Technology: A Dynamic, Human System

Learning Objectives

After studying this chapter, you will be able to:

- Define technology.
- List the basic features of technology.
- Describe technology as a dynamic process.
- List positive and negative features of technology.
- Explain the difference between science and technology.
- Name the major divisions in the evolution of technology.
- Cite technological developments in each period of technological history.
- Identify characteristics of the Information Age.

Do you know what technology is? Is it mainframe and desktop computers, industrial robots, laser scanners at the supermarket checkout, fiber optic communications, space shuttles at Cape Canaveral, and satellites circling the earth? Is it people using tools and machines to make their work easier and better? Is it an organization devoted to operating agricultural and related biotechnology, communication and information, energy and power, medical, manufacturing, construction, or transportation systems? See **Figure 1-1.**

Chances are that you are not sure whether it is one, two, or all three of these things. You are not alone in this lack of understanding. Throughout the world, people use technology every day but understand little about it. We go to school for years to learn how to read and write and to study mathematics, science, history, foreign languages, and other subjects. Few people, however, spend time learning about technology and its impacts on everyday life.

This book has been developed to aid in understanding this vast area of human activity. It will help you become more technologically literate. After this study of technology you will be able to find, select, and use knowledge about tools and materials to solve problems. This ability leads to an increased understanding of technology as it affects your life as a citizen, consumer, and worker, **Figure 1-2.**

Figure 1-1. We encounter technology in many areas of our lives. (Apple Computers)

Figure 1-2. Technology affects our daily lives.

Technology Defined

Almost everyone uses the word "technology," but what does it really mean? To some people, it means complicated electronic devices and hard-to-understand equipment. To others, it means the source of the radical changes that are happening in all phases of life. Some people fear it, whereas others see it as the source of longer and more complete lives. Some people believe it to be a development of the twentieth century. Each of these views is partly correct.

Technology is not necessarily a complex or space-age phenomenon. It can be primitive and crude, or it can be complex and sophisticated. Technology has been here as long as humans have been on earth. *Technology is humans using objects (tools, machines, systems, and materials) to change the natural and human-made (built) environment.* Technology is conscious, purposeful actions designed by people to extend human ability or potential to do work.

Technology has four basic features. As shown in **Figure 1-3**, these are:

- Technology is human knowledge.
- Technology uses tools, materials, and systems.
- Application of technology results in ***artifacts*** (human-made things) or other outputs (pollution and scrap, for example).
- Technology is developed by people to modify or control the environment.

These four characteristics suggest that the products and services available in society are the result of technology. They were designed through technological innovation (development). They were produced by technological means (processes). These means are integrated with people, machines, and materials to meet an identified need (management). The products were distributed to customers by technology. They are maintained and serviced through technological actions. When the products fulfilled their function or became obsolete, they were (or should have been) recycled by technological means. As you can see, without technology the world as we know it today would not exist.

Nearly all technology is developed and produced by profit-centered businesses. These businesses are organized by people to make a product or perform a service. Through this action, the owners have a product to sell. They hope to pay their business expenses (wages, material costs, taxes, and so on) and have money left over—a ***profit***. To stay competitive, these businesses change and improve their products and services, **Figure 1-4.** For example, automobile companies produce new models that they hope will capture a large share of the

Technology...

...is human knowledge.

...involves tools, materials, and systems.

...produces artifacts (outputs).

...modifies or controls the environment.

Figure 1-3. Technology has four characteristics.

Career Corner

Industrial Engineering Technicians

Industrial engineering technicians study the efficient use of personnel, materials, and machines in production facilities. They prepare layouts of machinery and equipment, plan the flow of work, conduct statistical studies, and analyze production costs. Their work is more practical and limited in scope than the work of industrial engineers. Almost all industrial engineering technicians have completed high school with an emphasis in mathematics and science. Many also complete a postsecondary program in engineering technology, earning either an associate's degree or a bachelor's degree.

Figure 1-4. What we like and need changes over time, and companies must keep up with these new wants and needs if they are to stay profitable. (Library of Congress, General Motors)

market. Likewise, cereal companies bring out new products to increase their sales. A competitive edge is gained through employing technology. These technologies include standardizing products, using interchangeable parts on machines, and adopting labor-saving practices.

This competition also exists between nations. Each country's economy is directly dependent on its ability to produce products and services that can compete in the world market. The country that produces the best products will have a healthy economy and a high number of citizens working. Countries that fail to develop competitive products will experience high levels of unemployment and growing poverty.

Technology As a Dynamic Process

All these points suggest that, by its very nature, technology is a ***dynamic process.*** It is changing. It causes change. Technology, almost always, seems to improve on existing technology. See **Figure 1-5.** The typewriter was an improvement over writing by hand. The electric typewriter was faster than the manual typewriter. The word processor and laser printer outperform the electric typewriter. Now, voice-recognition software enables humans to talk to computers, providing even faster communication.

Figure 1-5. Technology is a dynamic process. People are always trying to improve existing artifacts and systems. (Keytec, Inc.)

Connections to Technology: **History**

The Presidential Election of 1960

As vice-president during the eight-year term of a popular president, Richard Nixon was expected by most observers to win the presidential election of 1960. He was also running against a young and somewhat inexperienced senator, John Kennedy of Massachusetts, which further improved his chances.

Still, in large part because of a relatively new technological device called the television, the perceptions of the two candidates changed dramatically as election day grew closer. In a series of debates that were televised for the first time in history, Kennedy came across as confident and mature, whereas Nixon was seen as nervous and self-conscious. Interestingly enough, according to most people who listened to the debates on the radio, Nixon had won the contests, but according to most of those who watched on television, Kennedy had won.

In the end, Kennedy won the election, and television had forever changed the political process.

Can you name another technological device that might have changed the course of history? In what way or ways did it do so?

Nor will this be the end. New devices will improve on this system.

Past technology improved productivity, and future technology will increase it even more. There is no turning back. We cannot feed the world's population using the walking plow and mules. Commerce cannot be maintained using covered wagons. People cannot build hand-hewn log cabins fast enough to provide housing for a rapidly expanding population. Technology is necessary for survival and the hope for a better future.

Positive and Negative Aspects of Technology

Technology increases human capabilities. Through technology we can see more clearly and further by using microscopes and telescopes. Technology helps us to lift heavy loads by using hoists, pulleys, and cranes. Technology allows us to communicate better and faster by radio, television, and telephone. Technology makes distant places close at hand when traveling by automobiles, trains, and aircraft. Technology makes life more enjoyable with video games, motion pictures, and compact disc recordings. Technology brings us new products and materials, such as microcomputers, acrylic fibers, and artificial human organs.

Technology, however, also has negative aspects. Poorly designed and used technology pollutes the air we breathe and the water we drink and can cause soil erosion, **Figure 1-6.** Calm sunsets and unspoiled wilderness areas are threatened by inappropriately used technology.

Technology may cause unemployment and radical changes in the ways people live. New technological devices may displace workers from traditional jobs. Families may have to move long distances to find employment. Their new homes may be in areas that have vastly different values and lifestyles.

The consequences of technology are now more feared than natural events. Many people are more concerned about nuclear winter, acid rain, and air pollution than they are of earthquakes and tornadoes.

This good news/bad news view of technology requires a new type of citizen. Technology must be developed and used wisely and appropriately. This challenge requires individuals who *understand and can direct* new technology. People who have this understanding and ability are called ***technologically literate.***

Technology and Types of Knowledge

Some people think technology is applied science. They are mistaken! Technology and science are closely related, but they are different. To see this difference, let us use your imagination. Assume that you plan to see your first glacier in Rocky Mountain National Park. You drive to the parking lot and leave your car. As you climb up the trail the trees are getting shorter. After a while there are no trees. The type of plant life is changing. Lichen is growing on the north side of the rocks. Finally you reach your destination. You notice that the small lake in front of the glacier appears to be green. A cold wind is blowing over the continental divide and across the glacier. It is midsummer but snow still covers the ground. If you want to know why these phenomena happen, you must consult ***science,*** the knowledge of the natural world, **Figure 1-7.** Scientific knowledge is gathered from detached observation. Scientists distance themselves from the phenomena as they try to develop their explanations. They try to explain why something exists or happens in a certain way. Their work can be described as ***research***.

Let's return to our make-believe hike. The long hike has made you tired. Now you must climb down the mountain. As you come around a curve in the trail you see a welcome sight—your car! You get

Figure 1-6. Technology has positive and negative aspects. Poorly used technology can cause such problems as soil erosion and air pollution. (U.S. Department of Agriculture)

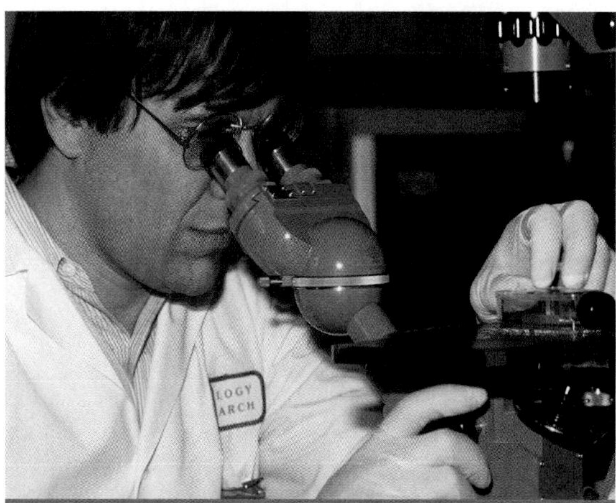

Figure 1-7. Scientific knowledge is different from technological knowledge. Scientists observe phenomena from a distance in an effort to understand how or why something exists.

into it, start the engine, and drive along a paved road. You pass a small hydroelectric dam with electric power lines leading in all directions. Along the road there are homes with lights and heating systems. Finally you reach town with its stores, gas stations, and a pizza shop. Food at last! All these things are a result of technology. They are part of the human-made world or the built environment.

These parts of the world have been developed by innovators through technological knowledge. These people work with materials and use machines and tools to make things happen. Their work can be described as ***development.*** They design and build products and structures to make our lives better. Those who work in technology are not observers; they become directly involved with the processes they develop and use.

Science and technology are two major types of knowledge. A third type is the ***humanities.*** This type of knowledge describes the relationships between and among groups of people. Those in the humanities study how people behave individually (psychology), how they work in groups (sociology), what they value (religion and ethics), and how they express themselves (art, music, and literature). People in the humanities also look at human behavior over time (history and anthropology).

You can begin to understand the world by using these three types of knowledge: the knowledge of the natural world (science), the human-made world (technology), and human actions within these environments (the humanities). Note that they all interact with one another and that knowledge from one area directly affects development in other areas, **Figure 1-8.** Leave out any one of them, and you will make errors in judgment and do things incorrectly.

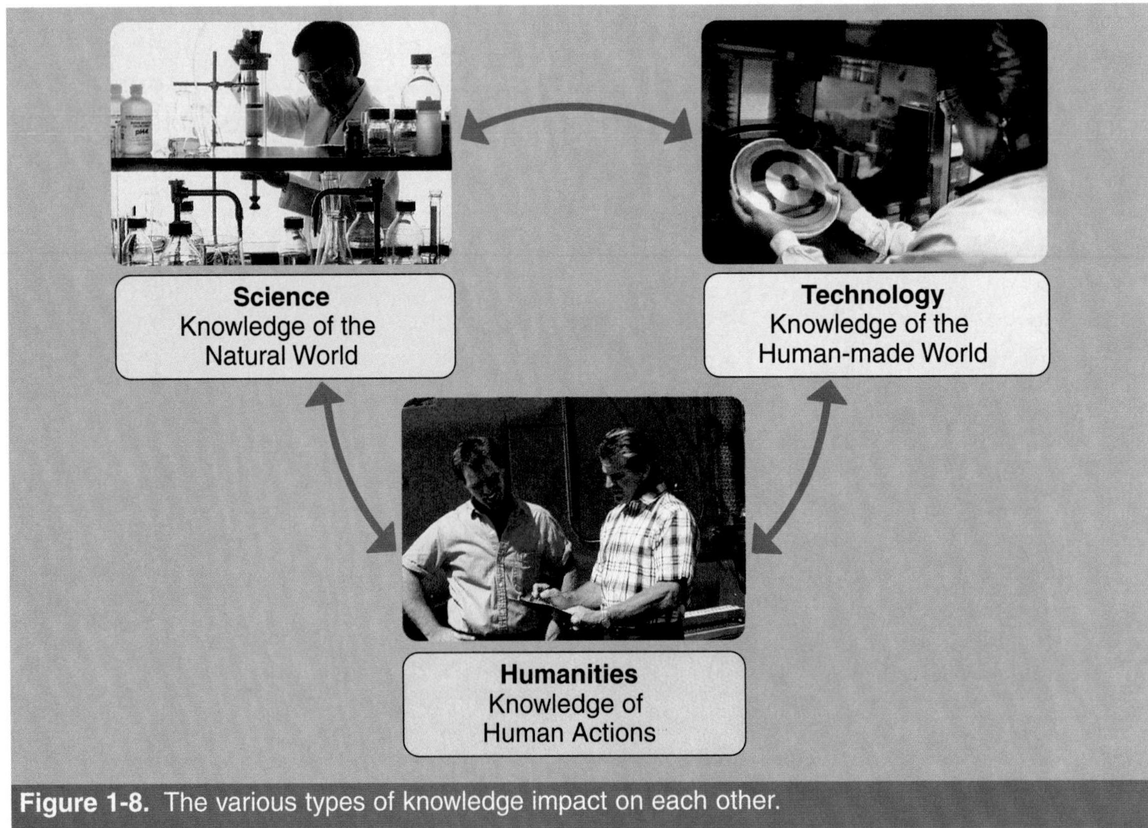

Figure 1-8. The various types of knowledge impact on each other.

Technology Explained

Smart Houses

smart house: A house that allows computers to control appliances and energy use.

Computer technology has been used to create an entirely new type of house called a *smart house.* The smart house system makes home automation possible. This system allows the electrical, telephone, gas, and television systems and appliances in a house to be interactive. It is being built into new homes across the United States, **Figure A.**

The heart of a smart house is found in the *system controllers,* **Figure B.** These units serve as the hubs for messages between products within the house. The system controllers constantly check the system for problems. Electrical power, telephone service, and cable television are brought into the house at a service center, where a system controller is installed.

The smart house system uses microprocessors, also known as chips, and a unique wiring system to allow the electrical appliances and products in a house to interact. Participating companies are developing products with built-in smart house chips. These products can then interact with the system controllers and the rest of the smart house system.

Smart houses allow home owners to change how their houses work. This is done by programming the system controllers. Reprogramming a switch changes what the switch controls. The heating system can be programmed to vary the home's temperature during the day. There can be cool bedrooms at night and warm bathrooms in the morning. The water heater can be set to turn on, so hot water will be available for a morning shower. The electrical range can be turned on to cook a meal in the oven. The dishwasher can be set to start at night, in order to take advantage of lower utility rates.

The wiring system in a smart house is unique. Smart houses use three types of cables, **Figure C.** Hybrid branch cables carry 120-volt AC power and control signals. These cables provide power for standard appliances, such as coffeemakers, VCRs, and lamps. Communications cables carry audio, video, telephone, and computer data signals. Applications cables carry 12-volt DC power and control signals. These cables provide power for low voltage devices, such as smoke detectors, security systems, and switches. Large appliances, such as ovens and clothes dryers, use conventional high amperage wiring.

Figure A. In the twenty-first century, many homes will use smart house systems to control energy use, security, communications, lighting, and entertainment.

Figure B. This is a series of smart house controllers in a recently built home.

Figure C. The wiring used in a smart house is special. The photo on the left shows the three types of cables: (from top to bottom) applications, communications, and hybrid branch. The photo on the right shows a convenience center. These special outlets can be changed to meet the needs of the home owner.

The Evolution of Technology

Many people see technology as new and dramatic. It is not. It is as old as humanity. The world is said to be about five billion years old. Humans have been on earth for the last 2.5 million years. During this time humans have been distinguished from other species by two major factors: their ability to make tools and their ability to use those tools. The level of this technology determines the type of life available.

Early humans were said to live in ***primitive conditions.*** These conditions are determined by nature. Primitive people tried to exist with nature. They did not attempt to control nature or improve the natural condition. For example, members of primitive cultures depended on harvesting natural vegetation and hunting game. Drought would severely affect their food supply, but they could not do anything about it. If nature did not provide ample food, the people starved. Likewise, primitive cultures would use only naturally occurring materials to build shelters and make clothing. Some groups of people on earth today still live in primitive conditions.

Civilized conditions are much different from primitive existence. In civilized societies humans exert their will on the natural scene. They make tools, grow crops, engineer materials, and develop transportation systems.

The evolution of civilization is directly related to the tools of the time. Many of the early major divisions of the history of civilization are named for the materials used to make tools, and others are defined by the products produced. See **Figure 1-9.** In each case, the technological developments during the period led to major changes in society and the creation of new needs and wants.

The Stone Age

The earliest period was the ***Stone Age,*** which began about two million years ago. In this period humans used pebbles as tools. These simple stone tools were used to cut and pound vegetables and cut meat from animal carcasses. From these modest beginnings came pointed stone hunting tools. These tools allowed people to obtain food more efficiently. Also during this period, humans learned how to harness fire for heating, cooking, and protection. The population became more productive, which meant that more people could live in a given area.

The Bronze Age

As populations grew, new technology was needed to support the demand for food and shelter. This need was coupled with the discovery of copper. People also learned that copper could be heated and then melted with other ores to produce bronze, a stronger metal than copper. These events allowed humans to enter the second major historical period, the ***Bronze Age,*** which began around 3000 B.C. During this period, humans used copper and copper-based metals as the primary materials for tools because of their hardness and durability. Through such developments as large-scale irrigation systems, humans transformed agriculture to the extent that they did not have to depend on native vegetation and animal life for survival. They also created better ways for storing food and developed writing, navigation, and other basic technologies.

The Iron Age

The next historical period is called the ***Iron Age,*** which began around 1200 B.C. Iron and steel became the primary materials for tools during this period because they were more plentiful and cheaper than copper and bronze. Now more

Stone Age (1,000,000–3000 B.C.)	Bronze Age (3000–1200 B.C.)	Iron Age (1200 B.C.–1300 A.D.)	Middle Ages (500 A.D.–1500 A.D.)	Renaissance (1300 A.D.–1600 A.D.)	Industrial Revolution (1750 A.D.–ongoing)	Information Age (Present)
Stone tools, fire, cave paintings, pottery	Copper and bronze tools, smelting, frescoes, writing, paper, ink	Iron tools, furnaces, aqueducts, body armor, ox-drawn plows, spinning wheel, windmill	Printing press, magnetic compass, paper money, waterwheel	Improved magnetic compass, telescope, hydraulic press, calculating machine, modern architecture	Steam engine, cotton gin, power looms, factories, electricity, automobile, airplane	Desktop computers, robots, lasers, solar energy, cell research, satellites

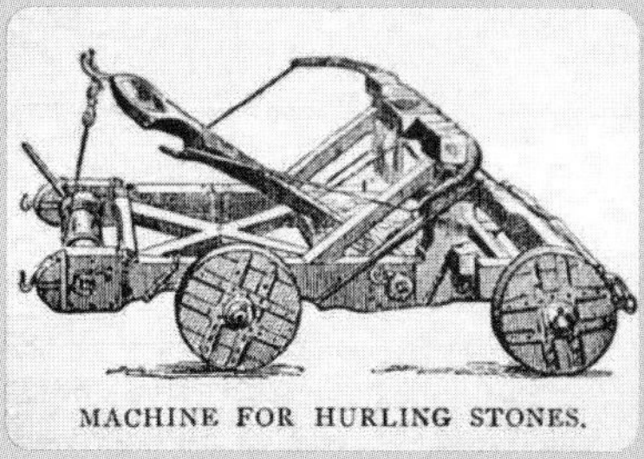

MACHINE FOR HURLING STONES.

Iron Age 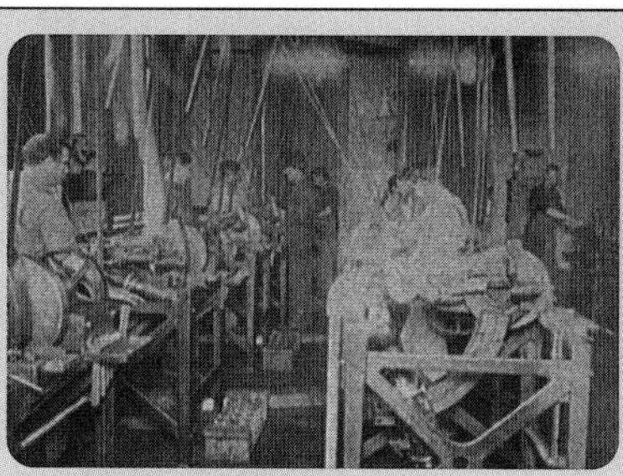 | Industrial Revolution 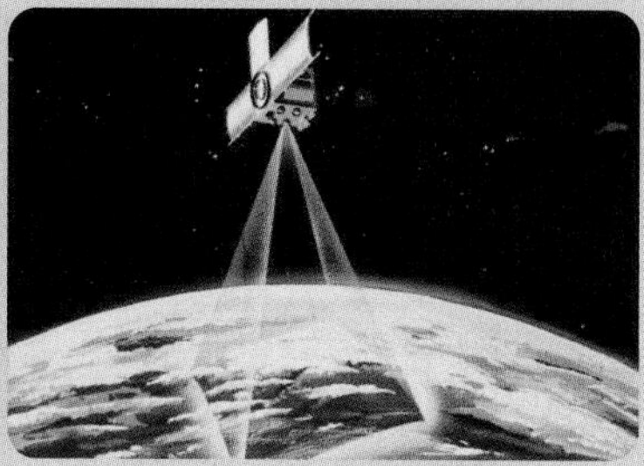 | Information Age

Figure 1-9. The major divisions in the history of civilization are identified by the tools and products produced and used. (WWW.PDImages.com, Library of Congress, NOAA)

people could afford tools, which created more opportunities for sustained technological advancement. Progress continued even during the Dark Ages (about 500 A.D. to 1000 A.D.), when many cultural and governmental institutions stagnated or declined in importance. For example, during the Iron Age, the alphabet came into general use, as did coins. Trade, transportation, and communication all improved, and civilization expanded.

The Middle Ages

The *Middle Ages* began about 400 A.D. when Germanic tribes invaded and conquered the Roman Empire. This period is known for its various upheavals, as tribes continually fought each other for territory, but technology still progressed. For example, one of the major inventions of this time was the printing press, developed by Johann Gutenberg in 1445.

This device changed people's lives forever. Previously, to make books people had to copy words and illustrations onto parchment by hand, a laborious process that made each book very expensive. With the printing press, books could be produced quickly and with less labor, thus making books less costly and available to more people.

Other items discovered and improved on during this time included the magnetic compass and the waterwheel, which changed the power of water into mechanized energy. The adoption and use of paper money increased the amount of goods bought and sold.

The Renaissance

The ***Renaissance***, the period that began in the early 1300s in Italy and lasted until 1600, was a time of great cultural advancement. The Renaissance is known as a period of new ideas in art, literature, history, and political science, but technological developments also occurred. For example, because of improvements in ships, European voyagers traveled to America. They brought with them new crops and farming tools and learned new methods of farming from Native Americans, thus improving agriculture for both groups.

In Italy, Leonardo da Vinci, the famous artist, kept a notebook in which he drew plans for many inventions, including a flying machine and a movable bridge. These drawings greatly influenced future inventors.

Other technologies developed during this period included the calculator and the telescope. Another new artifact, the screwdriver, was invented by gunsmiths because they needed an instrument to adjust gun mechanisms.

The Industrial Revolution

During the last 250 years, technology has dramatically changed the world. For example, humans applied technology to agriculture and went from cultivating food by hand to using a horse and plow to adopting the mechanical reaper. Because of these developments, a smaller percentage of the population could now grow the food and clothing fiber needed to sustain the population. This allowed large numbers of people to leave the farms and migrate to growing towns and cities. In these places, a new phenomenon was developing—the marvel of manufacturing.

This era is called the ***Industrial Revolution.*** It started in England about 1750 and moved to America in the late 1700s. Tremendous changes in technology occurred during this period. For example, Edmund Cartwright changed weaving from a manual process to a mechanical one with his power loom. Joseph Jacquard further revolutionized weaving by creating a series of punched cards with recorded instructions that could control the intricate patterns designed by the weaver, thus freeing up the weaver to spend more on managing. (These cards later became the inspiration for computer programming.) James Watt improved the steam engine and made it available for the driving of machinery, revolutionizing both the manufacturing process and methods of transportation. Eli Whitney, inventor of the cotton gin, developed the system of manufacturing standardized, interchangeable parts, which dramatically expanded manufacturing capabilities. As has been true throughout the history of technology, many of these developments were refinements of existing inventions and evolved after much testing and retesting. Here, however, scientific knowledge began to be applied to technological knowledge, greatly accelerating technical progress.

As manufacturing changed, so did the character of the workforce. Employees were divided into production workers and managers, and each was given specific tasks. Efficiency of production became an area of serious study. For example, in the early twentieth century, Frederick Winslow Taylor developed his "four principles of scientific management" to reduce waste and increase productivity. Revolutionary for the time, Taylor's ideas included studying production workers' motion and rearranging equipment to decrease the time the workers spent not producing goods.

The various developments in industry led to continuous manufacture, **Figure 1-10,** which is characterized by:

- Improved machine life because of interchangeable parts

- Division of the job into parts that are assigned to separate production workers
- Creation of material-handling devices that bring the work to the production workers
- Classification of management as a professional group

Later during the Industrial Revolution, enterprising people developed sophisticated transportation and communication systems to support the growing industrial activities. Dirt and gravel roads became paved highways. The diesel electric locomotive replaced the steam locomotive. The motor truck and the airplane successfully challenged the dominance of transportation by the railroad. The telegraph, telephone, radio, and television replaced pony express letter carriers.

Construction practices advanced to provide the factories, stores, homes, and other buildings needed to meet a growing demand for shelter. Mass-produced dwellings replaced log cabins. Metal buildings became the factory of choice. Shopping centers and shopping malls made downtown shopping areas less important.

Efficient production was coupled with rising consumer demand. Extended free time was available for the first time with the forty-hour workweek and annual vacations. Children could stay in school longer because they were not needed on the farm and in the factories. Universal literacy became a possibility, although it has yet to be reached.

The Information Age

The Industrial Revolution moved us from an agrarian era into the industrial era. We moved from a period when most people worked raising food and fiber to a period where people worked in manufacturing. Technology is now moving many nations into a new period. This stage of development is called many different names, including the ***Information Age.***

During the industrial age, the most successful companies processed material better than their competitors. The Information Age changes this emphasis. It places more importance on information processing and cooperative working relations between production workers and managers. The Information Age has several characteristics, including:

- Wide use of automatic machines and information-processing equipment
- High demand for trained technicians, technologists, and engineers
- Blurring of the previously sharp line between production workers and managers
- Constant need for job-related training and retraining of production workers

These factors promise change—but this is not something new. It takes place with every generation. One generation traveled west in covered wagons during their youth and saw an astronaut circle the globe in their later years. What will you see in your lifetime?

Figure 1-10. Some workers are involved with the continuous manufacturing process. (Maytag Co.)

Summary

Technology has always been a part of human life. It changes life and is changed as life progresses. It can be described as the use of tools, materials, and systems to extend the human potential for controlling and modifying the environment. It makes life better for many people. If it is improperly used, however, it can cause serious damage to people, society, and the environment. Throughout history, technology has had a powerful influence on the economic, political, social, and cultural characteristics of societies. Early technological development was based on tools, not scientific knowledge, but later inventors applied scientific principles to refine and advance existing technology. Only technologically literate people can properly develop, select, and responsibly use technology. This is the challenge for today's youth.

Key Terms

artifacts
Bronze Age
civilized conditions
development
dynamic process
humanities
Industrial Revolution
Information Age
Iron Age
Middle Ages
primitive conditions
profit
Renaissance
research
science
Stone Age
technologically literate

Test Your Knowledge

Write your answers on a separate sheet of paper. Please do not write in this book.

1. What is technology?
2. Which one of the following is *NOT* a feature of technology?
   a. Technology uses tools, materials, and systems.
   b. People develop technology to control their environment.
   c. Technology is found in nature.
   d. Technology results in artifacts and other outputs.
3. *True or false?* Competition causes businesses to develop new products.
4. *True or false?* Technology almost always improves on existing technology.
5. Name one positive and one negative aspect of technology.
6. _____ is knowledge of the natural world.
7. List the three major types of knowledge.

For Questions 8–18, match the historical period on the right to the correct technological development on the left. (Note: Some letters will be used more than once.)

Technological Development	**Historical Period**
8. _____ Airplane	a. Stone Age
9. _____ Printing press	b. Bronze Age
10. _____ Fire	c. Iron Age
11. _____ Solar energy	d. Middle Ages
12. _____ Copper tools	e. Renaissance
13. _____ Telescope	f. Industrial Revolution
14. _____ Ox-drawn plows	g. Information Age
15. _____ Paper money	
16. _____ Pottery	
17. _____ Calculating machine	
18. _____ Electricity	

19. James Watt's improvements to the _____ revolutionized both the manufacturing process and the transportation system.
20. *True or false?* The Information Age is characterized by the blurring of the line between production workers and managers.

Applying Your Knowledge

1. Design a simple technological device (a tool to make a job easier) that could be used in your daily activities.
2. Develop a chart like the one shown below. For each problem on the left, list the scientific and technological knowledge that would help solve it.

Problem	**Scientific knowledge**	**Technological knowledge**
Depletion of the ozone layer		
Rapid depletion of petroleum		
Rising cost of electricity		
Injuries on the football field		

3. Select three major problems in your community. Develop a chart like the one shown above. For each problem listed on the left, list the scientific and technological knowledge that would help solve it.

Chapter 2
Technology As a System

Learning Objectives

After studying this chapter, you will be able to:

- Explain how technology is a system.
- Identify the major components of a technological system.
- List the inputs to a technological system.
- List the steps in the problem-solving/design process.
- Cite the major activities involved in production processes.
- Cite the major activities involved in management processes.
- Name positive and negative technological outputs.
- Explain feedback and control.

Technology involves performing a task by using an object that is not part of the human body. For example, suppose that you want to crack the shell around a walnut. If you put it in your mouth and bite down, you are not using technology because the human body is doing the work. You may also find that your jaw is not strong enough to do the job or that you value your teeth and do not want to damage them. You need another way to open the nutshell. You see some rocks and realize their value in this situation. You place the shell on one rock and strike it with another rock. You have now employed technology. The rocks extended your potential, or ability, to do a specific task.

This example shows that technology is the development and application of knowledge, tools, and human skills to solve problems and extend human potential, **Figure 2-1.** Technology arises and moves forward out of human wants and needs. It is designed and evaluated by people and in time is modified or abandoned. Moreover, every technology, whatever the force behind it, is developed through a ***system.*** This means that each technology has parts and that each part has a relationship with all other parts and to the whole. The parts work together in a predictable way, and they are designed to achieve a goal. See **Figure 2-2.**

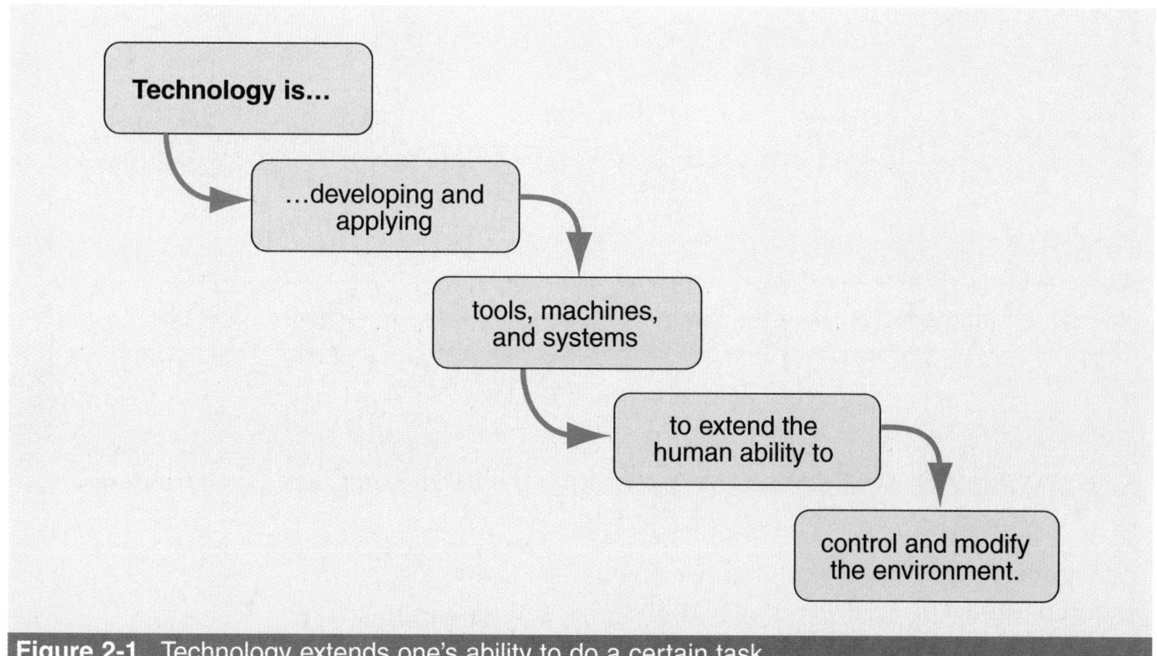

Figure 2-1. Technology extends one's ability to do a certain task.

System Components

As shown in **Figure 2-3,** a technological system has several major components. They are:

- Goals
- Inputs
- Processes
- Outputs
- Feedback and control

Figure 2-2. The parts that comprise this automobile engine are designed to work together in a predictable way. (Mazda Motor Corp.)

Goals

As we said above, humans develop technology to meet needs, which means, of course, that humans have a ***goal*** in mind when developing the technology. Note, however, that each artifact generally meets more than one goal For example, humans developed radar (*ra*dio *d*etection *a*nd *r*anging) to determine the position and speed of aircraft. Radar had a major impact on military strategies and actions during World War II. Today this technological innovation contributes to safe, reliable air transportation. We also use it to help predict the weather and to detect surface features on planets.

To examine this point a little further, let us look at another technology. Suppose that a company is developing a new coal-powered automobile. It is not enough to say that it is creating the automobile to *transport people.* This is certainly the primary goal. It might also be an opportunity for the innovators to *make money*, however.

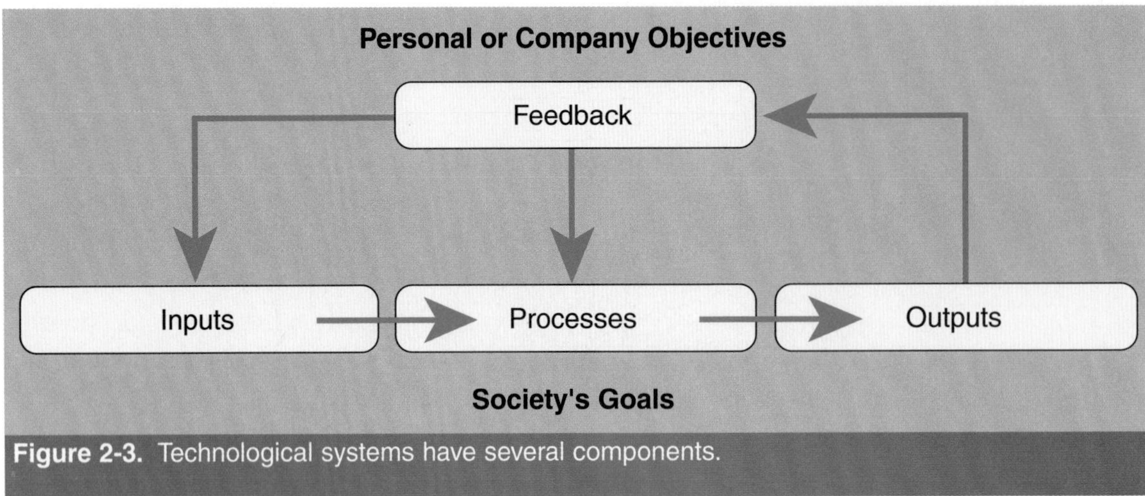

Figure 2-3. Technological systems have several components.

Local government leaders might look at its potential impact on *economic growth* in their city. The federal government might examine it in terms of whether it would help reaffirm *national technological leadership.* The general public might consider whether it would be a good alternative to petroleum-powered vehicles, thus reducing their *dependence on foreign oil.* Workers might look at the pros and cons in terms of whether it would *improve job security.* Coal-mining companies and oil companies might examine it with regard to whether it would *increase or decrease their market.* Environmental groups might comment on how the car *affects the environment.* Note the number of different goals and concerns that are highlighted here. As you can see, a technological development can meet a number of different goals and concerns that are important to different groups.

Inputs

All natural and human-made systems have inputs. ***Inputs*** are the resources that go into the system and are used by the system. Technological systems, as shown in **Figure 2-4,** have at least seven inputs:

- People
- Materials
- Tools and machines
- Energy
- Information
- Finances
- Time

People

People are the most important input to technological systems, **Figure 2-5.** Human needs and wants give rise to the systems. Human will and purposes decide the types of systems that will be developed. People bring to the systems specific knowledge, attitudes, and skills. They provide the management and technical know-how to design and direct the systems. Their labors make the systems function. Human ethics and values control and direct the systems. Through their local, state, and national governments, people make policies that promote or hinder the technological systems. Finally, people are the consumers of technological outputs. They use the products and services that the systems provide.

Materials

All technology involves physical artifacts, that is, objects made by humans, usually for a practical purpose. Tools, buildings, and vehicles are some examples of artifacts. ***Materials*** are the substances from which artifacts are made.

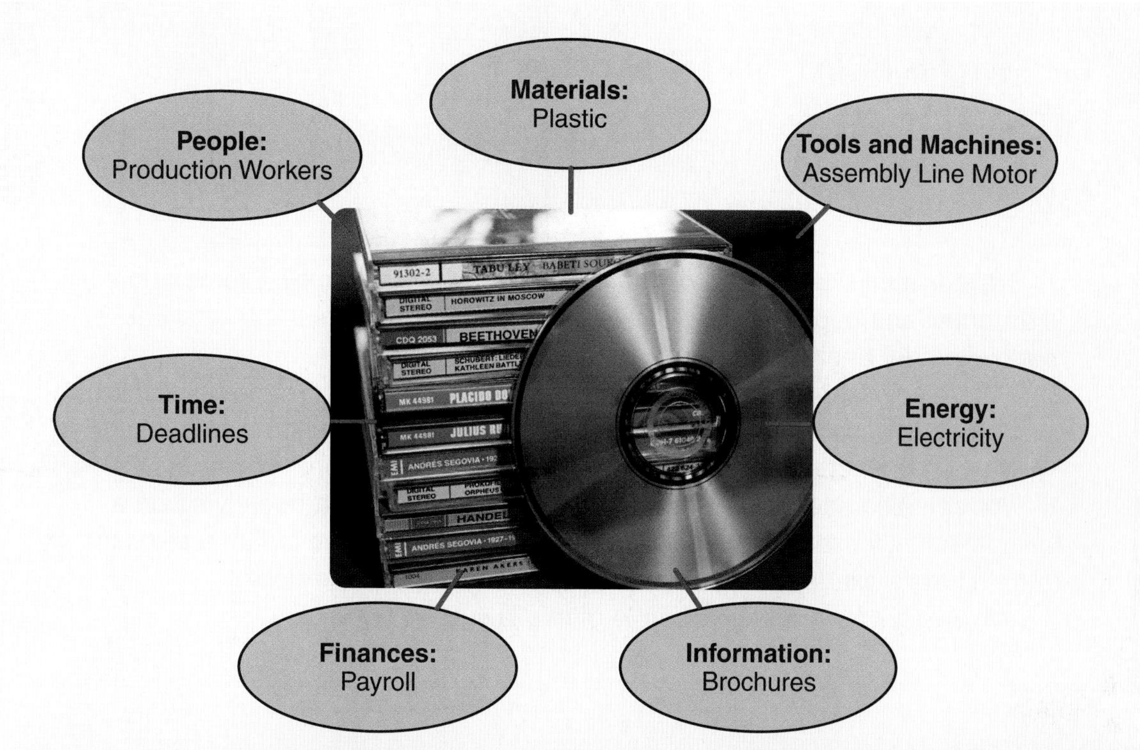

Figure 2-4. Technological systems have at least seven inputs.

Figure 2-5. People bring knowledge and skills to each technological system. For example, they operate machines, design devices, and provide management. (Apple Computers)

Among other classifications as discussed in Chapter 4, materials can be categorized as natural (found in nature, such as stone or wood), synthetic (made by humans, such as plastics or glass), or composite (natural products that have been modified, such as paper or leather), **Figure 2-6.**

Some of these materials provide the mass and structure for technological devices. Others support the productive actions of the system. For example, some materials lubricate machines; others contain data; still others package and protect products.

Connections to Technology: **Science**

Newton's Third Law of Motion

Newton's third law of motion states that for every action by a force, there is an equal and opposite reaction by another force. We can see this law at work with a jet engine.

The engine is designed to shoot out a mass of hot gas. The force of this gas in one direction creates an equally strong force in the opposite direction. Thus when the hot gas shoots backward, the jet plane shoots forward.

We can also see this law in other areas. For example, a book lying on a table is actually exerting a force on the table because of the pull of gravity. In turn, the table exerts an equal and opposite reaction on the book.

What might be another example of this law at work?

Tools and Machines

Technology is characterized by tools and machines. ***Tools*** are the artifacts we use to expand what we are able to do. Tools include both hand tools and machines. Hand tools expand muscle power and include such artifacts as hammers and screwdrivers. ***Machines*** amplify the speed, amount, or direction of a force. Early humans relied on power from animals and water to affect force, but now we use energy from such sources as electricity to power machines. We use tools and machines to accomplish such varied tasks as build structures, communicate information and ideas, convert energy, improve our health, and transport people and goods.

Figure 2-6. Materials can be found in nature. Here, for example, we see water that can be converted into energy, rocks that can be made into hand tools, and trees that can be made into paper.

Energy

Technology involves doing something, and all technological activities require energy. ***Energy*** is the ability to do work. Note that the form the energy takes varies. It can range from human muscle power to nuclear power, from heat energy to sound energy.

Technological systems require that energy be converted, transmitted, and applied. For example, the energy of falling water may be converted into electrical energy by a hydroelectric generator, **Figure 2-7.** The electricity may then be carried to a factory over transmission lines. At the factory, motors may convert the electrical energy into mechanical motion, halogen and fluorescent bulbs may convert it into light energy, or radiant heaters may convert it into heat energy.

Information

Everywhere we turn we encounter facts and figures that are called data. When we organize this data and group it according to its type, we create ***information.*** Information is essential for operating all technological systems.

An example will show the difference between data and information. You might measure the size and weight of everything you can find. This would be data because it is random and assorted. If you sort the data so the height and weight of all people are grouped together, however, you have information, that is, organized data. Note also that from this information you can see relationships and draw conclusions, for example that adults are generally taller than children and that men are generally taller than women. With this final step you have developed knowledge. Knowledge is people using information to understand, interpret, or describe a specific situation or series of events. We will examine the various aspects of information in more detail in Chapter 4.

Finances

We have seen that technology system inputs include people, materials, tools and machines, energy, and information. These resources have value and, therefore, must be purchased. Also, the outputs of technology systems have value and can be sold. Thus ***finances***, the money and credit necessary for the economic system to operate, are another technology input.

For example, a construction company determines that people might need such an artifact as a new home. The company must purchase the land and materials to build the structure. They must then obtain plans for the dwelling. They rent or buy equipment, and they hire workers to build the house. All of these actions require money.

Figure 2-7. The energy of falling water can be converted into electrical energy. (Library of Congress)

They then sell the finished home to cover these costs, **Figure 2-8.** Generally money will be left after all expenses are paid. As we noted in the last chapter, this amount of money is called profit, which is the goal of most economic activities.

Time

All jobs or activities take ***time.*** Each person has only sixty minutes in an hour and twenty-four hours in a day. This time is allotted to the various tasks that need to be done. Likewise, time must be allotted to all technological endeavors. The most important ones will most likely be completed. If time is not available, less-critical tasks will be left undone or postponed to a later date. Therefore, not all technology that is needed can be developed immediately. Some activities will have to wait until time and other resources are available.

Figure 2-8. This newly built home is being sold by the construction company to cover the costs of creating it and to earn a profit.

Processes

All technological systems are characterized by action. A series of identifiable tasks must be completed. The steps needed to complete these tasks are called ***processes.*** Technological systems use three major types of processes: problem-solving or design processes, production processes, and management processes, **Figure 2-9.**

The Problem-Solving/Design Process

In Chapter 1, we learned that science involves activities that study and interpret the natural world. Scientists use research to develop their descriptions. They carry out their work through a set of procedures called the scientific method. This method structures the research so that valid results can be obtained. Through the scientific method, scientists attempt to understand the world that now exists.

On the other hand, technology, as described earlier, develops and uses tools and machines, systems, and materials to extend the human ability to control and modify the environment. One key word in this definition is *develop.* This part of technology requires creative action. The procedure used to develop technology is

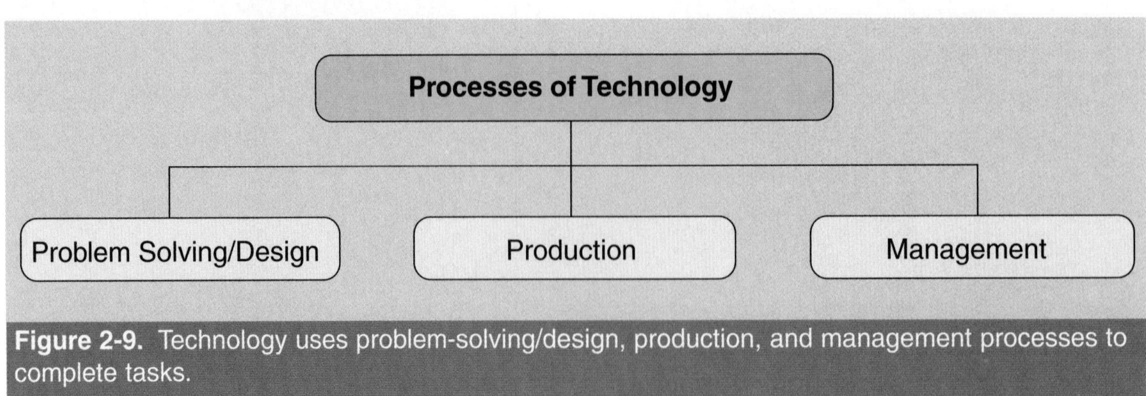

Figure 2-9. Technology uses problem-solving/design, production, and management processes to complete tasks.

called the ***problem-solving/design process.*** It involves five major steps, **Figure 2-10:**

1. Identifying the problem: A person or group develops basic information about the problem and the design limitations.
2. Developing solutions to the problem: A person or group develops several possible solutions to the problem and refines them through the formation of ideas and brainstorming (creative thinking) procedures.
3. Isolating, refining, and detailing the best solution: A person or group selects and refines the most promising ideas and then details the best solution.
4. Modeling and evaluating the solution: A person or group produces and tests physical, graphic, or mathematical models of the selected solution.
5. Communicating the final solution: A person or group selects a final solution and prepares documents needed to produce and use the device or system.

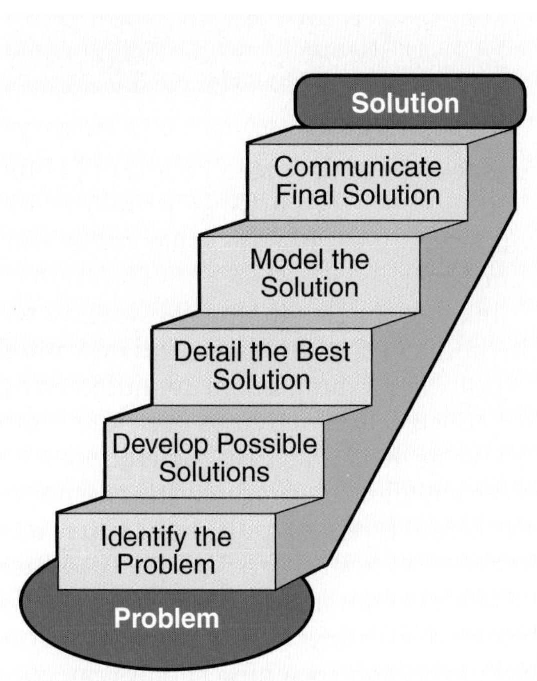

Figure 2-10. The problem-solving process involves five major steps to reach a solution.

Production Processes

Production processes are actions that are completed to perform the function of the technological system. For example, a company might use a series of production processes to produce an informational booklet, **Figure 2-11**. They write and edit the message or copy to be communicated. They produce photographs to illustrate the document. They put the photographs and copy together into a page layout. They convert the layout into printing plates from which copies of the booklet are printed.

Production processes are used to grow and harvest crops, change natural resources into industrial materials, prevent or treat illnesses, convert materials into products or structures, transform information into media messages, convert a form of energy, or use energy to power transportation vehicles to relocate people or goods. Each technological system has its own production processes, as we will see later in this book.

Figure 2-11. This production process converted page layouts (top) into booklets (bottom). (Graphic Arts Technical Foundation)

Management Processes

Management processes are all the actions people use to ensure that the production processes operate efficiently and appropriately. They use these processes to direct the design, development, production, and marketing of the technological device, service, or system. Management activities involve four functions, **Figure 2-12.**

- *Planning:* Setting goals and developing courses of action to reach the goals
- *Organizing:* Dividing the tasks into major segments so that the goals can be met and resources assigned to complete each task
- *Actuating:* Starting the system to operate by assigning and supervising work
- *Controlling:* Comparing system output to the goal

Individuals and groups use management processes to organize and direct their activities. For example, you might have a task to complete, such as writing a term paper. First, you must *plan* this activity by selecting a topic, establishing major steps to be completed, and setting deadlines for each task. You then must obtain and *organize* resources. You must secure reference materials, writing or word-processing equipment, and time. Then you *actuate* the work by reading and viewing reference material, taking notes, preparing a draft of the paper, and editing the draft into final form. Finally, your instructor compares your paper to established standards and gives you the results. This step is *control,* which includes evaluation, feedback, and corrective action.

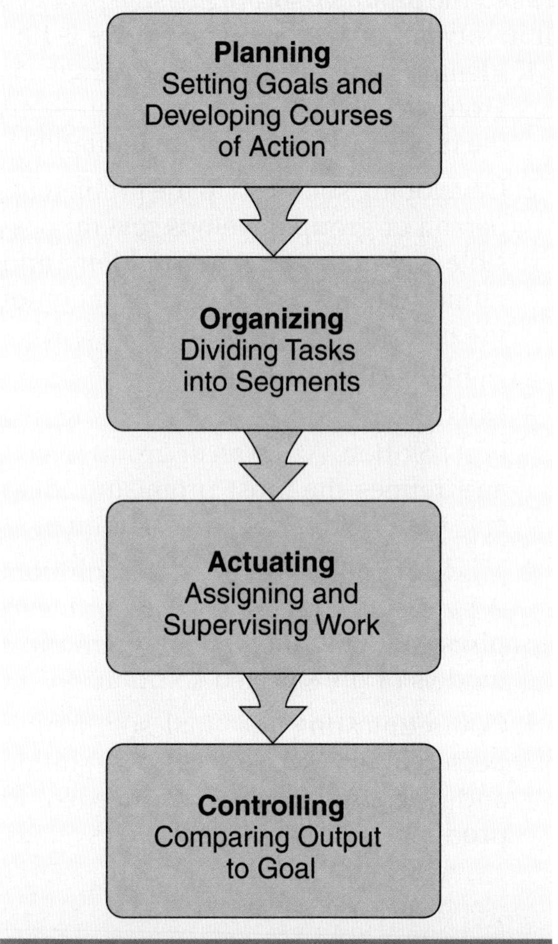

Figure 2-12. Management activities direct the operation of the production process.

Outputs

All technological systems are designed to produce specific outputs. These ***outputs*** might be manufactured products, constructed structures, communicated messages, or transported people or goods. Note, however, that these primary outputs are not the only ones that come from technology.

In operating technological systems, we can produce other less-direct and unwanted outputs. For example, our manufacturing and construction activities generate scrap and waste. Manufacturing operations also create chemical by-products, fumes, noise, and other types of pollution. Cars, airplanes, trains, and trucks often produce noise, air pollution, and congestion. Poorly designed housing developments, industrial parks, and shopping centers can contribute to soil erosion, as can careless farming practices.

Technology Explained

Solar Collectors

solar collector: A high temperature device heated by concentrated solar energy.

Solar collectors can be of two types: solar thermal and solar electric. Solar thermal collectors capture heat from sunlight by moving air or a fluid through the collector. The air or fluid absorbs the heat and carries it away for storage or use. Solar electric collectors use *photoelectric* materials. When light strikes a photoelectric material, electrons move and create an electric current.

Both solar thermal and solar electric collectors can be placed together in a solar orchard, **Figure A**. Solar thermal collectors are used to produce steam. This steam is fed to a generating plant to make electricity. Solar electric collectors can be connected directly to the electric utility grid. Both types of collectors use tracking systems to follow the sun across the sky.

A relatively new method of collecting and using solar energy is the solar furnace. The solar furnace uses the reflected light from mirrors called *heliostats*. Each device tracks the sun and reflects the solar energy onto a parabolic mirror. The mirror, in turn, reflects the light from all the heliostats onto a single point. Temperatures can reach 5400°F (3000°C). **Figure B** shows a drawing of a solar furnace built for research in France.

Figure A. This solar orchard is in Phoenix, Arizona. These collectors use solar electric cells with concentrating lenses to increase the power output.

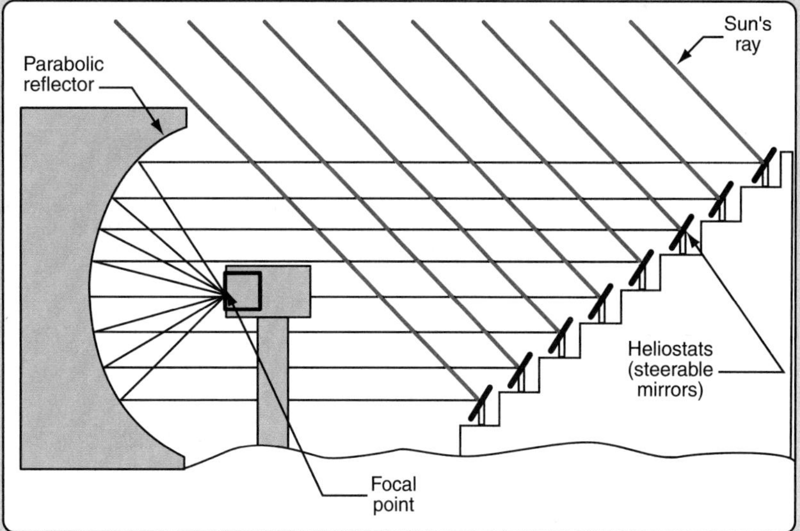

Figure B. The solar furnace in France uses 63 flat mirrors to collect and reflect the sun's energy.

Technological systems also have social and personal impacts. The products of technology shape society and are shaped by society. For example, the automobile was a novelty in the early 1900s. The wide acceptance of the automobile, however, has greatly changed where and how we live. We travel more and live farther from our place of work. The small compact town has been replaced by sprawling cities. Shopping is no longer close to the neighborhood. On the other hand, we have shaped the automobile. Rising gasoline prices and government fuel-economy standards have resulted in improved engine fuel efficiency with fewer exhaust products. Personal buying habits also dictate the types of cars that will be built.

Feedback and Control

All systems are characterized by ***feedback*** and ***control.*** This process involves using information about the outputs of a process or system to regulate the system. The process of feedback and control is used in many common situations. For example, many homes have heating systems that are controlled automatically with a thermostat, **Figure 2-13.** The occupant sets the thermostat at a desired temperature. The thermostat measures the room temperature. If it is too low, the thermostat closes a switch that turns on the furnace. As the room is heated, the thermostat monitors the temperature. It is using the output of the furnace to determine the necessary adjustments for the system. When this output (heat) warms the room to the proper temperature, the thermostat changes the system's operation. It opens the switch and stops the furnace. The cycle is repeated when the room cools to a specific temperature. Thus through feedback and control (in this case automatic control) the thermostat regulates the heating system.

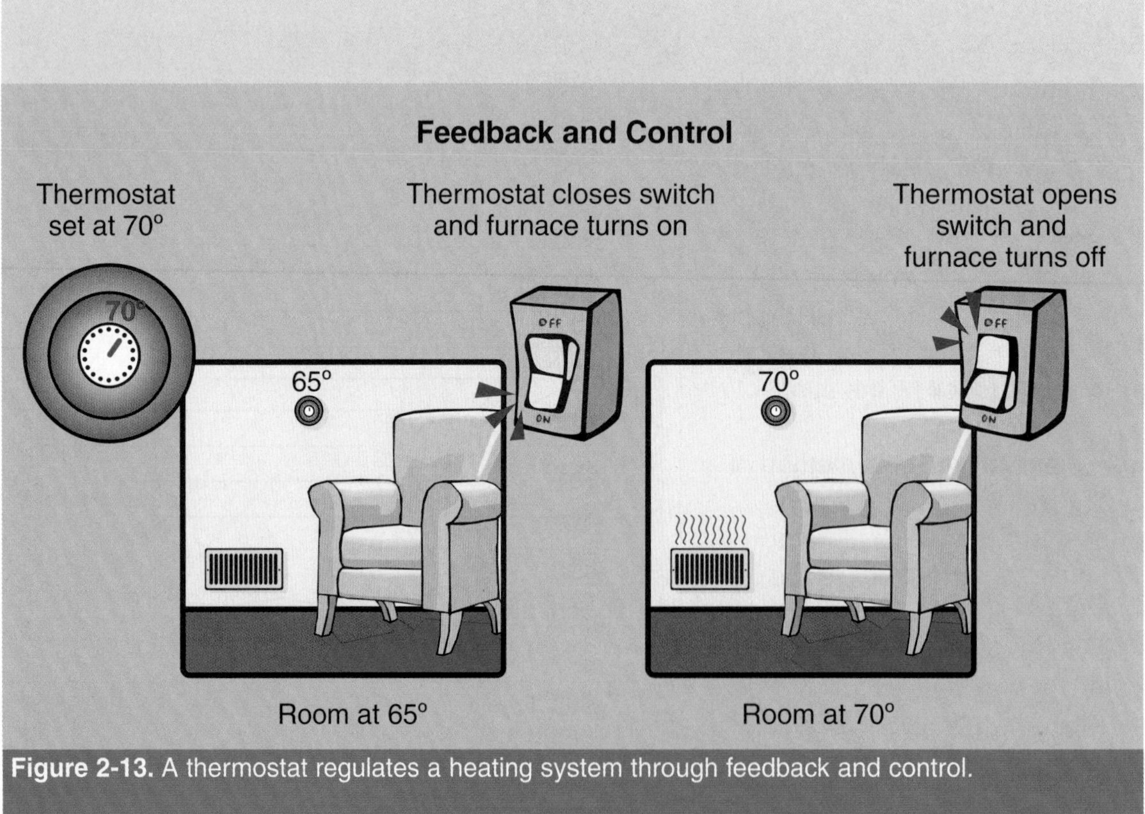

Figure 2-13. A thermostat regulates a heating system through feedback and control.

Summary

Technology involves human-made systems that use resources to produce desired outputs. The systems use seven major types of resources: people, materials, tools and machines, energy, information, finances, and time. These resources are used during the operation of the technological systems. The operation stage is called the process. It involves three major types of processes: problem solving/design, production, and management. The result of each system includes desired and unwanted outputs. For example, we use technological systems to produce products, construct structures, communicate information, and transport people and goods. These things improve our way of life. These are the desired outputs. We also get pollution, scrap and waste, altered lifestyles, and increased health risks, however. These are the negative impacts or outputs of technology. We use feedback and control to regulate the system. The challenge for each person developing or using technology is to use the feedback and control to maximize the desired outputs and minimize the negative impacts at the same time.

Key Terms

actuating
control
controlling
energy
feedback
finances
goals
information
inputs
machines
management processes
materials
organizing
outputs
planning
problem-solving/design process
processes
production processes
system
time
tools

Test Your Knowledge

Write your answers on a separate sheet of paper. Please do not write in this book.

1. In what way(s) is technology a system?
2. List the five major components of a technological system.
3. *True or false?* Each artifact is developed to meet one goal.
4. List the seven inputs to a technological system.
5. What is the most important input to a technological system?
6. *True or false?* Materials can be classified as natural, synthetic, or composite.
7. *True or false?* Energy needs to be converted before it can be used in a technological system.

8. Explain the difference between data and information.
9. *True or false?* Technology primarily uses the scientific method to develop new artifacts and services.
10. List the five steps in the problem-solving/design process.
11. What are the four functions of management activities?
12. List three desired outputs of a technological system.
13. List three undesired outputs of a technological system.
14. Give one example of how a technological development has affected society.
15. The process of using information about the outputs of a system to regulate it is called _____.

Applying Your Knowledge

1. Select a technological device that you use and list its inputs, processes, and desired and undesired outputs. Organize your answer in a form similar to the one shown below.

   Device:

   Major inputs:

   Production processes used to make the device:

   Desired outputs:

   Undesired outputs:
2. Select an early technological advancement (invention) and prepare a short report that includes the inventor's name, nation from which it came, events that led up to the invention, its impact on life at the time, and its later refinements. Include a sketch of the item.

Career Corner
Purchasing Agents

Purchasing agents find quality materials, supplies, and services at the lowest costs possible. They determine which items are best, select a supplier for the material or service, negotiate a low price, establish delivery dates, and release purchase orders. Most purchasing agents work in offices and often work more than 40 hours a week.

Many companies prefer to hire purchasing agents who have a college degree and some experience. Newly hired purchasing employees often attend specialized training programs. They generally spend time learning about their company's purchasing practices, while working with experienced purchasers.

Chapter 3
Types of Technological Systems

Learning Objectives

After studying this chapter, you will be able to:

- List the four general ways technology systems can be categorized.
- Identify obsolete, current, and emerging technologies.
- Describe how technology evolves in a private enterprise system.
- Cite examples of government-sponsored technology.
- Explain why cooperation is important in technological development.
- Identify tasks of each major area of technology.
- Explain how technological systems are related.

Technological systems are everywhere. They support our daily lives, produce the artifacts that constitute the human-made world, and are vital to the economic and political health of every nation. Technological systems, as shown in **Figure 3-1,** can be grouped in four general ways:

- Level of development
- Economic structure
- Number of people involved
- Type of technology developed and used

Level of Development

Technology is a constantly changing phenomenon. Simple technologies are constantly being replaced by more sophisticated ones. For example, the sail was used on boats as early as 5000 B.C. in Mesopotamia. This device made water transportation more efficient and took less human effort than rowing the boat. In 1807 Robert Fulton used the steam engine to power a ship down the Hudson River. This advancement soon replaced the sail as the primary power source for commercial shipping. The steam engine did not last as long as the sail, however. It was replaced by the diesel engine in the 1900s. In 1954 nuclear power was first used in the submarine USS *Nautilus,* and it is still in use in naval vessels around the world. At any stage of history, then, there are obsolete, current, and emerging technologies, **Figure 3-2.**

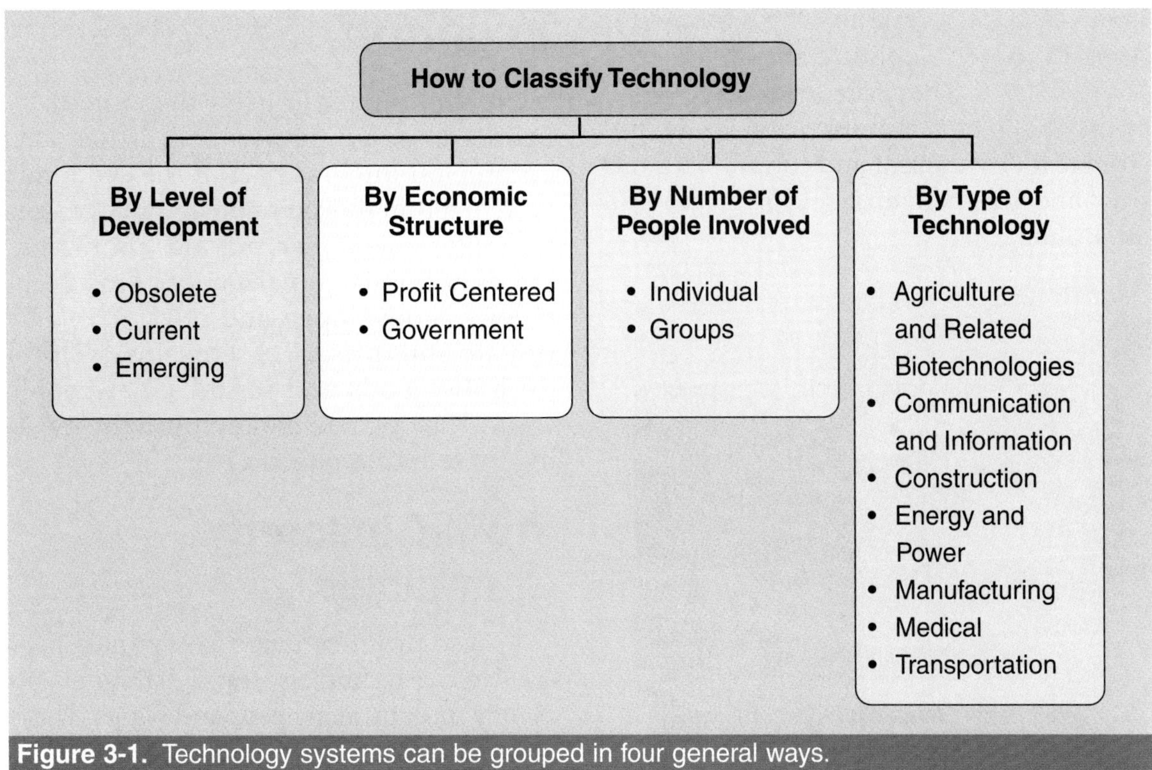

Figure 3-1. Technology systems can be grouped in four general ways.

Obsolete Technologies

Obsolete technologies are those that can no longer efficiently meet human needs for products and services. For example, hand spinning and weaving of cloth are obsolete. Some artists and home hobbyists still use this technology to produce works of art and personally designed items. Likewise, artists use the skills of the old-time blacksmith to produce decorative, forged steel items. This technology is seldom used to produce products for the mass market, however. Instead, machines are used to produce these items more quickly and efficiently.

Current Technologies

Current technologies include the range of techniques used to produce most of the products and services today. You can see these technologies everywhere you look. Trucks, trains, and aircraft are the vehicles we usually use to transport the majority of goods in this country. The print and electronic media deliver the majority of the information to the general public. We also use a series of common technologies to produce products and construct structures.

Emerging Technologies

Emerging technologies are the new technologies that are not widely employed today. They might, however, be commonly used in a later period of time. Today, laser machining, space travel, and electronically delivered magazines and books are emerging technologies. We often call these technologies ***high technology*** or ***high tech.*** This term is somewhat misleading because many of yesterday's high technologies are now current technologies, **Figure 3-3.** Today's high-tech processes may be tomorrow's current technology. For example, most Americans do not remember a world without television. Yet television was considered high technology in the early 1950s. Fiber-optic communication was unknown to most people until the 1980s. Today it is used in most telephone systems in the United States. Therefore, we should think of high tech as a moving focus: new and strange today, current and common tomorrow, and obsolete and seldom used at a later date.

Figure 3-2. At any stage of history, technologies are obsolete, current, or emerging. (Library of Congress, United Airlines, NASA)

Economic Structure

Technology has been developed by people to serve humankind. It has also become an integral part of the economic system. From this point of view, we can see technology as either private (profit centered) or public (government). Most technology is developed and put into use by private enterprise. These businesses produce the goods and services that people want. They plan to make a profit by meeting these human needs.

Profit-Centered Companies

These profit-centered companies generally move technology through three stages as they meet human needs and wants. These stages, as shown in **Figure 3-4,** are:

Figure 3-3. Many people remember a time without cellular telephones. Today, cell phones can be seen everywhere.

1. ***Research and development:*** Designing, developing, and specifying the characteristics of the product, structure, or service
2. ***Production:*** Developing and operating systems for producing the product, structure, or service
3. ***Marketing:*** Promoting, selling, and delivering the product, structure, or service

These stages are supported by two major activities that provide the resources needed to operate the systems. These are:

- ***Industrial relations:*** Setting up and managing programs to ensure that the relations between the company and its workers and the public are positive
- ***Financial affairs:*** Obtaining the money and physical resources and keeping the financial records needed to manage the system

Government

Some technology that is important to the general public is too expensive or risky to be developed by private enterprise. In this case the government may address what it sees as the general welfare of the citizens by developing a specific technology. The goal is not to make a profit from the undertaking. The space program with its specific hardware is an example of this type of public-funded technological development. Out of this type of program come many products and systems that can later be used by individuals and companies. Examples of NASA's (National Aeronautics and Space Administration) space program technological innovations that have become commonplace include the following:

Figure 3-4. Technology generally progresses through three stages: research and development, production, and marketing. Here, people created and sold books by designing pages (top), printing and binding the pages into books (middle), and displaying the books in a manner to attract customers (bottom).

- Mylar from the early Echo space program is now used for clothing, packaging tape, and decorative balloons.
- High-temperature composite materials from the Lewis Research Center are used in aircraft engines and aircraft structures.
- Satellite communications systems from many space programs now make it possible to communicate around the globe instantly, **Figure 3-5.**
- Film with a v-groove design, which was developed for increasing fuel efficiency of jet aircraft, reduces water and wind resistance. It was used on the America's Cup winning yacht *Stars and Stripes.*
- A corrosion-resistant coating that was developed for launch pads was applied to the refurbished Statue of Liberty.

Figure 3-5. Technology that benefits all people, such as this communications satellite, is sometimes developed by the government. (NASA)

Career Corner

Controllers

Controllers manage workers who prepare financial reports that summarize and forecast a company's financial position. Common reports these workers prepare include income statements and balance sheets. The workers also help predict future earnings or expenses for a company. Controllers generally provide overall management for the accounting, audit, and budget departments. Many controllers oversee the preparation of reports required by governmental regulatory agencies.

For most controller positions, a bachelor's degree in finance, accounting, economics, or business administration is required. Often, preference is given to applicants who have a master's degree in one of these areas. Controllers work in offices close to the departments that develop the financial data for top management. They typically have direct access to advanced computer systems. Many controllers work up to 50 or 60 hours per week.

Number of People Involved

We can also categorize technology systems by the number of people involved and how they work with the technological system. In isolated cases, individuals might develop technology for their own use. A cattle rancher might create an automatic feeding system for his or her livestock. A home gardener might develop a unique hoe that meets his or her personal needs, **Figure 3-6.** A home craftsperson may design and build a desk for his or her own use. This type of personal technology is not common, however. It accounts for only a small portion of all the technology in use today.

Most technology is developed and applied by groups or teams of people. The key to this activity is cooperative attitudes. People often say that we live in a competitive society. This belief is partly true. We compete to get on a sports team or to get a job. Companies compete to sell products and gain market share. Nations compete for prestige. Within each of these groups, however, is cooperation. The winning sports team is a study in cooperative attitudes. The most profitable companies exhibit high levels of management and worker cooperation, **Figure 3-7.** Nations cooperate in organizations such as the United Nations. Once a person gets a job, cooperation helps to ensure success.

This spirit of cooperation in developing and operating a technological system can be described as ***corporate participation.*** This does not mean that the corporate form of ownership must be present. It means that people are united and combined into one body. They share a common vision, work toward similar goals, and share in the success of the enterprise.

Type of Technology Developed and Used

The three ways of looking at technology discussed so far are useful in a general way. They give only part of the picture, however. A more meaningful approach to understanding technology categorizes the area by the type of technology developed and used. It looks at technology as human actions and groups them accordingly. This grouping, as shown in **Figure 3-8,** includes seven types of technology, all of which have been used by humans throughout history to modify and control their environment:

- *Agricultural and related biotechnologies:* Used in growing food and producing natural fibers
- *Communication and information technologies:* Used in processing data into information and in communicating ideas and information

Figure 3-6. A gardener developed this home-built garden hoe for his own use.

- ***Construction technologies:*** Used in building structures for housing, business, transportation, and energy transmission
- ***Energy and power technologies:*** Used in converting and applying energy to power devices and systems
- ***Manufacturing technologies:*** Used in converting materials into industrial and consumer products
- ***Medical technologies:*** Used in maintaining health and curing illnesses
- ***Transportation technologies:*** Used in moving people and cargo from one place to another

In each technological system, inputs are processed and transformed into outputs. For example, in agricultural technology, materials (seeds and fertilizers, for instance) are changed into food and fibers. In medical technology, information and materials are changed into devices that aid in maintaining health. In transportation technology, energy is changed into a form of power that moves vehicles containing goods.

These systems can be looked at individually, but such a focus gives an inaccurate view. These systems are all closely related and are part of a single effort: to help humans live better. They work

Figure 3-7. Most technology requires a high level of cooperation between people for its development and efficient operation. (Apple Computers)

Types of Technology

Communication

Construction

Manufacturing

Transportation

Agriculture

Medicine

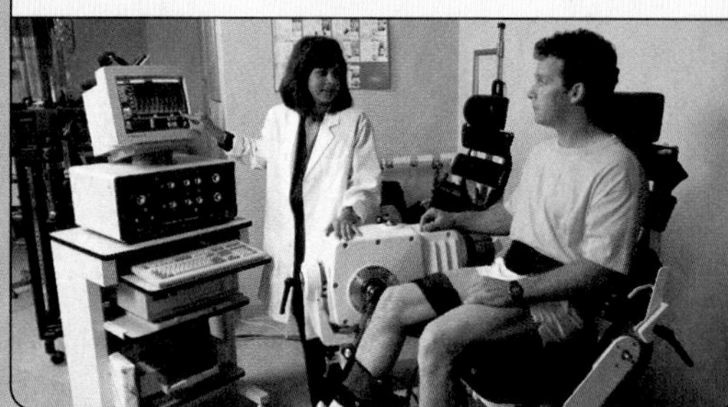

Energy

Figure 3-8. Technology can be categorized by the type of product developed.

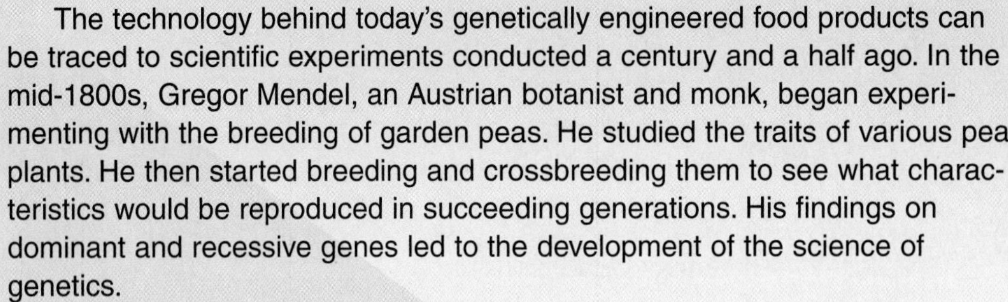

Connections to Technology: Science

Genetic Engineering

The technology behind today's genetically engineered food products can be traced to scientific experiments conducted a century and a half ago. In the mid-1800s, Gregor Mendel, an Austrian botanist and monk, began experimenting with the breeding of garden peas. He studied the traits of various pea plants. He then started breeding and crossbreeding them to see what characteristics would be reproduced in succeeding generations. His findings on dominant and recessive genes led to the development of the science of genetics.

We now use the principles developed by Mendel to produce better and more cost-effective foods. For example, because of genetic engineering, dairy cows injected with a growth hormone developed from genetically engineered bacteria produce greater amounts of milk. Beef cattle injected with the hormone have leaner meat. Genetically altered bacteria have also been used to make crops more insect resistant.

Genetic engineering has been used to great benefit, but some people have expressed concerns about the practice. Can you think of a problem that might result from genetic engineering? If one does not immediately come to mind, you might want to check your local newspaper. The pros and cons of various types of genetic engineering are in the news frequently.

together to support one another. In one case manufacturing may be the focus, with the other systems in a supporting role. In another situation, manufacturing may be the supporting technological system. To see this relationship, look at **Figure 3-9.** Let us follow one material from its natural state to a finished product that is designed to make life better for us. We will follow iron ore on its journey to becoming a stainless steel cookie sheet.

Our story starts with *manufacturing technology.* With regard to the cookie sheet, this technology system involves three major activities:

- Locating and extracting the raw materials (iron ore, limestone, and coal) to make steel
- Producing strips of steel from the raw materials
- Making (stamping) the cookie sheet from the steel strips

These manufacturing activities could not exist without the other technological systems. For example, *communication and information technologies* played a role at every point in the product's development. The need for the new product was communicated through sales orders. The specifications for the steel and the cookie sheet were communicated through engineering drawings and specification sheets. The availability of the product was communicated to potential customers through advertising. Sales reports communicated product success to the company's management.

Figure 3-9. The production of each manufacturing product involves a variety of steps. Each step depends on agricultural and related biotechnologies and communication and information, construction, energy and power, manufacturing, medical, and transportation technologies.

Energy and power technologies helped to power the melting furnaces and stamping presses, light the work areas, and heat the offices and control rooms. *Agricultural and related biotechnologies* facilitated the growth of trees, which were used to make the pallets that contained the boxes of finished products. *Medical technologies* helped to ensure that the workers remained in good health.

Likewise, *construction technologies* were essential. Constructed roads created access to the iron ore and coal mines and to the limestone quarries. Workers traveled to the steel mill and the production factory on constructed roads and worked in constructed buildings. Constructed power lines brought electricity to the various manufacturing sites. Constructed dams and pipelines made water available.

Transportation technologies also played a major role. They moved the raw materials to the steel mill and the steel strips from the mill to the product manufacturer. They delivered finished products to stores. Customers used private cars or public transportation systems to visit a store to purchase the product.

Throughout the rest of this book we will focus on these seven major technological systems. We will examine the components for each of these systems and explore the productive processes in greater depth.

Summary

Technology has always been a part of human existence. As civilization has become more advanced, technology has evolved. Older technologies are often replaced with newer, more efficient technologies; the high technologies of today often become the current technologies of tomorrow.

Most current technology is developed and applied by industrial and business enterprises. These companies hire research and development personnel to design and develop these technologies, which are then put into use through production activities and sold through marketing efforts. The system is driven by a desire to profit from the development and application of technologies. Some technology is also supported by government agencies. Most technology is developed by groups of people working together in a cooperative manner. Technologies are commonly developed to help people grow food, communicate information, construct structures, convert energy, manufacture products, improve health, and transport people and cargo. The systems are closely related and work together to provide better lives for everyone.

Key Terms

agricultural and related biotechnologies
communication and information technologies
construction technologies
corporate participation
current technologies
emerging technologies
energy and power technologies
financial affairs
high technology (high tech)
industrial relations
manufacturing technologies
marketing
medical technologies
obsolete technologies
production
research and development
transportation technologies

Test Your Knowledge

Write your answers on a separate piece of paper. Please do not write in this book.

1. List the four general ways technology systems can be categorized.
2. Give one example each of an obsolete technology, a current technology, and an emerging technology.
3. *True or false?* Emerging technology is also referred to as high technology.
4. Name the three stages a technological development goes through in a profit-centered company.
5. The two resources needed to support the stages of technological development are industrial relations and _____.
6. Give an example of a government-sponsored technology.
7. *True or false?* Most technology is developed by individuals attempting to improve their personal situation.
8. Define the term *corporate participation*.

For Questions 9–19, match the technological system on the right to the correct area in which it is used. (Note: Some letters will be used more than once.)

Area of Use

9. _____Used to maintain health
10. _____Used to build structures
11. _____Used to convert materials into products
12. _____Used to grow food
13. _____Used to move people
14. _____Used to process data
15. _____Used to apply energy
16. _____Used to produce natural fibers
17. _____Used to communicate ideas
18. _____Used to cure illness
19. _____Used to transmit information

Technological System

a. Agricultural and related biotechnologies
b. Communication and information
c. Construction
d. Energy and power
e. Manufacturing
f. Medical
g. Transportation

20. Give a specific example of how one technological system is related to another.

Applying Your Knowledge

1. Develop a chart similar to the following one and list five technological devices from each type of technology that you use daily.

Technology	Technological device
Agricultural and related biotechnologies	
Communication and information	
Construction	
Energy and power	
Manufacturing	
Medical	
Transportation	

2. Identify a new technological device that you could use and sketch what it would look like and how it would work.
3. Ask an older person (a grandparent or neighbor) to tell you about five technological devices that they used in their lifetime that are no longer around. List them (obsolete technologies) and the devices that have replaced them (current technologies) and ones that you think will replace the current ones (emerging technologies).

Section 1 Activities

Activity 1A

Design Problem

Background

Technology is the application of knowledge to create machines, materials, and systems to help us make work easier, make life more comfortable, or control the natural or human-made environment.

Situation

The Easy-Play game company has developed a new board game for two players. The game uses five red marbles and five green marbles as the playing pieces. The company has found that counting out the marbles one by one is too costly.

Challenge

Design a technological device (machine) that will count five marbles at a time from a box containing a large number of marbles.

Activity 1B

Fabrication Problem

Background

All products of technology have been developed to meet needs or opportunities. Most early technological products answered functional needs. Some of the earliest products of technology were clay containers for transporting liquids and storing food.

Challenge

Divide your class into two groups. Each group will use different clay-forming techniques to produce a clay pot and lid. Group 1 will form the two parts without the aid of any external devices, and Group 2 will use a form to aid in the production process. The form Group 2 will use is a frozen orange juice concentrate can. See **Figure 1B-1.** The product should be the diameter of the orange juice can and 1 1/2" tall. The lid should be 1/2" tall.

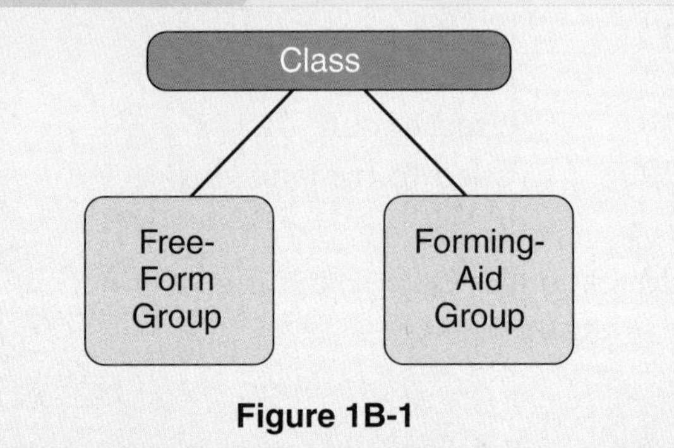

Figure 1B-1

Materials and Equipment

- Potter's clay
- Frozen orange juice concentrate cans
- Sponge
- Table knife

Procedure

Produce a clay pot and lid using the procedure for your group.

Free-Form Group

1. Obtain the supplies.
2. Separate a portion of clay.
3. Roll several 3/8" diameter strips of clay. See **Figure 1B-2.**
4. Coil a strip of clay flat on the tabletop until a disk the diameter of the juice can is produced. This is the bottom of the pot.
5. Build the wall of the pot by layering strips of clay until the 1 1/2" height is

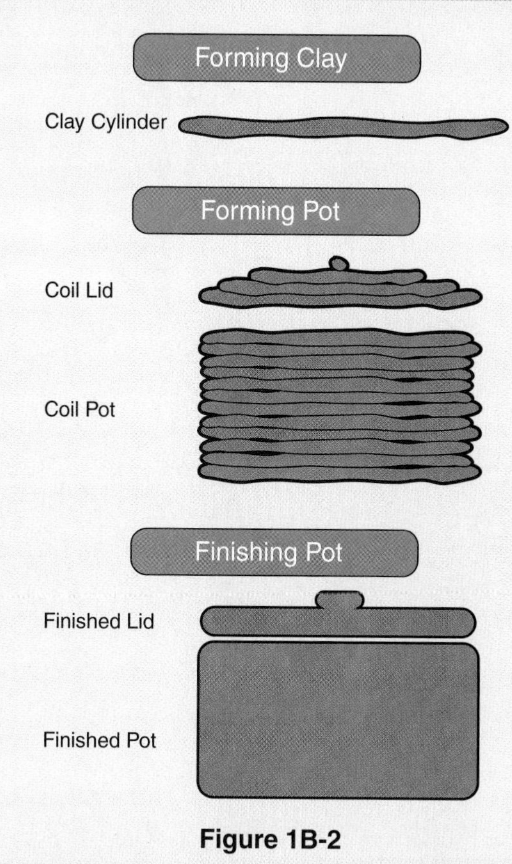

Figure 1B-2

reached.

6. Use a damp sponge and your fingers to smooth the clay and form the finished pot.
7. Coil a strip of clay flat on the tabletop until a disk the diameter of the juice can is produced. This is the lid for the pot.
8. Gently shape the disk into a concave shape 1/2" tall.
9. Form a handle (1/2"ball) and attach it to the center of the lid.
10. Use a damp sponge and your fingers to smooth the clay and form the finished lid.
11. Allow the pot to air dry.
12. Fire the pot in a kiln. (optional)

Forming-Aid Group

1. Obtain the supplies.
2. Separate a portion of clay.
3. Roll several 3/8" diameter strips of clay.
4. Coil a strip of clay flat on the tabletop until a disk just a bit larger than the diameter of the juice can is produced. This will become the bottom of the pot.
5. Use the can like a cookie cutter to cut a perfectly round disk.
6. Place the can on the tabletop with the open end down.
7. Build the wall of the pot by loosely coiling strips of clay around the can until the 1 1/2" height is reached. Do not push the clay against the can because it will stick.
8. Carefully slide a table knife between the can and pot to separate them.
9. Remove the can.
10. Place the pot wall on top of the bottom.
11. Use a damp sponge and your fingers to smooth the clay and form the finished pot.
12. Coil a strip of clay flat on the tabletop until a disk the diameter of the juice can is produced. This will become the lid for the pot.
13. Gently shape the disk into a concave shape 1/2" tall.
14. Form a handle (1/2" ball) and attach it to the center of the lid.
15. Use a damp sponge and your fingers to smooth the clay to form the finished lid.
16. Allow the pot to air dry.
17. Fire the pot in a kiln. (optional)

Analysis

Meet as a class and analyze the two clay-forming techniques. Answer the following questions:

1. Do you think the free-form method is a common technique for producing modern ceramic products? Why or why not?
2. Which method was easier to use?
3. Which method produced a higher quality pot?

Section 2

Technological System Components

Tomorrow's Technology Today
Self-Cleaning Windows

If you have ever cleaned windows, you know it is much harder than it looks. The windows streak if you wipe them dry and spot if you let them air dry. It seems the highest windows are always the dirtiest, and as soon as the windows are clean, it inevitably rains, depositing a new layer of dust on the glass. Now, imagine cleaning the windows of a 90-story high-rise.

In a recent survey *Better Homes and Gardens* conducted, homeowners chose self-cleaning windows as their favorite "fantasy product." Several manufacturers have heeded the call and developed windows that clean themselves. Although each manufacturer gives its self-cleaning glass a different name, the properties of each glass and the techniques used to create them are very similar. Titanium oxide is the key to transforming traditional glass into self-cleaning glass.

When self-cleaning glass is manufactured, an ultrathin coating of titanium oxide is applied to the surface of the molten glass. After the glass cools, the ultrathin (0.00005 mm) titanium oxide coating is perfectly transparent. When the coating is exposed to UV light (sunlight), any organic substances (such as dirt) in contact with it begin oxidizing, breaking down, and loosening from the glass.

The titanium oxide coating is also hydrophilic, which means it attracts water. Normal glass tends to repel water, which causes the water to bead and eventually run in rivulets down the glass. Since the self-cleaning glass is hydrophilic, the water spreads out evenly across the glass's surface. The thin film of water on the glass dries quickly and evenly. This eliminates the spotty deposits of dust you find on a regular pane of glass after a rain shower.

Attempts have been made in the past to create self-cleaning windows. Unfortunately, the early attempts at creating such glass involved applying films to sheets of glass. After several years of service, the films would deteriorate and peel away from the glass. The titanium oxide film used in modern self-cleaning glass is far more durable because it is applied while the glass is still in its liquid state. Because of this, the coating becomes an integral part of the glass, not simply a thin film adhered to it.

Also, the titanium oxide coating is a catalyst, rather than a reactant, in the oxidation of the dirt on its surface. This means the coating causes the dirt to break down, but the process does not affect the coating. For this reason, the coating should last for the life of the window without wearing out.

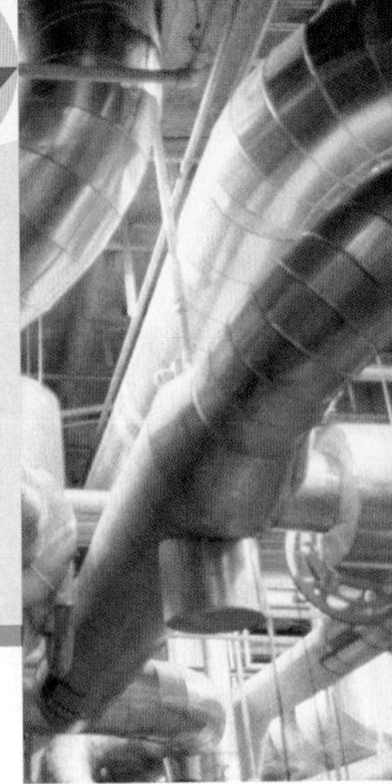

Chapter 4
Inputs to Technological Systems

Learning Objectives

After studying this chapter, you will be able to:

- Identify inputs to technological systems.
- Describe the types of skills and knowledge various groups of people bring to technological systems.
- Identify the types of tools and machines used as inputs to technological systems.
- Describe the types and properties of materials that are inputs to technological systems.
- Identify the types of information that are inputs to technological systems.
- Identify the major types and sources of energy as inputs in technological systems.
- List the sources of finances that are used as inputs to technological systems.
- Explain the importance of time with regard to technological systems.
- Explain why we need to use and conserve intelligently the resources that are the inputs to technological systems.

Humans have lived on the earth about 2.5 million years. This might seem like a long time, but it is relatively short considering the earth is about five billion years old. In the short span of human history people have developed a special ability. They have learned how to build and use tools, machines, systems, and materials to change their environment. You learned in Chapter 1 that this process is called technology. This ability to develop technology has led to many kinds of technological systems. You learned in Chapter 2 that all systems include several components. These components are inputs, processes, outputs, feedback and control, and goals. This chapter will explore in more depth the *inputs* that are common to all technological systems.

As you might recall from Chapter 2, all inputs can be grouped into seven major categories, as shown in **Figure 4-1.** These are:

- People
- Tools and machines
- Materials
- Information

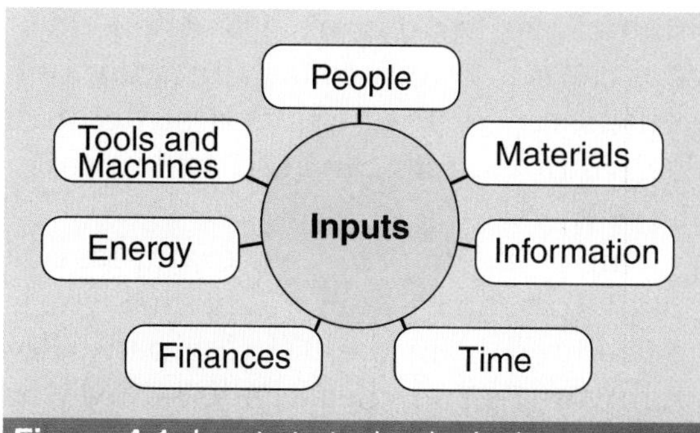

Figure 4-1. Inputs to technological systems can be grouped into seven major categories.

- Energy
- Finances
- Time

These inputs are the resources that are used to make the system operate. They are the elements that are changed by technological processes or are used by technology to change other inputs. Let us look at each one separately.

People

As stated in Chapter 2, people are the most important input to technological systems. People produce these systems. Their minds create and design the systems and their outputs. They use their skills to make and operate the systems. Their management abilities make systems operate efficiently. Their needs and wants are satisfied by the systems. Therefore, it is easy to see why people are fundamental to technology.

Various groups of people bring different knowledge, skills, and abilities to technological systems. For example, ***scientists*** generally develop the basic knowledge needed to help create products and processes, **Figure 4-2.** Fields of knowledge used by scientists include physics, materials science, geology, and chemistry.

Engineers apply scientific and technological knowledge in designing products, structures, and systems, **Figure 4-3.** They determine appropriate materials and processes needed to produce products or perform services. For example, civil engineers determine the correct structure for a bridge to carry vehicles across a river. Electrical engineers design circuits for computers.

Still other people work at such tasks as building products and structures, producing communication messages, and transporting people and cargo. These people can be referred to as ***production workers.*** Production workers in manufacturing and construction are often categorized as unskilled, semiskilled, and skilled workers. The specific designation depends on one's training and work experience. Skilled workers in laboratories and product-testing

Figure 4-2. Scientists are involved in various fields of knowledge when creating technological products and processes. (Northern Telecom)

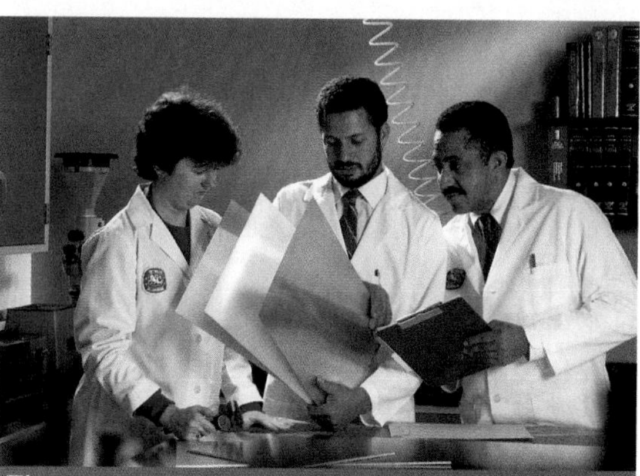

Figure 4-3. These engineers are developing product applications for a new composite material. (Reynolds Metals Co.)

facilities are usually called ***technicians,*** **Figure 4-4.** Skilled workers in service operations are often called ***mechanics.***

Other groups of people establish and/or help direct businesses. Those who create businesses are called ***entrepreneurs.*** They have vision of what can be done and are willing to take risks to see it happen. ***Managers*** organize and direct the work of others in these businesses. They set goals, structure tasks to be completed, assign work, and monitor results. Nonmanagerial ***support staff*** bring other knowledge to the businesses. They carry out such tasks as keeping financial records, maintaining sales documents, and developing personnel systems.

The final group of people involved in technological systems are ***consumers,*** **Figure 4-5**. They are the reason for the system in the first place. Their attitudes about styling, price, and service directly affect the systems' outputs. Their money is spent on the products or services. Therefore, they financially support the system.

We should note that one element common to the development of technology within all these groups of people is ***creativity.*** Creative people have the ability to see a need or a way of making life easier and design systems and products to meet the need or desire. In some instances, they develop ideas for a product and decide what size, shape, and color the product will be, **Figure 4-6.** Their decisions can add beauty to the world through well-designed products and structures. Their innovations can lead to improvements in existing technology.

One of the more famous examples of a creative person is Art Fry, a chemist who is now retired from the 3M Company. Fry was aware that another scientist at 3M was attempting to invent a stronger adhesive for some of their tapes by experimenting with various molecules. This scientist instead discovered a new adhesive, one that would adhere to paper but was not as sticky on tape. He was not sure what to do with this discovery but did mention it to other employees at 3M. A short time later, Mr. Fry, who was a member of his church's choir, was in church and, as usual, was trying not to get frustrated as he marked his place in the songbook with little pieces of paper

Figure 4-4. This skilled worker is involved with medical technology.

Figure 4-5. Consumers are fundamental to all technological systems.

that invariably fell out of the book. As the service continued, he suddenly thought—he could make a bookmark with the new adhesive, one that would stick to the page but could then be removed.

He prepared samples of his idea and distributed them to people at 3M. In that process, he and his manager, as they used the samples to make comments about their utility, realized they had something more than a bookmark. They had a new way of communication and organizing information more quickly and efficiently: self-attaching notes, which we now call Post-it Notes.

At this point other workers at 3M, with the support of management, joined the effort to design and produce this new item and, after methodically testing various forms and solving a series of manufacturing and packaging problems, came up with a product that today people probably cannot imagine being without.

We can see from this example that various groups of people contributed to this invention in different ways. We also can see that some products create their own demand. In this example, too, we can see that some changes in society lead to even more needs and demands. In this case, 3M has expanded its product line in this area to include such items as flip charts, tape flags, and room decorator kits. More items are sure to come.

Figure 4-6. This person used her creative abilities to design a greeting card. (American Greetings)

Tools and Machines

Humans are the only species on earth that can develop and use technology because, as we said earlier, humans are tool builders and users. ***Tools*** are the artifacts that expand what humans are able to do. From early humans on, people have used tools to increase their ability to do work, **Figure 4-7**. Today, tools include such diverse artifacts as milking machines used on dairy farms, machine tools used in factories, hammers and saws found on construction sites, and automobiles used to transport people and cargo.

One way to look at tools is by their area of activity, **Figure 4-8.** For example, microscopes and telescopes are used in scientific activities. Tennis rackets and pitching machines are used in recreational activities.

Figure 4-7. Humans have long used tools. This museum representation shows a person using a tool to start a fire.

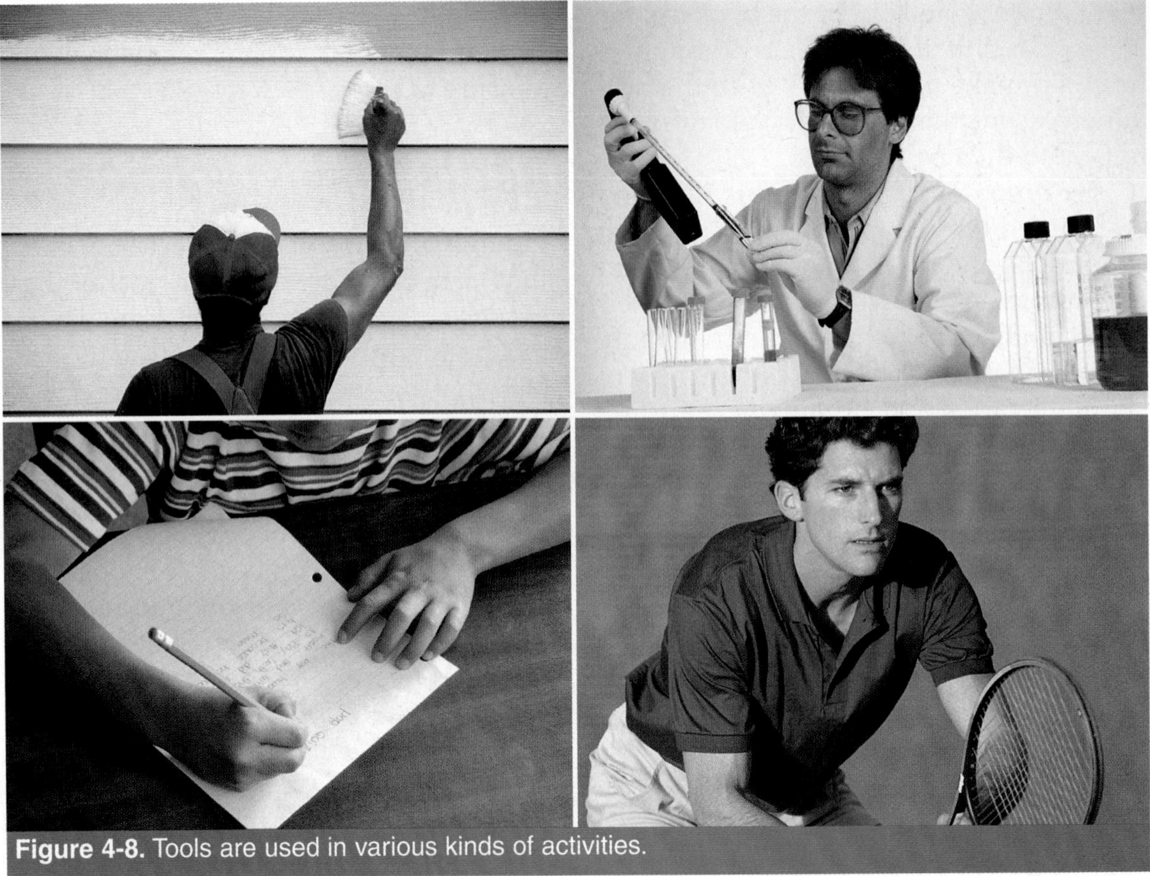

Figure 4-8. Tools are used in various kinds of activities.

Another way to group tools is by the technological system in which they are used. For example, some people in agriculture use plows; some people in communication use radio transmitters; and some workers in the field of transportation use automobiles or trucks.

As we noted in Chapter 2, however, the most common way we classify tools is to divide them into the two categories of hand tools and machines. Each is described in more depth below.

Hand Tools

Almost every technology uses a common set of hand tools to produce, maintain, and service products and equipment. ***Hand tools*** are those simple handheld artifacts that require human muscle power or air or electric power to make them work. These tools can be classified by their purpose, as shown in **Figure 4-9.** These categories are:

- Measuring tools: Used to determine the size and shape of materials and parts
- Cutting tools: Used to separate materials into two or more pieces
- Drilling tools: Used to produce holes in materials
- Gripping tools: Used to grasp and, in many cases, turn parts and fasteners
- Pounding tools: Used to strike materials, parts, and fasteners
- Polishing tools: Used to abrade and smooth surfaces

Machines

Machines are artifacts that transmit or change the application of power, force, or motion. Machines can be simple or complex, with complex machines being comprised of simple machines. Simple machines work on two basic principles, the

Figure 4-9. Hand tools can be grouped by the action they perform. (L. S. Starrett Co., Ridge Tool Co., Cooper Tools)

principle of the lever and the principle of the inclined plane, and can be grouped under six categories. As shown in **Figure 4-10,** they are:

- Lever
- Wheel and axle
- Pulley
- Inclined plane
- Wedge
- Screw

The first three mechanisms—the lever, the wheel and axle, and the pulley—operate on the basic principle of the lever. The second three mechanisms—the inclined plane, the wedge, and the screw—operate on the principle of the inclined plane. Each is described in turn below.

Lever Mechanisms

Almost everyone has used a lever. If you have ever pried open a crate with a crowbar or pulled a nail with a claw hammer, you have used a lever. A ***lever*** has a rod or bar (the ***lever arm***) that rests and turns on a support called a ***fulcrum,*** **Figure 4-11.** You apply a force of effort at one end of the rod or bar to lift a load at the other end. The purpose is to help lift weight more easily.

Levers are grouped into three categories: first class, second class, and third class, **Figure 4-12.** Each class of lever applies force differently to move the load. In ***first-class levers,*** the fulcrum is between the load and the effort. A pry bar is an example of a first-class lever. In ***second-class levers,*** the load is between the effort and the fulcrum. The wheelbarrow uses the principle of a second-class lever. In ***third-class levers,*** the effort is placed between the load and the fulcrum. A person moving dirt with a shovel applies the principle of a third-class lever.

Another aspect of levers we should note is that, like all other simple machines, a lever can be either a ***force multiplier*** or a distance multiplier. When it increases the force applied to the work at hand, it is a force multiplier. The fulcrum is close to

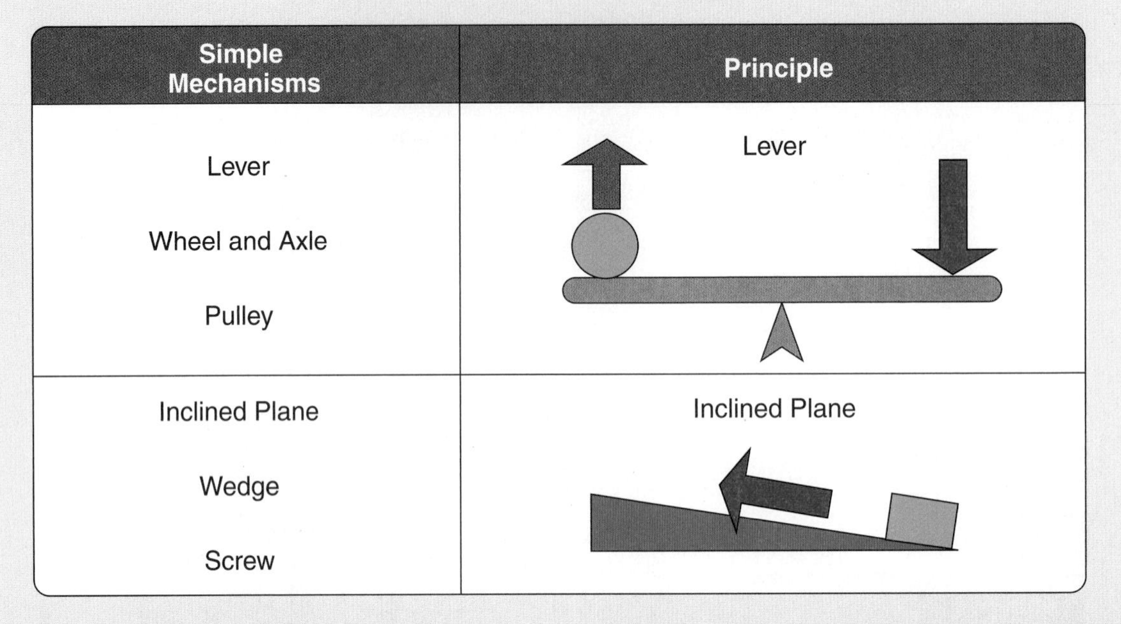

Figure 4-10. The six types of mechanisms work on two basic principles.

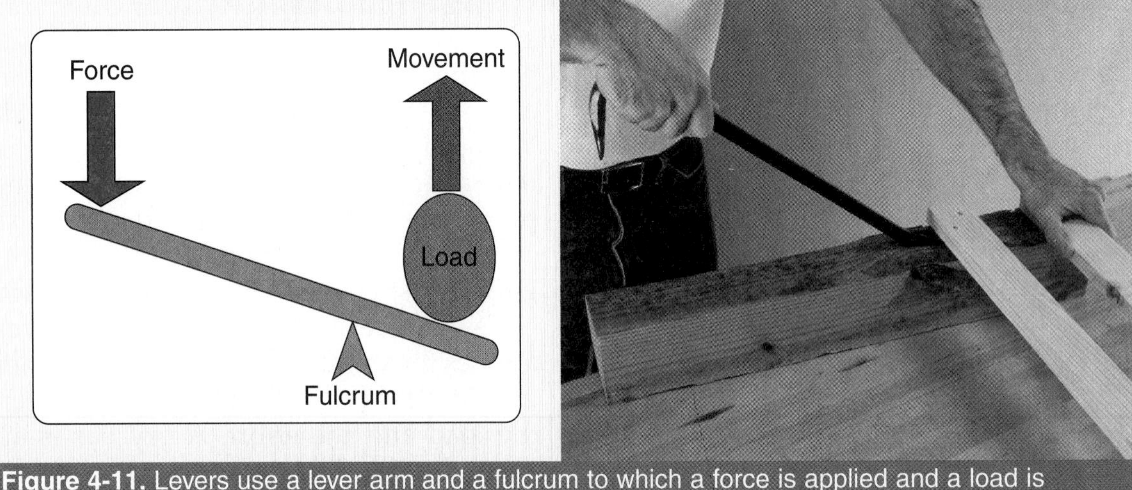

Figure 4-11. Levers use a lever arm and a fulcrum to which a force is applied and a load is moved. The crowbar uses the principle of the lever.

the load, and the force is applied at the other end. We thus can move a heavy load with a light force.

A ***distance multiplier*** lever is just the opposite. Here the fulcrum is close to the force, and the load is at the other end. The load will move a greater distance than the force, but note that a large force is required to move a light load. These two applications of levers are shown in **Figure 4-13.**

Wheel-and-Axle Mechanisms

A ***wheel and axle*** is a shaft attached to a disk. This mechanism acts as a second-class lever. The shaft or axle acts as the fulcrum. The circumference of the disk acts as the lever arm. If the load is applied to the shaft, the wheel-and-axle mechanism becomes a force multiplier, **Figure 4-14.** Automotive steering wheels use this

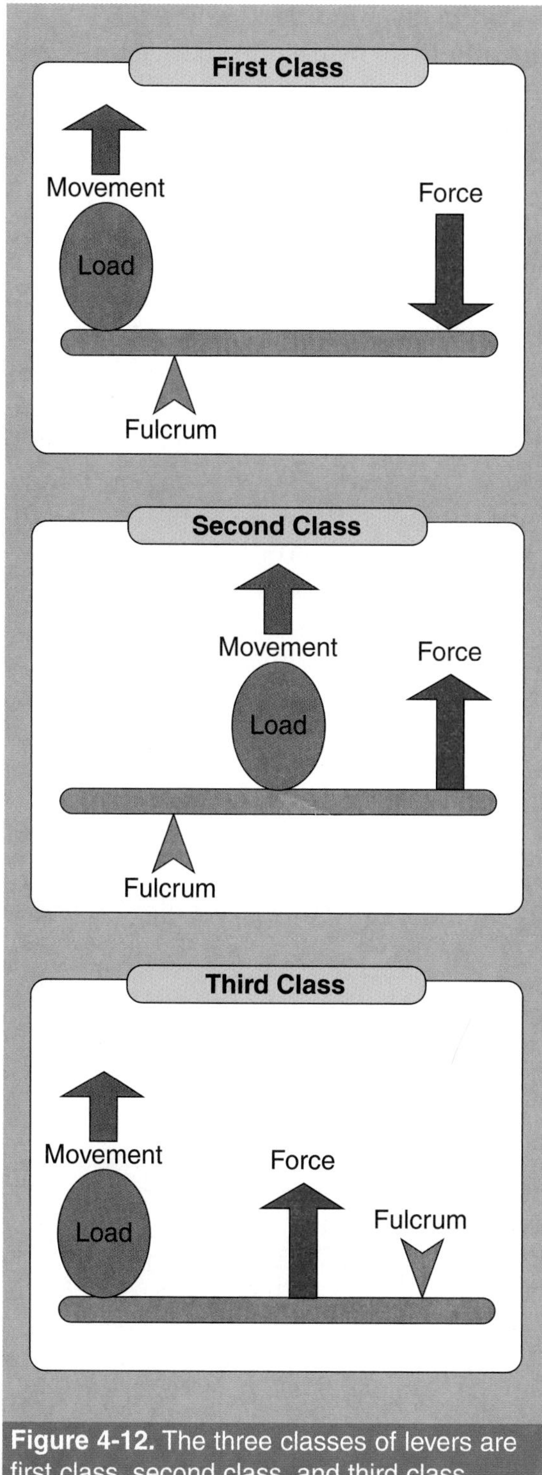

Figure 4-12. The three classes of levers are first class, second class, and third class.

principle. A 15-inch wheel attached to a 1/4-inch shaft will multiply the force by sixty times.

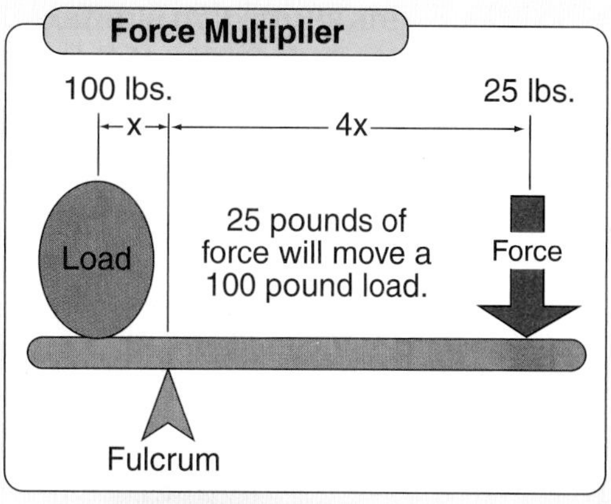

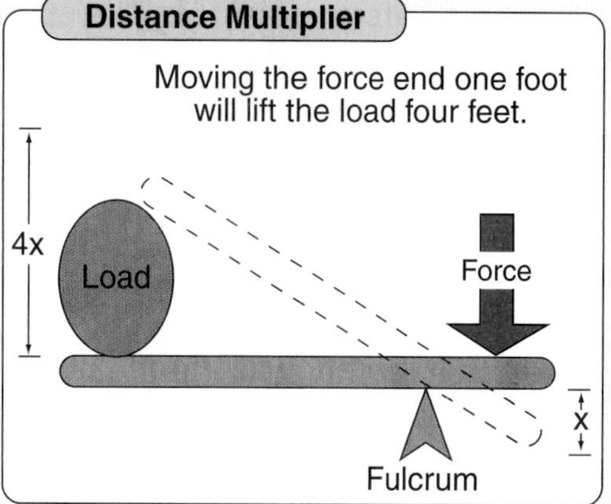

Figure 4-13. Levers can be used to multiply force (top) or distance (bottom).

If the load is applied to the disk, the wheel-and-axle mechanism becomes a distance multiplier. Automobile transaxles use this type of wheel-and-axle mechanism. One revolution of the axle will cause the wheel to revolve one time. The circumference of the wheel is many times that of an axle, however. Therefore, the vehicle moves a considerable distance down the road for each revolution. For example, a 20-inch wheel attached to a 1/2-inch shaft will multiply the distance for each revolution of the shaft forty times.

Pulley Mechanisms

Pulleys are grooved wheels attached to an axle. They also act as second-class levers. Pulleys can be used for three major

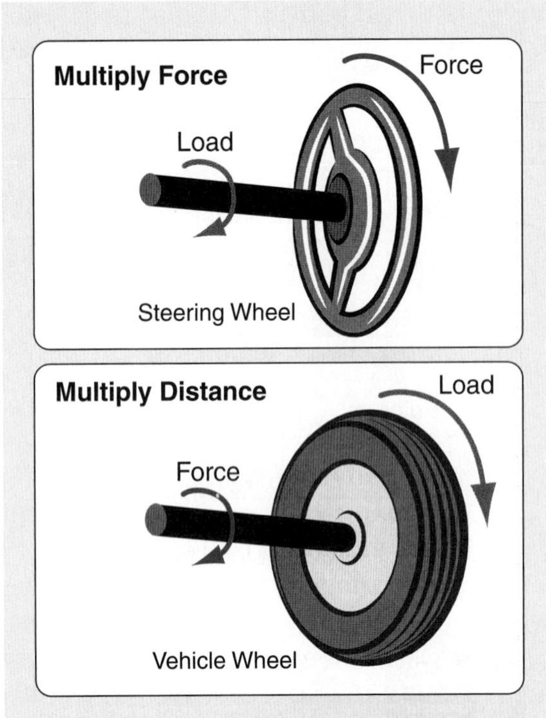

Figure 4-14. A wheel and axle can multiply force or distance.

purposes, as shown in **Figure 4-15.** A single pulley can be used to change the direction of a force. Two or more pulleys can be used to multiply force or to multiply distance. The number and diameters of the pulleys used will determine the mechanical advantage (force multiplication) of a pulley system.

Inclined Plane Mechanisms

Inclined planes are sloped surfaces used to make a job easier to do. The principle of the three mechanisms in this category (inclined plane, wedge, and screw, **Figure 4-16**) is that it is easier to move up a slope than up a vertical surface. The simplest application of this principle is the ***inclined plane.*** It is used to roll or drag a load from one elevation to another. Common examples of inclined planes are roadways in mountains and ramps to load trucks.

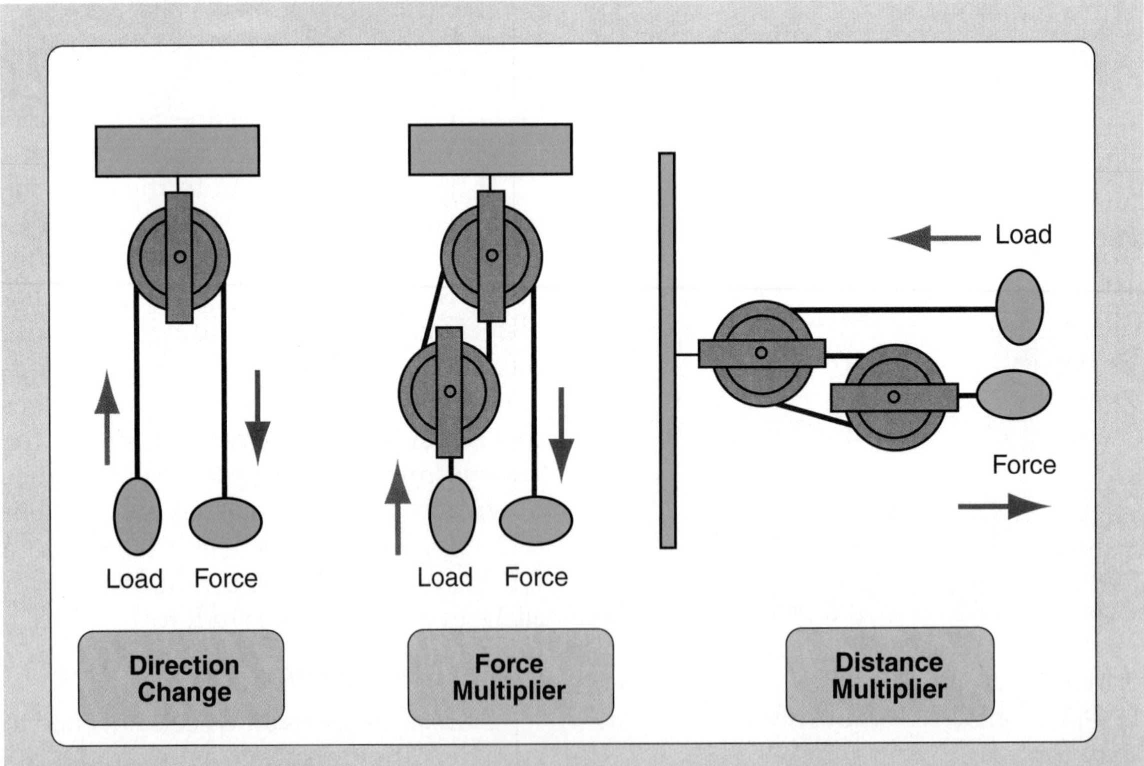

Figure 4-15. Pulleys can be used to change the direction of a force, multiply force, or multiply distance.

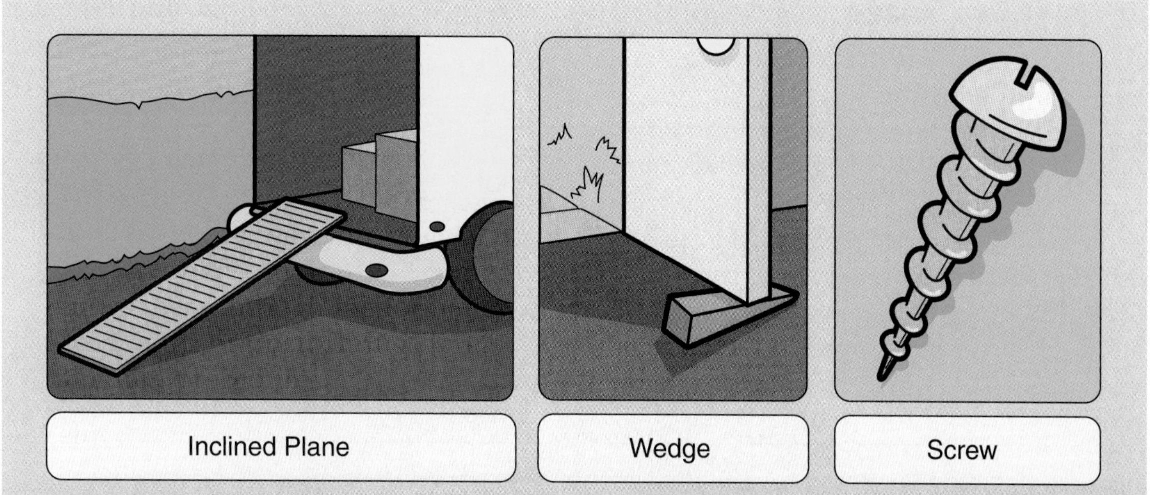

Figure 4-16. The three types of inclined plane mechanisms are the inclined plane, the wedge, and the screw.

Wedge Mechanisms

A second application of the inclined plane principle is the ***wedge.*** This device is used to split and separate materials and to grip parts. A wood chisel, a firewood-splitting wedge, and a doorstop are examples of this mechanism.

Screw Mechanisms

The ***screw*** is the third mechanism using an inclined plane. A screw is actually an inclined plane wrapped around a shaft. The screw is a force multiplier. Each revolution of the screw moves it into the work only a short distance. For example, a 1/2-inch x 12 machine screw is 1/2 inch in diameter and has 12 threads per inch. With one revolution of the screw, the circumference moves about 1-1/2 inches but into the work only 1/12 of an inch. In this example, the force is multiplied about eighteen times.

Materials

We are living in a material world. Everywhere you look you will see materials. They come in all sizes, shapes, and types. They possess a number of specific properties. All material is made up of one or more of the elements that occur naturally on earth. These elements, which number less than one hundred, combine to produce literally thousands of compounds. To understand materials and to use them effectively and efficiently, a person must know about the types of materials and the properties that materials exhibit.

Types of Materials

As we noted in Chapter 2, materials can be classified as natural, synthetic, or composite. They can also be classified by their origin, their ability to be regenerated, and their physical state, **Figure 4-17.** Each is described in turn below.

Classification of Materials as Natural, Synthetic, or Composite

Many materials occur naturally on earth. These materials are often called natural resources. Iron, carbon, petroleum, and silica are examples of ***natural materials.*** They can be refined and combined to make products.

Other materials are human made or ***synthetic materials.*** The most common synthetic materials are plastics. They are

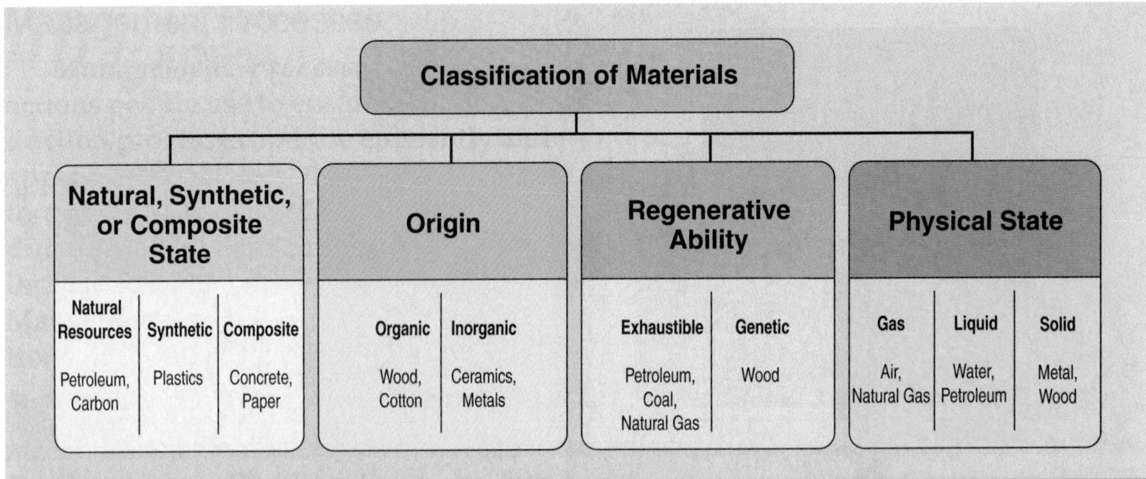

Figure 4-17. Materials can be grouped as being natural, synthetic, or composite and by their origin, ability to be regenerated, and physical state.

developed and produced from cellulose (vegetable fibers), natural gas, and petroleum.

Composite materials are combinations of natural and synthetic materials that are mixed to create items with other desirable properties. For example, concrete is a mixture of water, cement, sand, and gravel. It molds to almost any shape and then hardens into a long-lasting material that requires little care.

Classification of Materials by Origin

Materials that come from living organisms are called ***organic materials.*** For example, wood, cotton, and flax are products of plant fibers. Wool and leather are products of animals. Petroleum, coal, and natural gas are the products of decayed and fossilized organic materials. All materials that do not come from living organisms are called ***inorganic materials.*** For example, metals and ceramic materials are inorganic.

Classification of Materials by Ability to be Regenerated

Some materials naturally occur on earth in a specific amount. These materials cannot be replaced by human action or by nature. The quantity of these materials on earth is finite (limited). Once they are used up, there will be no more. They are also called ***exhaustible materials.*** Metal ores, coal, petroleum, and natural gas fall in this category of materials.

Other materials have a life cycle and can be regenerated. They can be produced by human action or by nature. They are called ***genetic materials*** because they are produced by living things. These materials are the result of farming, forestry, and fishing activities, among others. Wood, meat, wool, cotton, and leather are all genetic materials. The technology associated with producing these materials is called biorelated technology.

Other materials are the product of natural reactions. For example, carbon dioxide and oxygen are the by-products of natural processes. Water is purified through natural processes in lakes, wetlands, and rivers. Salt is produced around salt water by natural evaporation.

Classification of Materials by Physical State

Another important way of grouping materials is by their physical state, that is, as gases, liquids, or solids. ***Gases*** are materials that easily disperse and expand to fill any space. They have no physical shape but do occupy space and have volume. Gases can be compressed and put into containers. They meet many human needs. For example, they form the air we breathe, the fuel we use for rockets, the carbonation we add to beverages, and the compressed air we use to inflate tires.

Liquids are visible, fluid materials that will not normally hold their size and shape. They cannot be easily compressed. Common liquids include drinking water, fuels used for transportation vehicles, seawater for boats and ships, and coolants for industrial processes.

Solids are materials that hold their size and shape. They have an internal structure that causes them to be rigid. They can support loads without losing their shape.

Solids can be divided into four categories, as shown in **Figure 4-18:**

- Metallic materials (metals): Inorganic substances that have a crystalline structure, that is, their molecules are arranged in a boxlike framework called crystals. They are the most widely used of all engineering materials. They are generally used as an alloy, that is, a mixture of a base metal and other metals or nonmetallic materials. For example, steel is primarily an iron-carbon alloy, and brass is a copper-zinc alloy.
- Polymeric materials (plastics): Synthetic materials that contain complex chains of hydrogen-carbon (hydrocarbon) molecules. These materials are either thermoplastic (soften when heated) or thermosetting (made rigid by heat).
- Ceramics: Nonmetallic, mostly inorganic crystalline materials, such as

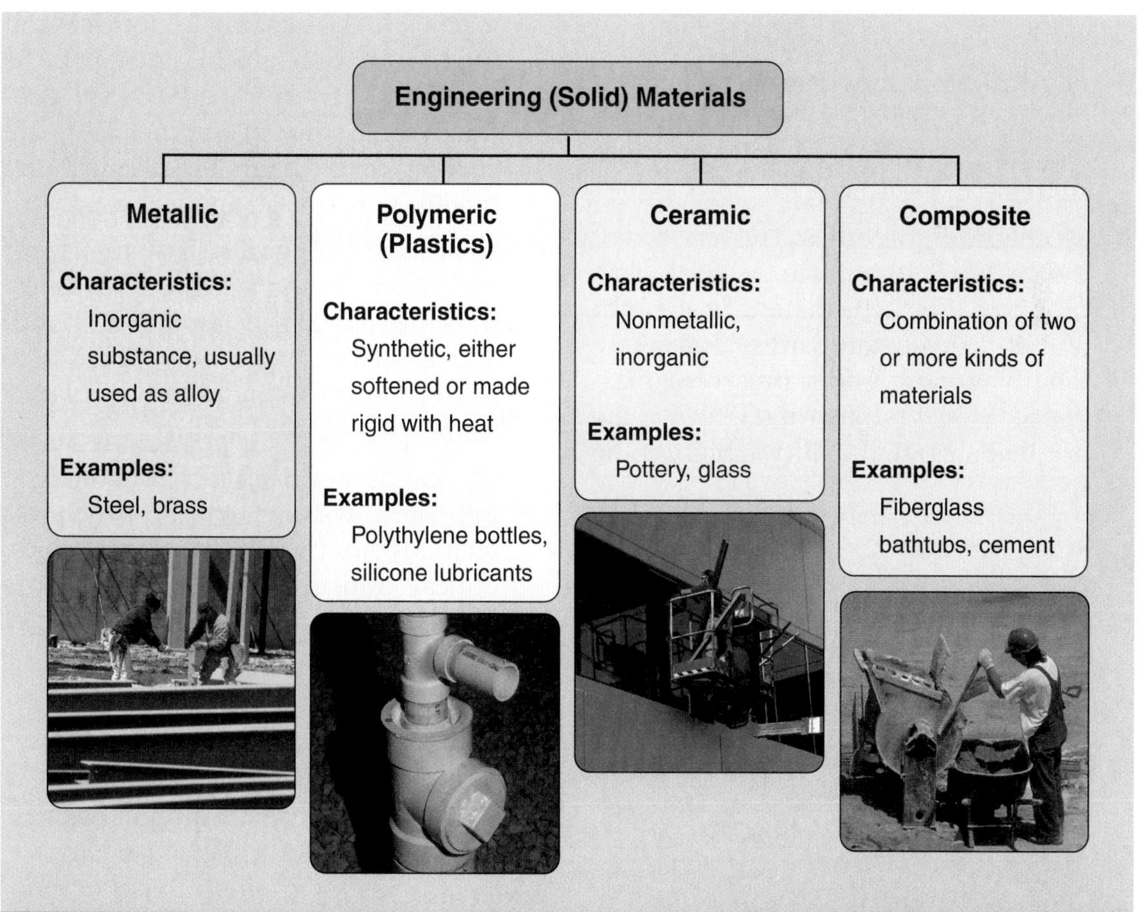

Figure 4-18. Solids can be grouped under the four categories of metallic, polymeric, ceramic, and composite.

clay, cement, plaster, glass, abrasives, or refractory material.

- Composites: A combination of two or more kinds of materials. One material forms the matrix or structure. The other material fills the structure. Fiberglass is a composite with a glass fiber structure filled with a plastic resin. Wood is a natural composite with a cellulose fiber structure filled with lignin, a natural glue, which bonds the structure together.

Properties of Materials

All materials exhibit a specific set of properties. For example, the properties of iron are different from those of oak. These properties are considered when materials are selected for specific uses. The common properties can be grouped under seven categories, as shown in **Figure 4-19.**

- Physical properties: The characteristics due to the structure of the material, including size, shape, density, moisture content, and porosity.
- Mechanical properties: The reaction of the material to a force or load. This property affects the material's strength (ability to withstand stress), plasticity (ability to flow under pressure), elasticity (ability to stretch and return to the original shape), ductility (ability to be bent), and hardness (ability to withstand scratching and denting).

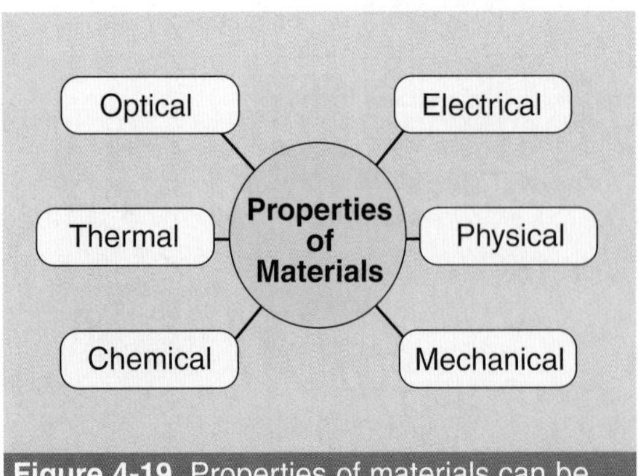

Figure 4-19. Properties of materials can be grouped under six categories.

- Chemical properties: The reaction of the material to one or more chemicals in the outside environment. This property is often described in terms of chemical activity (degree to which the material will enter into a chemical action) and corrosion resistance (ability to resist attack from other chemicals).
- Thermal properties: The reaction of the material to heating and cooling. This property is expressed as thermal conductivity (ability to conduct heat), thermal shock resistance (ability to withstand fracture from rapid changes in temperature), and thermal expansion (change in size due to temperature change).
- Electrical and magnetic properties: The reaction of the material to electrical and magnetic forces. This property is described in terms of electrical conductivity (ability to conduct electrical current) and magnetic permeability (ability to retain magnetic forces).
- Acoustical properties: The reaction of the material to sound waves. Acoustical transmission (ability to conduct sound) and acoustical reflectivity (ability to reflect sound) are measures of acoustical properties.
- Optical properties: The reaction to visible light waves. Optical properties include color (waves that are reflected), optical transmission (ability to pass light waves), and optical reflectivity (ability to reflect light waves).

Information

In contrast to other living species, humans have the ability to think, reason, and enter into articulate speech. They therefore can observe what is happening around them, make judgments about those

Connections to Technology: **Mathematics**

The Law of Equilibrium

A Greek mathematician who lived more than two thousand years ago actually proved the law behind the workings of the lever. Archimedes had observed how a small force can move a great weight. From this observation, the following law of equilibrium was created: A lever is in equilibrium when the product of the weight (w_1) and distance (d_1) on one side of the fulcrum [the center of gravity] is equal to the product of the weight (w_2) and distance (d_2) on the other side of the fulcrum. The mathematical formula is: $w_1 \times d_1 = w_2 \times d_2$.

Thus, if Roberto, who weighs 150 pounds, is 2 feet from the fulcrum of a seesaw, how far from the fulcrum would Becky, who is 60 pounds, have to sit to achieve balance?

Using the above formula, where w_1 is 150, d_1 is 2, and w_2 is 60, we can calculate the distance (d_2) as follows:

$$60 \times d_2 = 150 \times 2$$
$$60 \times d_2 = 300$$
$$d_2 = 300 \div 60$$
$$d_2 = 5$$

Thus Becky would have to sit 5 feet from the fulcrum.

If Isaac, who weighs 120 pounds, is 3 feet from the fulcrum, how far from the fulcrum would Josie, who is 90 pounds, have to sit to achieve balance?

observations, and explain them to other people. This unique human ability requires knowledge, which is derived from data and information, **Figure 4-20.**

Data are all the raw facts and figures that are collected by people and machines. ***Information*** is data that has been sorted and categorized for human use. Data processing involves collecting, categorizing, and presenting data so that humans can interpret it.

We can group information into three areas:

- Scientific information: Organized data about the laws and natural phenomena in the universe. Scientific information describes the natural world.
- Technological information: Organized data about the design, production, operation, maintenance, and service of human-made products and structures. Technological information describes the human-made world.
- Humanities information: Organized data about the values and actions of individuals and society. Humanities information describes how people interact with society and the values held by individuals and groups of people.

Information that is learned and applied by people is called ***knowledge.*** It is the result of reasoned human action. Knowledge guides people as they determine which course of action to take. Although knowledge can be described in terms of being derived from scientific, technological, or humanities information,

Figure 4-20. Data are processed into information so that humans can gain knowledge. (AC-Rochester)

in reality all these types of knowledge must be brought to bear on any problem needing a solution. Knowledge derived from science might provide a theoretical base for the solution. Knowledge derived from technology is used to implement the solution. Knowledge derived from the humanities will tell us if the solution is acceptable to society. We saw all these types of knowledge interacting at 3M when the people there created and produced Post-it Notes.

Energy

Energy is key to our survival, but it cannot be created or destroyed. It can only be converted from one form to another. After it is converted, energy powers our factories, heats and lights our homes, propels our vehicles, drives our communication systems, and supports our construction activities.

Various types of energy exist. For example, everyone uses human energy to complete tasks. Human energy falls short of meeting all our needs, however. We have a limited supply, and we do not want to use it all on work. Also, some tasks cannot be done with human energy alone. For example, we cannot heat a house with human energy because the heat radiated from the human body is not enough. Therefore, people throughout history have used other sources of energy. We can look at energy from two vantage points: type and source, **Figure 4-21.**

Types of Energy

Energy can be grouped into six major types. These are:

- Chemical energy: Energy stored in a substance and released by chemical reactions
- Electrical energy: Energy created by moving electrons
- Thermal (heat) energy: Energy that comes from the increased molecular action caused by heat
- Radiant (light) energy: Energy produced by the sun, fire, and other matter, which includes light, radio waves, X rays, and ultraviolet and infrared waves
- Mechanical energy: Energy produced by moving water, animals, people, and machines
- Nuclear energy: Energy produced by splitting atoms or uniting atomic matter

Sources of Energy

Energy is available from three major sources. These sources are grouped in terms of the supply. The first source of

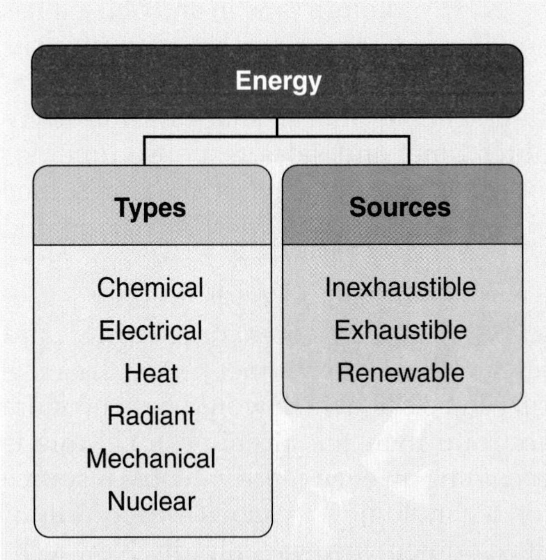

Figure 4-21. The two main factors in understanding energy are the type of energy and the source from which it comes.

energy is the sun. The sun is the fundamental source of energy for our solar system. Its actions warm the earth, cause the wind to blow, create weather, generate lightning, and, indirectly with our moon, create the ocean tides. This form of energy has always been with us and will continue to be so. It is said to be ***inexhaustible.***

A second source of energy comes from living matter. Human and animal muscle power can provide valuable sources of energy. Wood and other plant matter can be burned as a fuel. These sources are called ***renewable.*** They are used up but can be replaced with the normal life cycle of the energy source. For example, new trees can be grown, although this takes many years.

The third source of energy is termed ***exhaustible.*** These sources are limited by the quantity found on earth. Neither the sun nor human action can create additional supplies. Coal, petroleum, and natural gas are exhaustible energy sources.

The goal of wise energy utilization is to maximize the use of inexhaustible sources, recycle the renewable sources, and use the minimum of exhaustible energy.

Finances

As we have seen, technological systems require people; tools and machines; materials; and energy. These resources are generally purchased. For example, we pay people wages or salaries for their labor and knowledge. We buy or lease machines and purchase materials and energy. All these actions require money, which provides the financial foundation of the technological activity.

Money to develop and operate technological systems can be obtained in two ways, equity financing and debt financing, **Figure 4-22.** In equity financing, a person or people buy a technological system or company. If one person owns it, the operation is called a ***sole proprietorship***. If two or more people own it, they might form a ***partnership*** in which each person owns a portion of the company. In another circumstance, people might form a legal entity called a ***corporation*** to own the operation. They then sell shares, which are certificates of ownership in the corporation. In all three cases, the person or people raised the money by selling equity—a portion of the company. Therefore, raising money in this way is called ***equity financing.***

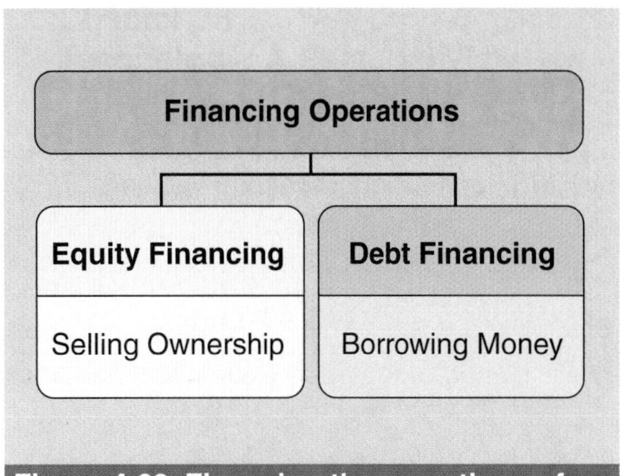

Figure 4-22. Financing the operations of technological systems and companies is achieved through equity and debt financing.

People also raise money by borrowing from other sources. Banks, insurance companies, or investment groups might loan money to support the activities of a company. This loan constitutes a debt that must be repaid. This type of financing is called ***debt financing.***

Time

Humans have always been aware of time, but our measurement, perception, and use of it have changed in a variety of ways, **Figure 4-23**. For example, early people measured time by way of the rising and setting of the sun. They knew they had only so much daylight in which to hunt and gather food, and they acted accordingly. Later farmers were very aware of particular months of the year because certain periods were more conducive to growing healthy crops.

Now, although time in and of itself has not changed, technology has accelerated its use and changed our standards of measurement. We allocate machine time, computer time, and sales response time, for example. At one point we measured time in years, months, and days. Then hours and seconds were observed. Now engineers worry about nanoseconds (billionths of a second) in computer processing, and researchers are experimenting with a mercury ion clock, which would be even more accurate than the atomic clock. Time is becoming an even more valuable resource for technological systems as the rate of development increases rapidly. New technologies build on old ones at a rapid rate, and the resulting products are made and disseminated quickly. For example, today most homes have a microwave oven, the result of research into streamlining the operations of a standing oven.

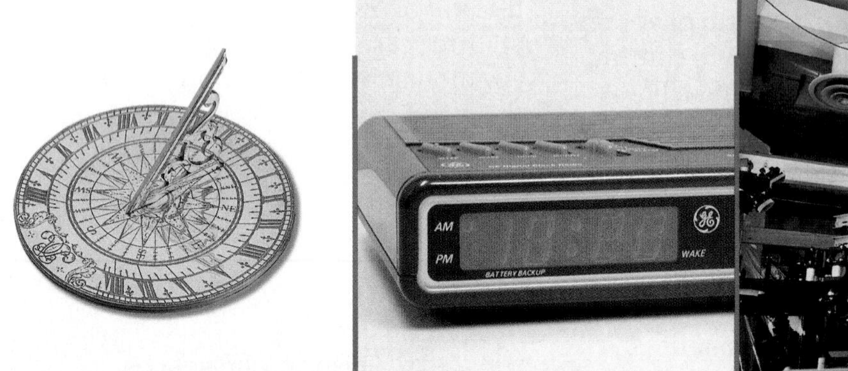
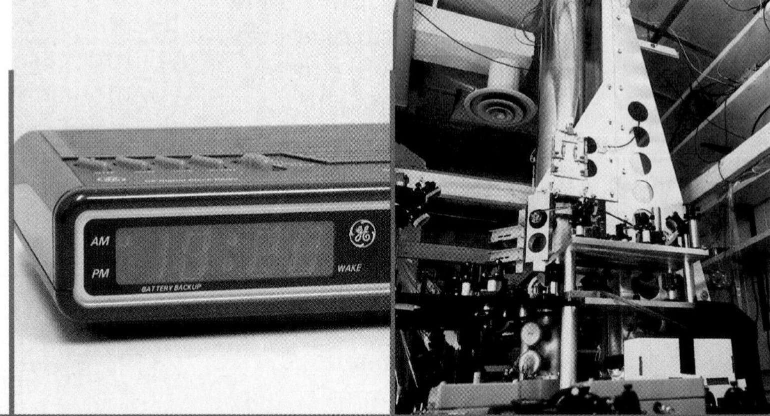

Figure 4-23. Time has been measured in different ways throughout human history. Early Egyptians used a sundial, which measured the time of day by the sun's shadow. We currently use various forms of electric and battery-operated clocks and watches. The new NIST-F1 is the eighth generation of atomic clocks at the National Institute of Standards and Technology. It keeps time with an uncertainty of about one second in twenty million years. (OutdoorDecor.com, National Institute of Standards and Technology Boulder Laboratories, U.S. Department of Commerce)

Summary

All technological systems involve inputs that are processed into outputs. People are the most important input, and they contribute to technological systems through their various individual knowledge and skills. They use tools and machines to increase their ability to do work. The various materials in the world comprise the forms that different technologies take, and our understanding of their types and properties is key to continuing advancement. Different kinds of information help create knowledge, which is needed to solve problems and implement solutions. The various forms of energy support the resources necessary to complete technical tasks. We use finances to purchase other inputs. Finally, time affects the amount and kinds of technology we produce.

The quality and quantity of these inputs often determine the type of technology a society can use. Many of these inputs are in limited supply. Therefore, everyone designing, building, or using technology should use inputs appropriately and recycle as many of them as possible.

Key Terms

composite materials
consumers
corporation
creativity
data
debt financing
distance multiplier
engineers
entrepreneurs
equity financing
exhaustible
exhaustible materials
first-class levers
force multiplier
fulcrum
gases
genetic materials
hand tools
inclined plane
inexhaustible
information
inorganic materials
knowledge
lever
lever arm
liquids
machines
managers
mechanics
natural materials
organic materials
partnership
production workers
pulleys
renewable
scientists
screw
second-class levers
sole proprietorship
solids
support staff
synthetic materials
technicians
third-class levers
tools
wedge
wheel and axle

Test Your Knowledge

Write your answers on a separate piece of paper. Please do not write in this book.

1. *True or false?* Energy is the most important input to technological systems.
2. Explain the statement, "Consumers are the reason for technological systems."
3. *True or false?* The drilling tool is used to abrade and smooth surfaces.
4. The three types of mechanisms that use the principle of the lever are the lever, the wheel and axle, and the _____.
5. Explain the difference between first-class, second-class, and third-class levers.
6. The three types of mechanisms that use the principle of the inclined plane are the inclined plane, the screw, and the _____.
7. Give an example of a genetic material.
8. In their physical state, materials can be classified as solids, gases, and _____.
9. *True or false?* The mechanical properties of a material deal with the material's reaction to a force or load.
10. Information can be classified into three areas: technological, the humanities, and _____.
11. *True or false?* Energy that comes from increased molecular action caused by heat is called thermal energy.
12. Give an example of an exhaustible source of energy.
13. The two ways to raise money to develop and operate technological systems are equity financing and _____.
14. *True or false?* Our perception of time has remained constant throughout history.
15. Explain why we need to use all resources carefully.

Applying Your Knowledge

1. Look around the room you are in. List the human abilities that were used to design, construct, and decorate it and the products within it. List three ways each ability was used.
2. Design and sketch a simple device that uses at least three of the six simple machines (mechanisms) to do a job.
3. Select a product that is made from exhaustible materials or a task that uses an exhaustible energy source. Describe how that product could be made or how the task could be completed using renewable materials or renewable energy sources.

Career Corner
Construction Laborers

Construction laborers perform a wide range of tasks involving building and maintaining constructed works. These works can include buildings, highways, tunnels, and dams. Construction laborers prepare construction sites, dig holes and trenches, mix concrete, unload buildings, and tend machines. They often help concrete workers, carpenters, plasterers, and masons.

Most laborers do physically demanding work that requires lifting and carrying heavy objects. They often work outdoors in all weather conditions. Construction laborers generally work eight-hour shifts, although longer shifts are common. Also, their work is often seasonal, working when the weather permits construction activity. Many construction laborer jobs do not require experience or training related to the occupation. Most laborers learn skills by observing and learning from experienced workers.

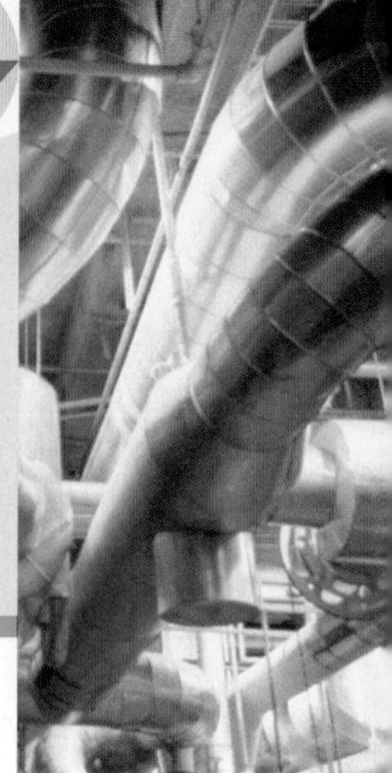

Chapter 5
Technological Processes

Learning Objectives

After studying this chapter, you will be able to:

- Describe the three major types of technological processes and their relationship to one another.
- Identify the steps used in the problem-solving/design process.
- Explain the major production steps used in the agricultural and related biotechnological, communication and information, construction, energy and power, manufacturing, medical, and transportation processes.
- Identify the steps involved in management processes.

You have learned that a technological system has inputs, processes, and outputs. Inputs are the resources that go into the system. ***Processes*** are what happen within a system. They take inputs and change them into outputs. Processes are used to design products, structures, energy conversion systems, communication messages, and transportation systems. They are also used to:

- *grow* and *process* crops
- *produce* products and structures
- *treat* medical conditions and illnesses
- *convert* and *apply* energy
- *communicate* information and ideas
- *transport* people and cargo

In addition, managerial processes see that the technological system runs efficiently and produces quality products and few unwanted outputs.

All of the above processes can be classified under three major headings, as shown in **Figure 5-1:** ***problem-solving/design processes, production processes,*** and ***management processes.*** Both the problem-solving/design process and the management process can be seen as generic in that the same steps within each process can be used no matter what the technology, **Figure 5-2.** The steps in the production process, however, vary according to the technological system under consideration. Each process is still related to the others, of course, in that actions that occur in one area directly affect the actions and reactions that occur in the other areas. We will now examine each process separately to understand them in more depth.

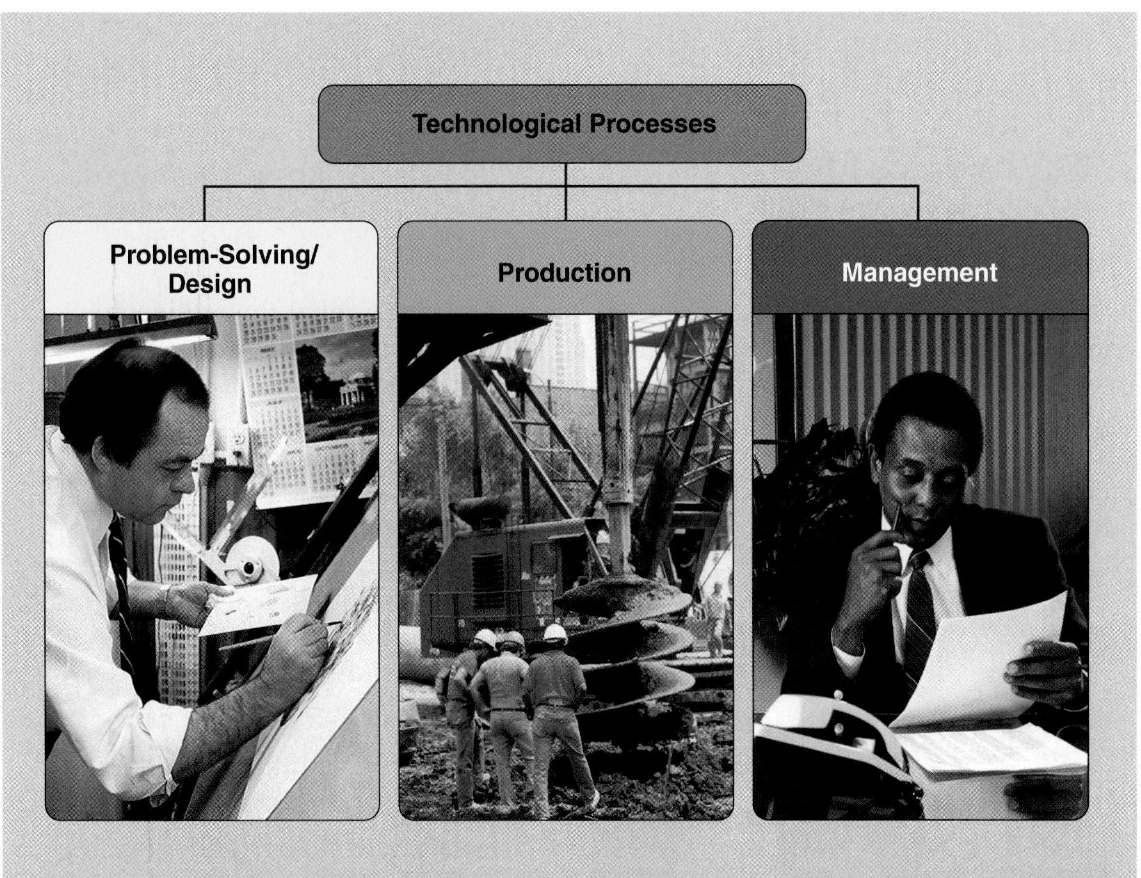

Figure 5-1. The problem-solving/design, production, and management processes are used to develop and operate technological systems. (Inland Steel Co.)

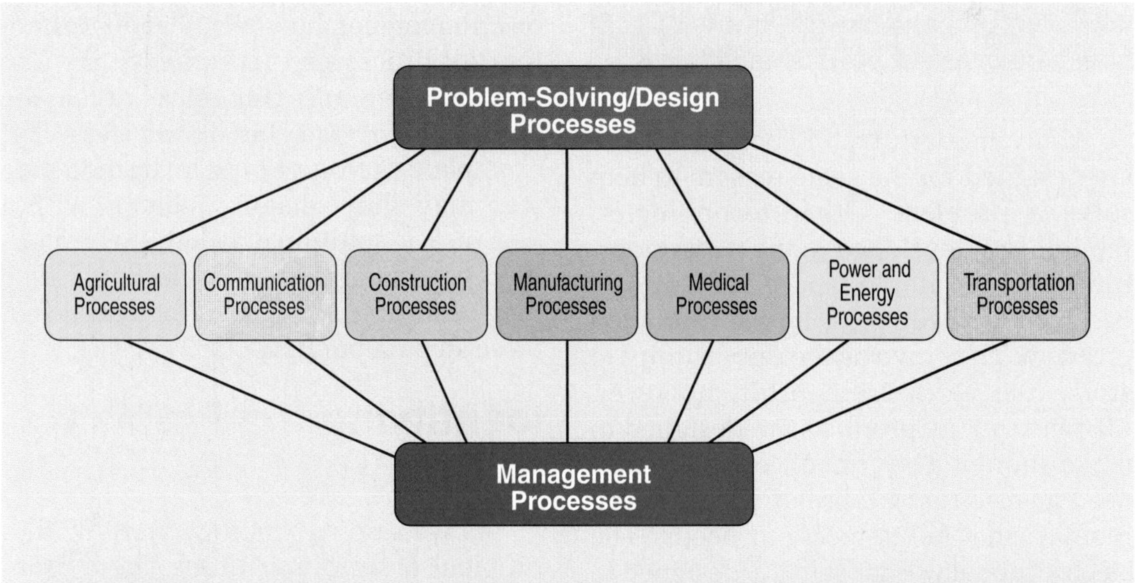

Figure 5-2. The same problem-solving/design processes and management processes are used for all seven technological areas.

Problem-Solving or Design Processes

All technology has been developed to meet human needs and wants. Each device or system is designed to solve a problem or to meet an opportunity, **Figure 5-3.** Early technology was almost always problem oriented. Tools were designed to make work easier. New housing technology made living more comfortable. New transportation devices helped humans move loads from place to place. New communication methods made information exchange easier.

Figure 5-3. Technology is designed to solve a problem or meet an opportunity. For example, new medical scanning devices have led to improved diagnoses.

Many modern technological devices are designed for the same reason: to help solve a problem. Other technologies, though, are developed to meet an opportunity. For example, spacecraft are not designed to solve a pressing transportation problem. They give us the opportunity to study the earth from orbit, however. Likewise, many products are designed to make money. How many people really need an electric toothbrush? Do standard toothbrushes fail to meet our needs? The electric model was most likely designed to give the inventors and manufacturers an opportunity to profit from their idea. There is nothing wrong with this motive. In fact, the drive to make money has brought us many things that we now take for granted. The change simply suggests that our civilization has progressed a great deal. We now have time to apply technology to luxuries. We are beyond the survival mode of our early ancestors.

No matter what the motive, however, designers and problem solvers follow a common procedure in designing and developing technology. The procedure is the same for solving problems or meeting opportunities and involves working within a systematic, ***iterative*** (repetitious), and creative process that leads to innovations and solutions. The five most important steps in the problem-solving/design process are shown in **Figure 5-4.** They are:

1. Identifying the problem or opportunity
2. Developing multiple solutions to the problem or opportunity
3. Isolating, refining, and detailing the best solution
4. Modeling and evaluating the selected solution
5. Communicating the final solution

Each of these steps leads to the next one. Remember, however, that all technological processes have feedback. The results from one step may cause the designer to retrace his or her steps. For example, a prototype built in the modeling step may show major problems with a design. The solution may have to be modified by changing it and selecting a new "best" solution. Now, let us look at each of these steps separately.

Identifying the Problem or Opportunity

As we said, technology starts with a problem or an opportunity. These problems and opportunities, however, are seldom clearly seen or felt. The first step

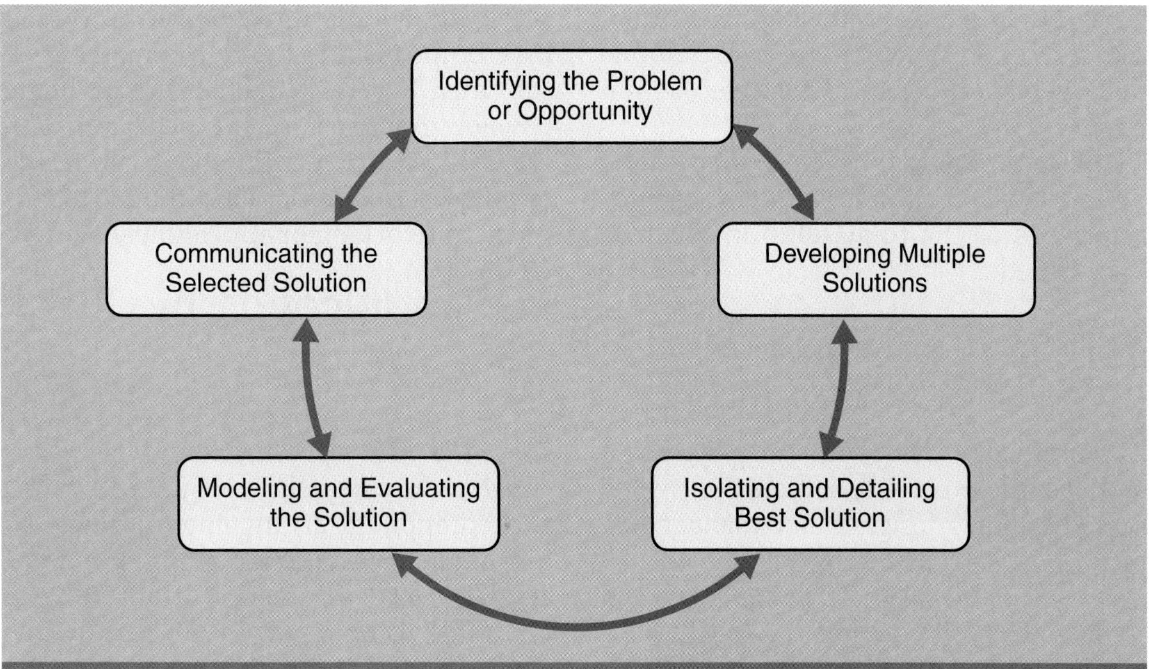

Figure 5-4. The problem-solving/design process has five steps, each of which affects the others.

in the design or problem-solving process is to describe the task that is being undertaken. The problem or opportunity must be identified.

This process involves defining the problem or opportunity. Included in the definition will be a list of limitations or requirements. These requirements include the ***criteria*** and ***constraints*** regarding such characteristics as appearance, operations, manufacturing, marketing, and finances. Criteria include a listing of the features of the technological product or system and focus on such aspects as the purpose of the technology and any problems that might result from various environmental, political, ethical, and social issues. Constraints deal with the limits on the design, such as size and cost. These criteria and constraints communicate the expectations for the solution.

Developing Multiple Solutions

Once the problem or opportunity is defined, a designer seeks solutions. Designers create many possible answers by letting their minds create solutions (a process called ***ideation***) and by brainstorming with others, usually in teams. They continually check and critique their answers, redefining their solutions and improving the designs. Their sketches show ways to meet the challenge. They first make rough sketches to capture thoughts. They then refine the sketches to mold thoughts into more specific solutions, **Figure 5-5.**

Isolating, Refining, and Detailing the Best Solution

Designers then isolate the most promising ideas and refine their work. They analyze ideas to select the most promising, realizing that criteria and constraints sometimes compete with one another and that trade-offs must occur. They know that this best solution may not meet all the criteria and constraints perfectly. They do understand, however, that the solution must be safe, efficient, and functional (work properly) and that it must be one that can be produced and marketed within cost limits.

Once the solution is chosen it must be detailed. Specifications and general characteristics of the final solution must be established.

Modeling and Evaluating the Selected Solution

Frequently, designers will produce a model of the expected solution, **Figure 5-6.** A model allows designers and managers to review the solution's performance. Models may be physical, graphic, or mathematical. Often, designers first use a computer to develop graphic models based on mathematics. Next, they refine these models to optimize the solution. Then they construct physical (working or appearance) models.

Figure 5-5. Designers use sketches as they explore possible solutions for design problems or opportunities.

Communicating the Final Solution

The final solution must be carefully specified for production. Designers develop drawings showing its size, shape, and component arrangement. They also formulate a material list or bill of materials and produce specifications for the materials to be used. These documents communicate the characteristics of the product, structure, media, or system. Designers also prepare written and oral reports to gain approval for the solution from decision makers.

The design/problem-solving process ends with a specified solution. Section 4 of this book contains a more detailed discussion of this process. For now we should note that the solution indicated by the design or problem-solving process then moves to the next step: the production process.

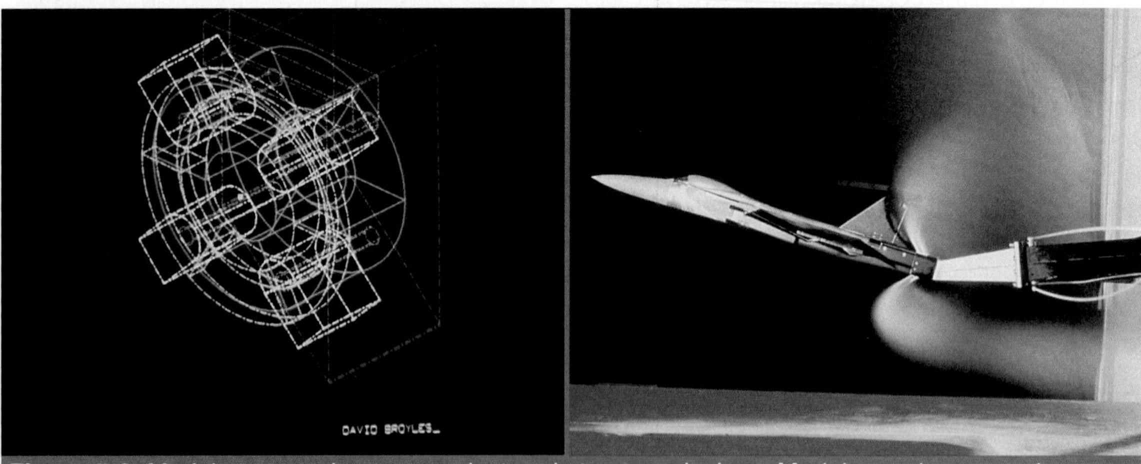

Figure 5-6. Models are used to test product and structure designs. Models can be created on computers or made physically, as is this airplane model in a wind tunnel (right). (DaimlerChrysler)

Production Processes

Production processes are actions that create the physical solution to the problem or opportunity. For example, they grow and harvest crops, construct structures, and generate communication messages. As noted earlier, the problem-solving/design process is used across system lines. This process generates the description of the solution. The same processes or sequence of events can be used to design agricultural and related biotechnological, manufacturing, construction, energy and power, medical, communication and information, and transportation systems.

Each of the technological systems has its own unique production processes, however. Manufacturing processes are different from medical processes. Agricultural and related biotechnological processes are different from construction processes. Communication and information processes are different from transportation processes. These processes move the solutions described by the design process into tangible solutions. Let us now examine each technological area with regard to its unique production processes.

Agricultural and Related Biotechnological Processes

Since our early history, people have raised crops and domesticated animals. This activity is called farming or agriculture. It is the art of cultivating the soil, growing crops, and raising livestock. Broadly speaking, this area includes farming, fishing, and forestry. Each of these areas involves a crop with a biological cycle that includes birth, growth, maturity, and death.

Farming is growing plants and animals for commercial use, **Figure 5-7.** Farming includes growing both crops and animals. Typical crops are fruits, vegetables, grains, and forage for animals. Individuals apply farming processes as they grow plants for landscapes and home gardens.

Fishing is harvesting fish from lakes and oceans for commercial use. Fish may also be raised and harvested in controlled areas called fish farms. Individuals also use fishing processes for recreational purposes.

Forestry is growing trees for commercial use. The trees may be used for lumber and veneer (thin sheets of wood), paper, or other products.

Agricultural practices involve the following major steps, **Figure 5-8.**

1. Propagation
2. Growth
3. Harvesting
4. Conversion/processing

Agriculture produces food and fiber for people to use, and it starts with the birth of a crop. Types of births include planting a seed, rooting a cutting from a plant, or allowing animals to breed. This step is called ***propagation.*** It allows a biological organism to reproduce.

After the new animal or plant appears, the next step is ***growth.*** Agriculture provides

Figure 5-7. These trees are a crop that will be sold as Christmas trees.

Figure 5-8. Agricultural practices that use technology involve four major steps.

a proper environment for this to happen. Growth could involve providing feed and water for animals or cultivating and watering (irrigating) crops.

When the animal or plant has reached maturity, ***harvesting*** occurs. Harvesting involves such actions as removing edible parts of plants from trees and stocks and butchering animals to produce meat and other products for consumption.

Finally, the food product undergoes ***conversion*** or ***processing*** to create a foodstuff. For example, wheat is ground into flour. Meat is cut into steaks and roasts. Trees are cut into lumber or chipped for papermaking.

We have enhanced agricultural practices by the application of biotechnological processes. For example, through combining biological processes with physical technology, we have genetically altered bacteria. These bacteria are used to protect crops from insects. More complete discussions of agricultural and related biotechnological processes are contained in Chapters 29 and 30.

Communication and Information Processes

Humans have always exchanged information and ideas. This process is called communication. The simplest communication processes involve spoken language. Note, however, that spoken language is not technology because no technical means are used in the process. Technology became involved as civilization grew and people developed additional techniques to help them communicate better. For example, we now use technology to produce printed, graphic, and photographic media to impart information and express ideas. We group these techniques under the general heading called graphic communications. We have also developed telecommunications technology, which allows us to communicate using electromagnetic waves. Each process is described in more detail below.

Graphic Communications

In ***graphic communications*** processes, messages are visual and have two dimensions. This category includes the printed messages commonly found in books, magazines, owners' and service manuals, and promotional flyers. Graphic communications media also include technical graphic messages such as technical illustrations and engineering drawings. The final graphic communications medium is photographic communication. This group includes the film and print media coming from photographic processes.

Telecommunications

Telecommunications processes depend on electromagnetic waves to carry their

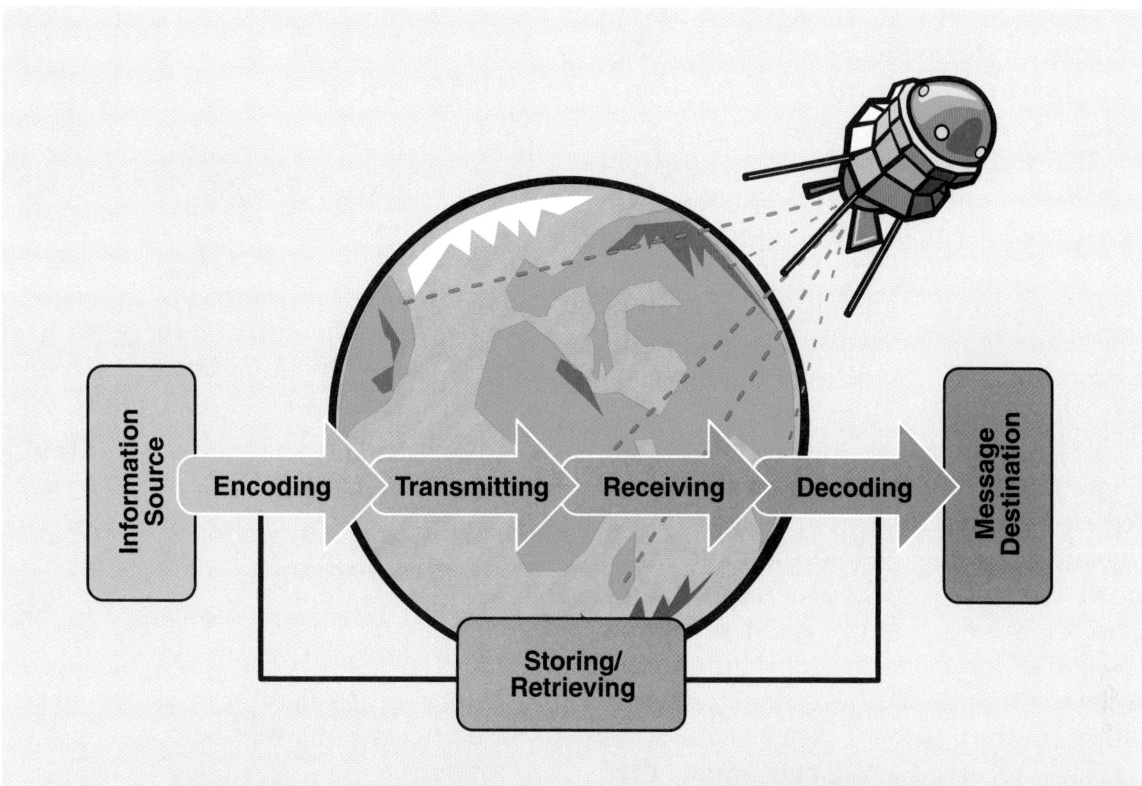

Figure 5-9. All communication technologies involve five major steps.

messages. Telecommunications techniques include broadcast (television and radio), hard-wired (telephone and telegraph), and surveillance (radar and sonar) systems.

All communication technologies, which will be discussed in depth in Section 6 of this book, involve five major steps, **Figure 5-9:**

- Encoding
- Transmitting
- Receiving
- Decoding
- Storing/retrieving

Communication technology organizes information so that people can receive it, **Figure 5-10.** The communication process starts with an information source. Most often this source is the human brain, although in some cases machines are used. The first step in the communication process is ***encoding*** the information. Encoding involves placing the information in a format or pattern that the communication

Figure 5-10. Communication technology allows us to send, receive, and store data and information.

technology can use. This process might involve producing images on film, pulses on a light wave, electrical charges on a tape, or graphics on paper.

Transmitting the coded message from the sender to the receiver is the next step. Transmission can include moving printed materials, broadcasting radio and television programs, transmitting telephone messages along wires or fiber-optic strands, and using pulses of light to send messages between ships at sea.

Transmitting a message is not enough. Someone or something must receive it. ***Receiving*** the message requires recognizing and accepting the information. A radio must recognize the coded radio waves. The telephone receiver must recognize the pulses of electricity. The human eye must see the semaphore flags (ship-to-ship communication).

The received message must then undergo ***decoding,*** which means that the coded information must be changed back into a recognizable form. The decoded information is presented in an audio or a visual format that humans can understand. The radio receiver changes radio waves into sound waves. The human brain places meaning relative to the position of the semaphore flags. The telephone changes electrical impulses to sound.

Throughout the communication process information undergoes ***storing*** and ***retrieving.*** Storage processes allow the information to be retained for later use. Books can be shelved in libraries, recorded music can be stored on tapes or compact discs, television programs can be stored on tapes, data can be stored on computer disks or tapes, and pictures can be stored in file folders. Later this information may be retrieved (brought back). It can be selected and delivered back into the communication process. Storage and retrieval can happen at any time in the communication system. Information is commonly stored at the source or at the destination.

Construction Processes

Humans first used construction technology to produce shelter. This effort allowed early humans to move out of caves and into crude homes. This humble beginning took the form of huts and tents. We now construct many types of buildings. Note, however, that construction technology does not stop with buildings. We also construct many types of structures to support other activities. These include:

- Roads, canals, and runways for transportation systems
- Factories and warehouses for manufacturing activities
- Studios, transmitter towers, and telephone lines for communication
- Dams and power lines as a part of our electricity generation and distribution systems

Constructed works are everywhere.

Constructed works can be grouped into two major categories: buildings and civil or

Figure 5-11. Construction technology is used to build buildings (top) and civil structures (bottom).

heavy engineering structures, **Figure 5-11.** ***Buildings*** are all the structures that are erected to protect people and machines from the outside environment. Buildings can be used for three major purposes. First, they are used as ***residential structures,*** that is, the places where people live. These structures can be homes, town houses, condominiums, and apartment buildings. Buildings are also used as ***commercial structures.*** They are the stores and offices that are used to conduct business. Government buildings such as schools, city halls, and state capitols are also placed in this category. Finally, buildings can be ***industrial structures.*** They are the power plants, factories, transportation terminals, and communication studios used by major companies.

The second major type of construction produces ***civil engineering structures*** or ***heavy engineering structures.*** These structures are primarily designed with the knowledge of the civil engineer. Common civil structures are roads, dams, communication towers, railroad tracks, pipelines, airport runways, irrigation systems, canals, aqueducts, and electricity transmission lines.

Most construction projects include several steps, which will be presented in more depth in Chapters 17 and 18. These, as shown in **Figure 5-12,** include:

1. Preparing the site
2. Setting the foundation
3. Building the superstructure
4. Installing the utilities
5. Enclosing the superstructure
6. Finishing the project
7. Completing the structure and site
8. Servicing the structure

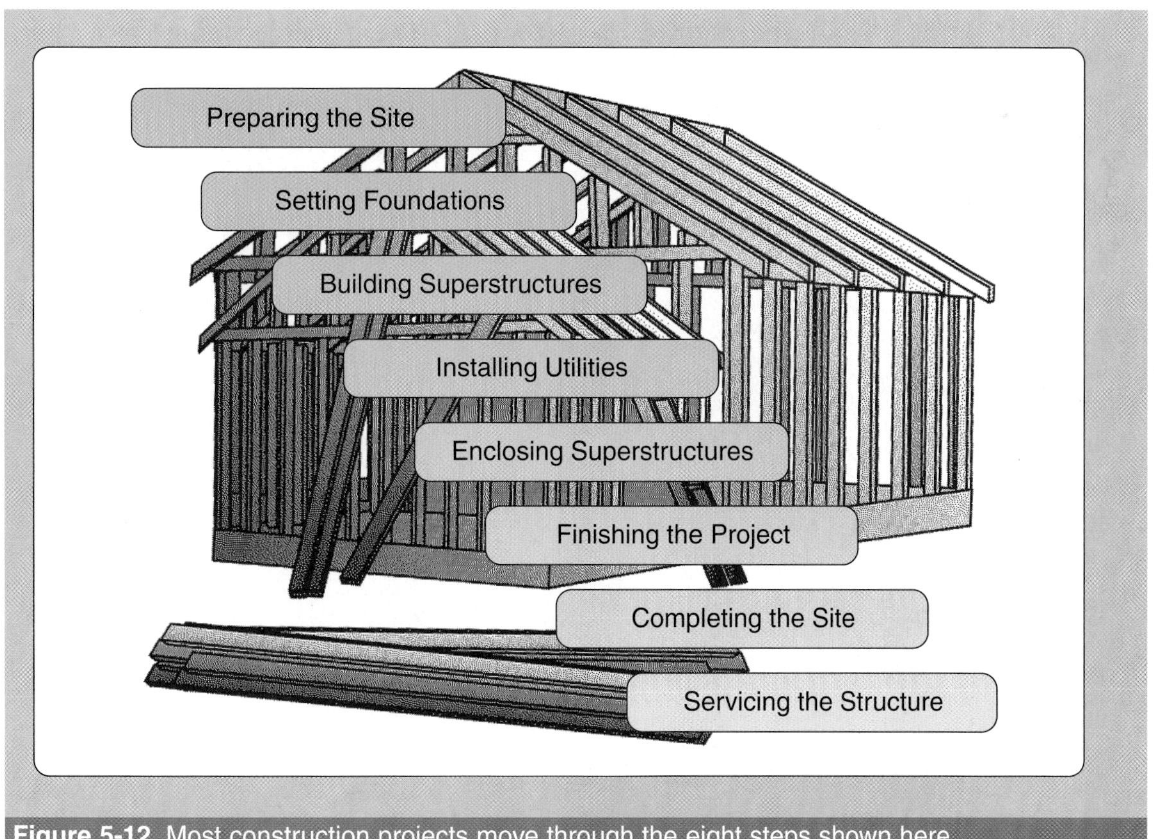

Figure 5-12. Most construction projects move through the eight steps shown here.

Technology Explained

Hybrid Vehicles

hybrid vehicle: A vehicle combining two or more sources of power.

With the cost of gasoline at high levels, people are looking for more efficient automobiles. One alternative to the standard car is the hybrid vehicle. What is a hybrid car? It is any vehicle that combines two or more sources of power.

A hybrid vehicle is not a new concept. A commonplace hybrid vehicle in American life is the diesel-electric locomotive, **Figure A.** It combines an internal combustion engine and electric generators and motors. Many submarines are also hybrids. They combine nuclear power with electric generators and motors. Some city buses are hybrids. They use diesel engines for part of their routes and overhead electric lines to power motors for other parts.

A recent development is the hybrid car. It combines a gasoline-powered car with an electric car. In a standard car, an engine uses gasoline as fuel. The gasoline turns a transmission, which drives the wheels. In an electric car, batteries are the power source (fuel). The electricity powers a motor that drives the wheels through a transmission.

Figure A. A common hybrid vehicle is the diesel-electric locomotive.

A hybrid vehicle uses both types of power, **Figure B.** It uses a gasoline engine smaller than one in a traditional car. The engine may be as small as 1.0 liter and have as few as three cylinders. It uses advanced technologies to increase its efficiency and reduce its emissions. The engine can provide power to the wheels or to the motor or generator to recharge the batteries.

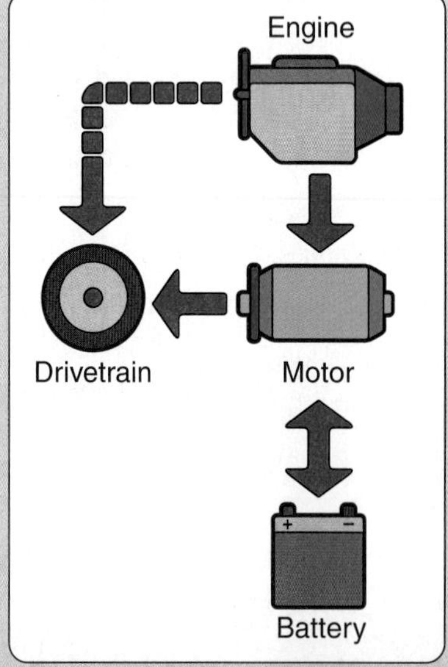

Figure B. This is a diagram showing the major parts of a hybrid vehicle power system.

The vehicle also has an advanced electric motor. This power source has special electronic controls allowing it to be both a motor and a generator. At times, it draws energy from the batteries to accelerate the car. At other times, it functions as a generator. It recharges the batteries by regenerative braking. During this phase, the energy from forward momentum is captured during braking. Therefore, no external power supply is needed for recharging the batteries.

The operation of the vehicle changes under different driving conditions. During initial acceleration, the electric motor is the primary source of power. The gas engine will start up under heavy acceleration or to turn the generator, which in turn charges the battery.

During city driving, the electronic control system controls both power sources. The gas engine and electric motor are used equally. The engine starts and stops, depending on the situation. At high speeds, the gas engine is the primary source of power. The electric motor will provide some power.

Preparing the Site

Not all construction projects use all the steps just listed. Most projects, however, start with ***site preparation.*** This task includes removing existing buildings, structures, brush, and trees that will interfere with locating the new structure. Workers then roughly grade the site and establish the desired slope. Then surveyors locate the exact spot for the new structure.

Setting the Foundation

The base or ***foundation*** for the structure is then constructed. This step generally involves digging pits or trenches. Then concrete or rock is placed in the holes. If the hole extends to solid rock, additional foundations might not be needed. Foundations provide a stable surface onto which the building can be constructed.

Building the Superstructure

The ***superstructure*** of the project is constructed on the foundation and includes the framework of the building or tower. Superstructures also include the pipes for pipelines, surfaces for roads and airport runways, and tracks for railroads.

Installing the Utilities

In many cases the constructed structure will include ***utilities,*** which means the installation of such items as water, gas, and waste pipes; electrical and communication wire; and heating and cooling ducts. Many runways, railroads, and highways also need lighting and communication systems installed.

Enclosing the Superstructure

After the utilities are installed, the superstructure must be enclosed. Walls will need interior and exterior skins. Roofs, ceilings, and floors must be covered. Doors and windows must be installed to close openings.

Completing the Structure and Site

Finally, the structure must be completed. Walls, ceilings, doors, windows, and building trim must be painted. Likewise, communication towers must be coated to protect them from the natural elements. Runways and roadways must have traffic stripes painted on them and signs installed.

Also, the site must be completed. It must be graded to reduce erosion and increase its beauty. Shrubs, grass, and trees are added to landscape the site and keep the soil in place.

Servicing the Structure

The result of this sequence of steps is a new constructed structure. During its life it will require ***servicing,*** which means it will undergo maintenance, repair, and reconditioning. For example, various people will periodically paint the surfaces, repair the roofs, and replace utility systems components. Servicing will attempt to keep the structure in good working order.

Energy and Power Processes

We noted earlier that energy is the ability to do work and that all technological systems require energy. Power, which is sometimes confused with energy, is actually the rate at which the work is accomplished or the energy is changed from one form to another. Power is calculated by dividing the work done by the time taken to do it.

We will explore energy and power processes in more detail in Chapter 27. For now, we should note that six basic forms of energy exist and that technology aids us in transforming and applying energy. Each is discussed in turn.

The six basic forms of energy are as follows:

- Mechanical energy: Energy found in moving objects
- Chemical energy: Energy contained in molecules of a substance
- Radiant (light) energy: Energy in the form of electromagnetic waves
- Thermal (heat) energy: Energy associated with the movement of atoms and molecules
- Electrical energy: Energy associated with moving electrons
- Nuclear energy: Energy released when atoms are split or united

The process of transforming and applying energy involves gathering or collecting the resources, converting the energy into a new form, transmitting the energy, and applying the energy to do work, **Figure 5-13.**

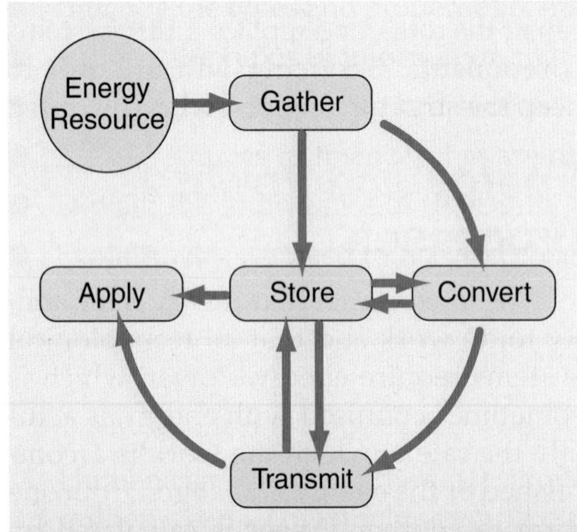

Figure 5-13. The process of transforming energy involves four major steps: gathering the resources, converting the energy, transmitting the energy, and applying the energy.

Gathering Resources

Energy resources include such items as petroleum, coal, wind, sunlight, and moving water. The type of gathering techniques used will depend on the resource to be used. We obtain coal through mining, whereas we obtain natural gas and petroleum through drilling. We capture sunlight with collectors. We collect water through the use of dams, and we capture wind with turbines.

Converting Energy

Energy resources are converted from one form to another through a variety of means. Internal combustion engines convert fuels (gasoline, diesel fuel, and natural gas, for example) into mechanical energy. Hydroelectric projects use turbines and generators to change the mechanical energy in flowing water into electricity. Home furnaces change the chemical energy in fuels such as coal, fuel oil, or natural gas into heat (radiant energy). Hundreds of energy converters exist that are designed to perform specific tasks.

Transmitting Energy

Energy must often be moved from one place to another. For example, the energy from an engine must be transmitted to the wheels that drive a car. Heat energy must be moved from a furnace to the appropriate rooms in a house. Electrical energy must be moved from generating stations to homes, businesses, and factories.

Applying Energy

Energy, in itself, is of little use to people until it is applied to a task. The energy must power a device or power a process. For example, energy is useful when it causes a vehicle to move, a machine to perform a task, or a device to light a space. See **Figure 5-14.**

Note that not all energy is used immediately after it is gathered. We often want to store it for later use. The reservoir behind a dam stores the energy in moving water. A battery stores the chemical energy of its parts. Tank farms store petroleum for later energy applications.

Manufacturing Processes

You use manufactured products for everything you do. You ride to school in a manufactured car or bus. You look out of the building through a manufactured window. You wear manufactured clothing. You are reading a manufactured book and sitting on a manufactured chair in a room lighted with manufactured fixtures. It is difficult to imagine a world without manufactured products. Note, however, that no matter what the product, all manufacturing activities involve three stages, as shown in **Figure 5-15:** obtaining resources, producing industrial materials through primary processing, and creating finished products through secondary processing.

Obtaining Resources

Manufacturing activities start when we obtain material resources. This effort first involves searching for or growing materials that can be harvested or extracted from the earth. For example, farmers and foresters grow the trees, plants, and animals needed to support the manufacturing processes. Exploration companies locate petroleum, coal, and metal ore reserves.

Figure 5-14. Energy is used to do work. Here, light enables a physician to examine X-ray film.

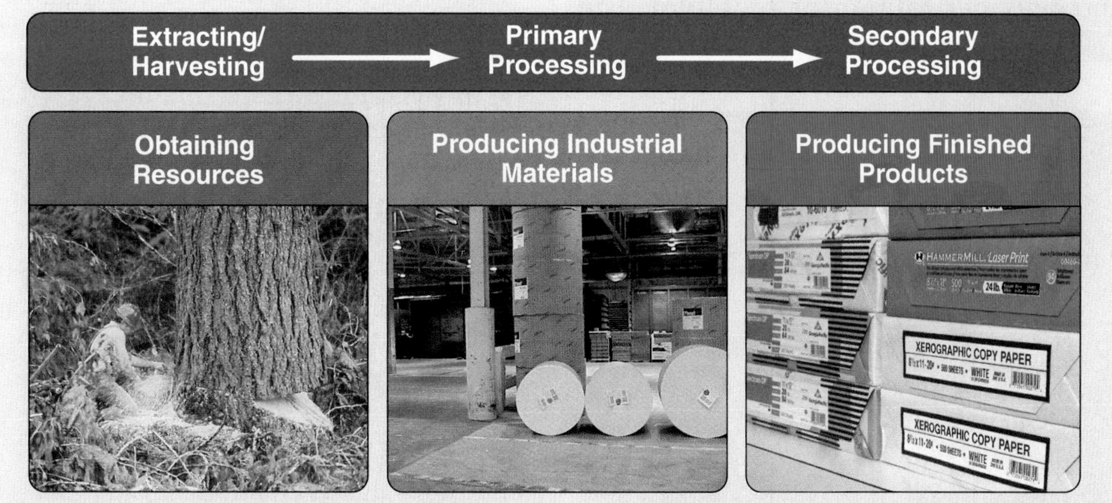

Figure 5-15. Manufacturing obtains material resources, which are changed into industrial materials. These materials are used to make products. (Weyerhauser Co., MeadWestvaco)

Once we grow or find the resources, we must gather or obtain them. These activities use one of three major processes:

- ***Harvesting:*** Gathering genetic materials (living materials) from the earth or bodies of water at the proper stage of their life cycle
- ***Mining:*** Obtaining materials from the earth through shafts or pits
- ***Drilling:*** Obtaining materials from the earth by pumping them through holes drilled into the earth

Producing Industrial Materials through Primary Processing

Material resources are changed to industrial materials in ***primary processing*** plants. For example, steel mills change iron ore, limestone, coke, and other materials into steel sheets and bars. Forest products plants change trees into paper, lumber, plywood, particleboard, and hardboard. Smelters change aluminum and copper ore into usable metal sheets, bars, and rods.

Typically these materials are refined using heat (smelting and melting, for example), mechanical action (crushing, screening, sawing, and slicing, for example), and chemical processes (oxidation, reduction, and polymerization, for example). The result is an industrial material or standard stock. Examples include a sheet of plywood, glass or steel, a plastic pellet, and metal rods and bars.

Creating Finished Products through Secondary Processing

Most industrial materials have little value to the consumer. What would you do with a sheet of plywood, a bar of copper, or a bag of plastic pellets? These materials become valuable when they are made into products. They are the material inputs to the ***secondary processes*** of manufacturing. These processes change industrial materials into industrial equipment and consumer products, **Figure 5-16.**

Secondary processing changes materials in six basic ways. Casting and molding processes, forming processes, and separating processes give materials size and shape. Materials also undergo conditioning, finishing, and assembling.

- ***Casting and molding*** introduce a liquid material into a mold cavity. There the material solidifies into the proper size and shape.
- ***Forming*** uses force applied from a die or roll for reshaping materials.
- ***Separating*** uses tools to shear or machine away unwanted material. This

Figure 5-16. Secondary manufacturing produces industrial equipment and consumer products. (Minster Machine Co., Northern Telecom)

activity shapes and sizes parts and products.

- *Conditioning* uses heat, chemicals, or mechanical forces to change the internal structure of the material. The result is a material with new, desirable properties.
- *Finishing* coats or modifies the surface of parts or products to protect them, make them more appealing to the consumer, or both.
- *Assembling* brings materials and parts together to make a finished product. They are bonded or fastened together to make a functional device.

The results of obtaining resources, producing industrial materials, and manufacturing products are the devices we use daily. More information about these manufacturing activities is in Chapters 14, 15, and 16.

Medical Processes

People in the industrialized nations are living longer, which is partly the result of better medical care. Technological and scientific advances have given health care professionals new tools to treat diseases and the aging process. These tools are used in three major areas, as shown in **Figure 5-17:**

- *Prevention:* Using knowledge, technological devices, and other means to help people maintain healthy bodies. Prevention focuses on such health aspects as proper nutrition, exercise, immunizations against diseases, and proper actions at work and recreation.
- *Diagnosis:* Using knowledge, technological devices, and other means to determine the causes of abnormal body conditions. This is a type of problem solving that matches symptoms to possible causes. For example, a person who experiences back pains might have vertebrae damage, muscle strains, or cancer of the spine. By using various tests, health care professionals can determine the actual cause of the pain.
- *Treatment:* Using knowledge, technological devices, and other means to fight diseases, correct body malfunctions, or reduce the impact of a body condition. Treatment includes such things as surgery, chemicals (drugs), physical therapy, or other medical actions.

You will have the opportunity to learn more about medical technology in Chapter 31.

Transportation Processes

Humans have always had the desire to move around. They frequently move themselves and their possessions from one place to another. This movement is called transportation. One way to understand transportation is to look at the media in which

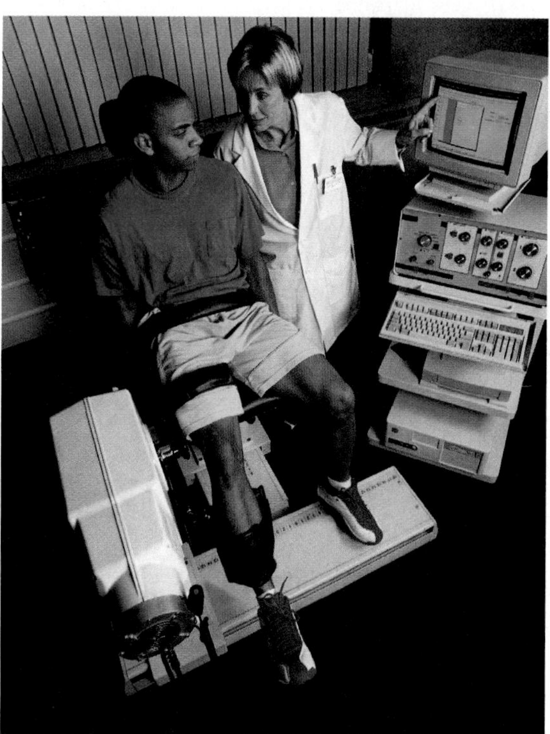

Figure 5-17. Medical technologies help people stay well through disease prevention, diagnosis, and treatment. (Corbis)

the various transportation systems have been developed. The three basic media are land, water, and air; and a fourth one, space, is on the horizon, **Figure 5-18.** Another way to understand transportation is to look at its subsystems, which are vehicular systems and support systems. Both the media and the subsystems are discussed as follows.

Transportation Media

The earliest movement occurred on land. People first walked the land. Then they tamed animals to help them move from one place to another. Finally they developed various land vehicles, which range from the horse-drawn wagons of old to the magnetic levitation trains now in development.

The second transportation medium is water. Again, water transportation has a

Figure 5-18. Transportation can be conducted on land, over or under water, and through the air. Later, we will probably use space transportation systems more. (Norfolk Southern Corp., OMI Corp., and NASA)

Connections to Technology: **History**

The Tennessee Valley Authority

The American experience with the Great Depression of the 1930s was of course not a pleasant one. The United States, along with most of the other countries in the world, went through a period of high unemployment, declining sales, falling prices, and rising debt.

One result of these conditions was a new willingness on the part of the American people to allow the federal government to become involved in improving social conditions. With this being the situation, the Roosevelt administration saw a way to use technology to help some of the poorest areas of the country.

In 1933 Congress authorized the administration to create a federal program called the Tennessee Valley Authority (TVA). Under this program, the federal government became involved in the production of electric power. The TVA covered the seven states of Tennessee, Kentucky, Alabama, Mississippi, Georgia, Virginia, and North Carolina. Some of the residents of these states did not even have electricity in their homes yet. The TVA built and operated hydroelectric plants and constructed dams to improve transportation and control floods. Under the TVA, the residents of this region were able to purchase cheap power. Thus although the depression created distressing conditions, some people's lives actually improved in some ways because of the application of technology.

The TVA is still in effect today. In what areas is it now involved?

long history. Means of transportation in this area have ranged from dug-out log canoes to nuclear submarines.

The third major transportation medium is air. This medium is of recent vintage. It can be traced back to early balloon travel and now extends to supersonic airplanes.

We are starting to develop a fourth transportation medium, space. Today we explore only the near reaches of space, **Figure 5-19.** The future holds the promise of space transportation systems, however.

Figure 5-19. Space travel is presently used for experimental work and study. In this rendering, astronauts test a construction system. (NASA)

Figure 5-20. Most vehicles are made up of five systems.

Transportation Subsystems

Most transportation systems contain two major subsystems: vehicular systems and support systems. ***Vehicular systems*** are the onboard technical systems that make a vehicle work. Most vehicles are made up of five systems, as shown in **Figure 5-20:**

- ***Structure:*** Provides spaces for devices. The structure, or framework, includes passenger, cargo, and power system compartments.
- ***Propulsion:*** Generates motion through energy conversion and transmission.
- ***Guidance:*** Gathers and displays information so that the vehicle can be kept on course.
- ***Control:*** Makes changes in speed and direction of the vehicle possible.
- ***Suspension:*** Keeps the vehicle held in or onto the medium being used (land, water, air, or space).

Support systems include pathways and terminals, **Figure 5-21:**

- ***Pathways:*** The structures along which the vehicles travel. Pathways include roads, railways, waterways, and flight paths.
- ***Terminals:*** The structures that house passenger and cargo storage and loading facilities.

A complete discussion of transportation technology is included in Section 7 of this book.

Management Processes

The third major type of process used in technology is management. ***Management processes*** are those designed to guide and direct the other processes. Management provides the vision for the activity. Specifically, management includes four steps, **Figure 5-22.** The first step is ***planning,*** which involves developing goals and objectives. The goals can be broad or specific and focus on such areas as production, finance, and marketing.

Once plans are developed, the activity must be structured. Procedures to reach the goal must be established. Lines and levels of authority within the group or enterprise must also be drawn. These actions are called ***organizing.***

After the activity is organized, actual work must be started, which is called ***actuating.*** Workers must be assigned to tasks. They must be motivated to complete the tasks accurately and efficiently.

Finally, the outputs must be checked against the plan, which is termed ***controlling.*** Control is the feedback loop that causes management activities to be adjusted.

Sometimes we think of management only in terms of companies, when in reality every human activity is managed. Goals are set. A procedure for finishing the task is established. Work is started and completed. Results are evaluated. For example, a coach plans, directs, and evaluates the plays a football team makes. A Girl Scout troop leader manages cookie sales. You develop a plan to mow a lawn.

Note that we manage not only personal but also group activities. Some group activities take place within social, religious, and educational organizations. Others occur in technological enterprises. The management of technological activities and companies will be explored in more detail in Section 9 of this book.

Figure 5-21. Transportation support systems include pathways such as roads (top) and railway terminals (bottom) (Amtrak)

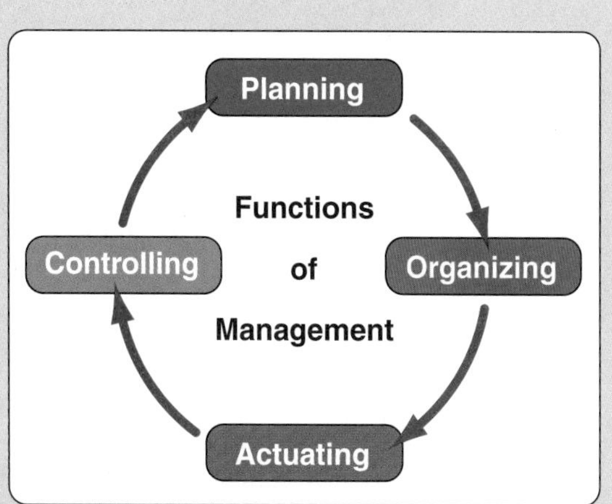

Figure 5-22. Management involves setting goals (planning), establishing a structure to meet them (organizing), assigning and supervising work (actuating), and monitoring the results (controlling).

Summary

All technological systems have processes. They include processes to design and engineer the outputs and the systems themselves. Often these processes use problem-solving methods. These methods guide the creative activities of the field.

Once we design the system and outputs, we use production processes. These processes transform inputs into outputs. Materials, human abilities, machines and tools, information, energy, finances, and time are used to do such things as grow crops, and construct structures.

Finally, all technological systems are managed. They are planned, organized, actuated, and controlled.

Key Terms

actuating
assembling
buildings
casting and molding
civil engineering structures
commercial structures
conditioning
constraints
control
controlling
conversion/processing
criteria
decoding
diagnosis
drilling
encoding
finishing
forming
foundation
graphic communications
growth
guidance
harvesting
heavy engineering structures
ideation
industrial structure
iterative
management processes
mining
organizing
pathways
planning
prevention
primary processing
problem-solving/design processes
processes
production processes
propagation
propulsion
receiving
residential structure
retrieving
secondary processes
separating
servicing
site preparation
storing
structure
superstructure
support systems
suspension
telecommunications
terminals
transmitting
treatment
utilities
vehicular systems

Test Your Knowledge

Write your answers on a separate piece of paper. Please do not write in this book.

1. The three major types of technological processes are problem solving/design, production, and _____.
2. *True or false?* The three major technological processes can be applied generically across technological systems.
3. Give three examples of criteria that might need to be considered when identifying a problem in the problem-solving/design process.
4. *True or false?* The best solution to a problem resolves all the criteria and constraint requirements.
5. The agricultural production process includes propagation, growth, harvest, and _____.
6. *True or false?* The first step in the communication process is encoding the information.
7. Give one example of a heavy engineering structure.
8. The part of a constructed work that supports the superstructure is called a(n) _____.
9. _____ energy is energy found in moving objects.
   a. Thermal
   b. Chemical
   c. Mechanical
   d. Electrical
10. List the three major processes used to gather or obtain manufacturing materials.
11. *True or false?* Secondary processing changes industrial materials into equipment and products.
12. The three areas in which health care professionals use technological tools are prevention, diagnosis, and _____.
13. Transportation media include water, air, space, and _____.
14. List the five systems common to most vehicles.
15. List the four steps in the management process.

Applying Your Knowledge

1. Select a product or structure about which you are familiar. Develop a chart similar to the one below and list five considerations that were used in each of the technological processes used to produce the item.

Product or structure name:	
Process	**Factors considered**
Design/problem solving	
Production	
Management	

2. Select a problem that exists in the world that can be partially solved through technology. List the role each technology would play in solving the challenge.
3. List and describe how you have used the production processes in one of the technological areas (agriculture and related biotechnology, communication and information, construction, energy and power, manufacturing, medicine, or transportation). Use a chart similar to the one below.

Product or structure name:	
Technology	**Contributions to solving the problem**
Agriculture and related biotechnology	
Communication and information	
Construction	
Energy and power	
Manufacturing	
Medicine	
Transportation	

Career Corner

Production Managers

Production managers coordinate the people, materials, equipment, and processes needed to manufacture industrial and consumer products. Their major responsibilities include scheduling production activities, requisitioning and maintaining equipment, staffing operations, and maintaining quality and inventory control. Production managers work closely with sales and purchasing managers.

Most industrial production managers work more than 40 hours a week. In plants that have more than one shift, managers often work late shifts. Many companies seek candidates with college degrees in industrial management or business administration. Companies also want managers who have good interpersonal and communication skills.

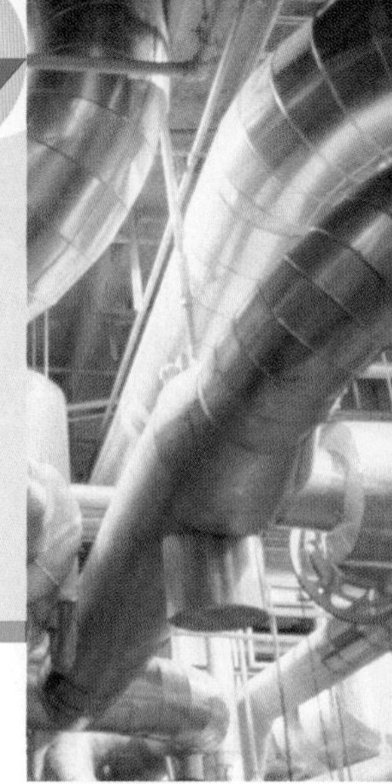

Chapter 6
Outputs and Feedback and Control

Learning Objectives

After studying this chapter, you will be able to:

- Describe the relationship between outputs and feedback and control in technology systems.
- Identify the major types of outputs of various technology systems.
- Differentiate between desired and undesired, intended and unintended, and immediate and delayed outputs of technological systems.
- Define feedback as used in technology systems.
- Differentiate between the two major types of internal control systems.
- Describe the major components of internal control systems.
- Describe the difference between manual and automatic control in internal control systems.
- Identify external controls in technology systems.

All technological systems have purposes. They are designed to meet specific needs and wants of people. Therefore, a direct relationship exists between the need (the reason for the system) and the output (the satisfying of the need). These outputs can be categorized in three ways, as shown in **Figure 6-1:**

- Desirable and undesirable
- Intended and unintended
- Immediate and delayed

The people designing and using technology should strive to maximize the desirable results. The undesirable outputs can never be totally eliminated, but they need to be held to a minimum. To achieve this goal, people have created control systems that compare the results of the outputs with the goals. As we saw in Chapter 2, they use the feedback to regulate the inputs. We will now examine in more depth both the outputs and the feedback and control systems essential to understanding and improving technological systems.

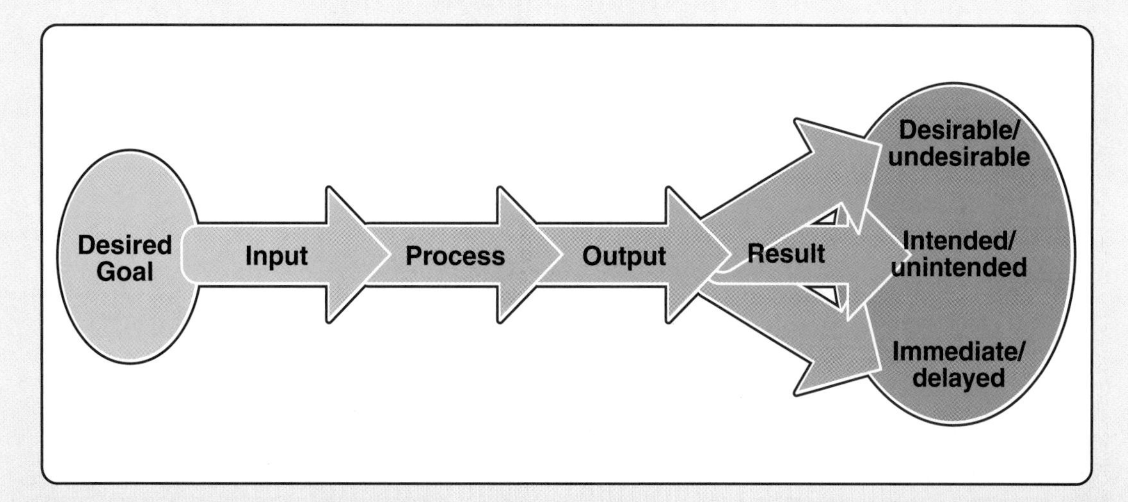

Figure 6-1. The outputs of a technological system can be categorized in three ways.

Outputs: Desirable and Undesirable

The needs for and outputs of technology are countless. Each of the technological systems we have already classified (agriculture and related biotechnology, communication and information, construction, energy and power, manufacturing, medicine, and transportation) has a general type of output, **Figure 6-2.** For example, agricultural and related biotechnology helps people satisfy their needs for nourishment. Its outputs are plants and animals that can be processed into food. Communication technology helps satisfy the need for information. Its outputs are media messages. Construction technology satisfies the need for shelter and support structures for other activities. Its outputs are buildings and civil structures. Energy and power technology helps satisfy the needs for heat, light, and motion. The output is mechanical, thermal, radiant, chemical, electrical, or nuclear energy. Manufacturing technology satisfies the need for tangible goods. Its outputs are consumer and industrial products. Medical technology helps satisfy our need for health and well-being. Its outputs are devices and treatments for medical ailments and for promotion of healthy bodies. Transportation technology satisfies a desire to move humans and things. Its output is the movement of people and goods (cargo). These are the ***desirable outputs*** of the systems.

Note, however, that in addition to desirable outputs, ***undesirable outputs*** also exist, **Figure 6-3.** For example, some manufacturing activities produce fumes and toxic chemical by-products. If improperly treated these outputs poison the air and water around us. Poorly planned agricultural and construction projects can cause soil erosion. The result may be the loss of valuable topsoil and increased flooding. Communication technology can use billboards, which some people find unsightly. Also, people differ on the value of some communication messages. Every communication message, to some people, is unsightly or is considered noise. Transportation systems create outputs we do not want. Exhaust fumes from cars and trucks pollute the air. Airport and highway noise impinge on residential areas. New rail lines and roads can separate neighborhoods or divide farms. All of these are examples of the undesirable outputs of technology.

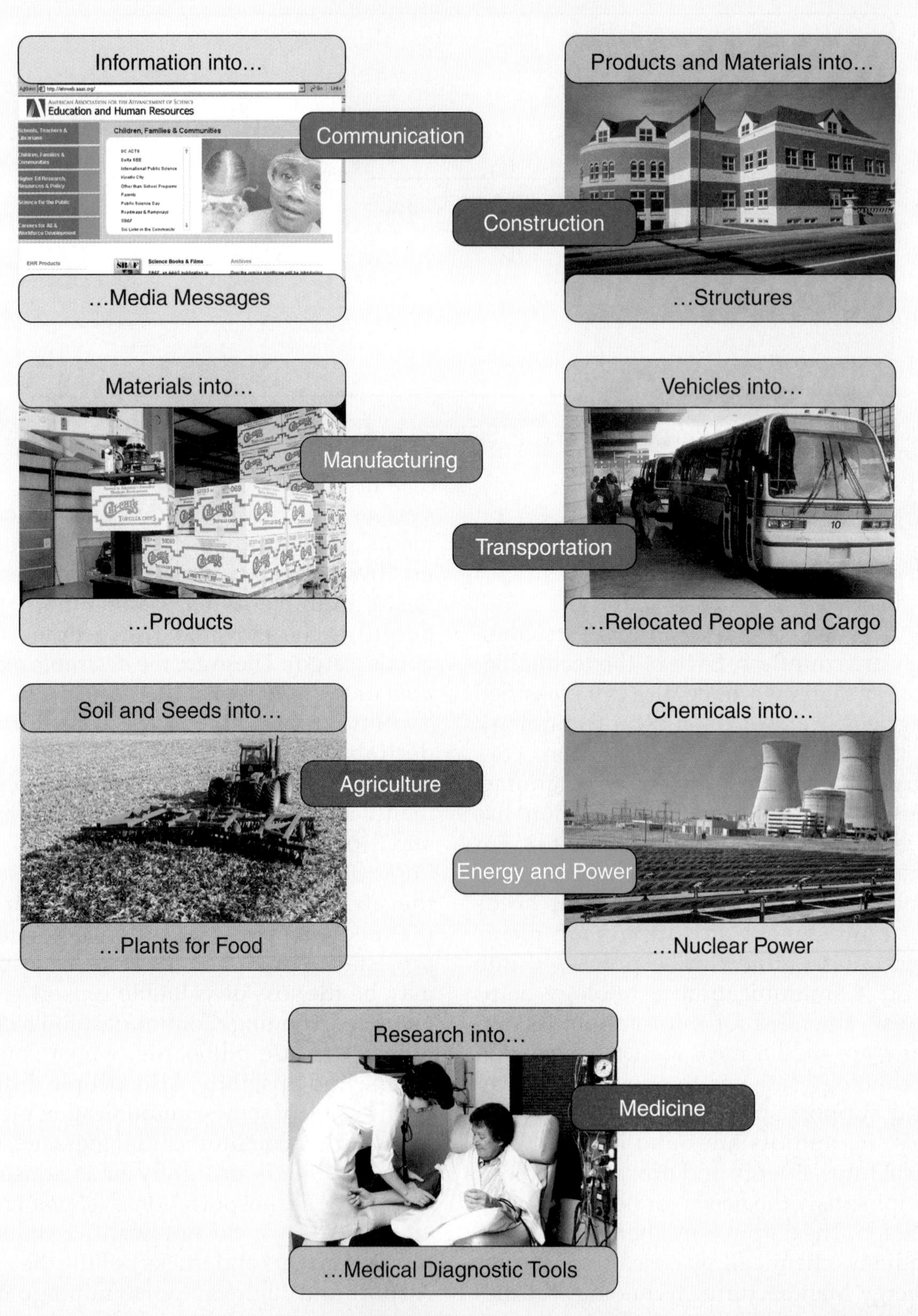

Figure 6-2. Each major type of technological system has a general type of output. (America Online, Kolbe and Kolbe Millwork Co., Inc., FANUC Robotics North America, Inc., John Deere and Co., U.S. Department of Energy)

Figure 6-3. Outputs of technological systems can be seen as desirable or undesirable. (John Deere and Co.)

Outputs: Intended and Unintended

As we said, technological systems are designed to produce specific outputs. These outputs include such varied items as refrigerators, weekly magazines, high-rise apartments, municipal bus transportation, or any of thousands of other products and services. These are the ***intended outputs.***

Depending on your point of view, the intended outputs may be desirable or undesirable. For example, suppose you planned to make 10"-diameter plywood disks. You would have to cut them from rectangular sheets of plywood. Cutting circles from rectangular material will produce waste material. You can plan for the waste, consider it in pricing your product, and reduce it to a minimum. You cannot totally eliminate it, however. Both the disks and the waste were intended. They were in the plan for the system. The disks are the intended *desirable* output, and the waste is the intended *undesirable* output.

Sometimes, technological systems produce ***unintended outputs,*** that is, outputs that were not considered when the system was designed. For example, some heating systems produced in the 1940s through the 1960s used asbestos to insulate pipes. The material was considered an excellent insulator. Later, people who worked with asbestos developed lung cancer. This was an undesirable, unintended output of the technological system.

Outputs: Immediate and Delayed

Most technological systems are designed to produce a product or service for use now, and thus we have ***immediate outputs.*** For example, we do not produce steel for use in the year 2051. We produce it for use this year. This practice reduces inventory costs and loss due to theft and spoilage. The production of steel is an example of an immediate, intended, and desirable output.

We also create ***delayed outputs,*** however. The chemicals used as propellants in aerosol (spray) cans and as refrigerants are now affecting the ozone layer around the earth. The accumulation of this matter has taken decades to reach a dangerous level. The same is true of the sulfur dioxide produced by coal-fired power plants. The resulting acid rain is killing forests in the United States and Canada. These examples show some delayed, unintended, and undesirable outputs of technology.

Technology Explained

Integrated Circuits

integrated circuit (IC): A piece of semiconducting material in which a large number of electronic components are formed.

We are living in the Information Age that began with the invention of the transistor. A transistor is a tiny electronic device that allowed engineers to eliminate bulky electronic devices, such as the vacuum tube. Technology has made even greater advancements in the field of electronics, expanding transistor technology to create the integrated circuit (IC), or microchip.

An IC is a thin piece of pure semiconductor material, usually silicon, **Figure A.** Tens of thousands of electronic devices and their interconnections can be produced in one IC, **Figure B.**

ICs must be manufactured in very clean environments. Often, the individual electronic devices on the chip are very small, **Figure C.**

These circuits are very complex and require careful design. This requires a high level of skill and can be very time-consuming.

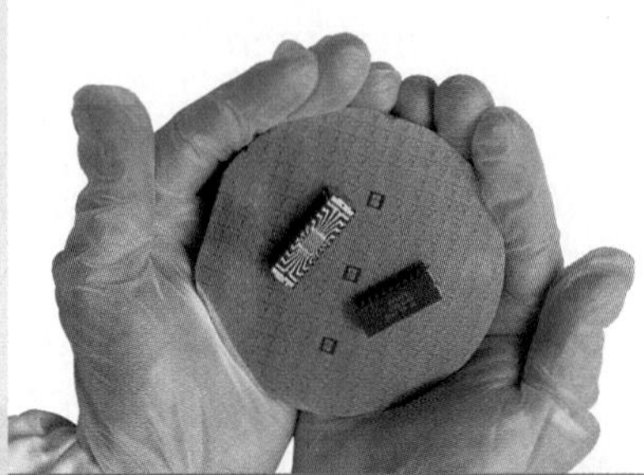

Figure B. Microchips are contained inside these cases. The electrical connections extend out of the case.

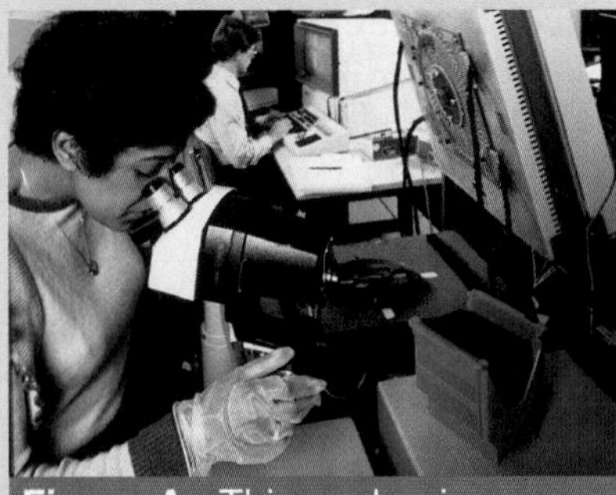

Figure A. This worker is inspecting an integrated circuit.

Methods used to manufacture ICs include the following steps:

1. Oxidizing the surface. This prevents impurities from entering the silicon.
2. Coating the surface with a photoresist, a light-sensitive chemical. When developed, it prevents the oxidized layer of the IC from being etched away.
3. Placing a pattern over the photoresist and exposing it to light. The unexposed areas are then washed away.
4. Etching the IC with acid. This creates windows of unprotected areas on the silicon.
5. Introducing impurities into the silicon (*doping*). These impurities create the tiny transistors, diodes, capacitors, and resistors in the IC, giving the chip its electronic properties.
6. Connecting the components with small aluminum leads.
7. Placing the IC in a protective casing.

The IC is responsible for the enormous growth of the computer industry in the twentieth century.

Figure C. Close-up of a silicon sensor chip. (BAE Systems)

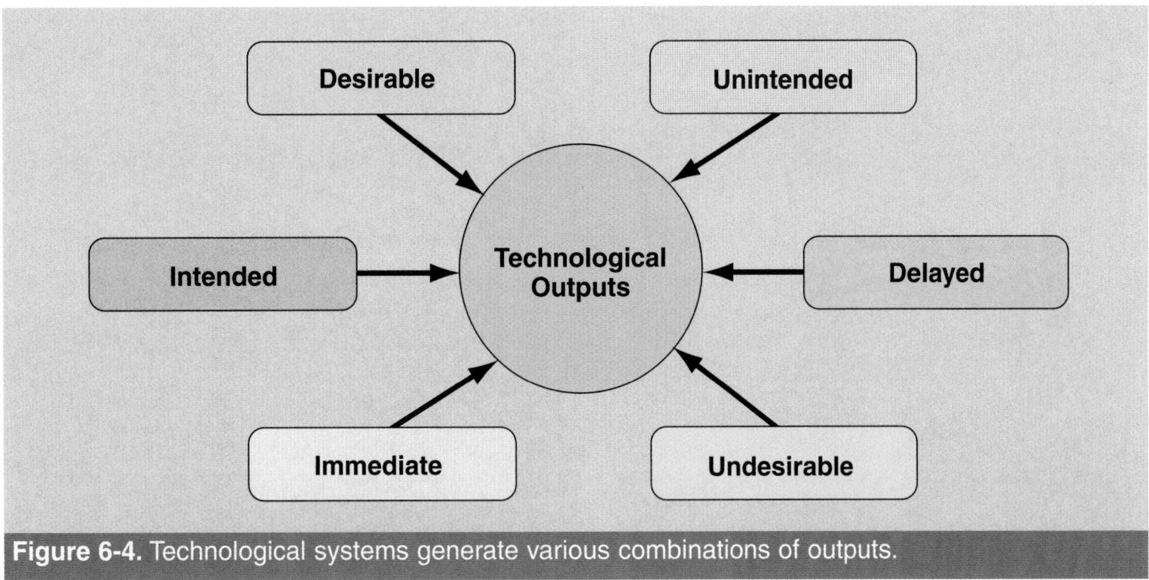

Figure 6-4. Technological systems generate various combinations of outputs.

Thus, although we design technological systems to produce specific outputs, many kinds of outputs can occur. The various combinations are shown in **Figure 6-4.**

Feedback and Control

As we noted earlier, humans have created control systems, most of which use feedback, so that they can reach their desired goals. Feedback can be defined as using information from the output of a system to control or regulate the system or its inputs to affect the output. We often view control systems as internal, that is, we look at how operating processes are controlled so that outputs meet specifications. We can view internal control systems from three different angles: the types of control systems, the components of control systems, and the operations (manual versus automatic) of control systems. To understand technological system control completely, however, we also need to consider other kinds of control, those external to the system. For example, political actions and personal value systems also exercise control. We will explore both internal and external areas of control in the rest of this chapter.

Internal Controls: Types of Control Systems

We use two major types of control systems in technology, the open-loop system and the closed-loop system, **Figure 6-5.** The open-loop system is the simpler of the two. This system uses no feedback to compare the results with the goal. The system is set and then operates without the benefit of output information. For example, suppose you decide to drive a car to an adjoining town. You start the car and then hold the gas pedal at a specific point throughout the trip. The car will accelerate to a certain speed. It will stay at that speed until you reach a hill. It will slow down as it takes more power to move the vehicle. As you crest the hill, the car will speed up as it coasts down the other side. As long as you do not adjust the gas pedal the car will change speed as it reaches each new road condition. This type of control is ***open-loop control.*** Output information is not used to adjust the process in this type of control system.

Most technological systems, however, use feedback in their control systems, which means that output information is used to adjust the processing actions.

Figure 6-5. Technology uses open-loop and closed-loop control systems. The food processing equipment on the left runs at a set speed (open loop), whereas the power generators shown on the right automatically adjust to increased demand for electricity (closed loop). (FMC Corp., American Electric Power)

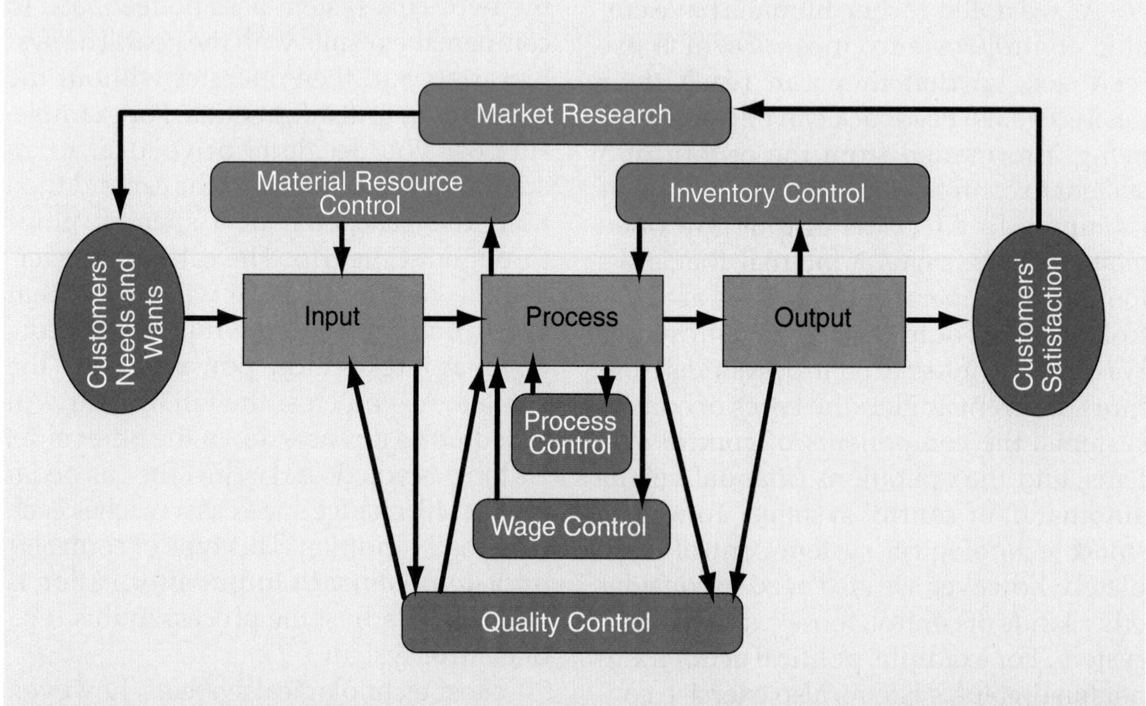

Figure 6-6. With feedback from market research, manufacturers use various types of control to meet customers' needs and wants. Information enters the control system, is processed, and is then fed back into the system.

Feedback is used to control various stages or factors within a system or the entire system. Systems that use feedback are called ***closed-loop control*** systems.

Let us look at closed-loop control in one technology system, manufacturing, **Figure 6-6.** After using *market research* to gather and analyze customers' positive and negative reactions, manufacturers evaluate present and future product designs. Using a process called *quality control,* manufacturers compare parts, assemblies, and finished products with engineering standards, **Figure 6-7.** Quality control ensures that the product performs within an acceptable range. Using a process termed *inventory control,* manufacturers compare finished goods in warehouses and products in process with sales projections, **Figure 6-8.** The goal is to match as closely as possible the product production with the product demand. With *process control,* manufacturers closely monitor the operation of machines and equipment. The goal is to see that each manufacturing process produces the proper outputs. Using *material resource control* (or planning), manufacturers monitor the need for materials and the quantity on order. The goal of this activity is to reduce the raw material inventory. Finally, using a process called *wage control,* manufacturers monitor the hours worked by employees producing products. The goal of this activity is to keep the labor content of a product within planned limits.

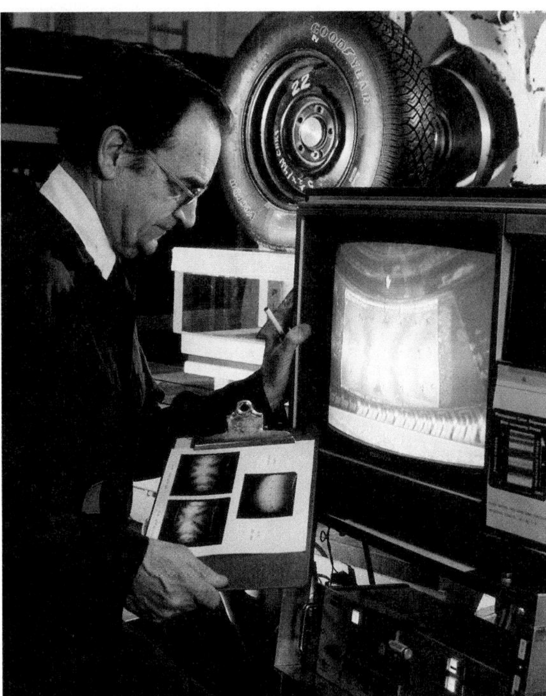

Figure 6-7. Quality control personnel gather data about product characteristics and compare them with standards. (Goodyear Tire and Rubber Co.)

Control is not limited to manufacturing, of course. Automobiles have emission control systems. Engines have governors to control their operating speeds. Some magazines are sent to people who need only that kind of information. Control systems monitor the ink being applied to paper during the printing process. Thermostats control the temperatures in buildings, ovens, and kilns. Fuses and circuits breakers control the maximum amount of electric current in a circuit. In other words, control is everywhere.

Internal Controls: Components of Control Systems

We want technological services, products, information, and structures to meet our needs. To ensure that this happens we design control systems with three major components, **Figure 6-9.** These components are:

- Monitoring devices
- Data-comparing devices
- Adjusting devices

Monitoring Devices

The first component of a control system, the monitoring device, gathers information about the action being controlled. It monitors inputs, processes, outputs, or reactions. The information gathered can be

Figure 6-8. This worker is checking the inventory of tread plates used in making truck bodies. (Reynolds Metals Co.)

of many types. For example, information can include such varied data as material size and shape, temperatures of enclosures, speed of vehicles, and impacts of advertising. The information can also vary in that it can be very specific and mathematical or more general. In addition, information can differ in that it might be gathered and recorded in short intervals or over long periods of time.

The type of information gathered during the monitoring step depends on how the data will be used. Process performance data will be very narrow and focused. Data about the impacts of technological systems will be much more general. For example, data needed to determine the emission levels of a motorcycle engine are more specific than the data needed to determine the impact of the air transportation system on business.

Process-operating information can be gathered using several types of monitoring devices. These include:

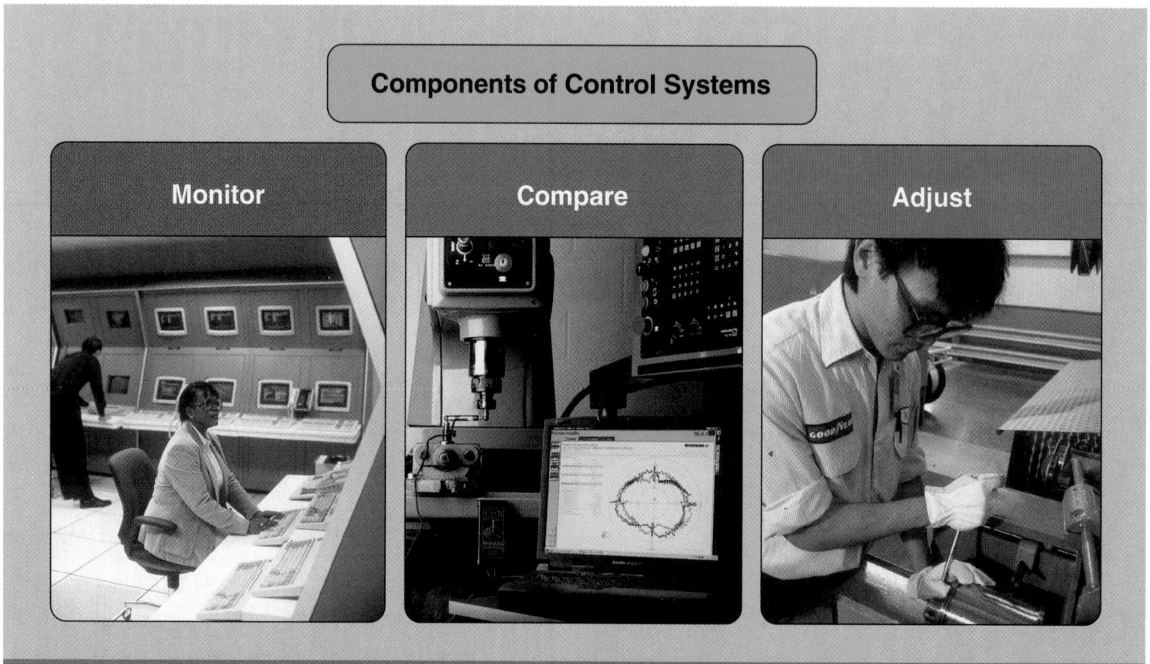

Components of Control Systems

Monitor

Compare

Adjust

Figure 6-9. Control systems monitor performance, compare it against standards, and adjust the system to ensure the output meets the goal. (General Motors, Renishaw, Inc., Goodyear Tire and Rubber Co.)

- *Mechanical sensors,* which can be used to determine position of components, force applied, or movement of parts.
- *Thermal sensors,* which can be used to determine changes in temperature.
- *Optical sensors,* which can be used to determine the level of light or changes in the intensity of light, **Figure 6-10.**
- *Electrical or electronic sensors,* which can be used to determine the frequency of or changes in electric current or electromagnetic waves, **Figure 6-11.**
- *Magnetic (electromagnetic) sensors,* which can be used to determine whether changes are occurring in the amount of current flowing in a circuit.

Data-Comparing Devices

The information gathered by monitoring devices is matched against expectations. The results are compared to the intent of the system through the use of data-comparing devices. This process is much like a balance scale in that the results should balance with the goal.

Comparisons can be analytical or judgmental. ***Analytical systems*** mathematically or scientifically make the comparison. Statistics might be used to see whether the output meets the goal. For example, the temperature of an oven can be compared with the setting on the knob. The size of a part can be checked with lasers and compared with data in a computer file. The position of the ship can be compared with the location of radio beams. The analytical type of comparison is commonly used in closed-loop control systems.

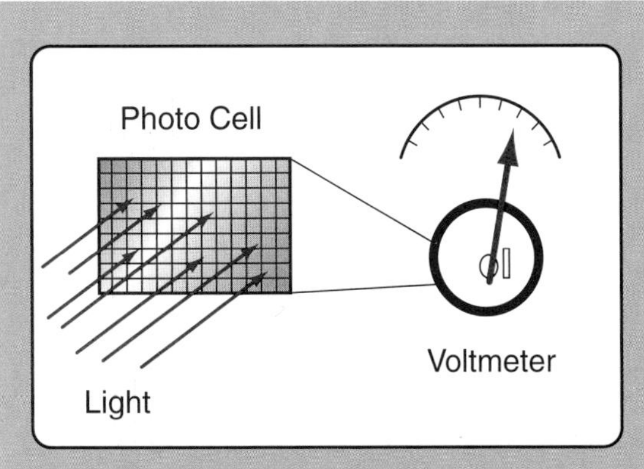

Figure 6-10. A photographic light meter is a common optical sensor. Light striking the photocell generates electricity, which is measured by a voltmeter. Strong light produces more voltage than weak light.

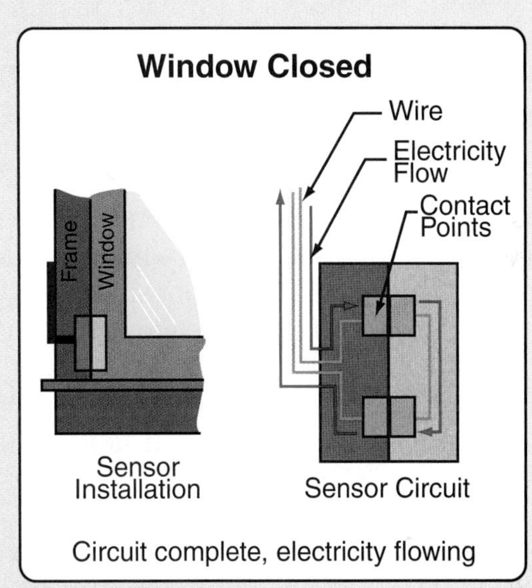

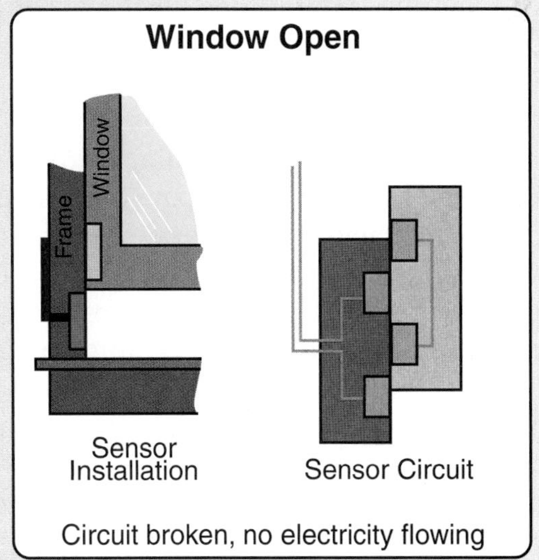

Figure 6-11. This electrical sensor is part of a home security system. It senses when windows are open.

Analytical systems can also be used for managerial control systems. For example, the number of products produced might be directly compared with the production schedule. The number of passengers buying bus tickets might be compared with sales projections. The number of people listening to a radio station might be compared with the total listeners for all stations. Analytical systems try to remove human opinion from the evaluation process. The information is reduced to percentages, averages, or deviation from expectations.

Judgmental systems use human opinions and values to enter into the control process. Open-loop control systems require human judgment. For example, the speed at which you drive a car or the movement you use to cut on a line is based on your judgment. You do not apply mathematical formulas to decide when to accelerate and when to brake as you drive on a mountain road.

Likewise, human opinion is used in determining value of system outputs. Judgments are used in the design phase of technological systems. It is almost impossible to select a design by analytical means. For example, people decide the styling, color, and shape of most products and buildings. They select the content of magazines and telecommunications programs. After the output is available, analytical means may be used to compare the results. After a television program is produced, the number of viewers watching the program can be measured. The success of the program can be subjected to mathematical analysis. Likewise, a furniture manufacturer can mathematically compare the sales of Early American tables with the sales of French Provincial tables. Note, however, that the decision to produce the show or manufacture the particular style of table was a judgment.

Adjusting Devices

A comparison of outputs to expectations might indicate that the system is not operating correctly. The output might not meet specifications, and thus the system might need modifying. Controllers do this modifying through the use of ***adjusting devices.*** They might speed up motors, close valves, increase the volume of burning gases, or perform a host of other actions. The goal is to cause the system to change and, therefore, produce better outputs.

Several types of adjusting devices, or system controllers, exist. These, as shown in **Figure 6-12,** include:

- ***Mechanical controllers:*** Using cams, levers, and other types of linkages to adjust machines or other devices
- ***Electromechanical controllers:*** Using electromagnetic coils and forces to move control linkages and operate switches to adjust machines or other devices
- ***Electrical and electronic controllers:*** Using electrical (switches, relays, and motors, for example) and electronic (diodes, transistors, and integrated circuits, for example) devices to adjust machines or other devices
- ***Fluidic controllers:*** Using fluids, including oil (hydraulic) and air (pneumatic), to adjust machines or other devices

Internal Controls: The Operation (Manual versus Automatic) of Control Systems

Technological control systems can also be classified as manual and automatic. ***Manual control systems*** require humans to adjust the processes, whereas ***automatic control systems*** can monitor, compare, and adjust the system without human interference.

Manual Control

A good example of a manual control system is the automobile. The driver uses various speed and direction controls to guide the vehicle to the destination (goal). As the vehicle enters a city the driver eases up on the accelerator to reduce the speed to the new speed limit. Brakes might be applied to meet the new conditions. The steering wheel is turned as the road curves. In this example, the human eye gathers information. It sees a new speed limit sign. The eye also scans the speedometer to determine the speed of the vehicle. The brain compares the posted speed limit with the speedometer reading. The brain then commands the foot to change its position to allow the speed to be adjusted until it matches the speed limit. In this example, the system is *monitored* by the human eye. The brain *compares* its actual and desired performance. The foot physically *adjusts* the system. Manual control systems are everywhere. We use them in such simple actions as riding a bicycle, adjusting the temperature of a gas stove when cooking, setting the focus on a camera, or checking

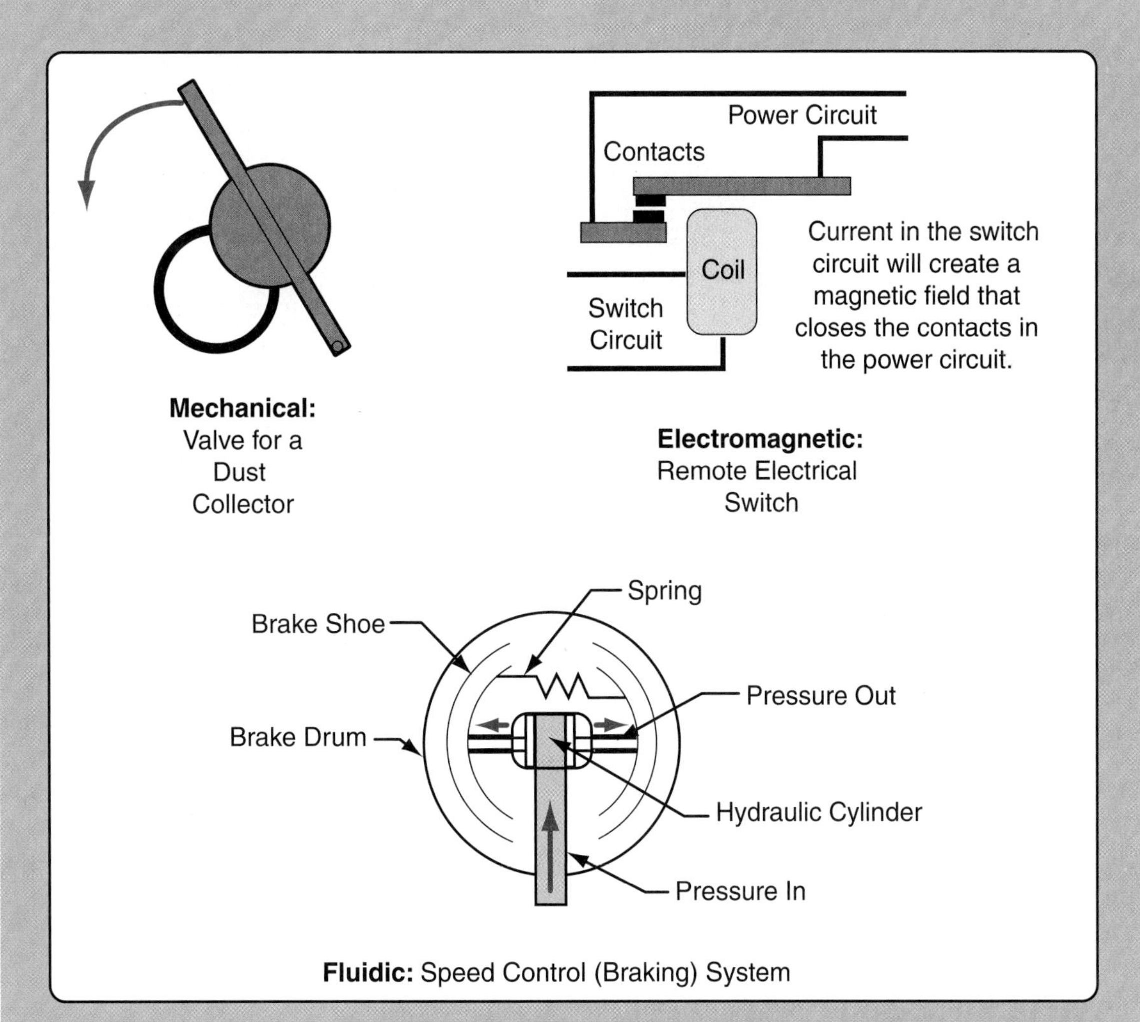

Figure 6-12. System controllers, or adjusting devices, modify the system to produce better outputs.

Figure 6-13. This pilot is using a manual control system to land an airplane. His eyes monitor the approach to the runway. His mind compares the actual flight path with the desired one. His hands and feet change control surfaces and speed to land the plane safely. (United Airlines)

a fence post for plumb. We also use them in performing more complicated actions, such as landing an airplane, **Figure 6-13.**

Automatic Control

Many technological systems have automatic control. A simple example of automatic control is the thermostat in a heating system, **Figure 6-14.** The device measures the room temperature. It compares the temperature of the room with the desired temperature. As long as the temperatures are within a preset range, no action is taken. When the room temperature drops below the set temperature range, a switch is activated. It turns on the heating unit. The unit will operate until the upper limit of the temperature range is reached. At this point the thermostat turns off the unit.

Automatic doors in stores, electronic fuel injection systems in vehicles, and programmable VCRs are other examples of automatic control systems you see every day. More sophisticated automatic control systems help pilots land the new generation of jet airliners, assist workers in controlling the output of electric generating plants, and help contractors lay flat concrete roads.

External Controls

We have looked at control primarily as an internal system component. As we said earlier, though, control also comes from

Connections to Technology: **Science**

Chlorofluorocarbons

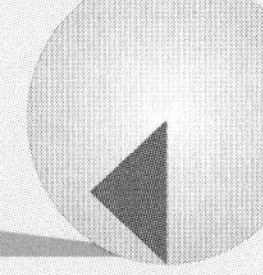

As this chapter indicates, we sometimes produce new technologies that turn out to have undesirable consequences. Such was the case with a group of synthetic organic compounds called chlorofluorocarbons (CFCs).

CFCs are a group of compounds that contain carbon and fluorine and often include chlorine and hydrogen. They change easily from liquid to gas or from gas to liquid and are used as refrigerants.

The development of a group of CFCs known as Freon in the late 1920s was originally considered to be a miracle find. Until that time refrigerators contained toxic gases, and sometimes fatalities occurred when the gases leaked from the refrigerators. Freon seemed to solve the problem; it is odorless, colorless, nonflammable, and noncorrosive. Within a few years Freon became the standard refrigerant for refrigerators and air conditioners.

As time went on, however, scientists discovered that CFCs damage the environment because they break down ozone molecules in the upper atmosphere of the earth. In 1978 the United States started to ban the use of some CFCs, and twelve years later they and other countries agreed to discontinue any production of CFCs by the year 2000.

What do we now use as a refrigerant in most air conditioners and refrigerators?

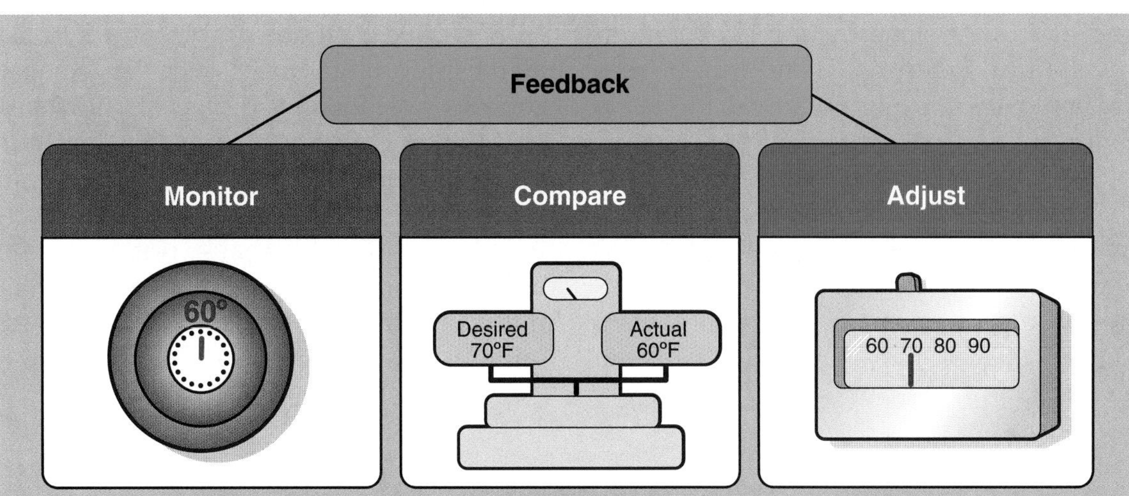

Figure 6-14. A thermostat is a common automatic control system found in many homes and buildings.

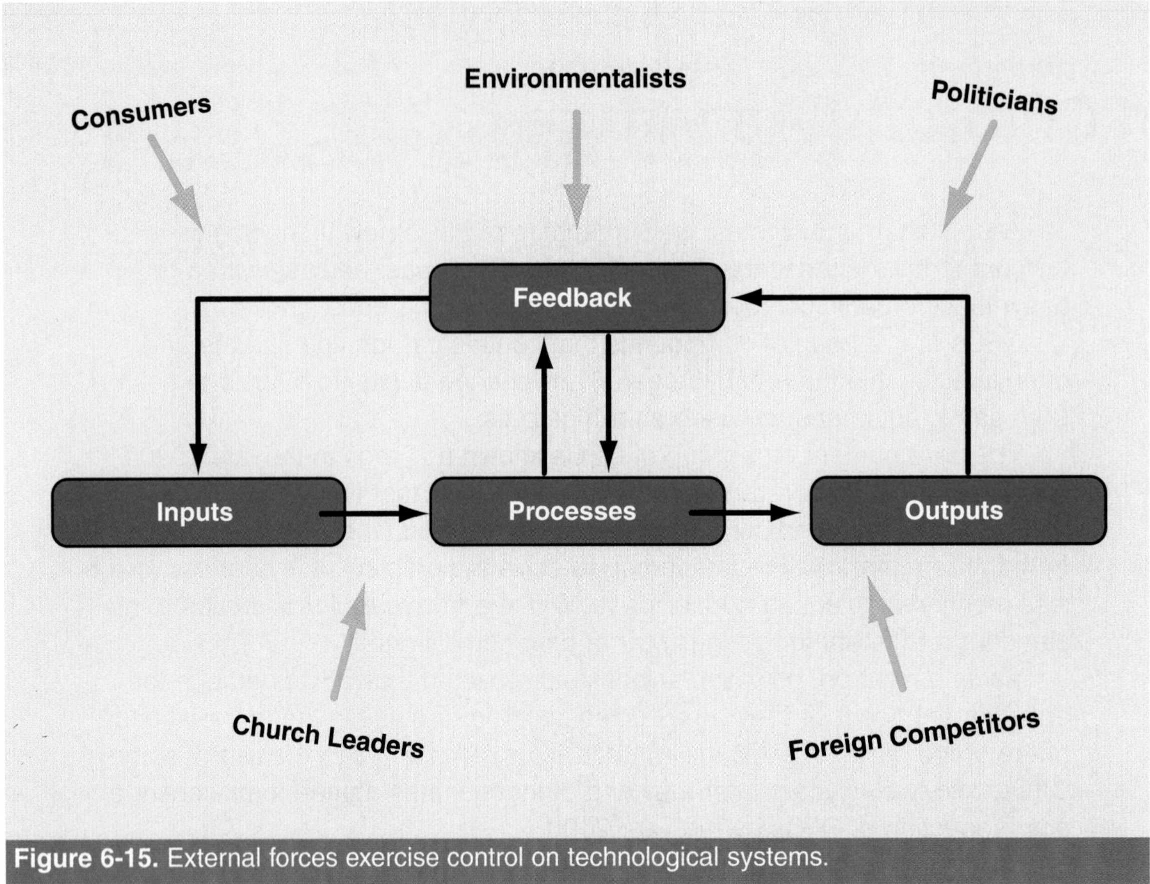

Figure 6-15. External forces exercise control on technological systems.

outside the system, **Figure 6-15.** For example, environmental responsibility is a societal goal that technological systems must address. Systems therefore are designed to reduce air, water, and noise pollution. Likewise, working conditions in factories and offices, on construction sites, and at transportation terminals and warehouses are controlled through political and labor union actions.

Public opinion is also a control factor. The opinions of civic groups, churches, labor unions, and various associations affect public policies. Our news media constantly report on issues that will control technological systems. The debate between environmentalists and forest products companies over timber cutting is one example. Another is the dialogue between the pro-nuclear and anti-nuclear power groups. The controversy over the types of nets that should be used in fishing is still another example of external control.

This discussion is not designed to list all the important public issues that exercise control on technology. It suggests, however, that looking only at the physical controls built into technological systems gives an incomplete view. We need to see that primarily through feedback, both internal and external control systems affect technology by creating new inputs, processes, and outputs, **Figure 6-16.**

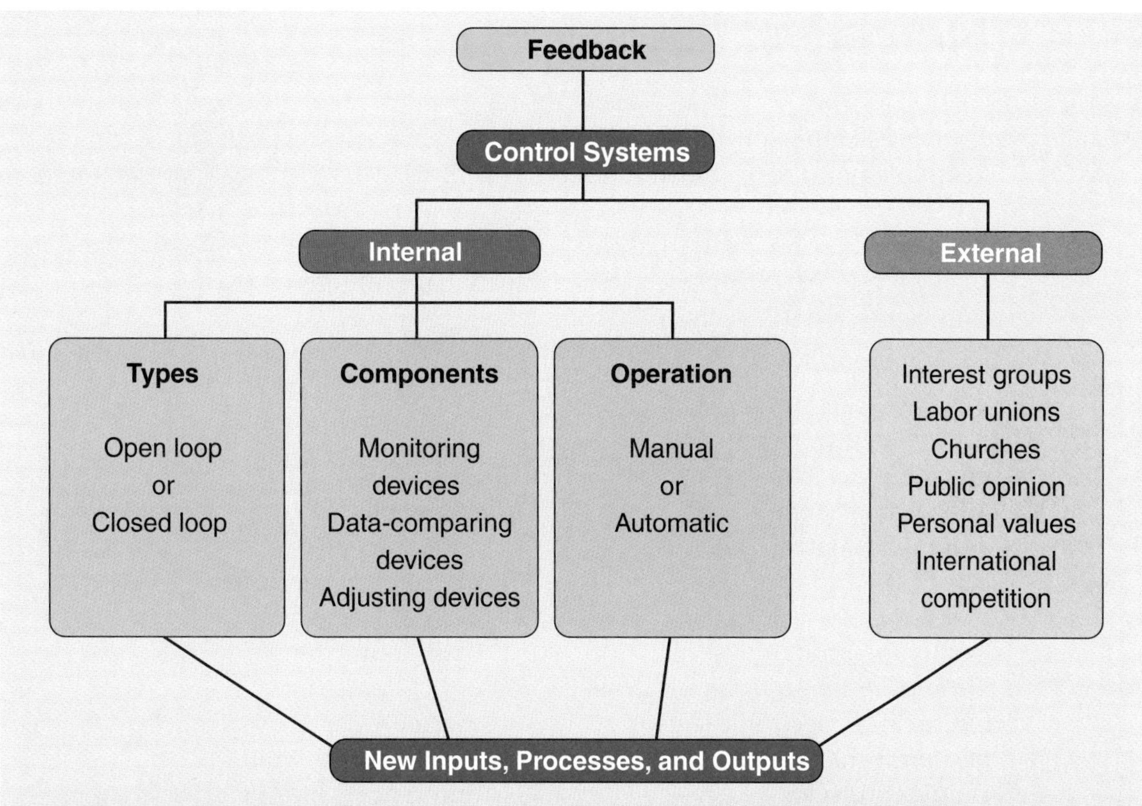

Figure 6-16. Primarily through feedback, both internal and external control systems lead to new inputs, processes, and outputs.

Career Corner
Dental Hygienists

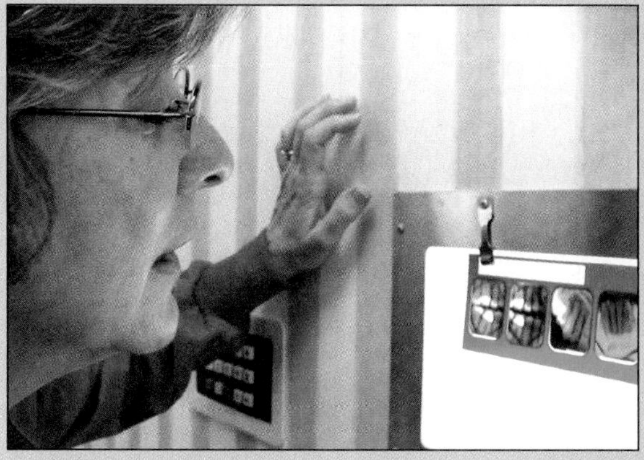

Dental hygienists remove deposits from teeth, teach good oral hygiene practices, and provide preventive dental care. They generally work in dentists' offices, where they examine patients' teeth and gums and record the presence of diseases. Hygienists remove plaque from teeth, take dental X rays, and apply cavity-preventive agents. They use a variety of tools and machines, including instruments and ultrasonics to clean teeth and X-ray machines to take dental pictures.

Dentists frequently hire hygienists to work only part of each week, so many hygienists have part-time jobs or work in more than one dental office. Dental hygienists must be licensed by the State and, therefore, must graduate from an accredited dental hygiene school. Also, they must pass a written and clinical examination.

Summary

Technology is a series of complex human-made systems that change resources into useful items and services. The outputs of the systems can be desirable or undesirable, intended or unintended, and immediate or delayed. Humans have created control systems, most of which use feedback, to increase the probability that the outputs will be desirable. These control systems range from being parts of the internal operating process to outside forces that influence the system.

Key Terms

adjusting device
analytical system
automatic control system
closed-loop control
delayed output
desirable output
electrical and electronic controller
electrical or electronic sensor
electromechanical controller
fluidic controller
immediate output
intended output
judgmental system
magnetic (electromagnetic) sensor
manual control system
mechanical controller
mechanical sensor
open-loop control
optical sensor
thermal sensor
undesirable output
unintended output

Test Your Knowledge

Write your answers on a separate piece of paper. Please do not write in this book.

1. Describe the relationship between outputs and feedback and control.
2. Name one major output for each of the classified technology systems.
3. *True or false?* Communication messages can be seen as an undesirable output.
4. Give one example of a delayed, unintended, and undesirable output.
5. Feedback can be defined as using information about the _____ of a system to control or regulate the system or its inputs.
6. What is the major difference between open-loop control systems and closed-loop control systems?
7. _____ sensors are a type of monitoring device that can be used to determine the level of light or changes in the intensity of light.
   a. Mechanical
   b. Thermal
   c. Optical
   d. Electrical

8. _____ controllers are a type of adjusting device using cams, levers, and other types of linkages to adjust machines.
   a. Electromechanical
   b. Mechanical
   c. Electrical
   d. Fluidic
9. Give one example of an automatic control system.
10. *True or false?* Public opinion is an external control factor.

Applying Your Knowledge

1. Select a technology that you use every day. Prepare and complete a chart like the following.

Outputs	
Desired	Undesired
Intended	Unintended
Immediate	Delayed

2. For the same technology, diagram the control system used and list the ways information is gathered, comparisons are made, and adjustment techniques are used.
3. For the same technology, list the outside influences that helped determine its design and that help control its operation.

Section 2 Activities

Activity 2A

Design Problem

Background

Technology uses tools to process materials to create products. These tools often use one or more of the six basic mechanisms. These mechanisms are the lever, the wheel and axle, the pulley, the inclined plane, the wedge, and the screw.

Situation

The Science Experiments Company markets instructional kits for use in elementary and middle school science programs. It has hired you as a designer for its physics kits.

Challenge

Design a kit containing common materials, such as tongue depressors, mousetraps, rubber bands, thread spools, small wooden blocks, and string. Develop a technological device (machine) using these materials, containing a power source and using three or more of the basic mechanisms to lift a tennis ball at least 6" off a table.

Activity 2B

Fabrication Problem

Background

Technological devices are often built to increase human ability. Some devices multiply our ability to lift loads, while others help us see better. Still other devices improve our ability to hear, move people and products from place to place, or communicate over long distances.

Challenge

Construct the device shown in the drawings below. Test the device to determine how well it helps you lift a load. Then, analyze how well the device worked.

Equipment

- Test stand
- 1/2" diameter plastic syringe
- 3/4" or 1" diameter plastic syringe
- 12" clear tubing
- Baseball or other load

Procedure

1. Obtain the supplies.
2. Assemble the hydraulic system. See **Figure 2B-1.**
3. Place the load on the small syringe.
4. Press the large syringe plunger to cause the load to rise.
5. Reverse the load, placing it on the large diameter plunger.
6. Press the small syringe plunger to cause the load to rise.
7. Complete the following analysis of the activity.

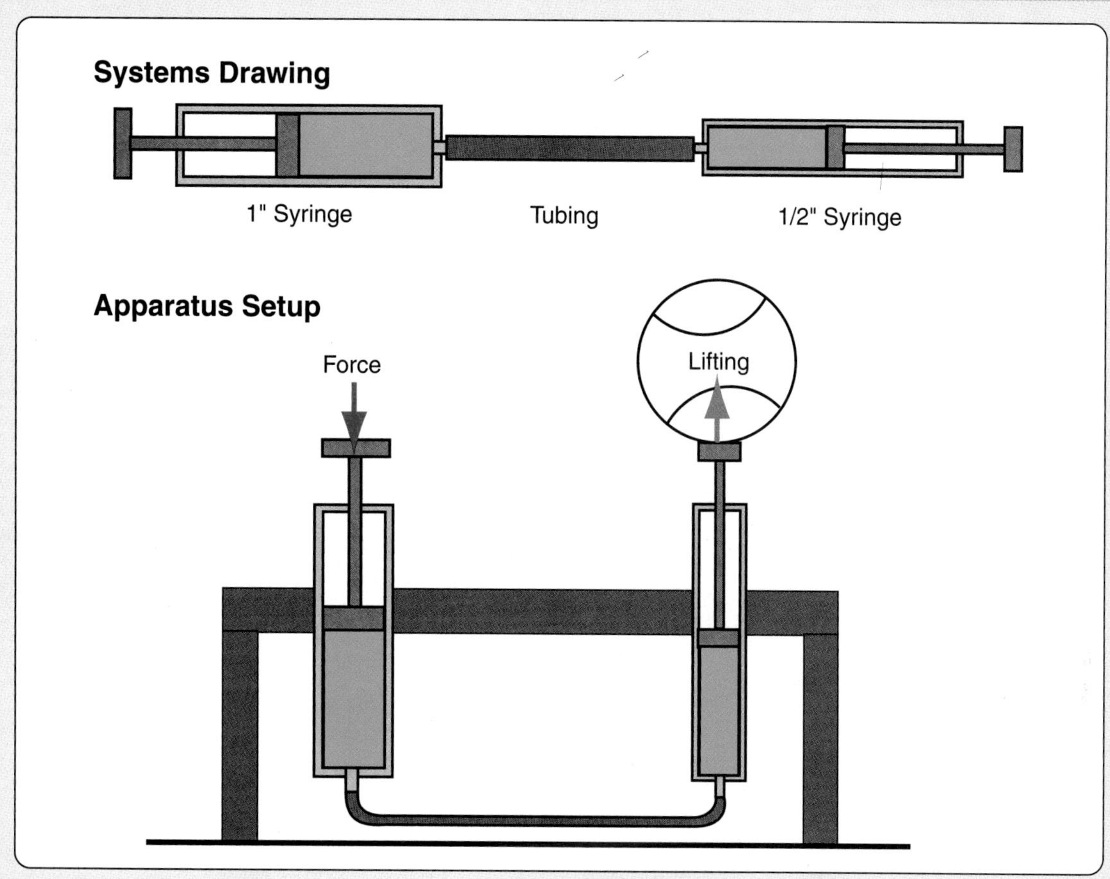

Figure 2B-1

Analysis

Analyze the two circuits by answering the following questions:

1. Which arrangement of cylinders lifted the load with the least effort?
   A. Force on the large cylinder and the load on the small cylinder.
   B. Force on the small cylinder and the load on the large cylinder.
2. Which arrangement of cylinders moved the load the greatest distance when the force cylinder was depressed?
   A. Force on the large cylinder and the load on the small cylinder.
   B. Force on the small cylinder and the load on the large cylinder.
3. Which arrangement is a force multiplier?
4. Which arrangement is a distance multiplier?
5. Which type of multiplier would you use as an automobile lift in a service station?
6. Which type of multiplier would you use for a hydraulic elevator in a four-story building?

Section 3

Tools of Technology

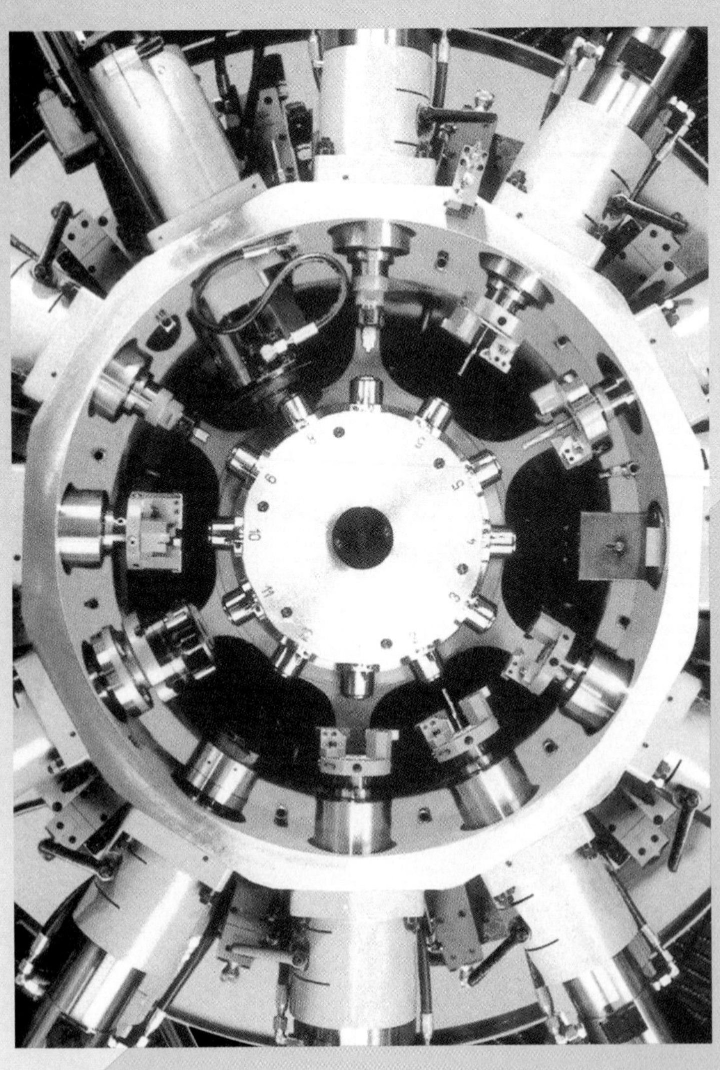

Tomorrow's Technology Today
Satellite Radio

Imagine a beautiful spring day. You jump into your car, roll down the windows, switch to your favorite radio station, and take off for the open road. Just as you reach the main stretch of highway, your favorite song begins to play. As you reach to turn up the stereo, the music begins to fade, and the big band sound from a nearby station encroaches on the classic rock you were enjoying. It is then that you realize you have traveled beyond the range of your local radio station.

Now imagine getting into your car and driving from Florida to California while listening to the same radio station. While this may seem like wishful thinking, it is within the realm of possibilities with the satellite radios being offered as optional equipment on many new vehicles. Satellite radios are also available for home use.

Most satellite radio receivers will pick up AM and FM signals by conventional means, but they will also allow you to choose from over 100 stations offering a variety of music genres, as well as news, talk, sports, and entertainment programming. In addition to being broadcast over a wide geographic area, satellite radio signals are free from static and distortion.

Today, there are three satellite-based radio broadcasters: XM Satellite Radio, Sirius Satellite Radio, and WorldSpace. While all three have the same goals in mind, each achieves them in a slightly different manner.

XM Satellite Radio uses two satellites in geostationary orbit to broadcast signals across the United States. A satellite in geostationary orbit will remain directly above a fixed point on the equator at a distance of approximately 22,223 miles above Earth's surface.

XM Satellite Radio's Earth station transmits a signal to the two satellites. The satellites bounce the signals back to radio receivers on the ground. The receivers unscramble the signals. In addition to the encoded sound, the signal contains information about the broadcast (such as song title and artist), which the radio receiver displays. In urban areas where buildings can interrupt the satellite signals, ground repeaters pick up the satellite signals and transmit them to the radio receivers.

Sirius Satellite Radio uses three satellites that form a high-angle, elliptical orbit over the United States. Radio signals are sent from Earth stations to one of the satellites, which then beams the signal back to the ground, where a radio receiver picks it up. Like the XM system, Sirius uses ground repeaters to pick up and distribute signals in urban areas.

The third satellite radio broadcaster, WorldSpace, does not broadcast in the United States. Instead, it offers satellite radio service to parts of Asia and Africa. Like the XM system, WorldSpace uses two satellites in geostationary orbit. Radio signals are broadcast to these satellites. The satellites then send the signals back to radio receivers on the ground. WorldSpace, which plans to launch a third satellite soon to service South America and parts of Mexico, hopes to send radio signals to parts of the world that would otherwise be unable to receive radio signals.

Chapter 7
Production Tools and Their Safe Use

Learning Objectives

After studying this chapter, you will be able to:

- List the categories of production tools used in technology.
- List the characteristics of machine tools.
- Identify the major types of machine tools.
- Describe the major types of energy-processing tools and machines.
- Describe the major types of information-processing tools and machines.
- List guidelines for using tools and machines properly and safely.

The human ability to design and use tools provides the foundation for technology. You might recall from Chapter 2 that tools range in kind from simple hand tools to complex machines. Because of tools we have such varied items as buildings, vehicles, and communications media and energy-conversion machines, **Figure 7-1.**

Tools can make our lives better, but they can also threaten our existence. For example, we travel in comfort in modern automobiles, but they pollute the atmosphere. High-rise buildings allow us to live and work in cities, yet social problems and crime in many of these cities frighten people. Television allows us to be instantaneously touched by events around the globe, but it threatens family interaction.

Our challenge is to design and use the tools of technology wisely. To do this we must first understand the tools and machines around us. We use thousands of different tools and machines today. We cannot discuss them all here because of space limitations, of course. Instead, this chapter will introduce you to selected tools and machines used in three major types of processing:

- Material processing
- Energy processing
- Information processing

Note, however, that understanding the tools and machines around us involves knowing not only their particular features but also how they should be handled. The final section of this chapter will cover the rules of safety for tools and machines.

Figure 7-1. Many different tools are required to build a complex product such as this airliner. (McDonnell Douglas)

Material-Processing Tools and Machines

The world around us is full of artifacts. Each was made using ***material-processing*** tools and machines. Each was the result of an action or actions that changed the form of materials. Tools and machines were used to cast, form, and machine materials into specific shapes. They also helped assemble products and apply protective coatings.

Fundamental to all material processing is a group of tools called ***machine tools***. These are the machines used to make other machines. To understand them better we will now look at their characteristics and types.

Characteristics of Machine Tools

Machine tools have some common characteristics. As shown in **Figure 7-2,** these include:

- A method of cutting materials to produce the desired size and shape. The new size and shape are achieved with what is called a cutting tool.
- A series of motions between the material and the tool. This movement causes the tool to cut the material.
- Support of the tool and the workpiece (the material being machined).

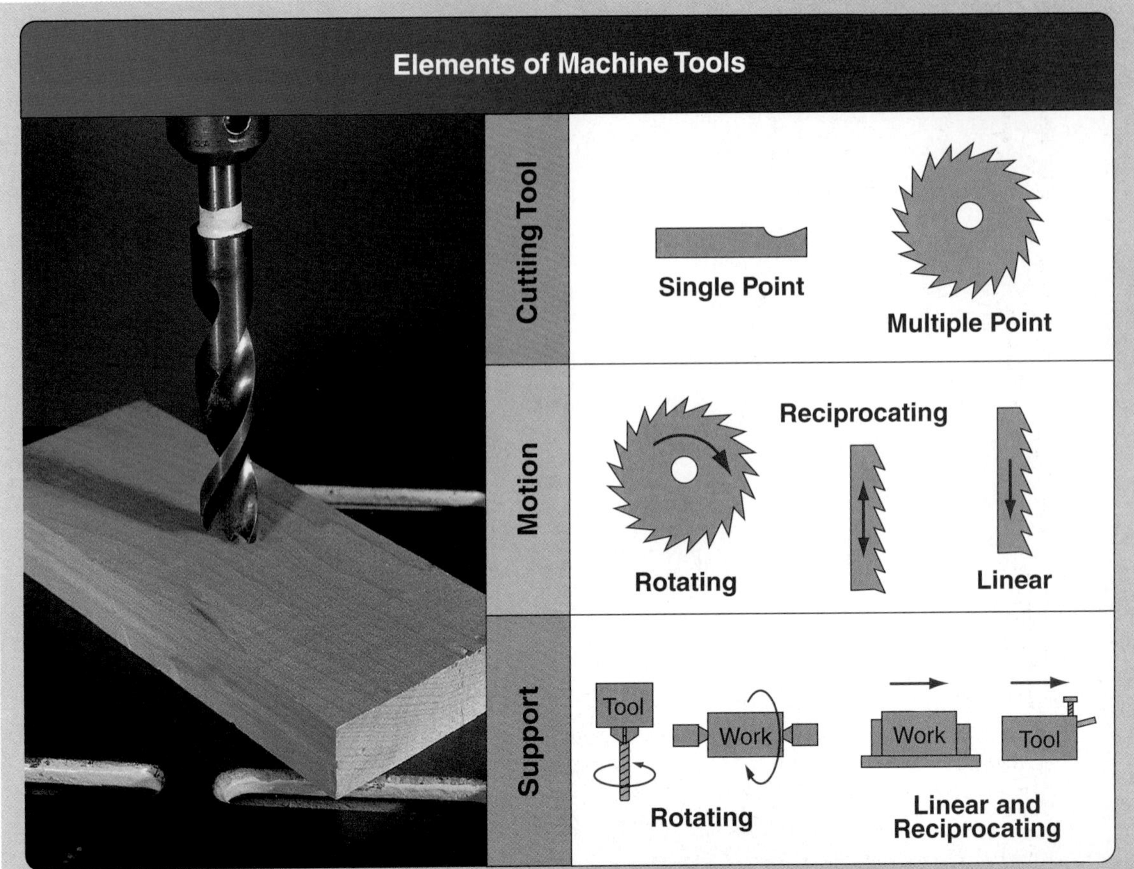

Figure 7-2. All machine tools have three basic elements: a cutting tool, a method of moving the tool or workpiece, and a method of supporting the tool and the workpiece.

The Cutting Tool

Most cutting actions require a ***cutting tool.*** Cutting tools come in many sizes and shapes, but they must meet certain requirements. First, they must be harder than the material they are cutting. For example, a diamond will cut steel, steel will cut wood, and wood will cut butter. Second, the cutting tool must have the proper shape. That is, it needs a sharpened edge and relief angles, as shown in **Figure 7-3.** The sharpened edge allows the tool to cut into the material. The relief angles keep the sides of the tool from rubbing against the material as it is cut. The rake angle also helps to create a chip as the material is cut. This action allows waste material to be carried away efficiently.

Two basic types of cutting tools are used in all hand tools and machines. These are single-point and multiple-point tools.

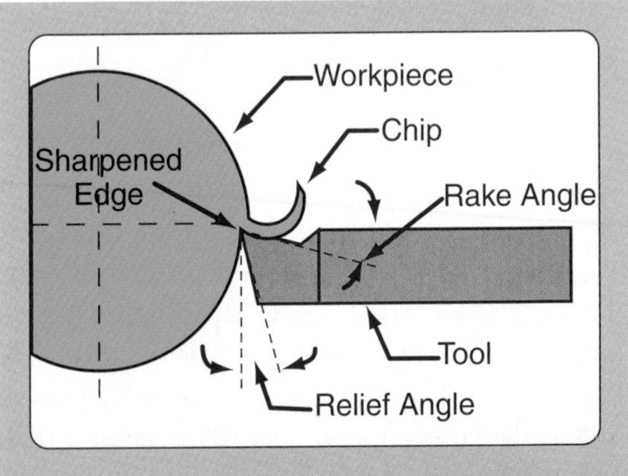

Figure 7-3. Cutting tools have a sharpened edge, a relief angle, and a rake angle.

The ***single-point tool*** is the simplest cutting device available. It has a cutting edge on the end or along the edge of a rod, bar, or strip. Common hand tools that use single points are knives, chisels, and wood planes, **Figure 7-4.** Most lathe-turning processes (where work is turned on a horizontal axis; see "Turning Machines" later in this chapter) use single-point tools.

With a ***multiple-point tool*** a series of single-point tools are arranged on a cutting device, **Figure 7-5.** Most often, these single points are arranged in a set pattern. For example, the teeth of a circular saw are evenly spaced around the circumference of the blade. Likewise, the teeth on band saw and scroll saw blades are spaced along the edge of a metal strip. In some cases there is no pattern for the arrangement of cutting points. The cutting edges or points have a random arrangement. Abrasive papers and grinding wheels are good examples of this type of cutting tool.

Figure 7-4. Single-point tools have a cutting edge on the end or along the edge. Among the tools pictured here are a hand plane, a utility knife, and various types of chisels.

Motion

Movement between the tool and the workpiece must occur before cutting can take place. All cutting operations have two basic motions: cutting motion and feed motion.

Cutting motion is the action that causes material to be removed from the work. It causes the excess material to be cut away. ***Feed motion*** is the action that brings new material into the cutter. It allows the cutting action to be continuous.

To understand these movements, consider a band saw, **Figure 7-6.** Imagine a piece of wood placed against the blade. When the machine is turned on, the first tooth will cut the material. This is the cutting motion. The next tooth follows the path of the first tooth. It will not produce

Figure 7-5. Multiple-point tools have a series of single-point tools arranged on a cutting device. Among the tools pictured here are circular, band, and scroll saw blades; grinding wheels; milling cutters; hand files; abrasive paper; and drill bits.

Career Corner

Machinists

Machinists interpret blueprints or specifications, set up machines, and produce precision parts. They use many different machine tools, including lathes, milling machines, grinders, and machining centers. Machinists must be able to calculate cutting and feed speeds, select proper cutting tools, plan the sequence operations, monitor machine operations, and use measuring tools to check the quality of the parts. Many machinists must be able to write and modify computer numerical control (CNC) programs. After a machinist has developed and tested operations and setups, repetitive operations generally are done by less-skilled machine operators. Most machinists are trained through apprenticeship programs, on-the-job training, vocational schools, or community colleges.

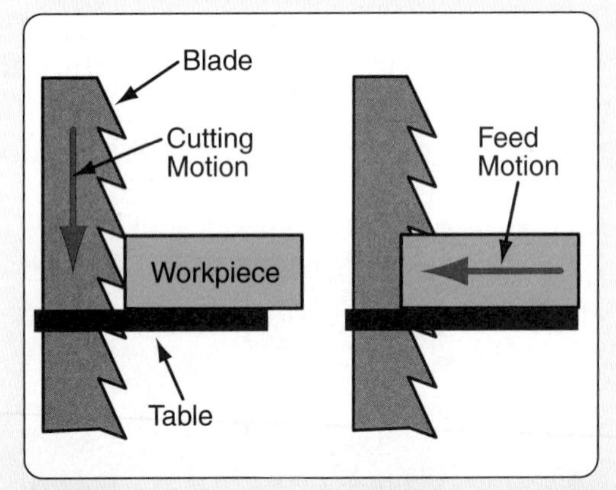

Figure 7-6. All cutting operations include cutting motion and feed motion.

a chip. The material in the *kerf* (the slot cut by a blade) has already been removed by the first tooth. The cutting action will continue only if the wood is pushed into the blade. This makes new material available for the next tooth to cut. Moving the wood is the feed motion.

Cutting and feed motions can be rotating, linear, or reciprocating. ***Rotating motion*** uses round cutters or spins the work around an axis. ***Linear motion*** moves a cutter or work in one direction along a straight line. ***Reciprocating motion*** moves the tool or the work back and forth. Look back to Figure 7-2 for a diagram of these three types of motion.

Support

The final element present in all machine tools is support for the tool and the workpiece. The types of cutting and feed motions used determine the support system needed. Rotating motion requires a holder that revolves around an axis. ***Chucks*** are attachments used to hold and rotate drills and router bits. ***Arbors*** are spindles or shafts that are used to hold table saw blades and milling cutters. Parts are placed between centers on lathes and cylindrical grinders.

Linear motion is produced in several ways. A tool may be clamped or held on a rest. It can then be moved in a straight line. This is a common practice for wood and metal lathes and hand wood planes. Band saw blades travel around two or three wheels to produce a linear cutting motion. Material may be pushed through a saw blade while it is supported by the machine table.

Reciprocating motion is common with scroll saws and hacksaws. The blade is clamped at both ends into the machine. Then the blade is moved back and forth. Likewise, a workpiece can be clamped to a table that reciprocates under a cutter. This action is used in a surface grinder. The operator moves a workpiece back and forth across the face of a grinding wheel. This action is used to sharpen hand tools, lawn mower blades, and machine tool cutters and knives.

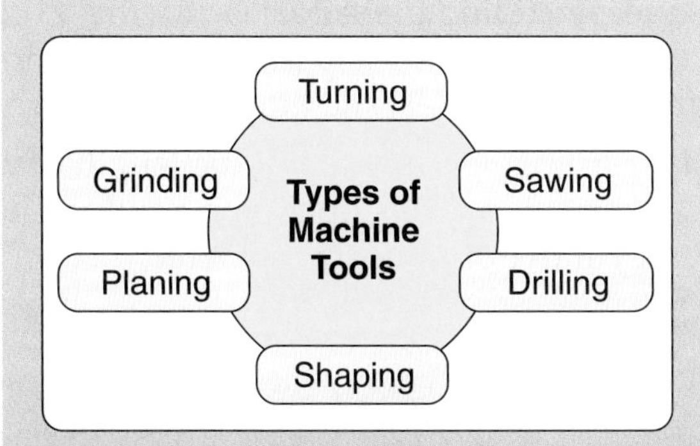

Figure 7-7. The six categories of machine tools are turning, sawing, drilling, shaping, planing, and grinding.

Types of Machine Tools

Hundreds of different machine tools exist. If you look at any machine tool catalog you can find many of them. All machine tools can be grouped into six categories, however, as shown in **Figure 7-7** These are:

- Turning machines
- Sawing machines
- Drilling machines
- Shaping machines
- Planing machines
- Grinding machines

Turning Machines

Turning machines use a process in which a workpiece is held and rotated on an axis. As noted earlier, this action is produced on machines called ***lathes,*** **Figure 7-8.** All lathes produce their cutting motion by rotating the workpiece. Linear movement of the tool generates the feed motion.

Lathes are primarily used to machine wood and metal. Plastics can be machined

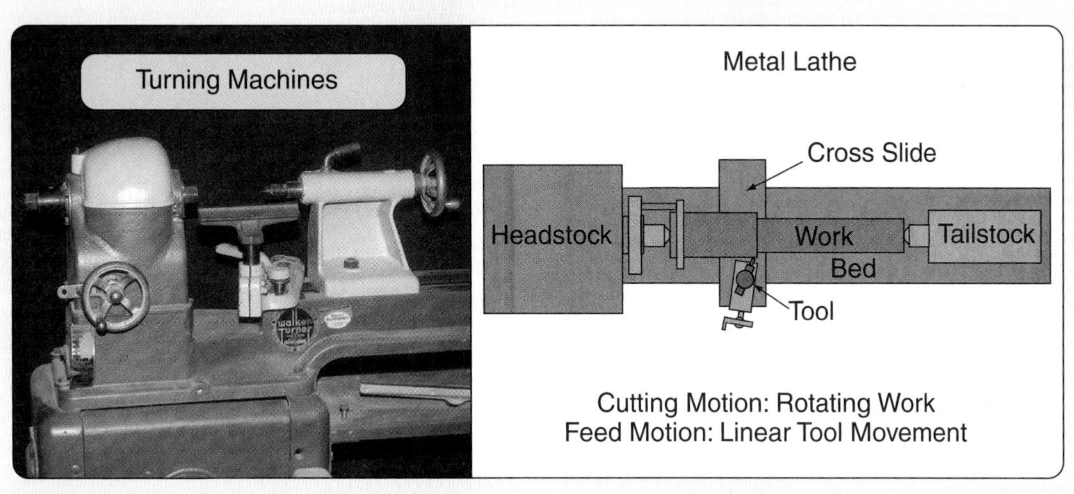

Figure 7-8. Lathes are used for turning operations.

on lathes as well. Wood and metal lathes contain four main parts or a system. A *headstock,* which contains the machine's power unit, is the heart of the lathe. This unit has systems to rotate the workpiece and to adjust the speed of rotation. At the opposite end of the machine is a *tailstock.* This unit supports the opposite end of a part that is gripped at the headstock. The headstock and tailstock are attached to the *bed* of the lathe. Finally, a *tool rest* or *holder* is provided to support the tool. Tool rests on metal lathes clamp the tool in position and feed it into or along the work. Wood lathes commonly have a flat tool rest along which the operator moves the tool by hand.

The work can be held in a lathe in two basic ways. It can be placed between centers. One center is in the headstock, and the other is in the tailstock. These centers support the workpiece and can be of two types. *Live centers* rotate with the workpiece, whereas *dead centers* are fixed as the work rotates around them.

Both wood lathes and metal lathes have centers, but they differ in the way the turning force is applied to the workpiece. Wood lathes apply the force through the headstock using a spur center (a center with cross-shaped extensions that seat into the wood). Metal lathes use a device called a dog to rotate the workpiece. The dog clamps onto the work and has a finger that engages the faceplate attached to the drive spindle.

A small workpiece may be rotated using only the headstock in the device called the chuck. Three-jaw, four-jaw, or collet chucks are commonly used to grip and rotate the part. Chucks are found on metal lathes. These chucks have jaws that squeeze and hold the part.

Lathes are used to perform many operations, **Figure 7-9.** The most common are:

- Turning: Cutting along the length of a workpiece. This operation produces a cylinder of uniform diameter.

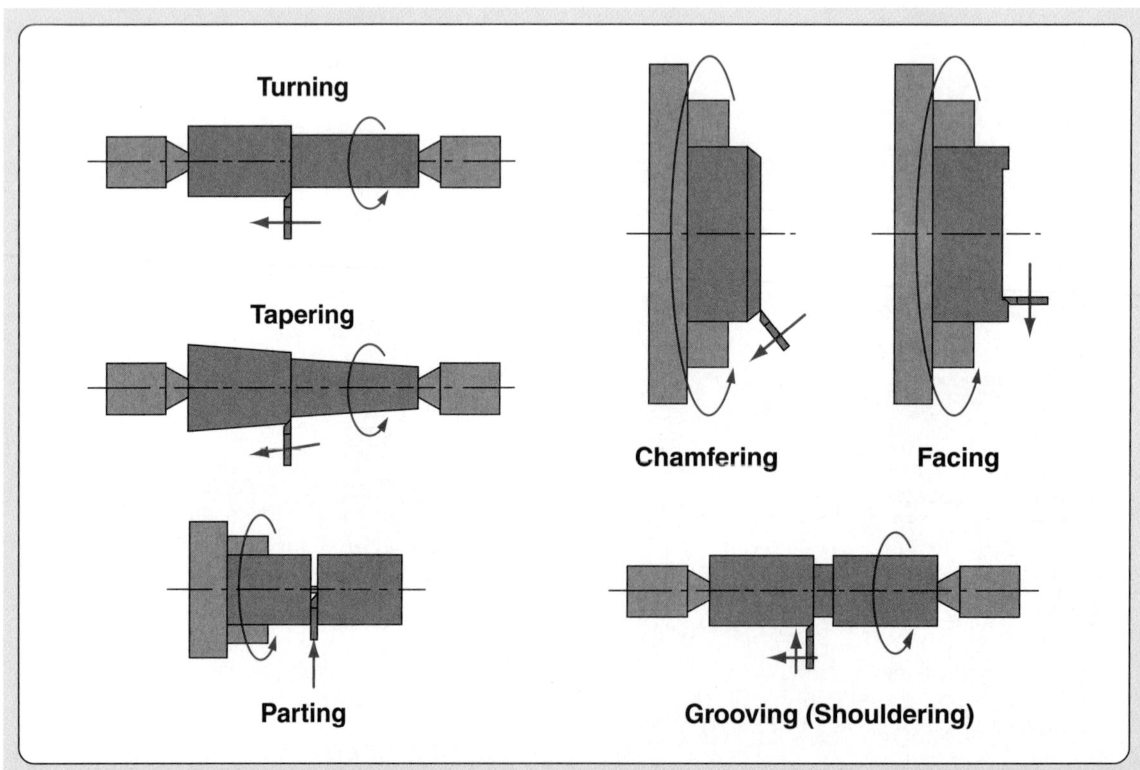

Figure 7-9. Shown here are common lathe operations.

- Tapering: Cutting along the length of a cylinder at a slight angle to produce a cylindrical shape with a uniformly decreasing diameter.
- Facing: Cutting across the end of a rotating workpiece. This operation produces a true (or square) end on the workpiece.
- Grooving (shouldering): Cutting into a workpiece to produce a channel with a diameter less than the main diameter of the workpiece.
- Chamfering: Cutting an angled surface between two diameters on the workpiece.
- Parting: Cutting off a part from the main workpiece.
- Threading: Cutting threads along the outside diameter or inside a hole in the workpiece.

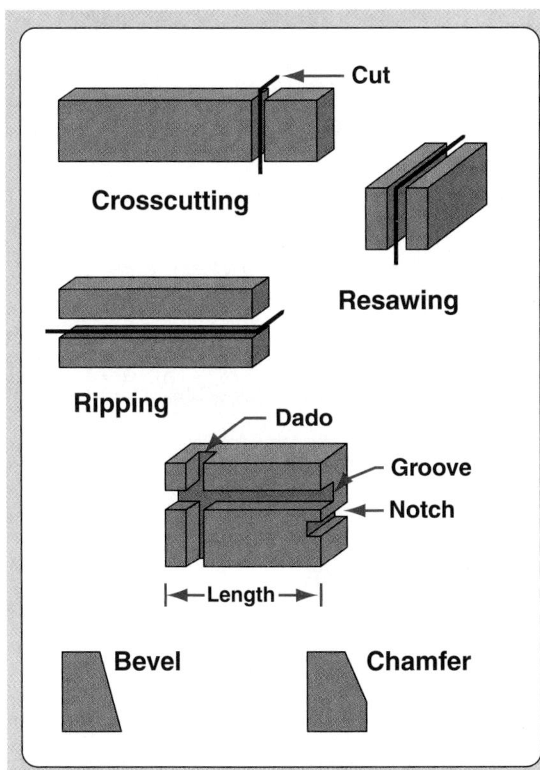

Figure 7-10. Typical sawing operations are crosscutting; ripping; resawing; grooving, dadoing, and notching; and chamfering and beveling.

- Knurling: Producing a diamond pattern of grooves on the outside diameter of a portion of the workpiece. This produces a surface that is easier to grip and turn.

In addition, drills and reamers can be placed in the tailstock to produce and finish holes in the workpiece.

Sawing Machines

Sawing machines use teeth on a blade to cut material to a desired size and shape. These machines are designed to perform a number of different cutting actions. These actions, as shown in **Figure 7-10,** include:

- Crosscutting (or cutoff): Reducing the length of a material
- Ripping (or edging): Reducing the width of a material
- Resawing: Reducing the thickness of a material
- Grooving, dadoing, notching: Cutting rectangular slots in or across a part
- Chamfering and beveling: Cutting an angled surface between two primary surfaces of a material

Sawing machines can be grouped according the type of blade they use and the methods used to produce the cutting action. This grouping identifies three basic types of saws, **Figure 7-11.**

- Circular saws
- Band saws
- Scroll saws

Circular Saws

Circular saws use a blade in the shape of a disk with teeth arranged around the edge. These teeth vary in shape and arrangement depending on the operation to be performed. Common blades are available for crosscutting, ripping, and combination cutting (crosscutting or ripping).

Technology Explained

Flexible Manufacturing

flexible manufacturing: Using a set of automatic, programmable machines to produce products in low volume production runs.

Computer control has greatly changed how manufacturing systems work. The Industrial Revolution was based on producing large numbers of uniform products using a *continuous manufacturing* system. Any change in the design of a product necessitates changes in the machines that make the product. These changes are very expensive to make.

The advent of computers has changed how products are designed. Products can be designed, tested, and refined on a computer. The design for a product is expressed in a *computer language.* This language can be used to control machines and inspect product quality, **Figure A.** Computers are also used to control how products move through a sequence of operations. This level of computer control has led to computer-integrated manufacturing (CIM), or flexible manufacturing systems.

These systems use a number of technologies, including computer-aided design, to develop and specify a product's size and shape. Robots are used to load and unload machines and move products from machine to machine, **Figure B.** Computer numerical control is used to guide machine operation. Electronic measurement and computer data processing help to improve inspection and quality control, **Figure C.**

Figure A. Flexible manufacturing is based on computer control. (Hitachi Seiki)

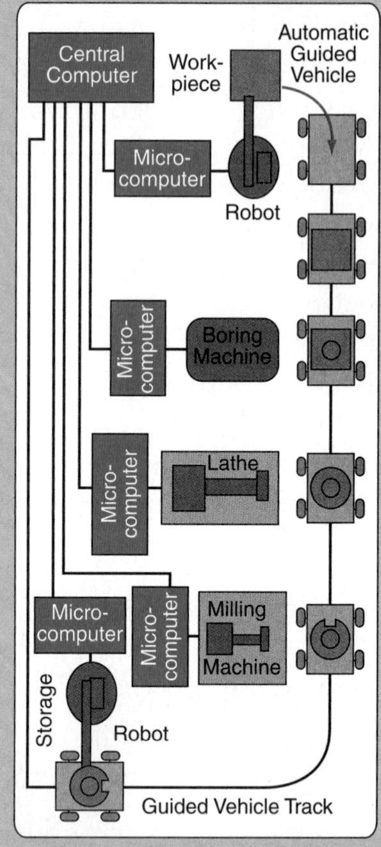

Figure B. This is a diagram of a simple flexible manufacturing system. Each part of the system is under computer control.

Figure C. Computers are used to direct the operation of machine tools and assembly machines (left), material handling systems (center), and inspection systems (right).

Three basic types of circular saws exist: the table saw, radial saw, and chop saw, **Figure 7-12.** All three of these machines generate the cutting motion by rotating the blade. Their feed motions are different, however. The ***table saw*** uses a linear feed of the material. The workpiece is pushed into the rotating blade to generate the cut. The machine operator manually feeds the material in low-volume production settings. Automatic feeding devices are used to increase the speed, accuracy, and safety of many high-volume sawing operations.

Many other machines use cutting and feed motions that are like the circular saw. These include metal milling machines and wood shapers, routers, jointers, and planers. These machines feed the workpiece into a rotating cutter.

The ***radial saw*** moves the rotating blade across the workpiece. The workpiece is positioned on a stationary table. The rotating blade produces the cutting motion. The rotating blade is manually drawn across the material for the feed motion. The direction of the blade travel is determined by the position of the arm of the machine. Generally the machine can cut a square angle and any angle up to 45° on each side of square.

The ***chop saw*** is used to cut narrow strips of material to length. The blade is attached to a pivot arm. The material is placed on the table of the machine. The blade assembly is pivoted down to generate the feed motion. Many cutoff saws have saw blade units that can be rotated so that the material can be cut off at a specified angle. This type of saw, called a miter saw, is widely used to cut trim moldings for homes and the parts of picture frames.

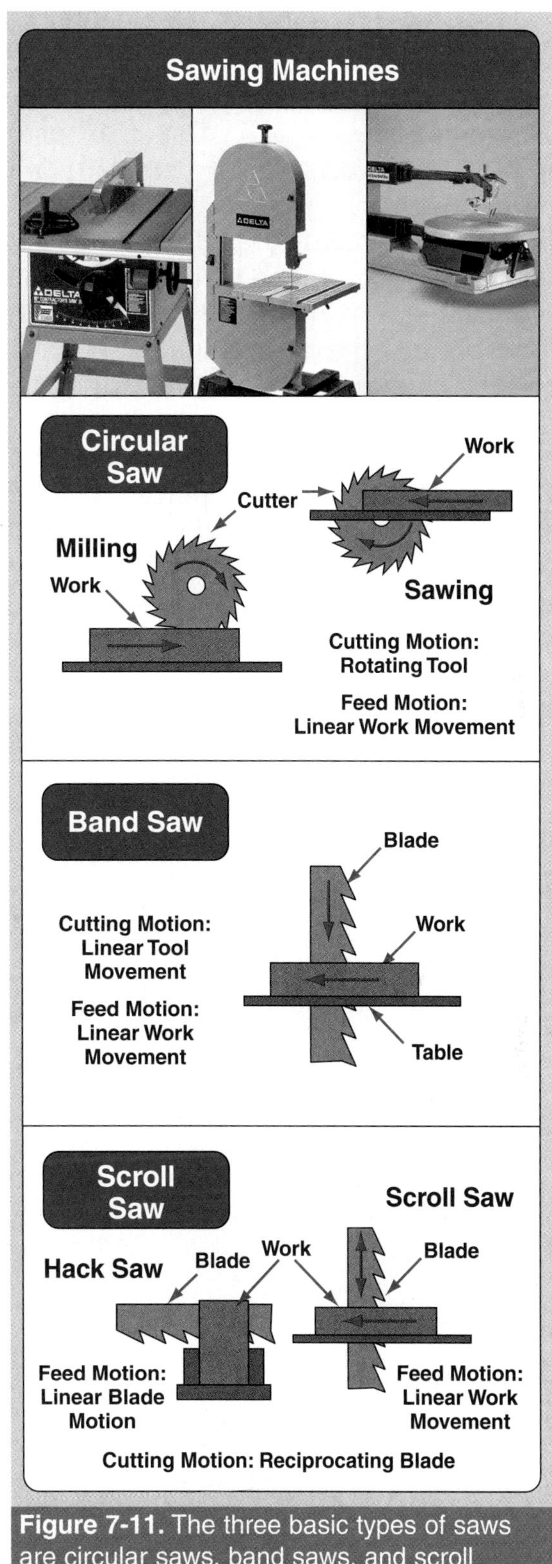

Figure 7-11. The three basic types of saws are circular saws, band saws, and scroll saws. (Delta)

Band Saws

Band saws use a blade made of a continuous strip or band of metal. Most of these bands have teeth on one edge. Large band saw blades used in lumber mills have teeth on both edges, however.

The band generally travels around two wheels, which gives a continuous linear cutting action. The band may be vertical or horizontal. Horizontal band saws are usually used as cutoff saws for metal rods and bars. They have replaced hacksaws as the primary method of cutting metal stock to length. Horizontal machines hold the material stationary in vises or clamps. They produce the feed motion by allowing the blade unit to pivot into the work.

Vertical band saws are widely used to cut irregular shapes from wood or metal sheets. The material is placed on the machine table. Then it is fed into the blade to produce the cut. Most often, the operator manually feeds the material to produce the desired cut.

Scroll Saws

Scroll saws use a straight blade that is a strip of metal with teeth on one edge. The blade is clamped into the machine. The machine then moves the blade up and down to produce a reciprocating cutting motion. The material is placed on the table and manually fed into the blade.

Portable scroll saws grip only one end of the blade. Then the saw is fed into the work to produce the feed motion.

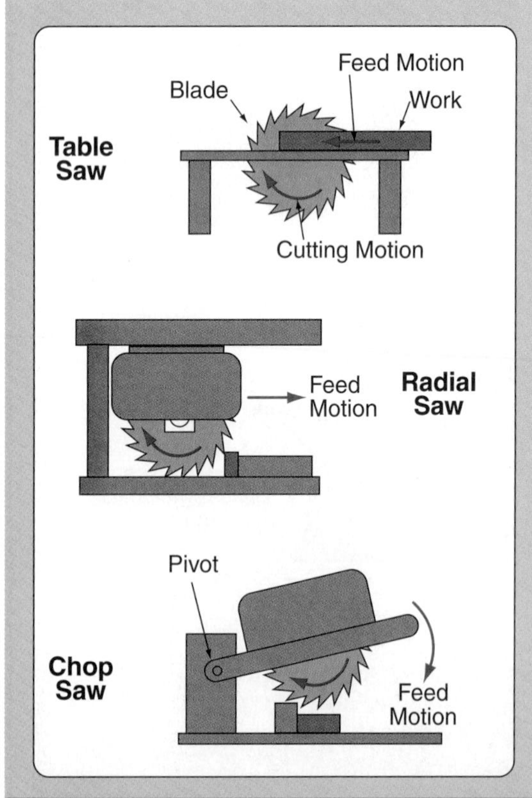

Figure 7-12. The three types of circular saws are the table saw, the radial saw, and the chop saw.

Drilling Machines

Holes in parts and products are very common and can be produced in various ways. Holes can be punched into sheet metal using punches and dies. Holes can be burned into plate steel using an oxy-acetylene cutting torch. Holes can be produced in the part directly as it is cast from molten material. Holes can be produced with powerful beams of light (laser machining) or electrical sparks (electro-discharge machining).

Many cylindrical holes, however, are produced through drilling operations. ***Drilling machines*** produce or enlarge holes using a rotating cutter. Generally, the cutting motion is produced as the drill or bit rotates. The drill is moved into the work to produce the feed motion. The drill press, **Figure 7-13,** is the most common machine using these cutting and feed motions.

Drilling may also be done by rotating the work to produce the cutting motion. Then the stationary drill is fed into the work. This practice is common with drilling on a metal lathe or a computer-controlled machining center.

Many operations can be completed on drilling machines. The most common are shown in **Figure 7-14.** These operations are:

- Drilling: Producing straight, cylindrical holes in a material. These holes can be used to accommodate bolts, screws, shafts, and pins for assembly. They can also be a functional feature of a product. For example, holes are essential for the functioning of furnace burners, automobile carburetors, and compact disc recordings.

- Counterboring: Producing two holes of different diameters and depths around the same center point. Counterbores are used to position shafts and recess heads of fasteners, among other purposes.
- Countersinking: Producing a beveled outer portion of a hole. Most often, countersinking is used with flathead wood and metal screws. Countersinking holes allows screw heads to be flush with the surface of the part.
- Reaming: Enlarging the diameter of a hole. This action is generally performed to produce an accurate diameter for a bolt hole.

Drilling operations use various drilling tools. These tools include twist drills, spade bits, and forstner bits. ***Twist drills*** are shafts of steel with points on the end to produce a chip. These chips are carried from the work on helical flutes that circle the shaft. ***Spade bits*** are flat cutters on the end of a shaft. The bottoms of the cutters are shaped to produce the cut. ***Forstner bits*** are two-lipped woodcutters that produce a flat-bottomed hole. Hole saws, which use sawing machine action, and fly cutters,

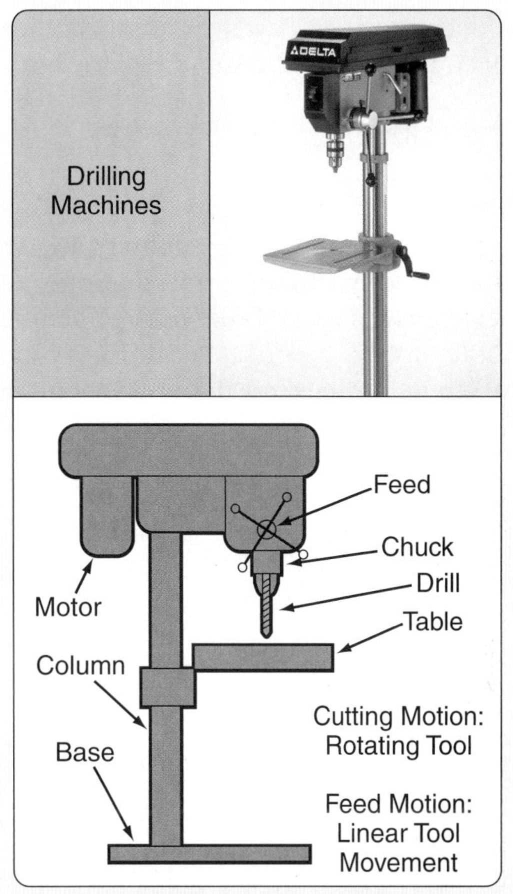

Figure 7-13. A drill press is the most common drilling machine. (Delta)

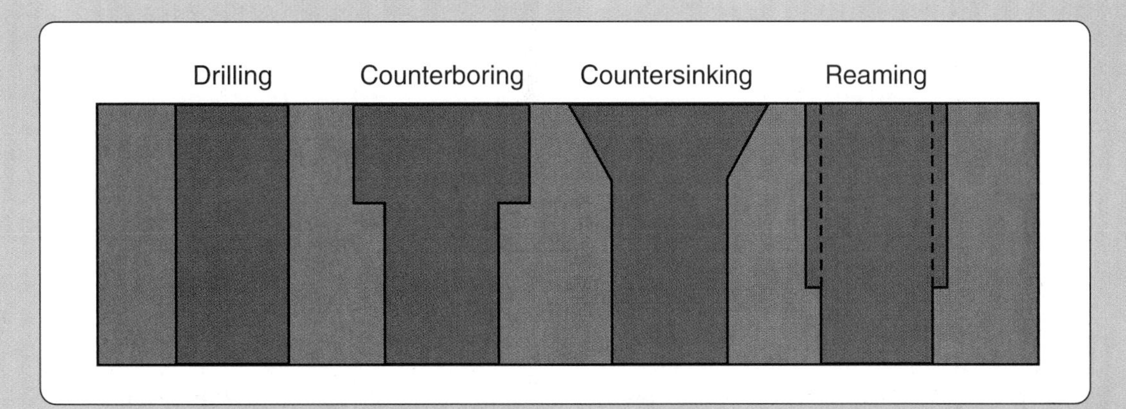

Figure 7-14. The most common operations performed on drilling machines are drilling, counterboring, countersinking, and reaming.

which use lathe-type tools, can also be used to produce holes. Several of these bits and drills are shown in **Figure 7-15.**

Shaping and Planing Machines

Shaping and *planing machines* are two metal working machine tools that produce flat surfaces, **Figure 7-16.** These machines should not be confused with the woodworking shaping and planing machines, which operate on the same principles as sawing machines. Wood shapers and planers use a rotating cutter into which the wood is fed.

Both the metal shaper and the metal planer use single-point tools and reciprocating motion to produce the cut. Their difference lies in the movements of the tool and the workpiece. The metal shaper moves the tool back and forth over the workpiece to produce the cutting motion. The work is moved over after each forward-cutting

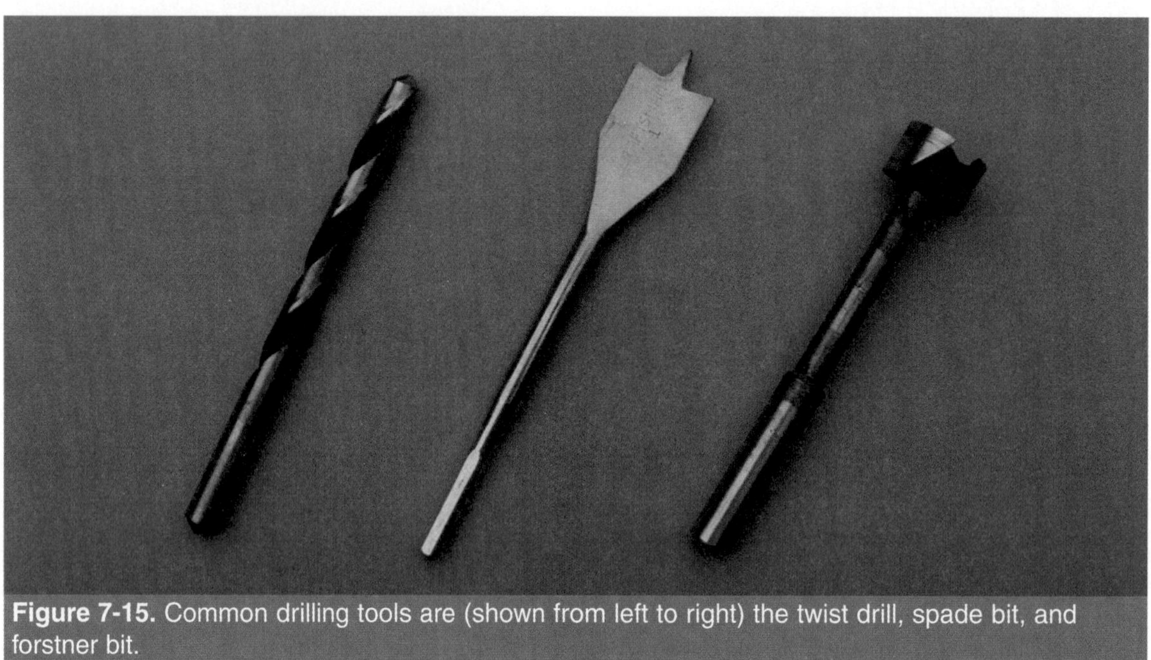

Figure 7-15. Common drilling tools are (shown from left to right) the twist drill, spade bit, and forstner bit.

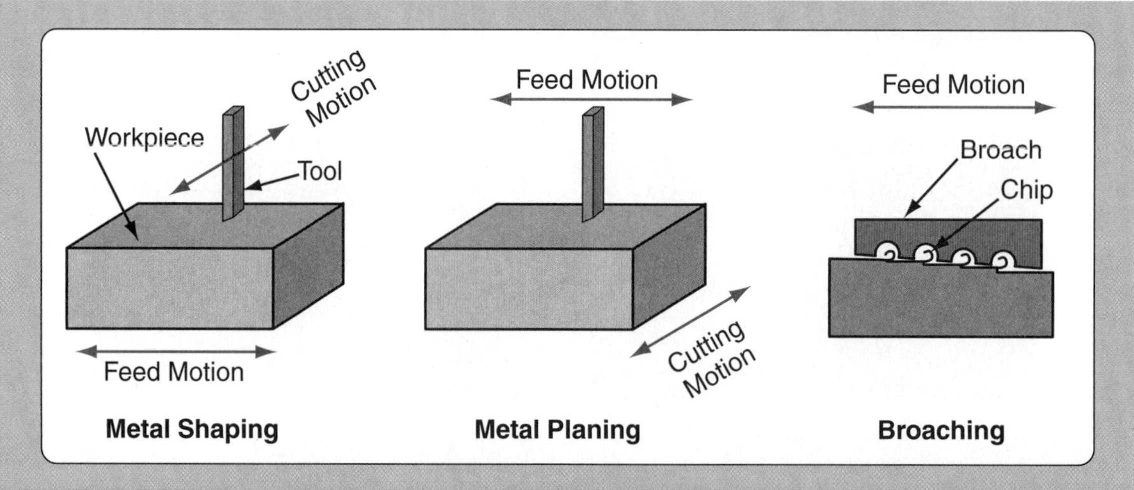

Figure 7-16. Shaping and planing machines use different feed and cutting motions.

stroke to produce the feed motion. The metal planer reciprocates the workpiece under the tool to generate the cutting motion. The tool is moved one step across the work for each cutting stroke.

Both shapers and planers can cut on the face or side of the part. They can also be used to machine grooves into the surface. Both machines have limited use in material processing.

A machine that is closely related to these machines is a broach. This machine uses a tool with many teeth. Each tooth sticks out slightly more than the previous tooth. As the broach tool is passed over a surface, each tooth cuts a small chip; however, when all the teeth pass over the work, a fairly deep cut is possible. Broaches are often used to machine a keyway (rectangular notch) in a hole. Keyways are widely used to assemble wheels and pulleys to axles.

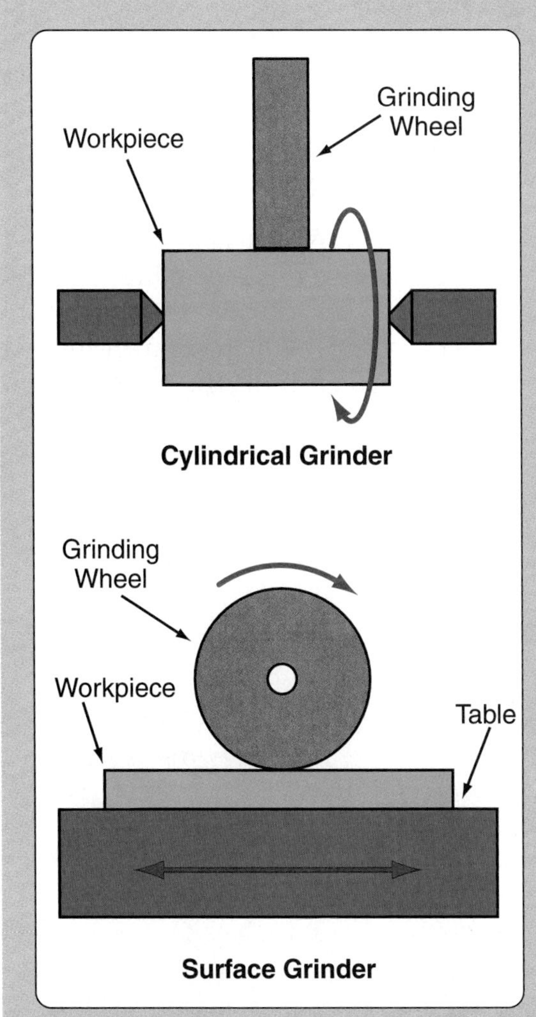

Figure 7-17. Two common grinding machines are the cylindrical grinder and the surface grinder.

Grinding Machines

Grinding machines use bonded abrasives (grinding wheels) to cut the material. These wheels have random cutting surfaces that remove the material in the form of very small chips.

Grinders are basically adaptations of other machine tools. The two most common are cylindrical grinders and surface grinders, **Figure 7-17.** ***Cylindrical grinders*** use the lathe principle to machine the material. The workpiece is held in a chuck or between centers and rotated. A grinding wheel is rotated in the opposite direction. The opposing rotating forces produce the cutting motion. The grinding wheel is fed into the work to produce the feed motion.

Surface grinders work on the metal planer principle. A rotating grinding wheel is suspended above the workpiece. The work is moved back and forth under the wheel to produce the cutting motion. The work is moved slightly, or indexed, after each grinding pass to produce the feed motion.

A third type of grinder used is the pedestal grinder. The rotating grinding wheel produces the cutting motion. The operator manually moves the workpiece across the face of the wheel to produce the feed motion.

Energy-Processing Converters

At the heart of all technological systems is energy. It powers most devices that process materials and information.

As we learned earlier, energy can be neither made nor destroyed, but it can be converted and applied to do work. This is the task of ***energy conversion*** or ***energy-processing converters.*** Literally hundreds of these converters exist, and they process energy in various ways. For example, energy-processing converters:

- Convert mechanical energy into electrical energy (example: electric generator)
- Convert electrical energy into mechanical energy (example: electric motor)
- Convert radiant energy into thermal energy (example: solar heater)
- Convert heat energy into mechanical energy (example: internal combustion engine)

Through a study of human history we can quickly identify three major energy-processing converters that have helped to shape life as we know it. They are the steam engine, the electric motor, and the internal combustion engine. We will focus on these converters in this chapter to gain an understanding of the importance of energy conversion. Other energy converters will be discussed in Chapter 27.

The Steam Engine

The steam engine was the foundation of the Industrial Revolution. Earlier converters simply translated one form of mechanical power to another. For example, animal, water, and wind power reduced human labor. These power sources were still mechanical power devices, however. The steam engine was the first device to convert energy effectively from one form to another. It converted heat energy to mechanical energy.

The invention of the steam engine dramatically altered people's lives. For example, the steam engine changed our early dependence on moving water to power our factories. Because of the steam engine, we could now locate manufacturing plants away from fast-moving rivers. We could expand both the numbers and the locations of these plants. Steam power was also applied to river, ocean, and rail transportation, **Figure 7-18.** This application allowed us to move more cargo over longer distances in a shorter period of time.

The first practical steam engine was developed by Thomas Newcomen in the early 1700s. It applied available knowledge about pistons, steam condensation, atmospheric pressure, vacuums, and mechanical linkages. The result was a large engine that was widely used in Europe to pump water from mines.

The engine was based on the fact that water takes up less space than does steam (hot water vapor). The engine, as shown in **Figure 7-19,** operated through the following steps:

1. All valves were closed.
2. The steam valve was opened to allow steam to move from the boiler into the cylinder. This caused the piston to move upward, which, in turn moved the pump end of the rocker arm downward.

Figure 7-18. Steam engines were widely used to power locomotives until the 1940s. (Norfolk Southern Corp.)

3. The steam valve was closed, and a catch caught the piston in the upward position.
4. The cold water valve was opened. This allowed a jet of cold water to enter the cylinder. The steam quickly condensed as the cylinder cooled. This caused a vacuum inside the cylinder.
5. The catch holding the piston was released, allowing the piston to be drawn downward by the vacuum. With this movement, the rocker arm was rotated, causing the pump end to rise.

This five-step cycle was then repeated, causing a continuous pumping action. The Newcomen atmospheric steam engine was very inefficient, however. It took a large amount of fuel to pump a small amount of water. As we noted in Chapter 1, James Watt developed a more efficient design. His engine used condensing steam to drive the piston in both directions. This resulted in a reciprocating steam engine that was widely used to power industrial machinery and transportation vehicles.

The motion of the Watt double-acting steam engine is shown in **Figure 7-20.** As shown in the first drawing, the steam

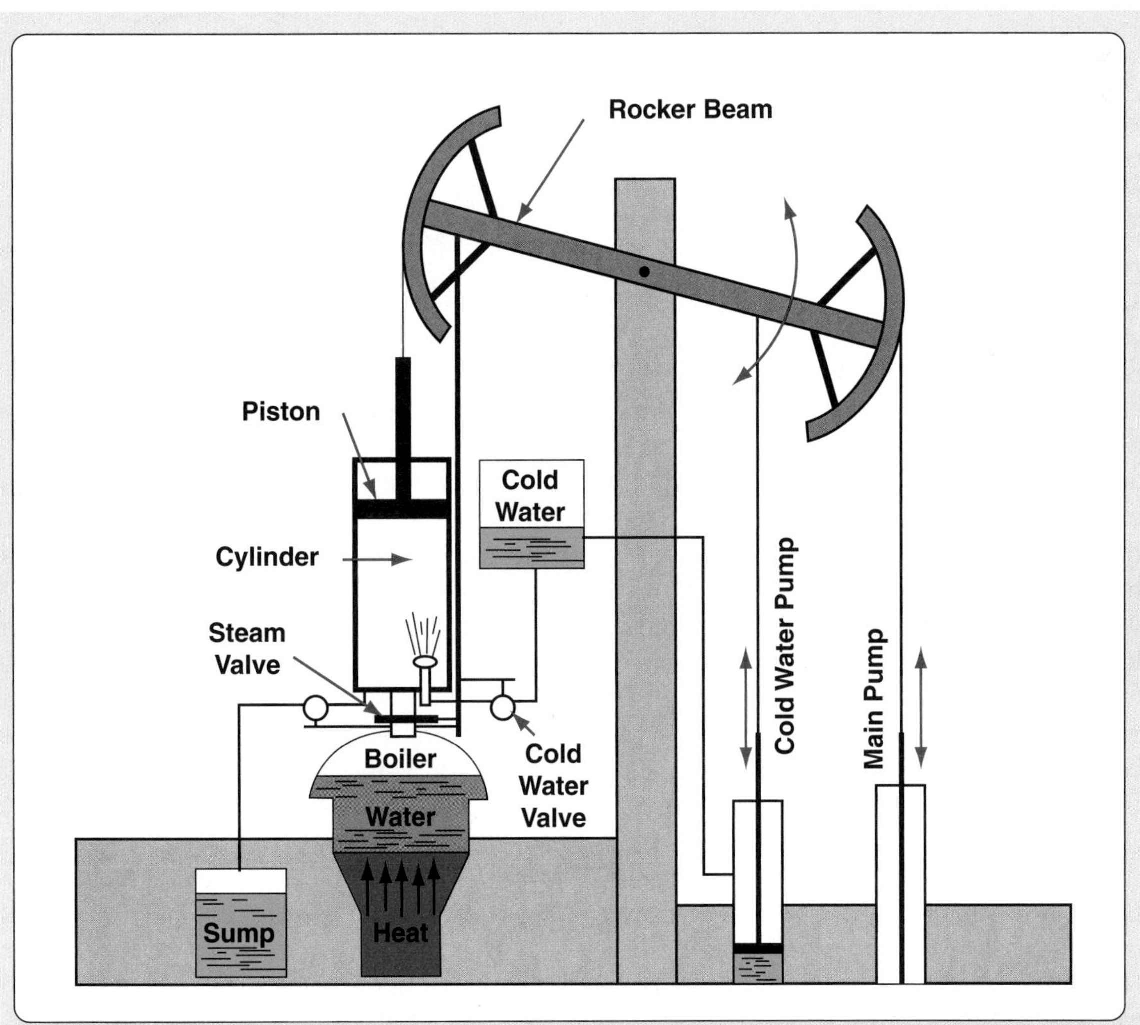

Figure 7-19. The Newcomen steam engine used steam and cold water to cause a piston and rocker beam to move, thus changing heat energy to mechanical energy.

enters the engine from the boiler. It flows to the left, causing the piston to move to the right. This action moves the piston rod, which turns the flywheel. As the flywheel turns it moves the slide valve to the left. The second drawing shows its new position. Now the steam flows to the right, forcing the piston to the left. Again the piston rod moves the flywheel. The slide valve is moved back to the position shown in the first drawing. The cycle starts over again. The reciprocating motion of the piston continues to turn the flywheel. This wheel can be connected to any device that needs a rotating power source.

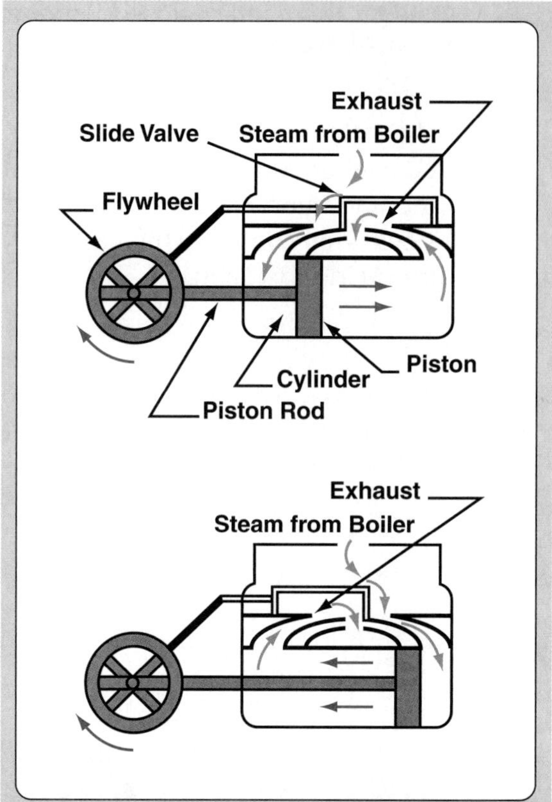

Figure 7-20. The Watt steam engine was a tremendous improvement on the Newcomen steam engine. This engine used condensing steam to drive the piston in both directions.

The Internal Combustion Engine

Because of technological advances, the steam engine has been almost totally replaced by other energy converters. The internal combustion engine, which was developed in the late 1800s, now powers most land transportation vehicles. Researchers continue work on energy-converter improvements, including the replacement of the internal combustion engine with fuel cells. Fuel cells use stored hydrogen and oxygen and offer many advantages over the internal combustion engine. For example, if we are able to use fuel cells to power our automobiles and trucks, we will have a cleaner environment, vehicles with fewer moving parts and thus less maintenance, and less dependence on foreign oil. The full-scale use of this type of energy conversion is still a few years away, but various manufacturers are working on operating models now.

The common internal combustion engines are gasoline and diesel engines. Both of these engines change heat energy into mechanical energy. They drive a reciprocating piston by igniting a fuel. The piston's reciprocating motion is changed into rotary motion by a crankshaft, **Figure 7-21.**

Gasoline engines can be either two-stroke cycle engines or four-stroke cycle engines. The four-stroke cycle engine is the engine most people come into contact with and will be the focus of this discussion.

A ***stroke*** is the movement of a piston from one end of a cylinder to another. A ***cycle*** is a complete set of motions needed to produce a surge of power. A two-stroke cycle engine moves the piston up and back once (two strokes: one up and one down) to produce a power stroke. A four-stroke engine moves the piston up and back twice (four strokes) to produce a power stroke.

Look at **Figure 7-22** to see these four strokes: intake, compression, power, and exhaust. During these strokes the following actions take place:

- Intake stroke: The piston moves downward to create a partial vacuum. The intake valve opens. Atmospheric pressure forces a fuel and air mixture into the cylinder.
- Compression stroke: The intake valve closes, and the piston moves upward. As the piston moves up, the fuel-air mixture is compressed in the small cavity at the top of the cylinder. The area forms a combustion chamber for the power stroke. The compression ratio of the engine will tell you how much the fuel-air mixture has been compressed. The standard compression ratio is about 15:1. This ratio tells you that the cylinder volume at the beginning of the compression stroke is fifteen times larger than at the end of the compression stroke. Therefore, the fuel-air mixture has been reduced to about 6.6 percent of its original volume.
- Power stroke: An electrical spark is produced between the two points of the

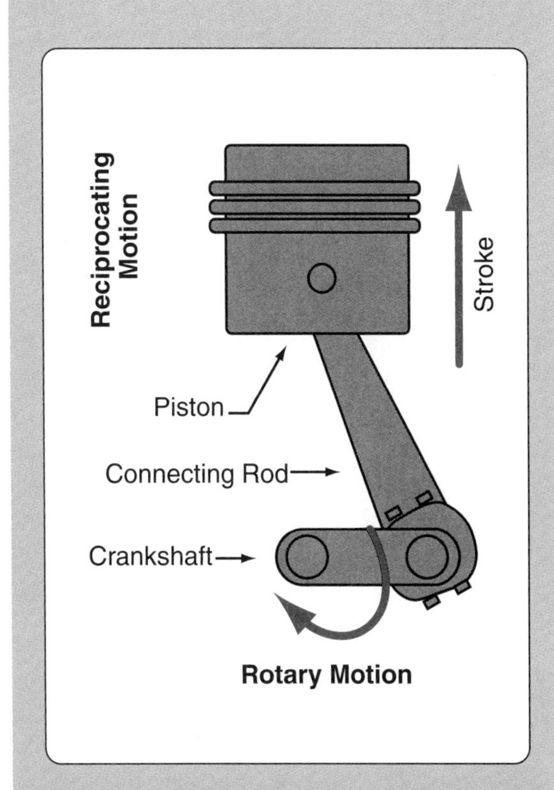

Figure 7-21. The reciprocating piston in an internal combustion engine turns a crankshaft.

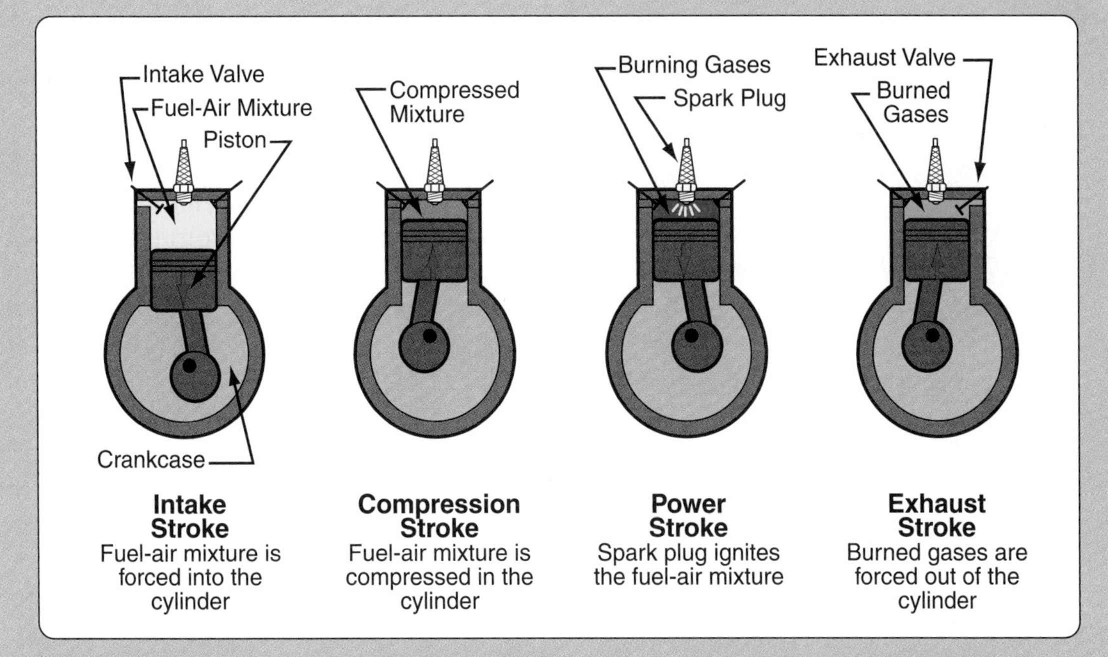

Figure 7-22. A four-stroke engine moves the piston up and down twice to produce a power stroke. The four actions are intake, compression, power, and exhaust.

spark plug. This action ignites the compressed fuel-air mixture. The burning gases produce temperatures as high as 4000°F (2200°C). This temperature expands the gas and generates pressure up to 1000 pounds per square inch (psi) or 7000 kilopascals (KPa). The pressure forces the piston downward in a powerful movement.

- Exhaust stroke: When the piston reaches the bottom of the power stroke, the exhaust valve opens. The piston moves upward to force exhaust gases and water vapor from the cylinder. These are the products of the combustion during the power stroke. At the end of this stroke the cylinder is ready to repeat the four strokes.

Single-cylinder engines are common for low-horsepower applications such as lawn mowers, cement mixers, and portable conveyors. For more demanding applications, several cylinders are combined into one engine. The cycle of each cylinder is started at a different point of time so that the engine has a series of closely spaced power strokes. Four-cylinder, six-cylinder, and eight-cylinder engines are common.

Figure 7-23. All the equipment in this metal stamping line is powered by electric motors (Minster Machine Co.)

The Electric Motor

The last energy-processing device that will be discussed in this section is the electric motor. This device is probably the most universally used source of power. The average home has more than forty motors in it. You can find motors in clocks, refrigerators and freezers, video recorders, compact disc players, furnaces and air conditioners, clothes washers and dryers, shavers, hair dryers, and many other appliances. They are also on construction sites and in factories, **Figure 7-23.** They play a vital part in every transportation and communication system.

The electric motor is based on the laws of magnetism and electromagnetism. These laws, as shown in **Figure 7-24,** state that:

- Like poles of a magnet repel one another, and unlike poles attract one another.
- Current flowing in a wire creates an electromagnetic field around the conductor.

Look at **Figure 7-25** to see how these laws are applied. You will note that there are two magnets. The outer magnet is stationary and is called the ***field magnet.*** The inner magnet is an electromagnet that can rotate and is called the ***armature.***

In View A, electrical current is allowed to pass through the armature. The current induces magnetism in the armature core. A north pole will be on one end and a south pole on the other. The direction the wire is wrapped around the core and the direction of the current will determine which end is the north pole.

View A shows the north field pole and the north armature pole next to each other. These are like poles and will repel each other. Therefore, the armature spins one-fourth turn to the position shown in View B. Now the unlike poles will attract each other, and the armature will continue rotating another one-fourth turn. When the armature is in the position shown in View C, the direction of the current is reversed. This action occurs mechanically in direct current motors. Alternating current motors use the existing direction changes in the line current. The current direction change reverses the poles on the armatures as shown in View D. Now the entire action is repeated. Like poles repel and turn the armature another one-fourth of a turn. Then unlike poles attract, turning the armature the final one-fourth turn to complete a full revolution of the armature.

Information-Processing Machines

Information processing and exchange are as old as human existence. Early results of these actions can be seen in carvings found on cave walls and rocks. Today the products of information processing are in books and photographs and on tape and film.

Modern information processing is built on a number of important technological advancements. Four of these will be explored here, and others will be introduced in Section 6 of this book.

The communication methods basic to all modern communication systems are printing, telecommunications, and the computer. Printing is based on movable type, developed in the 1400s. Telecommunications has its roots in Alexander Graham Bell's telephone and Marconi's radio. Computers are based on semiconductor circuits on microchips that process data. Each will be discussed in turn.

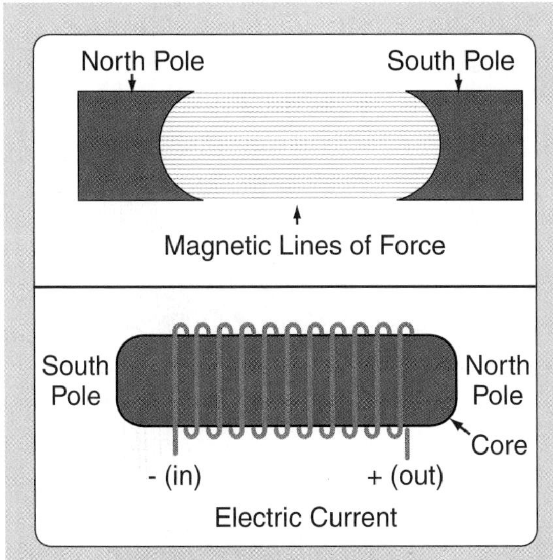

Figure 7-24. Motors take advantage of two principles of physics: (1) Magnetic lines of force travel between opposite poles of a magnet; and (2) Current flowing through wires wrapped around an iron core produces magnetic poles at the ends of the core.

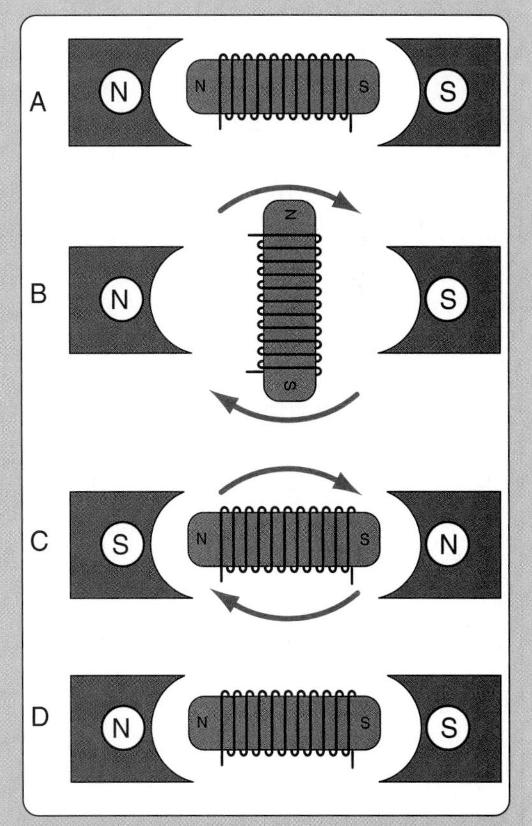

Figure 7-25. Electric motors operate by using a stationary magnet and an armature that switches its magnetic poles.

Printing

We have become accustomed to having printed materials at our fingertips. Books, magazines, newspapers, pamphlets, and brochures are everywhere. The abundance of printed materials is of recent history, though. Until the last five hundred years, most people could not read and had never seen a book. Those books that did exist were handwritten on parchment paper by monks and scribes. As civilization grew, so did the demand for information. In the 1440s this demand was addressed with a new printing process developed by Johann Gutenberg, a German goldsmith. It used movable type, as shown in **Figure 7-26.**

Before this time the Chinese printed entire pages from blocks of wood with the message carved in them. Each page required a separate block, however. Also, the block could not be changed or used again. Gutenberg used his metal-casting knowledge to produce standard-sized metal blocks. On the end of each block was a letter of the alphabet in reverse image. These blocks, called type, were all exactly the same height and on a body that was the same thickness. This type became part of the new way to produce a printed page, as shown in **Figure 7-27.**

Gutenberg assembled the type in groups to form words, sentences, and paragraphs. He then clamped the type in a frame to make a *page form.* He spread ink on the type and laid paper on top of the page form. He slid the form into a press that was adapted from a press that squeezed juice from grapes. He applied pressure by turning a screw that moved a pressure pad downward. The resulting force transferred the ink from the type to the paper. He then opened the press and removed the printed page from the type form. He was then ready to repeat the process until the desired number of pages were printed. Then, he could take apart the type form and use the individual pieces of type again.

The first book Gutenberg printed was the Bible. He and his helpers completed an unknown number of copies. Forty-seven books from this printing still exist.

From this humble start came the printing industry and a number of different printing processes. They will be explored in more depth in Chapter 20.

Telecommunications: The Telephone

Almost everyone uses a telephone. It allows us to communicate information, ideas, and thoughts to another person quickly and inexpensively. It is the simplest form of telecommunication in use today. It is called *hardwired* communication because a separate channel connects each telephone. These channels may be copper wires or fiber-optic strands.

The telephone converts sound into electrical energy. The electrical energy is then transmitted from the sender to the receiver. At the receiver, the electrical

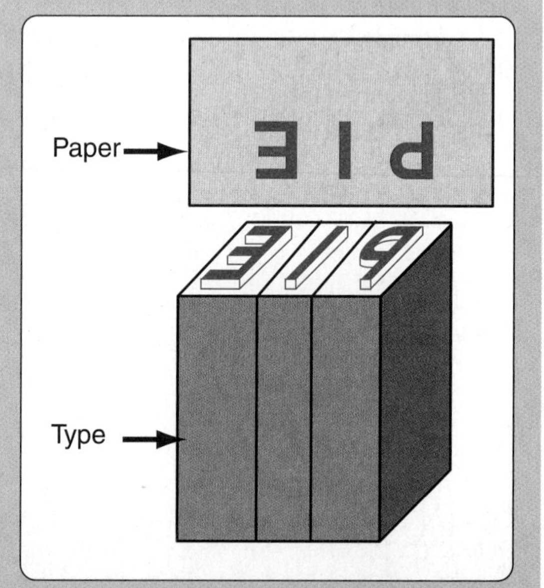

Figure 7-26. Gutenberg developed movable type, which revolutionized book production and started the printing industry.

energy is converted back into sound waves. In short, the telephone changes information contained in spoken messages into electrical signals.

The first information conversion occurs at the telephone microphone (mouthpiece), **Figure 7-28.** Carbon granules are placed between two wires in an electric circuit. A diaphragm is placed over the particles. Sound waves cause the diaphragm to vibrate. The movement compresses the carbon and changes its ability to conduct electrical current. Each tone vibrates the diaphragm at a different rate and, therefore, allows a specific amount of electricity to be conducted through the circuit. The pulse of electricity, which is coded information, is carried to the telephone receiver (earpiece). Here an electromagnet is in the circuit. The electric pulses cause a magnet to vibrate a diaphragm. This vibration produces sound almost identical to the sound that created the electrical impulse in the mouthpiece. The electrical impulse is converted back to information in the form of audible sound.

The use of the telephone has been expanded from a personal communication device to a data-transmission instrument. Telephone circuits are now used to transmit coded electronic messages through fax machines and computer modems. The cellular telephone integrates the telephone with the radio, which will be discussed next. It allows people to be connected to the phone system from almost any location.

Telecommunications: The Radio

The radio also sends messages from one place to another. Unlike the telephone, which uses wires to connect the sender and receiver, the radio uses radio waves. Radio waves are invisible electromagnetic energy that travels through the atmosphere and space. The radio was originally developed by Guglielmo Marconi for ships to send messages when they were in danger. These wireless sets were first used in 1901. The radio is now used to carry

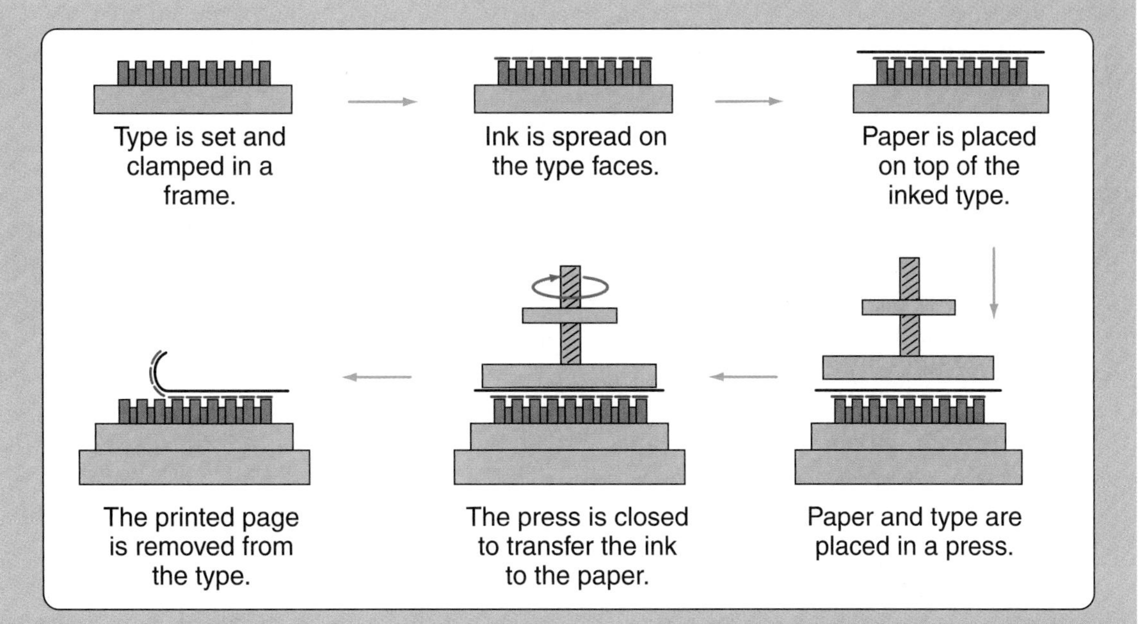

Figure 7-27. To create a printed page, Gutenberg set pieces of type, spread ink on them, put paper on the ink, and used a press to transfer the ink to the paper. The pieces of type could then be used again.

Connections to Technology: **Communication**

Computer Bugs

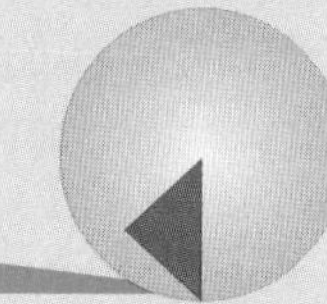

We have all probably heard or used the term "debug" to describe the repair of a computer problem. How many of us are aware, however, that the word came into common usage as the result of an actual insect?

In the early 1940s, a pioneer in the new world of computers, Grace Murray Hopper, was working for the U.S. Navy as a computer programmer. In fact, she was the third programmer of the Mark I, the world's first programmable digital computer. The navy used the computer to make quick calculations of such tasks as determining where to lay minefields.

One day as Ms. Hopper and others were checking the computer to see what had caused a breakdown, she discovered a dead moth inside one of the areas. As she was removing the moth with tweezers, someone asked her what she was doing. "I'm debugging the machine," she reportedly said. The term stuck and is now even defined in the dictionary as a word meaning the elimination of computer malfunctions.

Grace Murray Hopper continued to be one of the foremost computer programming experts in the world until her death in 1992. What was one of her other major accomplishments?

information and entertainment to all parts of the world.

A radio communication system contains a transmitter to produce the radio waves and a receiver to collect them. The transmitter changes sound (voice and music) into radio waves. It imposes the sound information onto a carrier wave. This action is called *modulation.* Modern transmitters use either frequency modulation (FM) or amplitude modulation (AM) to place the information onto the carrier wave. Frequency modulation changes the rate of the cycles on the carrier wave (or its frequency). Amplitude modulation changes the strength (height, or amplitude) of the carrier wave.

Figure 7-29 shows how radio transmitters and receivers work. You will see that the microphone changes sound waves into electrical impulses or signals.

These signals are carried into the transmitter. A carrier wave generator in the transmitter produces the carrier signal. A mixer merges the sound signals with the carrier wave through frequency or amplitude modulation to make a radio signal. The radio signal is then amplified (made stronger) and sent to a broadcast antenna. From the antenna, radio waves radiate through the atmosphere. A receiving antenna at a distant location collects the waves. A tuner circuit separates the desired radio signal from all other radio signals. The selected signal is then amplified and demodulated (the carrier wave is removed from the signal). The sound signal remains and is amplified. Much like a telephone, the amplified signal powers a speaker to produce sound waves that people can hear and understand.

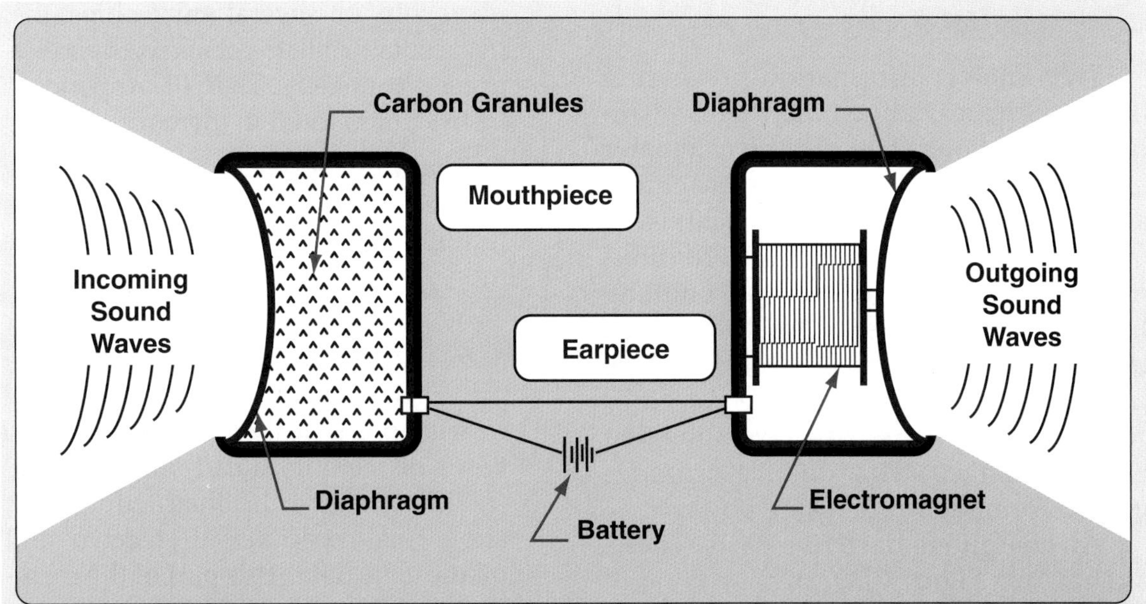

Figure 7-28. Telephones operate by changing sound waves into electrical impulses and then back into sound waves.

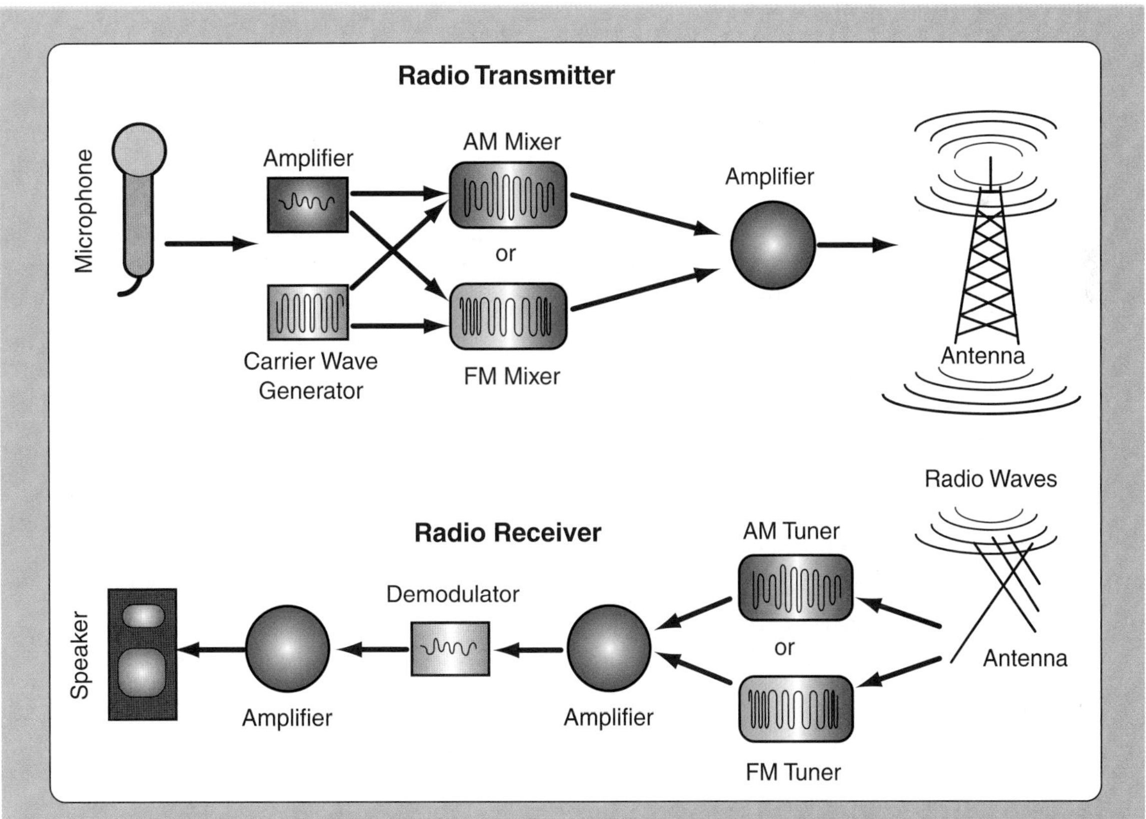

Figure 7-29. A radio communication system has a transmitter to produce radio waves and a receiver to collect them.

Computers

Two major information-processing machines have changed the way we handle information. They are the calculator and the ***computer.*** Both of these machines are different from all other machines in one important way: they can store information. We will focus only on the computer here, however, because of a feature unique to them. Unlike calculators, which have fixed programs that cannot be erased or changed, computers have programs that can be changed. This quality provides great information-processing power.

All computers have five parts. These, as shown in **Figure 7-30,** are:

- ***Input unit:*** A device used to enter data (numbers and letters) into the system. Input devices include keyboards, tape drives, disk drives, telephone modems, mice, and other computers.
- ***Processing unit:*** The part of the computer, called a central processing unit (CPU) or microprocessor, that manipulates the data. It is made up of a microchip or several microchips that may contain millions of microscopic electronic components. The CPU responds to instructions stored in memory.
- ***Output unit:*** A device used to display and record the results of the processing unit's actions. These devices include monitors, printers, plotters, and disk drives.
- ***Memory unit:*** The section of the computer that holds information and instructions. The memory unit has two types of memory. Read-only memory (ROM) contains the instructions that allow the computer to receive and manipulate data. This part of the memory cannot be changed. Random-access memory (RAM) temporarily stores data and feeds it to the central processing unit on command. RAM is constantly changing as the computer processes data. Also, it usually is erased when the computer is turned off.
- ***Program:*** The instructions the computer uses to process the data and to produce output. Often the program is

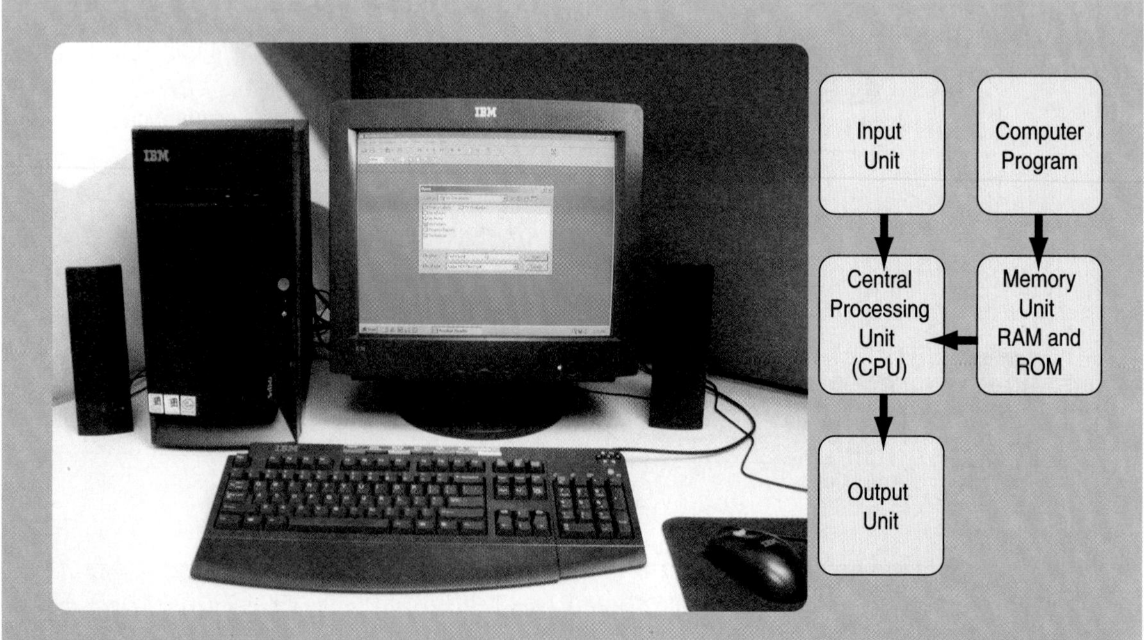

Figure 7-30. A computer has five parts: an input device, processing unit, memory, output device, and program.

called software because it cannot be seen or handled once it is loaded in the computer. The other four units described above are hardware. Hardware is the physical parts of the computer system. Programs are typically used to make mathematical calculations, maintain records, prepare drawings and other graphics, and handle text.

Computers are at the heart of most modern information-processing systems. They are at every level of business life. They process financial information and prepare financial reports, maintain schedules and ticket records for airlines, operate point-of-purchase units at supermarket check-out stands, guide spacecraft to distant planets, maintain the fuel-air mixture in automobiles, guide washing machines through their wash-rinse-dry cycles, help prepare layouts for advertising, and control industrial machines.

Using Technology Safely

Tools and machines make each of our lives more comfortable. They also injure or kill thousands of people each year. Every day local newspapers report injuries and deaths caused by careless use of such tools and machines as home appliances, factory machinery, and automobiles.

As we observed earlier in the chapter, if we want to understand technological devices and have them affect our lives positively, we must use them in a safe manner. Note that these devices include not just tools and machines that process materials but also those that process energy and information.

Operators should follow some basic guidelines when they select and use technological devices. These guidelines include:

- Select the correct tool for the job.
- Read about the proper operation of the device in owner's manuals and instruction books.
- Seek instruction and advice on the proper use of the device.
- Use the device only in the way described in operation manuals and for applications for which the device was designed.
- Never use a device when personal ability is impaired by medication, lack of rest, or distractions (loud music or people talking to you, for example).
- Never work alone or where help is not available in case of an accident.

Most people working in business and industry receive proper training on the equipment they use. Their challenge is to follow the safe procedures they are shown and expected to learn. Many people, however, also use technological devices in personal and educational settings. For example, they drive cars and work in home workshops and school laboratories.

These settings require special attention to safety. Remember that safety is both a state of mind and a series of actions. You must be concerned about your safety and the safety of those around you. Likewise, you must complete tasks using safe actions. Working safely requires you to follow the above general rules. The following rules are provided to help you work with laboratory materials and tools safely. These rules cover working safely with people, materials, and tools and machines.

Safety with People

Personal Safety

- Concentrate on your work. Watching other people or daydreaming can cause accidents.

- Dress properly. Avoid loose clothing and open shoes. Remove jewelry and watches.
- Control your hair. Secure any loose hair that may get caught in moving machine parts.
- Protect your eyes with goggles, safety glasses, or a face shield.
- Protect your hearing by using hearing protectors when working around loud machines or where high-pitched noises are present.
- *Do not* use compressed air to blow chips and dirt from machines or benches. Use a brush to sweep them away gently.
- Think before acting. Always think of what will happen before starting a task.
- *Ask* your teacher questions about any operation of which you are unsure.
- Seek first aid for any injuries.
- *Follow* specific safety practices demonstrated by your teacher.

Safety around Others

- Avoid horseplay. What you think of as harmless fun can cause injury to other people.
- *Do not* talk to anyone who is using a machine, except when necessary. You may distract him or her and cause an injury.
- *Do not* cause other people harm by carelessly leaving tools on benches or machines.

Safety with Materials

- Handle materials properly and with care.
  - Be careful when moving long pieces of material. Do not hit people and machines with the ends.
  - Use *extreme* caution when handling sheet metal. The thin, sharp edges can easily cut you and others.
  - Grip hot materials with pliers or tongs.
  - Wear gloves when handling hot or sharp materials.
  - Use extreme caution in handling hot liquids and molten metals.
- Check all materials for sharp burred edges and pointed ends. Remove these hazards when possible.
- If a material gives off odors or fumes, place it in a well-ventilated area, fume hood, or spray booth.
- Lift material properly. Use your legs to do the lifting, not your back. *Do not* overestimate your strength. If you need help, ask for it.
- Dispose of scrap material properly to avoid accidents.
- Clean up all spills quickly and correctly.
- Dispose of hazardous wastes and rags properly.

Safety with Tools and Machines

- Use only sharp tools and well-maintained machines.
- Return all tools and machine accessories to their proper places.
- Use the *right* tool or machine for the *right* job.
- *Do not* use any tool or machine without permission or proper instruction.

Casting and Molding Processes

- *Do not* try to perform a process that has not been demonstrated to you.
- Always wear safety glasses.
- Wear protective clothing, gloves, and face shields when pouring molten metal.

- *Do not* pour molten material into a mold that is wet or contains any water.
- Carefully fasten two-part molds together.
- Perform casting and molding processes in a well-ventilated area.
- *Do not* leave hot castings or mold parts where they could burn other people.
- Constantly monitor material and machine temperatures during the casting and molding process.

Separating Processes

- *Do not* try to perform a process that has not been demonstrated to you.
- Always wear safety glasses.
- Keep hands away from all moving cutters and blades.
- Use push sticks to feed all material into woodcutting machines.
- Use *all* machine guards.
- Always stop machines to make measurements and adjustments.
- *Do not* leave a machine until the cutter has stopped.
- Whenever possible, clamp all work.
- Always unplug machines from the electrical outlet before changing blades or cutters.
- Remove all chuck keys or wrenches before starting the machine.
- Remove all scraps and tools from the machine before using it.
- Remove scraps with a push stick.

Forming Processes

- *Do not* try to perform a process that has not been demonstrated to you.
- Always wear safety glasses and gloves.
- Always hold hot materials with pliers or tongs.
- Place hot parts in a safe place to cool. Keep them away from people and from materials that will burn.
- Follow correct procedures when lighting torches and furnaces. Use spark lighters, *not* matches.
- *Never* place your hands or any foreign object between mated dies or rolls.

Finishing Processes

- *Do not* try to perform a process that has not been demonstrated to you.
- Always wear safety glasses and a respirator.
- Apply finishes in properly ventilated areas.
- *Never* apply finishing materials near an open flame.
- Always use the right solvent to thin finishes and clean finishing equipment.

Assembling Processes

- *Do not* try to perform a process that has not been demonstrated to you.
- Always wear safety glasses.
- Wear gloves, protective clothing, and goggles for all welding, brazing, and soldering operations.
- Always light a torch with a spark lighter; *never* use a match.
- Handle all hot material with gloves and pliers.
- Perform welding, brazing, and soldering operations in well-ventilated areas.
- Use proper tools for all mechanical fastening operations. Be sure screwdrivers, wrenches, and hammers are the proper size and are in good condition.

Summary

Technological devices are designed to process materials, energy, and information. Tools and machines change the form of materials to make them more useful to people. They are also used to change the form of energy so we can do such things as heat buildings, power vehicles and machines, and light up dark areas. Information-processing tools and machines aid us in gathering and ordering data so that we can better understand it.

Each one of these processing activities can make our lives better, but each can also harm us and the environment. Processing activities must be carefully selected and properly and safely used.

Key Terms

arbors
armature
band saw
chop saw
chucks
circular saws
computer
cutting motion
cutting tool
cycle
cylindrical grinders
drilling machines
energy conversion
energy-processing converters
feed motion
field magnet
forstner bits
grinding machines
information processing
input unit
lathes
linear motion
machine tools
material processing
memory unit
multiple-point tool
output unit
planing machines
processing unit
program
radial saw
reciprocating motion
rotating motion
sawing machines
scroll saws
shaping machines
single-point tool
spade bits
stroke
surface grinders
table saw
turning machines
twist drills

Test Your Knowledge

Write your answers on a separate piece of paper. Please do not write in this book.

1. Give an example of a multiple-point tool.
2. What is the difference between feed motion and cutting motion?
3. *True or false?* Lathes are primarily used to machine wood and metal, but plastics can be machined on lathes also.
4. The three basic types of saws are the circular saw, the band saw, and the _____ saw.

5. Define the term "counterboring."
6. Describe the difference between shaping machines and planing machines.
7. The _____ was the first energy converter that did not simply change the form of mechanical energy.
8. _____ is an energy-converting device that uses the laws of magnetism and electromagnetism.
9. The radio uses _____ to connect the sender and receiver.
10. List the five major parts of a computer.
11. *True or false?* Safety is both a state of mind and a series of actions.
12. Give one example of what *NOT* to wear when working with tools and machines.
13. List one rule with regard to working safely around other people.
14. Give one example of how to handle materials properly.
15. *True or false?* When working with tools and machines, wear safety glasses until you reach the assembling process.

Applying Your Knowledge

1. Select a product that has been manufactured using material-processing technology. List one to three operations that were used to make the products that used each of the following machine tools:

Machine Tools	Process or Operation
Turning machines	
Sawing machines	
Drilling machines	
Planing or shaping machines	
Grinding machines	

2. Select a major development in material processing, energy processing, or information processing. Write a two-page report on the development.
3. Select a major development in material processing, energy processing, or information processing. Build a model of the technological device.
4. Produce a simple product that uses material-processing technology. Write a summary of the procedure you used to make the product and list the tool or machine you used for each step of the procedure.
5. Communicate data or information using an information-processing device. Write a one-page report on the processes you used to complete the task.
6. Select a household device that you use and develop a set of safety rules for its use.

Chapter 8
Measurement Systems and Tools and Their Role in Technology

Learning Objectives

After studying this chapter, you will be able to:

- Define measurement.
- Compare the U.S. customary and the metric measurement systems.
- Describe the major physical qualities that can be measured.
- Name types of measurements.
- Identify common measuring tools to measure linear distances, diameters, and angles.
- Relate measurement to quality control.

When you travel from your home to school, how far is the trip? How long does the journey take? Do you travel by car or bus? If so, how much fuel does that vehicle's fuel tank hold?

This book has some physical qualities. How heavy is it? What size are the pages? How thick is it?

Think of how you compare with your classmates. How much do you weigh? How tall are you? How fast can you run the one-hundred-yard dash?

These questions ask you to describe physical qualities. Physical qualities are *characteristics* of an object or event that can be described. If we say a tree is tall, but have nothing to compare it to, saying the tree is tall has no meaning. The tree could be a dwarf apple tree or a giant redwood. If it takes a long time to do something, what do we mean by a long time? Hours? Weeks? Years? Centuries?

To describe something to someone, we must have a common reference or standard. We use measurement to describe objectively the physical qualities of an item. ***Measurement*** is the practice of comparing the qualities of an object to a standard, **Figure 8-1.** To describe objects using measurement, we have created systems of standards for comparison.

Systems of measurement have evolved over time, with two major systems in use today. In this chapter, we will first look at a short history of measurement systems. We will then discuss the systems used today, the qualities we measure, and the types of measurement we commonly use. Finally, we will look at the kinds of measurement tools we use and how measurement affects technology.

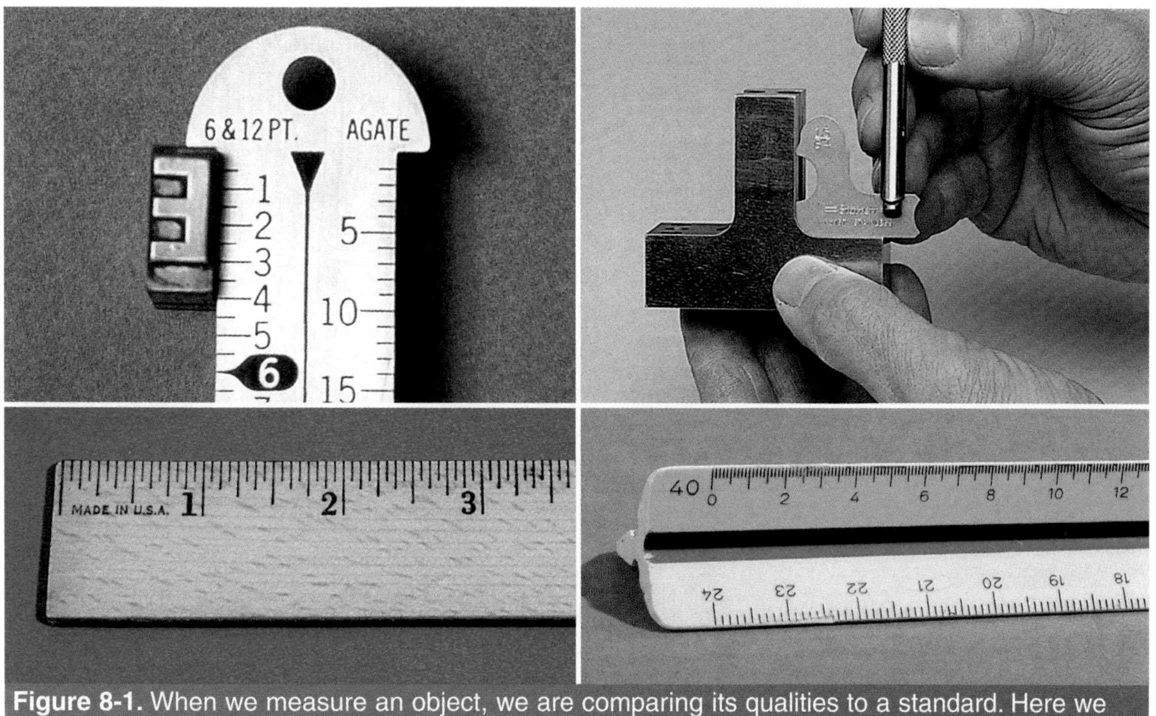

Figure 8-1. When we measure an object, we are comparing its qualities to a standard. Here we see four different standards of measurement.

Measurement Systems: Past and Present

All measurements compare the quality being described against a standard. In early times people used the sizes of human body parts as standards for measurement. The biblical story of Noah records that he built an ark that was three hundred cubits long, fifty cubits wide, and thirty cubits high. These are strange measurements to us. Early Egyptians, Romans, and Greeks used these terms, however. The standard was derived from the human arm, as shown in **Figure 8-2.** A cubit was the distance from the tip of the middle finger to the elbow. A shorter measurement was the palm, which was the width of the four fingers. The width of a single finger was called a digit.

Human arms vary in size, however. As long as technology was simple, measurements did not have to be highly accurate. Now that technology is more complicated, measurements have been standardized. Each major physical quality is compared to a standard measurement that has been set by governments and international agreements.

Two measurement standards are in use today. One is the customary or English system, here referred to as the ***U.S. customary*** or conventional system. The more widely used system is called the *International System of Units.* This system is abbreviated *SI* (from the French name *Système international d'unités*). SI is more commonly known as the ***metric system.*** See **Figure 8-3.** The United States is the only industrialized country that has not adopted the metric system for everyday use.

U.S. Customary System

The U.S. customary system of measurement is based on the system that developed in England from approximately the 1100s to the 1500s. As in earlier times, some of the system was based on the sizes

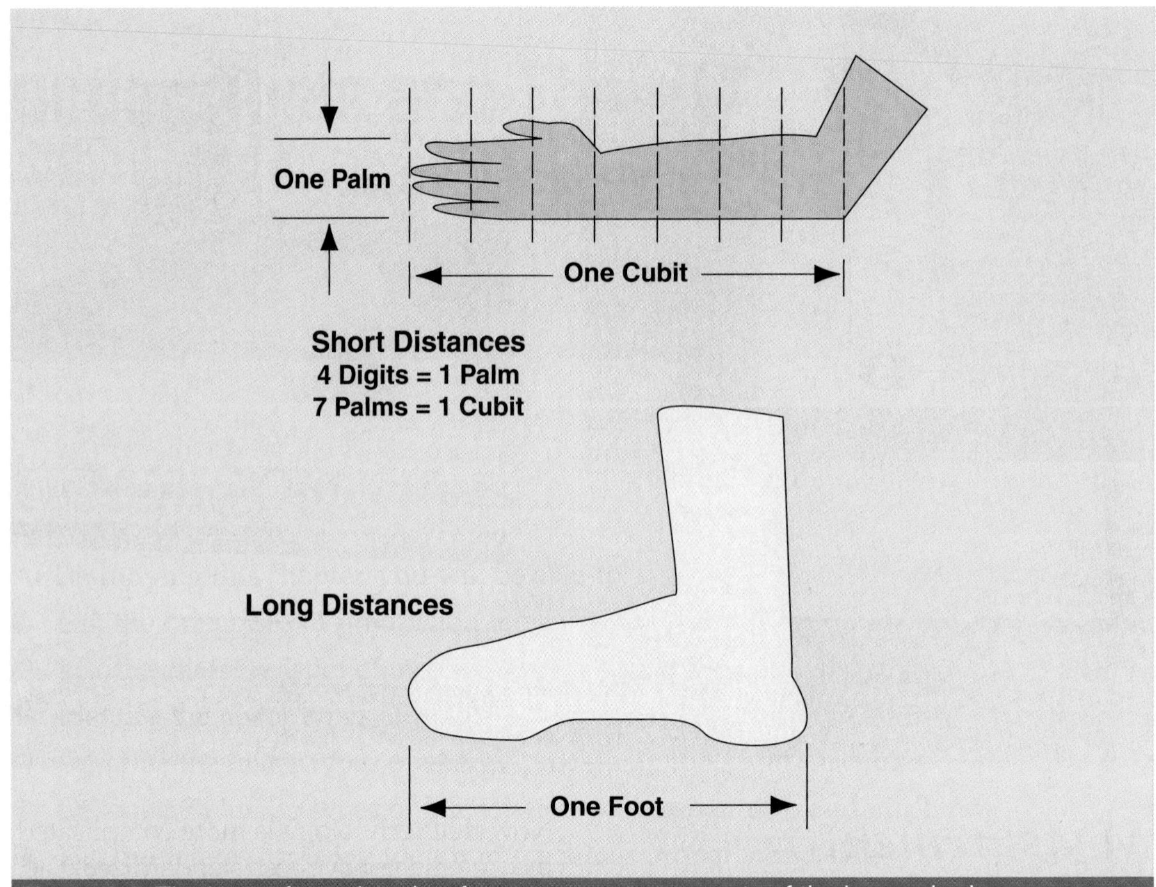

Figure 8-2. The bases for early units of measurement were parts of the human body.

of human body parts. For example, the term *inch*, a word meaning *thumb* in some languages, was used to represent the width of a human thumb. The *foot* was originally based on the length of a human foot. Other terms in the system were developed from Roman measurements. For example, the word *mile* originally meant the length of one thousand paces of a Roman legion.

Although it has some logic to it in that it developed from common understandings of various kinds of measurement, this system also creates a great deal of confusion. For example, numerous terms are used to describe the same kind of measure. Twelve inches equals one foot, but thirty-six inches equals either three feet or one yard.

The system is also confusing because of inconsistencies in computing fractions and multiples of different base measurements. For example, one inch is 1/12 of a foot, whereas one quart is 1/4 of a gallon.

SI Metric System

In the 1790s a group of French scientists assembled to create a measurement system that was more logical and exact. The SI metric system is based on decimals, or values of the number ten. There is a logical progression from smaller units to larger ones throughout the system because the base unit always decreases or increases by values of ten. Smaller units are decimal fractions (1/10, 1/100, 1/1,000, and so on) of the base unit. Larger units are multiples of ten (10, 100, 1,000, and so on) times the base unit.

The metric system is also simpler because it uses a prefix to show us how the base unit is being changed. For example,

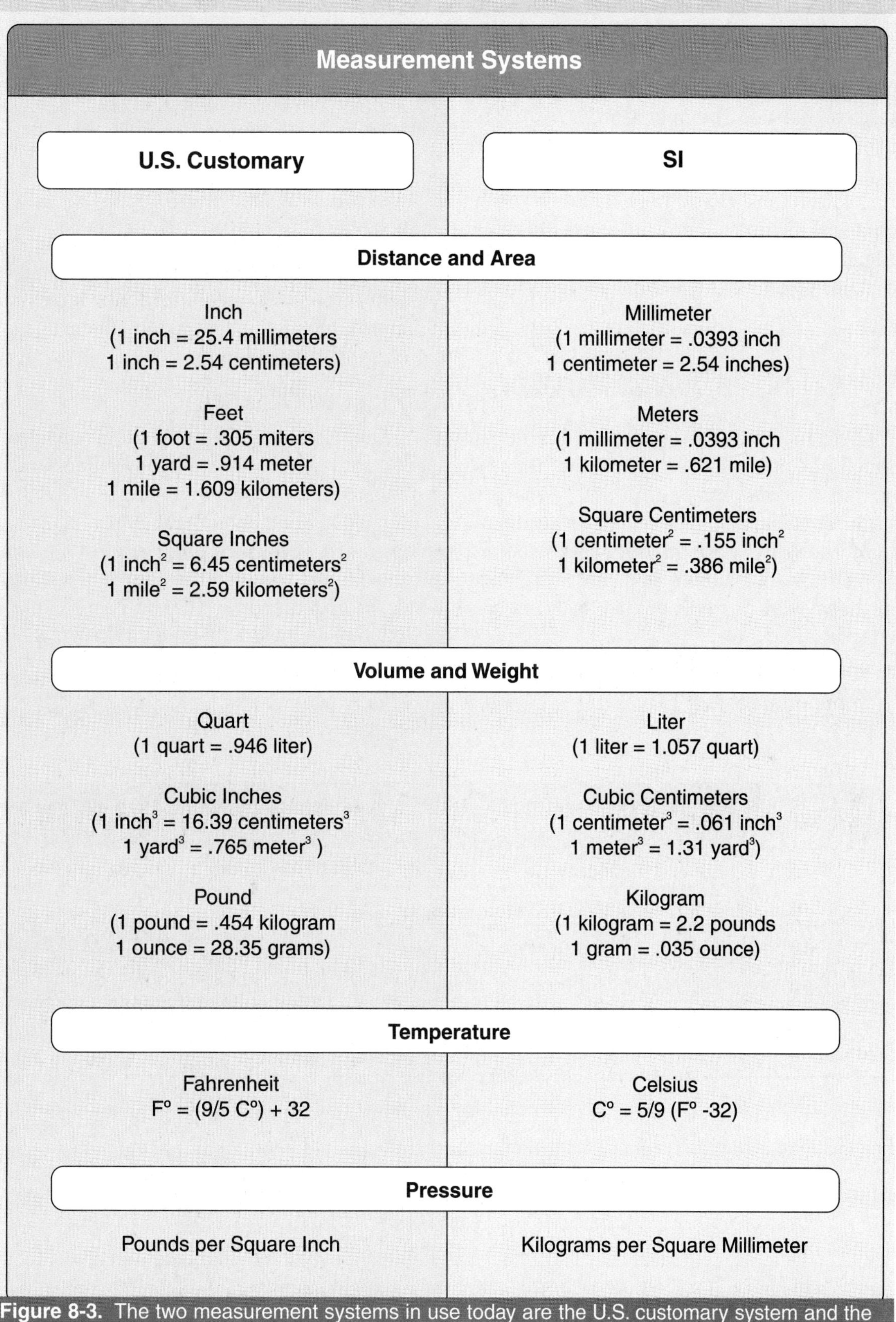

Figure 8-3. The two measurement systems in use today are the U.S. customary system and the metric system. Some of the more common measurements are shown here.

the unit for distance is the meter. For large distances, we use the word *kilometer.* The prefix *kilo* means 1,000, so seven kilometers is equal to 7,000 meters. For small distances, we use the word *millimeter.* The prefix *milli* means 1/1,000th. Twelve millimeters are 12/1,000 of a meter. The metric system lends itself to easy use in mathematical formulas. Look at **Figure 8-4** to see the most common metric prefixes. The metric system uses the same prefixes for all base units.

Qualities Measured

We have developed standards for seven physical qualities. These qualities are size and space, mass, time, temperature, number of particles, electrical current, and light intensity. These qualities are measured as technology is developed and used. Four of these qualities will be discussed in this chapter. They are:

- Size and space: The length, area, or volume of an object
- Mass: The amount of matter in an object
- Temperature: How hot or cold an object or place is
- Time: How long an event lasts

Size and Space

When we measure the size and space of an object, we may encounter three different but related measurements: length or distance, area, and volume. Each is discussed in turn below.

Length or Distance

Length or ***distance*** is the separation between two points. Lengths and distances are generally given in a single measurement from Point A to Point B. For example, you might find that a page in a book measures eleven inches from top to bottom. You might discover that it is two miles (just over three kilometers) from your home to school.

When we describe parts or products, however, we often give two or three

SI Measurement Prefixes			
Factor	**Magnitude**	**Prefix**	**Symbol**
10^9	1,000,000,000	giga	G
10^6	1,000,000	mega	M
10^3	1,000	kilo	k
10^2	100	hecto	h
10^1	10	deka	da
10^{-1}	1/10	deci	d
10^{-2}	1/100	centi	c
10^{-3}	1/1000	milli	m
10^{-6}	1/1,000,000	micro	μ
10^{-9}	1/1,000,000,000	nano	n
10^{-12}	1/1,000,000,000,000	pico	p

Figure 8-4. The metric system uses the same prefixes for all base units.

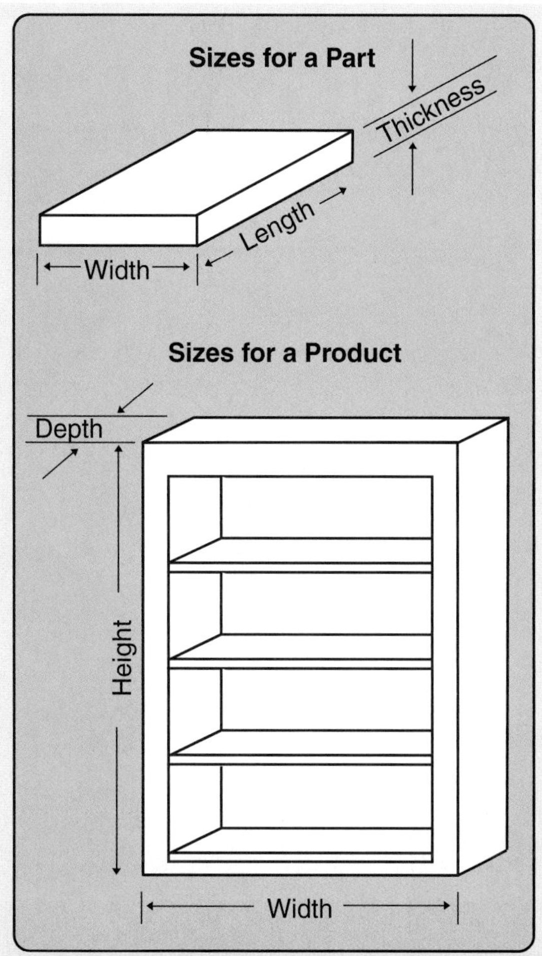

Figure 8-5. We use various distance measurements when describing parts and products.

different distance measurements. Many parts and products are described as having thickness, width, and length. For example, you could describe a block of wood as being 3/4″x 4″ x 12″. This means that the block is three-fourths of an inch thick, four inches wide, and twelve inches long. Round parts are said to have thickness and diameter. A piece of pipe may be described as being 3/4″ in diameter and 24″ long. Finished products such as furniture, television sets, and cabinets are described by giving their height, width, and depth, **Figure 8-5.**

The common units for length or distance in the U.S. customary system are (in increasing size) inch, foot, yard, and mile. The common units for length or distance in the metric system are millimeter, centimeter, decimeter, meter, and kilometer.

Area

The measurement of the separation between two points does not indicate the size, or ***area,*** of an object, however. Determining the area of something involves measuring both the length and the width and then multiplying the two figures to see how much surface the object covers, **Figure 8-6.** For example, you might find that the length of this page is eleven inches; still, that measurement does not tell you how much area the page covers because the page also has a width. To determine the size of the page, you also measure the width and then multiply that number by the length. Here, if the width of the page is eight inches, the surface area of the page would be eighty-eight square inches.

In the customary system we express area measurements in such units as square inches, square feet, and square yards. The metric system uses such units as square centimeters and square meters.

Volume

Area does not indicate total size, however. A third consideration when measuring size and space is volume. ***Volume*** means the amount of space that an object occupies or encloses, **Figure 8-7.** Volume is calculated by multiplying the length, width, and depth of an object. The measurement is expressed in cubic units.

For example, say we wanted to know how much space a room in a house will take up. We would measure the length of the room (say ten feet), the width of the room (say eight feet), and the depth (the measurement from the floor to the ceiling) of the room (say eight feet). The amount of space the room takes up in the house would then be 640 cubic feet (ten feet

Figure 8-6. The area of an object is the measurement of how much surface the object covers. It takes into account the length and width.

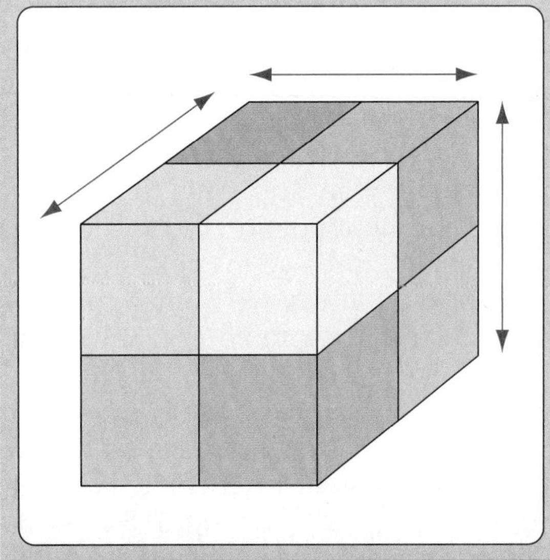

Figure 8-7. Volume is the amount of space an object occupies. It takes into account the length, width, and depth.

multiplied by eight feet multiplied by eight feet).

Related to volume is ***capacity,*** which is the amount of a substance that an object can hold. Capacity is also measured in cubic units. We perhaps encounter the idea of capacity most often when we purchase such items as milk and soft drinks. An item that is described as a two-liter bottle of a soft drink, for example, indicates that the bottle can hold two liters of the liquid.

In the U.S. customary system, cubic units for volume are expressed in such terms as cubic inch, cubic foot, and cubic yard. In the metric system, volume is measured in such terms as cubic millimeter, cubic centimeter, and cubic meter. The cubic units for capacity are somewhat different. In the U.S. customary system we find such terms as pint, quart, and gallon. In the metric system we find such terms as milliliter, deciliter, and liter.

Mass

Mass describes the quantity of matter present in an object. Matter is anything that has mass and takes up space.

Note that mass is not the same as weight, although they are related. ***Weight*** is the force of the earth's pull on the mass. Weight therefore can vary depending on gravity. For example, by the time astronauts reach outer space, their weight has decreased dramatically because of the lack of gravity. They are thus able to "float." Their mass, however, stays the same. Mass does not vary with temperature or location, **Figure 8-8.**

In most cases measurement of mass and weight are close enough that we use the terms interchangeably. For example, when we purchase an apple, we are buying the matter of the apple, but we weigh the apple on a scale to determine its size. The scale weighs the apple but also gives us an idea of its mass because everywhere on earth the apple will weigh the same. We buy products by the pound (fruit, vegetables, and sugar, for example), by the gram (prescriptions and precious metals, for example), or by the ton (hay, grain, sand, and gravel, for example).

The U.S. customary system measures the weight of materials and uses such terms as ounces and pounds. The metric system measures mass and uses such terms as milligram and gram.

Temperature

Temperature is the measurement of how hot or cold a material is. The terms *hot* and *cold* are relative, though, and therefore must be compared to a standard. For example, the term *hot* has different meanings for human comfort and for melting metals. Temperatures that are hot enough to make you uncomfortable would be considered cold for melting steel.

The range of temperature is broken into units. The U.S. customary system and the metric system use water as the basis to divide the temperature range into units called degrees. The range from the boiling point of water to the freezing point of water is divided into degrees. The customary *Fahrenheit* scale has 180° between the freezing point and boiling point. The freezing point of water at sea level is 32°F, and the boiling point of water is 212°F. The metric *Celsius* scale divides this range into 100°, with the freezing point at 0°C and the boiling point at 100°C.

Time

Time is how long an event lasts. The *second* is the basic unit of time in both the

Figure 8-8. The mass of this boulder will remain the same no matter what gravitational pull exists.

metric and the customary systems. Short-term measurements are given in hours, minutes, seconds, or fractions of a second. These terms measure everything from the time it takes to complete a race (minutes and seconds) to computer speeds (nanoseconds, or billionths of a second). Time may be measured in very long terms such as a millennium (one thousand years), a century (one hundred years), a decade (ten years), or a number of years. These terms are used to describe the age of civilizations, people, buildings, and vehicles.

Types of Measurement

The measurement of part and product size is important in technological design and production activities. Generally this type of measurement can be divided into two levels of accuracy, as shown in **Figure 8-9.**

- Standard measurement
- Precision measurement

Standard Measurement

Many production settings do not require close measurements. The length of a house, the width of a playing field, and the angle of the leg on a playground swing set need not be accurate to the utmost degree. If the product is within a fraction of an inch or a degree of angle, it will work fine. This kind of measurement is called ***standard measurement.*** Standard measurement is often given to the foot, inch, or fraction of an inch in the customary system or to the nearest whole millimeter in the metric system.

The kind of production is only one factor that affects the type of measurement used. The material being measured is also important. For example, wood changes (expands or shrinks) in size with changes in its moisture content and the atmospheric humidity. Measurements closer than 1/32 of an inch or 1 millimeter are not useful. Wood can change more than that amount in one day.

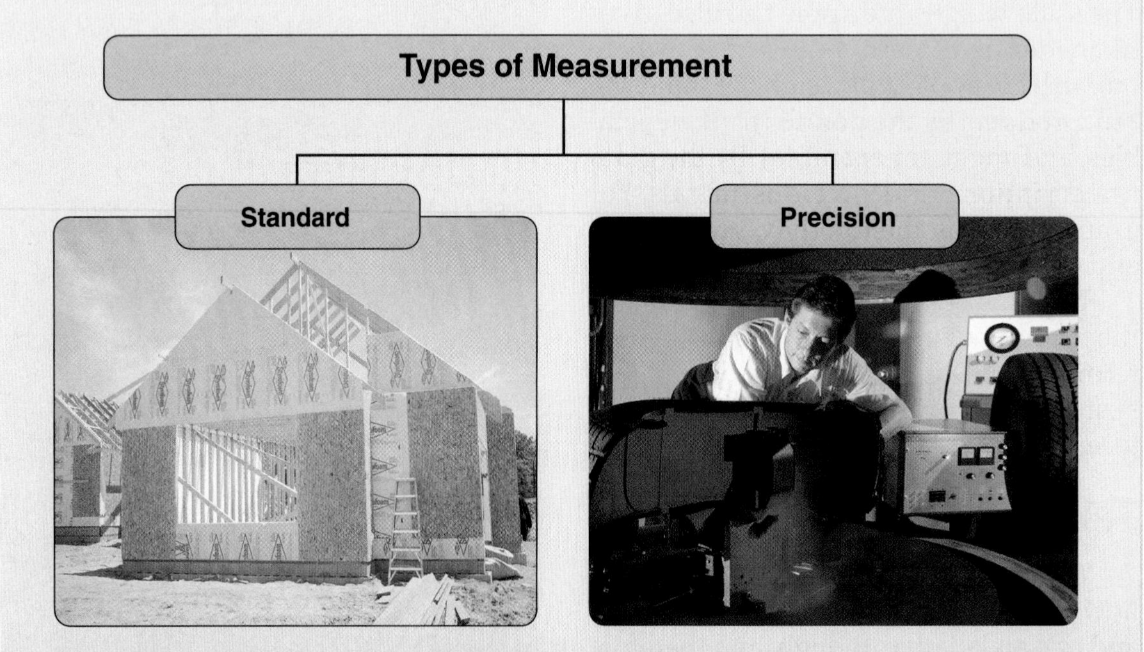

Figure 8-9. Standard measurement is widely used in the construction industry, as shown on the left. Precision measurement is found in many manufacturing applications, as shown on the right. (Goodyear Tire and Rubber Co.)

Figure 8-10. This device uses a laser for measurement. (FMC Corp.)

Standard measurements are common in cabinet and furniture manufacturing plants, construction industries, and printing companies. The printing industry uses it own system of standard measurement based on the pica (1/6″) and the point (1/72″).

Precision Measurement

Standard measurement is not accurate enough for many production applications. Watch parts and engine pistons would be useless if they varied by as much as 1/32 of an inch (0.8 millimeter). These parts must be manufactured to a more precise size. For this type of production, ***precision measurement*** is required. Precision measurement measures features to 1/1,000″ (one one-thousandth of an inch) to 1/10,000″ (one ten-thousandth of an inch) in the customary system. Metric precision measurement will measure features to 0.01 mm (one one-hundredth of a millimeter).

Workers in manufacturing activities that involve metals, ceramics, plastics, and composites use precision measurements. Those who deal with laboratory testing, material research, and scientific investigation also use precision measurements.

Measurement Tools

People who work in technological areas have many types of measurement tools available, **Figure 8-10.** The measurement tools that an operator manipulates and reads are called ***direct-reading measurement tools.*** In recent years, new measurement tools and machines have been developed. They bring together sensors and computers to automate measurement. These systems are called ***indirect-reading measurement tools.*** Each is discussed as follows.

Direct-Reading Measurement Tools

Three common uses of measurement are finding linear dimensions, diameters, and angles. Each of these three features can be measured using standard or precision devices, **Figure 8-11.**

Linear Measurement Devices

The most common linear measurement device is the ***rule,*** **Figure 8-12.** A rule is a rigid or flexible strip of metal, wood, or plastic with measuring marks on its face. Two types of rigid rules are used in linear measuring. These are the machinist's rule and the woodworker's or bench rule. A bench rule is generally divided into fractions of an inch. The most common divisions are sixteenths (1/16″). Metric bench rules are divided into whole millimeters. Machinist's rules are designed for finer measurements. Customary machinist's rules are divided in sixty-fourths (1/64″) or into tenths (1/10″) and hundredths (1/100″). Metric machinist's rules are divided in 0.5-mm increments.

The part is measured with a rule by first aligning one end of the part with the zero mark. The linear measurement is then taken by reading the rule division at the other end of the part.

Flexible rules are often called tape rules. They are used in woodworking and carpentry applications. The catch at one

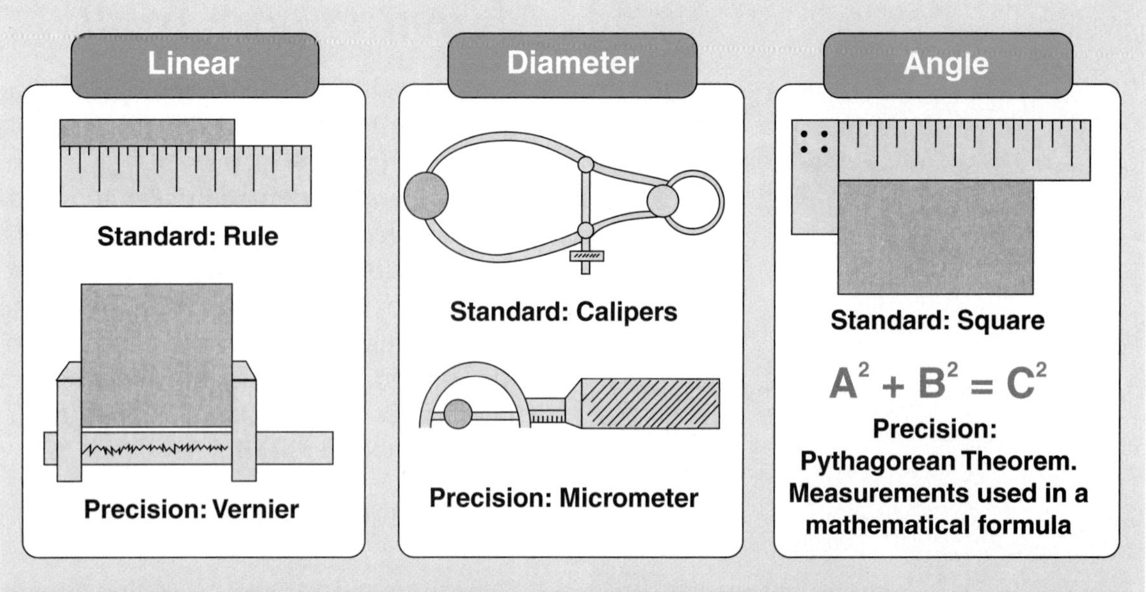

Figure 8-11. Both standard and precision measurement tools are used to determine linear sizes, diameters, and angles.

end of the rule is hooked to the end of the board or structure. The tape is pulled out until it reaches the other end of the board or structure. A measurement is then taken.

Metric tape rules are divided into 1-mm increments. Commonly, the smallest division on a customary tape rule is 1/16″. Some tape rules highlight every sixteen inches, which is the common spacing of studs in home construction. Tape rules are generally available in lengths from eight feet to one hundred feet or from two meters to thirty meters.

Various machinist's tools can make precision linear measurements. The most common are inside and outside micrometers, depth gages, and vernier calipers, **Figure 8-13.** These tools measure such physical qualities as length and thickness.

Diameter Measuring Devices

A common measurement task involves determining the diameter of round material or parts. A simple, rough measurement may be determined with hole gages or circle templates. These devices have a series of holes into which the stock can be inserted. The smallest hole into which the material will fit establishes the approximate diameter of the item.

More precise diameters can be established using a ***micrometer,*** as shown in **Figure 8-14.** The part to be measured is placed between the anvil (fixed part) and the spindle (movable rod). The spindle is

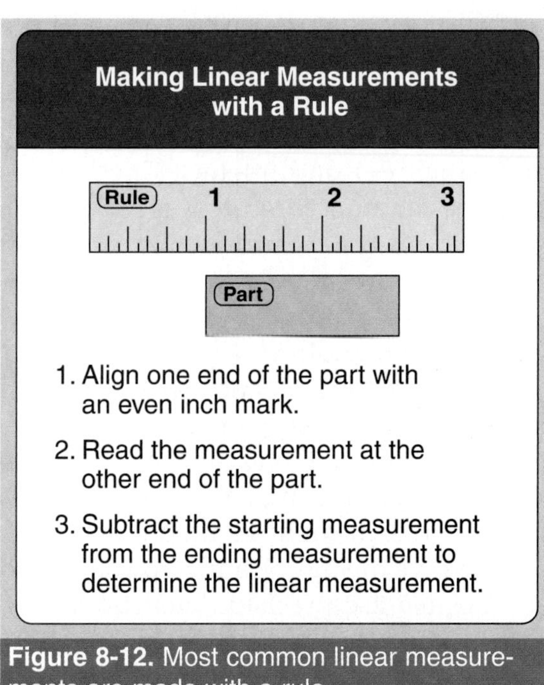

Figure 8-12. Most common linear measurements are made with a rule.

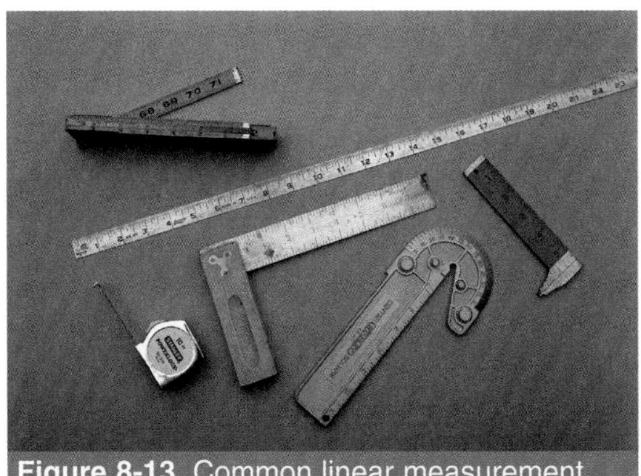

Figure 8-13. Common linear measurement tools include rigid rules, tape rules, micrometers, and calipers.

moved forward by turning the barrel. When the spindle and the anvil touch the part, a reading is taken on the barrel. Most customary micrometers will measure to within 1/1,000 of an inch; metric micrometers measure to 0.01 (1/100) of a millimeter.

Angle Measuring Devices

The angle between two adjacent surfaces or intersecting parts is important in many situations. The legs of a desk are generally square (at a 90° angle) with the top. The ends of picture frame parts must be cut at a 45° angle to make a square frame.

Measurements at 90° angles are commonly done with ***squares,*** **Figure 8-15.** These tools have a blade that is at a right angle to the head. The head is placed against one surface of the material. The blade is allowed to rest on an adjacent surface. If the blade touches the surface over its entire length, the part is square. Parts that are not square allow light to pass under the blade.

Some squares have a shoulder on the head that allows the square to be used to measure 45° angles. This angle is important in producing mitered corners on furniture, boxes, and frames.

The most common squares are the rafter or carpenter's square, the machinist's square, the try square, and the combination (90° and 45°) square.

Career Corner
Surveyors

Surveyors measure distances, directions, and angles between points and elevations along lines. They establish property boundaries and locate buildings on properties. Also, they write descriptions of land for deeds and leases. Surveyors use surveying instruments and electronic distance-measuring equipment to measure horizontal and vertical angles and linear distances. On some projects, surveyors use the global positioning system (GPS) to locate points on the land.

They generally work eight-hour days, five days a week. Seasonal demands may require them to work longer hours. Most licensed surveyors become trained by combining postsecondary school courses in surveying with extensive on-the-job training. Presently, technological advancements have caused some employers to require a four-year college degree.

Measuring with a Micrometer

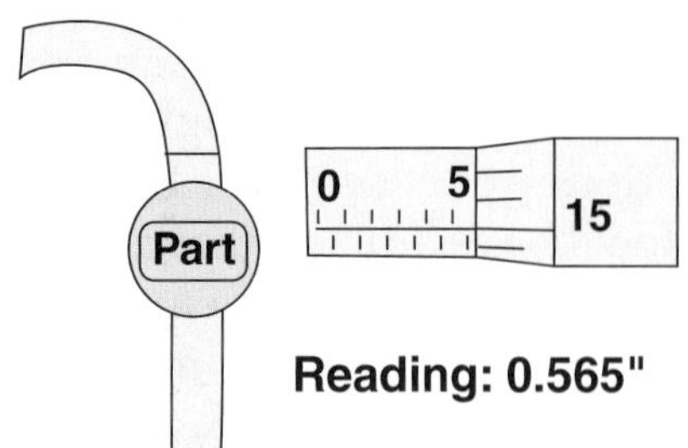

1. Align the part between the spindle and the anvil of the micrometer.
2. Gently move the spindle against the part.
3. Read the measurement on the barrel.
   a. Read to the nearest .050" (each mark on the barrel).
   b. Add the fraction from the thimble –each mark is .001".

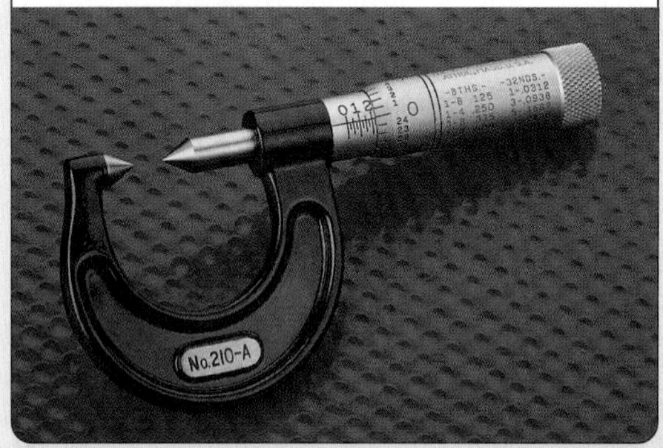

Figure 8-14. Micrometers are used to obtain precise measurements of diameters. (L.S. Starrett Co.)

Protractors and sliding T-bevels may be used to measure angles that a square cannot. A protractor allows for direct reading of angles. You place the protractor over the angle to be measured and then read the angle. The sliding T-bevel has an adjustable blade. You place the head on one surface and then clamp the blade along the angle of the second surface. You can then measure this angle with a protractor or by using a mathematical formula.

Indirect-Reading Measurement Tools

In many modern measuring systems, humans no longer take measurements. Sensors gather the measurement data, which is processed by computers or other automatic devices. The final measurement can be displayed on an output device such as a digital read-out, computer screen, or printout. These new systems include laser measuring devices, optical comparators, and direct-reading thermometers **Figure 8-16.** If you have weighed yourself on a digital bathroom scale, then you have used an indirect-reading measuring device.

Measurement and Control

So far you have learned that measurement is used to describe physical qualities. You might measure size and shape, mass, temperature, or other qualities. Note, however, that you also have a reason for doing the measuring. Measurement is the key to the production process and to quality control, as discussed below.

Measurement and Production Processes

All technological processes produce products or services. These products or services may be goods, buildings, or communication media. Measurement is necessary in designing an artifact. Its size, shape, or other properties are communicated through measurements. Processing equipment is set up and operated using these design measurements. Materials needed to construct the item are ordered using measurement systems. All personal or industrial production is based on measurement systems.

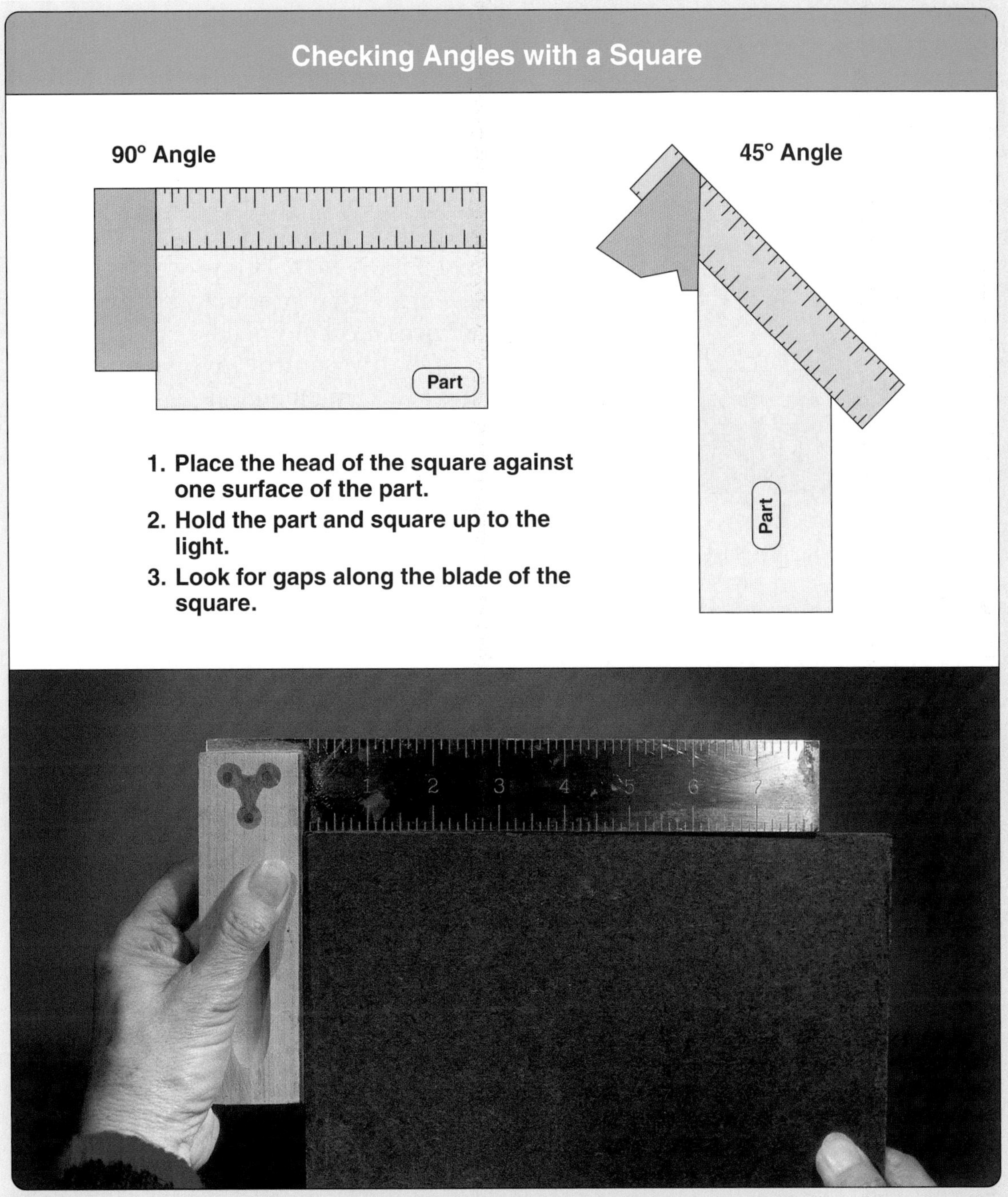

Figure 8-15. Ninety-degree angles can be measured with a square. When an angle is not 90°, the blade of the square does not touch the surface of the work, as shown in the photo.

Measurement and Quality Control

Measurement can also be used to compare the present condition with a desired condition. For example, say you are building a bookcase and need a shelf that is 24"(610 mm) long. You would probably cut it from a longer board. First, you would measure and mark the location for the cut. Then, you would saw along the line. If the board is too long, it will not fit into the case. A short board will fail to rest on the shelf supports.

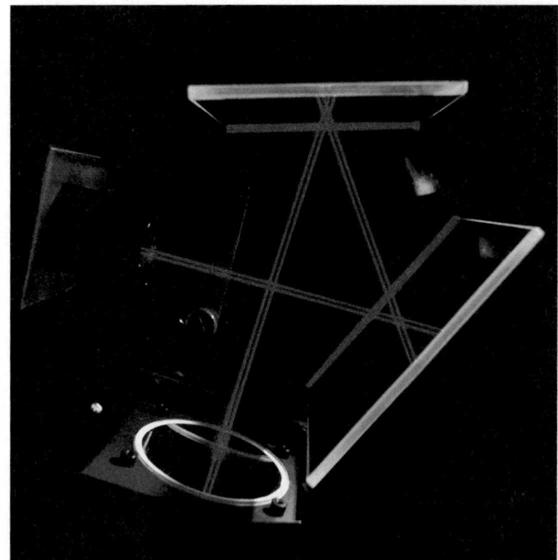

Figure 8-16. Machine operators use manual measurements for routine work. Indirect measurement systems, such as this laser measurement system, are often built into continuous processing and assembling operations. (GM-Hughes)

Measurement tells you if you have produced a 24″-long shelf.

A board of the correct length would meet the intended purpose. It would meet your quality standards. The process of setting standards, measuring features, comparing them to the standards, and making corrective actions is called ***quality control***, **Figure 8-17.** This process is designed to ensure that products, structures, and services meet our needs.

Quality control involves measuring and analyzing materials entering the system, work in process, and the outputs of the system. It is an on-going process designed to ensure that resources are efficiently used and that customers receive functional products.

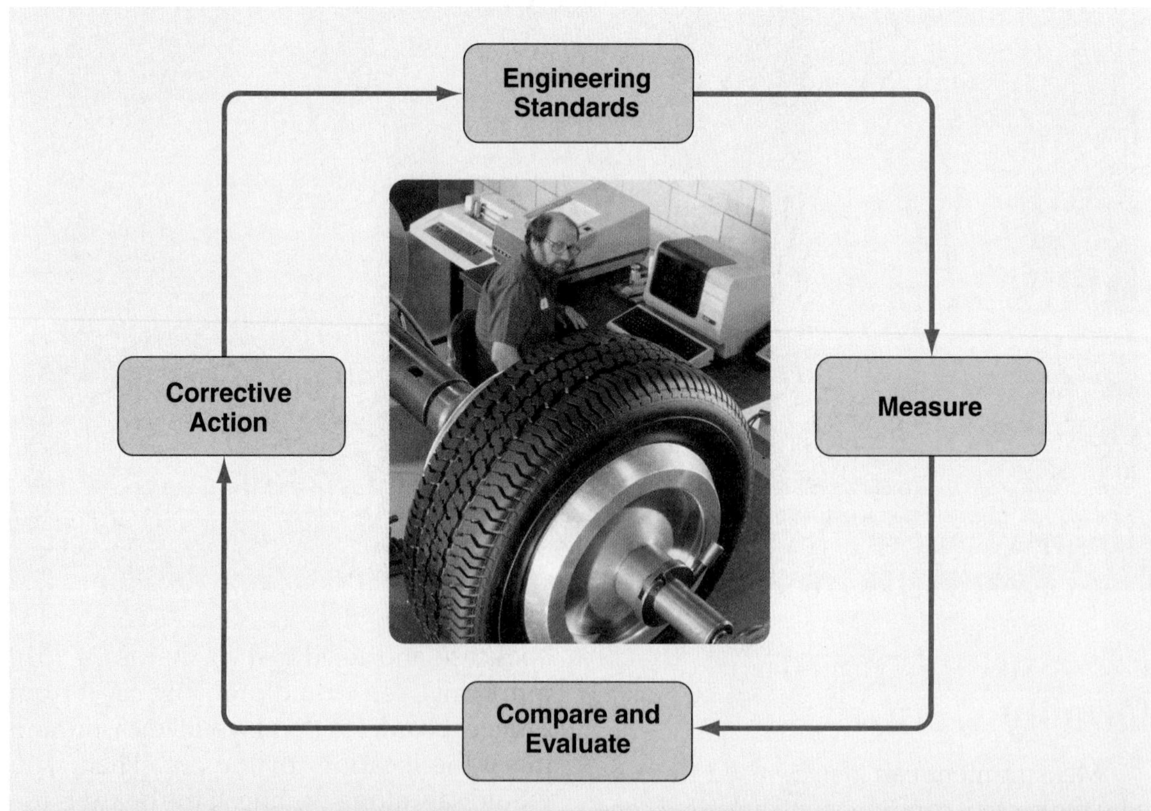

Figure 8-17. The foundation of a quality control system is measurement and analysis. (Goodyear Tire and Rubber Co.)

Connections to Technology: Mathematics

Measuring Area

Why do we need to use measurement systems in technology anyway? One answer is that measurement systems provide us with a common terminology when we wish to change or improve our environment.

Suppose, for example, that you wanted to paint your bedroom a different color. How much paint would you need to buy to cover all four walls?

To calculate the amount of paint needed, you would need to determine the area to be covered. As noted in this chapter, in this case area equals length times width, and the resulting amount is expressed in square feet. Say that two walls of the bedroom are 9 feet long and 8 feet high and the other two walls are 11 feet long and 8 feet high. Also take into account that you have one window that measures 36 inches by 48 inches and two doors that measure 36 inches by 84 inches each. Your calculations to figure the area to be covered would thus look something like this:

Two walls at 9 feet long by 8 feet high =	
72 square feet × 2 walls =	144 sq ft
Two walls at 11 feet long by 8 feet high =	
88 square feet × 2 walls =	176 sq ft
Subtract:	
One window opening at 36 inches by 48 inches	
(3 feet by 4 feet) =	–12 sq ft
Two door openings at 36 inches by 84 inches	
(3 feet by 7 feet) × 2 doors =	– 42 sq ft
Total =	266 sq ft

When you go to the store to purchase the paint, you find that one gallon of the paint you want will cover 250 square feet. Therefore, you would have to buy one gallon plus another smaller can to cover the 266 square feet.

How much paint would you need to purchase if the dimensions of your bedroom were as follows: two walls that are 10 feet long and 8 feet high, two walls that are 13 feet long and 8 feet high, two windows each measuring 32 inches by 48 inches, and two door openings that are 30 inches by 84 inches?

Summary

Measurement describes size and shape, mass, time, temperature, number of particles, electrical current, and light intensity. It involves comparing a physical characteristic to an established standard. The common standards are the metric system and the U.S. customary system. These systems allow people to communicate designs, order materials, set up machines, fabricate products, and control quality.

Key Terms

area
capacity
direct-reading measurement tools
distance
indirect-reading measurement tools
length
mass
measurement
metric system
micrometer
precision measurement
quality control
rule
squares
standard measurement
temperature
time
U.S. customary
volume
weight

Test Your Knowledge

Write your answers on a separate piece of paper. Please do not write in this book.

1. What is measurement?
2. Name one advantage of the metric system as compared with the U.S. customary system of measurement.
3. The three measurements related to size and space are length or distance, area, and _____.
4. Name one unit of measurement for length in the metric system.
5. Define the term *capacity* as it relates to measurement.
6. What is the difference between mass and weight?
7. The unit of measurement for temperature in both the U.S. customary and the metric system is the _____.
8. The basic unit of measurement for time in both the U.S. customary and the metric system is the _____.
9. Name one industry where standard measurements are commonly used.
10. Name one industry where precision measurement is usually required.
11. The most common linear measurement device is the _____.
12. *True or false?* Precise diameters can be obtained by measuring with a micrometer.

13. *True or false?* You can measure 90° angles with tools called squares.
14. Give one example of an indirect-reading measurement tool.
15. What is the relationship between measurement and quality control?

Applying Your Knowledge

1. On a form similar to the one below, write the metric unit and the U.S. customary unit of measurement for each feature listed.

Measurement	Metric	U.S. customary
Distance from Los Angeles		
Temperature on a hot day		
Weight of a loaf of bread		
Length of a pencil		
Capacity of a container		

2. Select an object, such as a desk or a bookcase. Describe its size using both the metric and U.S. customary measurement systems.
3. Check the squareness of a piece of furniture using a rafter, try, or combination square. Describe what you found and how any out-of-squareness could be corrected.

Section 3 Activities

Activity 3A

Design Problem

Background

All technology involves a machine or device to process materials, energy, or information. These machines are used to change the form of one or more of these resources into a new, more usable form.

Situation

You have been selected as the public relations director for a local citizens group. Your group is concerned about issues local politicians and business leaders are not adequately addressing. It wants to inform the public about these important issues.

Challenge

Choose an issue you feel political and business leaders are not addressing. Some examples of this type of issue are the greenhouse effect, fossil fuel dependence, and nuclear power. Use the library for research, if necessary. Develop a one-page flyer to communicate your group's position on the issue. If your school has computers available, use one or more computer software programs to help you produce the flyer.

Activity 3B

Fabrication Problem

Background

We live in a material world. All around us are products that have been developed using material processing technology.

Challenge

Work with a partner to make a game that can be given to a local charity. See **Figure 3B-1.** This figure shows the layout for the game

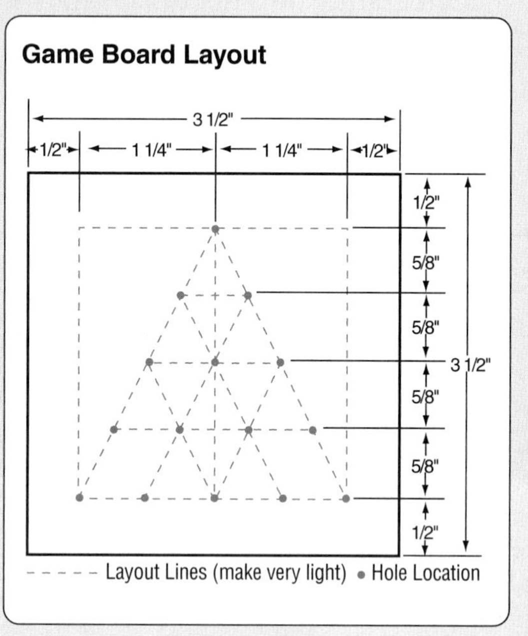

Figure 3B-1

Materials

- One piece 3/4" x 3 1/2" x 3 1/2" clear pine, redwood, or western red cedar
- Twelve 1/8" diameter x 1" wood pegs or golf tees

Procedure

Carefully watch the demonstration your teacher gives, showing the proper use of the tools and machines needed to

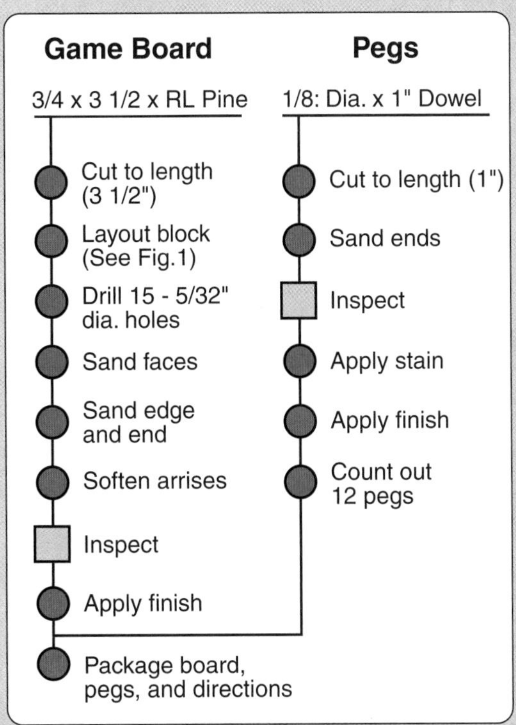

Figure 3B-2

make the product. Follow the procedure given on the operation process chart to construct the game. See **Figure 3B-2**.

Follow *all* safety rules your teacher discusses during the demonstration. Photocopy the directions shown below. These will be packaged with the game.

Analysis

Meet as a class and analyze the material processing activities you completed by answering the following questions:

1. Which material processing tools did you use?
2. Which measuring tools did you use?
3. How could you have increased the speed of the manufacturing process?
4. What changes in the material processing actions would you make to improve the quality of the product?

Mind Challenge

Directions

- Put a peg in each hole except one.
- Select one peg and jump another adjacent peg, ending in an empty hole.
- Remove the jumped peg.
- Continue jumping pegs until only one remains or no more jumps are possible.

Scoring

- One peg remains—Terrific—30 points.
- Two pegs remain—Good—20 points.
- Three pegs remain—Fair—10 points.
- Four pegs remain—Poor—0 points.

Section 4

Problem Solving and Design in Technology

Tomorrow's Technology Today
Nanotechnology

Many of today's technological products seem to be getting smaller and smaller—from machines to computer chips. It is estimated that within the next 50 years, they will continue to decrease in size, so much that thousands of machines will be able to fit on the point of a pencil! In the coming decades, nanotechnology, or "the science of the small," will use these tiny machines for many amazing applications. One nanometer equals one-millionth of a millimeter, so we are talking about very, very small machines.

All matter in the universe is made up of atoms. Cells are nature's tiny machines. We still have a lot to learn about constructing materials on such a small scale, but we are getting closer with new developments in nanotechnology. Many scientists believe we will be able to manipulate each individual atom of an object in the near future.

Nanotechnology combines engineering with chemistry. Its goal is to manipulate atoms and place them in a specific pattern. These patterns will produce a certain kind of structure. Depending on the atoms used and the patterns created, nanotechnology can be used in several different areas. The most promising uses for this technology are in medicine, construction, computers, and the environment.

This branch of technology is expected to greatly impact the medical industry. Scientists are working on creating nanorobots. These cells will be programmed to attack cancer cells and viruses and rearrange their molecular structures to make them harmless. These tiny robots may also be able to slow the aging process, which could increase life expectancy. Since these machines will be able to perform work a thousand times more precisely than current methods, they can also be programmed to perform delicate surgeries.

The construction industry will benefit from nanotechnology as well. The tiny machines will be used to make much stronger fibers. There is hope that one day we will be able to replicate anything, from diamonds to food.

The parts of a computer continue to get smaller all the time. They will, however, soon reach their limit. Nanotechnology will be used to make much smaller storage devices with the capability to hold much more information.

Finally, nanotechnology can be used to better the environment. We are currently running out of many natural resources. With nanotechnology, we will be able to construct more of them. Manufacturing with nanotechnology produces much less pollution than traditional manufacturing. In addition, nanorobots could potentially be programmed to rebuild the ozone layer and clean up oil spills.

It is clear that nanotechnology has many exciting potential uses. Research is just getting underway in this new branch of technology, but it certainly will continue. Researchers expect that we will see the benefits of nanotechnology within the next century.

Chapter 9
The Problem-Solving and Design Process

Learning Objectives

After studying this chapter, you will be able to:

- Describe the general problem-solving/design process.
- Define a technological problem or opportunity.
- Describe the technological problem-solving/design process.
- List the major steps involved in solving technological problems and meeting technological opportunities.
- Describe the steps followed in identifying a technological problem or opportunity.
- List the types of criteria and constraints to be considered in the problem-solving/design process.
- Describe the types of information gathered as a foundation for technological development projects.
- List the general methods of gathering information for the problem-solving/design process.

All technology is created for a purpose. It is designed to meet human needs or wants. All technology has come from people who wanted to solve a problem or address an opportunity. For example, some people have developed technological artifacts that protect us from the physical environment. Others have developed artifacts that have helped us become more informed. Still others have created artifacts that allow us to travel with ease.

Most modern technological artifacts and systems are not developed or discovered by accident. Instead, as we noted in earlier chapters, people use specific problem-solving/design, production, and management processes to meet needs and wants. In this section we will focus on the activities that occur in the first stage, the problem-solving/design process. In the initial chapter we will first learn about the problem-solving/design process in general and how that process applies to technology. We will then focus on the initial step in the process, that is, identifying the technological problem or opportunity.

Figure 9-1. Technology is born from a problem or an opportunity, which can be as simple as how to clear snow from a road or as complex as how to land a person on the moon. (NASA)

The Problem-Solving/Design Process in General

People usually realize that a problem exists when they encounter a difficulty and are not sure how to resolve it. A problem can be as simple as opening a walnut when tools are unavailable. It can be as complex as moving five hundred people from Los Angeles to Tokyo in two hours, **Figure 9-1.** These problems have two things in common: (1) a goal exists (cracking a walnut or moving people), and (2) no clear path is apparent from the present state (an intact walnut or people who need to be moved) to the goal.

Note that not all problems are technological, however. You might have a problem getting along with a classmate or a family member. This is a social problem. You might have a problem identifying and describing the weather produced by some types of clouds. This is a scientific problem. You might be trying to decide whether to keep some money you found on the street. This is an ethical problem.

Problem solving is a common human activity. We all use a fairly universal process as we approach problems. In short we do the following:

- Develop an understanding of the problem through observation and investigation
- Devise a plan for solving the problem
- Implement the plan
- Evaluate the plan

Within this general framework, however, those in technological areas have created a specific approach to help solve problems and meet opportunities. This approach is called the design process or, as we call it here, the problem-solving/design process.

The Problem-Solving/Design Process in Technology

The ***problem-solving/design process*** in technology is used in cases of attempting to satisfy people's technological needs and wants. We should note here that those in technology use this problem-solving/design process not only where they see a problem that they might be able to solve but also where they see opportunities for improvement. The difference between these two kinds of situations is described in more detail as follows.

Technological Problems

You face a ***technological problem*** if you need to develop tools, machines, or systems to help you do work. An example within the agricultural area will highlight this point.

Until the 1940s, most farmers raised their own hay and used it as feed for the animals on the farm. In recent years, however, the self-sufficient family farm has been disappearing. Family farms are being replaced with larger farms that specialize in a specific crop. Some farmers operate large dairy farms that do not raise hay. Other farmers specialize in raising and selling hay to these dairy farmers and beef cattle feedlots. Hay now had to be shipped to other farms and to the feedlots, but having people gather and ship loose hay was not very productive.

Added to this problem was the fact that farm labor was becoming more expensive, and many people did not want to do heavy manual labor. The solution to this problem was to develop farm implements that allowed one person to do more work with the assistance of machines, **Figure 9-2.** People learned how to collect hay into bales for easier transportation. This process included the following steps:

1. The field of hay was cut with a mowing machine and allowed to dry on the ground.
2. A machine called a side delivery rake was used to collect the hay into a narrow pile called a windrow.
3. A tractor pulled the baler along the windrows, where it collected, compacted, and tied the hay into bales.
4. The bales were then pushed out of the machine onto the ground, where farm workers picked them up and loaded the bales onto a truck.

More recent technology has further improved the machines used in harvesting hay. Mowing machines and side delivery

Career Corner
Engineers

Engineers apply science, mathematics, and technology knowledge to develop solutions to technical problems. They develop ways to extract and process natural resources. Engineers design products, machines to build products, production systems, and production facilities. They also design buildings, highways, and transit systems. Engineers develop ways to harness, convert, and use energy.

Some branches of engineering are aerospace, agricultural, biomedical, chemical, civil, computer hardware, electrical and electronics, environmental, industrial, mechanical, mining and geological, and nuclear. Most engineers work in office buildings, laboratories, and industrial plants or at construction sites, mines, and oil and gas production sites. A bachelor's degree in engineering is required for almost all entry-level engineering jobs. Most engineering education programs include study in an engineering specialty and courses in mathematics and science.

rakes have been replaced by a machine called a swather, which cuts, conditions, and windrows hay in one operation. The hand loading and hauling of baled hay is being replaced with bale wagons that automatically pick up and stack bales into cubes. Hay is also gathered into large rolls. These cubes or rolls are loaded with forklifts onto trucks that haul the hay to customers. Thus, harvesting hay, which was once *labor intensive,* has become *equipment intensive* through the use of technological problem solving.

Figure 9-2. Technology has changed farming from a labor-intensive activity to an equipment-intensive activity. A single farmer can now do the work of many people.

An examination of other technological areas reveals the same pattern. For example, in manufacturing operations, computer-aided design systems have increased the productivity of drafters.

Robots perform many routine manufacturing tasks without human interference or monitoring, **Figure 9-3.** In transportation, computer-controlled "driverless" people movers speed people between terminals at airports. Guided bus systems, such as the one shown in **Figure 9-4,** promise to increase the efficiency of urban transport. Automated equipment on new commercial aircraft has reduced the number of flight deck officers from three to two.

Technological Opportunities

Not all technology is developed to solve a problem, however. Many conditions exist that people do not view as a problem. Still, if a "nonproblem" condition can be improved with technology, a ***technological opportunity*** has been discovered. People may want to use the new device even though an old device was adequate for their needs. The automatic

Figure 9-3. Machines such as this robot can perform repetitive tasks with high accuracy. (Cincinnati Milacron)

Figure 9-4. New transportation systems, such as this guided bus, use fewer drivers to move more people and with greater safety. (DaimlerChrysler)

coffeemaker is an example of meeting a technological opportunity. There was no consumer cry for a new way to make coffee. The percolator and drip coffeemakers in use were adequate for the job. Nevertheless, the first Mr. Coffee® automatic-drip coffeemaker was an almost instant success. Advertising helped many people develop a "need" for the new product. An opportunity was identified, and a new product was born.

As we noted earlier, though, no matter whether the situation is a problem or an opportunity, those in technology use a specific problem-solving/design approach. That approach involves a series of steps that follow one from the other. They are described as follows.

Steps in Solving Technological Problems and Meeting Opportunities

Solving technological problems and meeting technological opportunities require specific action. As shown in **Figure 9-5,** these challenges involve four major phases. These four phases and the basic activities they include are:

- Identifying or recognizing a technological problem: This phase of the technology development includes:
  - Defining the problem: Describing the situation that needs a technological solution and establishing the criteria and constraints under which the device or system must operate.
  - Gathering information: Obtaining background information needed to begin developing solutions to the problem or situation.
- Developing a solution: This phase of the technology development involves:
  - Developing possible solutions: Originating a number of different solutions that could solve the problem or meet the opportunity.
  - Isolating, refining, and detailing the best solution: Picking out the most promising solution; integrating, modifying, and improving it; and creating detailed sketches of best solution.
- Evaluating the solution: This phase of the process involves:
  - Modeling the best solution: Testing and evaluating the proposed solutions through graphic, mathematical, and/or physical modeling techniques.
  - Selecting the best solution: Comparing the design solutions in terms of economic, market, technical, production, and environmental criteria to determine the best solution to the problem.
- Communicating the solution: This phase of the process includes:
  - Interpreting the solution: Communicating the final solution through such detailed documents and reports as engineering drawings, bills of materials, and specification sheets.
  - Presenting the solution for approval: Presenting written and oral reports to obtain appropriate approval (from management or government, for example) for implementing the solution

As stated earlier, the remainder of this chapter will explore the first phase of the technology development activity: identifying or recognizing a technological problem. The other phases will be discussed in the following three chapters.

Identifying a Technological Problem or Opportunity

We have learned that all technology starts with a problem or an opportunity. We want to create a new technological artifact or system to solve that problem or meet that opportunity. To achieve this goal, however, we must first identify what the problem or opportunity really is.

Such a task is not always simple. For example, we know that holes have developed in the ozone layer over the Antarctic and Artic regions. The holes are not the real problem, however; instead, their presence indicates that a problem might exist. In this situation, we look further by gathering more information and discover that the holes are the consequence of our use of such items as polluting vehicles and aerosol paint and personal care products. The problem, then, is to create such items as vehicles with cleaner fuels and products without the aerosol component.

Thus if designers are to solve a technological problem or meet an opportunity, they must first define what the problem or opportunity really is and then gather information needed to begin developing solutions to the situation. In each of these steps, designers follow certain procedures, which are described in more detail as follows.

Defining the Problem or Opportunity

The definition of a problem or opportunity should complete two major tasks:

- Describe the situation that needs a technological solution
- Establish the criteria and constraints under which the artifact or system must operate

Describing the Situation

Defining a problem or opportunity first involves describing it in a clear and concise statement. In this statement, we must take care that we do not confuse the problem or opportunity with the solution. For example, a bookend is a solution, not a problem. The problem is to hold books in an organized manner on a flat surface. Likewise, a chair is a solution. A chair can be described

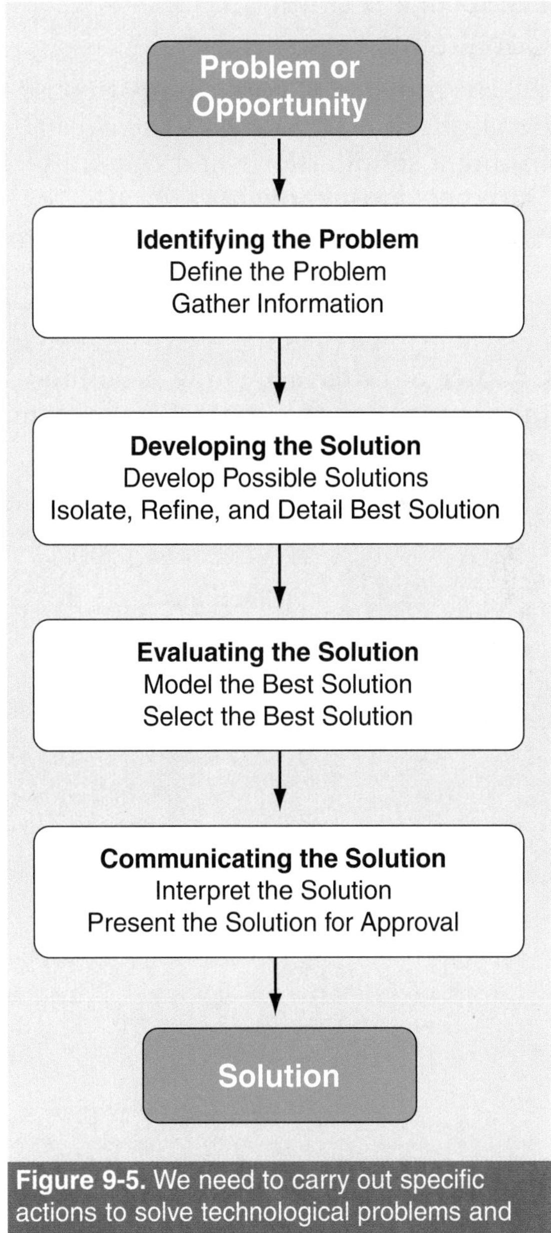

Figure 9-5. We need to carry out specific actions to solve technological problems and meet technological opportunities.

as a device with a seat, back, and four legs. With this description, a designer starting with the problem statement "design a chair" would never end up with a beanbag or a wicker basket suspended from a chain. Descriptions can restrict creativity or open doors to a variety of solutions.

Look back at our ozone layer "problem." To close the hole is the solution. The problem is to develop technological devices that emit fewer pollutants into the atmosphere.

Establishing Criteria and Constraints

The problem or opportunity description leads directly to the next step, establishing a set of criteria and constraints. Designers must know how the effectiveness of the new technology will be evaluated. To do this, they establish criteria and constraints under which the product or system must operate. As we learned in Chapter 5, ***criteria*** include a listing of the features of the artifact or system; ***constraints*** deal with the limits on the design. These criteria and constraints, as shown in **Figure 9-6,** can be grouped into the following general categories:

- *Technical or engineering criteria and constraints* describe the operational and safety characteristics that the device or system must meet. These criteria and constraints are based on how, where, and by whom the product will be used. An example of an engineering criterion for a new windshield wiper might be: Must effectively clear the windshield of water when the vehicle is traveling at highway speeds.
- *Production criteria and constraints* describe the resources available for producing the device or system. These criteria and constraints are based on the natural, human, and capital (machine) resources available for the production of the device or system. A production constraint for a product might be: Must be manufactured using existing equipment in the factory.

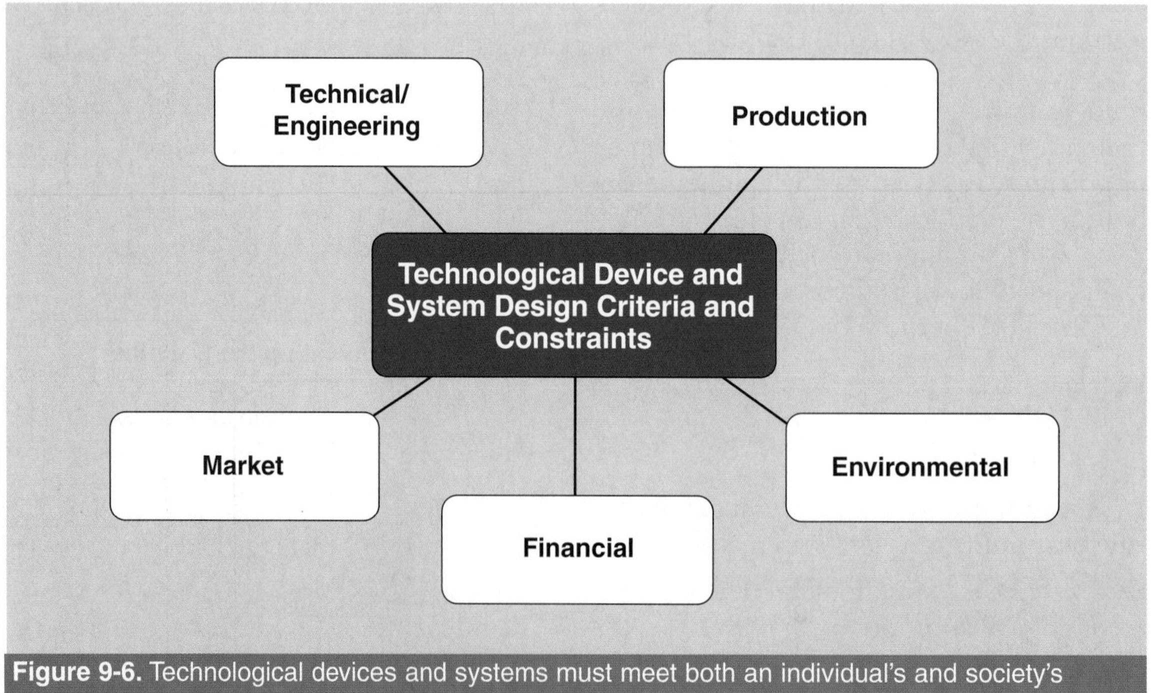

Figure 9-6. Technological devices and systems must meet both an individual's and society's expectations. These expectations can be communicated through five types of criteria and constraints.

- *Market criteria and constraints* identify the function, appearance, and value of the device or system. These criteria and constraints are derived from studying what the user expects from the device or system. An example of a market constraint might be: Must be compatible with Early American decor.
- *Financial criteria and constraints* establish the cost-benefit ratio for the device or system. These criteria and constraints address the amount of money required to develop, produce, and use the technological device or system and the ultimate benefits from its use. An example of a financial constraint might be: Must be priced at 5 percent less than the major competitor's product.
- *Environmental criteria and constraints* indicate the intended relationship between the device or system and the natural and social environments. These criteria and constraints deal with the impacts of the technological device or system on people, societal institutions, and the environment. An environmental criterion might be: Must remove 95 percent of all sulfur dioxide emitted from the electric generating plant smokestacks.

The problem or opportunity description and set of criteria and constraints provide designers with direction and a focus for their work. The description and set of criteria and constraints allow the designers to start the next step in the identification process, gathering information.

Gathering Information

Designing technological devices and systems requires knowledge. This knowledge is derived from obtaining and studying information. A wide variety of information may be needed, **Figure 9-7.** Typically this information includes:

- ***Historical information*** about devices and systems that were developed to solve similar problems
- ***Scientific information*** about natural laws and principles that must be considered in developing the solution
- ***Technological information*** about materials and energy-processing techniques that could be used to develop, produce, and operate the device or system
- ***Human information*** that will affect the acceptance and use of the device or system. This information might include such factors as ergonomics, body size, consumer preferences, and appearance
- ***Legal information*** about the laws and regulations that will control the installation and operation of the device or system
- ***Ethical information*** that describes the values people have toward similar devices and systems

Designers can obtain this information through three major activities. First, designers can use ***historical methods,*** that is, gather information from existing records. For example, they might consult books, magazines, and journals housed in libraries, **Figure 9-8.** They can review sales records, customer complaints, and other company files. They can check judicial codes that describe laws that may affect the development project.

Designers also use ***descriptive methods*** to record observations of present conditions. For example, they might survey people to determine product preference, opinions, or goals. They might observe and describe the operation of similar devices. They might measure and record physical qualities, such as size, weather conditions, and weight.

Finally, designers use ***experimental methods*** to compare different conditions. One condition is held constant while the other is varied. Designers use this method

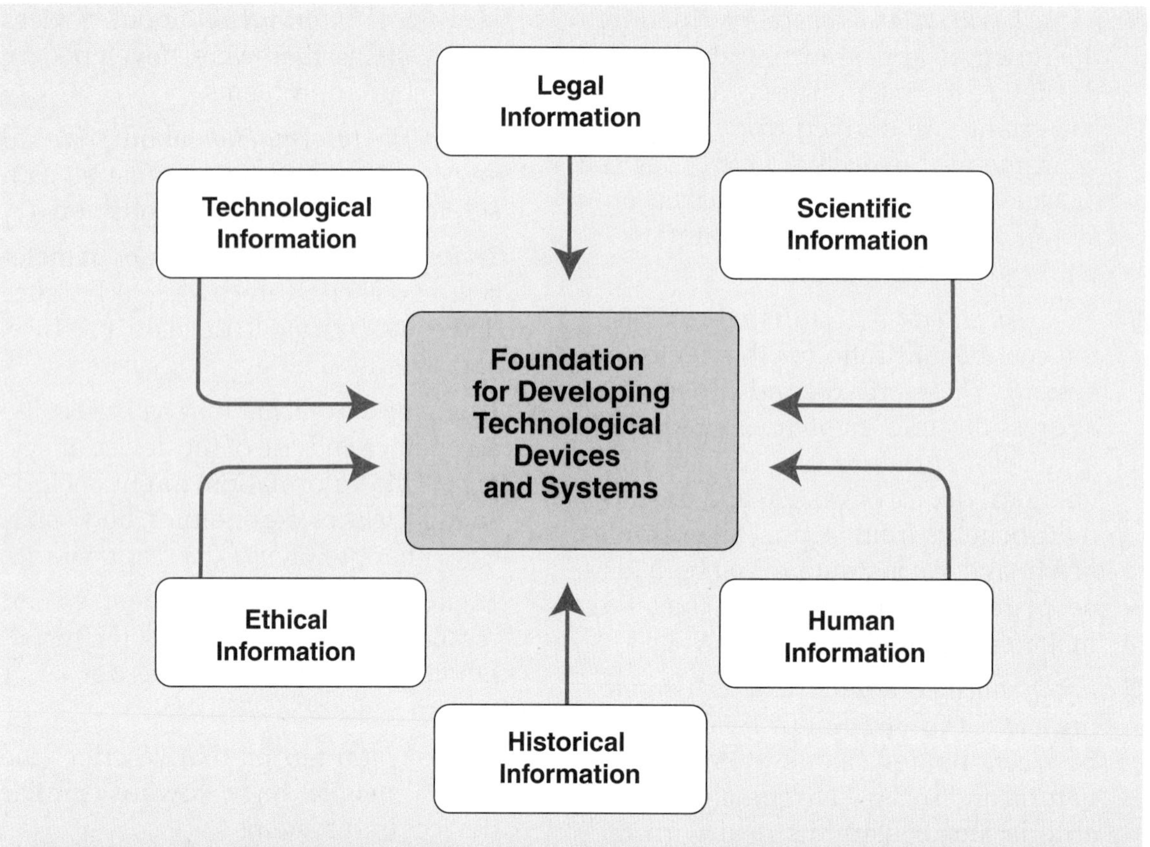

Figure 9-7. Many types of information are used as a foundation for technological development activities.

Figure 9-8. Books, magazines, and journals provide a valuable source of information for solving technological problems.

Figure 9-9. Laboratory experiments provide valuable information that can be used to solve technological problems. (Goodyear Tire and Rubber Co.)

Connections to Technology: History

The Origin of Radar

As indications of an impending world war grew in the late 1930s, various countries realized they had a major problem. Although scientists had discovered that radio echoes could detect ships and planes, the early devices made for this purpose were not reliable. Knowing that keeping track of an enemy's air and sea movements would be critical in a war, the various leaders kept urging scientists to improve the technology termed radar (*ra*dio *d*etection *a*nd *r*anging).

In the late 1930s and early 1940s, American and British researchers worked with each other to improve radar. They knew their problem—increasing the reliability and accuracy of radar—and developed and refined solutions to solve the problem. For example, the British developed an improved vacuum tube that had enough microwave energy to be used in radar systems. The Americans then took it over and developed units small enough for airplanes and boats. As the war continued, this improved radar indeed became a key element in the Allies' eventual victory.

During the last half-century, researchers have further developed the capabilities of radar. In what areas is it used today?

to determine scientific principles, assess the usefulness of technological processes, or gauge human reactions to situations, **Figure 9-9.**

All these methods are used to gather information that can provide the background needed to begin developing solutions for the problem or opportunity. The procedures for developing these solutions will be the focus of the next chapter in this book.

Summary

People develop technology in response to problems and opportunities. Development activities involve identifying the problem or opportunity, developing solutions for the problem or opportunity, evaluating the chosen solution, and communicating the solution.

The base for all technology development programs is a clear description of the problem and a set of criteria and constraints that describe the goals for the solution. These two steps are followed by efforts to collect information needed to start developing appropriate devices or systems.

Key Terms

constraints
criteria
descriptive methods
ethical information
experimental methods
historical information
historical methods
human information
legal information
problem-solving/design process
scientific information
technological knowledge
technological opportunity
technological problem

Test Your Knowledge

Write your answers on a separate piece of paper. Please do not write in this book.

1. *True or false?* Technology is often discovered by accident.
2. List the four universal steps used by people to solve problems.
3. Describe the difference between a technological problem and a technological opportunity.
4. List the four major steps used in technology to solve problems and meet opportunities.
5. What is the first step in defining a problem or opportunity?
6. *True or false?* The way a problem is described can restrict creativity.
7. Give one example of a production constraint in developing a technological artifact or system.
8. Give one example of a market criterion in developing a technological artifact or system.
9. List the six types of information that can help designers solve technological problems and meet technological opportunities.
10. List the three methods used to obtain information.

Applying Your Knowledge

1. Select a problem or opportunity that technology can address. Use a form similar to the one shown below to help you:
   a. Write a clear definition of the problem or opportunity.
   b. List the criteria and constraints that you would consider in solving the problem or meeting the opportunity.

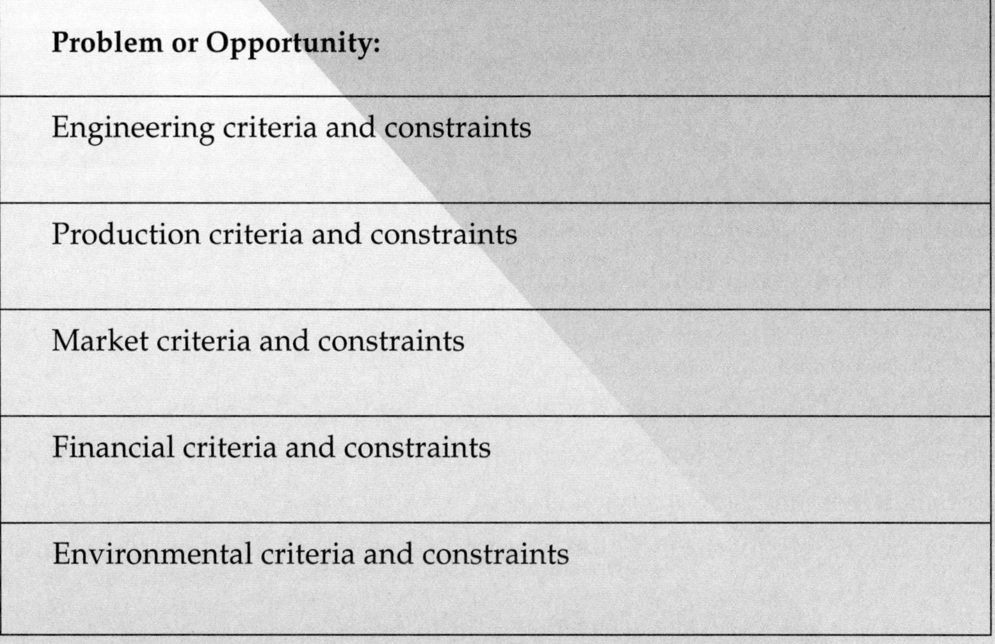

Problem or Opportunity:
Engineering criteria and constraints
Production criteria and constraints
Market criteria and constraints
Financial criteria and constraints
Environmental criteria and constraints

2. Suppose you were given the following problem to solve: Design a device that would allow you to generate enough electricity to power a television set while riding an exercise bicycle. Where would you go to gather information to start the design process?

Chapter 10
Developing Design Solutions

Learning Objectives

After studying this chapter, you will be able to:

- Name the two major areas in which designers develop solutions to technological problems and opportunities.
- List the three main steps followed in developing technological designs.
- Identify the ways in which ideas can be stimulated when developing design solutions.
- Differentiate between kinds of design sketches.
- List the three major types of information required for detailed sketches when building models.
- Identify the three types of pictorial sketches used in product design.

Technological devices are designed to meet identified problems and opportunities. These problems and opportunities, as shown in **Figure 10-1,** can be divided into two major areas:

- System design problems and opportunities
- Product design problems and opportunities

No matter what the area, though, developing design solutions involves three major steps. We will discuss those steps after we differentiate between system design and product design.

System Design

System design deals with the arrangement of components to produce a desired result. For example, automotive braking systems are a result of system design efforts. Look at the drum brake system shown in **Figure 10-2.** This design brings together mechanical and hydraulic components into a speed-reduction system. The brake pedal unit is a mechanical linkage. When the pedal is depressed, a plunger in the master cylinder is moved. This motion causes the fluid to move in the hydraulic system that connects the master cylinder to the wheel cylinders. The fluid movement pushes the pistons outward in the wheel cylinders. These pistons are attached to the brake shoes. The piston movement causes the shoe to be forced against the brake drum. This mechanical action creates friction between the shoe and the drum, which slows the automobile.

System design can be used in all technological areas. For example, it is an important part of construction technology, **Figure 10-3.** Electrical, heating and cooling, plumbing, and communication systems are designed for buildings. In manufacturing, the methods of production, warehousing, and material handling must be designed. Messages are carried over fiber optic and microwave communication systems. Transportation systems combine manufactured vehicles and other components to move goods and passengers from place to place. Irrigation systems are used to water crops. Pipelines are part of natural gas distribution systems. Doctors and hospitals provide patient care in health care systems.

Figure 10-1. Designers create both systems and products. (Harris Corp., Zenith Data Systems)

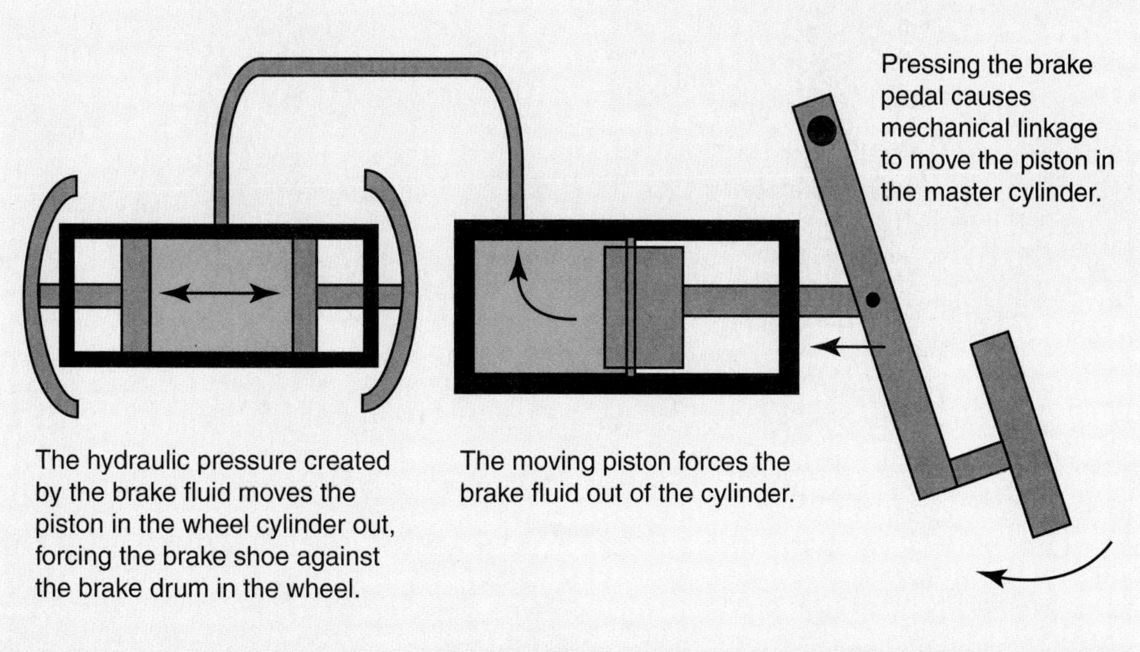

Figure 10-2. This brake system was the result of system design efforts.

Product Design

Product design deals with two areas: manufactured products (involving designers) and constructed structures (involving architects), **Figure 10-4.** The goal of both activities is to develop a product or structure that meets the customer's needs. This task means that the product or structure must function well, operate safely and efficiently, be easily maintained and repaired, have a pleasant appearance, and deliver good value.

In addition, products and structures must be designed so that they can be produced economically and efficiently. They must also be sold in a competitive environment. In short, the product or structure must be designed for:

- Function: Easy and efficient to operate and maintain
- Production: Easy to manufacture or construct
- Marketing: Appealing to the end user

Developing Design Solutions

System and product designs start with a clear definition of the situation or opportunity. We learned the procedures for developing the definition in the previous chapter. This problem definition leads to the next step in product design—developing design solutions. These solutions often evolve through three steps, as shown in **Figure 10-5:**

- Developing preliminary solutions
- Isolating and refining the best solution
- Detailing the best solution

Figure 10-3. System designers devise heating and cooling, plumbing, communication, and electrical systems for buildings. (General Electric Co.)

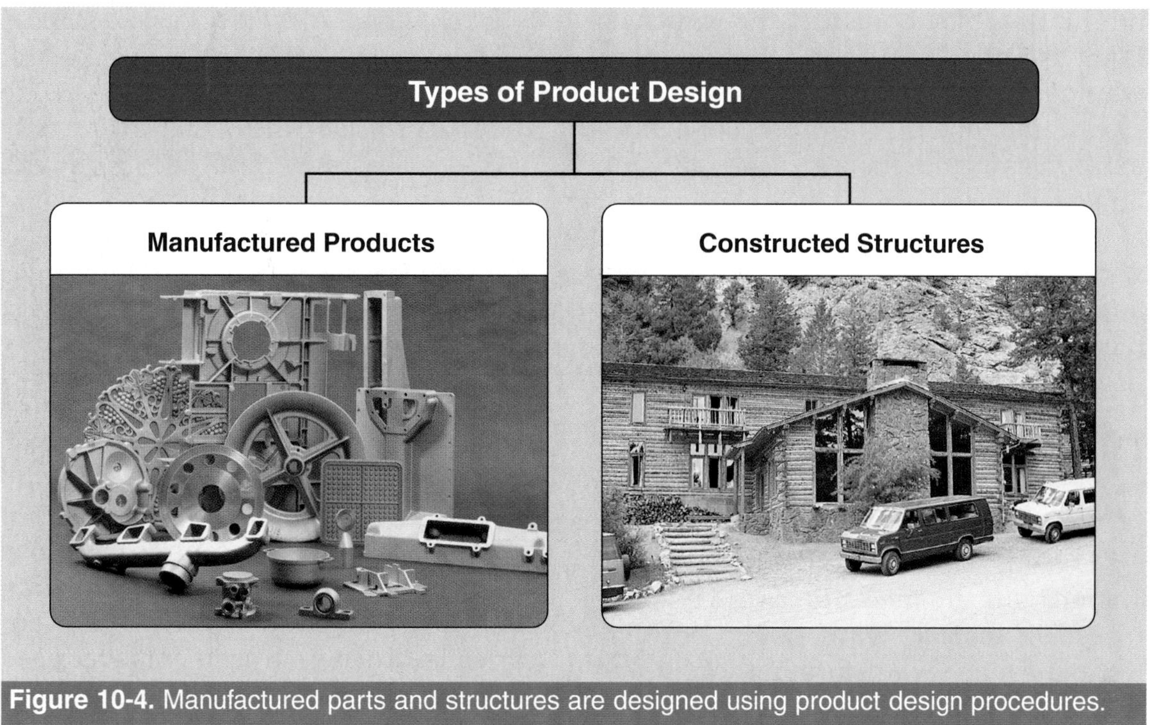

Figure 10-4. Manufactured parts and structures are designed using product design procedures.

This process can be described as *"imagineering."* First, the designer uses his or her imagination to develop a number of unique solutions or designs. These solutions are then engineered back to reality through design refinement and detailing activities. The first step starts with broad thinking. This kind of thinking is called ***divergent thinking.*** It seeks to think of as many different (divergent) solutions as possible. The most promising solutions are then refined and reduced until one "best" answer is found. The refinement of ideas requires ***convergent thinking.*** The goal is to narrow and focus (converge) the ideas until the most feasible solution is found.

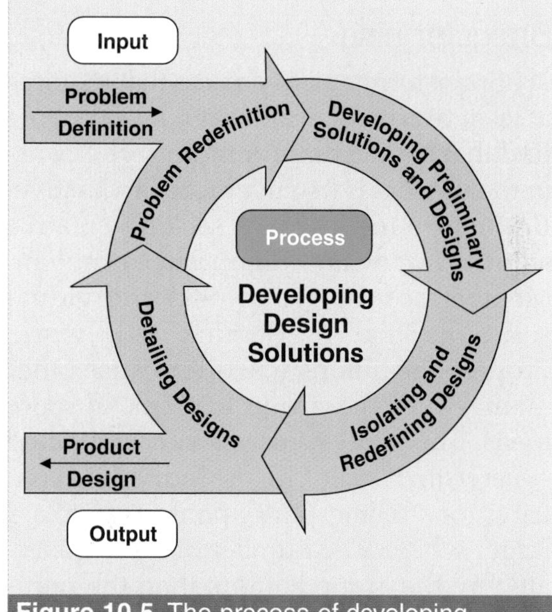

Figure 10-5. The process of developing design solutions evolves through three steps.

Keep in mind that the best solution may not be the one that works best or is the least expensive. As we noted earlier, criteria and constraints can compete with one another; trade-offs often occur among appearance, function, and cost. Many of us cannot always afford the very best answer. Our budget and the length of time we expect to keep the product enter into our product choices. For example, you might find it unwise to purchase a $900 racing bicycle because you are only going to use the bicycle to ride occasionally to the park. If you regularly ride a bicycle to work or go on cycling vacations, however, you might be able to justify the expense. Likewise, the "snapshot" photographer would probably not need the most expensive digital camera. The professional photographer would choose such a camera, however.

Product design activities produce a wide range of products. This variety allows the consumer to select one that meets his or her performance needs and financial resources.

Developing Preliminary Solutions

Designs start in the minds of designers, engineers, or architects. Ideas can be stimulated in various ways. Three popular techniques are brainstorming, classification, and "what if" scenarios. Then, after people have put forth their ideas, they record the ideas in rough sketches.

Brainstorming

Brainstorming is a process that requires at least two people, although most people find that having three or more participants in the process is usually more productive. ***Brainstorming*** involves seeking creative solutions to an identified problem. Members of the group offer individual solutions that they think will work. Proposed solutions will often cause other members of the group to think of more ideas. The strategy uses a concept called ***synergism***. It builds on the individual contributions of the participants to make a larger whole. The number of ideas generated by the group is more than the number they could develop if everyone worked alone.

Brainstorming activities work best when the group accepts some basic rules. These rules include:

- *Encourage wild, far-out ideas.* There are no "bad" or "stupid" ideas. Wild but promising ideas can always be engineered back to reality.
- *Record the ideas without reacting to them.* Many people will stop offering some of their ideas if they are criticized. To avoid criticism, they will provide only those ideas they think the group will like.
- *Seek quantity, not quality.* The chances of good ideas emerging are increased as the number of ideas increases.
- *Keep up a rapid pace.* A rapidly paced session will keep the mind alert and reduce the chance of judging the ideas.

Classification

Classification can be conducted by one person or a group of people. ***Classification*** involves dividing the problem into major segments. Then each segment is reduced into smaller parts. For example, buildings could be classified as business and commercial, homes, and industrial. Homes could be further classified as houses, apartments, and condominiums, for example. A house could then be classified by its major features: foundations, floors, walls, ceilings, roof, doors, windows, and so on. Foundations could then be classified as poured concrete, concrete block, wood posts, timber, and so on. This process might result in a classification chart. This type of chart is often developed as a tree chart with each level having a number of branches below it, as shown in **Figure 10-6.** It would end up looking much like a family tree that people use to trace their ancestors.

"What If" Scenarios

"What if" scenarios start with a wild proposal. Then its good and bad points are investigated. The good points can be used to develop solutions. For example, peeling paint is a problem for house painters. They must remove the old paint from a house before repainting. A wild solution would suggest mixing an explosive material with the paint before it is applied. Whenever the building is ready to be repainted the old

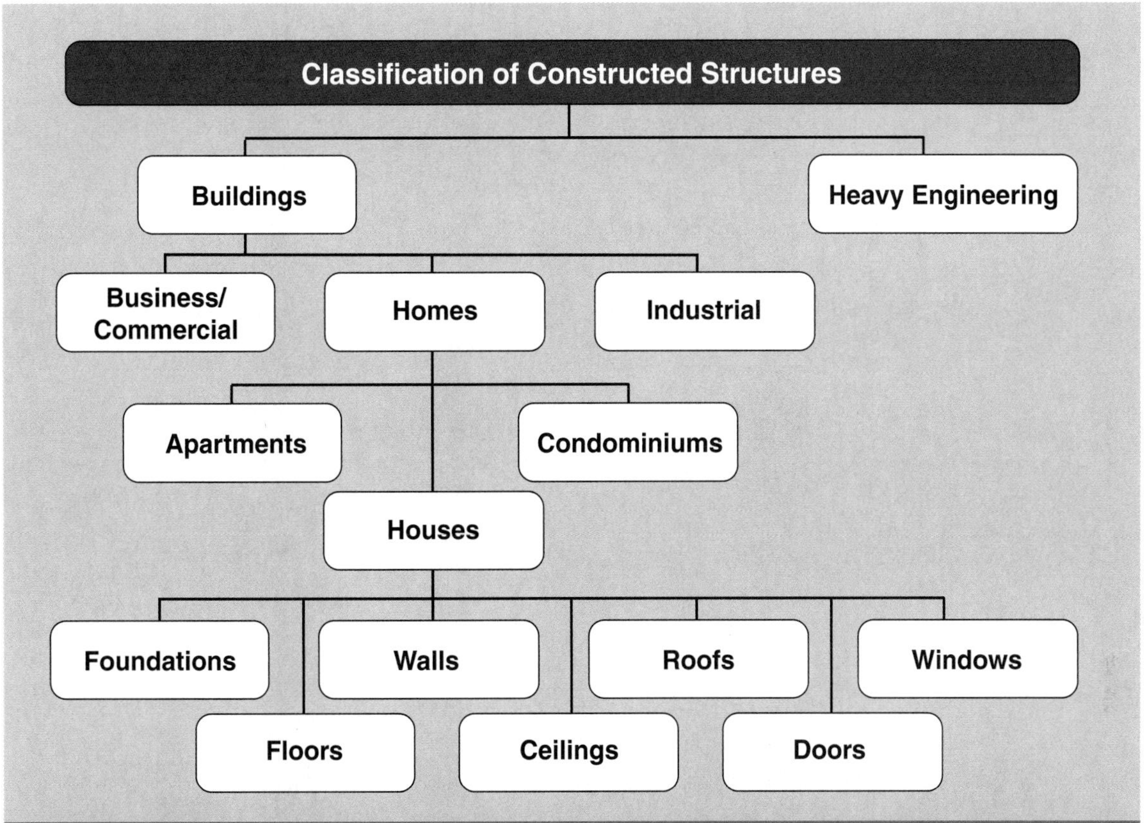

Figure 10-6. A classification chart divides the design problem or opportunity into segments to develop solutions more easily. This type of chart is called a tree chart.

paint could be blown off the building. Obviously, exploding house paint is ridiculous. The proposal, however, could lead to a solution. Paint sticks to a house through the adhesion between the paint and the siding. Maybe a material could be mixed with the paint that would cause it to lose adhesion when a special chemical is applied. At repainting time, the chemical could be sprayed on to loosen the paint. The paint could then be easily removed from the siding.

Rough Sketching

Once designers have conceived of a number of ideas in their minds they must record the ideas. The most common recording method is to develop ***rough sketches*** of the products, structures, or system components, as shown in **Figure 10-7.** These sketches are as much a part of the thinking process as they are a communication medium. Designers are forced to think through concepts such as size, shape, balance, and appearance. The sketches then become a library of ideas for later design efforts.

The term *rough* is not used to describe the quality of the drawing. Rough sketches are not necessarily crude. They often represent good sketching techniques. The term *rough* describes the state of the design ideas. It suggests that the designs are incomplete and unrefined.

Isolating and Refining Design Solutions

The rough sketches allow designers to capture a wide variety of solutions for the design problem or opportunity. The sketches are like books in a library: they

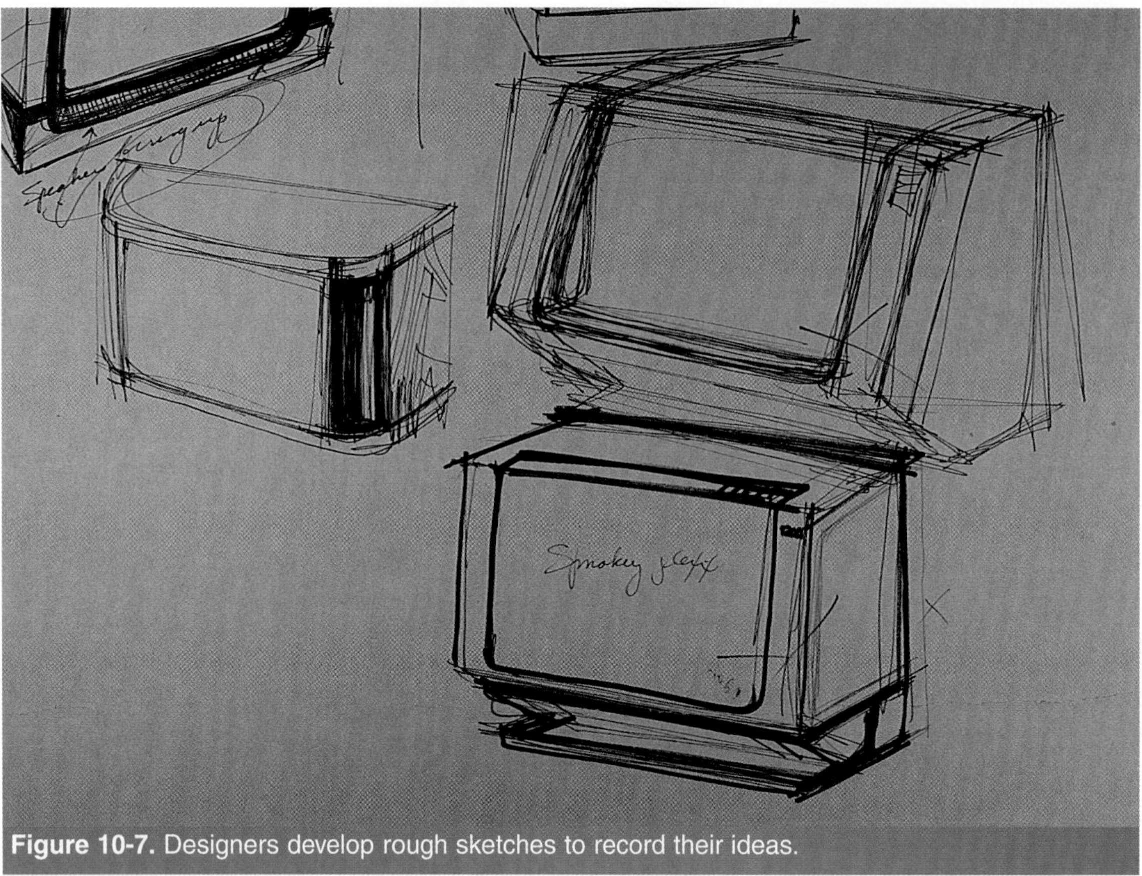

Figure 10-7. Designers develop rough sketches to record their ideas.

contain a number of different thoughts, views, and ideas. They can be selected, refined, grouped together, or broken apart.

Isolating and refining original designs in the "library of ideas" is the second step in developing a design solution, **Figure 10-8.** Promising ideas are chosen and then studied and improved. This process might involve working with one or more good rough sketches. The size and shape of the product or structure might be changed and improved. Details might be added, and the shape might be reworked. In short, the design is becoming refined as problems are worked out and the proportions become more balanced.

Refined design ideas might also be developed by merging ideas from two or more rough sketches into a ***refined sketch.*** The overall shape might come from one sketch, and specific details might come from others. This approach is one of integration, which blends the different ideas into a unified whole. The new idea may not look anything like the original rough sketches.

Detailing Design Solutions

Rough and refined sketches do not tell the whole story. Look back at the sketches shown in Figure 10-8. What size is the product in the sketch? You cannot tell. The sketches communicate shape and proportion. They do not communicate size. For this task, we need to add more details, and thus we need a third type of sketch, called a ***detailed sketch.*** It communicates the information needed to build a model of the product or structure. Detailed sketches can also be used as a guide to prepare engineering drawings for manufactured products and architectural drawings for

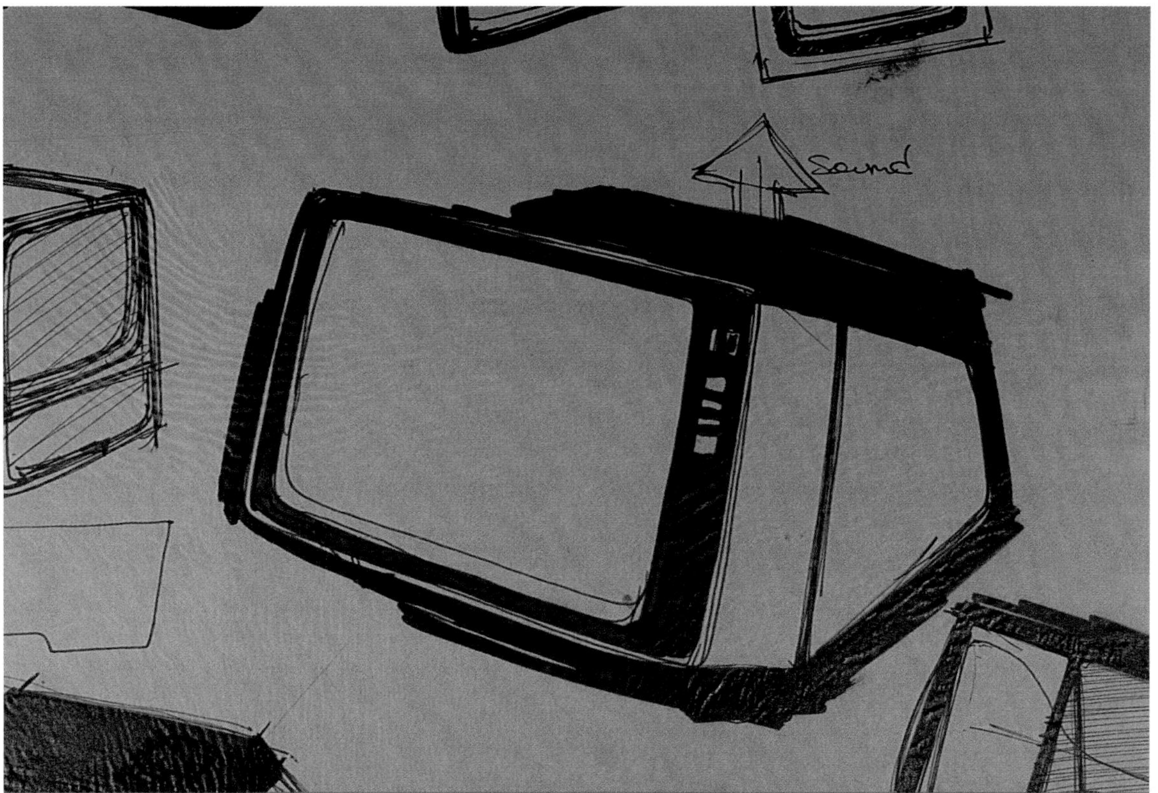

Figure 10-8. In this refined product sketch, the shape has been worked out and shading has been added to improve the looks of the sketch. (RCA)

constructed structures. Engineering and architectural drawings will be discussed in Chapter 12.

Detailed sketches are helpful when models of products or structures are made. Building models requires three major types of information. These types, as shown in **Figure 10-9,** are:

- *Size information:* Explains the overall dimensions of the object or the size of features on an object. This information might include the thickness, width, and length of a part, the diameter and depth of a hole, or the width and depth of a groove.
- *Location information:* Gives the position of features within the object. This information may establish the location of the center of a hole, the edge of a groove, or the position of a taper.
- *Geometry information:* Describes the geo-

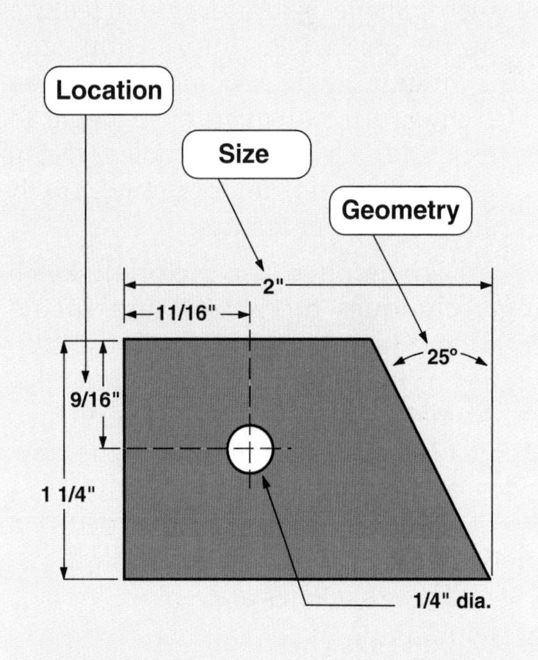

Figure 10-9. The types of information provided on detailed sketches are size, location, and geometry.

Connections to Technology: **Mathematics**

Solid Geometry

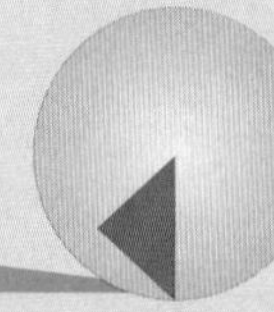

Designers should be familiar with some basic geometric concepts in order to create effective pictorial sketches. Designers use the concepts of solid geometry, for example, when drawing such three-dimensional images as pyramids, cones, cylinders, and cubes.

A pyramid, to use the first example, is a solid figure with a *polygon* as a base. A polygon, as you might recall from earlier geometry lessons, is a closed plane figure bounded by three or more straight lines. (The word *polygon* comes from the prefix poly-, meaning many, and the suffix *-gon,* meaning angle. Thus a polygon is a many-angled figure.) The faces (surface) of the pyramid are triangles with a common vertex, or point where they intersect.

In a regular pyramid the base is a regular polygon, and the faces are *congruent* triangles. Again, as you might remember from earlier geometry, congruent means equal in size and shape.

The other three-dimensional images share some characteristics but differ in others. For example, in what way or ways is a cone like a pyramid, and in what way or ways is it different?

metric shape or relationship of features on the object. This information could communicate the relationship of intersecting surfaces (square or 45° angle, for example), the shapes of holes (rectangular or round, for example), or the shapes of other features.

Designers often use pictorial sketching techniques to capture and further refine product design ideas. These techniques try to show the artifact much like the human eye would see it. Therefore, a single view is used to show how the front, sides, and top would appear.

Designers produce three different kinds of pictorial sketches when refining ideas. These sketches are:

- Oblique sketches
- Isometric sketches
- Perspective sketches

Oblique Sketches

Oblique sketches are the easiest pictorial sketches to produce. ***Oblique sketches*** show the front view as if a person was looking directly at it. The sides and top extend back from the front view. They are shown with parallel lines that are generally drawn at 45° to the front view.

To produce an oblique sketch, the designer completes steps like those shown in **Figure 10-10.**

1. Lightly draw a rectangle that is the overall width and height of the object.
2. Lightly extend parallel lines from each corner of the box back at 45°.
3. Lightly mark the extension lines at a point equal to the depth of the object.
4. Lightly connect the depth lines to form a box.

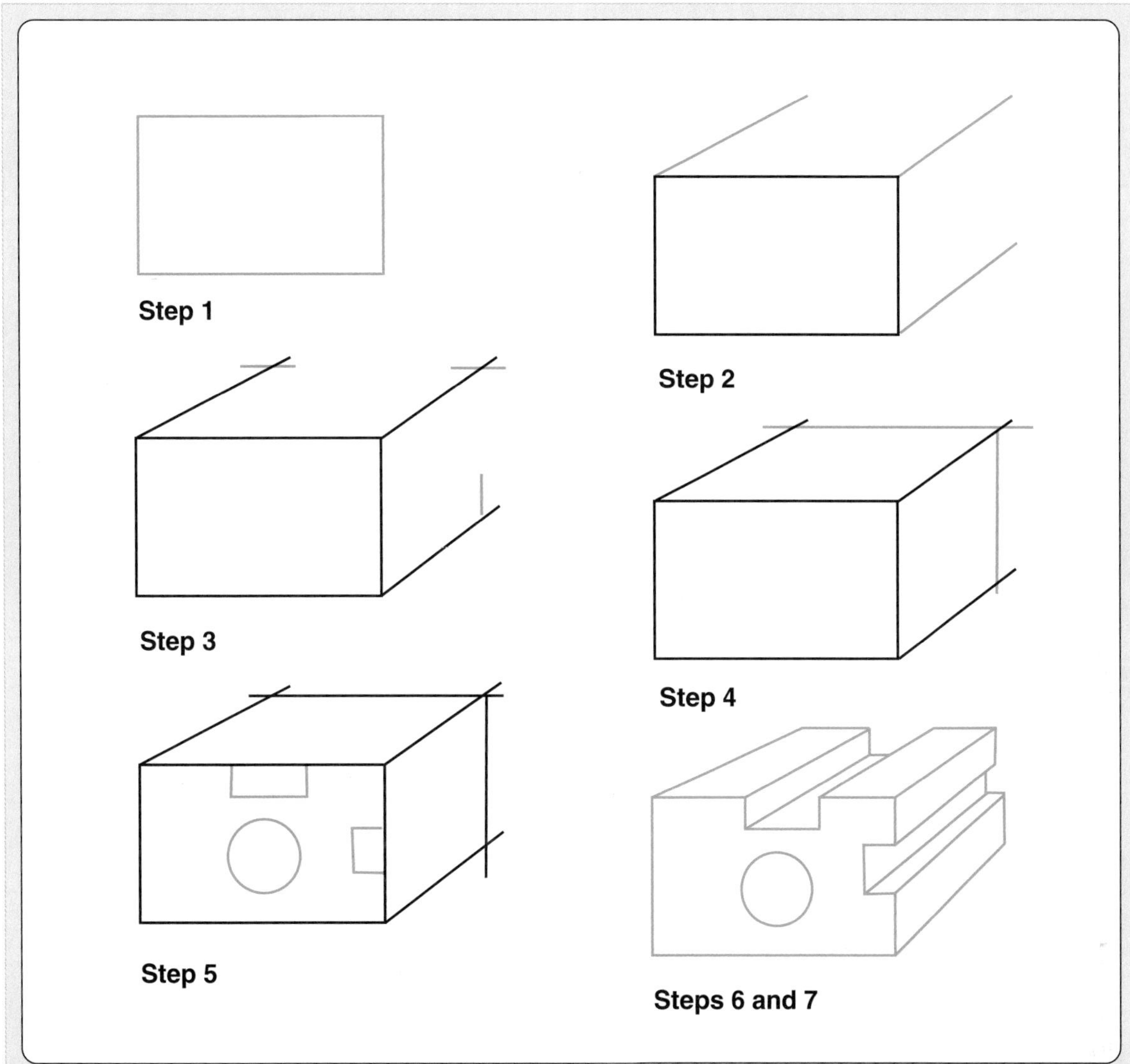

Figure 10-10. Designers create oblique sketches through a series of steps.

5. Add any details such as holes, notches, and grooves onto the front view.
6. Extend the details the depth of the object.
7. Complete the sketch by darkening in the object and detail outlines.

The procedure listed above will produce a cavalier oblique drawing. This type of drawing causes the sides and top to look deeper than they are. To compensate for this appearance designers often use cabinet oblique drawings, as shown in **Figure 10-11.** This type of drawing shortens the lines that project back from the front to one-half their original lengths.

Isometric Sketches

Isometric sketches are the second type of pictorial drawings used to produce refined sketches. The word *isometric* means equal measure. ***Isometric sketches*** get their name from the fact that the angles formed by the lines at the upper-right corner are equal—each is 120°.

Designers use isometric sketching when the top, sides, and front are equally important. The object is shown as if it were viewed from one corner.

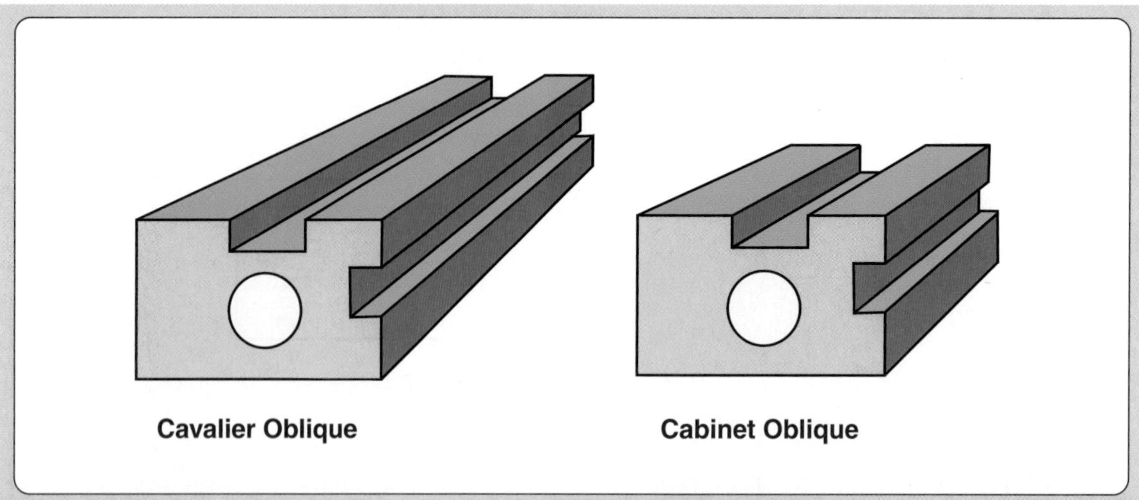

Figure 10-11. The two types of oblique drawings are cavalier oblique and cabinet oblique. Cabinet oblique drawings use one-half the depth of the object for a more natural appearance.

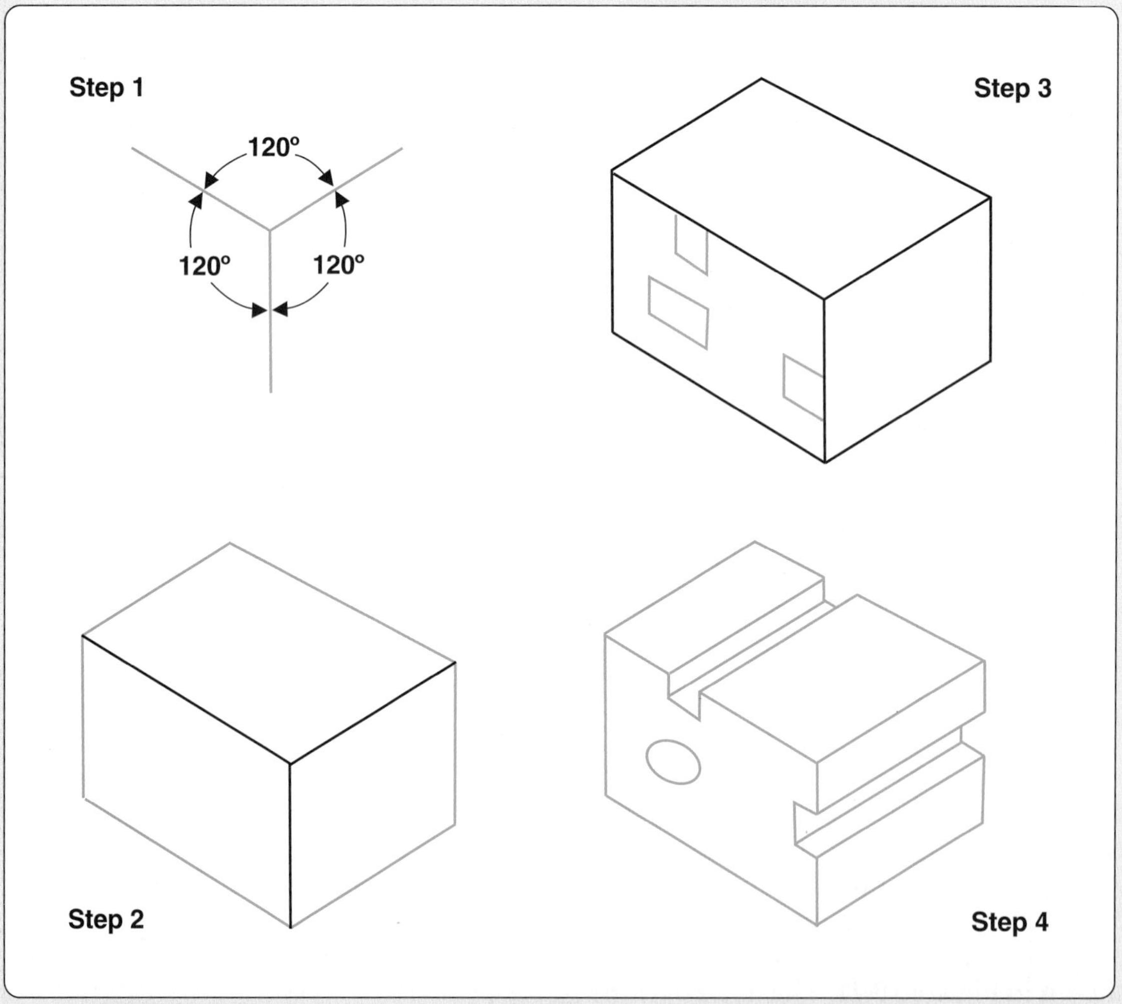

Figure 10-12. Designers follow four major steps when creating isometric sketches.

Designers follow four major steps when producing an isometric drawing. These steps, as shown in **Figure 10-12,** are:

1. Lightly draw the upper-right corner of an isometric box that will hold the object.
2. Complete the box by lightly drawing lines parallel to the three original lines.
3. Locate the major features such as notches, tapers, and holes.
4. Complete the drawing by darkening the features and darkening the object outline.

Perspective Sketches

Perspective sketches show the object as the human eye or a camera would see it. This realism is obtained by having parallel lines meet at a distant vanishing point. If you look down a railroad track you will see a similar effect. The rails remain the same distance apart, yet your eye sees them converge (come together) in the distance.

Three major types of perspective views exist: one point, two point, and three point. The difference between these types is determined by the number of vanishing points used, **Figure 10-13.**

A one-point perspective shows an object as if you were directly in front of it. All the lines extending away from the viewing plane converge at one point. The one-point perspective is like an oblique drawing with tapered sides and top.

A two-point perspective shows how an object would appear if you stood at one corner. It is constructed much like an isometric drawing. Again, the sides are tapered as the lines extend toward the vanishing points.

A three-point perspective shows how the eye sees the length, width, and height

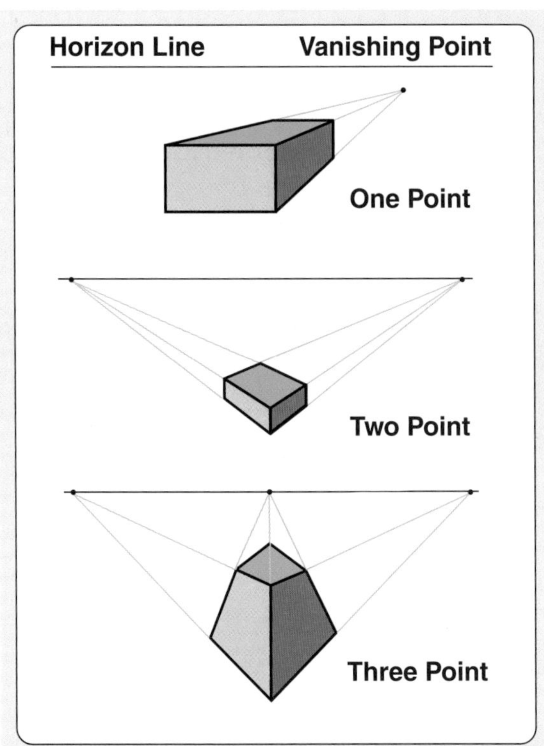

Figure 10-13. The three types of perspective drawings are one point, two point, and three point.

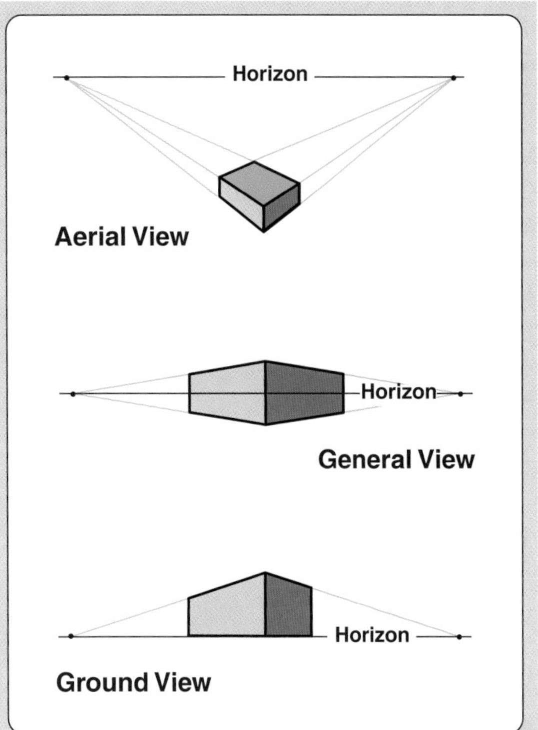

Figure 10-14. Changing the location of the horizon changes the appearance of a perspective drawing.

of an object. All lines in this drawing extend toward a vanishing point.

The appearance of a perspective drawing changes as the horizon changes, **Figure 10-14.** Changing the position of the horizon line can cause the object to be seen as if the observer was looking down on the object (aerial view), directly at it (general view), or up at it (ground view). The designer must decide which of these views best suits the object and the audience who will see the sketch.

When developing the basic structure for one-, two-, or three-point perspective sketches, designers follow the same basic steps. These steps, as shown in **Figure 10-15,** are:

1. Establish the horizon line, vanishing point(s), and the front of the object. Connect the front line(s) to the vanishing point(s).
2. Establish the depth of the objects along the lines that extend to the vanishing point(s).
3. Connect the depth lines to the vanishing point(s). Darken in the object.

Designers then add details to complete the sketch. Perspective sketches are often shaded to add to their communication value.

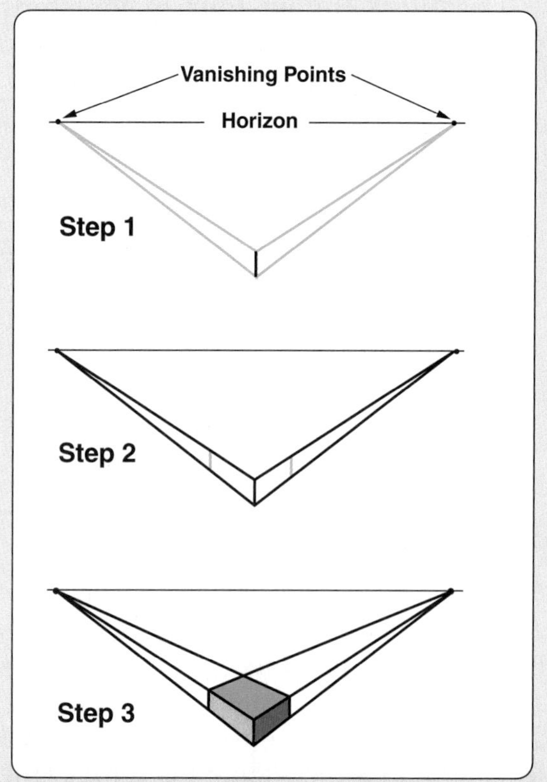

Figure 10-15. Designers follow three basic steps in developing perspective sketches.

Developing the perspective, or "human eye" view is more difficult than developing the oblique or isometric views. Note, however, that perspectives are the most realistic of the three pictorial sketches.

Summary

Design problems and opportunities in technological areas can be divided into two major areas: systems and products. System design deals with the arrangement of components to produce the desired result. Product design deals with manufactured products and constructed structures. In both system and product design, though, designers must first study the definition of the problem or opportunity before developing design solutions. Then, they must generate a number of possible solutions and create rough sketches of their ideas. These sketches become a library of design ideas. From this library, designers select specific ideas and refine them to bring the solution into focus. Finally, they describe the refined ideas through detailed sketches.

Key Terms

brainstorming
classification
convergent thinking
detailed sketches
divergent thinking
isometric sketches
oblique sketches
perspective sketches
refined sketch
rough sketches
synergism
"what if" scenarios

Test Your Knowledge

Write your answers on a separate piece of paper. Please do not write in this book.

1. What are the two major areas in which designers develop solutions to technological problems and opportunities?
2. List the three steps followed in developing design solutions.
3. What is the difference between divergent and convergent thinking?
4. *True or false?* The best solution is always the one that is the least expensive.
5. Name one of the rules used for effective brainstorming.
6. Define the term "classification" as it is used in developing design solutions.
7. What is a "what if" scenario?
8. Describe the difference between rough sketches, refined sketches, and detailed sketches.
9. List the three types of information required in detailed sketches when building models.

For Questions 10–18, match the type of pictorial sketch on the right to the correct definition on the left. (Note: Some letters will be used more than once.)

Definition

10. _____Used when the top, sides, and front are equally important
11. _____Shows the object as the human eye would see it
12. _____Sides extend back at 45°
13. _____The angles formed by the lines at upper-right corner are equal
14. _____Shows front view as if you were looking at it
15. _____Most difficult of three sketches to develop
16. _____Parallel lines meet at distant vantage point
17. _____Always has one corner made up of three 120° angles
18. _____The two types are cavalier and cabinet

Pictorial Sketch

a. Oblique

b. Isometric

c. Perspective

19. What is the primary difference between one-point, two-point, and three-point perspective sketches?
20. *True or false?* The appearance of a perspective drawing will change as the horizon changes.

Applying Your Knowledge

1. Develop a set of rough sketches for the following design definition:

> Problem or opportunity: The director of the school cafeteria would like a holder that would contain a salt shaker, a pepper shaker, twenty rectangular (1″ x 1 1/2″) packages of sugar, and a bottle of ketchup. The holder should be easily removable from the table at the end of the lunch period.

2. Refine the best sketch that you produced for the lunchroom table organizer.
3. Develop a detailed sketch for your lunchroom table organizer.
4. Select a device in the technology laboratory and develop a perspective sketch of it.

Career Corner
Drafters

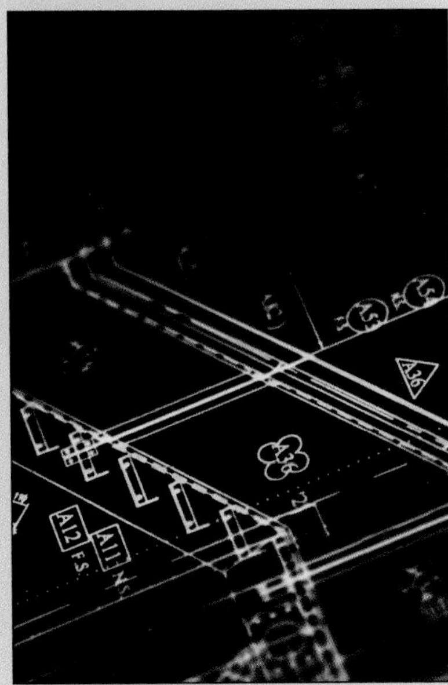

Drafters prepare engineering and architectural drawings and plans that manufacturing and construction workers use. They communicate technical details using rough sketches, drawings, and specifications. Most drafters use computer-aided design (CAD) systems to prepare their drawings. There is a number of different types of drafters, including aeronautical, architectural, electrical and electronic, and mechanical drafters.

Many drafting positions require postsecondary school training in drafting, which can be acquired at technical institutes, community colleges, and four-year colleges and universities. Entry-level drafters usually do routine work under close supervision. More experienced drafters do difficult work with less supervision.

Chapter 11
Evaluating Design Solutions

Learning Objectives

After studying this chapter, you will be able to:

- Define modeling.
- Describe how models are used in the design of technological artifacts and systems.
- List the three major types of models.
- Differentiate between types of models.
- Describe the types of analysis used to evaluate designs.
- Identify the role of redesigning.

We use the products of technology every day. Each product started with a problem or an opportunity that could be defined. Then, designers were challenged to solve this problem or meet the opportunity. They explored various ideas and developed solutions. These solutions started in the designers' minds, and then the designers recorded them on paper. They developed rough, refined, and detailed sketches. These sketches contained design ideas, not product plans. At some point the designers had to change two-dimensional drawings into three-dimensional models that could be evaluated, **Figure 11-1.** This conversion from drawings to models is the focus of this chapter. We will discuss the three basic activities of this process:

- Modeling design solutions
- Analyzing the design
- Redesigning the product or structure

Modeling Design Solutions

Everyone is familiar with models. Children play with dolls, model cars, and toy trains. They build towns out of wooden blocks, convert cardboard boxes into frontier forts, and have grand prix races with their tricycles. They pretend that they are dealing with real-life situations.

Likewise, museums use models to show how things worked in the past, **Figure 11-2.** These models show a slice of

Figure 11-1. The design process changes product sketches (left) into models (right) that can be evaluated. (Pontiac)

historical life so we can better understand the present day.

This activity of imitating reality is widely used in product and system design. People simulate expected conditions to test their design ideas. This process is called ***modeling*** or ***simulation.***

Simply stated, modeling is the activity of simulating actual events, structures, or conditions. For example, architects may build a model of a building to show clients how it will look. They may use structural models to test a building's ability to withstand an earthquake and the forces of the wind. Economists may devise a model to predict how the economy will react to certain conditions. Weather forecasters use models to show the public the location and movement of storms. They also use sophisticated computer models to predict the intensity and movement of hurricanes, **Figure 11-3.**

A model allows us to reduce complex mechanisms and events into an easily understood form. They allow us to focus on important parts of the total problem. This focus permits us to build understanding one part at a time. For example, an automobile is a very complex artifact. It is almost impossible to study and understand

Figure 11-2. This museum model shows how yarn was spun from wool in the colonial period.

Figure 11-3. This forecaster is viewing a computer model of a storm. (Harris Corp.)

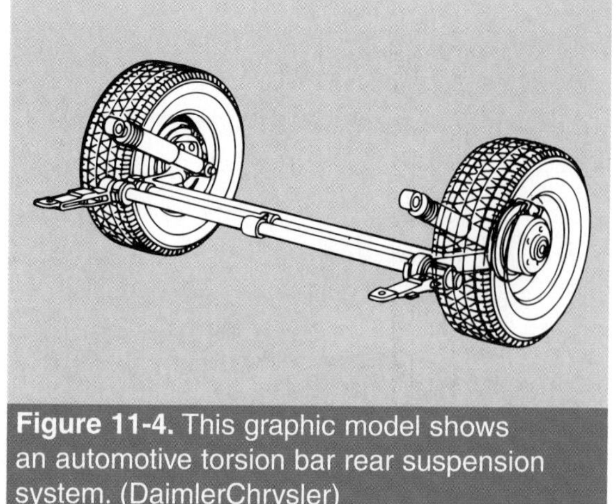

Figure 11-4. This graphic model shows an automotive torsion bar rear suspension system. (DaimlerChrysler)

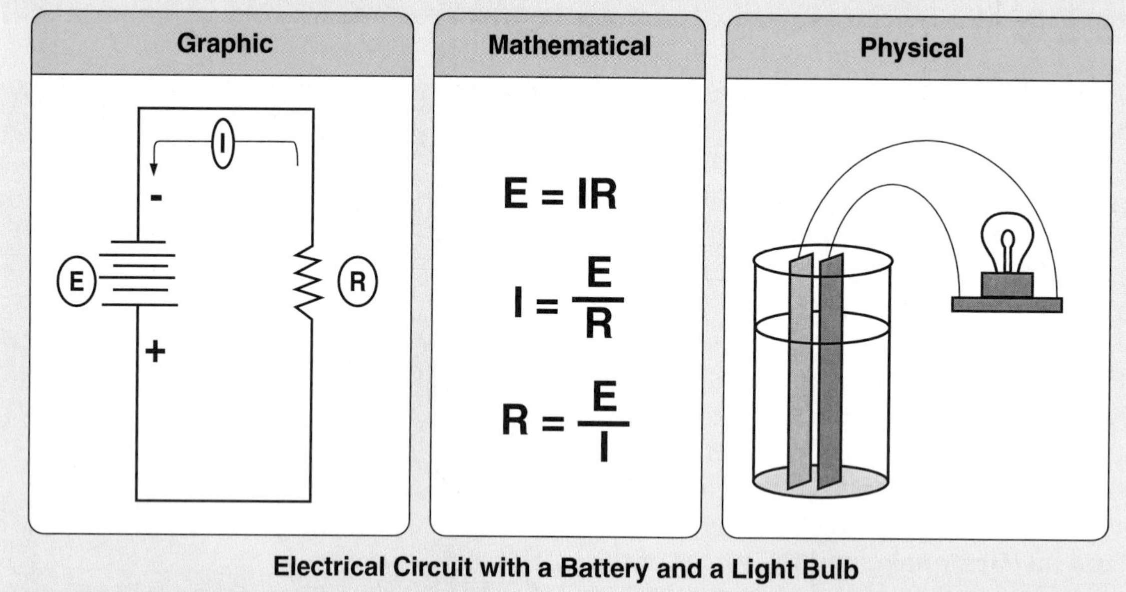

Figure 11-5. Models may be graphic, mathematical, or physical.

it as a whole. If you look at and understand the systems that make up the automobile one at a time, however, the whole becomes clear. For example, you can model and study its power train, cooling system, electrical system, lubricating system, or suspension system, **Figure 11-4.**

Literally thousands of models are used each day. Each of them starts as a conception, an idea in the human mind. This idea is often communicated through verbal descriptions. These written or oral descriptions provide the foundations for the development of models. Each model is designed to help people understand what an artifact looks like and how it functions or how various events take place.

All models can be grouped into three types, as shown in **Figure 11-5.**

- Graphic models
- Mathematical models
- Physical models

We should note here that the advent of the computer has made modeling much easier. In many cases designers no longer need physical models; they can develop and analyze the structure or product with computer modeling and simulation. This aspect of modeling will be discussed in more detail after the three main types of models are described.

Graphic Models

Designers cannot make physical models early in the product and structure development process. They do not have enough information about the design to construct a physical model. They must explore ideas for components and systems, however. One way they do this is by creating ***graphic models.*** Typical graphic models are conceptual drawings, graphs, charts, and diagrams. Each of these graphic models serves a specific purpose.

Conceptual models capture the designer's ideas for specific structures and products. They show a general view of the components and their relationships, **Figure 11-6.** Conceptual models are often the first step in evaluating a design solution. Relationships and working parameters of systems and components can be studied, modified, and improved using conceptual models. The refined and detailed sketches discussed in the previous chapter could serve as conceptual models.

For example, conceptual models could be developed for a toy train. These models would explore ways to connect the cars together, fabricate wheel-and-axle assemblies, and attach the car bodies to chassis assemblies.

Graphs allow designers to organize and plot data. Graphs display numerical information that can be used to design products and assess testing results.

For example, a graph can be developed that shows vehicle speed and braking distance for different types of brakes. The data can be charted on a line graph. This information will help designers select the type of braking system to be used in a specific vehicle. Likewise, plotting data on the colors of shirts purchased during a specific period can help designers select colors for next year's products. This type of data can be shown on a line graph, a bar graph, or a pie graph, **Figure 11-7.**

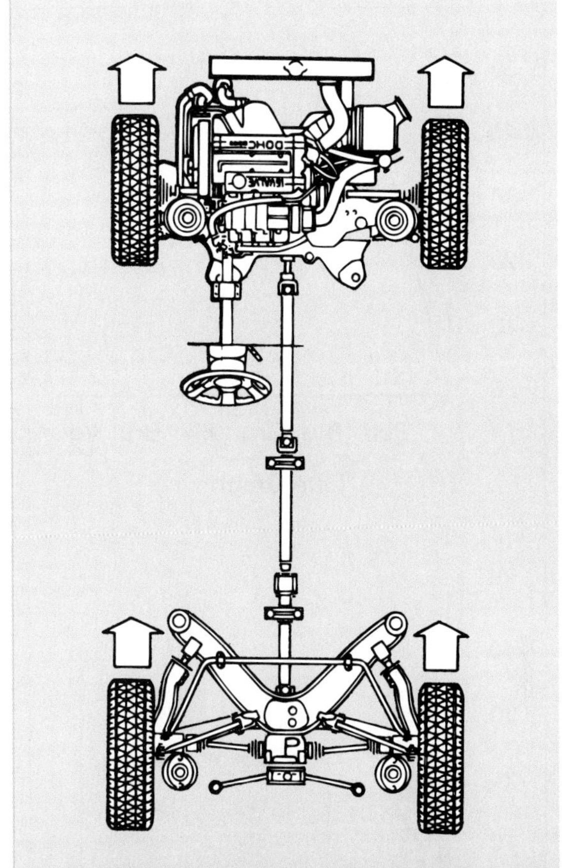

Figure 11-6. This conceptual model shows a designer's ideas for an automotive four-wheel drive system. (DaimlerChrysler)

Charts show the relationship between people, actions, or operations. They are useful in selecting and sequencing tasks needed to complete a job. Various charts are used for specific tasks. Flow-process charts help computer programmers write logical programs. Flow charts help the manufacturing engineer develop efficient manufacturing plants. Organization charts show the flow of authority and responsibility within a company, school, or business, **Figure 11-8.**

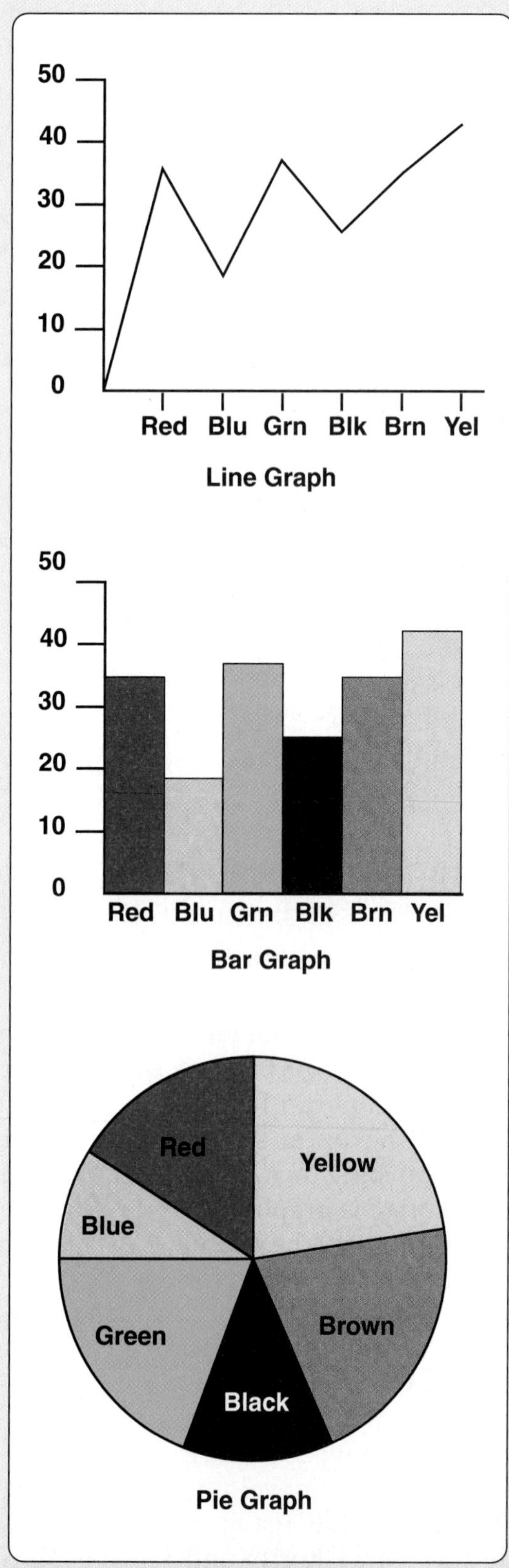

Figure 11-7. We use graphs to show the relationship among numerical data gathered about specific factors.

Diagrams show the relationship between components in a system. A schematic diagram can be used to indicate the components in electrical, mechanical, or fluidic (hydraulic or pneumatic) systems, **Figure 11-9.** Schematics do not show the specific location of the parts. Their position in the system is the important information that is communicated.

Flow diagrams show how parts move through a manufacturing facility. Lines and arrows indicate the path the material takes as it moves from operation to operation.

Another type of diagram with which many people are familiar is a play diagram. For example, football coaches develop them to show how the players (components) should interact during an offensive or defensive play (system). Analysts on television often use this type of diagram to show what happened during the game. The play diagram summarizes the purpose of all diagrams: show how something is designed to happen or how an event took place.

Mathematical Models

Mathematical models show relationships in terms of formulas. For example, the relationship among voltage, amperage, and resistance in an electrical circuit is shown by the formula E = IR, where E = electromotive force measured in volts, I = electrical current measured in amperes, and R = electrical resistance measured in ohms.

In a similar way, the relationship between the force needed to move an object and the distance it is lifted is shown in the formula for work: Work = Force x Distance.

The formulas used to explain chemical reactions are also mathematical models. The formula that follows shows the chemical reactions that take place as an automotive lead-acid storage battery is charged and discharged:

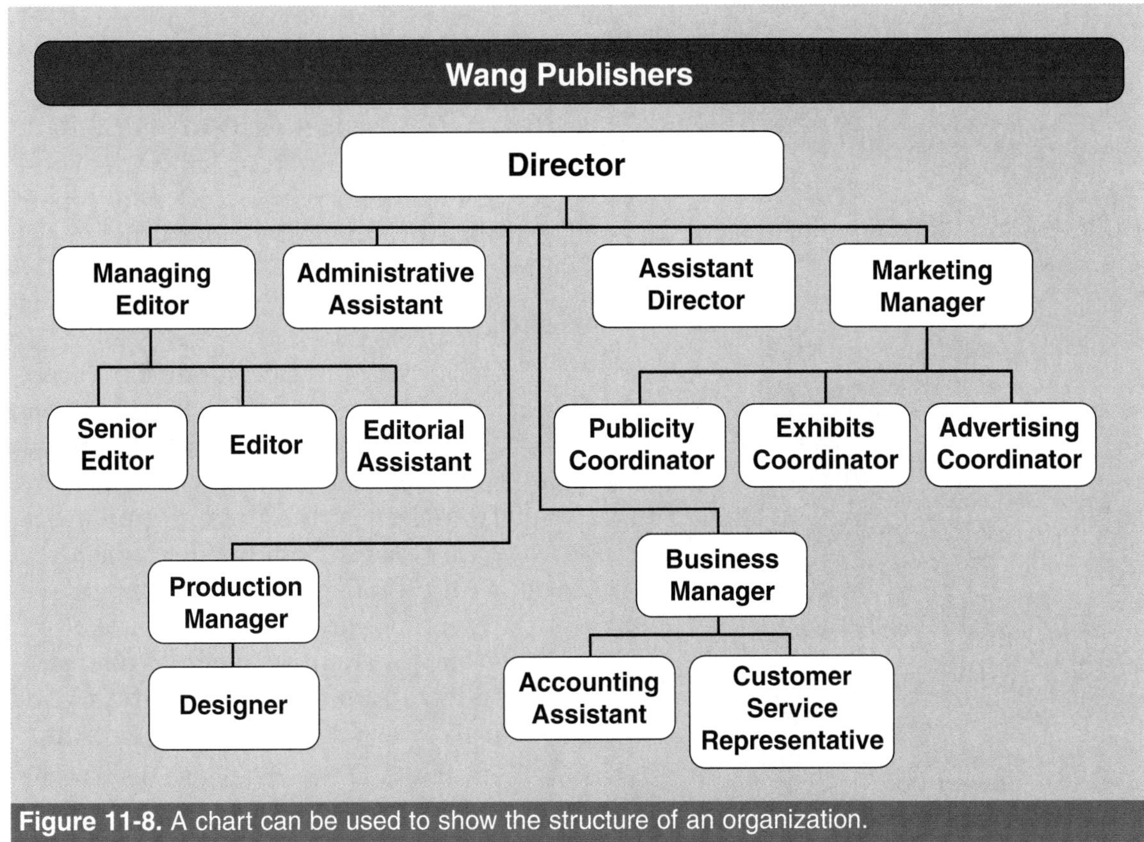

Figure 11-8. A chart can be used to show the structure of an organization.

$$PbO_2 + Pb + 2H_2SO_4 \underset{\text{charge}}{\overset{\text{discharge}}{\rightleftharpoons}} 2PbSO_4 + 2H_2O$$

The examples given so far are for simple mathematical models. Individuals use more complex models with thousands of formulas to predict the results of complex relationships. For example, these formulas can be part of economic models that predict the economic growth for a period of time. Also, complex mathematical models track storms and space flights, predict ocean currents and land erosion, and help scientists conduct complex experiments.

Physical Models

Physical models are three-dimensional representations of reality. Two types of physical models exist: mock-ups and prototypes.

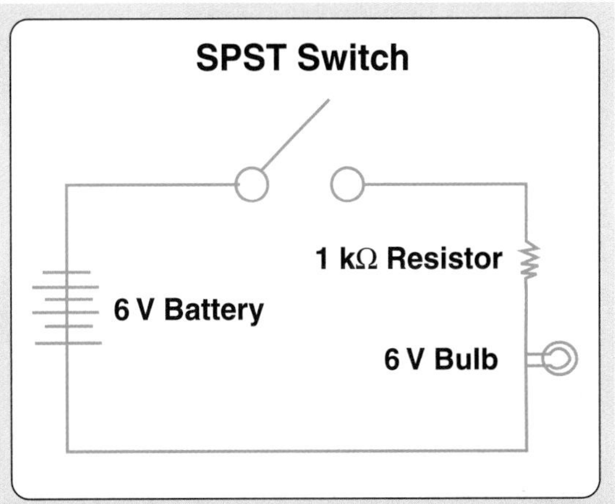

Figure 11-9. This schematic diagram shows the relationship among components in an electrical system.

Mock-ups

The first type of physical model is designed to show people how a product or structure will look. This type of model is an appearance model or ***mock-up.*** It is used to evaluate the styling, balance, color,

Figure 11-10. These designers are fabricating a clay mock-up of a new automobile. (DaimlerChrysler)

or other aesthetic feature of a technological artifact, **Figure 11-10.**

Mock-ups are generally constructed of materials that are easy to work with. Commonly these materials include wood, clay, Styrofoam, paper, and paperboard (cardboard and poster board, for example), **Figure 11-11.**

Prototypes

The second physical model is a ***prototype.*** A prototype is a working model of a system, assembly, or a product. Prototypes are built to test the operation, maintenance, and/or safety of the item, **Figure 11-12.** They are generally built of the same material as the final product. In some cases, however, substitute materials are used. For example, some automobile manufacturers have found that a specific plastic reacts to external forces in the same manner as steel does. Because the plastic

Figure 11-11. These designers are developing mock-ups of new toys. (Hasbro Inc.)

prototypes are easier to fabricate, they are used in place of steel ones.

Both types of physical models may be built full-size or to scale. Full-size models are needed to test the product's operation. For example, a full-size model is needed for people to evaluate to the comfort of a new bus seat.

In other cases full-size models are impractical. Building a full-size model of a new skyscraper would be a waste of money. A scale model is used when the product or structure is too large to construct in full size just for a test. A scale model is proportional to actual size. This means that the model's size is related to actual size by a ratio. A ratio of 4 to 1 (written as 4:1) means that four units in actual size are equal to one unit on the model. A scale model of a new building, for example, is used to show a client how the structure will look, how it fits on the site, and how it will be landscaped.

Computer Models

As we noted earlier, computers, because of their ability to produce ***computer models,*** have affected the way we work with models. Computers can be used to develop three types of three-dimensional models: wire frame, surface, and solid. Two of these are shown in **Figure 11-13.** The top view is a ***wire frame model.*** It is developed by connecting all the edges of the object. The process produces a structure made up of straight and curved lines.

The bottom view is a ***surface model.*** It can be thought of as a wire frame with a sheet of plastic drawn over it. Surface models show how the product will appear to an observer. Surface models can be colored to test the effects of color on the product's appearance and acceptance.

Surface models are widely used in developing sheet metal products. The surface model is first developed. Then the

Figure 11-12. This prototype is being used to test the aerodynamics of an aircraft. (DaimlerChrysler)

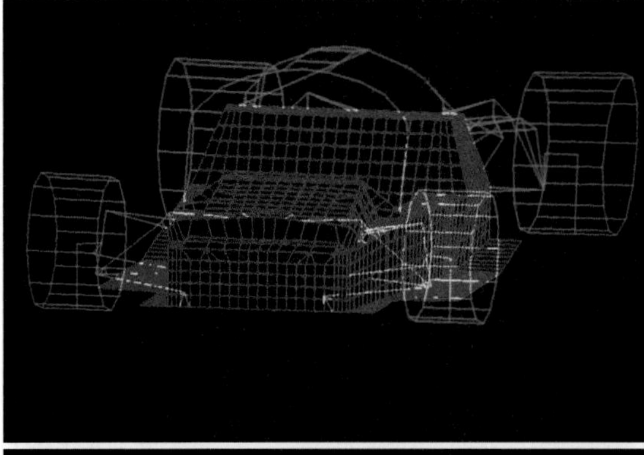

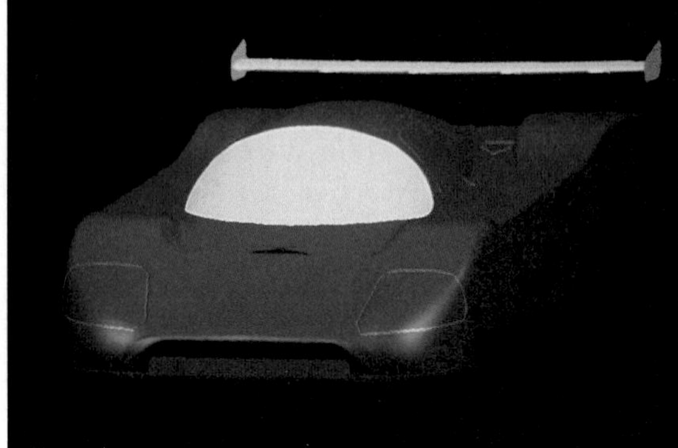

Figure 11-13. Two types of computer models are the wire frame (above) and surface models (below). They are used here in the design of a race car. (DaimlerChrysler)

Figure 11-14. This computer program is plotting the stress placed on a structure.

computer unfolds the model to produce a cutting pattern for the metal.

A ***solid model*** is the most complex of these models. Solid models look like surface models except the computer "thinks" of them as solids. This allows the designer to direct the computer to cut away parts, insert bolts and valves, and rotate moving parts, for example. Solid models can also be used to establish fits for mating parts and to set up the procedures needed to assemble parts into products.

Other computer simulations allow designers to test strengths of materials and structures, **Figure 11-14.** The designers create the structure on the computer and then apply stress. They might simulate forces created by weight, operating conditions, or outside conditions such as wind or earthquakes.

Finally, designers might use computer models to observe the product during normal operation. For example, computer programs simulate the flow of metals and plastics into molds. Other programs, such as the one shown in **Figure 11-15,** test air circulation in buildings and vehicles. This information is essential for the designers of heating and cooling systems.

Analyzing the Design

Modeling helps designers and decision makers enter an important design activity: design analysis. This stage requires people to evaluate a design carefully in terms of such criteria as its purpose and its likely acceptance in the marketplace. This evaluation helps designers and decision makers to choose the best design.

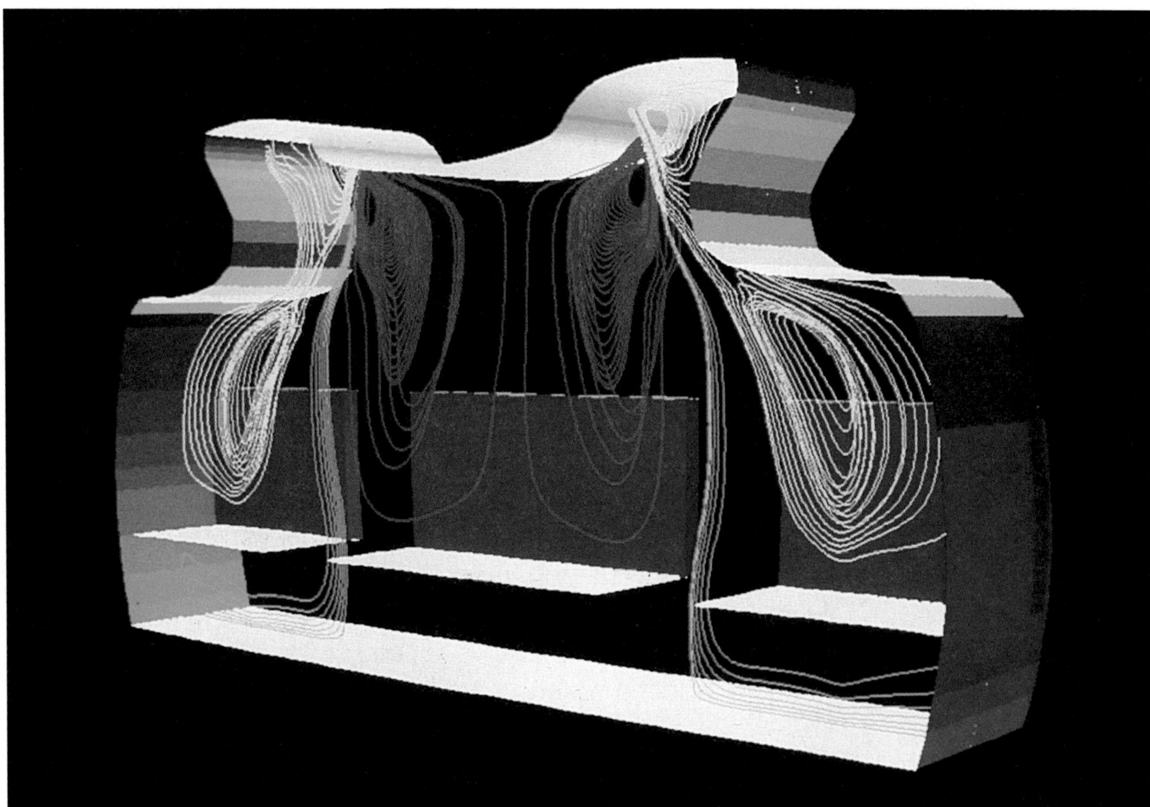

Figure 11-15. This computer program shows a simulation of the air circulation in the interior of an aircraft. Note the seats (in purple) and the shape of the overhead bins. (Boeing)

The five types of design analysis usually carried out **(Figure 11-16)** are:

- Functional analysis
- Specification analysis
- Human factors analysis
- Market analysis
- Economic analysis

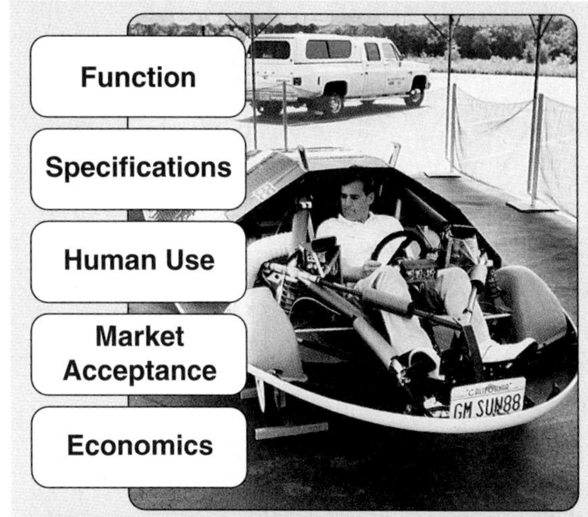

Figure 11-16. Product designs must be analyzed in five different areas. (GM-Rochester)

Functional Analysis

As we noted earlier, every product, structure, or technological system is designed to meet human needs or wants. Functional analysis evaluates the degree to which the product meets its goal. It answers the basic question, "Will the artifact operate effectively under the conditions for which it was designed?" These conditions may relate to the outside environment (weather, terrain, or water, for example), operating conditions (stress, heat, or gases produced during use, for example), human use and abuse and normal wear (material fatigue or part distortion, for example).

Technology Explained

Radar

radar: A system that can detect, locate, and determine the speed of distant objects with the use of radio waves.

Radar is an acronym for *ra*dio *d*etection *a*nd *r*anging. Practical radar systems were developed in the 1930s as a means of detecting enemy aircraft and ships. They were used widely during World War II. A radar system contains three separate subsystems: a transmitter producing and radiating radio signals, a receiver capturing the radio signals, and a display system presenting the data gathered.

The transmitter produces and sends out short bursts of powerful radio signals from an antenna, **Figure A.** These waves travel into space until they strike an object. When the waves strike an object, some of the signal is reflected back in the direction of the antenna. There, the signal is received and amplified.

The radar's signal is emitted in bursts. This allows a single antenna to both send and receive the signal. The radar system uses the elapsed time between the sending and receiving of the signal to determine distances of objects. Radio waves travel at the speed of light (186,000 miles per second). This allows elapsed time to be used as an accurate means to determine distances.

There are several different types of radar, **Figure B.** The intended use of the system determines the area the beam covers. Detection radar broadcasts a wide beam of radio waves. Ships use flat arcs sweeping the area just above the water, in order to avoid other objects in the water. Aircraft use radio beams sweeping up and down, as well as right and left.

Location radar uses a narrow, flashlight-type beam. The beam is focused on an object so accurate elevation, distance, and speed data can be obtained. This type of radar has many applications. Most people are familiar with its use by police officers to enforce speed limits.

The majority of radar antennas rotate to provide large viewing areas. Ground-based systems generally rotate in a 360° circle. Onboard radar systems often view only the areas in the direction of the vehicle's travel.

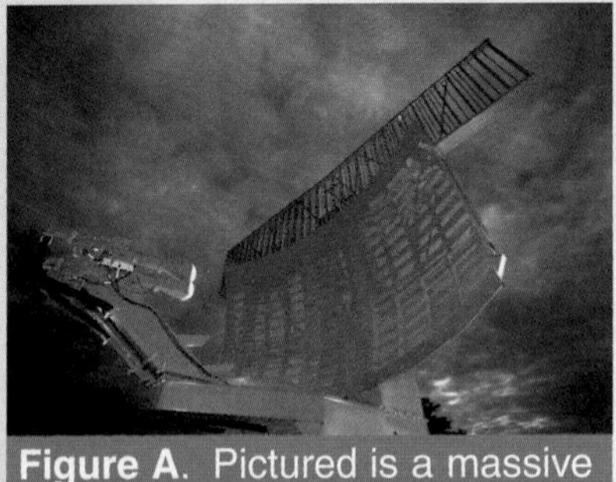

Figure A. Pictured is a massive radio antenna.

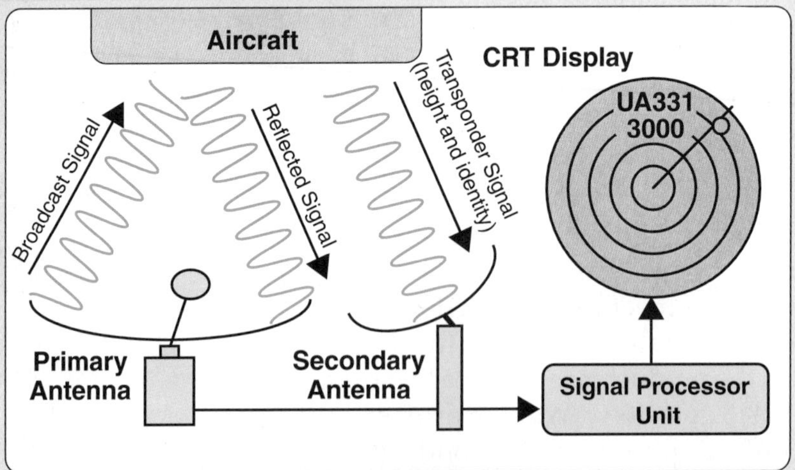

Figure B. This is a typical radar system used for commercial aviation.

Specification Analysis

Every product must meet certain specifications. These specifications may be given in terms of size, weight, speed, accuracy, strength, or a number of other factors. The specifications for all new products and structures must be analyzed. Note that the specifications must be adequate for the function of the product. Excessive specifications, however, will add to the cost of the product. Holding bicycle handlebar diameters to 1/1000″(.025 mm) is foolish. That level of precision is not needed for the part to function. Tolerances (the amount a dimension can vary and still be acceptable) must be close for spark plug threads, however. The goal is to produce an economical, efficient, and durable product that will operate properly and safely.

Specifications must also relate to the material and manufacturing processes to be used. For example, holding wood parts to tolerances smaller than 1/64″(.397 mm) is impossible. Normal expansion and contraction due to changes in humidity can cause this much change. Likewise, specifying aluminum for the internal parts of a jet engine is unwise. The temperatures inside the engine will melt aluminum parts. Similarly, specifying green sand casting as a process to produce precision parts is a mistake. Green sand casting produces low-cost parts, but they cannot be held to close tolerances.

Human Factors Analysis

To meet human needs, we design artifacts and structures for the people who will use them, travel in them, or live and work in them. Designing products and structures around the people who use them is the focus of human factors analysis, more commonly known as ***ergonomics.*** This science considers the size and movement of the human body, mental attitudes and abilities, and senses such as hearing, sight, taste, and touch.

Ergonomics also considers the type of surroundings that are the most pleasing and help people to become more productive. A good example of matching the environment to humans is an aircraft flight deck, **Figure 11-17.** All the controls are within easy reach. Dials and indicator lights are within the pilot's field of vision. Windows are located so that the pilots have a clear view of the sky ahead and above them.

Market Analysis

Most products of technology are sold to customers. These customers may be the general public, government agencies, or businesses. During design activities the market for the product must be studied. Then the designs must be analyzed in terms of that market. Market analysis will include finding customer expectations for the product's appearance, function, and cost. It will also include studying present and anticipated competition. Market analysis is often conducted by taking surveys of potential customers and analyzing competing products that are available on the market.

Economic Analysis

As we said earlier in this book, most technological artifacts and structures are developed by private companies. They risk money to develop, produce, and market the items. In turn, they hope to make a profit as a reward for their risk taking. To increase their chances of success, individuals will often conduct a financial analysis for the new products. The product is studied in terms of the costs of development, production, and marketing. These data are compared with expected sales income to determine the financial wisdom for producing the product. Often the product is judged on what is called its return on investment (abbreviated ROI). This figure indicates the percentage of return based on

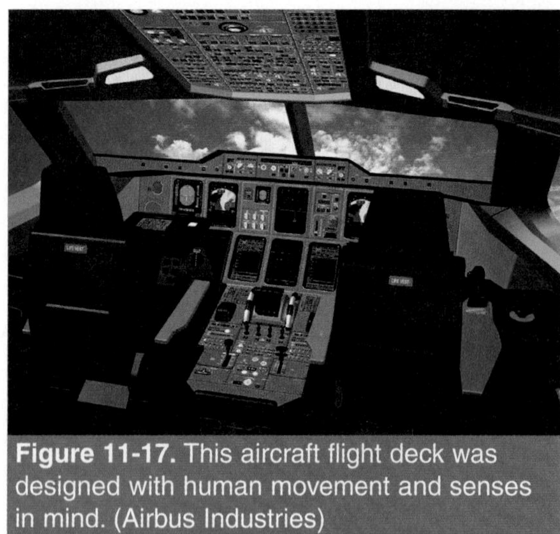

Figure 11-17. This aircraft flight deck was designed with human movement and senses in mind. (Airbus Industries)

the money invested in developing, producing, and selling the product. The higher the ROI, the better the anticipated financial returns to the company.

Redesigning Products and Structures

The final goal of all product and structure development activities is to design an artifact that will help people control or modify the environment. This goal requires that the problem-solving/design process be continual. Needs are identified and defined. Solutions are designed and modeled. The designs are analyzed, and flaws are identified. Redesign is then often required. Problems in the original design must be solved. New designs may be developed, or the original designs may be altered. New models will be built and evaluated until an acceptable product or structure emerges, **Figure 11-18.**

Nor is this the end of the process. Products have a life expectancy. New technologies and changes in people's needs and attitudes make some products obsolete. Old products will need redesigning, and new product definitions will appear. Product and structure design is a never-ending process.

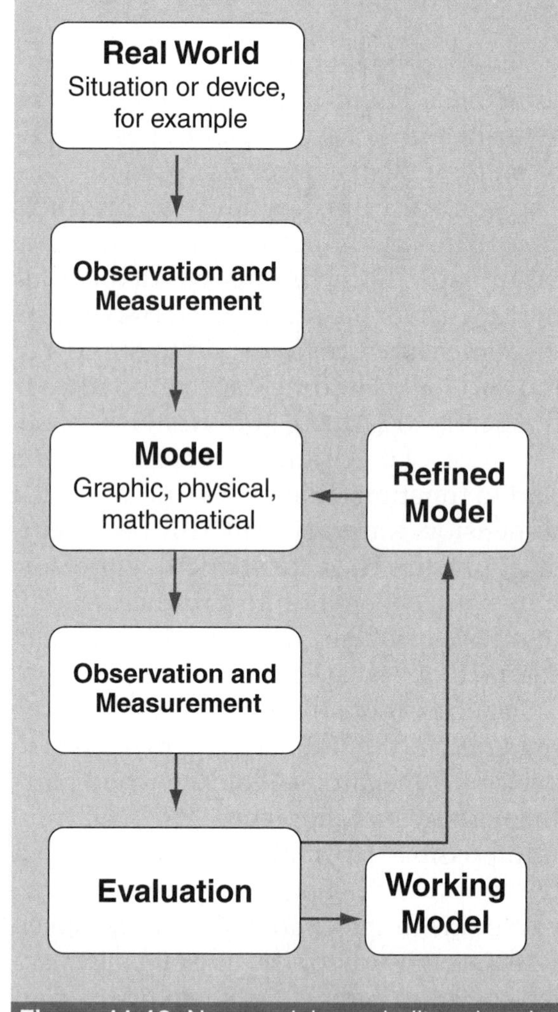

Figure 11-18. New models are built and evaluated until an acceptable product or structure emerges.

Summary

Designers develop product ideas to meet human needs and wants. These designs must be evaluated. Evaluation requires that graphic, mathematical, or physical models be built. The models allow designers to study and refine specific mechanisms and the total product.

The completed model is subject to careful analysis. Designers and decision makers evaluate its function, specifications, ergonomic qualities, market acceptance, and economic qualities. This is not the end of the design process, however. These individuals address shortcomings in the design through redesign activities. Finally, an acceptable design emerges.

Key Terms

charts
computer models
conceptual models
diagrams
ergonomics
graphs
graphic models
mathematical models
mock-up
modeling
physical models
prototype
simulation
solid model
surface model
wire frame model

Test Your Knowledge

Write your answers on a separate piece of paper. Please do not write in this book.

1. Define the term *modeling.*
2. Name one use of a model.
3. List the three types of models.
4. Name one difference between information shown on a graph versus that shown on a chart.
5. A working model of a system, assembly, or product is called a(n) _____.
6. List the three types of three-dimensional models that a computer can display.
7. Evaluating whether a product or system will operate effectively under the conditions for which it was designed is the goal of _____ analysis.
8. The analysis of how humans will be affected by a new artifact or structure is commonly called _____.
9. *True or false?* In market analysis, a product is often judged by its return on investment.
10. *True or false?* Product redesign is a common activity.

Applying Your Knowledge

1. Develop a graphic model, and build a physical model of a microwave communication tower. Assume the toothpick represents a piece of steel that is 15′ long. Build the tower that is 10″ square at the bottom and 69″ high.
2. Select a simple game that is on the market. Analyze it in terms of:
   - its specifications by developing a set of manufacturing specifications.
   - its function and market acceptance by playing it with classmates.

   Write a brief report that summarizes your analysis.

This activity develops the skills used in TSA's Computer-Aided Design, Engineering Animation event.

Computer-Aided Design, Engineering Animation

Activity Overview

In this activity, you will create an animation illustrating the assembly sequence for a product.

Materials

- Paper
- Pencil
- Computer with CAD and animation software

Background Information

Product selection. Use brainstorming techniques to develop a list of possible products to model in your animation. Part modeling will be easier if you have a sample of the product you can use for measurements or if you have actual part drawings. Some possible items to animate:

- A piece of self-assemble furniture, such as a computer desk
- A ballpoint pen or mechanical pencil
- A small construction project, such as a doghouse or shed
- Sports equipment, such as a swing set, tennis racquet, or weight bench
- A mechanical device, such as a wheel-and-axle assembly

Part modeling. After selecting the object, you must create models for each part.

Animation. Your animation is intended to illustrate the assembly sequence. You can begin the illustration with all parts shown, or you can have the parts appear as they are assembled. To help the viewer anticipate the action, highlight a part before it is moved into place in the assembly. You can highlight the part in several ways, such as changing its color, outlining it, momentarily enlarging the part, or momentarily stretching the part.

Output. Output the animation as an AVI or MOV file. Using a medium color depth (such as 16-bit) and low resolution (approximately 320 x 200 pixels) will speed the processing of the animation.

Guidelines

- Your assembly must include at least five components and there must be at least five steps in the assembly procedure.
- Your animation should clearly show the assembly procedure.
- Your animation must be at least 10 seconds long.

Evaluation Criteria

Your project will be evaluated using the following criteria:

- Originality and creativity in design
- Effectiveness of animation in illustrating assembly sequence
- Use of animation features
- Technical animation skills
- Smoothness and timing of animation

Career Corner
Market Researchers

Market researchers gather and analyze information about the potential sales of products and services. They use past sales data, information about competitors, and market trends to predict the market success of products. Market researchers develop methods for obtaining the data they need. They evaluate information they gather and make recommendations based on their findings.

People who hold bachelor's degrees in marketing and related fields might qualify as applicants for many entry-level positions. A master's degree is required for many private sector market research positions. The applicants might have degrees in business administration, marketing, statistics, or communications.

Chapter 12
Communicating Design Solutions

Learning Objectives

After studying this chapter, you will be able to:

- List the three basic documents through which designers communicate product information.
- Identify the three types of engineering drawings.
- Describe a computer-aided design system.
- Identify the information found on a bill of materials.
- Identify the information found on a specification sheet.
- Name the types of reports used to gain approval for designed products.

In the 1790s, Eli Whitney revolutionized the way products are made. He developed a system to mass-produce muskets for the army. The foundation of his new system was the concept of *interchangeable parts.* A part made for one gun would fit all other guns of the same make and model.

Today, we take interchangeable parts for granted. If you break a part of a product you expect to be able to buy a replacement. This ability requires a well-developed communication system between those who develop and engineer the product and those who make the parts and assemble the product, **Figure 12-1.**

Communicating design solutions is thus a key element in technology. Those engaged in communicating design solutions are involved with two areas of communication. First, they create documents and reports that specify all the details of the product. Second, they prepare documents and reports designed to obtain approval for the solution from various decision makers. Each type of communication is discussed in turn as follows.

Product Documents and Reports

The workers who make the parts and assemble them into products must be well informed. They must have knowledge of manufacturing processes. They must be able to set up and operate machines, apply finishing materials, and perform assembly operations. Still, this is only part of the manufacturing knowledge they need.

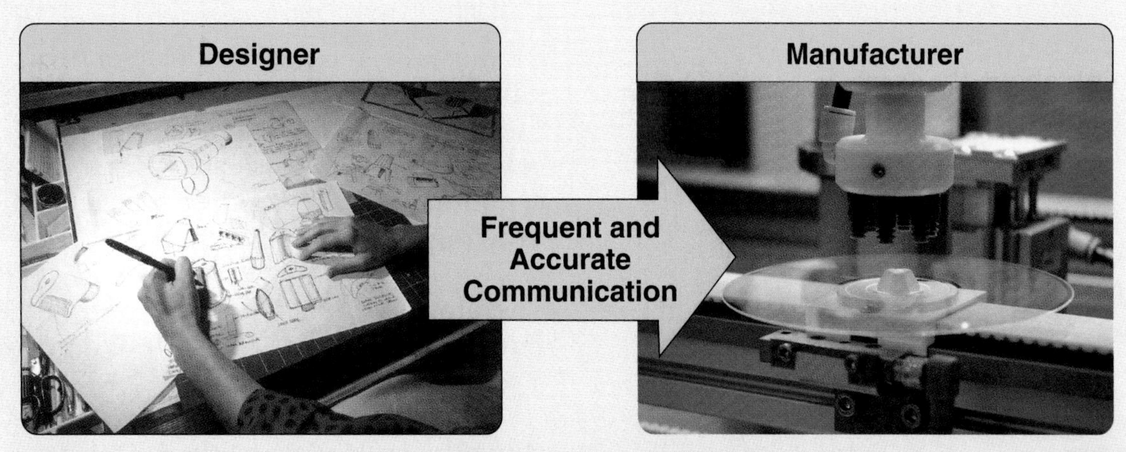

Figure 12-1. The design for the product must be communicated from the designers to the manufacturer. (Ohio Art Company)

They must "know" the product. This means they must have knowledge about the materials to be used in its manufacture. The workers must also know the size and shape of each part. Finally, they must know how the product is assembled from parts and fasteners (bolts, screws, and rivets, for example).

This knowledge of the product is delivered through three basic kinds of documents. They are, as shown in **Figure 12-2**:

- Engineering drawings
- Bills of materials
- Specification sheets

Let us look at how each of these kinds of documents is developed and the information each one communicates.

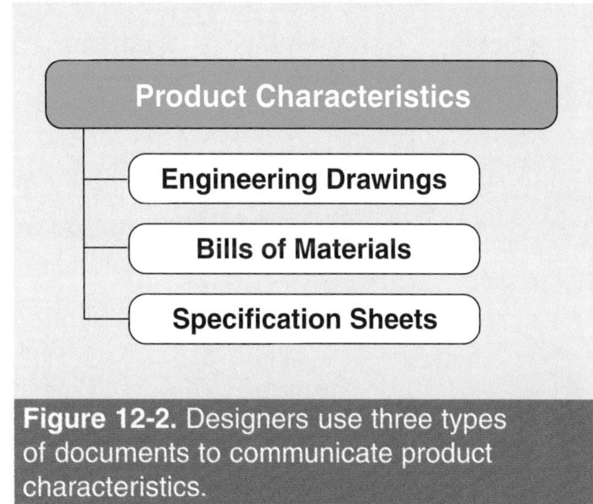

Figure 12-2. Designers use three types of documents to communicate product characteristics.

Engineering Drawings

Engineering drawings communicate basic information needed to construct the product or structure. In manufacturing they are called engineering or working drawings. In construction, however, they are called architectural drawings. Here, to keep things simplified, we will focus on engineering drawings. Keep in mind, however, that although architectural drawings are different from engineering drawings, the basic principles used to prepare both are similar.

Designers commonly use three types of engineering drawings to communicate product information. These are, as shown in **Figure 12-3.**

- Detail drawings: Drawings that show specific information needed to produce a part
- Assembly drawings: Drawings that show how parts go together to make a sub-assembly or product
- Systems drawings: Drawings that show the relationship between electrical, hydraulic, or pneumatic components

Before we go further here, however, we need to note the following. When we think about the term *drawing,* we might visualize someone with pencil in hand, illustrating the product on sheets of paper. Much drawing today, however, is accomplished with the use of computers. We will explore the role of computers in this area after we discuss the three basic types of drawing documents.

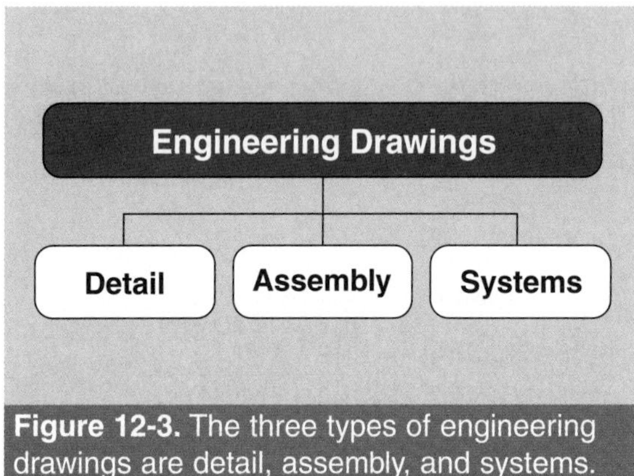

Figure 12-3. The three types of engineering drawings are detail, assembly, and systems.

Detail Drawings

Most products are made up of several parts. Each of these parts must be manufactured to meet the designer's specifications. These specifications are often communicated on ***detail drawings.*** Detail drawings commonly contain *all the information needed to manufacture one part.* Therefore, designers usually generate a number of different detail drawings for a complete product.

Most detail drawings are prepared using the ***multiview method.*** This drawing method places one or more views of the object in one drawing, as shown in **Figure 12-4.** The number of views will depend on how complex the part is. The most common multiview drawings, as shown in **Figure 12-5,** are:

- ***One-view drawings,*** which are used to show the layout of flat, sheet metal parts. The thickness of the material is listed on the drawing.

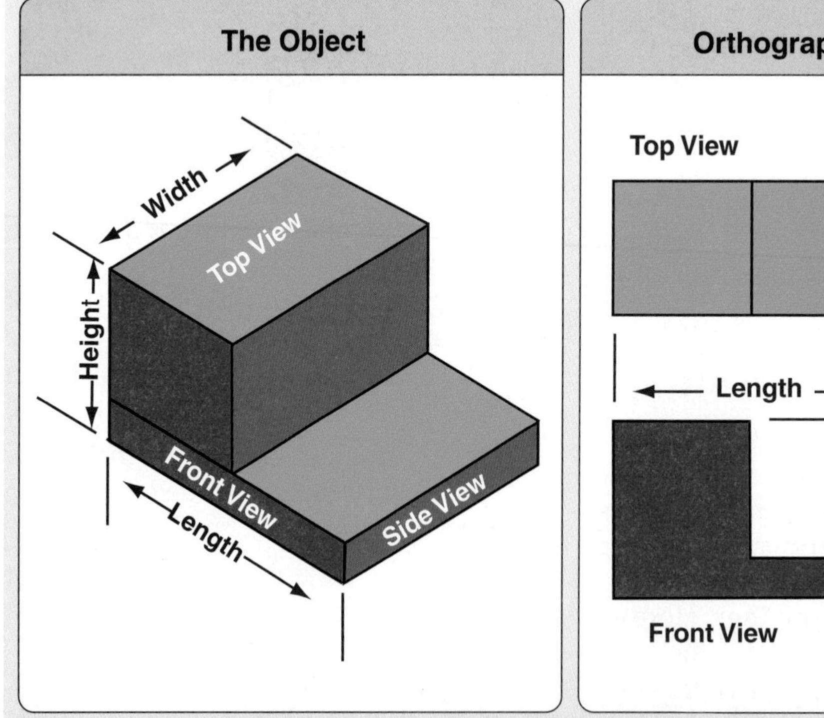

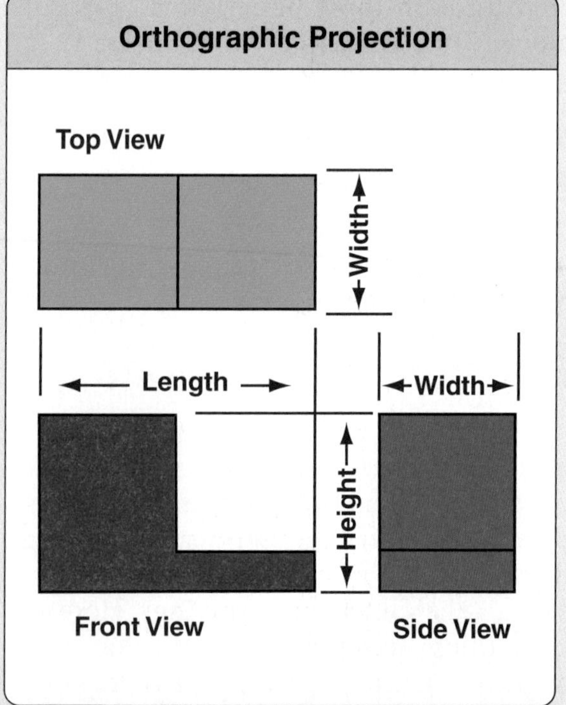

Figure 12-4. The orthographic projection shown on the right represents the object shown on the left.

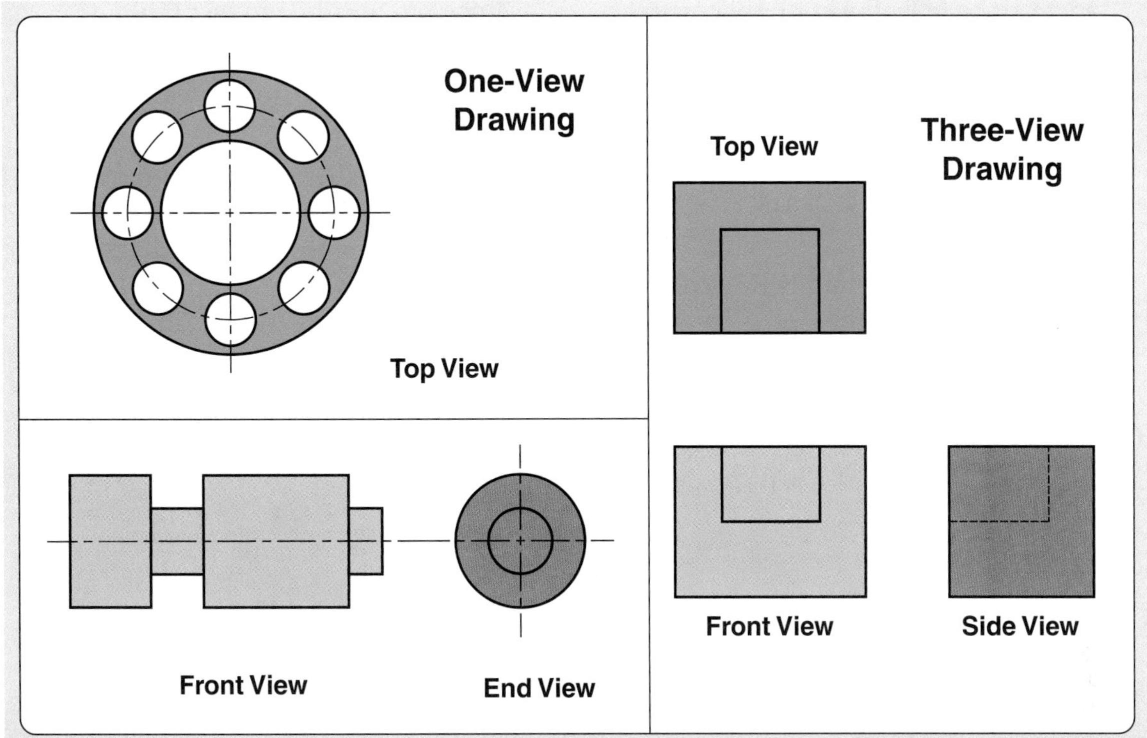

Figure 12-5. The most common multiview drawings are one-view, two-view, and three-view drawings.

- ***Two-view drawings,*** which are used to show the size and shape of cylindrical parts. The front and top views are generally identical. Therefore, only one is needed. The two views shown are the front view, which shows the features along the length of the part, and the end view.
- ***Three-view drawings,*** which are used to show the size and shape of rectangular and complex parts. Generally a top, right side, and end view are shown. This arrangement is called third-angle projection and is used in the United States, Britain, and Canada. Other countries use first-angle projection, which shows the top, front, and left side views. For very complex parts, designers may need to add more views, called *auxiliary views.*

In all cases, the least number of views are used. Creating unnecessary views costs time and money. Each drawing must meet the "acid test" of all engineering drawings. It should communicate all of the information needed to make the part. The designer and the drafter will not be on the manufacturing floor to answer questions or supply missing information.

Preparing Multiview Drawings

Multiview drawings use ***orthographic projection*** to project information at right angles to new views (see Figure 12-4). Before beginning a multiview drawing, drafters must select the surface of the object to be shown in the front view. The surface that has the most detail is chosen to be shown in the front view. The goal is to have as few hidden details as possible. Drafters complete several steps as they prepare a three-view drawing. The most important of these steps, as shown in **Figure 12-6,** are:

1. The front view is drawn in the lower-left quadrant of the paper. An accepted practice is to construct a box that will

enclose the object using light construction lines. Then details on the object are located and lightly drawn.

2. At the same distance above and to the right of the front view, lines are lightly drawn as the place to start the right side and top views.
3. Projection lines are lightly drawn above and to the right of the front view. These lines are used to project the size of the object to the other views.
4. The outline and details of the top view are lightly drawn.
5. A 45° line is drawn in the upper-right quadrant. The outline and features of the top view are projected to the side view. This is done by projecting sizes and features to the 45° line and then

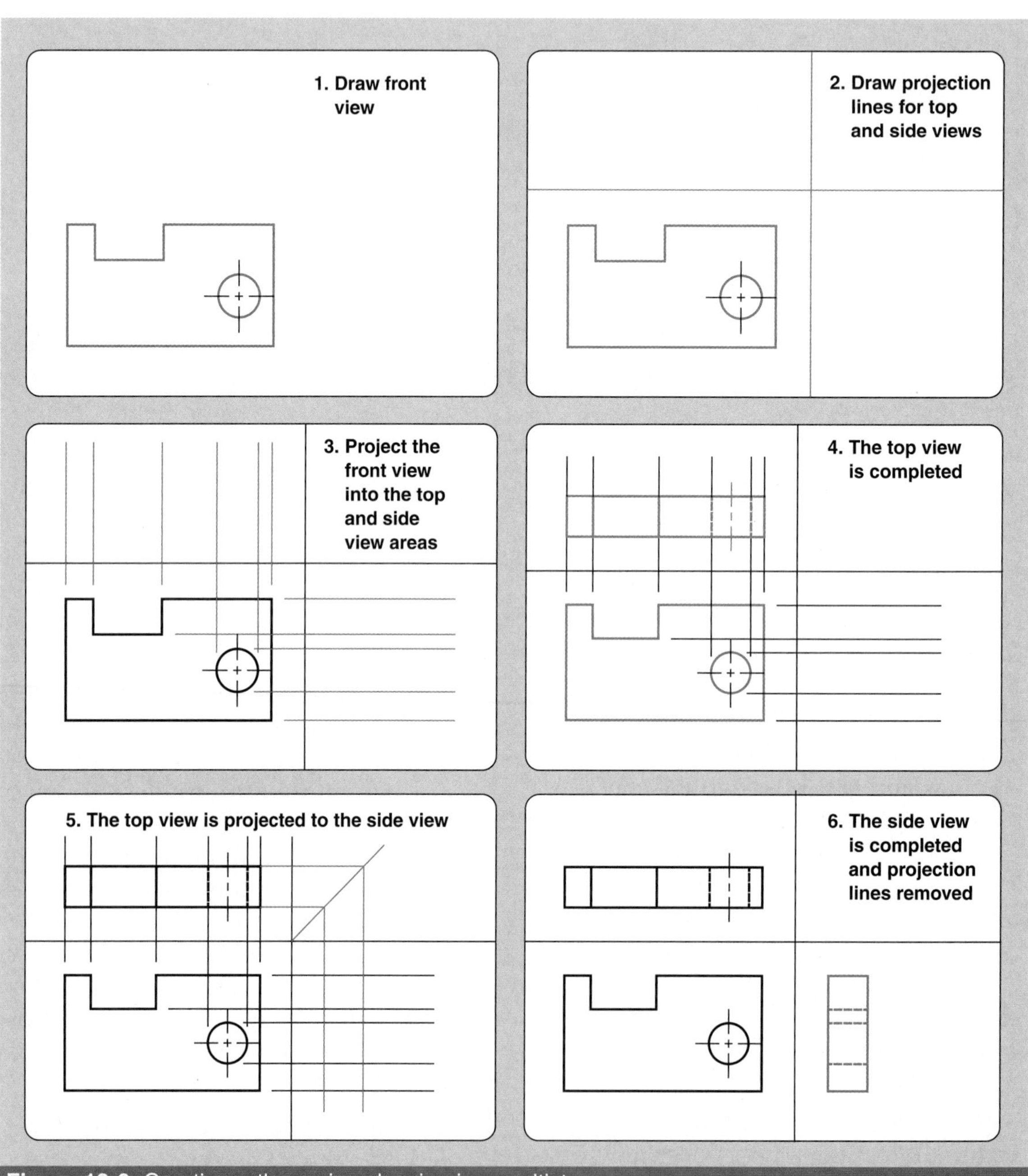

Figure 12-6. Creating a three-view drawing is a multistep process.

down to cross the projection lines drawn earlier from the front view. The side view will appear where these projection lines cross. Like those in step 2, these lines should be very light so they can be erased.

6. The side view is completed by constructing the overall shape from the front and top projections. Details are located, and dark object lines in all the views are made.

After drawing the three views of the object, the drafter adds dimension and extension lines, which will be discussed later. Finally, the projection lines are erased to complete the drawing.

Dimensioning Drawings

The size and shape of an object are communicated by detail drawings. To accomplish this task, designers include dimensions. The basics of dimensioning were presented in Chapter 10. At that time you were introduced to three important types of dimensions:

- ***Size dimensions:*** Indicate the size of the object (length, width, and height) and its major features (diameter and depth of holes and width and depth of notches, for example)
- ***Location dimensions:*** Indicate the position of features on the object, such as center points for holes, edges of grooves, and starting points for arcs
- ***Geometry dimensions:*** Indicate the shape of features and the angle at which surfaces meet (round holes or square corners, for example)

All these dimensions must be included on the drawings. One technique that ensures they are present suggests:

1. Dimension the size of the object first, followed by that of all major features
2. Dimension the location of all features next
3. Indicate any necessary geometric dimension last (angles not indicated are assumed to be 90°)

Dimensioning uses two kinds of lines. First, ***extension lines*** indicate the points from which the measurements are taken. Between the extension lines are ***dimension lines.*** These have arrows pointing to the extension lines that indicate the range of the dimension. The actual size of the dimension is shown near the center of the dimension line.

These dimensions may be given in fractions or decimals of an inch. In addition, a ***tolerance*** may be included with the dimension. This number indicates the amount of deviation in the dimension that will be allowed. For example a +/– (plus or minus) 1/64″ (0.4 mm) after the dimension indicates that the size may be 1/64″ larger or 1/64″ smaller than the dimension and still be acceptable.

Alphabet of Lines for Drawings

Users should be able to read the drawings easily. Therefore, a set of drafting standards has been developed so that all drawings communicate well. These rules are much like the rules of grammar that a writer uses. They allow each reader to interpret the prose (using grammar rules) or the drawing (using drafting standards) in a similar manner.

One set of essential drafting standards deals with lines and line weights, **Figure 12-7.** The shape of the object is of primary importance. Therefore, the lines that outline the object and its major details must stand out. These solid lines are called ***object lines*** and are the darkest on all drawings.

Some details are hidden in one or more of the views. Their shape and location, however, are important to understanding the drawing. Therefore, they are shown but with lighter, dotted lines called ***hidden lines.***

A third type of line locates holes in the part. These lines pass through the center of

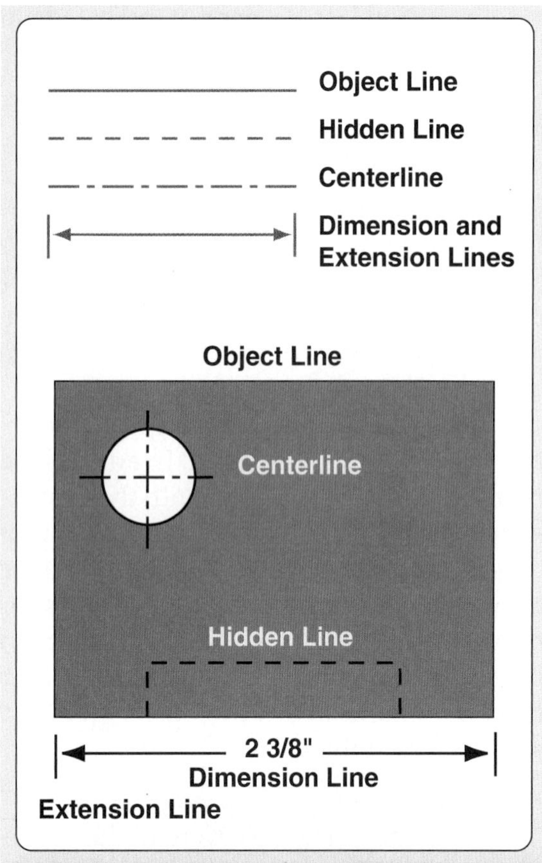

Figure 12-7. This "alphabet of lines" shows the types and weights of lines used in drawings.

the hole and are thus called ***centerlines.*** These are constructed of a series of light long and short dashes.

Dimension and extension lines, which were introduced earlier, are the same weight as hidden lines. They are important but should not dominate the drawing. Remember, at first glance the outline of the object and its details are the dominant feature that should "jump out" at the viewer.

Assembly Drawings

A second type of engineering drawing is the assembly drawing. ***Assembly drawings*** show how parts fit together to make *assemblies,* which are put together to make products. Two types of assembly drawings exist, as shown in **Figure 12-8.** They are orthographic assembly drawings and pictorial assembly drawings.

Orthographic assembly drawings use standard orthographic views to show parts in their assembled positions. ***Pictorial assembly drawings*** show the assembly using oblique, isometric, or perspective views like those discussed in Chapter 10.

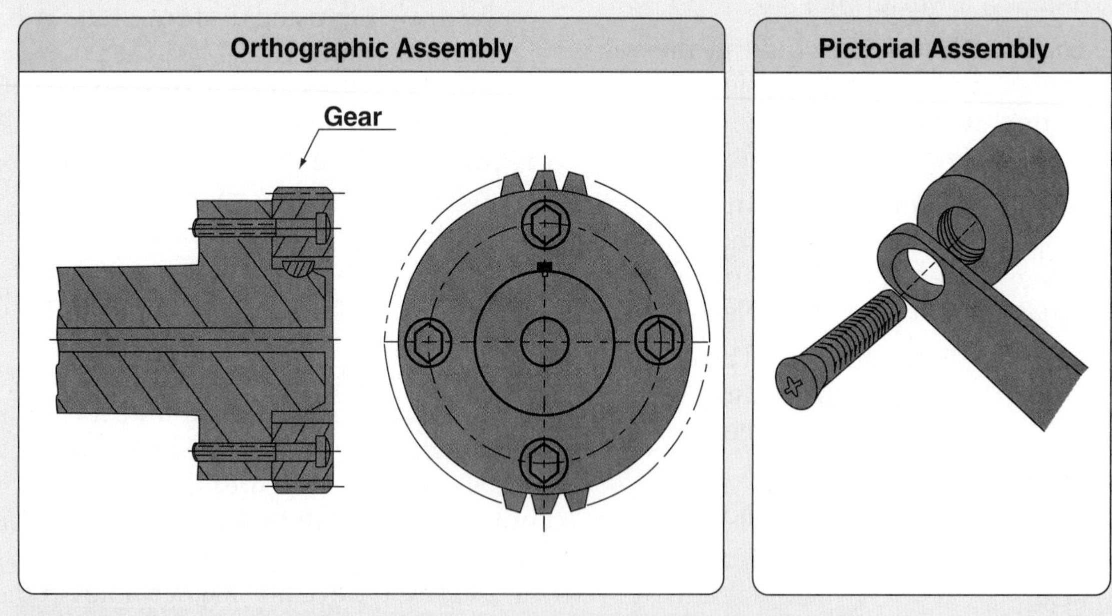

Figure 12-8. Assembly drawings can be orthographic or pictorial. (General Motors Corp.)

Connections to Technology: **Communication**

Principles of Design

This chapter described the documents and reports designers use to gain approval for their solution to a technological problem or opportunity. Also important is the *way* in which these documents and reports are presented to the various decision makers. Designers have found that specific principles of design are effective in communicating a clear and pleasing message, thus leading to a successful presentation.

The basic principles of design that a presenter needs to take into account are *balance, unity, rhythm, contrast,* and *proportion.* When the message is in proper balance, the images being presented are distributed evenly and one element dominates. The message has unity when all elements are harmoniously balanced. When the message has the proper rhythm, the viewer's eyes are drawn to the area the presenter wanted to emphasize. Proper contrast means that variations in the elements exist (although not to the extent that the image becomes confusing). When an image is proportioned correctly, each element is proportional to all the others, thus creating a unified design.

These basic principles of design are used every day in the world of advertising. Look at some ads in current issues of newspapers or magazines—can you find some examples of these design principles at work?

Either type of drawing may be a standard view or an exploded view, **Figure 12-9.** ***Standard views*** are constructed using the normal techniques for constructing orthographic or pictorial drawings. Standard views show the product in one piece, as it would be after it is assembled.

Exploded views show the parts that make up a product as if it were taken apart. The parts are arranged in the proper relationship to each other on the drawing. This type of drawing is often found in owner's manuals and parts books. They are used to show the parts that comprise a product. Each part on the drawing generally has a code that allows the owners or repair person to order a replacement part.

Most assembly drawings do not have dimensions. The exception would be when the assembler must manually position the parts for assembly.

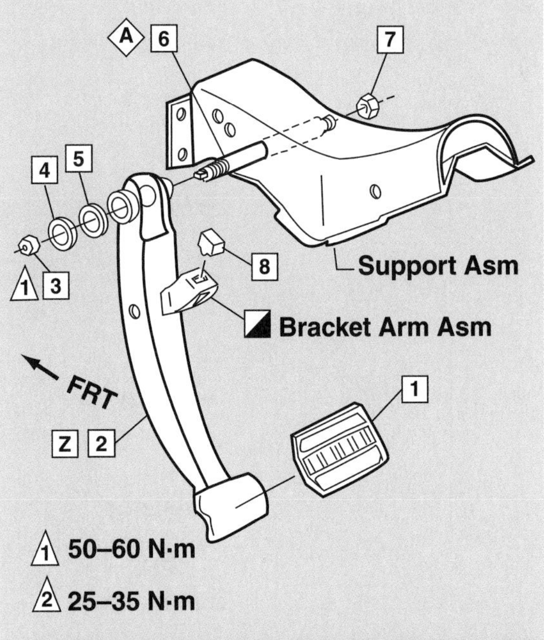

Figure 12-9. An exploded view assembly drawing shows the parts of a product as if it were taken apart. (General Motors Corp.)

Systems Drawings

Systems drawings are used to show how parts in a system relate to each other and work together. They are used for electrical, hydraulic (fluid), and pneumatic (gas) systems. They are often called *schematic drawings.* They do not attempt to show the actual position of the parts in a product. Assembly drawings would do this. They are designed to show the connections for wires, pipes, and tubes.

Systems drawings use symbols to represent the components. Standard symbols have been developed for electrical, pneumatic, and hydraulic parts. **Figure 12-10** includes some common symbols for electrical and electronic components.

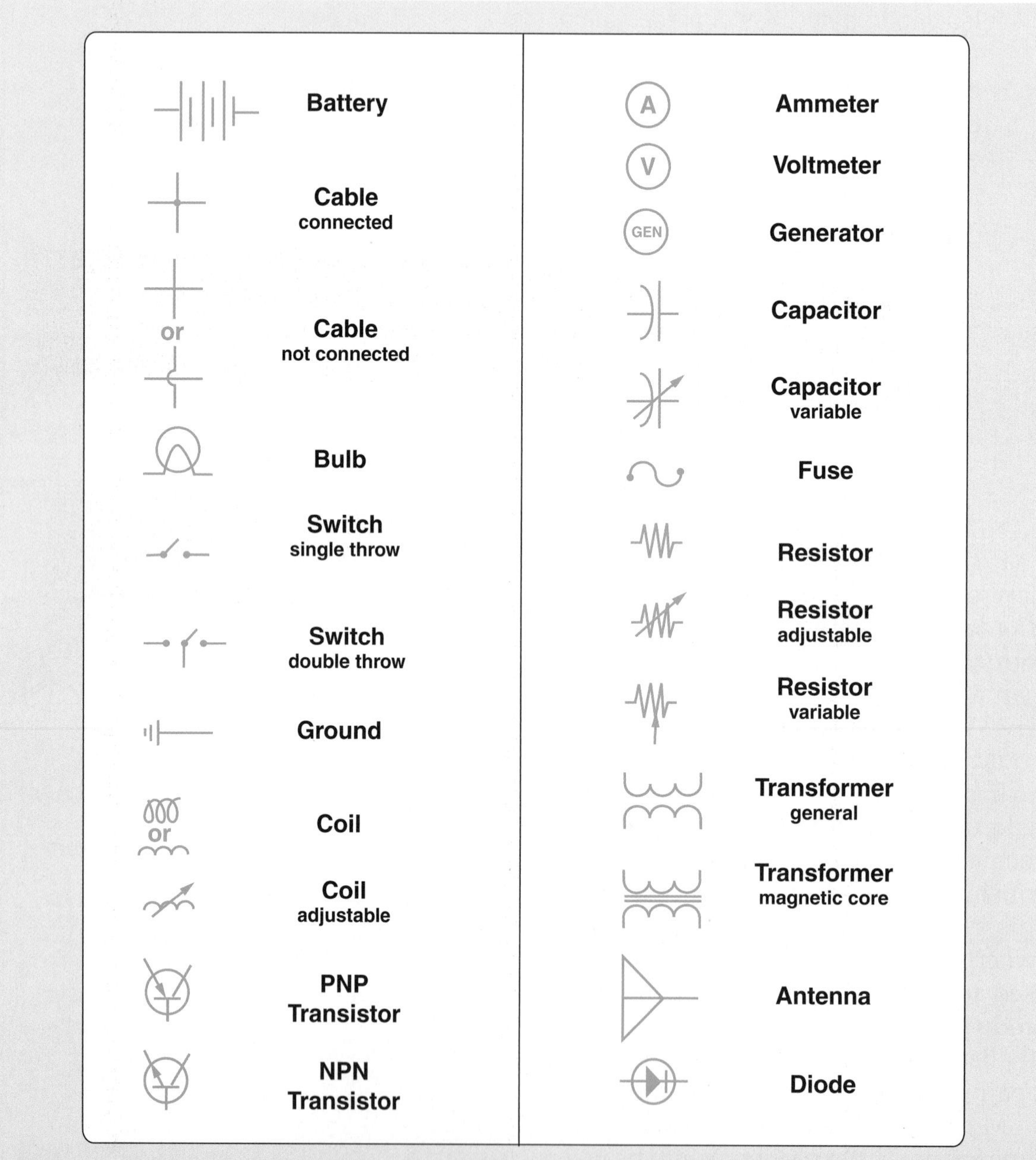

Figure 12-10. Shown here are some common symbols used on electrical and electronic systems drawings.

Systems drawings are developed by first arranging the major components on the sheet. Then connecting wires, pipes, and tubes are indicated. Special drawing techniques are used to indicate when the lines connect or simply cross each other.

Computer-Aided Design

A computer is a valuable technological tool. Computer systems can be applied to a number of tasks. A common industrial application for computer systems is in preparing drawings and models. This application uses a computer, a plotter or printer, input devices (keyboard, mouse, and graphics tablet, for example), and ***computer-aided design*** (CAD) software.

There is nothing magical about this system. It allows an operator to complete the steps of laying out and producing drawings following the methods that a drafter uses. CAD systems, however, do the job more quickly and uniformly. Also, computer drawings are easier to correct, store, and communicate. Computer drafters can send their drawings across the country or around the world in seconds using the Internet, **Figure 12-11.**

Figure 12-11. Designers can use graphic tablets to enter data.

Technology Explained

Computer-Aided Design

computer-aided design (CAD) system: A system using computers to create, change, test, and store drawings communicating design information.

Computers are used throughout industry to increase productivity and improve product quality. One area in which computer systems have made vast inroads is the drawing room. Computer-generated drawings and designs are commonplace in many industries. The systems used to develop these designs are known as computer-aided design (CAD) systems, **Figure A.**

Data can be entered into a CAD system using keyboard commands, a mouse, or a menu pad. A menu pad has a number of common drawing commands on its surface. The drafter simply points to the commands with a wand. This enters the command into the computer. These controls allow drafters to produce standard two-dimensional detail and assembly drawings. Additional features and text may then be placed on any engineering drawing, **Figure B.**

A growing use for CAD systems is in producing solid models (three-dimensional representations) of objects. This process uses a concept similar to weaving a rug. Strands are intertwined to produce a frame for the object. This produces a special drawing called a *wire frame representation,* **Figure C.** The frame is the skeleton required to produce the object. Such a skeleton can be covered to produce a solid representation, **Figure D.** This three-dimensional drawing is often used when considering styling and appearance. Solid models can also show how parts fit together in an assembly.

In addition to their display functions, CAD systems can be used to test and examine parts. Many CAD systems can calculate the mass, volume, reactions to stress, and other attributes of a designed part. CAD systems have become important tools for design engineers.

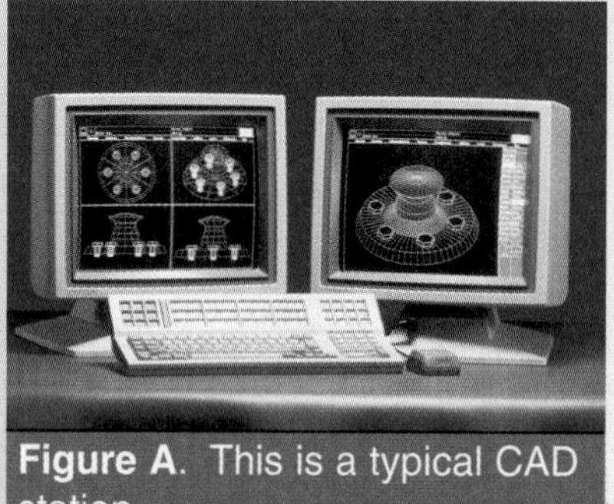

Figure A. This is a typical CAD station.

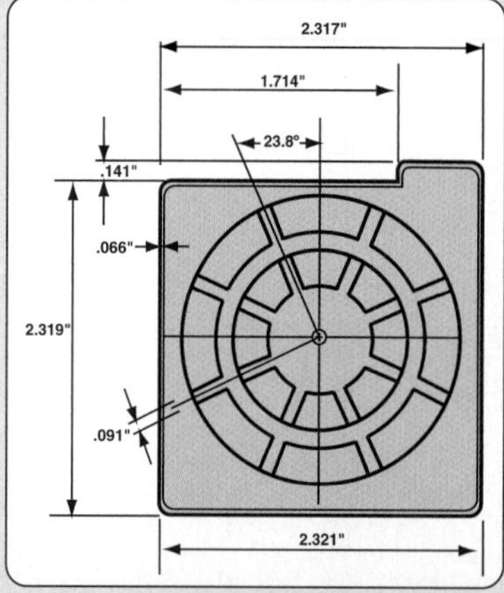

Figure B. Here is a simple drawing produced on a CAD sytem.

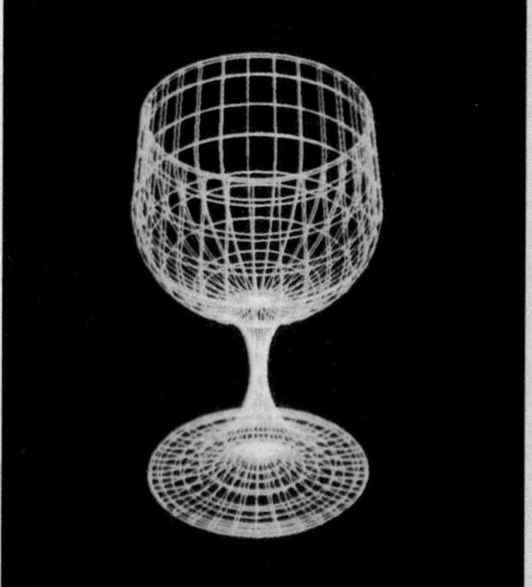

Figure C. Wire frame representations allow you to see through an object.

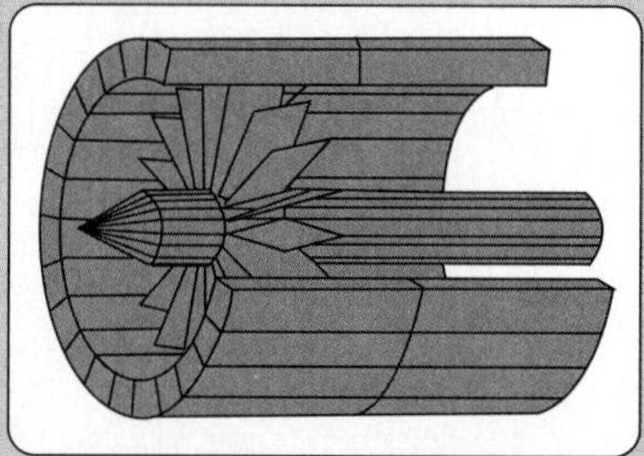

Figure D. Shown is a solid representation of a turbine.

Bills of Materials

Not all information needed to produce a product can be contained on detail, assembly, and system drawings. Additional documents are needed to provide complete production information. One important document is a ***bill of materials.*** The name of this document causes some confusion. We all pay bills for things we buy and use. A bill of materials does not contain cost information. Instead, it is a list of the materials needed to make one complete product. An example is shown in **Figure 12-12.**

Most bills of materials contain the following information for each part on the product:

- A *part number* that can be used on assembly drawings and for ordering repair parts.
- A descriptive *name* for the part.
- The number or *quantity* (abbreviated qty.) of parts needed to manufacture one product.
- The *size of the part,* indicating its thickness, width, and length (for rectangular parts) or its diameter and length (for round parts). Sizes are given in the order shown: T x W x L or Dia. x L.
- The *material* out of which the part is to be made.

The items on a bill of materials are listed in a priority order. Manufactured parts are listed first. Parts that are purchased ready to use and fasteners are listed after the manufactured parts.

Specification Sheets

Not all materials can be shown on a drawing. Can you make a drawing of engine oil, an adhesive, or sandpaper? If you did make a drawing of any of these items, it would be of little value. These and thousands of other items are not chosen for their size and shape. Other properties are important in their selection.

For example, some important factors in adhesives are the working time (time between application and clamping), clamping time (time the work will be held together for the glue to set), and shear strength. Window glass must be transparent. Insulating materials must stop heat from passing through them.

Specification sheets communicate the important properties that a material must possess for a specific application, **Figure 12-13.** These properties, which were introduced in Chapter 4, might include:

- Physical properties: moisture content, porosity, surface condition
- Mechanical properties: strength, hardness, and elasticity

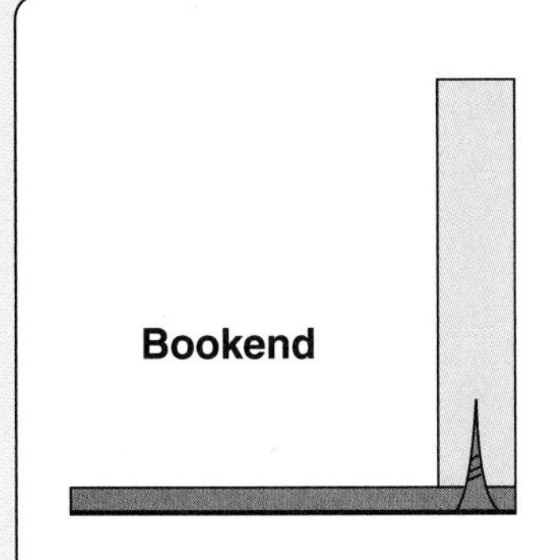

Bill of Materials

Part #	Part Name	Qty	Size T	Size W	Size L	Material
BE-1	End	1	3/4	6	8	Oak
BE-2	Bottom	1	28ga	6	6	Steel
BE-3	Protection	1	1/32	6	6	Felt
	FH Screws	3	#6		1	Steel

Figure 12-12. A bill of materials for a simple bookend lists the materials needed to make one product.

- Chemical properties: corrosion resistance
- Thermal properties: resistance to thermal shock, thermal conductivity, heat resistance
- Electrical properties: resistance, conductivity
- Magnetic properties: permeability
- Acoustical properties: sound absorption, sound conductivity
- Optical properties: color, transparency, optical reflectivity

Specifications are included on two types of sheets. In the first type, manufacturers prepare ***technical data sheets*** to communicate the specifications for products that they have on the market. These kinds of products are often called standard, or "off-the-shelf" materials and components. They are generally kept in stock by the manufacturer and are often listed in a supplier's catalog.

For example, you could write to a manufacturer regarding your need for an adhesive. You would probably receive technical data sheets on several adhesives the manufacturer makes that might meet your needs. You would study the specifications for each product and choose the one that meets your needs.

In the second type, large organizations might prepare their own specifications for materials and products they need. They send them to suppliers, who then compete to supply a specific item. One example is the Military Specification system, also known as the MILSPEC system, in which the government lists its specifications. Large manufacturing companies also have specification systems.

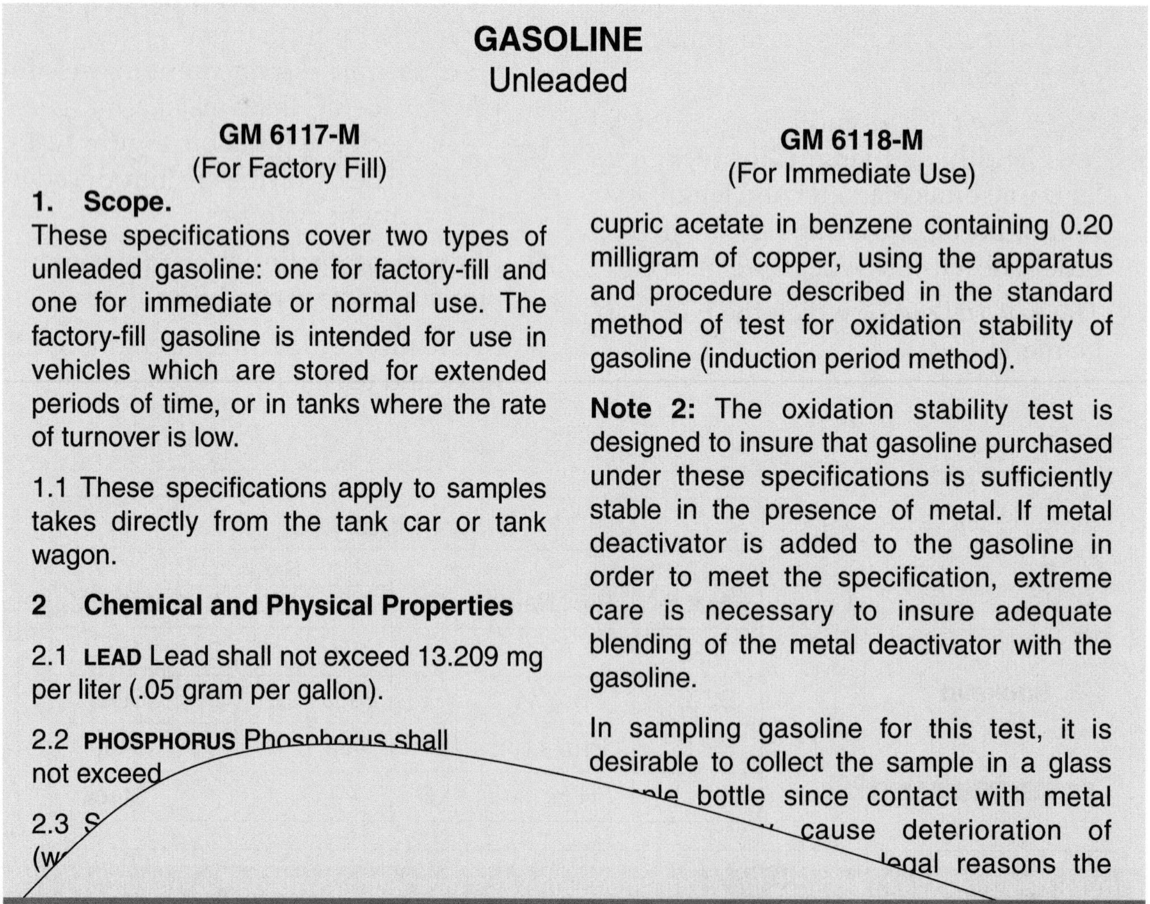

GASOLINE
Unleaded

GM 6117-M
(For Factory Fill)

1. Scope.
These specifications cover two types of unleaded gasoline: one for factory-fill and one for immediate or normal use. The factory-fill gasoline is intended for use in vehicles which are stored for extended periods of time, or in tanks where the rate of turnover is low.

1.1 These specifications apply to samples takes directly from the tank car or tank wagon.

2 Chemical and Physical Properties

2.1 **LEAD** Lead shall not exceed 13.209 mg per liter (.05 gram per gallon).

2.2 **PHOSPHORUS** Phosphorus shall not exceed

2.3 S
(w

GM 6118-M
(For Immediate Use)

cupric acetate in benzene containing 0.20 milligram of copper, using the apparatus and procedure described in the standard method of test for oxidation stability of gasoline (induction period method).

Note 2: The oxidation stability test is designed to insure that gasoline purchased under these specifications is sufficiently stable in the presence of metal. If metal deactivator is added to the gasoline in order to meet the specification, extreme care is necessary to insure adequate blending of the metal deactivator with the gasoline.

In sampling gasoline for this test, it is desirable to collect the sample in a glass ...ple bottle since contact with metal ... cause deterioration of ...egal reasons the

Figure 12-13. The characteristics of materials are described on specification sheets.

Career Corner

Engineering Technicians

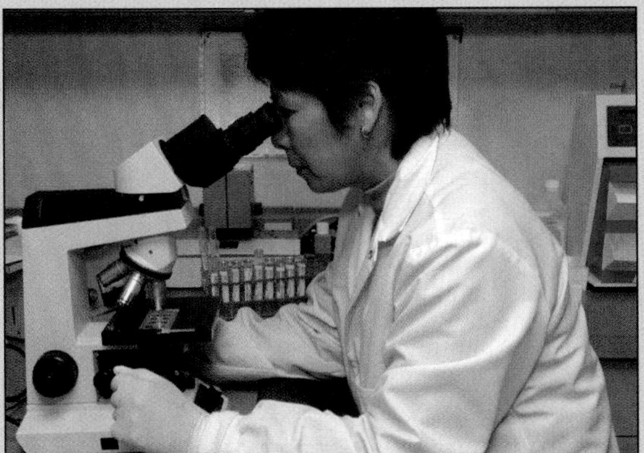

Engineering technicians use science, engineering, and mathematics to solve technical problems in research and development, manufacturing, and construction. Their work is more practical than that of scientists and engineers. Many engineering technicians work in quality control, assist in product development, and repair electrical, electronic, or mechanical equipment. Some common types of engineering technicians are aerospace, chemical, electronics, industrial, and mechanical engineering technicians.

Most engineering technicians have at least an associate's degree in engineering technology. Entry-level engineering technicians begin by performing routine duties under the supervision of an engineer, a scientist, or another employee. With more experience, they are given more challenging assignments.

Approval Documents and Reports

So far we have discussed methods of communicating designs for manufacture. We should note, however, that before anything can be built it must be approved by someone in charge. This "someone" might be company management, government agencies, or the customer.

Generally, the approval process requires two types of communication. First, designers prepared written reports. These reports might include need statements, proposed design solutions, cost estimates, marketing strategies, economic forecasts, and environmental impact statements.

The written reports are supported by oral reports to those who will give approval to the project. In oral reports designers present the highlights of the written report. They use graphs, illustrations, and other visual media to help communicate the design data, **Figure 12-14.** This presentation is designed to make the final "sale" of the design idea. The result is official approval to proceed with the manufacturing or construction project. The actual practices used to produce the product or construct the structure will be the focus of the next section of this book.

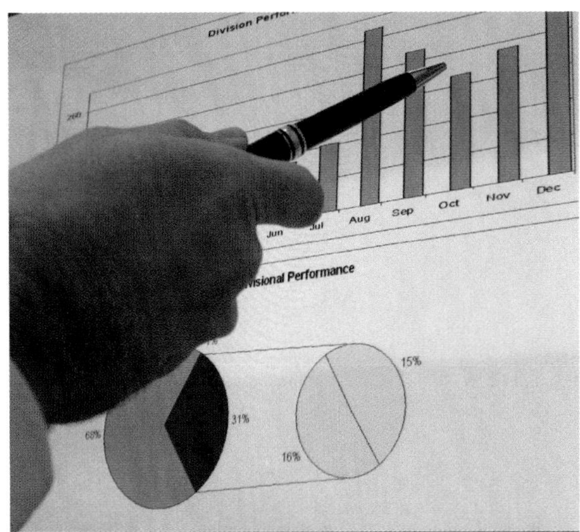

Figure 12-14. Graphic media are used to help communicate information about design solutions.

Summary

Communicating design solutions is a key element in creating successful technological systems. Designers must communicate vital information to manufacturing personnel and decision makers. They do this by preparing such documents as engineering drawings, bills of materials, and specification sheets. Engineering drawings communicate the details of each part, the way parts are assembled into products, and the arrangement of system components. Bills of materials list the parts needed to make one complete product. Specification sheets list the properties a material must possess for a specific application. Designers also prepare written and oral reports to gain approval for the product or structure.

Key Terms

assembly drawings
bill of materials
center lines
computer-aided design
detail drawings
dimension lines
engineering drawings
exploded views
extension lines
geometry dimensions
hidden lines
location dimensions
multiview method
object lines
one-view drawings
orthographic assembly drawings
orthographic projection
pictorial assembly drawings
size dimensions
specification sheets
standard views
systems drawings
technical data sheets
three-view drawings
tolerance
two-view drawings

Test Your Knowledge

Write your answers on a separate piece of paper. Please do not write in this book.

1. List the three basic kinds of documents through which designers communicate information about a product.

For Questions 2–10, match the type of engineering drawing on the right to the correct definition on the left. (Note: Some letters will be used more than once.)

Definition

2. _____Use symbols to represent components drawings
3. _____Size and shape dimensions included drawings
4. _____Two kinds are orthographic and pictorial drawings

Type of Engineering Drawing

a. Detail
b. Assembly
c. Systems

Definition	Type of Engineering Drawing
5. _____Usually do not include dimensions	a. Detail
6. _____Designed to show connections for wires, pipes, and tubes	b. Assembly
7. _____Also called schematic drawings	c. Systems
8. _____One is prepared for each part	
9. _____Use the multiview method	
10. _____Can include standard and exploded views	

11. Name one advantage of a computer-aided design system.
12. *True or false?* Bills of materials list the costs involved in making a product.
13. *True or false?* The important properties that a material must possess are listed on specification sheets.
14. What is a technical data sheet?
15. Give one example of a report designers need to prepare to gain approval for a project.

Applying Your Knowledge

1. Disassemble a simple product such as a ballpoint pen or a flashlight. Prepare:
    a. A detail drawing for one part.
    b. An assembly drawing for the product.
    c. A bill of materials for the product.
2. Select a simple game that is on the market. List the materials and parts that:
    a. Would need a drawing to produce.
    b. Would need specification sheets prepared for them.

This activity develops the skills used in TSA's Computer-Aided Design, 2D Architectural event.

Computer-Aided Design, 2D Architectural

Activity Overview

In this activity, you will create CAD drawings of an original residence design. The following drawings are required:

- Foundation plan with a section view of the typical footing and wall
- Floor plan for each level
- Front elevation
- One full section illustrating basic roof design
- Detail sheet with at least three details

Materials

- Paper
- Pencil
- Computer with CAD software
- Printer/plotter

Background Information

Basic layout. Begin by considering the type of residence you plan to design. Will you design a single-level ranch style or multiple-level residence? What type of foundation system will the residence have? Develop rough sketches of room sizes and locations. Consider which rooms should be grouped together and the traffic flow through the residence. Generally, bedrooms are grouped in one section of the residence. The kitchen and dining room are normally adjacent to one another. If the residence has multiple levels, the upper-level bathroom is normally located directly above a lower-level bathroom to simplify the plumbing. Will the residence include other rooms and features, such as an attached garage, home office, exercise room, or home theater?

Floor plan. Floor plans are normally drawn at a scale of 1/4" = 1'0". Use a scale that will allow your drawing to fit on the size of paper used for plotting. The floor plan should show the wall layout and dimensions, door and window locations, and room labels and sizes.

Foundation plan. The foundation plan can be created from the floor plan. Offset the outside wall line to locate the edges foundation wall and footing. Add square footings to support interior columns.

Front elevation. Copy the wall locations from the floor plan to begin the elevation. Add doors, windows, and other elements. Select an exterior finish for the home.

Full section. Determine the location of the section and identify it on the floor plan. Develop the section, which shows cross sections of the walls, floor, ceiling, and roof.

Details. Each of your details must be referenced on another drawing. Some common details are wall details (showing the interior and exterior components of the wall), kitchen cabinet details and elevations, and bathroom details and elevations.

Guidelines

- Create a separate CAD drawing file for each drawing.
- Plot each drawing.

Evaluation Criteria

Your project will be evaluated using the following criteria:

- Originality and creativity in design
- Accuracy of drawings
- Use of good drafting practice, including line type, lettering, and symbol usage

Section 4 Activities

Activity 4A

Design Problem

Background

All technological devices are the results of design efforts by people. People define problems and opportunities, think up many solutions, and model the solutions selected. Finally, designers communicate the design to production personnel.

Situation

Road Games Inc. is a company specializing in designing small, compact games people can take with them on trips. You have been recently employed in the creative concepts department of the company.

Challenge

Design a travel game using pegs and a 3/4" x 3 1/2" x 3 1/2" wood board. Dice can be used in the game, but are not required. Your boss expects you to produce rough sketches for five different ideas, a refined sketch for the best idea, a prototype, and a detailed drawing of the game board. Note: Activity 3B is an example of a game fitting these criteria.

Activity 4B

Fabrication Problem

Background

Product designers often work to improve an existing product. They also develop new and improved products to meet the changing demands and requirements of customers.

Challenge

You are a product designer for the Acme Bookend Company. The company makes bookends for different markets. Each market gets unique graphics and special shapes for the bookends. Your boss has asked you to modify an existing product to meet these new criteria:

- A new shape.
- New graphics that will appeal to high school students.

Materials

- Sketch paper
- Drawing paper
- Pencils and felt-tip pens
- Drafting ruler, T square, and triangles

Procedure

Redesign the bookend to meet the new criteria. See **Figure 4B-1.**

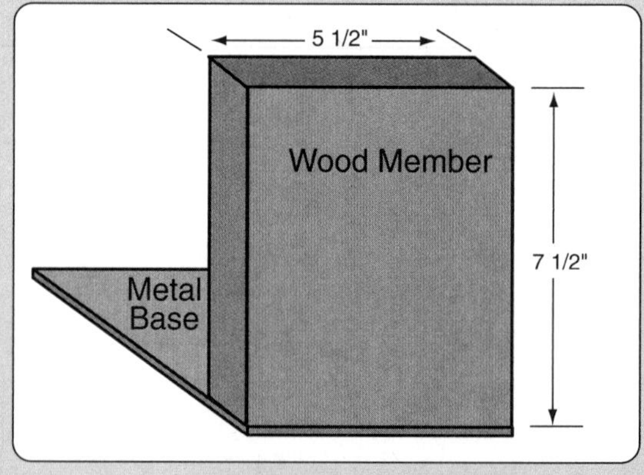

Figure 4B-1

Designing the Shape

1. Photocopy the bookend layout sheet. See **Figure 4B-2.**
2. Sketch four new shapes for the bookend.
3. Select the best shape and circle it with a colored marker.

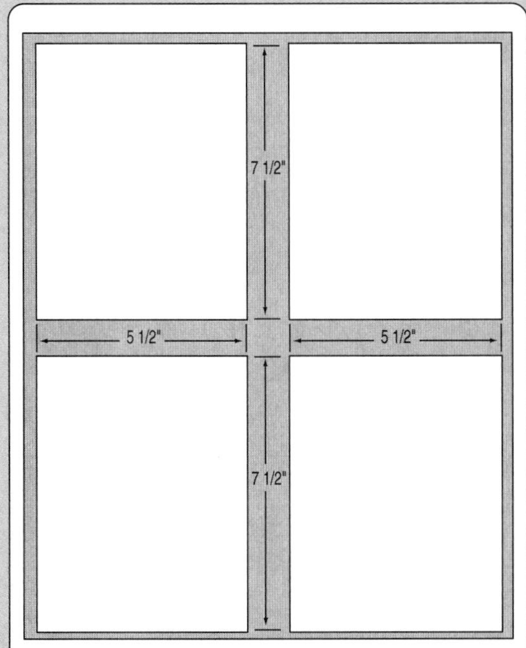

Figure 4B-2

Decorating the Product

1. Make four layouts, using the shape you chose.
2. Sketch four new graphic designs. Make sure the design will appeal to high school students.
3. Select the best design and circle it with a colored marker.

Communicating the Design

1. Obtain a piece of drawing paper.
2. Draw a border 1/2" in from all edges of the paper.
3. Draw a title box. See **Figure 4B-3.**
4. Produce a dimensioned two-view orthographic drawing of your design for the new bookend.

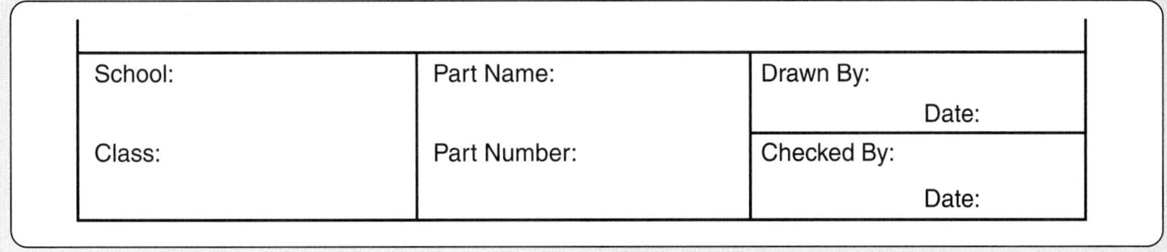

School:	Part Name:	Drawn By: Date:
Class:	Part Number:	Checked By: Date:

Figure 4B-3

Section 5

Applying Technology: Producing Products and Structures

Tomorrow's Technology Today
Gas-Plasma Displays

Dreaming of a big screen TV, but lost for space in your living room or den? Why not hang your big screen TV on the wall? With a gas-plasma display, you can do just that. The gas-plasma display is a flat-panel display about 3 to 6 inches thick, giving it the look of a picture or a work of art. Its thin frame frees precious space that a bulky cathode ray tube (CRT)–based TV would normally take.

Gas-plasma displays operate on the principle of charging an inert gas made of xenon and neon with a high voltage. The gas is normally in a balanced state, but when energized by a high-voltage, the particles inside the gas collide, exciting the gas atoms. When the gas atoms are excited, they release an ultraviolet light. This light strikes the phosphor coating of red, blue, or green in a sub-pixel causing the sub-pixel to release a light of corresponding color. By varying the pulses of current in a pixel area, the entire light spectrum of color can be duplicated.

The gas-plasma display has many advantages over traditional TVs. Unlike a typical TV, which has a curved surface, the gas-plasma's surface is flat, and does not distort the picture due to bending around the curved edges. The gas-plasma display allows for a wider viewing angle, up to 160°. This means you do not have to view the screen straight on to perceive the clearest picture, but can sit almost anywhere in the room. If you viewed an LCD at an extreme angle, you would not be able to see a picture. Also, a gas-plasma display can have a larger screen area, up to 60 inches diagonal, without the need to increase its depth.

Gas-plasma displays do have their disadvantages. With high voltage levels needed to excite the gas, the chances of screen burn increases. Just as a computer monitor is at risk of phosphor burn if left on with an unchanging image, so is the gas-plasma display. Also, because of high voltages, there is excessive heat, which is alleviated by cooling fans that sometimes hum. The biggest disadvantage yet is the price. However, the price is quickly decreasing as more and more manufacturers are producing gas-plasma displays.

As gas-plasma displays increase in popularity, you will find them in airports, railway stations, and sports arenas for informational purposes. You may also see them along city streets advertising products, or in the boardroom as you sit through a presentation. They can also serve as a replacement for bulky computer monitors when their price decreases. Gas-plasma TVs will lose their value as status symbols, but they will definitely remain a work of art as they hang on the walls of more and more homes across the world.

Chapter 13
Using Technology to Produce Artifacts

Learning Objectives

After studying this chapter, you will be able to:

- Identify the two major types of production activities.
- Differentiate between activities involved in resource processing, product manufacturing, and structure construction.
- Define servicing.
- Define repair.

We developed some of our earliest technology for production systems, **Figure 13-1.** These systems produce what is called *form utility*. They change the form of materials to make them more valuable. Look at **Figure 13-2.** Which is more useful and therefore more valuable to people: the lumber, shown in the left view, or the home that is built from it shown in the right view? You probably answered the home. Wood, in the form of a home, is more useful to you than in the form of lumber. Therefore, the actions that change lumber into houses provide form utility.

Production activities have two major goals: to produce a product or produce a structure. Then, after the product or structure has been produced, the focus of activities turns to service and repair. Each is discussed as follows.

Production Activities

As we stated, the goal of production activities is to produce either a product or a structure. Production activities that make products are called ***manufacturing*** activities. Production activities that produce structures such as buildings or roadways are called ***construction*** activities. Construction may be described as producing a structure on the site it will be used. It is different from manufacturing in that a manufactured product is produced in a factory and shipped to its point of use.

Literally thousands of different production activities fall under the categories of manufacturing and construction. These activities, however, can be grouped into three major systems:

- Resource processing systems (primary manufacturing)

Figure 13-1. These people are practicing early production activities using technology of the colonial period.

Figure 13-2. Lumber may be changed in form to become a home. (Weyerhaeuser Timber Co.)

- Product manufacturing systems (secondary manufacturing)
- Structure construction systems (construction)

Each of these systems plays a unique role in converting a material resource into a product or structure to meet human needs and wants.

Resource Processing Systems

Few materials occur in nature in a usable state. Generally, they must be converted into new forms before products and structures can be made. This conversion uses ***primary processing*** technology.

Typically, these processing systems involve two actions. First, the material must be located and obtained from the earth. This might involve growing and harvesting trees, crops, and domesticated animals. It might involve searching for minerals or hydrocarbons (petroleum and coal). Drilling or mining might be required to extract these resources from the earth.

Once the natural resource has been obtained, it must be transported to a processing mill. There the natural resource is changed into ***industrial materials.*** These materials are the inputs to secondary manufacturing activities. For example, iron ore, limestone, and coke might be changed into steel, **Figure 13-3.** The steel might then be processed into bars, rods, sheets, or pipes. This standard material becomes the raw inputs for systems that make products for industrial companies and retail consumers. It may end up as part of a lathe, automobile, broadcast tower, airline terminal, or one of thousands of other products and structures.

Specialized processing systems are used to manufacture food products, medicines, and chemicals. Other systems process petroleum into fuels and lubricants, natural gas into plastics, and coal into coke, **Figure 13-4.** More detailed information on obtaining material resources and processing them into industrial materials is contained in Chapters 14 and 15 of this book.

Figure 13-3. Primary processing activities change natural resources into industrial goods such as the steel rods shown here. (Inland Steel Co.)

Figure 13-4. Process systems produce such products as these tires. (Goodyear Tire and Rubber Co.)

Product Manufacturing Systems

Most standard materials have limited use to the average person. Suppose someone gave you ten sheets of plywood. You might think that this is a good gift. What are you going to use them for, however? You will probably have to ask the question, "What should I build out of the plywood?" You might build a bookcase or a storage locker. This action would use ***secondary processing*** activities.

Secondary processing systems change industrial materials into products, **Figure 13-5.** They cause the material to take on a desired size and shape. This might involve casting, applying force for forming, or machining using tools or other cutting devices. Secondary processing systems also assemble parts into products by welding, fastening, or gluing them together. Other secondary processing systems apply coatings to the material, part, or product. These coatings protect the item from the environment and improve its appearance. Finally, secondary processing systems might change the properties of the material. This change might cause the material to be harder, stronger, or more resistant to fatigue.

The outputs of secondary processing systems are either industrial or consumer products, **Figure 13-6.** ***Industrial products*** are items that are used by companies in conducting their businesses. For example, a computer terminal is an industrial product that may be used in the accounting activities of a company. Likewise, a furnace is an industrial product that becomes part of a building.

Consumer products are outputs that are developed for the end user in the product cycle: people such as homeowners, athletes, or students, just to name a few. Consumer products include such items as bath soap, television sets, lawn mowers, baseball bats, and furniture. The range of secondary manufacturing processes will be introduced and discussed in Chapter 16.

Structure Construction Systems

The three basic physical needs of all people are food, clothing, and shelter. Food comes to us from resource processing systems. Clothing comes from manufacturing

Figure 13-5. Secondary processing systems change industrial materials into products such as this finishing sander and this power nailer. (Hitachi, Bostitch)

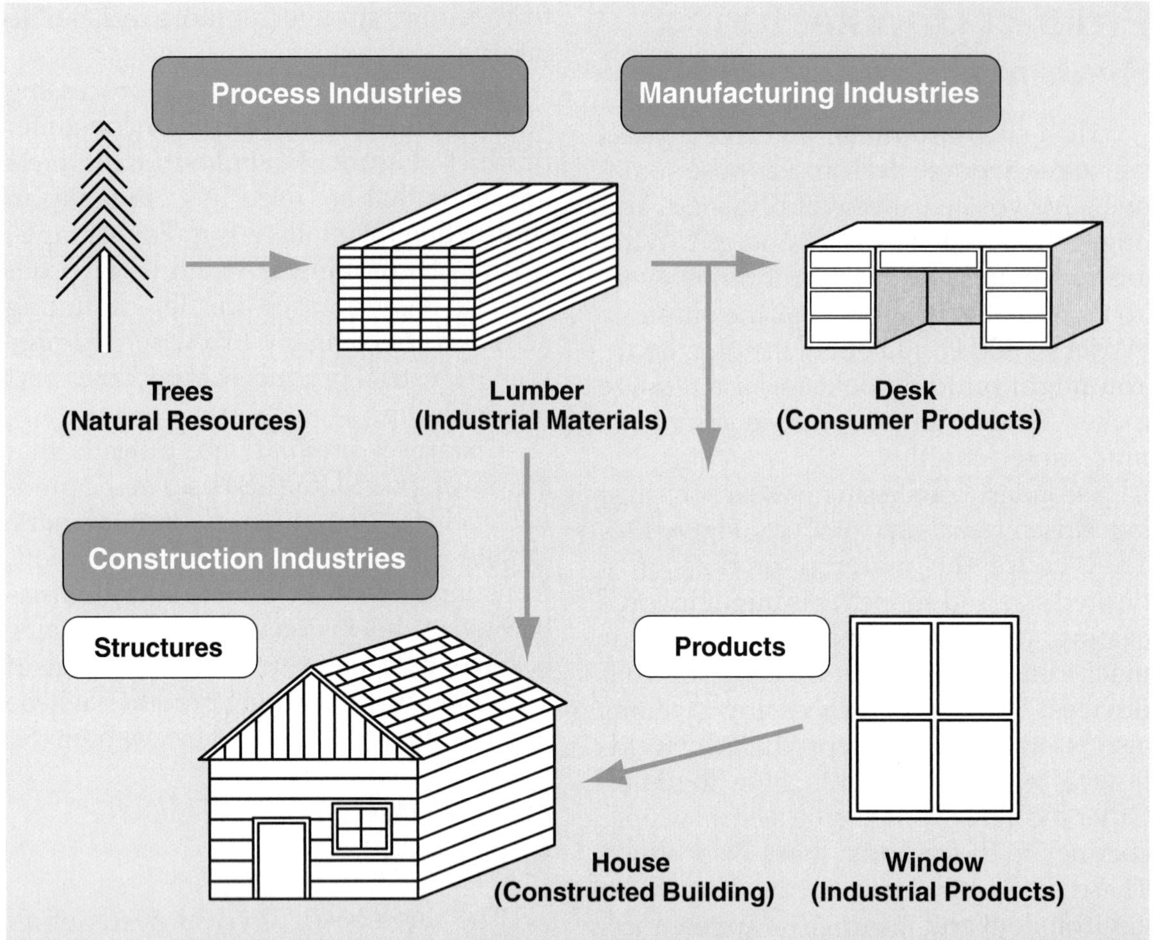

Figure 13-6. Production systems convert natural resources into industrial materials. These materials may be manufactured into industrial or consumer products. Materials and some manufactured goods are used in constructing structures.

industries. Shelter is the output of construction activities.

Constructed works can be grouped into the categories of buildings and civil structures. Homes, factories, stores, and offices are typical buildings. Common civil structures are roads, railways, canals, dams, power transmission lines, communication towers, and pipelines.

Each one of these structures is developed to meet a specific need, **Figure 13-7.** The development involves such activities as preparing the site, building foundations and superstructures, installing utilities, enclosing the superstructure, and landscaping the site. Construction activities will be presented in detail in Chapter 17.

Servicing and Repairing Products and Structures

All products and structures are subject to wear and tear. They can become damaged, worn, or outdated over time. Also, some products need attention during operation. All these conditions call for service and repair.

Servicing, or maintenance, is the scheduled adjustment, lubrication, or cleaning required to keep a product or structure operating properly. Automobiles require oil changes and engine tune-ups periodically,

Connections to Technology: Science

The Principles of Expansion and Contraction

Many scientific principles come into play when we use technology to produce artifacts. For example, the principles of expansion and contraction become major considerations when designing structures such as bridges.

Expansion is the increase in the size of a material, and *contraction* is a decrease. Each can be explained by the kinetic theory of matter. That theory states that the minute particles of matter in materials are always in motion. When an item such as a piece of steel is heated, the heat increases the vibrations of the particles. The increased vibrations create more space between the particles, thus increasing the size of the item. The opposite happens when the piece of steel is cooled. The particles move closer together, and the item contracts.

Because of the principles of expansion and contraction, designers will add what are called expansion joints when determining how a long steel bridge will be constructed. The joints have interconnected seams that move closer together in the summer and move further apart in the winter. Thus the bridge will not buckle in the extreme heat of summer.

Numerous other items would experience expansion and contraction problems except that allowances have been made in their construction. Can you name two of these items?

Figure 13-7. Construction activities produce such structures as these grain storage bins.

Career Corner

Plumbers

Plumbers install, maintain, and repair pipe systems. They work on systems that move water into and throughout residential, commercial, and public buildings. Plumbers also work on systems that dispose of waste, provide gas to stoves and furnaces, or supply air conditioning.

They work from drawings that show the planned locations of pipes, plumbing fixtures, and appliances. Plumbers cut holes for pipes, assemble pipe systems, install fixtures and appliances, connect systems to outside lines, and check the operation of completed systems. Almost all plumbers are trained through an apprenticeship training program, which local union-management committees administer. These four- to five-year programs include on-the-job training, supplemented by classroom instruction.

Figure 13-8. Buildings must be cleaned, and floors must be waxed. Machines need adjustment and cleaning. These are servicing acts and are usually performed at a specific point in time. The goal of servicing is to keep the product or structure in good working order.

Repair, on the other hand, involves fixing a broken, damaged, or defective product or building. It is designed to return a disabled product or structure to working condition. Repair requires diagnosing (determining) the problem and fixing or replacing defective parts or materials. Repairing a building may include replacing damaged ceiling tiles, applying new wall coverings, or fixing a hole in a wall. Product repair can involve replacing worn-out or broken parts and adjusting mechanisms. The actions involved in servicing products and structures are presented in Chapter 18.

Figure 13-8. Servicing returns products to their original operating conditions.

Summary

Production activities have two major goals: to produce a product or produce a structure. Although numerous kinds of production activities exist, they can be classified into three major groups or systems: resource processing, product manufacturing, and structure construction. After the produce or structure has been produced, we turn our attention to service and repair.

Key Terms

construction
consumer products
industrial materials
industrial products
manufacturing
primary processing
repair
secondary processing
servicing

Test Your Knowledge

Write your answers on a separate sheet of paper. Please do not write in this book.

1. *True or false?* Manufacturing activities are designed to produce products.
2. *True or false?* A roadway is an example of a construction activity.
3. Give one example of a resource processing system.
4. *True or false?* Iron ore is an industrial material.
5. *True or false?* Secondary processing systems change industrial materials into products.
6. *True or false?* An item used by a company can be considered an industrial product.
7. Buildings are outputs of _____ activities
8. Name a common type of civil structure.
9. Define the term *servicing* as used in this chapter.
10. Define the term *repair* as used in this chapter.

Applying Your Knowledge

1. Select two products or structures that you see around you. Complete a chart like the one that follows for each product.

Product		
Primary natural resource(s)		
Industrial material(s) used		
Product of manufacturing or construction		
Service or maintenance required		

Chapter 14
The Types of Material Resources and How They Are Obtained

Learning Objectives

After studying this chapter, you will be able to:

- List the types of natural resources that are used as inputs to production systems.
- Identify the types of genetic materials that are used in production systems.
- Identify the types of fossil fuel materials that are used in production systems.
- Identify the types of minerals that are used in production systems.
- Describe how genetic materials are obtained for use in production systems.
- Describe how fossil fuel materials are located and obtained for use in production systems.
- Describe how minerals are located and obtained for use in production systems.

You have learned that technology involves people designing and using tools and artifacts. This action extends human abilities to control or modify the environment. When no object or technical means exists, no technology is present. All technological objects (human-made objects, called artifacts) are made of materials. The materials that each object is made from can be traced back to one or more natural resources. For example, plastic materials are made from plant fiber (cellulose) or natural gas. Glass is made from silica sand and soda ash. Cotton is grown and harvested from plants. Plywood comes from trees in the forest. Steel is made from iron ore, limestone, and coal.

Materials form the foundation for all production activities. Without material resources, production is not possible. In this chapter we will explore the types of natural resources we have available for production and how we obtain them.

Types of Natural Material Resources

Production technology uses materials and energy as inputs and makes products and structures as outputs. Three types of natural resources can become the inputs to production systems. These materials, as shown in **Figure 14-1,** are:

Figure 14-1. Genetic, fossil fuel, and mineral materials are used as inputs to production systems. Here we see them in the form of trees, a coal mine, and a diamond, respectively. (U.S. Department of the Interior)

- Genetic materials
- Fossil fuel materials
- Minerals

Genetic Materials

Many resources come from living things. These resources, as you learned in Chapter 4, are called organic materials. Some organic materials are from organisms that have been dead for hundreds of years. This type of organic material includes fossil fuels that will be discussed later in this chapter. Other organic materials are obtained during the normal life cycle of plants or animals. These materials may be called ***genetic materials.***

We obtain genetic materials through three activities: farming, fishing, and forestry. Each of these activities works directly with nature as plants and animals move through the stages of life, **Figure 14-2.**

Typical genetic materials used in production systems are grains (wheat, oats, barley, and corn, for example), vegetable fibers (wood, flax, and cotton, for example), and animals or fish (meat, hides, and wool, for example).

The origin of all genetic materials is in birth or ***germination.*** The appearance of animal life is called birth. Plant life generally starts with the germination of seeds or spores.

Young plants and animals grow rapidly early in their life cycle. Their growth generally slows down as they reach older age. This period is called ***maturity***. The organism is still healthy at maturity, but it stays about the same size. The length of this maturity stage may be a matter of days, as with mushrooms, or centuries, as is the case with redwood trees. All organic life ends at some point, however. The plant or animal dies from old age or disease.

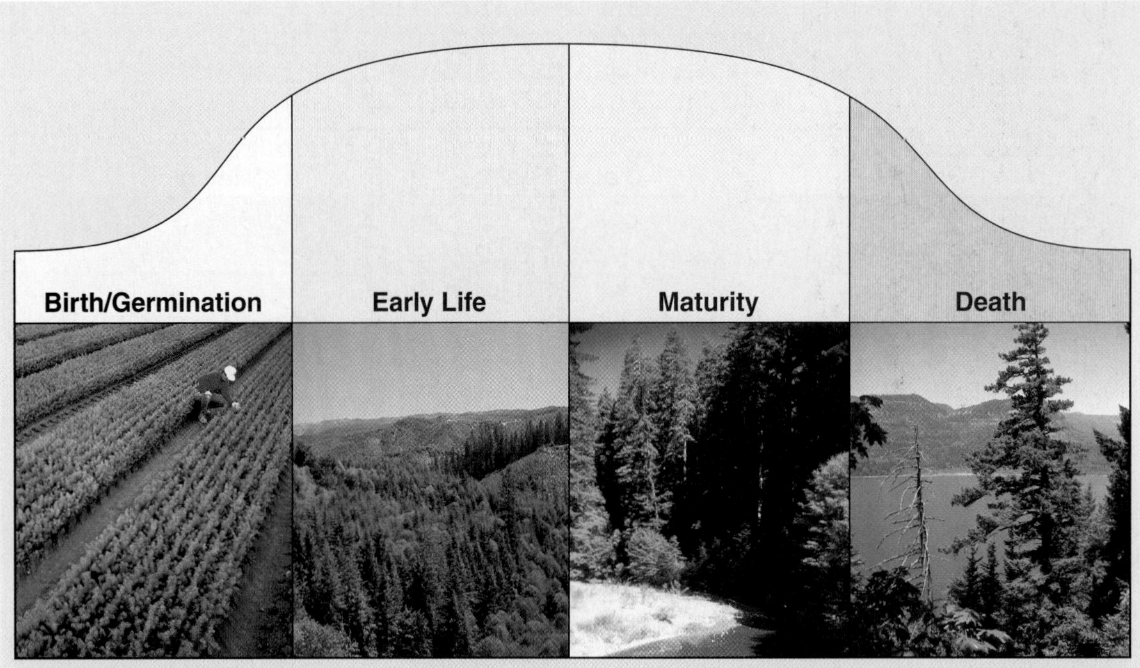

Figure 14-2. All genetic materials have a life cycle starting with birth, or germination, and ending with death.

Fossil Fuel Materials

Fossil fuels are mixtures of carbon and hydrogen. They are called hydrocarbons. Hydrocarbons include a vast number of products in use today, from fuels to medicines. Most hydrocarbon products, however, come from only three fossil fuel resources. These are:

- Petroleum
- Natural gas
- Coal

Petroleum is an oily, flammable mixture of hydrocarbons that has no specific composition. Instead, it is a mixture of a number of different solid and liquid hydrocarbons. The composition of petroleum will vary with where it is found on earth. Petroleum is the principal source of the liquids we use, such as gasoline, diesel fuel, heating oil, and lubricants, **Figure 14-3.** Lighter hydrocarbon products that come from petroleum are gases. They include methane, ethane, butane, and propane. These gases are widely used in producing plastic resins such as polypropylene and polyethylene.

Petroleum is the source for fuel most used by transportation vehicles. We should note, however, that exhaust from gasoline and diesel engines produces harmful air pollutants including nitrous oxides and carbon monoxide.

Natural gas is a combustible gas that occurs in porous rock. It is composed of light hydrocarbons. Typically, natural gas is about 85 percent methane. The rest is made up of propane and butane. Natural gas is used as a fuel for homes and industry. Natural gas is also used to make plastics, chemicals, and fertilizers.

Natural gas burns cleanly compared with other fossil fuels. It requires complex pipeline networks to distribute it to potential users. In some cases it is compressed and used as a fuel for vehicles. In this form it is called compressed natural gas, or CNG.

Coal is a combustible solid that is composed mostly of carbon. It started as plant

Figure 14-3. Petroleum is a vital natural resource that provides the fuels to power vehicles and other machines.

matter thousands of years ago. In moist areas the plant matter did not decay easily and layered up to make peat. Peat is brownish-black plant matter that looks like decayed wood. When dry, it will burn but gives off a great deal of smoke. The peat was buried by sediment, and pressure and heat changed the peat into coal.

The principal types of coal are:

- Lignite: A soft and porous material made from peat that has been pressed by natural action. It gives off more heat than peat and is used in electrical generating plants and for industrial heating.
- Bituminous coal: The most commonly found coal, which is harder than lignite. It is sometimes called soft coal because it can be easily broken into various sizes. Bituminous coal is widely used for power generation and heating. It can

Figure 14-4. Bituminous coal is sometimes called soft coal. (Jeffboat Shipyard)

also be used for coal gasification and chemical processes. See **Figure 14-4.**

- Anthracite coal: The hardest coal is a shiny black material that burns without smoke. Anthracite has the highest carbon content of all the types of coal. Anthracite coal is used for heating and to produce coke for steel making.

Coal does not burn cleanly, and its sulfur content is a source of chemicals that make acid rain. Also, its bulk makes it costly to ship.

Minerals

Minerals are any substance with a specific chemical composition that occurs naturally. They are different from fossil fuel resources, which are all chemical mixtures. Typical minerals are iron ore, bauxite (aluminum ore), and sulfur.

We can classify minerals in various ways. One way is to group them by their chemical composition. This grouping would include native elements (elements that occur naturally in a pure form), oxides, sulfides, nitrates, carbonates, borates, and phosphates.

We have a more useful way to group minerals available, however. This method groups the minerals that have economic value into families with similar features. These groups include:

- ***Ores:*** Minerals that have a metal chemically combined with other elements; can be processed to separate the metal from other elements
- ***Nonmetallic minerals:*** Substances that do not have metallic qualities, such as sulfur
- ***Ceramic minerals:*** Fine-grained minerals that are formable when wet and become hard when dried or fired
- ***Gems:*** Stones that are cut and polished and prized for their beauty and hardness

Locating and Obtaining Natural Resources

All natural materials are found on the earth. They are not all visible as we travel across the land and water or through the air, however. In fact, we need to perform two major actions to obtain natural resources. First, we must locate them, and then we must extract or gather them.

Obtaining Genetic Materials

Most genetic materials are easy to find. Trees and farm crops are on easily used plots of land. Domesticated animals and the fish raised on fish farms are contained in specific locations. Only those who fish commercially must seek genetic resources that are sometimes hard to find.

The major challenge for people dealing with genetic resources is to harvest the plant or animal at the proper stage of growth. This stage will vary with the growth cycle and growing habits of the organism. Many trees are harvested during their mature phase. Some young trees, however, will not grow in the shade of older trees. Thus sometimes all the trees in a single plot must be harvested at one time. Then new trees can be planted, which will grow in the cleared area.

Likewise, most farm crops are planted at one time in fields. The plants may be fertilized and irrigated to stimulate their growth, **Figure 14-5.** Most of the crop will mature at the same time. Therefore, the entire crop can be harvested at one time.

Let us look at harvesting a genetic material—trees. The forest management process requires that each stand of trees be evaluated and designated for a specific use. Wilderness areas are set aside. No roads or logging is permitted in a wilderness area. National parks, and many state

Figure 14-5. Intensive farm management, like this irrigation process, can increase crop yields.

and provincial parks, protect scenic beauty. Roads are allowed in the parks, but logging is not. National, state, and provincial forests, generally, are multiuse lands. Lakes, hiking trails, and camping areas are set aside for recreational use. Logging is permitted in selected areas to harvest mature trees. Private forests are generally designed and managed to maximize the amount of harvested material.

The logging process requires both planning and action. Planning for the removal of trees involves several steps. First, the forest is studied to determine whether it is ready for harvesting. This study involves a process called ***timber cruising.*** Teams of two or three foresters measure the diameter and height of the trees. Their task is to find stands of trees that can be economically harvested. They also prepare topographical maps showing the location and elevations of the features on the potential logging site.

Forest engineers must then plan the proper way to harvest the trees. They plan logging roads and loading sites. They also select the type of logging to match the terrain and the type of forest. Three logging methods exist:

- ***Clear cutting:*** All trees, regardless of species or size, are removed from a plot of land that is generally less than one thousand acres, **Figure 14-6.** This process allows for replanting the area with trees that cannot grow in competition with mature trees. Also, the number of tree species can be controlled.
- ***Seed-tree cutting:*** All trees, regardless of species, are removed from a large area except three or four per acre. These trees are used to reseed the area. Again, the number of species that will be reseeded is controlled by the type of seed trees left.

Figure 14-6. This clear cutting increased the yield of Douglas fir.

- ***Selective cutting:*** Mature trees of a desired species are selected and cut from the plot of land. This technique is used in many pine forests where tree density is limited.

These steps are the prelude for the main activity of ***logging,*** **Figure 14-7.** Workers move equipment into the forest to remove the trees. Loggers called ***fellers*** use a chain saw to cut down (fell) the appropriate trees. Smaller trees can be harvested by a machine that shears the trees off at ground level. Loggers are careful to drop the trees into a clear area so they are not damaged or cause damage to other trees.

Workers called ***buckers*** remove the limbs and top. These parts of the tree are called slash. The slash is piled for later burning or chipping so that nutrients are returned to the soil. The trunk is then cut into lengths called logs.

Workers gather the logs in a central location called the landing. This process of gathering is called ***yarding*** and can be accomplished in several ways. First, workers use chokers (cables) to bind the logs into bundles. Then they can use cables to drag logs using high-lead and skyline yarding, as shown in **Figure 14-8.** These systems use a metal spar (pole), cables, and an engine. High-lead yarding drags the logs to the landing. Skyline yarding lifts and carries the logs over rough or broken terrain.

On gentle terrain, ground yarding is used. This system uses tractors and an implement called an arch. These items drag the logs, bound together with a choker, to the landing. Very steep terrain may require helicopter yarding, as shown in Figure 14-7.

Once logs arrive at the landing workers load them on trucks and move them to the processing plant. This plant might be a lumber, plywood, particleboard, hardboard, or paper mill. The logs are often stored at the mill in ponds or stacked up and sprayed with water to prevent cracking and insect damage.

Harvesting genetic materials such as trees is much different from finding fossil fuels. We discuss fossil fuels next.

Career Corner

Forest and Conservation Workers

Forest and conservation workers perform tasks associated with replanting and conserving timberlands. Also, they work to maintain forest facilities, such as roads and campsites. Forest workers plant seedlings, remove diseased or undesirable trees, spray trees to kill insects, and apply herbicides on undesirable brush and trees. They might manually clear brush and debris from hiking trails and camping sites and help maintain recreational facilities and campgrounds.

Forest and conservation work is concentrated in the West and Southeast of the United States. Most forest and conservation jobs require little formal education. Many secondary schools and community colleges in timber states offer courses in general forestry, wildlife, conservation, and forest harvesting.

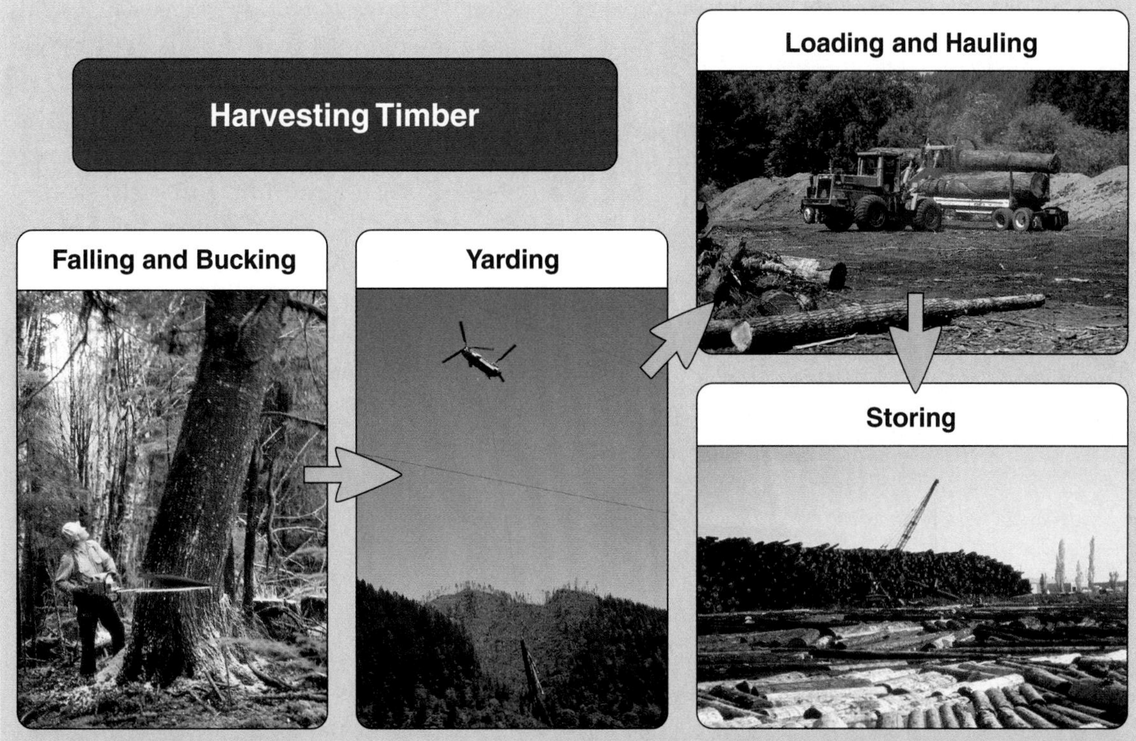

Figure 14-7. Logging starts with selecting and cutting trees. The felled trees are collected using yarding. Then, the trees are hauled to a mill for storage and further processing. (Boise Cascade)

Obtaining Fossil Fuel Resources

Most fossil fuel resources are buried under the surface of the earth. Fossil fuels can be pools of petroleum, pockets of natural gas, or veins of coal. Locating and extracting these resources is the major challenge. The techniques used for natural gas and petroleum are much different from those used for coal. Therefore, let us look at them separately.

Obtaining Petroleum and Natural Gas

People do not look for oil (petroleum) and gas directly. They look for rock formations that might contain deposits of oil and gas. Finding these rock formations is the job of geologists and geophysicists, **Figure 14-9.**

Most geologists believe that petroleum comes from decayed plant and animal

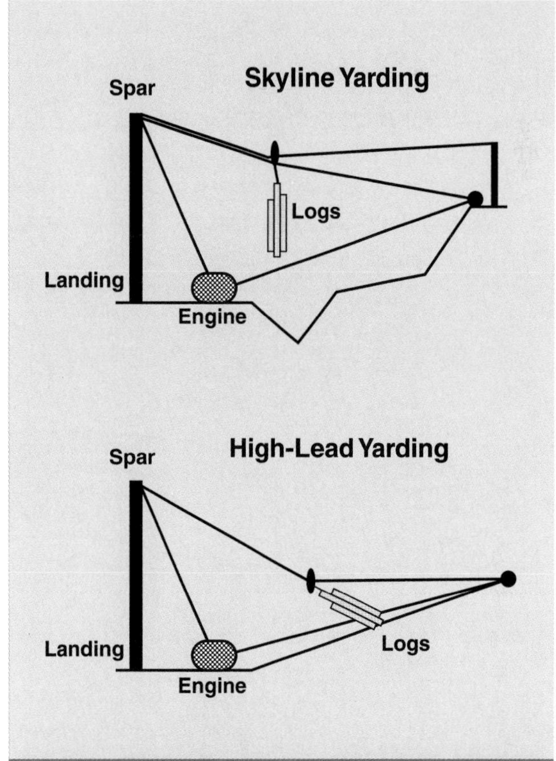

Figure 14-8. Two ways used to move logs to the landing are high-lead and skyline yarding.

Figure 14-9. This geologist is seeking deposits of petroleum. (BP)

matter. This organic matter was covered by layers of sediment from rivers. The layers built up and created great pressures and heat. Over millions of years the pressure and heat turned the organic matter into oil. Oil and gas are generally found in porous rock under a layer of impervious (dense) rock. The oil and gas collect under the dense rock. These deposits may be under oceans, mountains, deserts, or swamps. They may be near the surface, as they are in the Middle East, or several miles beneath the land or sea.

We have developed various ways to explore for petroleum and natural gas. One the most accurate ways is ***seismographic study,*** **Figure 14-10.** This technique uses shock waves such as those in an earthquake. A small explosive charge is

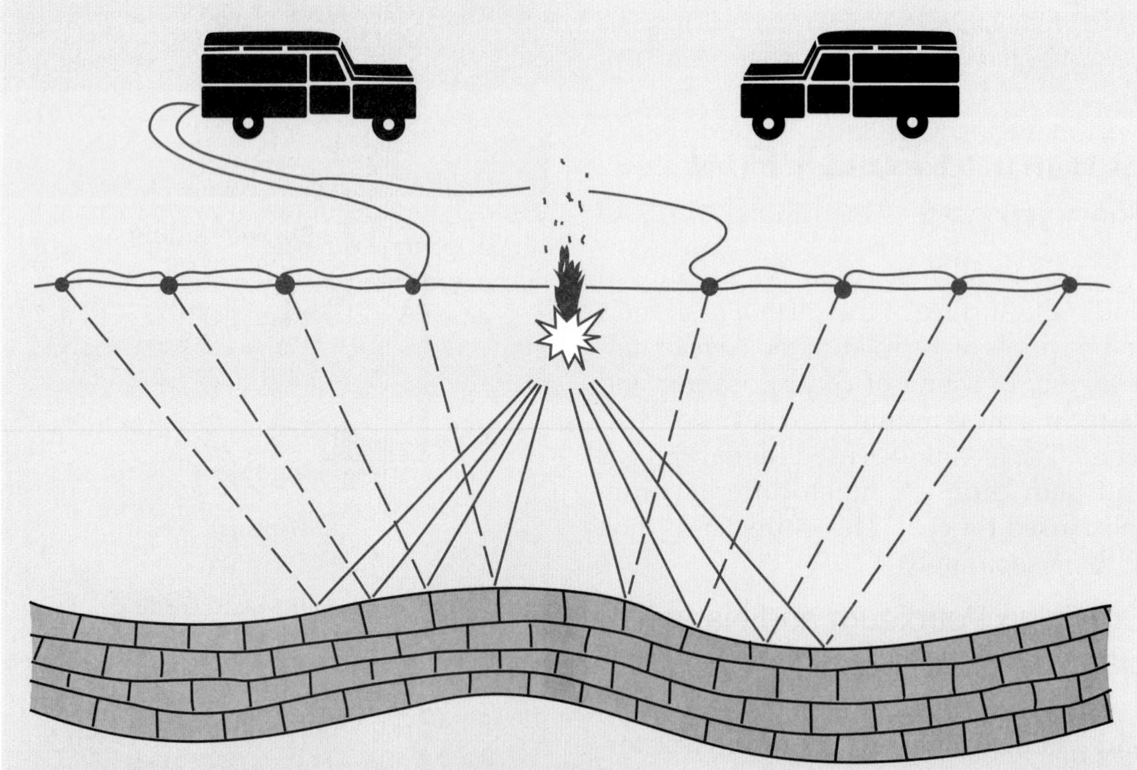
Figure 14-10. Seismographic surveys are used to determine the shape of underground rock formations that may hold oil and gas. An explosive charge is detonated, creating a shock wave that travels down into the earth. The boundaries between different types of rock reflect the shock waves back to the detectors on the surface. Geologists measure the time it takes for the waves to reach the detectors. The timing information gives the geologists an idea of the shape of a formation. (Shell Oil Co.)

detonated in a shallow hole. The shock waves from the explosion travel into the earth. When the waves hit a rock layer they reflect back to the surface. Seismographic equipment uses two listening posts to measure the shock waves that bounce off various rock layers. Measuring the time it takes the waves to go down to the layers and reflect back allows the geologist to construct a map of the rock formations.

Other methods use geological mapping, which measures the strength of magnetic forces, fossil study, and core samples from drilling to search for the deposits, **Figure 14-11.**

The geological study helps people select a good site for exploration. If an area has never produced oil or gas it is called a ***potential field.*** Producing fields are called ***proven reserves.***

A drilling rig is brought to promising sites. This rig may be either a land rig or an offshore drilling platform, **Figure 14-12.**

Figure 14-11. This scientist is studying satellite photographs to determine where to conduct seismographic surveys. (American Petroleum Institute)

Figure 14-12. Oil and gas rigs allow people to drill wells on land and under lakes and seas. (BP, Gulf Oil Co.)

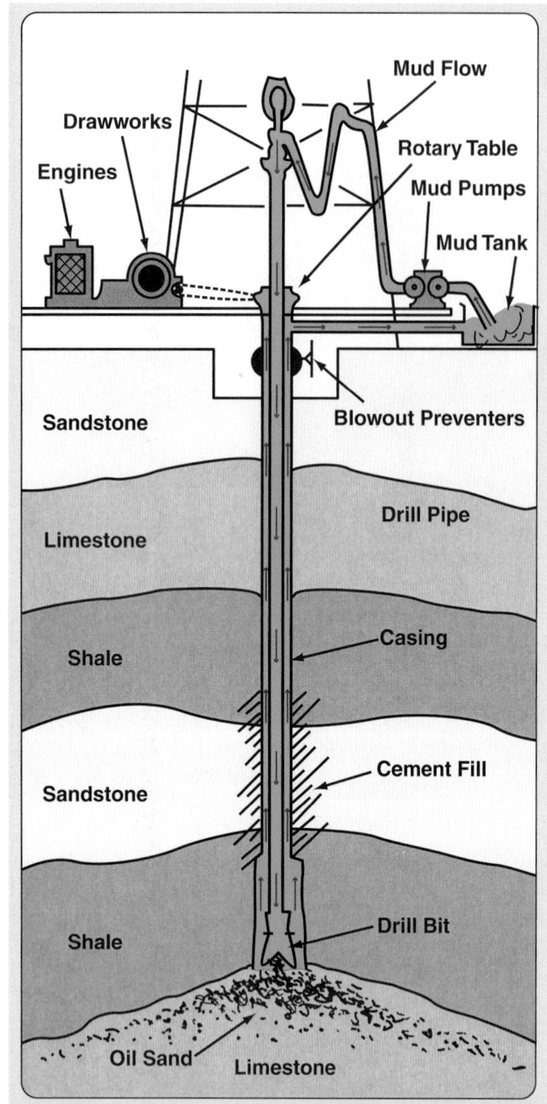

Figure 14-13. Drilling involves rotating a drill bit on the end of a drill pipe. (Mobil Oil Co.)

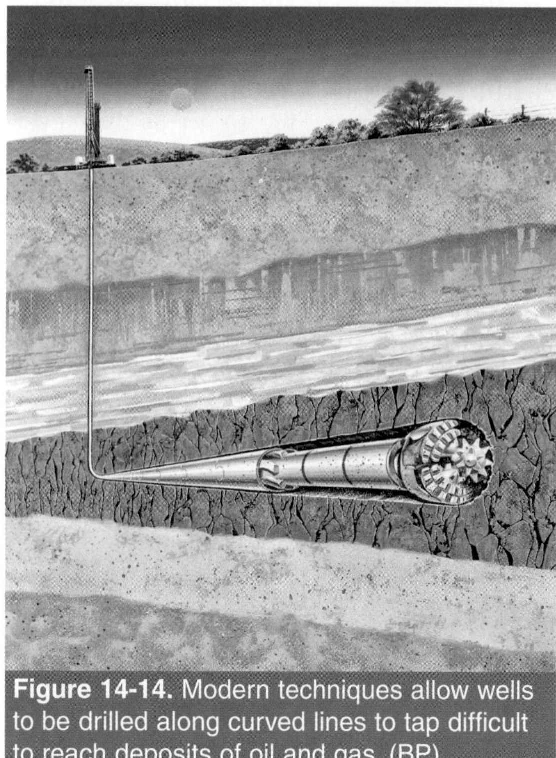

Figure 14-14. Modern techniques allow wells to be drilled along curved lines to tap difficult to reach deposits of oil and gas. (BP)

Drilling involves rotating a drill bit on the end of a drill pipe. Lengths of pipe are added to the drill pipe as the hole gets deeper, **Figure 14-13.**

During drilling, a mixture of water, clay, and chemicals is pumped down the drill pipe. This mixture is called ***mud.*** It flows out through holes in the drill bit to cool and lubricate. The mud also picks up the ground-up rock and carries it to the surface. Finally, it seals off porous rock and maintains pressure on the rock. Pressure is maintained to prevent a blowout. A ***blowout*** occurs when oil surges out of the well. Blowouts are very dangerous and waste large quantities of oil and gas.

Early oil wells were drilled straight down or at a specific angle. Modern techniques allow the well to be drilled along a curve to reach deposits that cannot be otherwise tapped, **Figure 14-14.**

Once an oil or gas deposit is found, the drilling rig is replaced with a system of valves and pumps, **Figure 14-15.** The recovered resource flows through pipes into storage tanks. From the well the petroleum is transported to refineries. Natural gas is compressed and sent to petrochemical plants. Pipeline companies sell natural gas to home heating and electric power customers.

Obtaining Coal

Coal is the most abundant fossil fuel and is found on every continent. Most of the known reserves are in the Northern Hemisphere, however. These reserves are generally recovered through mining

Connections to Technology: Science

Synthetic Fuels

As we have learned, the earth's supply of fossil fuels is limited. Because one day these materials will be exhausted, scientists and other researchers have been working on creating synthetic fuels, also called *synfuels.* These fuels result from the action of various chemical processes on such materials as coal and biomass (organic matter).

For example, coal can be turned into gas through a process called *gasification.* In this method, mined coal is combined with steam and oxygen and becomes a mixture of carbon monoxide, hydrogen, and methane. This mixture can then be used in place of natural gas.

In another process, organic matter is fermented to produce ethyl alcohol. This product is then mixed with gasoline and becomes a fuel called gasohol.

Although synthetic fuels might eventually be the answer to our limited supply of fossil fuels, many drawbacks still exist. For example, in the process of creating synfuels from coal, we use up a lot of coal, thus reducing our supply. Producing synfuels is also extremely costly at this time.

Scientists continue to work on these and other problems associated with synfuels. Can you name a synfuel currently being investigated and a problem that is being addressed in the research?

Figure 14-15. A valve system, called a Christmas tree, is set on top of an oil well. (AMAX)

operations. There are three types of mining: surface or open pit, underground, and fluid. The first two are used to mine coal. The third is used to obtain minerals and will be discussed in the next section.

Surface or ***open-pit mining*** is used when the coal vein is not very deep underground. This type of mining generally involves four steps:

1. The surface layers of soil and rock are stripped from above the coal. This material is called *overburden.* It is saved for later use in reclaiming the land.
2. The coal is dug up with giant shovels, **Figure 14-16.**
3. The coal is loaded on trucks or railcars to be transported to a processing plant.
4. The site is reclaimed by replacing the topsoil and replanting the area.

Figure 14-16. Surface mining uses giant shovels to dig up and load the natural resource. (American Electric Power)

Underground mining requires shafts in the earth to reach the coal deposits. The three types of underground mining are shaft, drift, and slope, **Figure 14-17.** ***Shaft mining*** requires a vertical shaft to reach the coal deposit. Workers then dig horizontal shafts to remove the coal, **Figure 14-18.**

Miners use ***slope mining*** when the vein is not too deep under the ground. They dig a sloping shaft to reach the coal. They then dig a horizontal shaft to follow and remove the coal vein.

Miners use ***drift mining*** when the coal vein extends to the surface of the earth. They then dig a horizontal shaft directly into the vein.

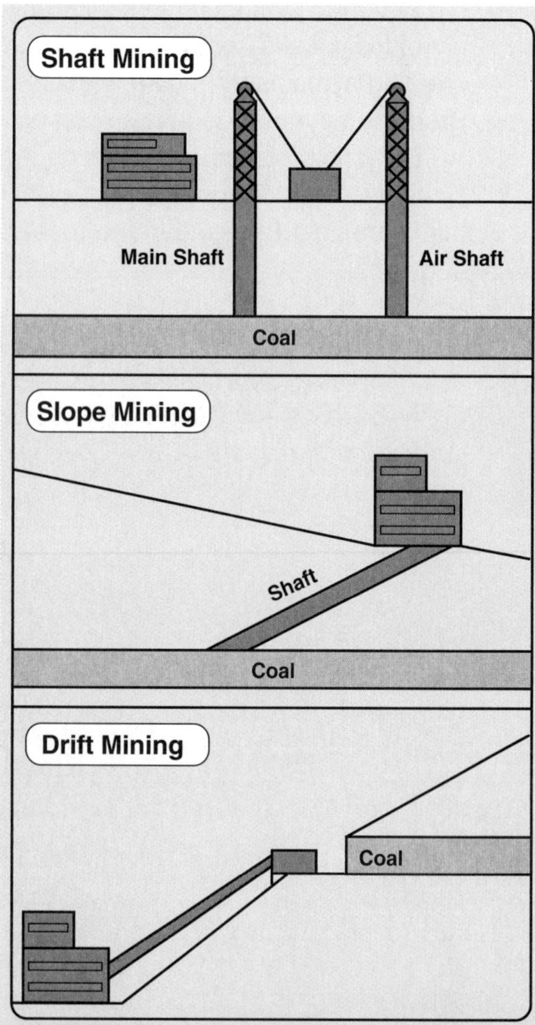

Figure 14-17. The three types of underground mining are shaft, slope, and drift.

Figure 14-18. This long-wall mining machine is recovering coal from an underground mine. (FMC Corp.)

Figure 14-20. Open-pit mineral mines can extend several thousand feet into the earth. (Brush-Wellman)

Figure 14-19. Coal is removed from slope and drift mines on tractors (top) or conveyors (bottom). (AMAX)

Those miners involved with underground mining use elevators to remove the mined coal. The same elevators move people and equipment in and out of the mine. Those working with slope and drift mines often use coal cars or conveyors to remove the coal, **Figure 14-19.**

Obtaining Minerals

We can extract minerals from the earth and oceans in a variety of ways. One way is through ***evaporation.*** We can pump seawater or water from salt lakes into basins. We can use solar energy to cause the water to evaporate. The mineral resource will be left behind. We recover a number of minerals in this manner from the Great Salt Lake in Utah.

We can also use mining and adopt the same techniques that were described in the section on recovering coal. Open-pit mines for minerals are generally much deeper, however. They can extend several thousand feet into the earth. They will appear as a giant inverted cone with ridges around the edge. The spiral ridge is the road used to move equipment into the mine and minerals out of it, **Figure 14-20.**

Another mining method we can use is called ***fluid mining.*** This technique uses two wells that extend into the mineral deposit. Hot water is pumped down one of the wells. The water dissolves the mineral and is forced up the other well. This mining process is often called the Frasch process. It is widely used to mine sulfur that is found in the limestone rocks that cover salt domes.

Summary

All production systems use materials. Common production materials come from genetic, fossil fuel, and mineral resources. To use these resources we must first locate them and then know how to extract them. Extraction can involve harvesting trees, farm crops, and animal and fish life. It also includes drilling oil and gas wells. Finally, we can extract material resources from the earth using mining techniques.

Key Terms

blowout
bucker
ceramic minerals
clear cutting
coal
drift mining
evaporation
feller
fluid mining
fossil fuels
gems
genetic materials
germination
logging
maturity
minerals
mud
natural gas
nonmetallic minerals
open-pit mining
ores
petroleum
potential field
proven reserves
seed-tree cutting
seismographic study
selective cutting
shaft mining
slope mining
surface mining
timber cruising
underground mining
yarding

Test Your Knowledge

Write your answers on a separate piece of paper. Please do not write in this book.

1. The three types of natural resources that are inputs to production systems are genetic materials, fossil fuel materials, and _____.
2. *True or false?* We obtain genetic materials through farming, fishing, and forestry.
3. *True or false?* Fossil fuels are mixtures of carbon and hydrogen.
4. List the three main fossil fuel resources.
5. What is the main difference between fossil fuel resources and minerals?
6. Name a typical mineral.
7. What is the major challenge involved in obtaining genetic materials?
8. The process of gathering logs is called _____.
9. What is the major challenge involved in obtaining fossil fuel materials?
10. Describe the method of seismographic study.

11. What is the difference between a potential field and a proven reserve?
12. *True or false?* Coal is the most abundant fossil fuel.
13. Name one type of mining used to obtain coal.
14. Name one way we can obtain minerals from water.
15. What is fluid mining?

Applying Your Knowledge

1. Select a resource from the categories of genetic, fossil fuels, and minerals. Then research how it is located (exploration) and recovered from the earth (production). Use a chart similar to the one below to record your findings.

Resource:
Exploration processes:
Extraction processes:

2. Build a model to show a selected mining or drilling technique.
3. Read an article and write a report on a controversial resource recovery issue. The report may deal with a number of issues, including protecting the environment, wildlife, or the supply of the resource.

Chapter 15
Processing Resources

Learning Objectives

After studying this chapter, you will be able to:

- List the types of primary processes.
- Identify materials that result from mechanical processing.
- Describe characteristics of materials resulting from mechanical processing.
- Identify materials that result from thermal processing.
- Describe characteristics of materials resulting from thermal processing.
- Identify materials that result from chemical and electrochemical processing.
- Describe characteristics of materials resulting from chemical and electrochemical processing.

You have learned that natural resources are used to produce products and structures. You also learned that the production process involves several steps. One of these steps comes between obtaining the resource and manufacturing the product or constructing the structure. This step can be called ***primary processing,*** **Figure 15-1.** The goal of primary processing is to convert material resources into industrial materials. Industrial materials are often called standard stock.

For example, primary processing converts wheat into flour; aluminum ore into aluminum sheets, bars, and rods; logs into lumber, plywood, particleboard, hardboard, and paper; natural gas into plastic pellets, film, and sheets; and silica sand and soda ash into glass. These industrial materials are used as inputs to further manufacturing or construction activities.

Primary processing uses many production actions. These actions can be grouped in a number of ways. One useful way to group processes is by the type of energy used. This grouping, as shown in **Figure 15-2,** includes:

- Mechanical processes
- Thermal (heat) processes
- Chemical and electrochemical processes

This system provides a general overview of most primary processing activities. Note, however, that many materials are produced using more than one type of process. For example, steel is made from iron ore, coke, and limestone using a thermal process. Some steel that is produced is formed into bars, rods, and sheets using mechanical processes. Some sheets of steel are coated with zinc to produce galvanized steel. This process uses an electrochemical process.

In this chapter you will be introduced to the manufacture of several materials that you come into contact with daily. These materials include lumber, plywood, steel, glass, petroleum products, and aluminum.

The primary processes used to produce these materials are viewed in terms of the first process used to change them from a raw material to an industrial material. The other processes used after the initial action will be briefly covered during each specific discussion.

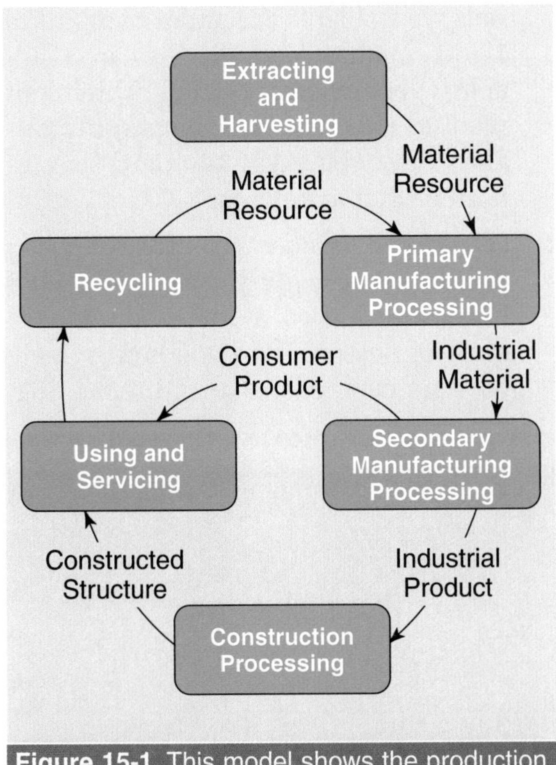

Figure 15-1. This model shows the production process. Note how primary processing fits into the cycle.

Mechanical Processes

Mechanical processes use mechanical forces to change the form of natural resources. They may use compression (pressure) to crush the material to reduce its size or change its texture. Other processes use shearing forces to cut and fracture the material. Still other processes run the material over screens to sort it by size.

A number of natural resources are first processed using mechanical means. The most common of these produces forest products from trees.

As you learned in Chapter 4, wood is a natural composite. Wood is cellulose fibers held together by a natural adhesive called lignin. The processing of forest products uses two methods. The material may be cut or sheared into new shapes. These mechanical actions are used to produce lumber, plywood, and particleboard.

Other wood-processing techniques use chemical action. They digest the wood so

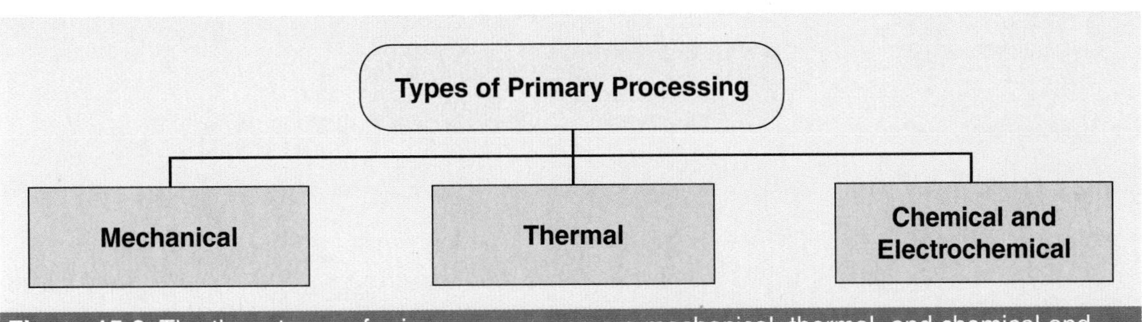

Figure 15-2. The three types of primary processes are mechanical, thermal, and chemical and electrochemical.

that the lignin releases the cellulose fibers. The cellulose fibers are then processed into new materials. Chemical processes are used to produce hardboard, cardboard, and paper.

Producing Lumber

Trees were one of the first natural resources used by humans. Trees provided the raw material for shelters and crude tools. Humans also used wood as the primary fuel for cooking and heating. In fact, even today wood is a major energy source in developing countries around the world.

By the time of ancient Egypt, wood had become a basic material for carpentry and boat building. This use of wood, a natural resource, continues in today's modern civilization.

One form of wood that is widely used is lumber. A piece of ***lumber*** is a flat strip or slab of wood. Lumber is available in two types:

- ***Softwood lumber*** is produced from *needle-bearing* trees such as pine, cedar, and fir. Softwood lumber is used for construction purposes, for shipping containers and crates, and for railroad ties. Softwood lumber is produced in specific sizes called nominal sizes. Typical sizes range from 1 by 4 (3/4″ x 3 1/2″ when finished [smoothed]) and the common 2 by 4 (1 1/2″ x 3 1/2″ when finished) to as large as 4 by 12 (3 1/2″ x 11 1/2″when finished). These materials are available in standard lengths in 1′ increments. Generally, they are available in lengths from 6′ to 16′. You might want to look at a building supply catalog to see the many standard sizes of lumber that are available.
- ***Hardwood lumber*** is produced from deciduous, or *leaf-bearing,* trees that lose their leaves at the end of each growing season. Hardwoods are widely used for cabinet and furniture making,

Figure 15-3. Logs are shipped to the mill on trucks or floated down rivers. Lumber production starts when the logs arrive at the mill. (Boise Cascade)

for making shipping pallets, and for manufacturing household decorations and utensils. Hardwood lumber is produced in standard thicknesses. These range from 5/8″ thick rough to 1″ thick rough boards, known as 4/4 (pronounced "four-quarter"), to as large as 4″ thick rough boards, or 16/4. The boards are available in random widths and lengths. Hardwoods are usually not cut to specific widths and lengths like softwood lumber is.

The largest quantity of lumber is produced from softwood trees. Therefore, let us look at how lumber is manufactured from logs.

In the previous chapter, you learned how trees are harvested and shipped to the mill, **Figure 15-3.** At the mill, logs are stored in ponds to prevent checking (cracking) and to protect them from insect damage. These logs are the material input for lumber manufacturing.

We follow some basic steps when we change the logs from a natural resource into the industrial material of lumber. These steps, as shown in **Figure 15-4,** include:

1. The log is removed from the pond and cut to a standard length. This length is established to give the mill a uniform input and maximum yield from the log.
2. The log is debarked. In this process the operator might use mechanical trimmers or high-pressure water jets to remove the bark. The bark is a by-product that can be used as fuel for the mill or sold as landscaping mulch.
3. The log is cut into boards and cants at the head rig. A ***head rig*** is a very large band saw that cuts narrow slabs from the log. When the square center section (called a ***cant***) remains, a decision is made. The cant might remain at the head rig to be cut into thick boards, or it might move to the next step.

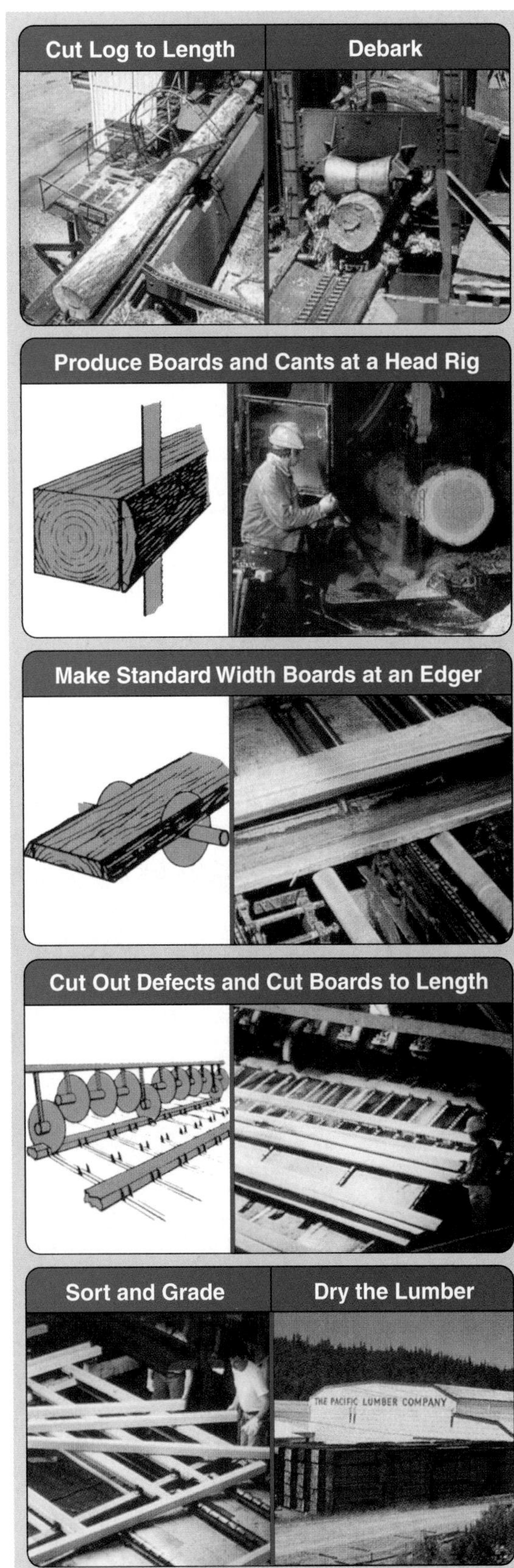

Figure 15-4. We follow certain steps when we change logs into lumber.

4. The cant is cut into thin boards at a ***resaw.*** This machine is a group of circular or scroll saw-type blades evenly spaced to cut many boards at once. Small logs often move directly from the debarker to the gang saw, bypassing the head rig.
5. The boards are cut to standard widths at an ***edger saw.*** This machine has a number of blades on a shaft. The blades can be adjusted at various locations to produce standard widths from 2″ to more than 12″.
6. The edged boards are cut to standard lengths at a ***trim saw.*** This machine has a series of blades spaced 2′ apart. The operator can actuate any or all the blades. This allows for cutting out defects and producing standard lengths of lumber. The boards may be 6′, 8′, 10′, 12′, 14′, or 16′ long. All the blades cut low-quality boards. This cuts the board into 2′-long scrap pieces. The scrap from the edgers and trim saws is used as fuel for the sawmill or becomes the raw material for board products and paper.

The processed lumber then moves onto the green chain. The boards move down a conveyor and are inspected and sorted by quality. The lumber is dried to make it a more stable product. The two kinds of drying are air and kiln (oven) drying.

The dried lumber is shipped as an industrial material. Some lumber receives special processing. Short boards can be processed into longer boards, **Figure 15-5.** Special finger joints are cut in the ends of the boards. The boards are then glued together to form a continuous ribbon of lumber. The ribbon is cut into standard lengths as it leaves a glue-curing machine.

Figure 15-5. These operations are changing short lengths of redwood lumber into long boards. Later, they are made into molding for homes.

Producing Plywood

Another common forest product is plywood. Plywood is a composite material made up of several layers, **Figure 15-6.** Plywood is more stable than solid lumber because cross-grained layers in the plywood reduce warping and expansion.

The outside layers are called ***faces.*** Between the faces are layers called crossbands. The grain of the ***crossbands*** is at a right angle (90°) to the face grain. The layer in the center is called the ***core.*** Its grain is parallel with the face grain. Plywood with only three layers does not have crossbands. Three-layer plywood has a core with its grain running at a right angle to the face layers.

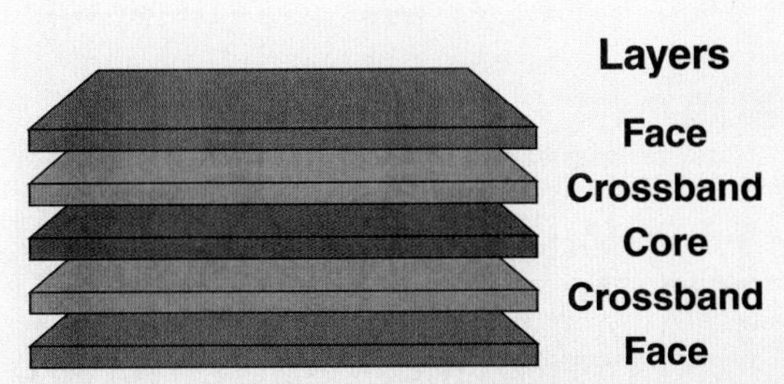

Figure 15-6. Plywood is made of layers of wood that are glued together under pressure. The number and the thickness of the layers are varied to make different types of plywood.

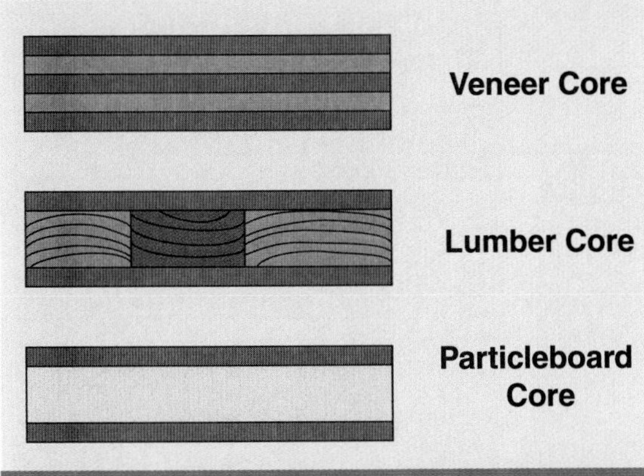

Figure 15-7. Plywood is available with three types of cores: veneer, lumber, and particleboard.

Three types of cores are used for plywood, **Figure 15-7.** The most common is ***veneer core plywood.*** A ***veneer*** is a thin sheet of wood that is sliced, sawed, or peeled from a log. Plywood used for cabinetwork and furniture usually has a lumber core or a particleboard core. ***Lumber core plywood*** has a core made from pieces of solid lumber that have been glued up to form a sheet. ***Particleboard core plywood*** has a core made of particleboard. Particleboard is made up of wood chips that are glued together under heat and pressure.

Plywood is typically available in 4′ x 8′sheets. Thicknesses from 1/8″ to 3/4″ are available.

Veneer core plywood is produced in two stages. The first stage makes the veneer. The veneer is sliced or peeled from the log and moves through a dryer. The dried veneer is sheared into workable size pieces. Defects are cut out or are patched.

Now the veneer is ready for the second stage: plywood production. Glue is applied between the layers. The layers of veneer are stacked up. Then, the sheet is placed in a heated press. The press is closed, and the glue cures under heat and pressure.

After pressing, the sheet is removed, trimmed to size, and sanded. The completed sheets are inspected and loaded for shipment. Look at **Figure 15-8.** This series of photos shows the steps used in manufacturing common plywood.

Thermal Processes

Many industrial materials are produced by processes that use heat to melt and reform a natural resource. These types of processes are called ***thermal processes.*** The thermal process is widely used to extract metals from their ores. This process is often called ***smelting.*** Other thermal processes are used to make glass and cement. For this discussion we will look at steel and glassmaking. Both of these use a

Log Is Loaded in a Veneer Lathe

Ribbon of Veneer Is Cut from Log

Veneer Is Cut and Spliced

Defects Are Patched

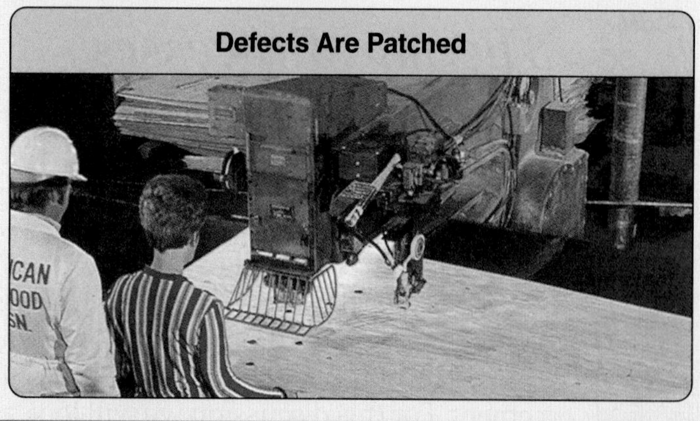

Veneer Is Glued and Pressed into a Sheet

Sheets Are Cut to Size and Sanded to Thickness

Finished Sheets Are Shipped to Customers

Figure 15-8. We follow certain steps in plywood manufacture.

Connections to Technology: **Mathematics**

Calculating Board Footage

As mentioned in this chapter, hardwood lumber is usually not cut to specific widths and lengths. Instead, this type of lumber is sold by volume. The unit of measurement for the volume of lumber is the board foot.

We generally use one of the two following formulas to calculate board footage:

(1) If the length is in feet:

$$\frac{\text{Length (in feet)} \times \text{Width (in inches)} \times \text{Thickness (in inches)}}{12}$$

(2) If the length is in inches:

$$\frac{\text{Length (in inches)} \times \text{Width (in inches)} \times \text{Thickness (in inches)}}{144}$$

Thus if a piece of lumber is 4′ long, 6″ wide, and 1″ thick, the board footage would be calculated using formula (1):

$$\frac{4' \times 6'' \times 1''}{12} = \frac{24}{12} = 2$$

The piece of lumber is thus 2 board feet.

What would be the board footage for a piece of lumber that is 96″ long, 12″ wide, and 2″ thick?

combination of thermal and chemical processes. The thermal energy melts the materials. During the melting process, chemical reactions take place to produce a new material. The new material is then shaped into standard stock. The shaping process uses mechanical techniques. The material is cast, drawn, rolled, or squeezed into new sizes and shapes.

Producing Steel

Steel is an alloy, a mixture of iron and carbon. Adding other elements in small amounts gives the steel specific qualities. These elements include manganese, silicon, nickel, chromium, tungsten, and molybdenum. Steel with nickel and chromium is called stainless steel. Adding molybdenum increases the hardness of the steel. Molybdenum steels are widely used for tools. Adding tungsten makes the steel more heat resistant.

The making of steel requires three steps, as shown in **Figure 15-9.**

- Iron smelting
- Steel making
- Steel finishing

Iron smelting produces pig iron, the basic input for steel making. ***Pig iron*** results from thermal and chemical actions that take place in a ***blast furnace.*** The blast furnace uses a continuous process. At no time is the blast furnace empty during the smelting process. A blast furnace operates 24 hours a day, 365 days a year. Every so often raw materials are added to the top and molten pig iron is removed from the bottom.

The operation of a blast furnace is simple. Alternating layers of iron ore, coke (coal with the impurities burnt out), and limestone are added to the furnace. The blast furnace is charged with four parts iron ore, two parts coke, and one part limestone.

Very hot air is blown into the bottom of the furnace. The coke burns, causing the iron ore to melt. The oxygen in the iron ore combines with the carbon to make carbon monoxide gas. During the melting, limestone joins with impurities to form slag. The slag floats on the molten iron and can

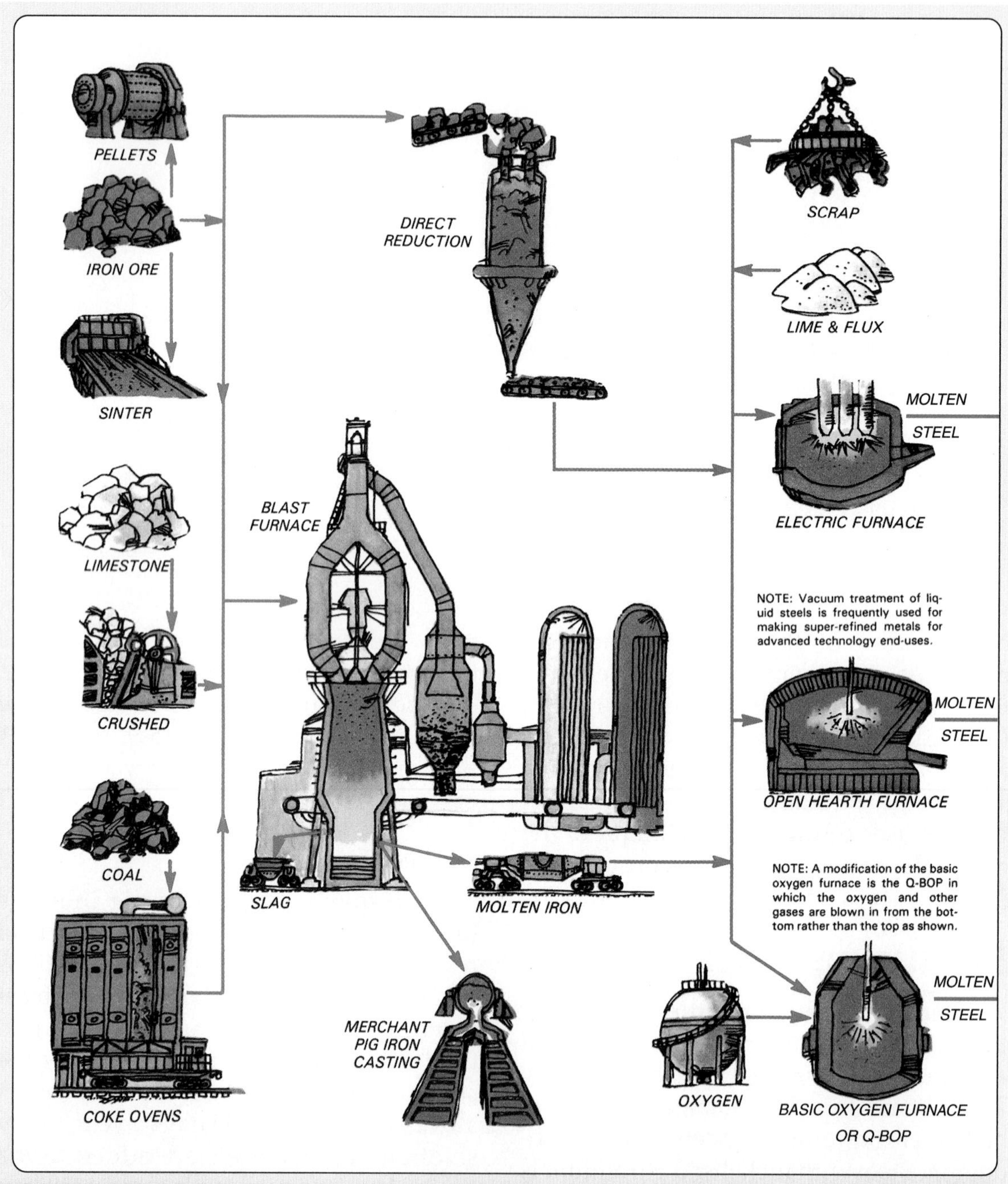

Figure 15-9. This illustration shows the steps it takes to turn iron ore into steel products. The three main stages are smelting, steelmaking, and steel finishing. (American Iron and Steel Institute).

be drawn off. This leaves molten iron with carbon dissolved in it. This material is called pig iron, which is iron with 3 to 4.5 percent carbon. Pig iron also has 1 to 2 percent other elements, including manganese, silicon, sulfur, and phosphorus.

The steel-making process starts with the pig iron produced in the iron-smelting step. Steel making actually removes some carbon from the iron. Heat and oxygen are used to take some of the carbon out of molten pig iron.

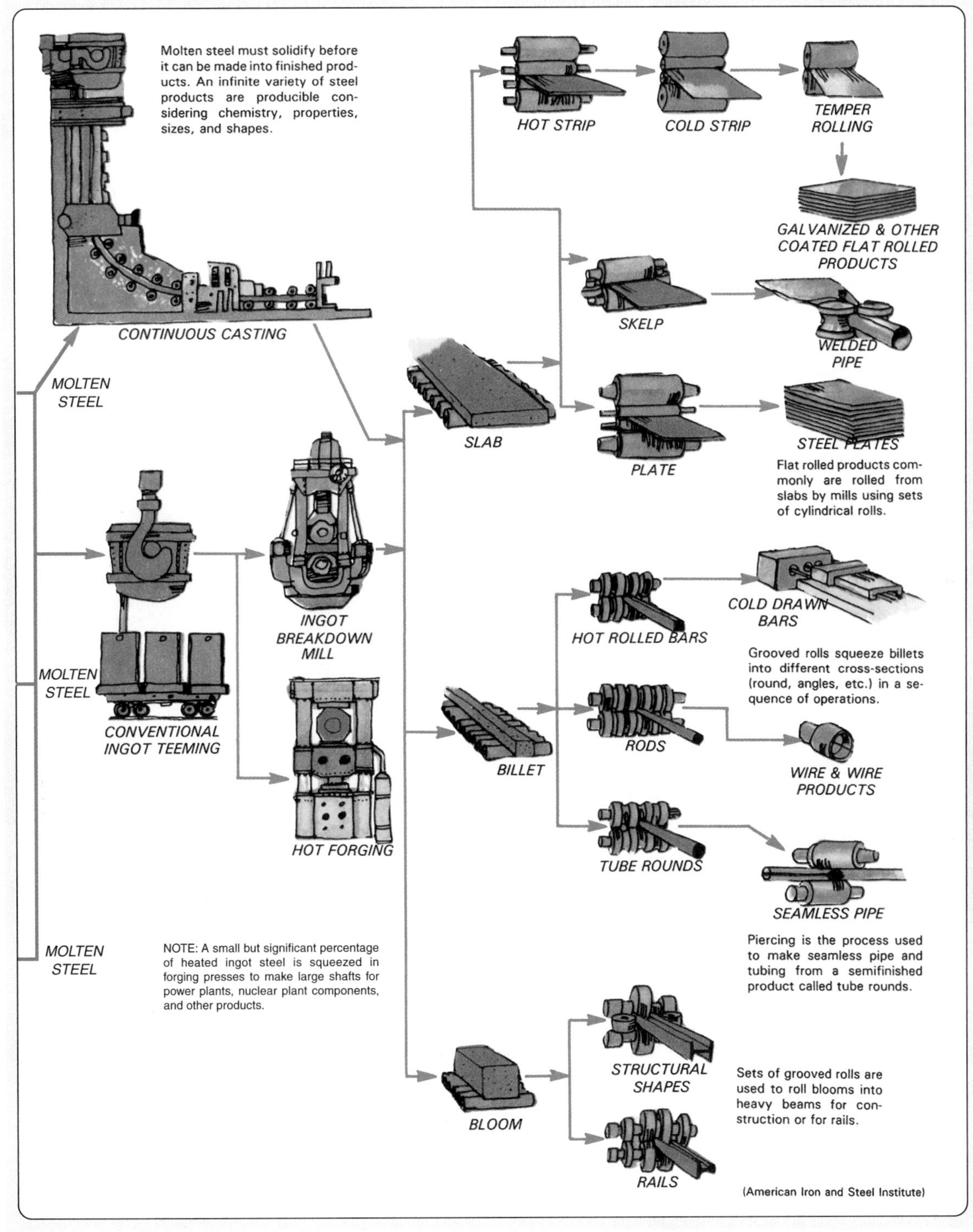

The most common steel-making process uses the basic oxygen furnace. Making steel in a basic oxygen furnace involves three steps. The first is ***charging.*** The furnace tilts to one side to receive pig iron, scrap steel, and flux, a material that combines with impurities to form slag. This charge provides the basic ingredients for steel.

The second step is ***refining.*** The furnace moves into an upright position, and the charge is melted. Then, a water-cooled oxygen lance is placed above the molten material. Pure oxygen is forced out of the lance into the iron at supersonic speeds. The oxygen causes the part of the carbon to burn away, producing steel and slag.

The final step is ***tapping,*** **Figure 15-10.** The floating slag is skimmed off the melt. The entire furnace tips to one side, and the steel is poured out. It is now ready to enter the steel finishing cycle.

Steel finishing changes molten steel into sheets, plates, rods, beams, and bars. The first step involves pouring the steel into ingots or into the head end of a continuous caster. A continuous caster solidifies the molten steel into shapes called slabs, billets, and blooms. ***Slabs*** are wide, flat pieces of steel, **Figure 15-11.** Sheets, plates, and skelp are produced from them. Sheets are wide, thin strips of steel, whereas plates are thicker. ***Skelps*** are strips of steel that are used to form pipe.

Billets are square, long pieces of steel. Bars and rods are produced from them. ***Blooms*** are short, rectangular pieces used to produce structural shapes and rails.

Steel is prone to rust if it is left exposed to the atmosphere. Therefore, many steel shapes are finished. They may receive a zinc or tin coating, **Figure 15-12.** Zinc-coated steel is called ***galvanized steel.*** It is used for such automobile parts as fenders and for such containers as buckets and trash cans. Tin-coated steel is called ***tin plate.*** Tin plate is widely used to make food cans.

Figure 15-10. Here a steel furnace is being tapped. The furnace is tilted and the steel pours out. (American Iron and Steel Institute)

Figure 15-11. These hot slabs are waiting to be rolled into sheets or plates. (Inland Steel Company)

Producing Glass

Glass is another material that is produced using thermal processes. ***Glass*** is made by solidifying molten silica in an amorphous state. An amorphous material has no internal structure like the regular, uniform lattice structures of metals.

Glass is made from sand (silica), soda ash (sodium carbonate), and lime (from limestone). These ingredients are weighed and mixed to form a batch. For sheet glass the mixture contains about 70 percent silica, 13 percent lime, and 12 percent soda.

The batch is moved into a melting furnace. A typical furnace for flat glass could be 30′ (9 m) wide, 165′ (50 m) long, and 4′(1.2 m) deep. It holds about 1200 to 1500 tons (1100 to 1400 metric tons) of glass at one time. The melting end of the furnace reaches temperatures of 2880°F (1580°C). This heat causes the material to melt and flow together.

Figure 15-12. This massive machine applies a tin coating to steel to make tin plate. (American Iron and Steel Institute)

From the furnace the glass may go to a secondary manufacturing process. Products such as jars, bottles, dishes, glasses, and cookware are manufactured using casting and forming techniques, **Figure 15-13.**

Primary manufacturing lines may change the glass into sheets for windows and similar products, **Figure 15-14.** Most of this glass is called ***float glass.*** Float glass is formed by floating the molten glass on a bed of molten tin. The glass flows out of the furnace onto the tin. The glass forms a ribbon as it cools and moves toward the end of the float tank. A typical float tank is 150′ (46 m) long and can form glass 160″(4 m) wide.

The formed glass moves from the float tank to an annealing oven, called a ***lehr.*** The temperatures in the lehr start at 1200°F (650°C) and fall to 400°F (200°C). This gradual temperature drop relieves internal stresses in the glass. From the lehr the ribbon of glass is cut and packed for shipment.

Other glass products are ***fiberglass*** and ***fiber-optic cables.*** These strands of glass are formed, cooled, and annealed. Fiberglass is used as the matrix for composite materials

Figure 15-13. Large quantities of glass are processed into food containers such as these jars. (Owens-Brockway)

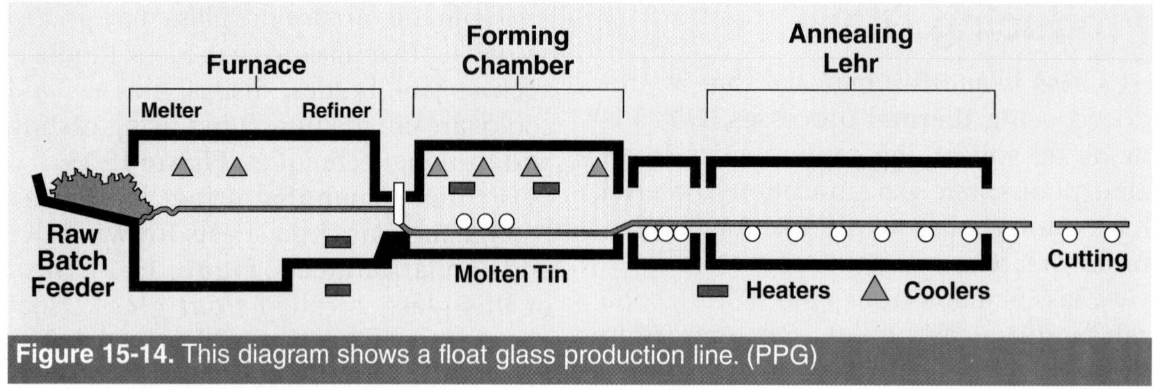

Figure 15-14. This diagram shows a float glass production line. (PPG)

and for insulation. Fiber-optic cables are used to transmit voice, television, and computer data at high speeds.

Refining Petroleum

Petroleum is another resource that is processed using thermal actions. As you learned earlier, ***petroleum*** is not a uniform material. Instead, it is a mixture of a large number of different hydrocarbons. The process used to isolate these hydrocarbons is called refining.

Most petroleum refineries use three processes: separation, conversion, and treating. ***Separation*** does exactly what its name says. Separation breaks petroleum into major hydrocarbon groups. The process used is called ***fractional distillation.*** The petroleum is pumped through a series of tubes in a furnace. There it is heated to about 725°F (385°C). The petroleum becomes a series of hot liquids and vapors. They pass into the bottom of a ***fractionating tower,*** **Figure 15-15.**

This tower can be a tall as 100′ (30 m). Within the tower are a series of pans or trays that can hold several inches of liquid. Each pan and its fluid are maintained at a specific temperature. The higher the level in the tower, the lower the temperature.

The hot vapors coming from the furnace rise in the tower. As they rise they are forced to bubble through the liquid in the pans. This action cools the vapors to the temperature of that pan. Hydrocarbons with a boiling point at or below the pan's temperature will condense and stay in the pan. Other vapors will continue to rise. The condensed liquids are continuously drained from the trays.

Typically, the lighter fractions (products) are taken off the top. These are gasoline and gases such as propane and butane. Other fractions in descending order of temperature are jet fuel, kerosene, diesel, fuel oil, and asphalt. **Figure 15-16** shows a diagram of two distillation towers.

The process of ***conversion*** changes hydrocarbon molecules into different sizes, both smaller and larger. For example, heavier hydrocarbons may be broken into smaller ones. This process is called ***cracking.*** Thermal cracking heats heavier oils in a pressurized chamber, **Figure 15-17.** The heat and pressure cause the hydrocarbon molecules to break into smaller ones. Catalytic cracking does the same by using a chemical called a catalyst. A catalyst is a chemical that helps a reaction to take place. The catalyst is not used up during the reaction. Both catalytic and thermal cracking are used to increase the amount and the quality of products produced from a barrel of petroleum.

A second conversion process is ***polymerization.*** It is the opposite of cracking. Polymerization causes small hydrocarbon molecules to join together. Refinery gases are subjected to high pressures and temperatures in the presence of a catalyst.

Figure 15-15. These fractionating towers are in Saudi Arabia. (American Petroleum Institute)

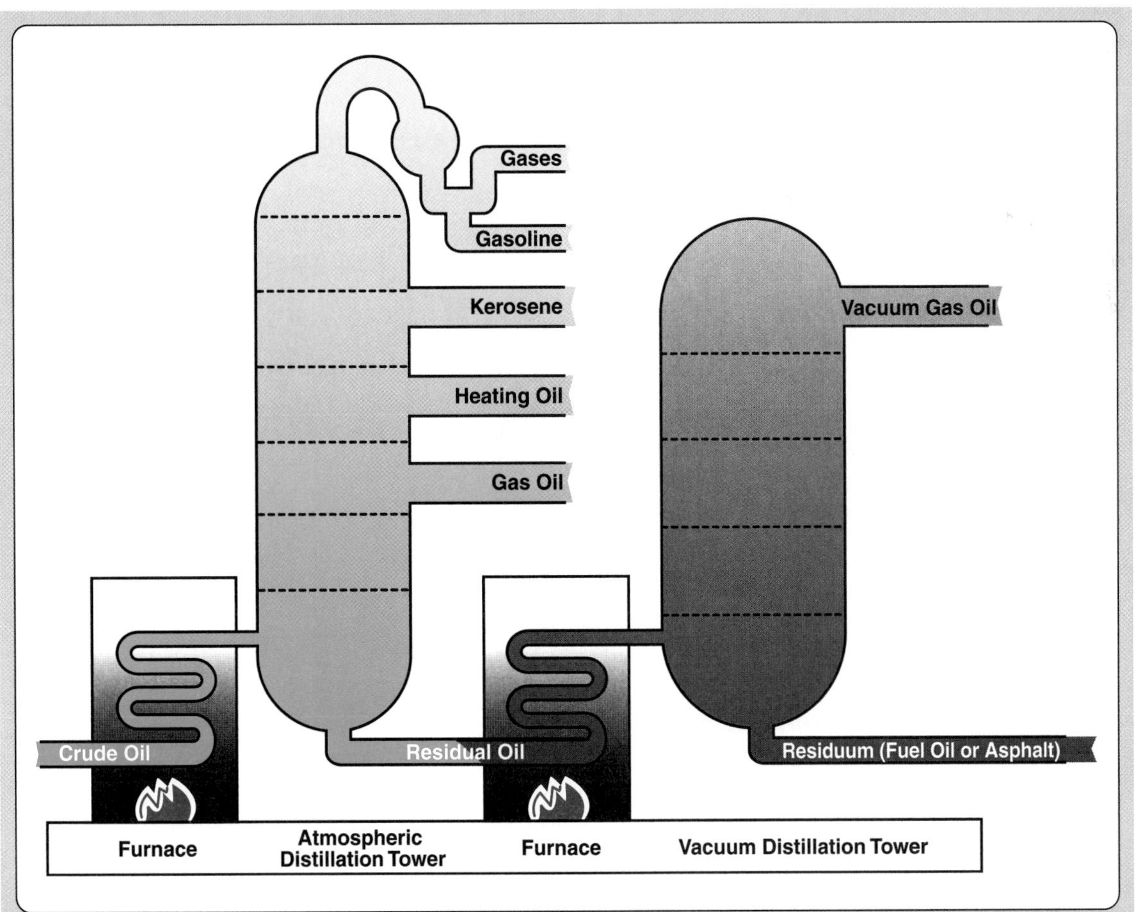

Figure 15-16. This diagram shows a two-stage distillation unit. (Chevron-USA)

They unite (polymerize) to form hydrocarbon liquids. This increases the yield of petroleum products.

Treating is the third petroleum-refining process. Treating adds or removes chemicals to change the properties of petroleum products. Sulfur may be removed so kerosene will burn cleaner and smell better. Additives improve the lubrication properties of oils. Other additives help fuels burn quickly and cleanly.

Chemical and Electrochemical Processes

Some primary processes use ***chemical*** or ***electrochemical*** processes. These are used to produce synthetic fibers, pharmaceuticals, plastics, and other valuable products. These processes break down or build up materials by changing their chemical composition. The most common chemical and electrochemical process in use is the one to make aluminum, which will be the focus of this section.

The process of making aluminum is carried out in two stages, **Figure 15-18.** The first stage chemically changes aluminum ore, known as ***bauxite,*** into aluminum oxide or alumina. The bauxite is crushed and mixed with a caustic soda, sodium hydroxide. The soda dissolves the aluminum oxide, forming a sodium aluminate solution that is called green liquor. This process leaves behind a residue containing iron, silicon, and titanium. This sludge is called red mud.

The sodium aluminate liquid is pumped into a digester. Some aluminum trihydrate, an aluminum oxide/water compound, is added to start or seed the process. The mixture is agitated with compressed air and cooled. During the process, a chemical

Figure 15-17. This cracking facility in Venezuela processes more than 50,000 barrels of heavy oil a day. (American Petroleum Institute)

Technology Explained

Nuclear Energy

nuclear energy: The energy holding atoms together, which is released by using controlled reactions.

Humans have always marveled at the Sun. It is a giant ball of hydrogen gas. The Sun's intense heat warms our planet and creates Earth's weather patterns. Reactions that turn millions of tons of hydrogen into helium every second create this heat. This process is called *nuclear fusion.*

Fusion combines the nuclei (the centers of atoms) of two atoms with low atomic numbers. It produces heavier nuclei and releases neutrons and great amounts of energy.

The most important fusion process unites atoms of hydrogen, **Figure A.** This element is the lightest of all. Most hydrogen atoms contain one proton in the nucleus and one orbiting electron. Other forms of hydrogen, however, do exist. These forms are called *isotopes.* The two isotopes used in fusion are deuterium and tritium. Every atom of deuterium has one neutron in its nucleus, while each tritium atom has two neutrons. Fusion of these two isotopes creates one helium nucleus. Helium, in its normal condition, has two protons and two neutrons. One neutron is left over. Also, over 17 million electron volts of energy are released.

Creating a controlled fusion reaction is very difficult. The atoms are very small, and the nuclei carry positive charges. Since like charges repel, the nuclei resist coming close together. Therefore, they must be accelerated to high speeds to overcome the repelling forces. Also, the extra neutron causes a serious problem. These particles can break free and damage the vessel in which the reaction is taking place.

Fusion is not the only way to use the energy of the atom. A second method is called *fission,* **Figure B.** This process breaks down radioactive materials into smaller molecules. Fission releases some of the energy holding the atom together.

The most common material used in fission is uranium. Uranium has three main isotopes: U-238, U-235, and U-234. U-235 is the isotope used in nuclear reactors. When a U-235 molecule is bombarded with neutrons, the nucleus breaks apart. It forms two new molecules. The extreme heat released is used to create steam, which turns turbines in electric generating plants. Fission reactors also provide power on submarines and ships.

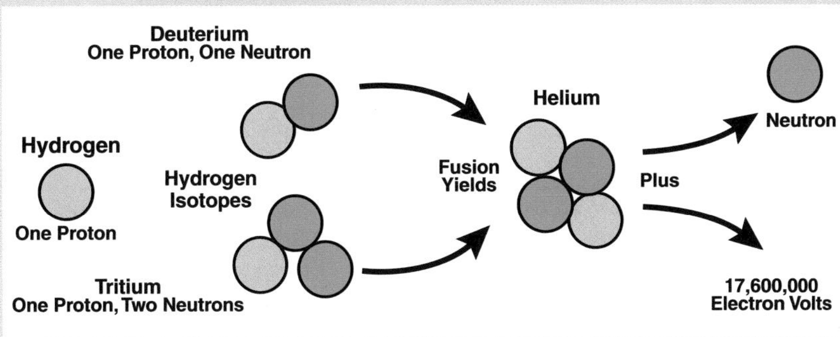

Figure A. Nuclear fusion forces the nuclei of atoms together. The extra neutron can be a problem to contain.

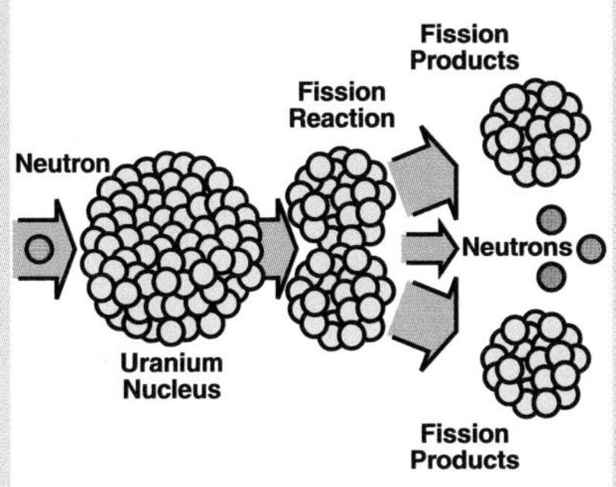

Figure B. Nuclear fission breaks apart atoms of radioactive materials. The waste products from fission reactors are, however, very harmful to living things.

action takes place. Aluminum trihydrate and caustic soda are formed. The aluminum compound settles out of the solution. It is removed and dried in a kiln. The drying drives the water out of the aluminum trihydrate, leaving pure aluminum oxide. This material is called ***alumina*** and is the input to the second phase of making aluminum.

In the second stage, alumina from stage one is converted into pure aluminum. This process takes place in large electrolytic cells called pots. The cell has a carbon lining, which is the cathode. The anode is also made of carbon.

The pot is partially filled with an electrolyte, a material that will conduct electricity. This material is a mixture of molten aluminum fluoride (a salt) and molten cryolite. The electrolyte is kept at 1650°F (900°C). The alumina is dissolved in the electrolyte. Large quantities of electrical current are passed between the anode and cathode. The electrical energy causes the alumina to break into pure aluminum and oxygen. The aluminum settles to the bottom of the pot, where it is gathered, purified, and cast into ingots.

The ingots are then formed into sheets, bars, rods, or other shapes. These become industrial materials used for numerous secondary manufacturing processes.

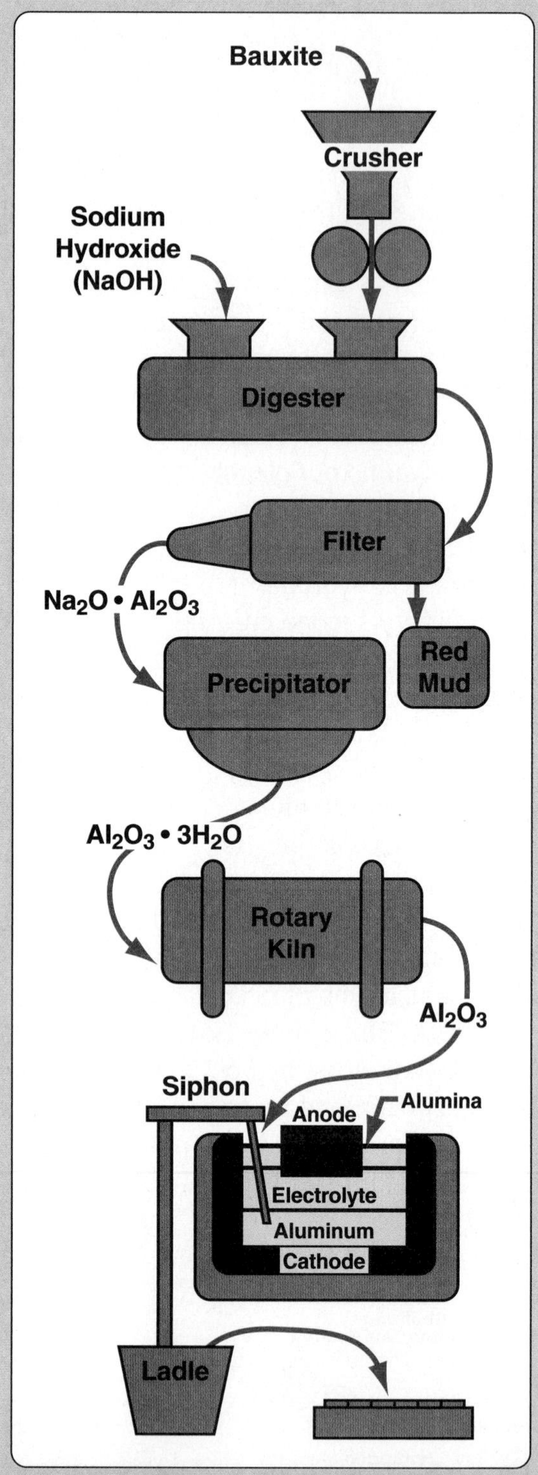

Figure 15-18. This diagram shows the stages of aluminum production from bauxite to pure aluminum ingots.

Summary

Primary manufacturing processes change natural resources into industrial materials. They accomplish this through mechanical, thermal, and chemical and electrochemical means. Mechanical processes change the form of natural resources through the use of mechanical forces. Thermal processes use heat, and chemical and electrochemical processes change materials by changing the chemical composition. The resulting materials are used to manufacture products and construct structures.

Key Terms

alumina
bauxite
billets
blast furnace
blooms
cant
charging
chemical processes
conversion
core
cracking
crossbands
edger saw
electrochemical processes
faces
fiberglass
fiber-optic cables
float glass
fractional distillation
fractionating tower
galvanized steel
glass
hardwood lumber
head rig
lehr
lumber
lumber core plywood
mechanical processes
particleboard core plywood
petroleum
pig iron
polymerization
primary processing
refining
resaw
separation
skelps
slabs
smelting
softwood lumber
steel
tapping
thermal processes
tin plate
treating
trim saw
veneer
veneer core plywood

Test Your Knowledge

Write your answers on a separate piece of paper. Please do not write in this book.

1. List the three types of primary processes.
2. What kind of primary processing produces lumber?
3. *True or false?* Most lumber comes from softwood trees.
4. List the three types of cores that are used in plywood manufacture.
5. What kind of primary processing changes iron ore into steel products?

6. *True or false?* Steel is an alloy.
7. Pig iron is produced in a(n) _____ furnace.
8. Zinc-coated steel is called _____ steel.
9. *True or false?* Glass is produced by electrochemical means.
10. What is float glass?
11. List the three processes most petroleum refineries use.
12. What is fractional distillation?
13. What kind of primary processing produces aluminum?
14. *True or false?* Bauxite is the raw material for producing aluminum.
15. What is alumina?

Applying Your Knowledge

1. Build a scale model of a plant to produce an industrial material from a natural resource.
2. Select a material that was not discussed in this chapter. Examples are paper, hardboard, copper, gold, nylon, or polyethylene. Research the processes used to produce the material you choose. Report your results in a chart similar to the one shown below.

Material:
Primary natural resource used to make the material:
Primary production process (steps):
Uses for the material:

3. Develop a list of materials that you see used in the room you are in. List the type of process used to produce each one.

Career Corner
Millwrights

Millwrights install and repair machinery used in many industries. They unload, inspect, and move new equipment into position. Millwrights assemble new machines by leveling the bases, fitting bearings, aligning gears and wheels, attaching motors, and connecting drive belts.

Also, millwrights work with maintenance workers to repair and maintain machines and processing equipment. Millwrights generally are trained through four-year apprenticeship programs, which combine on-the-job training with classroom instruction. Most employers want trainees with high school diplomas and some vocational training or experience.

Chapter 16
Manufacturing Products

Learning Objectives

After studying this chapter, you will be able to:

- List the basic steps involved with casting and molding.
- Identify methods used in casting and molding processes.
- Name safety rules involved with casting processes.
- Identify the basic concepts of forming actions.
- Name the types of separating processes.
- Identify methods used in separating processes.
- Name safety rules involved with using machine tools.
- List the three types of conditioning processes.
- Identify methods used in conditioning processes.
- Identify methods used in assembling processes.
- Identify methods used in finishing processes.
- Name safety rules involved in finishing processes.

In the last chapter you learned that primary processing produces industrial materials. These materials have little worth to the average person. What can you do with a sheet of steel, a 2 × 4 stud, a pound of polypropylene pellets, or an ingot of pure aluminum? These materials must be changed into products before they are useful to you. We use the term ***secondary manufacturing processes*** to describe the actions used to change industrial materials into products, **Figure 16-1.**

Types of Manufacturing Processes

Thousands of manufacturing processes exist. They are used to change the size and shape of materials, to fasten materials together, to give materials desired properties, and to coat the surfaces of products. It would be very difficult to study and comprehend all the individual ways to

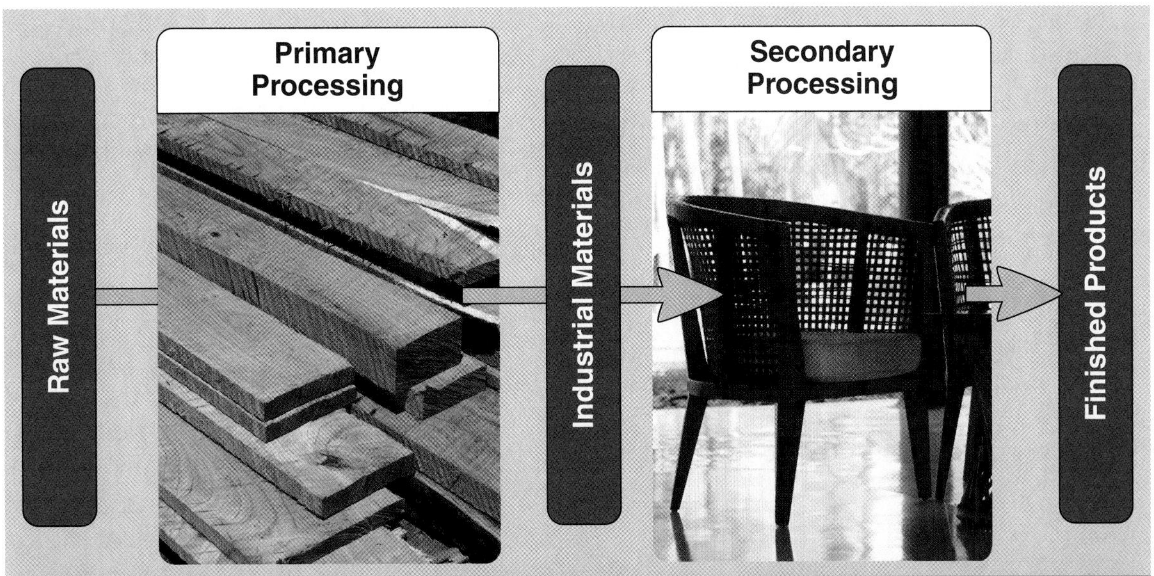

Figure 16-1. Secondary manufacturing processes turn industrial materials, such as lumber, into useful products, such as chairs.

process materials. Instead, we can understand secondary manufacturing processes more easily when we classify them into six groups, **Figure 16-2:**

- Casting and molding
- Forming
- Separating
- Conditioning
- Assembling
- Finishing

Each process has actions or concepts common to all the other processes in its group. Within these groups are specific processes or techniques. Each one differs in some way from the other processes in the group. In other words, these groups are like the members of a family. They do many things alike and look alike, but each member is unique.

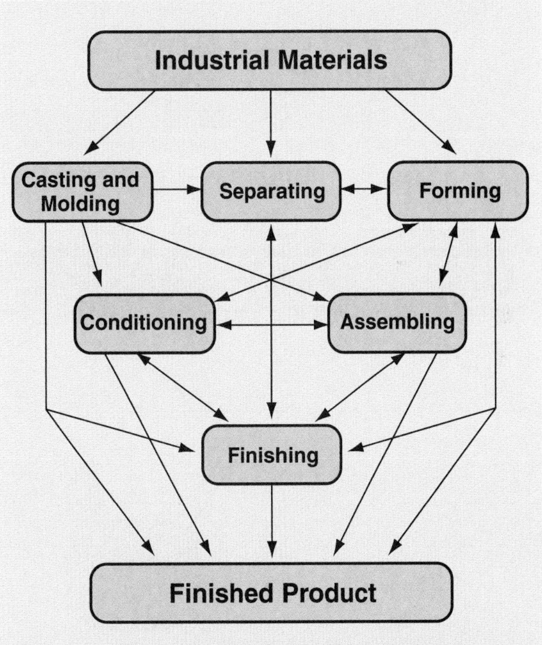

Figure 16-2. The six groups of manufacturing processes are all related.

Casting and Molding Processes

Three groups of processes give size and shape to pieces of material, **Figure 16-3.** The first of these is casting and molding. ***Casting and molding processes*** give materials shape by introducing a liquid material into a mold. The mold has a cavity of the desired size and shape. The liquid material is poured or forced into the mold. The material is allowed to solidify before being removed.

In this discussion, we will refer to molten and fluid materials as *liquids.* Molten refers to materials that are heated to a fluid state. These materials are normally solid at room temperature. Fluid materials, such as water and casting plastics, are liquid at room temperature.

All casting and molding processes involve five basic steps, which are shown in **Figure 16-4.** These steps are the following:

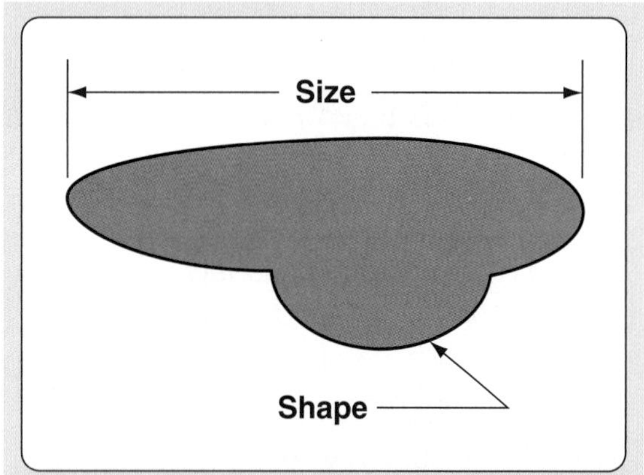

Figure 16-3. Size is the dimensions of an object, and shape is its form and outline.

- Producing a mold of the proper size and shape
- Preparing the material
- Introducing the material into the mold
- Solidifying the material
- Extracting the product from the mold

Producing a Mold

All casting and molding processes require a mold to hold the liquid material until it becomes solid. Inside the mold is a cavity of the proper size and shape. Channels, called gates, guide the liquid into the cavity. Two major types of molds are used in casting and molding processes. These are expendable molds and permanent molds.

Expendable Molds

Most cast products are made in molds that are used once. The mold is destroyed to remove the cast item. These molds are called ***expendable molds.*** Molds can have two or more parts. They are generally made in two steps. First, a pattern of the same shape as, but slightly larger than, the finished product is made. The extra size allows for shrinkage. Most materials

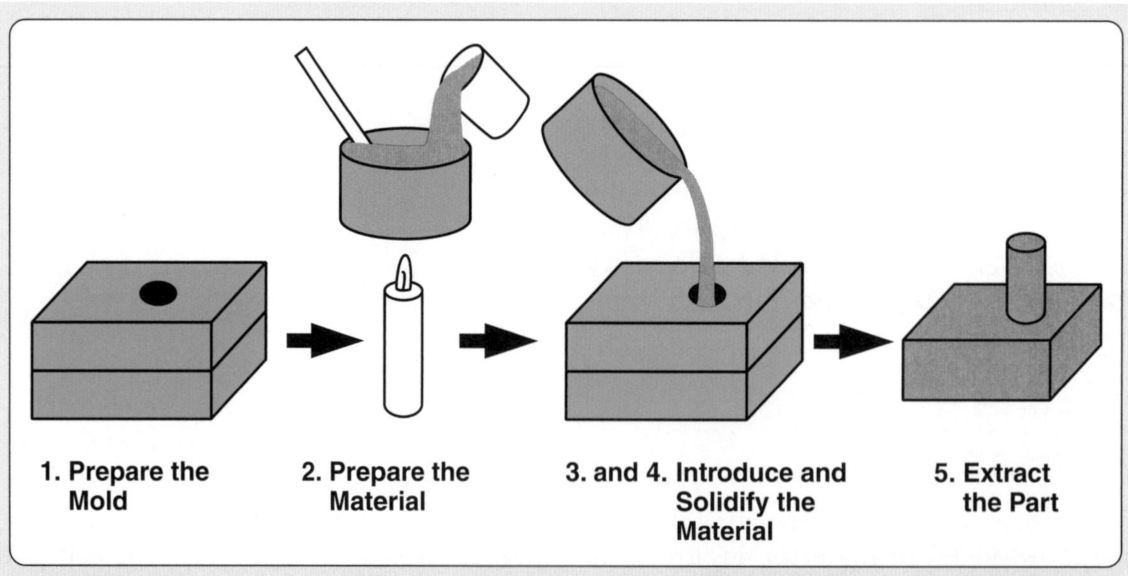

Figure 16-4. All casting and molding processes involve five basic steps.

shrink when they change from a liquid to a solid. An exception is water, which expands as it changes from a liquid to a solid.

The pattern is the foundation for making an expendable mold. It is surrounded with an inexpensive substance, such as sand or plaster. Sand is tamped into place, and plaster is allowed to dry around the pattern. When the mold is completed, the mold parts are separated, and the pattern is removed. This leaves a cavity of the correct shape to be filled by the liquid material. The mold parts are put back together to make a ready-to-use mold, **Figure 16-5.**

Safety in Casting Occupations

People who use casting processes need to observe some basic safety rules:

- Wear protective clothing and goggles when pouring a casting.
- Be sure everything to be used in the casting process is free of moisture.
- Place hot castings where they will not start a fire or burn someone.
- Keep the casting area orderly.
- Do not talk to anyone while pouring.
- Do not stand over the mold when pouring molten metal.

The key to safety in the technology education laboratory is a *safe* attitude.

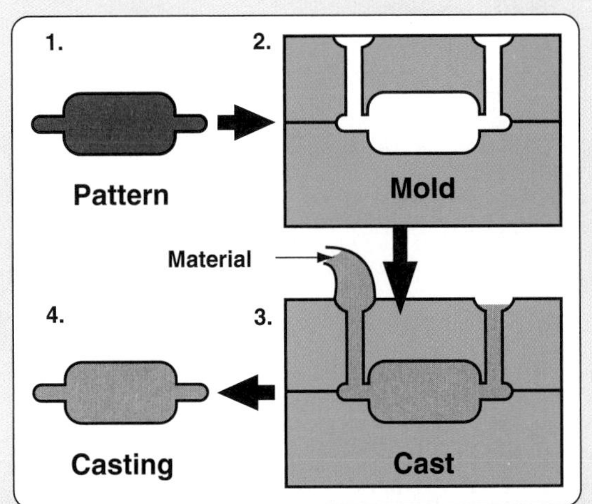

Figure 16-5. Four stages are followed in making an expendable mold casting: 1. Making the pattern. 2. Surrounding the pattern with sand or plaster. 3. Introducing the liquid material and allowing it to solidify. 4. Breaking up the mold and removing the casting.

Permanent Molds

A second type of mold is called a permanent mold. ***Permanent molds*** withstand repeated use. They are often made from steel, aluminum, or plaster, **Figure 16-6.** The mold must withstand temperatures above the melting point of the material being cast. Most metal molds are produced by machining out the cavity. Plaster molds are often produced by casting the mold material around a pattern.

Processes that use permanent molds are die casting, injection molding, and slip casting. Die casting is used to produce aluminum and zinc parts. Injection molding can produce a wide range of plastic parts in many different resins. Slip casting is used to produce clay products. These processes will be discussed later in this chapter.

Preparing the Material

All casting and molding processes require a material in a liquid state. A liquid can flow freely. Three types of liquids

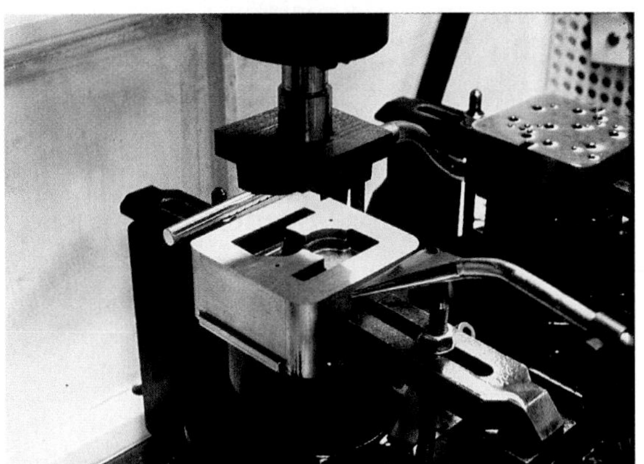

Figure 16-6. This permanent mold has been machined using the electrical discharge machining (EDM) process. (Agie Losome, Switzerland)

are used in casting: solutions, suspensions, and molten materials.

A *solution* is a uniform mixture of two substances. Casting plastics and frozen treats are examples of solutions. They are mixtures that will give desired properties when solid. The act of mixing the parts of a fluid for casting is called ***compounding.***

Suspensions are mixtures in which the particles will settle out. To keep the components mixed, you must shake the suspension. When you stop shaking the mixture, the particles begin to settle out. Slip, which is used in casting ceramics, is a suspension of clay and water.

Some materials, such as metals, must be heated to a *molten* state in order to be cast. At room temperature, steel is a solid and cannot be poured into a mold. Plastics that are injection molded are heated to a liquid state so they can be forced into a mold.

Introducing the Material

A liquid material can be introduced into the mold. Most expendable molding processes use gravity to fill the mold. The material is poured into the mold and allowed to fill the cavity.

Many permanent molding processes use force to introduce the material into the mold. An example of this technique is shown in **Figure 16-7.** This figure depicts the action of an injection-molding machine used with plastic resins. The plastic is heated to a liquid state. A ram then forces the resin into the mold cavity. After the resin cools and solidifies, the mold opens. The finished part is ejected. The mold then closes, and the process repeats. Die casting with metals uses the same procedure in a machine similar to an injection molder.

Solidifying the Material

Once the liquid material is introduced into the mold cavity, it must become a solid. We can make materials solid in three ways:

- Cooling
- Drying
- Chemical action

In cooling, the heat energy that caused the solid material to melt leaves. This makes the material return to its solid state. The mold can be cooled by letting the heat radiate into the air. Water can be pumped through passages in the mold to carry the heat away.

Water-based products are also solidified through cooling. Many food products, such as ice pops and ice cream bars, are given their shape in a mold by cooling. This action we call freezing.

Suspended materials are solidified by allowing the solvent to be absorbed into the mold. This action is caused by ***drying.*** An example of this technique is slip casting, **Figure 16-8.** Slip, a clay and water mixture, is poured into the mold. The plaster mold absorbs water from the slip. The action causes a layer of solid clay to build up on the mold walls. The longer the slip stays in the mold, the more moisture it loses, and

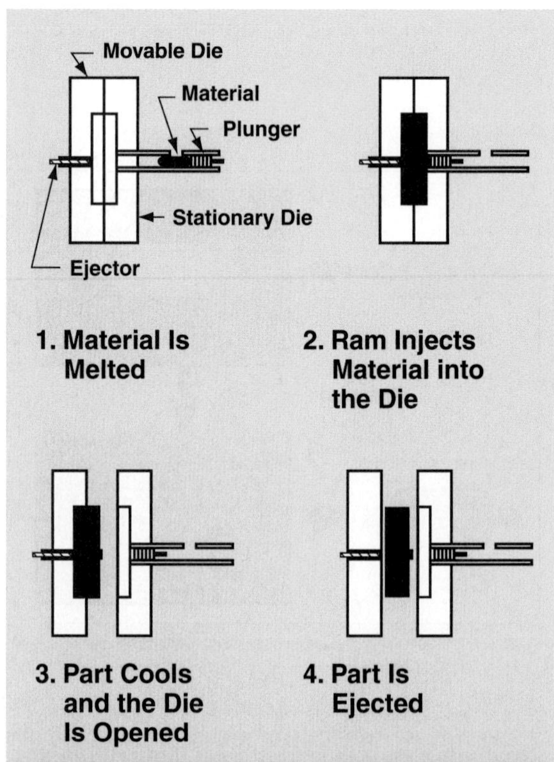

Figure 16-7. This diagram shows the action of an injection-molding machine.

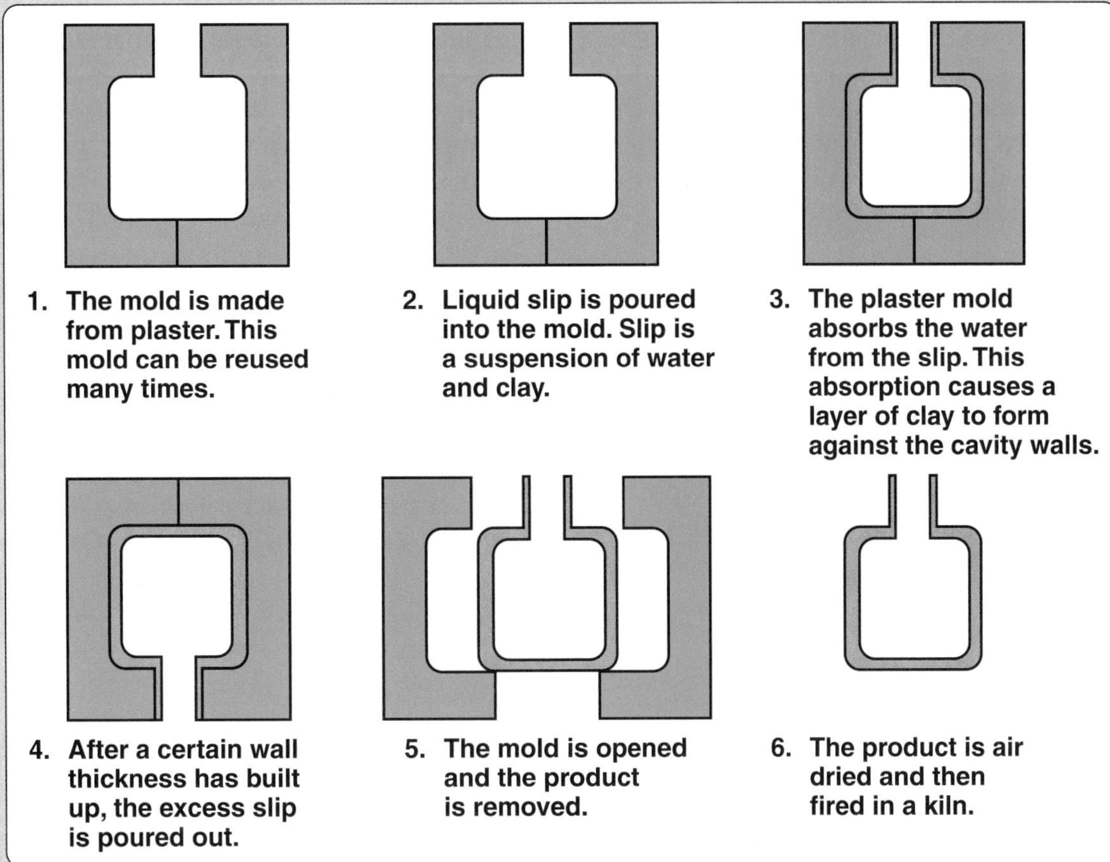

Figure 16-8. Shown here are the steps in making a slip cast product.

the thicker the walls become. When the proper wall thickness is reached, the remaining slip is poured out. The product is allowed to dry partially in the mold. The mold is then opened, and the product is removed. It is placed in a humidity-controlled cabinet for further drying.

The third method of making a material solid is by ***chemical action.*** For example, casting plastics have hardeners added. These chemicals cause the material to become solid, or set up. The short molecules in the plastic resin become longer, more complex molecules. This action, called *polymerization,* changes the liquid plastic into a solid.

Extracting the Product

The solidified product must be removed from the mold, **Figure 16-9.** Expendable molds are destroyed to

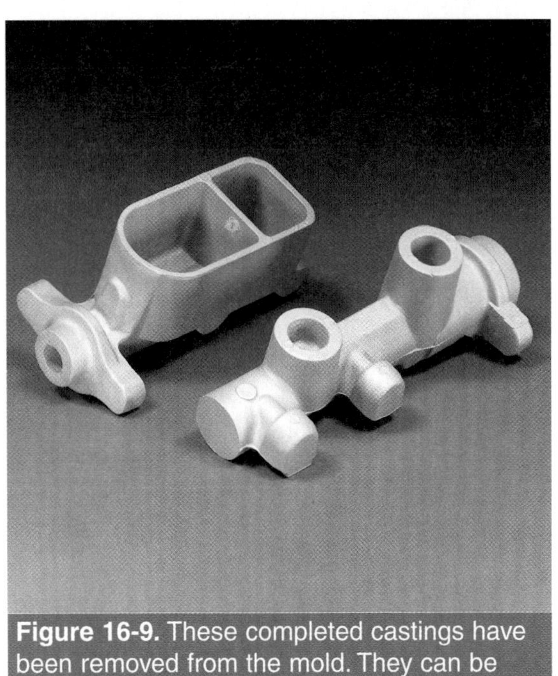

Figure 16-9. These completed castings have been removed from the mold. They can be machined and assembled with other parts to make useful products. (Stahl)

remove the part. Permanent molds are designed for easy opening. Often the molding machine automatically introduces the material, causes it to solidify, and then opens and ejects the finished part.

Forming Processes

Forming is the second family of processes that give materials size and shape. All *forming processes* apply force through a forming device to cause the material to change shape. This force must be in a specific range, above the material's yield point and below its fracture point.

All materials react to outside forces, **Figure 16-10.** Small forces cause a material to flex (bend). When the force is removed, the material returns to its original shape. If the force increases, a point is reached at which the material will not return to its original shape. This point is called the ***yield point.*** The range between rest and the yield point is called the material's ***elastic range.***

Above the yield point, the material will be permanently deformed. The greater the force, the more the material will be stretched, compressed, or bent. This range is called the material's ***plastic range.*** At a certain point, the material cannot withstand any more force, and it breaks. This point is called the ***fracture point.*** All forming processes operate in the plastic range of the material. These processes have three things in common:

- The presence of a forming device
- The application of force
- The consideration of material temperature

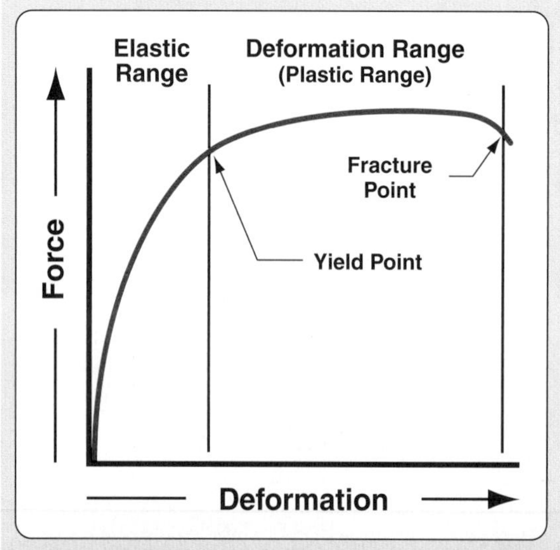

Figure 16-10. This stress-strain chart shows the elastic and plastic range for a material. Note the yield and fracture points.

Forming Devices

Forming processes are used to produce specific shapes in a material. They must have a way to ensure that the shape is correct and consistent. This goal is achieved by using one of two devices:

- Dies
- Rolls

Dies

Dies are forming tools made of hardened steel. They can be used to form any material softer than they are. The three types of dies, shown in **Figure 16-11,** are

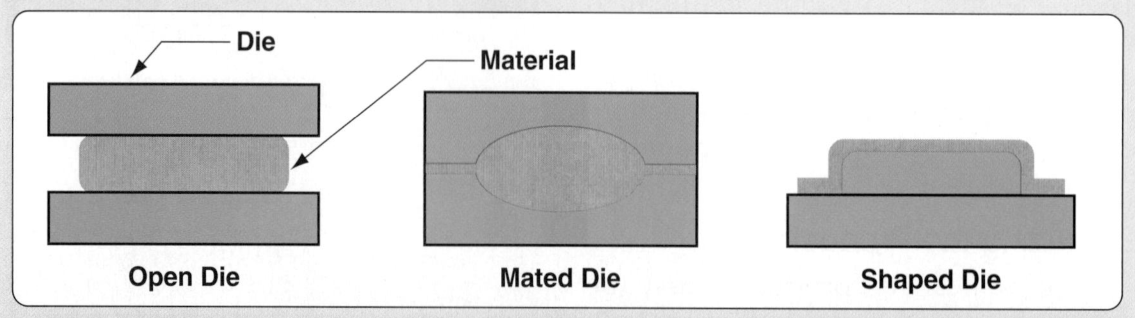

Figure 16-11. The three types of dies are open, mated, and shaped.

open dies, mated dies, and shaped dies. ***Open dies*** are the simplest of all dies. They consist of two flat die halves. The material is placed between the halves, and the dies are closed. This action presses the material into a new shape. ***Mated dies*** have the desired shape machined into one or both halves of the die set. The material is placed between the die halves. The die set is closed, and the material is forced to take on the shape of the die cavities. One-piece ***shaped dies*** are widely used to form plastic objects. A common process using this type of mold is vacuum forming. A sheet of heated plastic material is placed over the mold. The air between the mold and the sheet is sucked out. This allows atmospheric pressure to force the material tightly over the mold. When it cools, the product is removed, **Figure 16-12**.

Figure 16-14. This ice cube tray was molded using the mold shown at the top. (Brush-Wellman)

Rolls

Some processes use rolls to form the material. Two types of rolls exist: smooth and shaped. Smooth rolls are used to make curved shapes from sheets and shapes of metal. Shaped rolls have shapes machined into their surface. When material passes between shaped rolls, the material is squeezed into the shape of the roll. Shaped rolls are used to make pipe, tubing, and corrugated metal.

The Application of Force

The force needed to complete a forming action can be delivered in a number of ways. The four most common are the following:

- Presses
- Hammers
- Rolling machines
- Drawing machines

Presses and hammers are very much alike, **Figure 16-13.** They have a bed onto which a stationary die can be attached. The other die part is attached to a ram. A power unit lifts the ram, and the material to be formed is placed between the die halves. ***Presses***

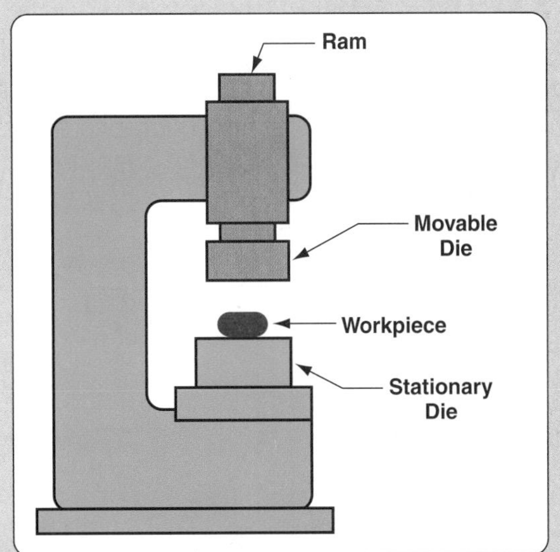

Figure 16-13. This diagram shows the parts of a forging hammer.

Technology Explained

Robots

robot: A programmable part-handling or work-performing device, often used to replace human labor in industrial settings.

Robots are industrial devices used to increase the productivity of operations. They also remove humans from dangerous and undesirable working conditions. The Czech author Karel Capek first coined the word *robot* in the play *Rossum's Universal Robots*. He derived the word *robot* from the Czech word *robota*, which means "work."

Some of the first robots were designed in the 1940s to handle radioactive materials. The first industrial robot was developed in 1962. The robot's functions were limited to picking up an object and setting it down in a new location. This simple type of robot is called a pick-and-place robot. Today, there are many different types of robots, **Figure A.** They are used in a wide variety of applications, including parts handling, welding, and painting.

A robot contains three important units: a mechanical unit for performing a task (the manipulator), a power unit to move the robot arm (the power supply), and a control unit to direct the robotic movement (the controller). Electric and hydraulic power supplies are the most common units used to raise, lower, and pivot robot arms into various positions. Hydraulic units are usually used to handle heavier objects, and they are generally considered faster than electric units. Electric units, though, take up less floor space and run more quietly.

Most robots operate in multiple planes, **Figure B.** The simplest robots rotate around a single axis. More complex units can produce motion in two directions—horizontally and vertically. Even more complex units add an in and out motion. These three basic motions, or degrees of freedom, can be combined with an additional three degrees of freedom (six total) in the end effector (the device at the end of the arm). The six degrees of freedom can place the robot's effector anywhere in its work area.

Figure A. Pictured is a common industrial robot.

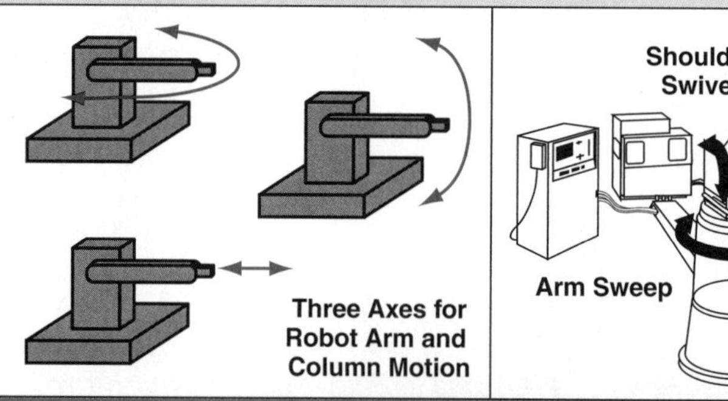

Figure B. These are typical motions of an industrial robot. (Cincinnati Milacron, Inc.)

slowly close the die halves by lowering the ram to produce a squeezing action. ***Hammers*** drop or drive the ram down with a quick action. This motion causes a sharp impact, which creates the forming force.

Rolling machines use two rolls that rotate in opposing directions to form the material. The rolls draw the material between them and squeeze or bend it into a new shape. Forming rolls may be smooth, or they may have a pattern machined into them. Smooth rolls are used to produce metal sheets and foils and to shape sheets and strips into curved products, such as tanks and pipes. Shaped rolls produce patterns or bend materials. Typical products produced with shaped rolls are corrugated roofing, bent metal siding, and metal trim.

Drawing machines pull or push materials through die openings. They make wire from rods and extrude other shapes. Wire drawing pulls the rods through a series of dies that have progressively smaller openings. At each die, the material is reduced in diameter and increased in length.

Extrusion works much like squeezing toothpaste from a tube. A shaped die is placed over the opening of the machine. The material is held in a closed cavity behind the die. A ram opposite the die forces the material out of the cavity, **Figure 16-14.** As the material passes through the die, it takes on the shape of the opening. Extrusion is used to produce complex shapes from plastic, ceramic, and metallic materials. Forming forces can be generated by air pressure, electrical fields, or explosive charges.

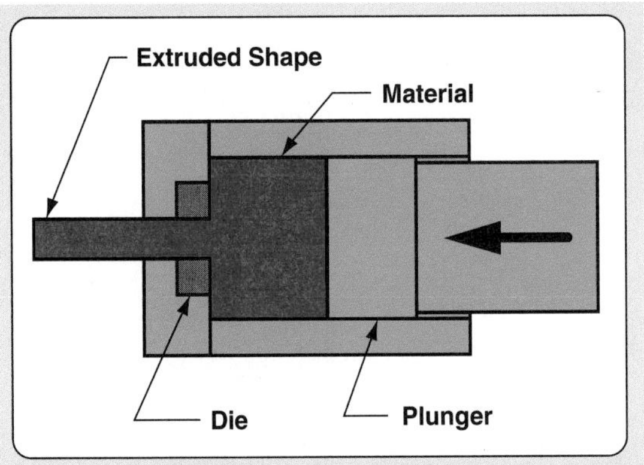

Figure 16-14. This diagram shows how extrusion works. The material takes on the shape of the die.

Material Temperature

Something to consider about forming processes is the temperature of the material while it is being formed. Metals can be formed hot or cold. Plastics and glass are formed hot, and ceramics are formed cold.

All materials have a temperature at which their properties change. The internal structure changes in all materials as they are heated. We must remember that the terms *hot* and *cold* are relative. A "cold" slab of steel at a steel mill might be hot enough to burn you.

When a metal is cold formed, internal stresses are built up. These stresses can cause the material to become brittle. This is called work hardening. Work hardening is relieved by heat treating the metal. Heat treating involves heating the metal and allowing it to cool slowly.

When a metal is hot formed, it is important that the metal be heated above its recrystallization point. Hot forming takes place above this point. Cold forming takes place below this point. Hot forming prevents work hardening of the metal. As the material cools, it forms a normal structure. Therefore, the minimum temperature for hot forming is different for each material. Also, material shaped by hot forming will be stress free, whereas cold forming builds internal stress in the material.

Separating Processes

Separating processes remove excess material to make an object of the correct size and shape. Casting and forming processes change the shape and size of materials without any removal. Separating processes remove material by either machining or shearing, **Figure 16-15.** ***Machining*** is based on the motion of a tool against a workpiece to remove material. ***Shearing*** uses opposing edges of blades, knives, or dies to fracture the unwanted material away from the work.

Machining

Machining removes excess material in small pieces. It cuts material away using three methods: These methods are chip removal, flame cutting, and nontraditional machining.

Chip Removal

Chip removal is the most common separating process. All chip-removal processes have three things in common:

- A tool or other cutting element is always present.
- Motion occurs between the tool and work.
- The work and the tool are given support, **Figure 16-16.**

Machining removes excess material through cutting actions. The most common cutting device is a chip-removing tool. Tools are based on the fact that a harder material will cut a softer material.

With regard to motion, note that two types of motion are needed to make a cut. These types are the cutting motion and the feed motion. ***Cutting motion*** moves a cutting tool through a material to make chips. ***Feed motion*** brings new material in contact with the cutting tool.

Safety with Machine Tools

People who work with machine tools use various machines. Some basic rules should be observed during the operation of these machines:

- Wear eye protection.
- Avoid loose-fitting clothing and keep long sleeves rolled up.
- Wear hearing protection when loud or high-pitched noises are present.
- Keep the laboratory clean. Wipe up any spills immediately.
- Wait until the machine has stopped to make adjustments, remove the workpiece, or clean up any scrap. Use a brush to clean chips off a machine.
- *Do not* talk to anyone while running a machine.
- Most of all, if you do not understand how to run a machine, *ask first!*

The key to safety in the technology education laboratory is a *safe* attitude.

Common separating machines can be grouped by the cutting and feed motions they use. These groupings include the

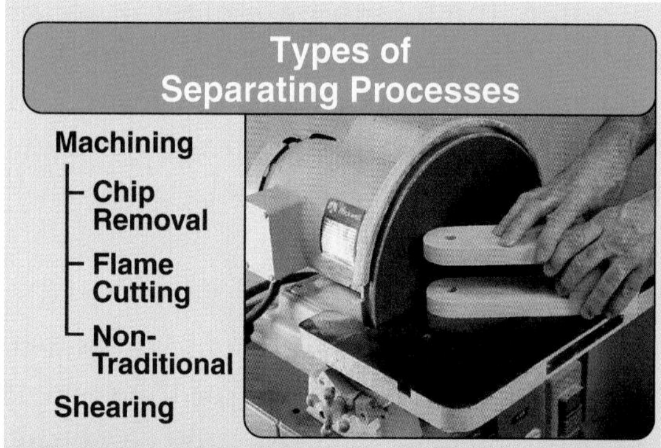

Figure 16-15. Separating processes include machining and shearing. The photo shows a sanding machine, which uses chip removal to take off unwanted material.

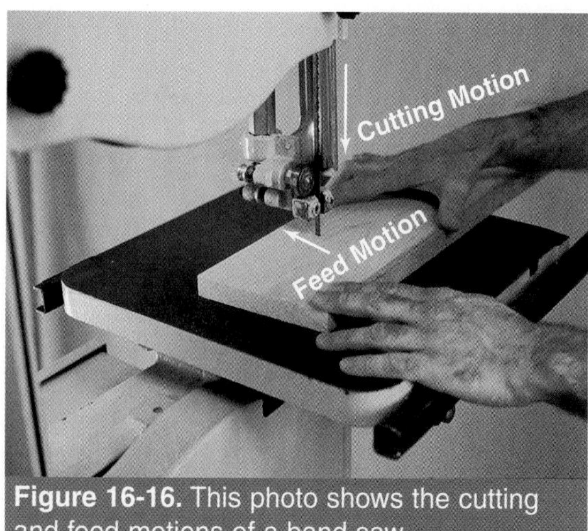

Figure 16-16. This photo shows the cutting and feed motions of a band saw.

following, **Figure 16-17.** ***Milling machines*** use a rotating cutter for the cutting motion. The feed motion is linear. The material is pushed into the cutter in a straight line. Machines that use the milling principle are horizontal and vertical milling machines; table saws; wood shapers, planers, and jointers; routers; disc sanders; and pedestal and bench grinders. The radial saw and the cutoff saw are variations of this group. Rotating multiple-point tools, called blades, are rotated. The blade, however, moves to produce the feed and cutting motions at the same time.

Drilling machines rotate a cutter for the cutting motion. The cutter is fed in a linear manner for the feed motion. Machines that use this principle are the drill press and the portable electric drill.

Turning machines rotate the work against a single-point tool to produce the cutting motion. The tool is fed along or into the work to produce the feed motion. Wood and metal working lathes are turning machines.

Shaping machines use a single-point tool that moves back and forth across the work to produce the cutting motion. The cut is usually made on the forward stroke. The tool is lifted slightly for the backstroke. The work is fed linearly under the tool. Metal-working shapers are the primary machines using this cutting action. Small saws, such as hacksaws and scroll saws, use a reciprocating motion also.

Planing machines move the work under the tool to make the cutting motion. Moving the tool across the work in small steps creates the feed motion. The metal planer is the machine that uses this technique. Both metal shapers and planers are used to make large machinery.

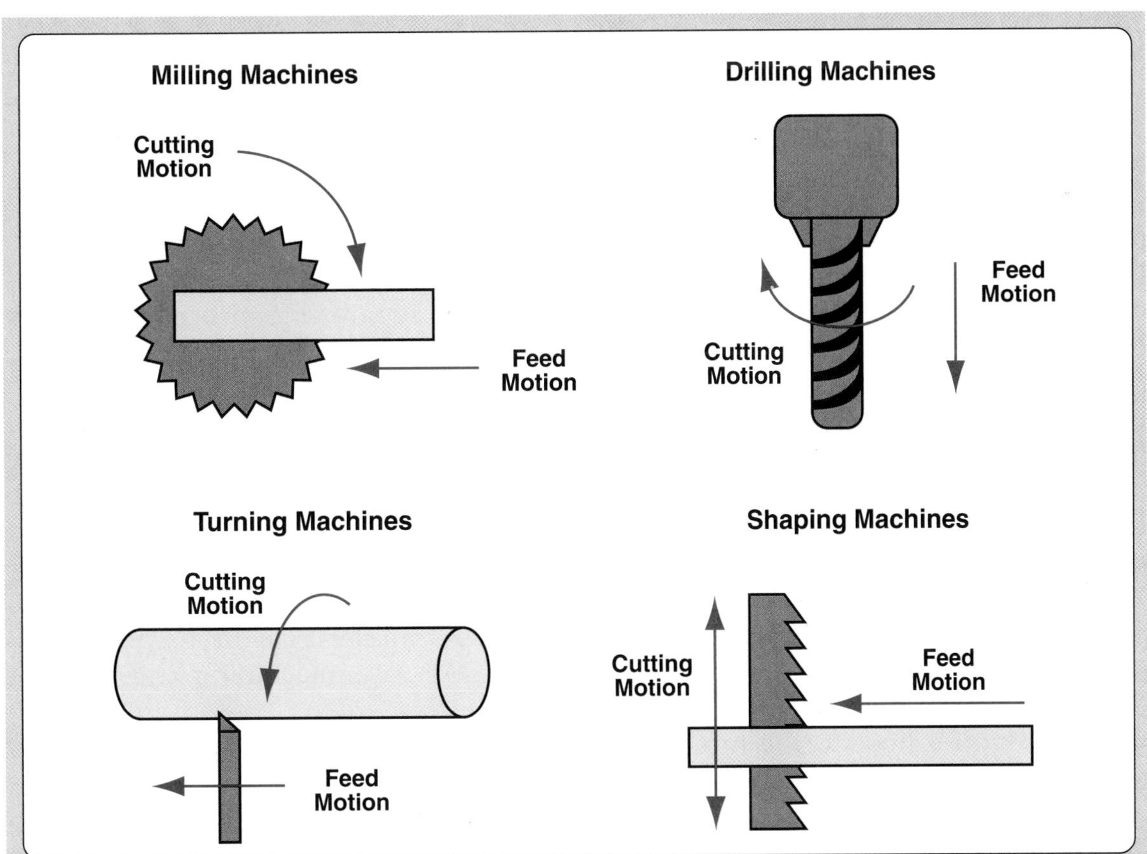

Figure 16-17. Cutting and feed motions are used to group machines into types. A few examples of these types are shown here.

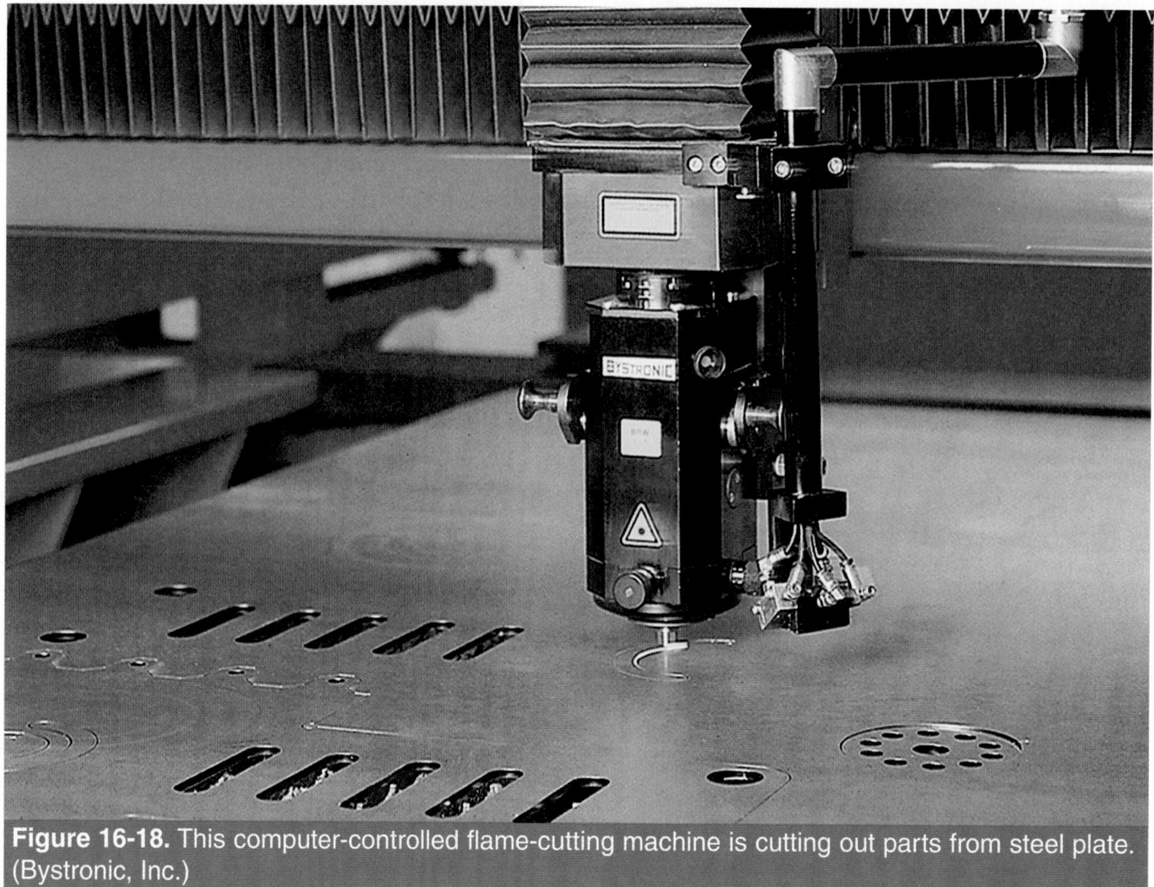

Figure 16-18. This computer-controlled flame-cutting machine is cutting out parts from steel plate. (Bystronic, Inc.)

Flame Cutting

Burning gases can be used to remove unwanted material from a workpiece, **Figure 16-18.** A fuel gas and oxygen are mixed in a torch, ignited, and allowed to heat the surface of the metal. When the metal is hot enough, a stream of pure oxygen is forced out of the torch. The stream of oxygen causes the metal to burn. The burning separates the workpiece from the scrap. Cutting material to size and shape using burning gases is called ***flame cutting.***

Acetylene gas and natural gas are used as fuel gases for cutting metals. The fuel gas and oxygen are fed through separate regulators to control the pressure. Each gas flows through a hose to the torch. The torch controls the volume and mixture of the gases. The mixture then moves out of the torch. The blast of pure oxygen is controlled separately from the fuel gas and oxygen mixture.

Nontraditional Machining

Chip removing and flame cutting have been used for a long period of time. In the 1940s, people started developing a series of new separating processes. These processes use electrical, sound, chemical, and light energy to size and shape materials. They are often called ***nontraditional machining*** processes. This term is not very accurate today because many of these processes are widely used. They are rapidly becoming traditional ways to machine materials. Three common nontraditional processes are electrical discharge machining (EDM), laser machining, and chemical machining.

Electrical discharge machining (EDM) uses electric sparks to make a cavity in a piece of metal. This process requires an electrode, a power source, a tank, and a coolant, **Figure 16-19.** The workpiece is connected to one side of the power supply

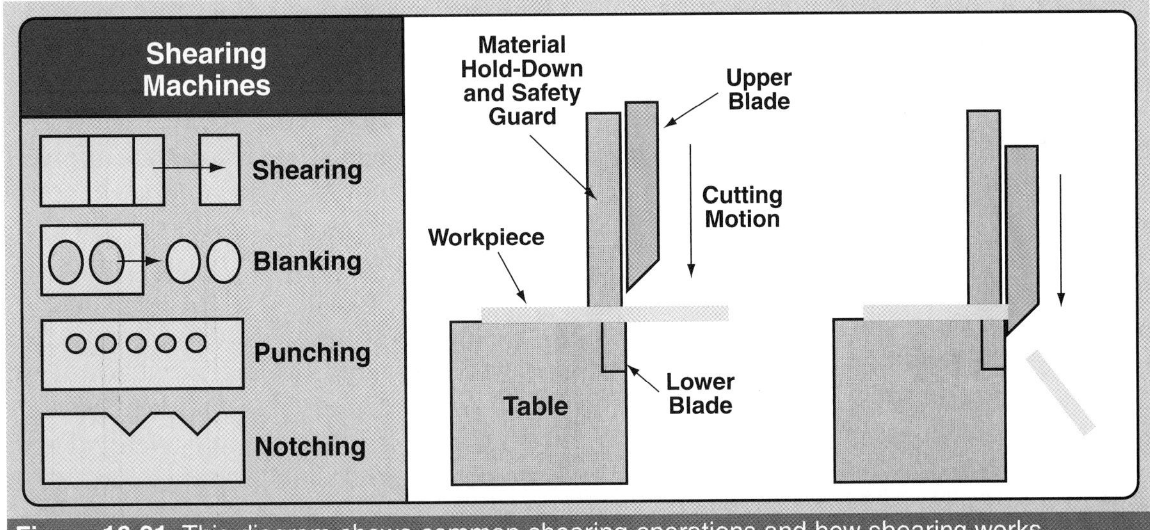

Figure 16-21. This diagram shows common shearing operations and how shearing works.

shatterproof. Likewise, the resins used in fiberglass layups undergo chemical action as they cure and become hard.

The most common conditioning processes use heat. These are called ***thermal conditioning*** and include heat treating, firing, and drying. ***Heat treating*** is a term used to describe the thermal conditioning processes used on metals. These processes include the following:

- ***Hardening*** is used to increase the hardness of a material, **Figure 16-22.** Hardening steels requires that the part be heated to a specific temperature. The part is allowed to "soak." This ensures that the entire part is at a uniform temperature. The part is then rapidly cooled in a tank of oil or water (quenched).
- ***Annealing*** is used to soften and remove internal stress in a part. The part is heated to a specific temperature and allowed to soak. It is then removed from the oven and allowed to cool slowly to room temperature.
- ***Tempering*** is used to relieve internal stress in a part. Hardening often creates internal stress, which will cause a part to crack under use. This stress is removed by heating the part to a specific temperature and allowing it to slowly cool. The tempering temperature is much lower than the annealing temperature. Tempering is used for metal parts and most glass products.

Firing is a thermal conditioning process used for ceramic products. Most ceramics are made from clay materials that are plastic when wet. After drying, the clay can be heated to a high temperature. The water is driven out of the clay particles, and the grains bond together to make a solid structure.

Likewise, certain coatings are fired. Porcelain enamels are fired to give a glasslike

Figure 16-22. These hot parts are leaving a heat-treating furnace. They will be quenched to harden them. (Bethlehem Steel Co.)

finish. The enamel is applied to the part by spraying or dipping. The product is heated, and the coating fuses with the part.

Drying is a common thermal conditioning process. It removes excess moisture from materials. Ceramic materials and wood products must be dried before they are useful. Drying can happen naturally or be helped by adding heat. For example, lumber is air dried or kiln dried. Air-dried lumber is stacked outdoors to dry after cutting. Kiln-dried lumber is carefully heated in special ovens called dry kilns.

Assembling Processes

Look around you. How many products with one part do you see? You might notice such items as paper clips or straight pins, but most of the things you see are made from more than one part. These products are assembled from two or more parts, **Figure 16-23.** The word *assemble* means "to bring together." Through ***assembling processes,*** for example, a simple product, such as a lead pencil, is created from five parts. The barrel is two pieces of wood glued around the graphite "lead." The eraser is held onto the barrel with a metal band. In fact, a lead pencil uses the two methods by which products are assembled: bonding and mechanical fastening.

Bonding

Bonding holds plastic, metal, ceramic, and composite parts to each other. Many different methods of bonding are used to attach parts together. ***Bonding*** uses cohesive or adhesive forces to hold parts together. *Cohesive forces* hold the molecules of one material together. *Adhesive forces* occur between different kinds of molecules. They cause some materials to be "sticky." The way in which parts are bonded together is affected by the following:

- The bonding technique.
- The bonding material.
- The type of joint.

Bonding Techniques

Five bonding techniques are used for assembly. The first is ***fusion bonding.*** This technique uses heat or solvents to melt the edges of the joint. The surfaces are allowed to flow together to create a bond. In some cases, more material is added to the joint to increase the strength. Oxyacetylene, arc, inert gas, and plastic welding are examples of fusion bonding methods.

A second method is flow bonding. ***Flow bonding*** uses a metal alloy as a bonding agent. The base metal is cleaned and then heated. The alloy is applied where the two parts meet. It melts, flows between the parts, cools, and creates a bond. Soldering

Figure 16-23. Products are assembled using both mechanical and bonding techniques. These washing machines have mechanisms bolted into place and sound insulation bonded inside.

and brazing are examples of flow bonding. The base metal is not melted.

Pressure bonding applies heat and pressure to the bond area. This method is used on plastics and metals. Resistance (spot) welding is an example of pressure bonding, **Figure 16-24.** Spot welding has four stages in its cycle:

1. The parts are held together between two electrodes.
2. Electrical current is passed between the electrodes. The current melts the metal between the electrodes.
3. The current is stopped, and the molten metal solidifies.
4. The electrodes are released, and the welded parts are removed.

Plastics can be pressure bonded using a similar process. No electric current passes between the parts. Heated jaws hold the parts and bond them together. The jaws release, and the parts are removed.

Cold bonding uses extreme pressure to squeeze the two parts to create a bond. This process is not used often. It is used only for small parts made of soft metals, such as copper and aluminum.

Adhesive bonding uses substances with high adhesive forces to hold parts together. These techniques are gaining wide use beyond their original use in woodworking. The advent of synthetic adhesives allows a wide range of plastics, metals, and ceramics to be joined together.

Bonding Materials

All bonding techniques use a bonding material. This material is referred to as a ***bonding agent.*** A bonding agent can be one of three kinds:

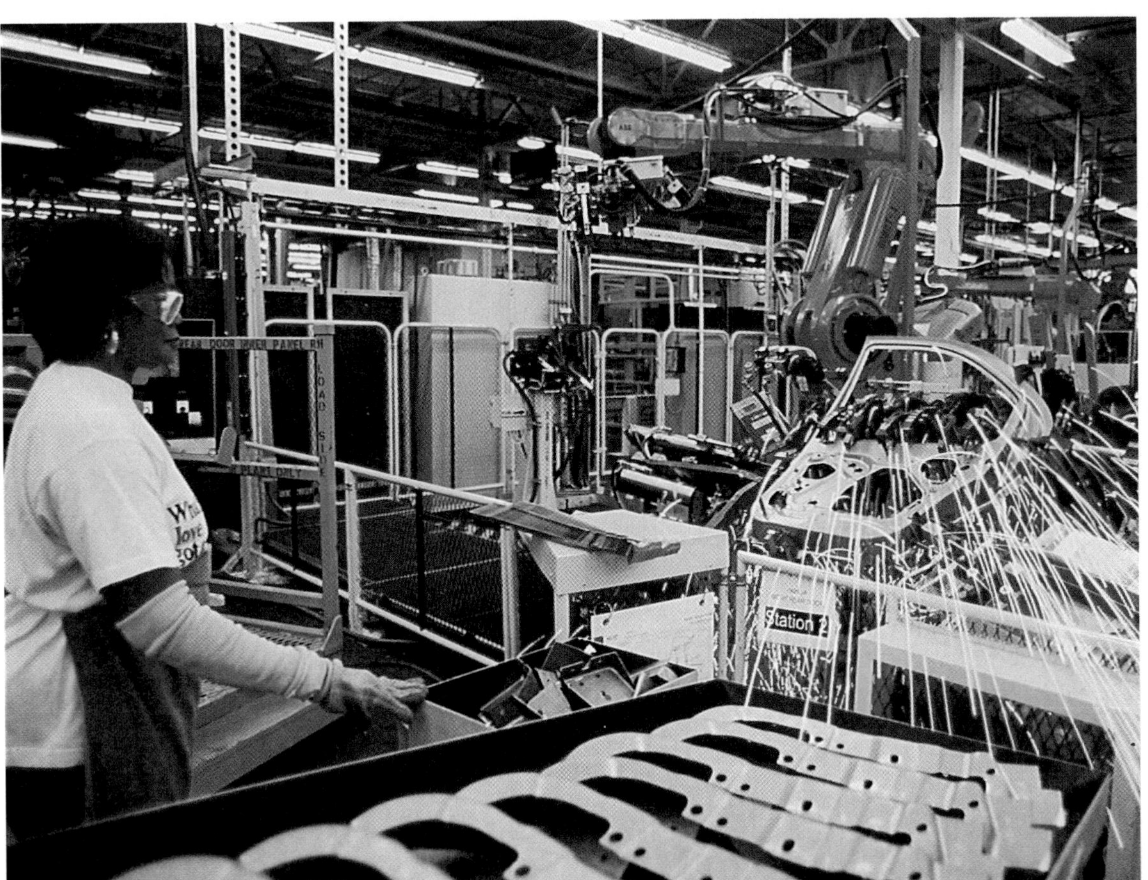

Figure 16-24. This automobile assembly is being put together using resistance welding. Resistance welding is a type of pressure bonding. (DaimlerChrysler)

- The same material: The same material as the base material holds parts together. For example, two metal parts might be welded together, or a steel welding rod might be used to strengthen the bond between two steel parts, **Figure 16-25.**
- Similar material: The same type of material holds parts in position. For example, a metal (solder) might be used to bond metallic parts (copper wires).
- Different material: A different type of material holds parts together. For example, white glue (a plastic) might hold wood together, or an epoxy (a plastic) might hold ceramic parts together.

Types of Joints

Joints are where parts meet. Pieces of material are fastened together at joints. Joints can occur on parts at the ends, sides, and faces. They are used to add length, width, or thickness and to make corners. The type of joint chosen will affect the type of bonding process used. A simple joint can be modified to increase its strength or improve its appearance, **Figure 16-26.**

Mechanical Fastening

Mechanical fastening uses mechanical forces to hold parts together. Friction between the parts can be used. For example, a part might be pressed or driven into a hole slightly smaller than it is. The friction between the parts will cause the parts to remain together. This type of fit is called a ***press fit.*** Press fits can be used to hold bearings in place on a shaft.

In other cases, the parts might be bent and interlocked to hold the parts together. This type of joint is called a seam. Many sheet metal parts are held together using seams.

Figure 16-25. This welding operation uses filler rods that are made from the same metal as the parts being welded.

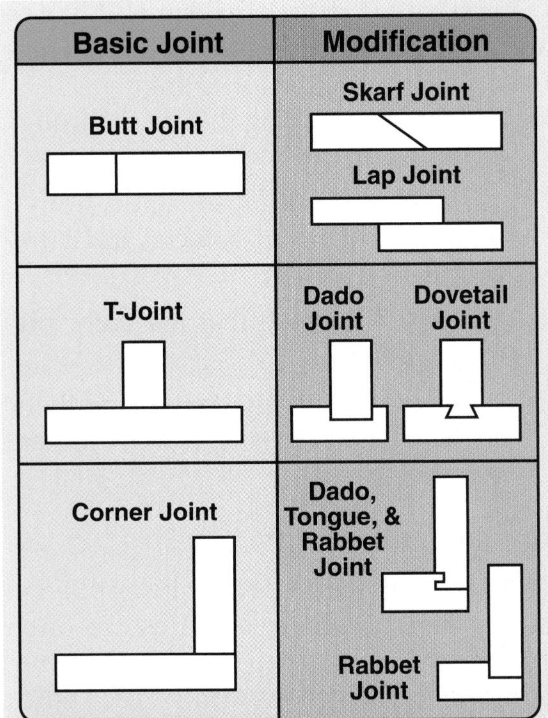

Figure 16-26. These drawings show simple joints can be changed for higher strength and better appearance.

The most widely used method to hold parts together is mechanical ***fasteners***, **Figure 16-27.** Examples of fasteners are staples, rivets, screws, nails, pins, bolts, and nuts. Mechanical fasteners can be described as permanent, semipermanent, or temporary. Permanent fasteners are not intended to be removed. Their removal will damage the fastener or the parts. Nails, rivets, and wood screws are good examples of permanent fasteners.

Semipermanent fasteners can be removed without causing damage to the parts. They are used when the product must be taken apart for maintenance and repair. For example, a furnace is held together with semipermanent fasteners so it can be serviced. Machine screws, bolts, and nuts are good examples of semipermanent fasteners.

Temporary fasteners are used when frequent adjustments or disassembly are required. They hold parts in position, but the parts can be taken apart quickly. A wing nut is a good example of a temporary fastener.

Finishing Processes

Finishing processes are the last of the secondary processes most products go through. These techniques protect products and enhance their appearance. Finishing processes can be broken into two types. One group changes the surface of the product. The other group applies a coating.

Most metals will begin to corrode if they are not protected in some way. Products made of metals need to be protected so they will last. Metals are easy to protect by changing the surface chemically. For example, anodizing converts the surface of aluminum products to aluminum oxide. This type of finish is called a ***converted surface finish.***

The other type of finishing applies a coating to the product. A film of finishing material is applied to the product or base material. These coatings protect the surface and can add color. Finishing processes involve cleaning the surface, selecting the finish, and applying the finish.

Safety with Finishing Materials and Equipment

People who work with finishing processes observe basic safety rules:

Figure 16-27. The lug nuts on this wheel are mechanical fasteners.

- Wear eye and face protection.
- Apply finishes in a well-ventilated area. Wear a respirator and protective clothing if solvents are toxic.
- Never apply a finish near an open flame.
- Keep the finishing area orderly, and clean up any spills immediately.
- Store all rags and chemicals in approved containers.
- Use a spray booth (an enclosed area with fans to remove fumes) to remove toxic fumes.
- Dispose of any waste solvents and finishes properly. Wash your hands to remove any finishes you have handled.
- Most of all, if you do not understand how to apply a finish, *ask first!*

The key to safety in the technology education laboratory is a *safe* attitude.

Cleaning the Surface

Finishing materials are applied to the surface of a product. This surface must be free of dirt, oil, and other foreign matter. Mechanical means or chemical cleaning can remove these unwanted materials. Chemical cleaning is often called pickling. ***Pickling*** involves dipping the material into a solvent that will remove the unwanted materials. The clean part is then rinsed to remove the solvent. Mechanical cleaning includes abrasive cleaning (sand blasting or sanding), buffing, and wire brushing.

Selecting the Finish

Hundreds of finishing materials exist. We can classify them as either inorganic or organic materials, however, to understand them better. Inorganic materials include metal and ceramic coatings. For example, steel is coated with zinc to produce galvanized steel or with tin to make tin plate, **Figure 16-28.** Clay products, such as floor and wall tile, are coated with a glaze. Electric range tops are coated with a ceramic material called porcelain enamel. All of us are familiar with the organic finishes. These materials include the following:

- ***Paints:*** Any coating that dries through polymerization (hardening)
- ***Varnish:*** A clear finish made from a mixture of oil, resin, solvent, and a drying agent
- ***Enamel:*** A varnish that has color pigment added
- ***Lacquer:*** A solvent-based, synthetic coating that dries through solvent evaporation

Applying the Finish

Finishing materials may be applied in many ways. Metallic coatings are often applied through dipping or plating. ***Dipping*** involves running the stock through a vat of molten metal. This

Figure 16-28. This roll of steel has been coated with tin to make tin plate. Tin plate is used to make cans for food. (American Iron and Steel Institute)

technique is widely used to produce galvanized steel.

Plating is an electrolytic process. The parts are hung on racks and lowered into a cell full of electrolyte, **Figure 16-29.** The plating metal is the anode, and the part is the cathode. Electrical current is run through the cell. The electrical current causes the metal to move from the anode into the electrolyte. From there, it moves across to the cathode and is deposited as a uniform coating of the product.

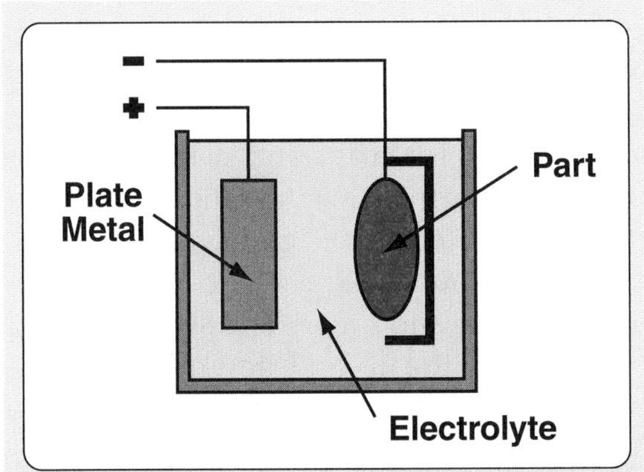

Figure 16-29. This schematic drawing shows an electrolytic cell. This cell is used to apply metal plating to parts.

Organic materials can be applied through brushing, rolling, spraying, flow coating, and dip coating. Most of us are familiar with brushing and rolling. The material is gathered in the brush or on the roller. It is then applied by wiping the brush or rolling the roller across the surface of the material.

Spraying uses air to carry fine particles of finishing materials to the surface of the product. Some of these processes are manual, in which a worker uses the spray gun. Automatic spraying systems, such as the one shown in **Figure 16-30,** are becoming widely used.

Flow coating passes the product under a flowing stream of finishing material. The material flows over the surface, and the excess runs into a catch basin to be used again. Dip coating is the same process as that used for galvanizing. In both cases, excess coating is allowed to drain off the product for reuse. After parts have a finish applied, they are allowed to air dry, or they can be run through a heated drying oven.

Figure 16-30. This automobile body is being sprayed with a coating. Automatic systems such as this one give good results and do not put workers at risk from paint fumes. (DaimlerChrysler)

Automating and Controlling Processes

Historically, skilled workers completed the primary and secondary manufacturing processes using simple tools. During the Industrial Revolution, much of this handwork was moved to machines in factories. Humans who had limited skills operated most of these machines. The operators loaded materials into the machines, adjusted tool feeds and speeds, monitored the machines' operation, and unloaded the finished work. In recent years, a new breed of machines has replaced many humans on the factory floor. These machines make up systems called *automation*. This term refers to the automatic systems used to operate or control equipment or a process. Automation has many parts, which include robots, programmable logic controllers (PLCs), computer-controlled machines, and artificial intelligence (AI) systems.

Robots are mechanical devices that can perform tasks automatically or with varying degrees of direct human control. Industrial robots are a very common type of robot. See **Figure 16-31.** Industrial robots can be described as programmable, multipurpose manipulators. They can operate using computer programs, and they can manipulate (hold and rotate) parts or devices. Other robots can mow your lawn, vacuum your house, gather samples on the moon, clean up toxic waste, and search the ocean's floor.

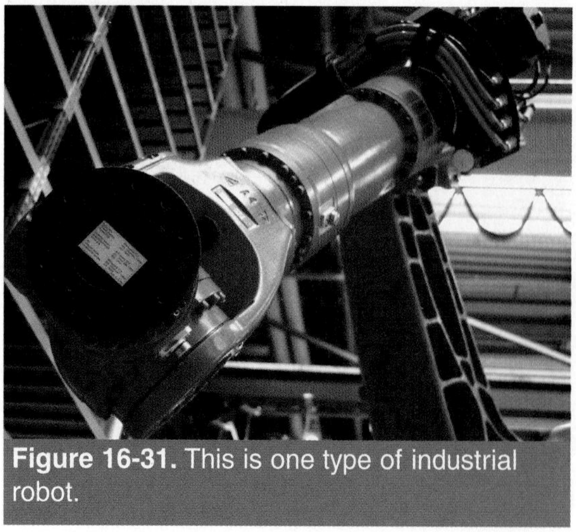

Figure 16-31. This is one type of industrial robot.

Some common applications of industrial robots are welding, painting, assembling, moving materials and parts (pick and place actions), and inspecting products. A typical industrial robot has five main parts: an arm, a controller, a drive, an end effector, and a sensor. See **Figure 16-32.**

The arm of the robot moves and positions the end effector to a desired position. Most arms resemble a human body and arm, in that they have a body, a shoulder, an elbow, a forearm, and a wrist. These parts of the arm allow the robot to move in three axes—left and right, up and down, and in and out. The base allows the robot to move left and right. The shoulder and elbow create the up-and-down and in-and-out motions. The wrist rotates to create the final positioning for the tool or material.

The controller is the "brain" of the robot. It controls the arm movements the motors or hydraulic systems create. Computer commands direct the motions.

The end effector is the "hand" connected to the wrist. It can be many different things, ranging from a gripping device

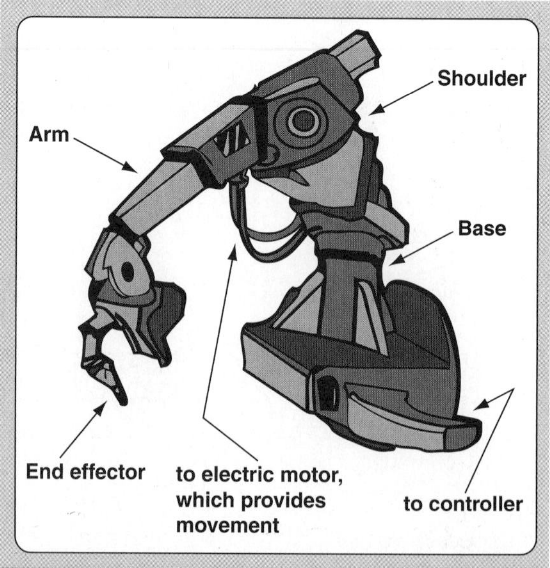

Figure 16-32. Parts of a common robot.

Career Corner

Automobile Assembly Workers

Automobile assembly workers put together parts to form subassemblies. Also, they assemble parts and subassemblies to build automobiles, trucks, and buses. They might mount tires, adjust brakes, and add fluids as the vehicles are assembled. These workers complete a number of tasks, including welding metal, cutting and molding plastic and glass, and sewing fabric. Many assemblers work with robots, computers, and other forms of automation as they complete vehicle assembly.

They often work more than 40 hours per week. Overtime is common during periods of peak demand. Some production work involves uncomfortable conditions, with heat, fumes, noise, and repetition. Newer plants are more automated and have safe working conditions.

to a paint spray nozzle. Sensors provide feedback so the robot can "understand" its position. They allow the robot to know where its end effector is and determine the moves needed to move the effector into a new position.

Programmable logic controllers (PLCs), or programmable controllers, are devices that use microprocessors to control machines or processes. A technician on the factory floor generally creates the program for the controller, rather than a computer programmer located elsewhere. PLCs have three basic functions, which are input, control, and output. The PLC works by analyzing its inputs and, depending on their state, turning on or off its outputs. For example, the controller might receive input such as temperature, liquid levels, or shaft speed. Based on this input and the logic written into the controller, it activates appropriate outputs. A flame might be reduced to lower the temperature, a valve might be opened to fill a tank, or the speed of a motor might be changed to improve a machine's operation.

Computer-controlled machining adds automatic control to any basic machine. See **Figure 16-33.** This automatic control is generally provided by a computer program that controls the motions of the machine. The control is called computer numerical control (CNC). The basic function of a CNC machine is to produce automatic, precise motion of the machine's tools or the material being processed. This motion is in two or more directions, called axes. The tool or workpiece can be precisely and automatically positioned along the length of travel on any axis.

For example, a person turning a handwheel attached to a lead screw positions the table holding a workpiece on a milling machine. In CNC milling machines, these motions are still created by turning a lead screw. A motor responding to computer-programmed commands, however, rotates the lead screw.

Figure 16-33. A computer numerical control (CNC) machining center can be programmed to do many different tasks.

Artificial intelligence (AI) is a field that is starting to be more widely used in manufacturing. It can be described as intelligence exhibited by a manufactured device or system. Most AI systems involve a computer to act as the "brain" for the system. These systems often use a type of AI called machine vision. The purpose of a computer vision system is to program a computer to "see and understand" a scene, using digital cameras or other sensors. The system then acts on this understanding to inspect a part, locate a cutter, or pick up a randomly positioned part. Other uses for AI include handwriting recognition for handheld communication devices, optical character recognition (OCR) programs for scanners, speech recognition systems for automated customer call centers, and face recognition for security systems. See **Figure 16-34.**

Figure 16-34. Many types of modern devices use artificial intelligence (AI).

Summary

Secondary manufacturing processes convert industrial materials into finished products. They add worth to materials by making them useful to consumers. The secondary processes can be grouped as casting and molding, forming, separating, conditioning, assembling, and finishing. The first three change the size and shape of the materials. Conditioning changes the properties of materials by altering their internal structures. Assembling processes attach parts together to make assemblies and products. Finishing produces coatings that protect and improve a product's appearance.

Key Terms

adhesive bonding
annealing
artificial intelligence (AI)
assembling process
bonding
bonding agent
casting and molding processes
chemical action
chemical conditioning
chemical machining
chip removal
cold bonding
compounding
computer-controlled machining
conditioning process
converted surface finish
cutting motion
die
dipping
drawing machine
drilling machine
drying
elastic range
electrical discharge machining (EDM)
enamel
expendable mold
extrusion
fastener
feed motion
finishing process
firing
flame cutting
flow bonding
flow coating
forming process
fracture point
fusion bonding
hammer
hardening
heat treating
joint
lacquer
laser machining
machining
mated dies
mechanical conditioning
mechanical fastening
milling machine
nontraditional machining
open die
paint
permanent mold
pickling
planing machine
plastic range
plating
press
press fit
pressure bonding
programmable logic controller (PLC)
robot
rolling machine
secondary manufacturing process
separating process
shaped die
shaping machine
shearing
spraying
tempering
thermal conditioning
turning machine
varnish
yield point

Test Your Knowledge

Write your answers on a separate piece of paper. Please do not write in this book.

1. List the five basic steps involved in casting and molding.
2. What is an expendable mold?
3. List two basic safety rules to observe when using casting processes.
4. Name the three ways we can make materials solid.
5. What is a material's plastic range?
6. Name the three things all forming processes have in common.
7. What are mated dies?
8. Name one product shaped rolls produce.
9. True or false? Plastics and glass are formed hot, but ceramics are formed cold.
10. The two kinds of separating processes are machining and _____.
11. What are the three things all chip-removal processes have in common?
12. The two types of motion needed to make a cut are cutting motion and _____ motion.
13. List two basic safety rules to observe when using machine tools.
14. True or false? Burning gases to remove unwanted material from a workpiece is a nontraditional form of machining.
15. What are the three types of conditioning processes?
16. Name two kinds of bonding techniques.
17. Give one example of a mechanical fastener.
18. List two basic safety rules to observe when working with finishing processes.
19. True or false? Chemical cleaning of product surfaces is often called pickling.
20. Name one type of organic finish.

Applying Your Knowledge

1. Build a simple product, such as a kite, using secondary manufacturing processes. List each step and the type of process used.
2. Select a simple product made from more than one part that you see in the room.
   A. List the parts it is made of.
   B. Select one part and list the steps you think were used to manufacture it.
   C. Complete a form like the following for one of the steps.

Product:
Part name:
Production process:
Step needed to complete the process:

Chapter 17
Constructing Structures

Learning Objectives

After studying this chapter, you will be able to:

- List the two types of construction.
- Describe the major types of buildings that are constructed.
- Identify the steps involved in constructing a structure.
- Identify the types of foundations.
- Describe the framework used in buildings.
- Describe how buildings are enclosed.
- Explain the types of utility systems used in buildings.
- Describe materials used in finishing buildings.
- Identify other types of buildings.
- Identify the characteristics of heavy engineering structures.

Human beings have three basic needs: food, clothing, and shelter. Each of these can be satisfied using technology. Agriculture and related biotechnologies help us grow, harvest, and process food. Manufacturing helps us to produce natural and synthetic fibers. These fibers become the inputs to clothing and fabric manufacture. Materials and manufactured goods can be fabricated into dwellings and buildings using construction technology. ***Construction*** uses technological actions to erect a structure on the site where it will be used.

Types of Structures

Construction builds two types of structures. These are buildings and heavy engineering structures, **Figure 17-1.** ***Buildings*** are enclosures to protect people, materials, and equipment from the elements. Buildings also provide security for people and their belongings. ***Heavy engineering structures*** help our economy function effectively.

Buildings

Buildings are grouped into three types: residential, commercial, and industrial, **Figure 17-2.** These groupings are based

Types of Construction Projects

Buildings

Heavy Engineering Structures

Figure 17-1. Construction erects buildings and heavy engineering structures.

on how the buildings are used. Note, however, that other types of buildings exist. These special buildings follow the same construction steps as the other buildings. This section will describe each of the main types, provide an overview of the other types, and discuss the general steps involved in constructing buildings.

Residential Buildings

Residential buildings are buildings in which people live. These buildings can be single-family or multiple-unit dwellings. The multiple-unit dwellings include apartments, town houses, and condominiums.

A residential building can be either owner occupied or rented from the owner. The owner of a dwelling is responsible for its upkeep. In some types of dwellings, such as condominiums, the costs of upkeep are shared between the owners. Each owner belongs to and pays fees to an association. This group elects officers who manage the maintenance of common areas such as entryways, garages, parking areas, and lawns. The association is also responsible for exterior repairs and insurance on the building. The individual owners maintain their own living quarters and insure their personal belongings against fire and theft.

Commercial Buildings

Commercial buildings are used for business and government purposes. These buildings can be publicly or privately owned. Commercial buildings range in size from small to very large. Retail stores, offices, courthouses, schools, libraries, and warehouses are commercial buildings.

Industrial Buildings

Industrial buildings house the machines that make products. These buildings are used to protect machinery, materials, and workers from the weather. The building supports the machines and supplies the utility needs of the manufacturing process. Many industrial buildings are specially built for one manufacturing process.

Other Types of Buildings

You see commercial, industrial, and residential buildings all around you. If you look around your town or city, however, you will probably see other types of buildings, **Figure 17-3.** These may include:

Types of Buildings

Residential

Commercial

Industrial

Figure 17-2. Construction is used to build residential, commercial, and industrial buildings. (Marvin Windows and Doors, Inland Steel Company)

- Monuments: These structures pay tribute to the accomplishments or sacrifices of people or groups.
- Cultural buildings: These buildings house theaters, galleries, libraries, performance halls, and museums. These buildings host musical, dramatic, and dance performances, literary activities, and art exhibits.
- Government buildings: These buildings house government functions. Examples include city halls, post offices, police stations, firehouses, state capitols, courthouses, and government office buildings.
- Transportation terminals: These buildings are used to aid in the loading and unloading of passengers and cargo from transportation vehicles. Examples are airports, train and bus stations, freight terminals, and seaports.
- Sports arenas and exhibition centers: These facilities are used for sporting events, concerts, trade shows, and conventions.
- Agricultural buildings: These structures include barns and storage buildings used to house livestock, shelter machinery, and protect farm products (grain and hay, for example).

As noted earlier, these special buildings are built using the same construction steps used for a single-family home.

A special type of building is the ***manufactured home.*** As you will remember, manufacturing produces products in a factory. The completed product is transported to its place of use. This is exactly how manufactured homes are produced, **Figure 17-4.** Most of the structure is built in a factory. This type of home is usually built in two halves. The floors, walls, and roof are erected; then the plumbing and electrical systems are installed. The structure's interior and exterior is enclosed and finished. This step includes installing flooring, painting walls, setting cabinets and plumbing fixtures, and installing appliances and electrical fixtures.

The two halves of the structure are transported to the site. The foundation is already in place. Each half is lifted from its transporter and placed on the foundation. The two halves are finally bolted together. The final trim that connects the halves is installed. The utilities are hooked up, and the home is ready for the homeowner.

Similar techniques are used to produce temporary classrooms, construction offices, and modular units that can be assembled into motels or nursing homes.

Figure 17-3. Buildings other than the three main types can be found in many forms.

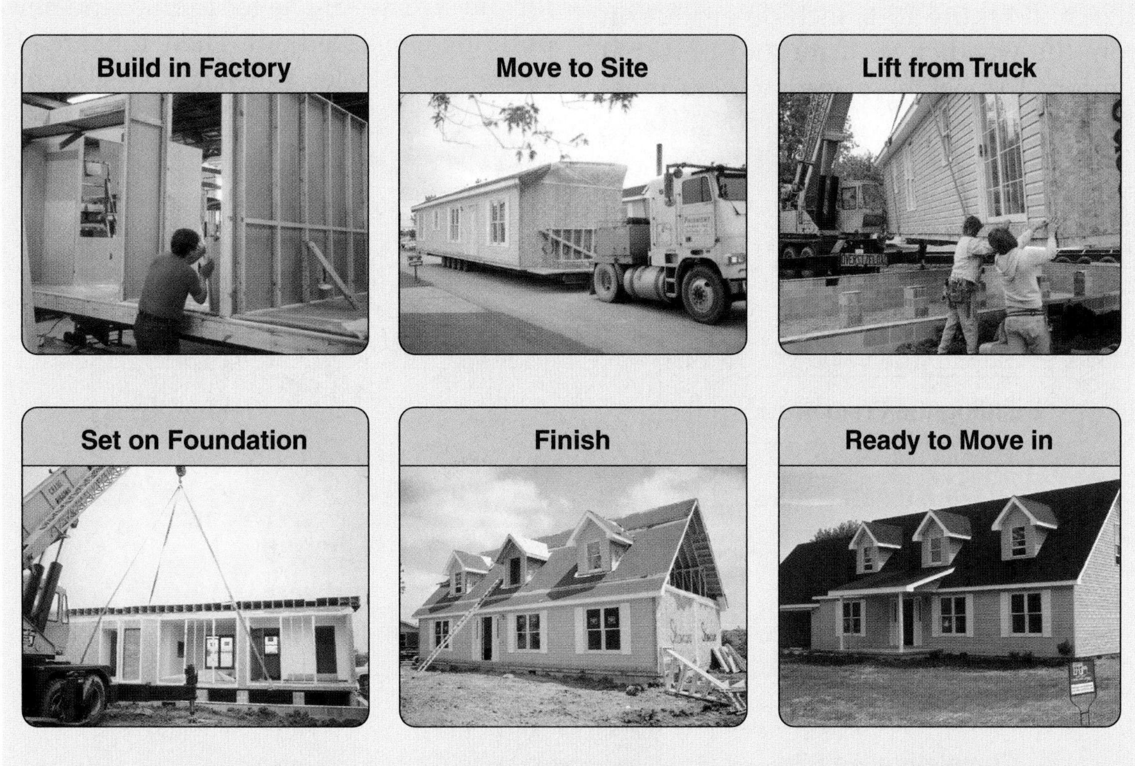

Figure 17-4. Shown here are the steps in building a manufactured home.

Constructing Buildings

Most construction projects follow the same basic steps. These steps, as shown in **Figure 17-5,** include the following:

- Preparing the site
- Setting foundations
- Building the framework
- Enclosing the structure
- Installing utilities
- Finishing the exterior and interior
- Completing the site

Each type of structure needs to have specific actions taken during each step. This helps complete the structure on time. We will look at the steps used to construct a small single-family home. Later in the chapter, other construction activities will be discussed.

A common type of building is a single-family home. It is designed to meet a number of needs of the owners. These needs, as shown in **Figure 17-6,** include protection from the weather, security, and personal comfort. To meet these needs a home must be properly designed and constructed. The construction process starts with locating, buying, and preparing a site.

Preparing the Site

The location for a home needs to be carefully selected. It should meet the needs of the people who will live there. For example, a family with children may think about the schools serving the area. The parents will consider the distance to work, shopping, recreation, and cultural facilities. The condition of other homes in the neighborhood, building codes, and covenants are other factors to consider.

Once the site is chosen, it is purchased from the original owner. This may require working with a real estate agent and obtaining a bank loan or other financing. The financing will probably include the money to erect the home. This is important because most banks will not loan money to build a house on land that is mortgaged.

Next, the site is cleared to make room for the structure. The location of the new building is marked out. Then, that area is cleared of obstacles. When it is possible, the

Preparing the Site

Setting Foundations

Building the Framework

Enclosing the Structure

Installing Utilities

Finishing the Exterior and Interitor

Completing the Site

Figure 17-5. Most construction projects follow the same basic steps. (Gehl Co.)

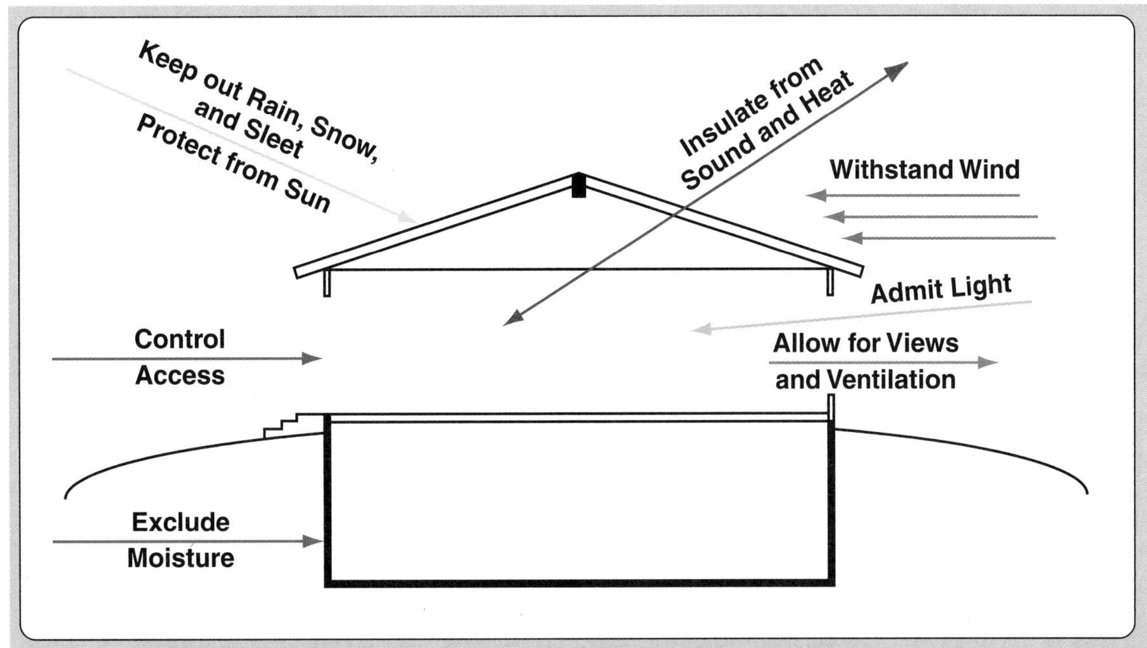

Figure 17-6. This diagram shows some of the needs a home must meet for its owner.

building should be located to save existing trees and other plant life. The site may require grading, **Figure 17-7,** to level the site. Grading prepares areas for sidewalks and landscaping and helps water to drain from the site. These preparations are needed for the next step, setting foundations.

Setting Foundations

The foundation is the most important part of any building project. The foundation serves as the "feet" of the building. Try to stand on just your heels. You will be unstable and wobble. Likewise, a building without a proper foundation will settle unevenly into the ground. It will lean, become unstable, and may fall to the ground. The Leaning Tower of Pisa in Italy is an example of a building that has a poor foundation. Over time, the tower has settled and is leaning several feet to one side.

Figure 17-7. Before a building can be built, the site must be cleared and graded.

A complete foundation has two parts: the footing and the foundation wall, **Figure 17-8.** The footing spreads the load over the bearing surface. The bearing surface is the ground on which the foundation and building will rest. This can be rock, sand, gravel, or a marsh. Each type of soil offers unique challenges for the construction project.

The type of foundation to use is selected to match the soil of the site. Three types, as shown in **Figure 17-9,** are:

- *Spread foundations:* These types of foundations are used on rock and in hard soils such as clay. The foundation walls sit on a low flat pad called a footing. On wide buildings, posts support the upper floor between the foundation walls. These posts also rest on pads of concrete called footings.

Figure 17-8. The foundation wall and footing spread the building's weight onto the bearing surface. The concrete foundation shown in the photo has been insulated to reduce heat loss.

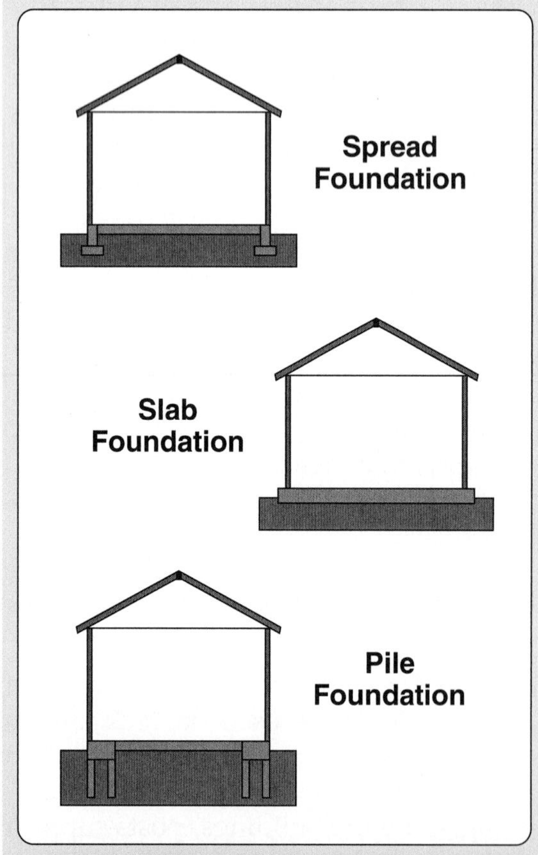

Figure 17-9. Three types of foundations used for buildings are spread, slab, and pile.

- *Slab foundations:* These types of foundations are used for buildings that are built on soft soils. They are sometimes called floating slabs. The foundation becomes the floor of the building. Such foundations allow the weight of the building to be spread over a wide area. This type of foundation is used in earthquake areas because it can withstand vibration.
- *Pile foundations:* These types of foundations are used on wet, marshy, or sandy soils. Piles are driven into the ground until they encounter solid soil or rock. Piles are large poles made of steel, wood, or concrete. They are widely used for high-rise buildings, marine docks, and homes in areas that flood easily.

Each type of foundation is built in a unique way. Let us consider a spread foundation. The site is surveyed to locate the foundation, **Figure 17-10.** Then the site is excavated in preparation for the footings and the walls. If the building is to have no footings, excavation does not go as deep. Buildings with basements require deeper

Figure 17-10. This worker is surveying a site for a new building. The survey will locate where the foundation will be placed. (Inland Steel Co.)

Figure 17-11. This worker is excavating a hole for a pool.

excavations. Footing forms are set up next. Forms are a lumber frame to hold the wet concrete until it cures (hardens). Forms give the footings or slabs height and shape. Concrete is poured and leveled off. When the concrete is cured, the forms are removed. Walls of poured concrete or concrete block are built atop footings. Slabs are ready for aboveground superstructures. **Figure 17-11** shows an excavation for a pool. Wooden foundations use no concrete for either footings or walls.

Career Corner
Carpenters

Carpenters help construct buildings, highways, bridges, factories, and other structures. Some carpenters do all types of work, while others specialize in doing a specific job, such as setting forms for concrete, framing walls and partitions, laying hardwood floors, or installing interior and exterior trim. Carpenters must know local building codes and be able to work from blueprints or instructions from supervisors. Those who remodel homes must be able to do all aspects of a job and, therefore, require a good basic overall training.

Carpentry work is somewhat strenuous and requires standing, climbing, bending, and kneeling. Carpenters must be competent in using tools and power equipment. Carpentry is considered a skilled trade. Most carpenters learn their trade through on-the-job training, vocational education, or apprenticeships.

Figure 17-12. The materials used for framework are lumber, steel, and reinforced concrete.

Building the Framework

The foundation becomes the base for the next part of the building, the framework. Erecting the framework gives the building its size and shape. The framework includes the floors, interior and exterior walls, ceilings, and roof. Also, the location of doors and windows are set up at this time.

The framework can be built out of three different materials, **Figure 17-12.** Small and low-cost buildings have frameworks made from ***lumber.*** Industrial and commercial buildings have either ***steel*** or ***reinforced concrete*** frameworks.

Building the framework involves three steps. First, the floor is constructed, **Figure 17-13.** Homes with slab foundations use the surface of the slab as the floor. Homes with basements or crawl spaces use lumber floors.

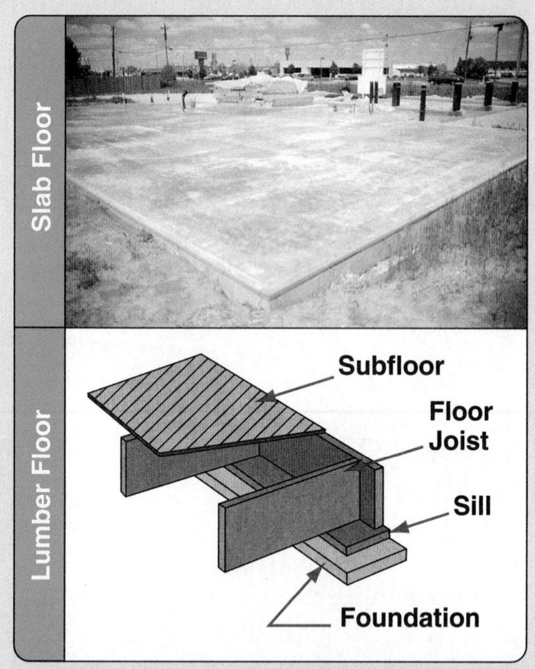

Figure 17-13. The floors in single-family homes are either concrete slabs or lumber.

Lumber floors start with a wood ***sill*** that is bolted to the foundation. ***Floor joists*** are then placed on the sill. They extend across the structure. Floor joists carry the weight of the floor. The size and spacing of the joists will be determined by the span (distance between outside walls) and the load on the floor. On top of the joists a ***subfloor*** is installed, usually made from plywood or particleboard. After the building is enclosed flooring material will be installed on top of the subfloor.

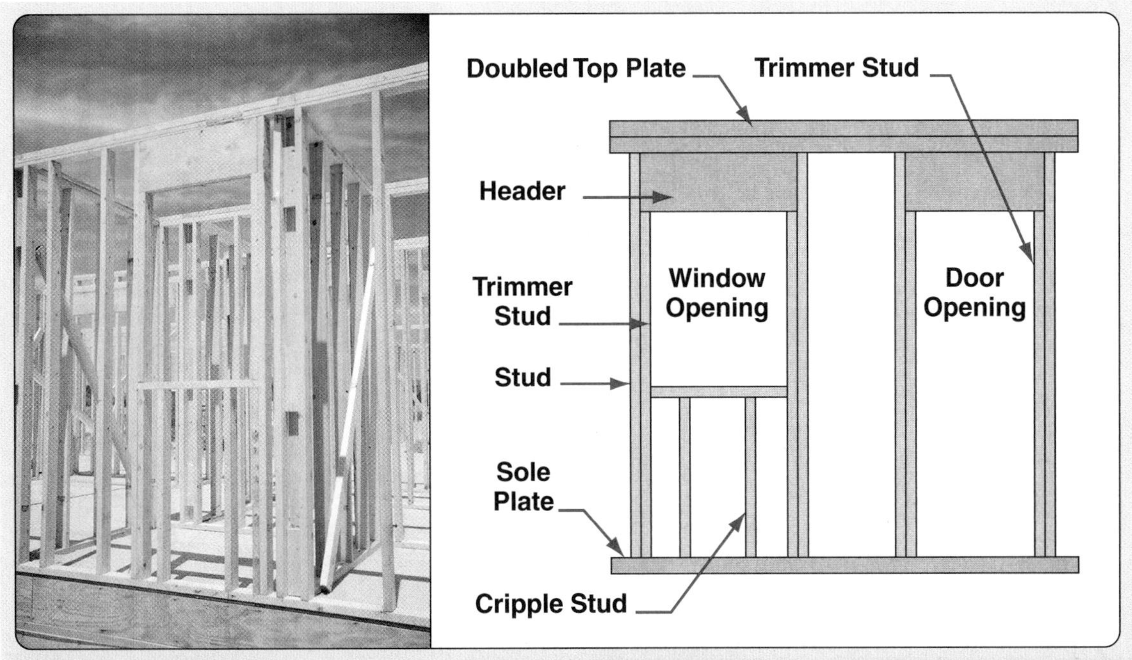

Figure 17-14. This photo shows many of the parts of a wood-framed wall.

The wall frames are placed on top of the floor. These frames support both exterior and interior walls. Wall framing is often made of 2 x 4 or 2 x 6 construction-grade lumber, **Figure 17-14.** A framed wall has a strip at the bottom called the ***sole plate.*** Nailed to the sole plate are uprights called ***studs.*** The length of the studs is set by how high the ceilings will be. At the top of the wall the studs are nailed to double ribbons of 2 x 4s called a ***top plate*** or wall plate. Door and window openings require headers above them. Headers carry the weight from the roof and ceiling across the door and window openings. ***Headers*** are held up by shorter studs called trimmer studs.

The walls support the ceiling and roof, **Figure 17-15.** The ***ceiling*** is the inside surface at the top of a room. The roof is the top of the structure that protects the house from the weather.

Ceiling joists support the ceiling. These joists rest on the outside walls and some interior walls. Interior walls that help support the weight of the ceiling and roof are called load-bearing or bearing walls.

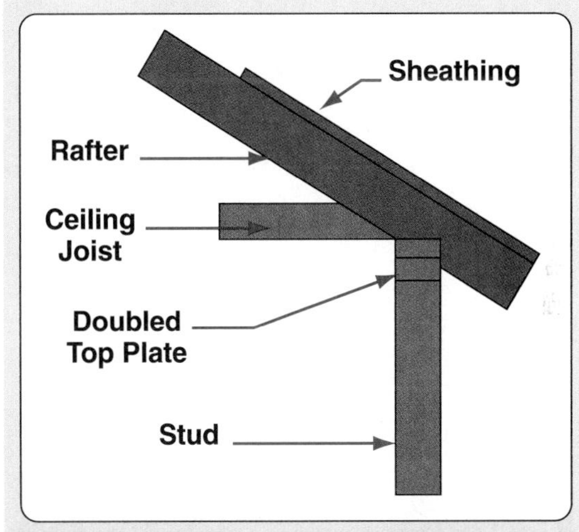

Figure 17-15. This illustration shows how the roof meets the wall frame.

The roof forms the top of the building. There are many types of roofs including flat, gable, hip, gambrel, and shed, **Figure 17-16.** The type of roof is chosen for its appearance and how it withstands the weather. For example, flat roofs are poor choices in areas with heavy snow. This type of roof cannot easily support the weight of deep snow. Likewise, a hip roof would look out of

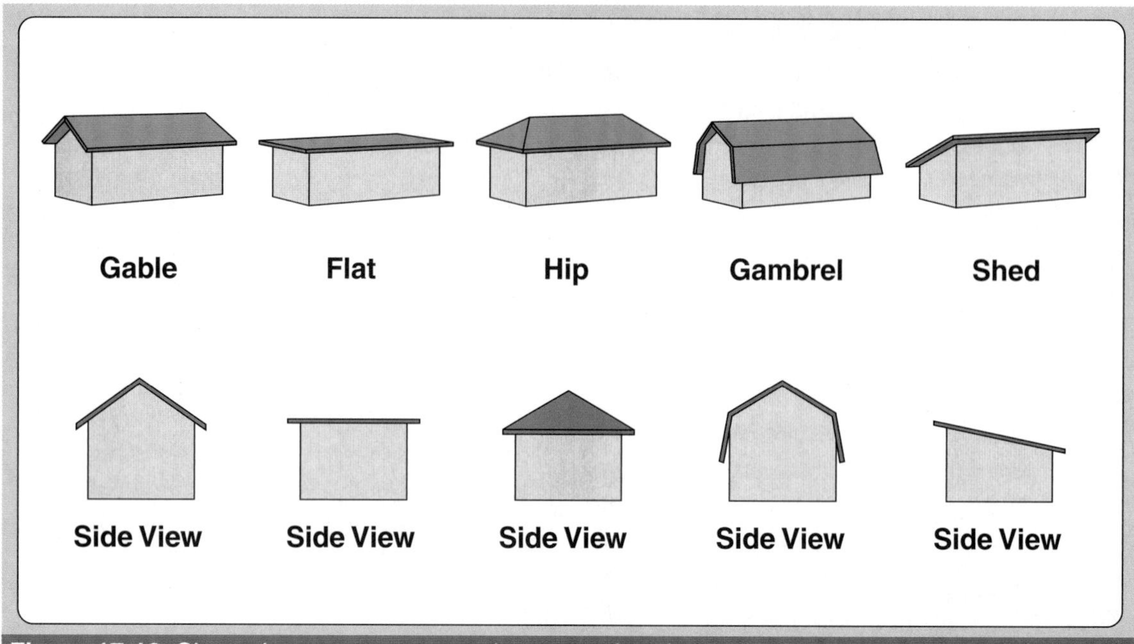

Figure 17-16. Shown here are some popular types of roofs used on homes.

place on Spanish-type homes. This kind of roof would not give the "Spanish-style" look.

Roof construction involves two steps. First, the roof frame is built with rafters. ***Rafters*** are angled boards that rest on the top plate of the exterior walls. Often a special structure called a truss is used. A ***truss*** is a triangle-shaped structure that includes both the rafter and ceiling joist in one unit. Trusses are manufactured in a factory and then shipped to the building site.

The rafters or trusses are covered with plywood or particleboard sheathing. This step completes the erection of the frame.

Enclosing the Structure

After the framework is complete the structure needs to be enclosed. The roof and wall surfaces need to be covered. This process has two steps: enclosing the walls and installing the roof.

With regard to enclosing the walls, we should note that all homes have both interior and exterior wall coverings. These coverings improve the looks of the building and keep out the elements (rain, snow, wind, and sun).

The first step is enclosing the exterior walls, **Figure 17-17.** This involves ***sheathing*** (covering) all the exterior surfaces. Plywood, fiberboard, or rigid foam sheets are used to sheath the walls. Most foam sheets have a reflective backing to improve the insulation value of the sheet. Most homes constructed today have a layer of plastic over the sheathing to prevent air from leaking in.

With regard to roof installation, normally the roof is put in place before the utilities are installed, **Figure 17-18.** The actual roof surface has two parts. Sheathing is applied over the rafters. This sheathing may be plywood or waferboard. Now the roofing material is installed. Builder's felt is often applied over the roof sheathing. Wood or fiberglass shingles, clay tile, or metal roofing is then installed over the sheathing and felt. Flat and shed roofs often use a built-up roof. A built-up roof starts with laying down sheets of insulation. Roofing felt is laid down, followed by a coat of tar, which is covered with gravel.

On many structures the overhang of the roof is also finished. A ***fascia*** board is used to finish the ends of the rafters and the

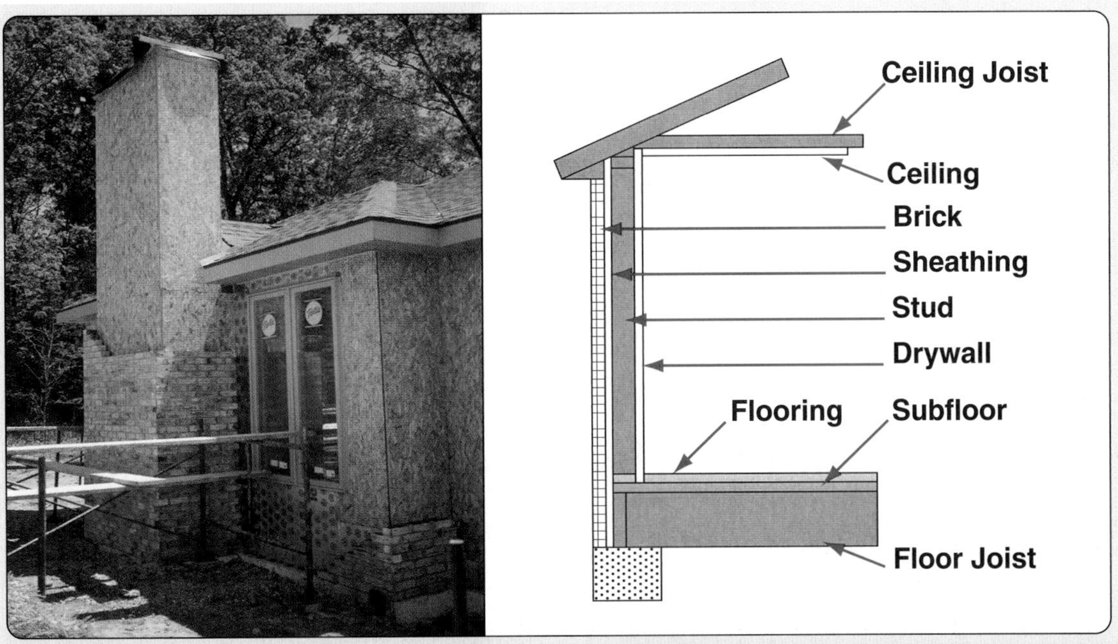

Figure 17-17. This is a cross-section diagram of a finished wall. The photo shows brick being applied as a siding material.

overhang. The ***soffit*** is installed to enclose the underside of the overhang. The soffit can be made of aluminum, vinyl, or plywood. Soffits must have ventilation holes or vents to prevent moisture and heat build up in the attic.

Once the sheathing and roof are installed, the openings for doors and windows are cut out. Then the doors and windows are set in place. Now the house is secure and weather tight.

Installing Utilities

Normally the utilities are installed after the building has been enclosed. This prevents theft and damage from the weather. Some parts of the utilities are installed earlier, such as large plumbing lines. The utility system includes four major systems:

- Electrical
- Plumbing
- Climate control
- Communications

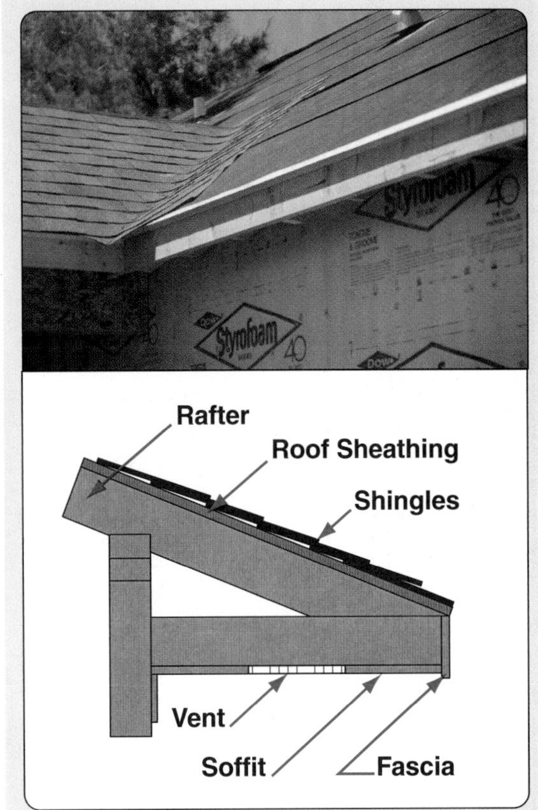

Figure 17-18. These are the parts of a finished roof. The photo shows asphalt shingles being installed on a new roof.

Electrical

The electrical system delivers electrical power to the different rooms of the home. The power is brought into the house through wires to a meter and distribution panel. This panel splits the power into 110-volt and 220-volt circuits. Each circuit has a circuit breaker to protect against current overloads.

Appliances such as clothes dryers, electric ranges, water heaters, and air conditioners require 220-volt power. Circuits for smaller appliances use 110 volts. Outlets may have power fed to them at all times. Outlets can also be controlled by switches. **Figure 17-19** shows a 110-volt circuit with wall (duplex) outlets and a ceiling light. You will note that the outlets will always have power. The circuit to the light has a switch, however.

Most 110-volt circuits are limited to 15 or 20 amps. Therefore, a number of different circuits are required to supply various parts of the home. A kitchen might have one or two circuits because of how many appliances are used there. One circuit might feed two bedrooms because there are few appliances in these rooms.

Plumbing

The plumbing system has two separate parts. One part supplies potable water, **Figure 17-20.** ***Potable water*** is safe for drinking. The other part of the system carries away wastewater. Plumbing fixtures and systems are designed to prevent mixing of potable water and wastewater and to stop sewer gas from leaking into the dwelling.

The potable water system starts with a city water supply or a well for the house. The water enters the house through a shutoff valve. The water may pass through a water conditioner. This device removes impurities such as iron and calcium.

The water line is split into two branches. One line feeds the water heater. The other line feeds the cold water system. Separate hot and cold water lines feed fixtures in the kitchen, bathrooms, and utility room.

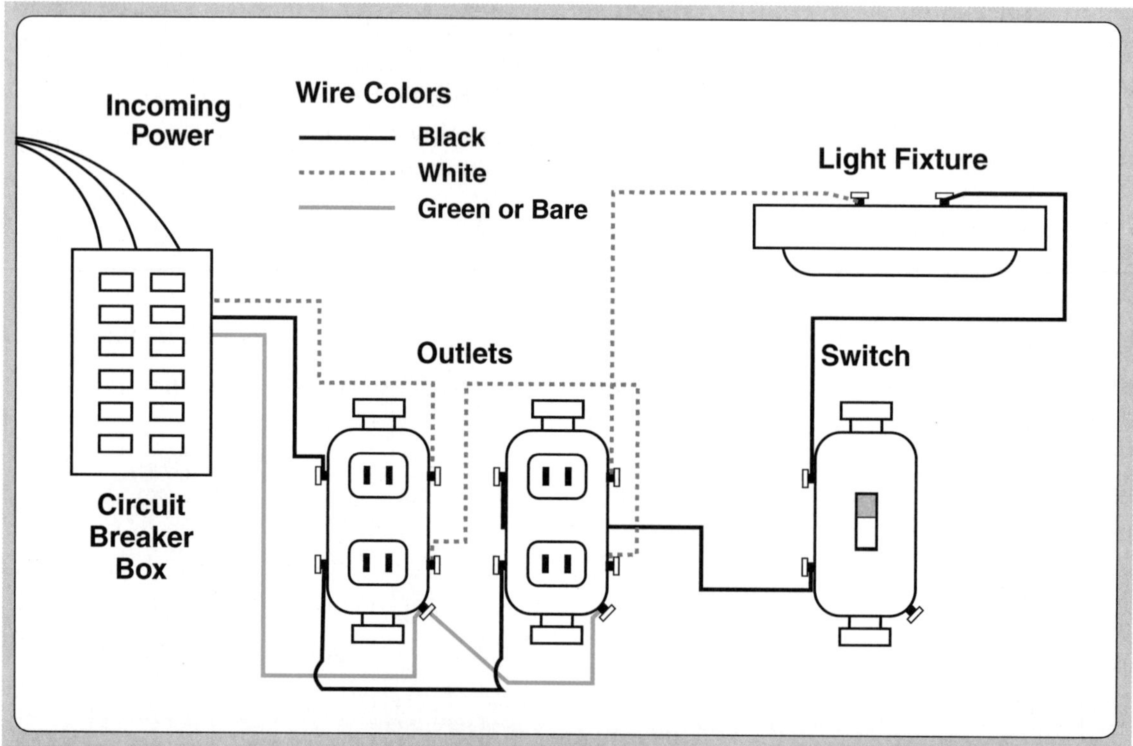

Figure 17-19. This diagram shows a 110-volt electric circuit. Note how the light is controlled by the switch but the outlets are not.

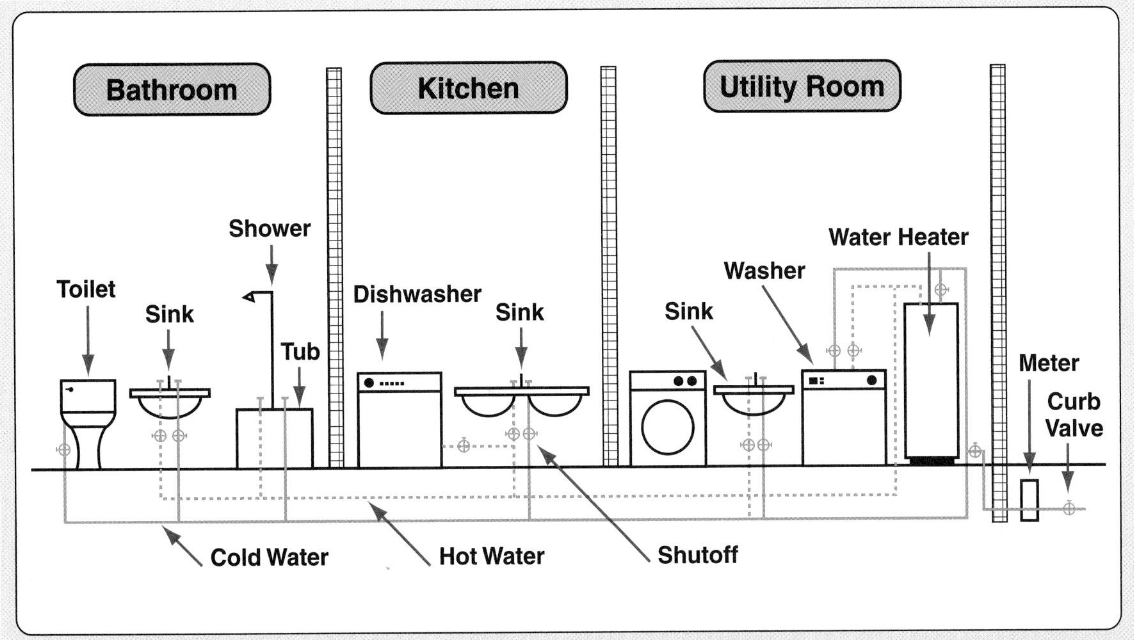

Figure 17-20. This schematic diagram shows the potable water system for a home. Each fixture has a shutoff valve.

Toilets, however, receive only cold water. Most water lines have shutoff valves before they reach the fixture.

For example, the water lines under a sink should have a shutoff valve. The valve allows repairs to be made without stopping the water flow to the rest of the house.

The second part of the plumbing system is the ***wastewater*** system. This system carries used water away from sinks, showers, tubs, toilets, and washing machines. The wastewater is routed to a city sewer line or to a septic system. At each of these fixtures and appliances, a device called a trap is provided. A trap is a U-shaped piece of pipe that remains full of water. The water in the line stops gases from the sewer system from leaking into the home. Wastewater systems have a network of vents to prevent the water from being drawn out of the traps. The vents also allow sewer gases to escape above the roof without causing any harm.

Homes that use natural gas have a third type of plumbing. Gas lines carry natural gas to furnaces, stoves, water heaters, and other appliances. Shutoff valves are installed at the entrance and at the major appliances.

Climate Control Systems

In many homes the climate control system is used to heat the building in winter and cool it in summer. This may be done with a single unit or with separate heating and cooling units.

With regard to heating systems, rooms in a home can be directly or indirectly heated. In a direct heating system, the fuel is used in the room to be heated. Direct heating may use a stove or a fireplace that burns wood or coal. Note, however, that burning wood or coal can be expensive and causes considerable air pollution.

Other direct heating methods use electrical power. These systems use resistance heaters that are installed in the walls or along the baseboards. Also, ceiling radiant wires or panels may supply the heat.

Indirect systems heat a conduction medium such as air or water. This medium then carries the heat to the rooms. The heat is then given off to the air in the room,

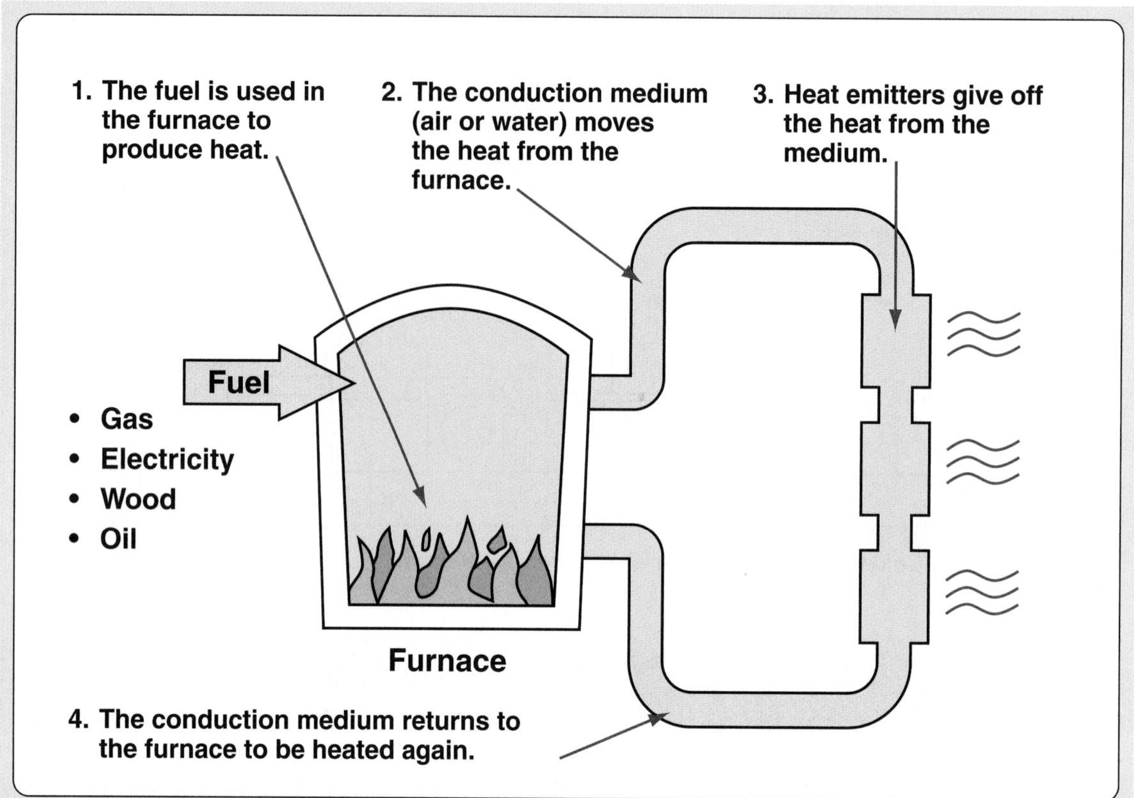

Figure 17-21. Indirect heating systems heat a conduction medium. The resulting heat is then put into the air in the room.

Figure 17-21. The energy sources for these systems are electricity, coal, oil, wood, natural gas, or propane.

Furnaces that heat air as a conduction medium are called *forced-air heating* systems. Forced-air furnaces draw air from the room. This air is heated as it moves through the furnace. A fan delivers the heated air through ducts to various rooms.

Hot water heating uses water to carry the heat. The water is heated in the furnace and pumped to various rooms. The water passes through room-heating units that have metal fins that surround the water pipe. The fins dissipate the heat into the room.

Some homes are heated with solar energy. Two types of solar homes exist: active and passive. Passive solar homes use no mechanical means to collect and store heat from the sun. Active solar homes use pumps or fans to move a liquid or air to collect solar heat. After the heat is collected the liquid or air is moved to a storage device. Solar heating systems are very effective in areas that have ample sunshine.

With regard to cooling systems, many buildings have cooling systems to cool the air during the warm parts of the year. Cooling systems use compressors, evaporators, and condensers much like a refrigerator. The system has a fan that draws the air from the room. The air passes over a cold evaporator. This is similar to a forced-air furnace. Instead of the air being heated, it is cooled. The cool air is returned to the room.

Another system used in climate control is a unit called a ***heat pump.*** A heat pump works as a cooling and a heating system. A heat pump can be operated in two directions. Operating in one direction the heat pump acts like an air conditioner. The heat pump takes warmth from inside the house and discharges it outside. In the winter

Connections to Technology: **Communication**

Word Origins

Probably most people would be happy to have their name become part of everyday language as a result of their invention. John McAdam, a Scottish engineer who experimented with road construction, might be doubly pleased; his work with roads has resulted in two words in common use today.

As described in this chapter, John McAdam developed the crushed stone road. This new type of road had three layers of crushed rock compacted into a solid mass. The road was also made slightly convex. McAdam's design improved roads tremendously because now the traffic load was spread, and rainwater ran off the surface. We are now more familiar with this type of roadway through the use of the term *macadam.*

The other word is even more familiar and is also related to roads. In an effort to improve roads even more, people used tar to bind the crushed rock together. This process was given the name tarmacadam or, as we now call it when we use it on runways, *tarmac.*

Modern roads are still built using John McAdam's principles. Can you find another common word we use today that is based on someone's name and invention?

the heat pump works in reverse. It takes warmth from the outside and brings it into the house. Heat pumps can use air, water, or the ground as a heat source. Heat pumps that use air work best in areas that do not get very cold, such as the southern and central United States. Groundwater heat pumps use well water as the heat source. The water is pumped from the well to the heat pump. The water has heat removed or the water receives excess heat, and then the water returns to the well. Ground coil systems are buried in the soil to take or give off heat. Groundwater and ground coil systems can be used in colder climates. Otherwise, heat pumps need a small furnace or other auxiliary heat source as a backup.

Communication Systems

Most homes have communication systems such as telephone, radio, and television. These systems require special wiring. Telephone wiring and television cables are normally installed during the construction of the building. Installing them after a building is finished is costly. It takes considerable work to feed the wires through attics, under floors, and inside walls. Some homes have intercom systems that allow two-way communication between rooms. Many intercom systems allow radio programs to be played throughout the house.

Finishing the Exterior and Interior

The final exterior finishing step is installing siding and trim. Siding is the finish covering used on a wood building. Many siding materials are in use. Wood shingles and boards, plywood, hardboard, brick, stone, aluminum, vinyl, and stucco are all used as siding. Look back at Figure 17-17. You will see bricks being installed over plywood sheathing. Trim is the strips of wood that cover the joints between window and door frames and the siding.

The interior walls are the next walls to be finished. Insulation is placed between the studs and around the windows and doors of all exterior walls. Insulation reduces heat loss on cold days and heat gain on hot days. The most common type of insulation is fiberglass. It is available in blankets or batts. A vapor barrier of polyethylene film is attached to the studs over the insulation. The vapor barrier prevents moisture from building up in the insulation.

Once insulation and utilities are in place, the interior wall surfaces can be covered. The most widely used interior wall covering is gypsum wallboard, commonly known as ***drywall.*** Drywall has replaced plaster in most applications. Drywall is a sheet material made of gypsum bonded between layers of paper. The sheets of drywall are nailed or screwed onto the studs and ceiling joists. The fastener heads and drywall seams are then covered with a coating called joint compound. The compound is applied in several thin coats. This is done to make smooth surfaces and joints between the sheets of drywall.

Figure 17-22. This site was finished by grading the lot (top) and planting landscaping (bottom).

The inside and outside of the house are now ready for the finishing touches. Interior wood trim is installed around the doors and windows. Kitchen, bathroom, and utility cabinets are set in place. Floor coverings such as ceramic tile, wood flooring, carpet, or linoleum are installed over the subflooring. Baseboards are installed around the perimeter of all the rooms. The exterior siding and wood trim is painted. Interior trim is painted or stained. The walls are painted or covered with wallpaper or wood paneling. Lighting fixtures, switch and outlet covers, towel racks, and other accessories are installed. The floors and windows are cleaned. Now the home is finished and ready to be occupied.

Completing the Site

Completing the building is the major part of the project. Other work remains to be done, however. The site must be finished. Earth is moved to fill in areas around the foundation. Sidewalks and driveways are installed.

The yard area needs to be landscaped. ***Landscaping*** helps to prevent erosion and improves the appearance of the site. Trees, shrubs, and grass are planted. Landscaping can divide the lot into areas for recreation and gardening. Landscaping can be used to screen areas for privacy, direct foot traffic, and shield the home from wind, sun, and storms.

Look at **Figure 17-22.** The top view shows dirt being moved onto the site for landscaping activities. The bottom view shows a finished landscaped area. Notice how the trees and lawn improve the appearance. Also, note that a grassy mound is used to guide people onto the sidewalk.

Heavy Engineering Structures

Construction activities do not always produce buildings. We need and use many other types of constructed structures. These structures are sometimes called civil structures, or heavy engineering structures. These structures include highways, rail lines, canals, pipelines, power transmission and communication towers, hydroelectric and flood control dams, and airports. They provide the paths for the movement of water, people, goods, information, or electric power. Each of these structures is the result of ***heavy engineering construction.***

These projects can be grouped in various ways. For this discussion, we will group them into transportation, communication, and production structures.

Transportation Structures

Transportation systems depend on constructed structures. These structures include railroad lines, highways and streets, waterways, and airport runways. Other constructed works help vehicles cross uneven terrain and rivers. These structures include bridges and tunnels. Pipelines are land transportation structures that are used to move liquids or gases over long distances.

Let us look at some examples of these constructed works. We will discuss roadways and bridges.

Roadways

Roads are almost as old as civilization. People first used trails and paths to travel. Later, they developed more extensive road systems. The Romans built the first engineered roads more than two thousand years ago. Their influence remained until the 1700s, when modern road building started. Today's roads have their roots in the work of the Scottish engineer John McAdam. He developed the crushed stone road. His roads were built of three layers of crushed rock that was laid in a ribbon about 10″ (25 cm) thick. Later, this roadbed was covered with an asphalt-gravel mix. Asphalt roads are very common today. A more recent development is the concrete roadway.

Building a road starts with selecting and surveying the route. Next, the route is cleared of obstacles such as trees, rocks, and brush. The roadway is graded so that it will drain. Drainage is important to prevent road damage from freezing and thawing. Also, a dry roadbed withstands heavy traffic better than a wet, marshy one. Another reason for grading is to keep the road's slope gentle. Elevation changes are described using the term *grade.* Grades are expressed in percentages. A road with a 5 percent grade would gain or lose 5′ of height for every 100′ of distance. Most grades are kept below 7 percent.

Once the roadbed is established, the layers of the road are built, **Figure 17-23.** The graded dirt is compacted, and a layer of coarse gravel is laid. This is followed with finer gravel that is leveled and compacted. Next, the concrete or asphalt top layer is applied. Concrete roads are laid in one layer. Asphalt is generally applied in two layers: a coarse undercoat and a finer topcoat. Finally, the shoulders or edges of the road are prepared. The shoulders can be gravel or asphalt.

Bridges

Another constructed structure vital for transportation is the bridge. Bridges provide a path for vehicles to move over obstacles. These obstacles include marshy areas, ravines, other roads, or bodies of water. Bridges can carry a number of transportation systems. These systems include highways, railroads, canals, pipelines, and footpaths.

Generally there are two types of bridges: fixed and movable. A fixed bridge does not move. Once the bridge is set in place it stays there. Movable bridges can change

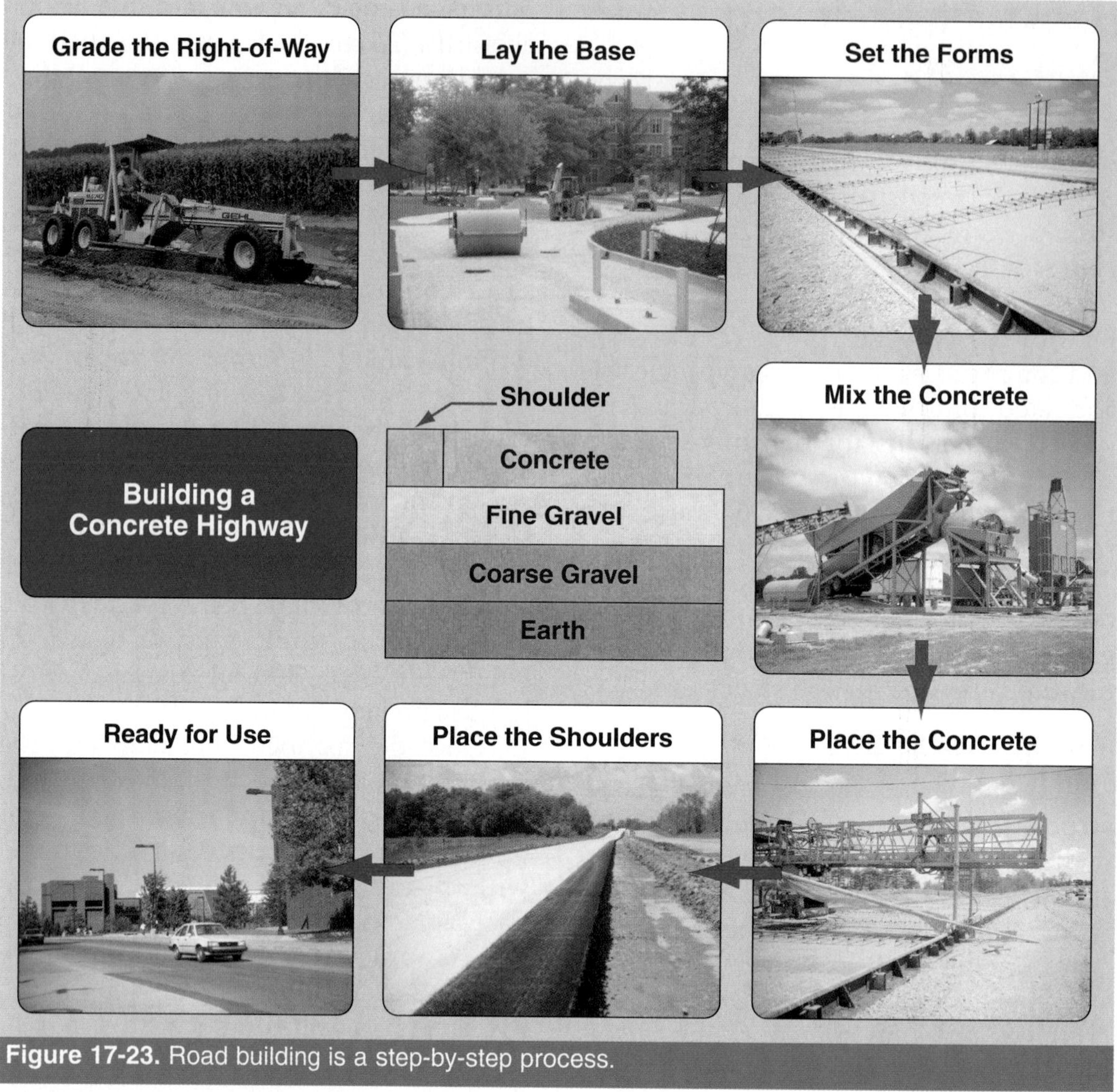

Figure 17-23. Road building is a step-by-step process.

their position to accommodate traffic below it. This type of bridge is used to span ship channels and rivers. The bridge is drawn up or swung out of the way so that ships can pass.

Bridges have two parts, **Figure 17-24.** The substructure spreads the load of the bridge to the soil. The abutments and the piers are parts of the substructure. The superstructure carries the loads of the deck to the substructure. The deck is the part used for the movement of vehicles and people across the bridge.

The kind of superstructure a bridge has indicates the type of bridge. The types of bridges are beam, truss, arch, cantilever, and suspension, **Figure 17-25.**

Beam bridges use concrete or steel beams to support the deck. This type of bridge is widely used when one road crosses another one. Beam bridges are very common on the interstate highway system.

Truss bridges use small parts arranged in triangles to support the deck. These bridges can carry heavier loads over longer spans than beam bridges. Many railroad bridges are truss bridges.

Arch bridges use curved members to support the deck. The arch may be above or below the deck. Arch bridges are used for longer spans. One of the longest arch bridges spans more than 1650′ (502 m).

Cantilever bridges use trusses that extend out like arms. The ends of the arms can carry small span or hook up to each other. The load is transmitted by the arms to the center. None of the load is carried by the ends of the arms.

Suspension bridges use cables to carry the loads. A large cable is suspended from towers. From the large cable, smaller cables drop down to support the deck. Suspension bridges can span distances as great as 4000′ (1220 m) and longer.

Communication Structures

Most telecommunication technology relies on constructed towers to support antennas. These towers are usually placed on a concrete foundation. A steel tower is built on top of the foundation. Once the tower is complete the signal wiring can be installed. Similar techniques are used to construct towers for power transmission lines, **Figure 17-26.**

Production Structures

Some structures that are used for production activities are not buildings. For example, petroleum refineries are a mix of machinery and pipelines. Irrigation systems are constructed to bring water to farms in dry areas. Evaporation basins are built to recover salt and other minerals from seawater.

Another important production structure is the dam. Dams are used for controlling floods, supplying water, making recreational lakes, or generating electricity.

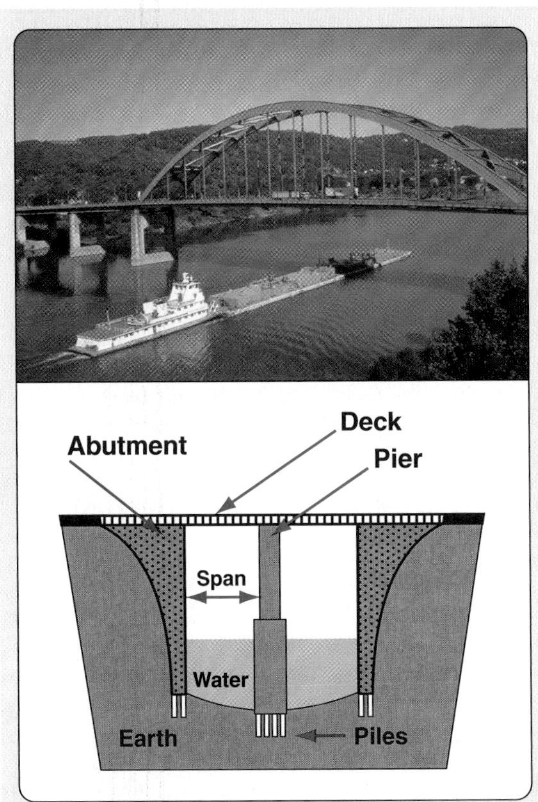

Figure 17-24. This diagram shows the parts of a bridge. An arch bridge is shown in the photo. (American Electric Power Co.)

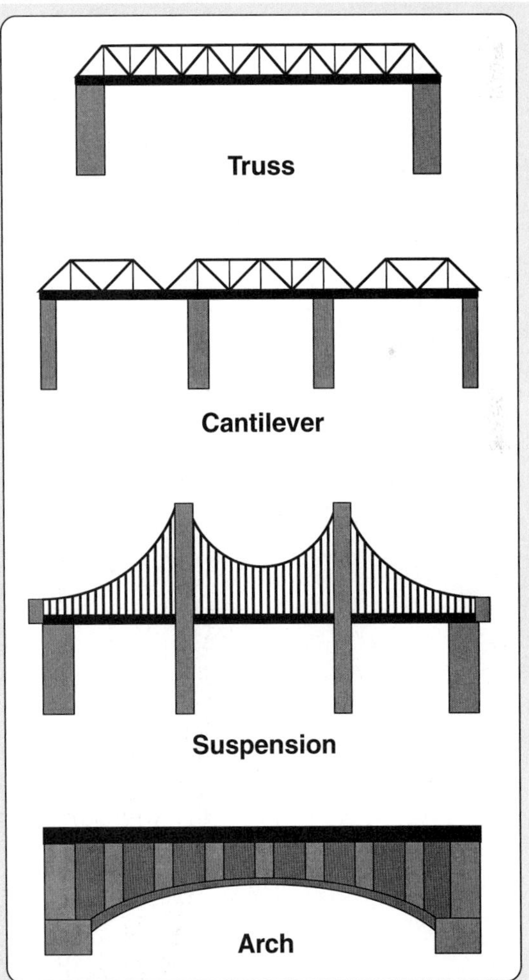

Figure 17-25. Four of the five types of bridges are the truss, the cantilever, the suspension, and the arch. The fifth is the beam bridge.

Figure 17-26. This helicopter is helping to construct a tower for an electricity transmission line. (American Electric Power Co.)

Several types of dams exist. One type is called a ***gravity dam.*** Its lakeside is vertical whereas the other side slopes outward. The sheer weight of the concrete the dam is made from holds the water back. The dam on the left in **Figure 17-27** is a gravity dam.

Two more types of dams are the rock dam and the earth dam. The earth dam is also shown in Figure 17-27. A rock dam looks like two gravity dams placed back-to-back. Both sides slope outward. Rock and earth dams must be covered with a waterproof material to prevent seepage. Clay is often used for this covering.

A ***buttress dam*** uses its structure to hold back the water. This type of dam is not solid. It uses walls of concrete to support a concrete slab or arches against the water.

Tall dams that hold back large quantities of water are called arched dams. The arched shape increases the strength of the dam. The arched shape also spreads the pressure onto the walls of the canyon where the dam is built.

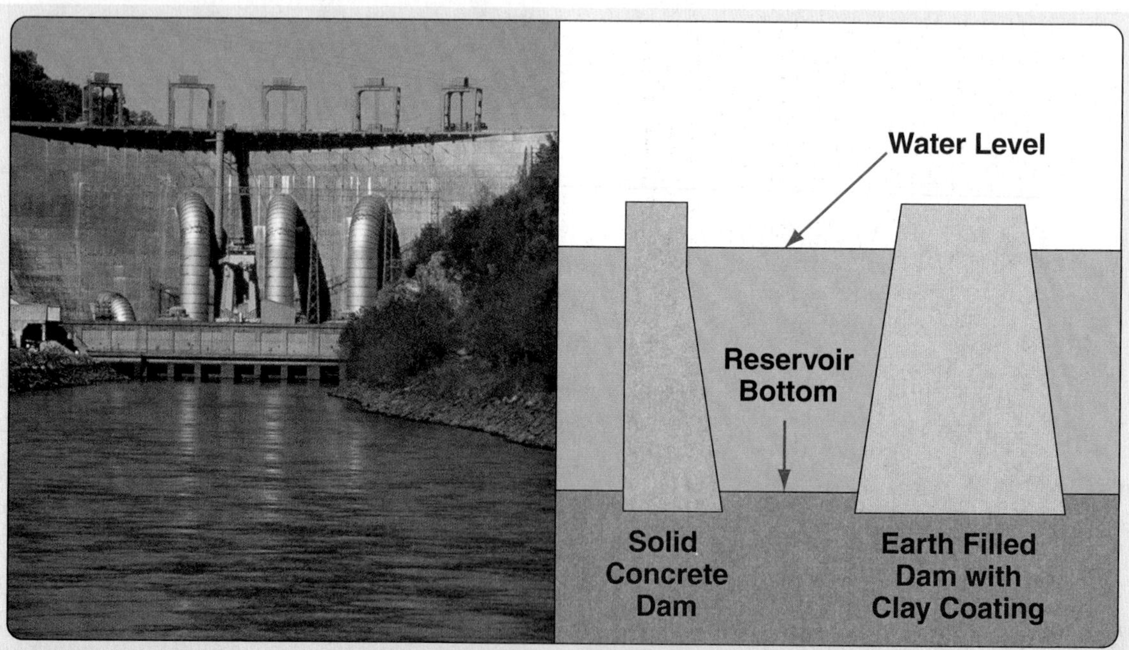

Figure 17-27. The drawing shows two common types of dams. The photo shows a gravity dam. (American Electric Power Co.).

Summary

Construction is a vital production activity. Construction provides us with homes, offices, factories, highways, railroads, pipelines, bridges, dams, and other structures. Construction can be divided into two kinds of projects: those that produce buildings and those that produce heavy engineering structures.

Most buildings are constructed using the same steps. The site is cleared and prepared for construction. The foundation for the structure is set. Then, the framework or superstructure is erected. Utilities are installed, and the structure is enclosed. The building is finished, and the site is landscaped

The construction of heavy engineering structures varies according to type. We can group these kinds of projects into three categories: transportation structures, communication structures, and production structures.

Key Terms

arch bridges
beam bridges
buildings
buttress dam
cantilever bridges
ceiling
ceiling joists
commercial buildings
construction
drywall
fascia
floor joists
forced-air heating
gravity dam
headers
heat pump
heavy engineering construction
heavy engineering structures
hot water heating
industrial buildings
landscaping
lumber
manufactured home
pile foundations
potable water
rafters
reinforced concrete
residential buildings
sheathing
sill
slab foundations
soffit
sole plate
spread foundations
steel
studs
subfloor
suspension bridges
top plate
truss
truss bridges
wastewater

Test Your Knowledge

Write your answers on a separate piece of paper. Please do not write in this book.

1. List the two kinds of constructed works.
2. *True or false?* A condominium is a residential structure that is generally owned by the people living in it.

For Questions 3–15, match the construction step on the right to the correct description on the left. (Note: Some letters will be used more than once.)

Description	Construction Step
3. _____ Sheathing the walls	a. Preparing the site
4. _____ Putting up drywall	b. Setting foundations
5. _____ Grading	c. Building the framework
6. _____ Putting in footings	d. Enclosing the structure
7. _____ Landscaping	e. Installing utilities
8. _____ Putting in heat pump	f. Finishing interior and exterior
9. _____ Installing a subfloor	g. Completing the site
10. _____ Installing the roof	
11. _____ Driving in piles	
12. _____ Marking the building site	
13. _____ Placing floor joists	
14. _____ Adding baseboards	
15. _____ Installing sidewalk	

16. *True or false?* You should use spread foundations on wet or sandy soils.
17. The two types of floors in single-family homes are concrete slab or _____.
18. What is a fascia board used for?
19. The two types of water systems that are part of a plumbing system are called _____ and _____ systems.
20. The most common type of insulation is _____.
21. A home built in a factory is called a(n) _____.
22. What does the term *grade* mean as used in this chapter?
23. *True or false?* Most railway bridges are beam bridges.
24. *True or false?* With cantilever bridges, none of the load is carried by the end of the arms.
25. Tall dams that hold back large quantities of water are called _____ dams.

Applying Your Knowledge

1. Use a chart like the one below to list and describe a few of the constructed structures you see as you travel from your home to school.

Structure	Type of Construction	Description–Use

2. Select one structure that you saw in completing the previous assignment. Make a drawing or model of the structure and label the major parts.

This activity develops the skills used in TSA's Structural Engineering event.

Structural Engineering

Activity Overview

In this activity, you will create a balsa-wood bridge and determine its failure weight (load at which the bridge breaks).

Materials

- Grid paper
- 20′ of 1/8″ x 1/8″ balsa wood
- 3″ x 5″ note card
- Glue

Background Information

General. There are several types of bridges: beam, truss, cantilever, suspension, and cable-stayed. The type of bridge used in a particular situation is generally determined by the length of the span and available materials. For this activity, a truss design is considered the most efficient.

Truss bridges. The truss bridge design is based on the assumption that the structural members carry loads along their axes in compression or tension. The members along the bottom of the bridge carry a tensile load. The members along the top of the truss carry a compressive load. The members connecting the top and bottom chords (members) can be in tension or compression.

Gussets. Gussets are plates connected to members at joints to add strength. Gussets are normally used in steel construction. The structural steel members are welded or bolted to the gusset. When designing your bridge, include a gusset at each joint, if possible.

Wood properties. Due to its molecular structure, wood can normally carry a larger load in tension than it can in compression. Also, a shorter member can carry a greater compressive load than a longer member.

Guidelines

- You must create a scale sketch of the bridge before building.
- Two pieces of balsa wood can be glued together along lengthwise surfaces. No more than *two* pieces of balsa can be glued together. You cannot use an excessive amount of glue.
- Gussets cut from the 3″ x 5″ card can be no larger than the diameter of a U.S. quarter coin. A gusset cannot touch another gusset. A gusset cannot be "sandwiched" between two pieces of balsa wood.
- The bridge design must take into account the loading device. Your teacher will provide specific guidelines for bridge length, width, and required details for attachment of a loading device.
- Your bridge will be weighed before being loaded.

Evaluation Criteria

Your project will be evaluated using the following criteria:

- Accuracy of sketch compared to completed bridge
- Conformance to guidelines
- Efficiency (failure weight ÷ bridge weight)

Chapter 18
Using and Servicing Products and Structures

Learning Objectives

After studying this chapter, you will be able to:

- Describe the factors to consider when selecting a product or structure.
- List the five steps used in installing a product.
- Describe factors involved in maintenance.
- List the steps used in repairing.
- Describe activities involved in altering products and structures.
- Explain the importance of recycling.
- Describe ways to dispose of materials.

People use manufactured products and constructed structures every day. They choose the items that meet their needs. People make their choices to live better. Sometimes, however, people misuse some of their belongings. People can also choose the wrong item, use it improperly, or maintain it poorly.

If technology is to help us live better, we must choose the correct product or structure for our needs or for the task at hand. This involves six steps: selecting, installing, maintaining, repairing, altering, and disposing of technological products, **Figure 18-1.**

Selecting Technological Products

The next time you go to a store, stop and look around you, **Figure 18-2.** You will see thousands of products on display. Each of these products is developed for a specific use. Some products help to fulfill basic needs or perform essential tasks. For example, stoves cook our food. Automobiles and buses transport us to work and school. Other products are designed to make life more pleasant. For example, games, videotapes, and recorded music help to entertain us. Colorful pictures and works of art brighten our homes and workplaces.

Choosing the best product to fill a need or want is quite a challenge. *Needs* come first because they are the things that are necessary for living. *Wants* are requirements that are not necessary for living or to complete a task but are nice to have. It is not always easy to determine what is a need and what is merely a want. Once you have separated needs and wants, you must decide how much money can be spent on meeting the needs. You probably cannot afford to pay for all of your wants at one time. Therefore, you will have to rank them by their importance to you. Such essentials as food and clothing will rank high on your list. Decorative items may rank low on your list.

Knowing what you need leads to the next step. This step is selecting the best of the many products that fit your needs. Some people buy products on impulse. A better way is to analyze the products. This allows you to think about the products you want to buy. You should consider three important factors:

- ***Function:*** How well will the product meet your needs? How well does it work? Is the product durable and easy to maintain? How well will it do the job you have in mind?

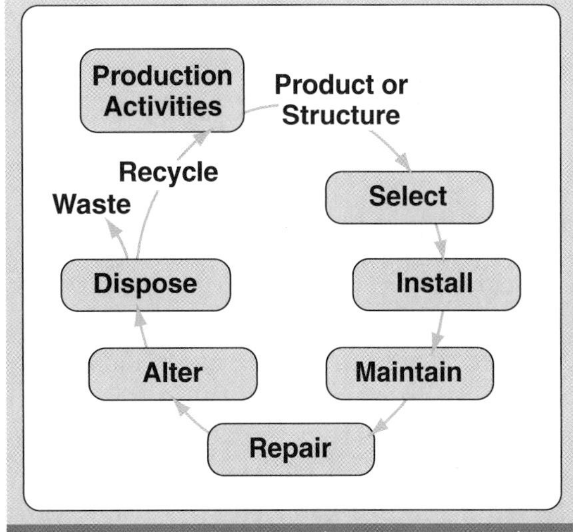

Figure 18-1. Using a product or structure in the proper way involves six steps.

Figure 18-2. Thousands of products are made to meet our needs and wants.

- ***Value:*** Does the performance of the product match its price? Do other products meet your needs as well but cost less? Is it worth the selling price?
- ***Appearance:*** Is the design and color of the product pleasing to you? Is it something you would be proud to own?

Installing Technological Products

Some products are ready to use when you buy them. Other products must be installed. For example, items such as cookware, clothing, and tools are sold ready to use. No set up is required. Other products require installation, such as the dishwasher shown in **Figure 18-3.** The purchaser or a service person must complete several tasks to make a product ready to use.

First, products must be *unpacked.* Products are shipped from the factory in protective crates and boxes. Boxes, crates, and rigid foam components protect the product from damage during shipment. The boxes and packing materials must be removed and properly discarded or recycled.

Second, some products require *utilities* in order to work. The products are attached to electrical, water, and wastewater lines. Natural gas or compressed air may also be needed.

Third, the product may need to be *positioned* and secured to the floor or inside a cabinet. It may also need to be leveled to operate properly.

Fourth, the product may require *adjustment.* Electrical meters may need to be set on zero. Clearances between doors may need to be adjusted. The space between moving parts should always be checked.

Finally, most products must be *tested.* For example, a sample roll of film may be shot to test a camera's operation and adjustments. A water conditioner may be cycled to see that its controls are operating.

Directions for the installation and use of many products are in the owner's manual, **Figure 18-4.** This information may be on one sheet of paper or in a large book. The complexity of the product determines the size of the owner's manual.

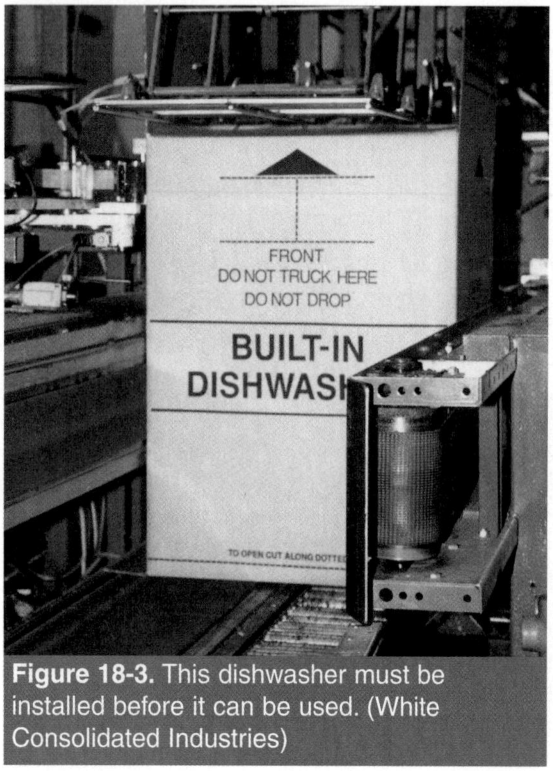

Figure 18-3. This dishwasher must be installed before it can be used. (White Consolidated Industries)

Legs at the front corners of the freezer should be set so the freezer is firmly positioned on the floor, and the front is raised just enough so the door closes easily when opened about halfway.

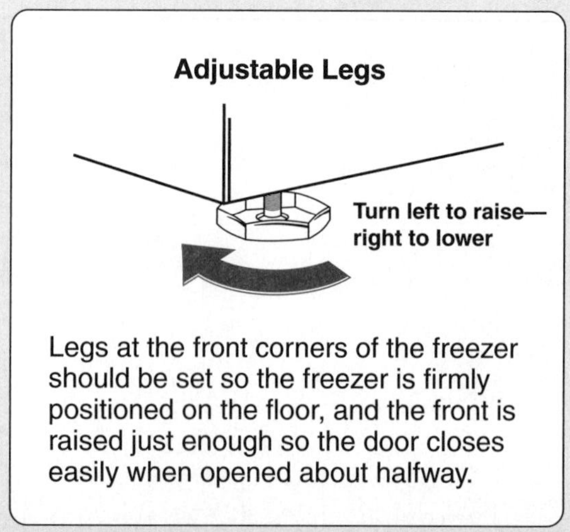

Figure 18-4. This is an example of installation instructions for a product. (General Electric Co.)

Maintaining Technological Products

We want products to work properly when we need them. This often requires a maintenance program. The goal of ***maintenance*** is to keep products in good working order, **Figure 18-5.** Clothing and dishes are washed to maintain their usefulness and extend their life. Filters in furnaces and air conditioning units are changed so the units function properly. The oil in automobile engines is changed to maintain the oil's ability to lubricate moving parts. Locomotives and buses are serviced to maintain their performance.

Figure 18-5. This mechanic is performing routine maintenance on an aircraft engine. The maintenance keeps the engine in good working order. (United Parcel Service)

Most maintenance is done on a schedule, **Figure 18-6.** The schedule is designed to keep the product working properly. Therefore, maintenance is sometimes called ***preventive maintenance.*** It is designed to *prevent* breakdowns.

Many products come with a maintenance manual. This document lists:

- The types of maintenance needed
- Methods for performing maintenance
- A time schedule for each maintenance task

Buildings and other constructed structures need maintenance just like manufactured products. Buildings must be cleaned and painted. Windows are cleaned. Roofs are sealed to prevent leaks. Bridges and communication towers are painted to prevent rusting. Railway tracks are leveled, and switches are lubricated. Streets and driveways receive periodic coatings to seal out water, **Figure 18-7.**

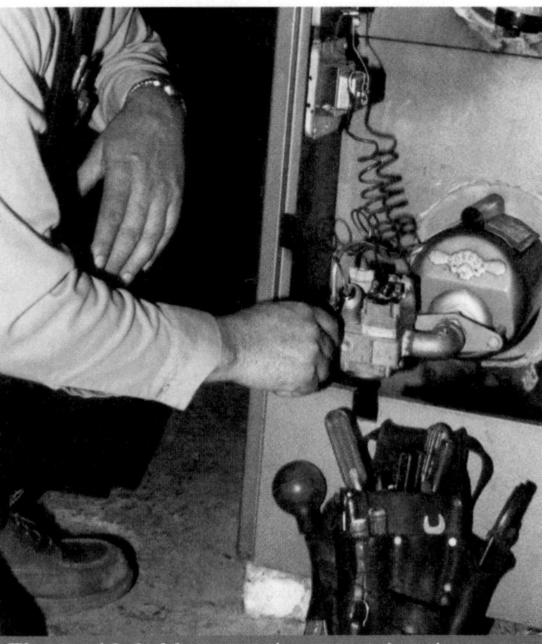

Figure 18-6. Many products need maintenance done periodically. This technician is performing a yearly check of a gas furnace.

Figure 18-7. Painting is an example of home maintenance.

Repairing Technological Products

No product or structure will work all the time or last forever. Some products are used until they stop functioning and are then discarded. For example, few of us try to salvage bent paper clips or bolts with stripped threads.

Many products, however, are too costly to discard the first time they stop working. Throwing away a bicycle or a car every time a tire goes flat would be very expensive. It costs less money to repair the product. ***Repair*** is the process of putting a product back into good working order.

Career Corner

Building Inspectors

Building inspectors examine the construction work done on various types of buildings. Their job is to make sure the structures meet building codes and zoning regulations. Almost half of all inspectors work for local governments. Other inspectors conduct home inspections as part of the home-purchasing process. Building inspectors generally work alone and typically work regular hours. They spend time inspecting construction work sites and in field offices, reviewing blueprints, answering letters, writing reports, and receiving telephone calls.

Inspectors should have technical knowledge, experience, and education. Many inspectors have formal training and experience in the aspect of construction they inspect. Generally, inspectors must have at least a high school diploma and might be required to have studied engineering or architecture or taken courses in building inspection. Most states and cities require some type of certification for employment. This requires proof of appropriate construction experience and education and passing an examination on code requirements, construction techniques, and materials.

This requires three steps, as shown in **Figure 18-8.**

- *Diagnosis:* The cause of the problem is determined.
- *Replacement* or *adjustment:* Worn or broken parts are replaced. Misaligned parts are adjusted.
- *Testing:* The repaired product must be tested to ensure that it works properly.

The information needed to repair a product is contained in the product's service manual. This manual provides a parts list, as shown in **Figure 18-9,** so that repair parts can be ordered. It also gives

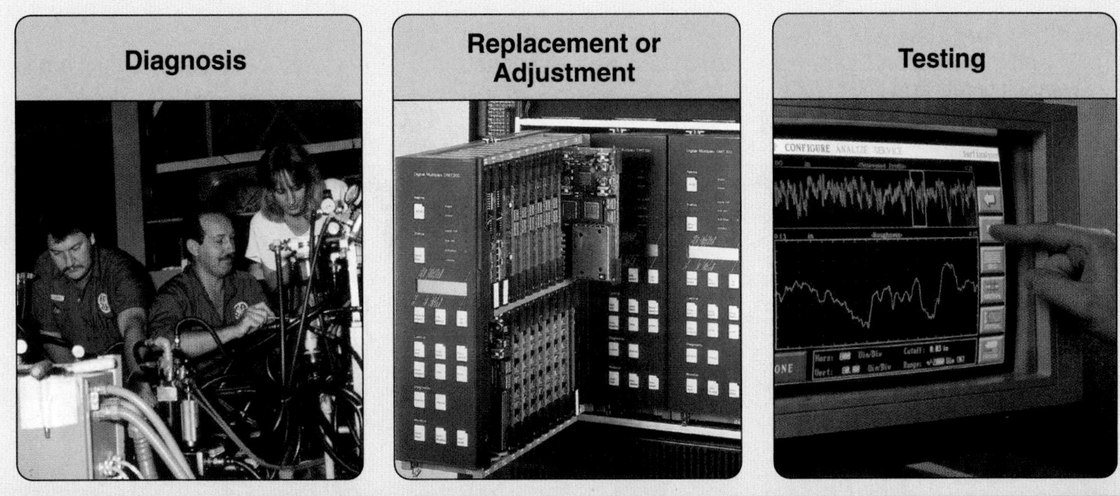

Figure 18-8. Three steps are followed when repairing a product. (General Electric, Hewlett-Packard, Federal Products Corp.)

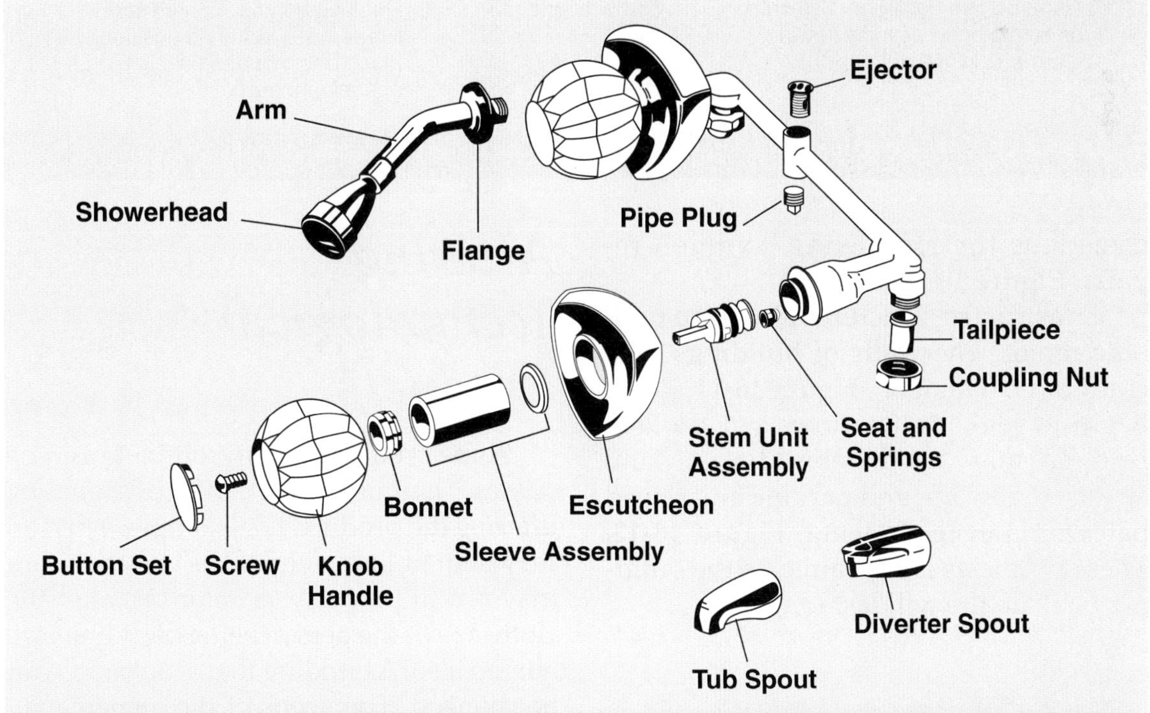

Figure 18-9. This exploded pictorial view provides a parts list. This helps a service person order repair parts.

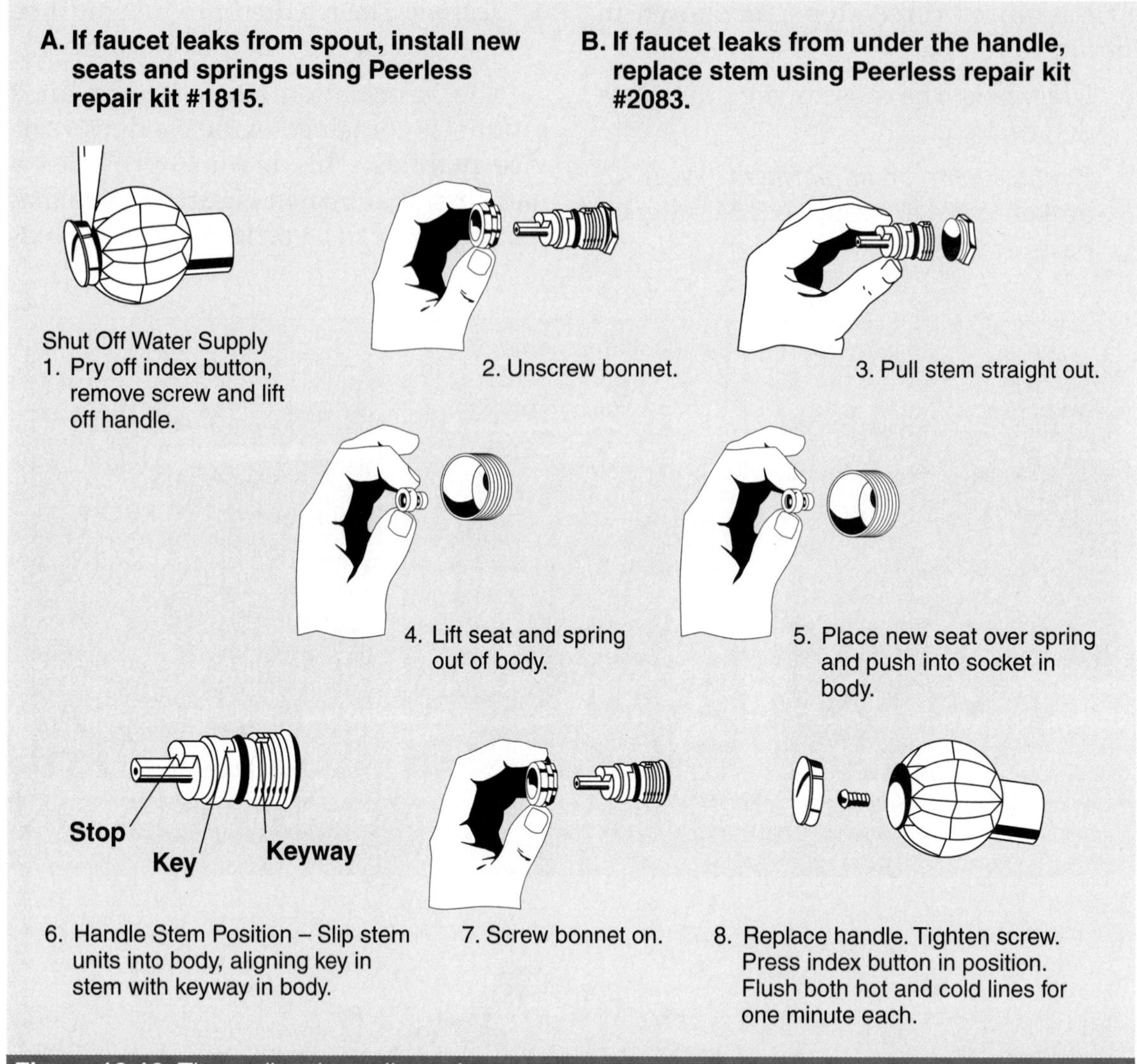

Figure 18-10. These directions allow a homeowner to repair a bathroom faucet.

directions for completing common repairs, **Figure 18-10.**

Constructed structures also need periodic repair. The walls of buildings may crack or be damaged, **Figure 18-11.** Roofs begin to leak, or windows get broken. Bridges may need rebuilding. Many streets, highways, and parking areas need patching and resurfacing, **Figure 18-12.** These problems are repaired so the building or structure will last longer.

Altering Technological Products

Some products become obsolete as time passes. Their useful life can be extended by ***altering*** the product. For example, a person may gain or lose weight. His or her clothes may not fit properly. A tailor can alter the clothes to fit the person better. Buildings are altered also. A building that is outdated can be changed. The needs of the owners may change. The rooms might be too small, or the windows might be too large. Contractors

Figure 18-11. Workers sometimes need to repair a faulty wall of a building.

Figure 18-12. The parking lot shown here is being repaired by applying a cold-mix asphalt (left) and compacting it (right).

can alter (remodel) the building to meet current needs, **Figure 18-13.** Remodeling can involve restoring or changing the appearance of a building.

Some altering is done to change the performance of a product or structure. Memory chips may be added to improve a computer's speed. A better set of tires and wheels can be installed on a car to improve its handling. New lighting fixtures and better heating and cooling systems can lower a building's energy use. Scrubbers are added to electrical generating plants to reduce sulfur emissions. These actions help products and structures to last longer and work better.

Figure 18-13. The exterior of this historic building has been preserved. Its interior has been altered to provide modern offices for a law firm.

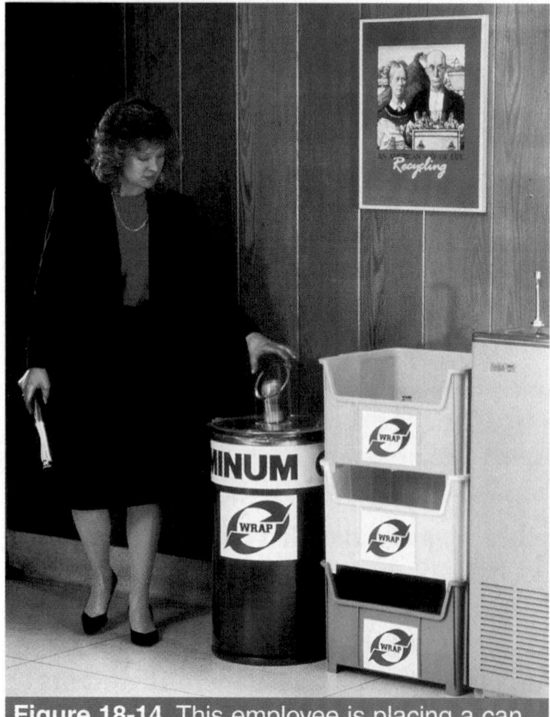

Figure 18-14. This employee is placing a can in a container as part of a company-sponsored recycling program. (Boise Cascade)

Disposing of Technological Products

All products and structures eventually reach the end of their useful life. They become so worn that they no longer can be repaired or altered. At that point, the owner has the responsibility to dispose of them properly, **Figure 18-14.**

The first choice for disposal should be ***recycling.*** This means that the materials in the product or structure are reclaimed. The reclaimed material is used to make new products.

Recycling can reduce the strain on both resources and landfill disposal sites. The average home produces an enormous amount of waste. Typically, household garbage contains the following materials:

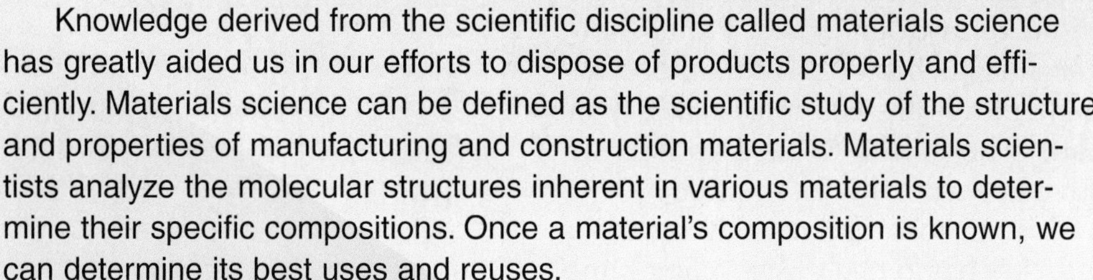

Knowledge derived from the scientific discipline called materials science has greatly aided us in our efforts to dispose of products properly and efficiently. Materials science can be defined as the scientific study of the structure and properties of manufacturing and construction materials. Materials scientists analyze the molecular structures inherent in various materials to determine their specific compositions. Once a material's composition is known, we can determine its best uses and reuses.

Scientists have placed these materials into four specific categories for study: metals, plastics, ceramics, and composites. Each exhibits a specific set of properties. For example, most glass (a ceramic) is composed primarily of silica, sodium oxide, and calcium oxide. These properties make glass easy to melt and reshape. Thus glass is a good material to recycle because it can be easily heated and shaped into new forms.

On the other hand, plastics are created from fractions of natural gas or crude oil changed with chemicals into solid form. These chemicals include benzene, cadmium compounds, and carbon tetrachloride. Plastics do not degrade easily and never completely decompose. Furthermore, because of their chemical properties, plastics, when burned, release many pollutants into the air. Thus recycling efforts are focused on reusing the material as is and in reducing its volume through such methods as pulverizing and dissolving the material with a solvent.

In what ways are the other two types of materials (metals and composites) disposed of, and why are these methods used?

- Paper and paperboard—42 percent
- Food and yard wastes—24 percent
- Glass—9 percent
- Metals—9 percent
- Plastics—7 percent
- Other—9 percent

Newsprint is the largest single component in landfills. Newsprint accounts for nearly 15 percent of the volume. The paper can be de-inked (the ink is removed) and used to make cardboard and new newsprint. Magazines and coated book paper are more difficult to recycle because of the coatings used to give the paper a glossy finish.

Some food wastes and most yard wastes can be placed in a compost bin. The waste will decompose and become excellent garden fertilizer. Glass is easily recycled into new glass. It readily melts and can be reformed into containers and sheet glass. Metal is also easily recycled. Nearly half of all aluminum is now coming from recycled beverage cans.

Plastics provide a unique recycling challenge. There is not just one plastic. Hundreds of types are in use today.

Household packaging is divided into seven categories for recycling. Each of these is identified by a code number inside a triangular symbol, **Figure 18-15.**

Some materials in household waste are very dangerous. These materials are called household hazardous waste. Paints, solvents, engine oil, batteries, and other hazardous chemicals are examples of this type of waste. They should be disposed of through a hazardous waste center.

In addition to consumers doing their part, industry has joined the recycling campaign. Steelmakers use recycled iron and steel as base material in the steelmaking process. Lumber mills use scraps from its processes to make paper, particleboard, and hardboard. Lumber mills also burn scrap lumber to provide power for the mills.

It is important that all companies dispose of their hazardous materials through licensed subcontractors. Paving contractors are using material ground from highway surfaces as aggregate for new blacktop. Old concrete becomes excellent flood control fill and is also used as aggregate.

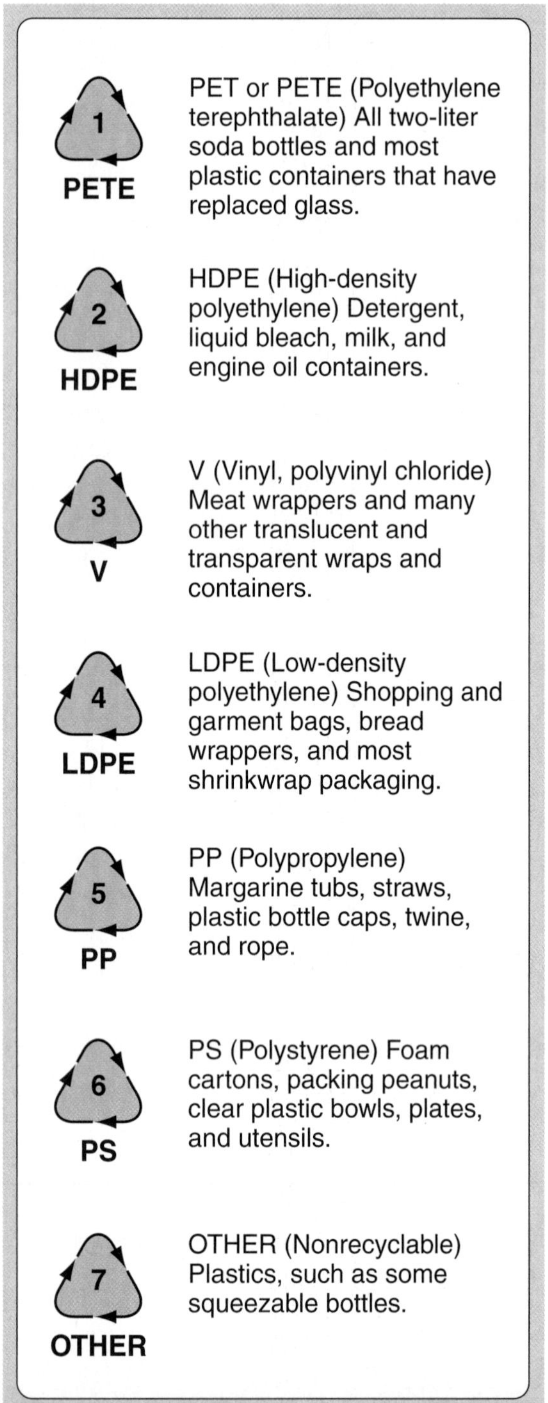

PET or PETE (Polyethylene terephthalate) All two-liter soda bottles and most plastic containers that have replaced glass.

HDPE (High-density polyethylene) Detergent, liquid bleach, milk, and engine oil containers.

V (Vinyl, polyvinyl chloride) Meat wrappers and many other translucent and transparent wraps and containers.

LDPE (Low-density polyethylene) Shopping and garment bags, bread wrappers, and most shrinkwrap packaging.

PP (Polypropylene) Margarine tubs, straws, plastic bottle caps, twine, and rope.

PS (Polystyrene) Foam cartons, packing peanuts, clear plastic bowls, plates, and utensils.

OTHER (Nonrecyclable) Plastics, such as some squeezable bottles.

Figure 18-15. These symbols are used on plastic containers and materials to aid in sorting for recycling.

Summary

Using products makes our lives better. However, we must use and dispose of products wisely. We should select products that meet our needs and our ability to pay for them. Instruction manuals should be carefully read and the instructions followed. All products should be used only for the purpose for which they were designed. Each product and structure should receive periodic maintenance and necessary repairs. Finally, each product should be disposed of properly after it has served its purpose. Whenever possible, products should be recycled. Recycling helps to reduce the strain on natural resources and disposal sites.

Key Terms

adjustment
altering
appearance
diagnosis
function
maintenance
preventive maintenance
recycling
repair
replacement
testing
value

Test Your Knowledge

Write your answers on a separate piece of paper. Please do not write in this book.

1. What are the three major factors to consider when analyzing whether to purchase a product?
2. List the five steps followed in installing a product.
3. Changing the oil in an engine is part of a(n) _____ maintenance program.
4. Name the three major items discussed in a maintenance manual.
5. List the three steps involved in repairing a product.
6. *True or false?* Service manuals contain parts lists.
7. Give one example of altering that could be done to change the performance of a product or structure.
8. Why is recycling important?
9. *True or false?* All products can be recycled.
10. Give one example of a specific material and a manner in which it can be disposed.

Applying Your Knowledge

1. Select one day in your life. List all the items that you throw away. Determine their type and whether they can be recycled.
2. Select a complex product that you use often. List:
   a. The preventive maintenance it requires.
   b. The repairs that it needs now or may need in the future.

Section 5 Activities

Activity 5A

Design Problem—Manufacturing Technology

Background

A major task in developing a manufacturing system is to select and properly sequence the production operations.

Situation

Balsum Manufacturing Company has purchased the rights to a game. See **Figure 5A-1.** Golf tees or 1" long, 1/4" diameter dowels are used as pegs. You will need five light colored and five dark colored pegs. This game is a modification of tic-tac-toe developed in central Africa. The rules are as follows:

Each player uses four pegs. The object of the game is to get three pegs in a horizontal, vertical, or diagonal row. Play begins with each player alternately placing a peg in a hole. If no one has completed a row when all eight pegs are in place, the second phase starts. Each player alternately moves a peg into an empty hole adjacent to one of his or her pegs until one player makes a row and wins.

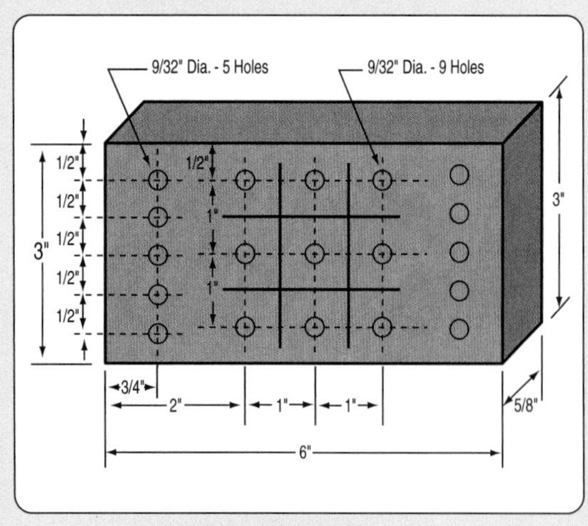

Figure 5A-1

Challenge

List the operations you would use to produce the product. Set up a simple production line to make one product for each member of the class.

Safety

Use the safe and appropriate procedures your teacher demonstrates.

Activity 5B

Fabrication Problem—Manufacturing Technology

Background

Manufacturing adds worth to materials by changing their form. This can be done by a number of different techniques.

Challenge

Study the drawing and operation sheet for the recipe card holder. See **Figure 5B-1.** The recipe cards hang from the dowel on split key rings. Build a holder individually or use a continuous production line.

Safety

Be sure to use the safe and proper practices your teacher demonstrates.

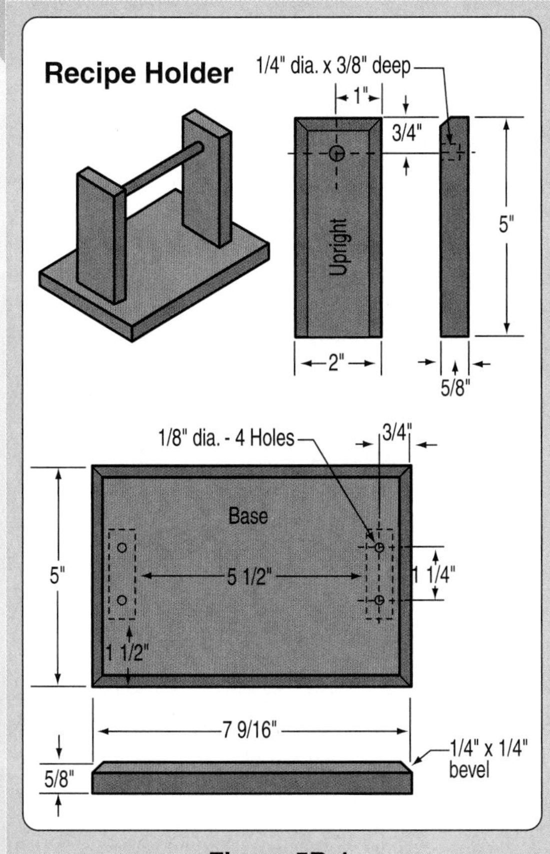

Figure 5B-1

Materials

- 5/8" thick x 5" wide random length pine (base)
- 5/8" thick x 2" wide random length pine (uprights)
- 1/4" diameter birch dowel
- 3" x 5" note cards
- Split key rings

Procedure

Use the procedure shown on the operation process chart, **Figure 5B-2.** Note that the parts are finished before assembly. This makes it easier to produce a high quality finish.

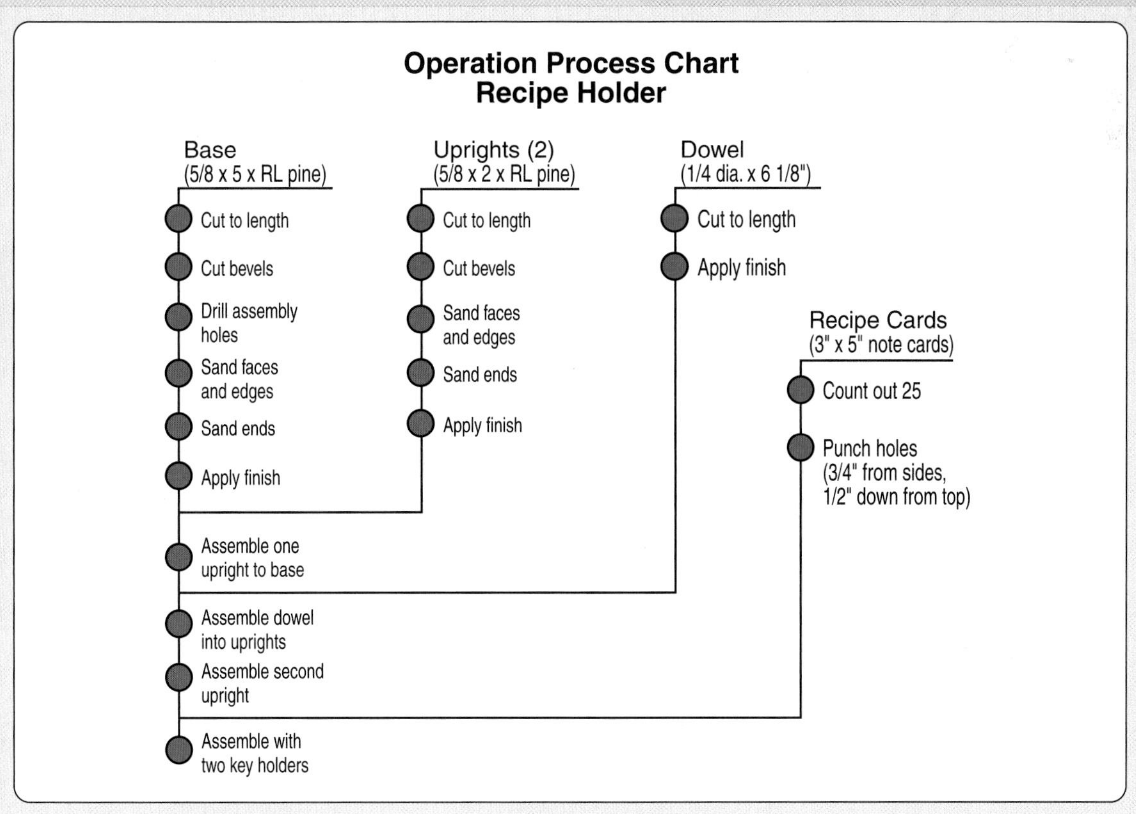

Figure 5B-2

Activity 5C

Design Problem—Construction Technology

Background

During harvest, farmers haul their grain to local and regional elevators. Often, more grain is brought in than can be processed. This extra grain is stored on the ground next to the elevator until it can be processed.

Situation

The managers of Southwest Grain Elevators expect to receive 20 percent more grain than they will be able to process during the month of August. In previous years, they stored the excess grain on the ground. Storing the grain in this manner results in a 10 percent reduction in quality.

Challenge

Design and construct a model of an inexpensive shelter for 8,000 ft^3 (equal to a pile 25' wide, 40' long, and 8' high) of grain.

Activity 5D

Fabrication Problem—Construction Technology

Background

More people are living in multiple family dwellings. Sound transmission between dwellings is a major concern. A number of ways have been developed to reduce the sound transmitted from one apartment to another.

Challenge

Select one of the high sound transmission class (STC) partition designs. See **Figure 5D-1.** Construct a model wall, using the design you selected.

Safety

Use the tools and techniques your teacher demonstrates.

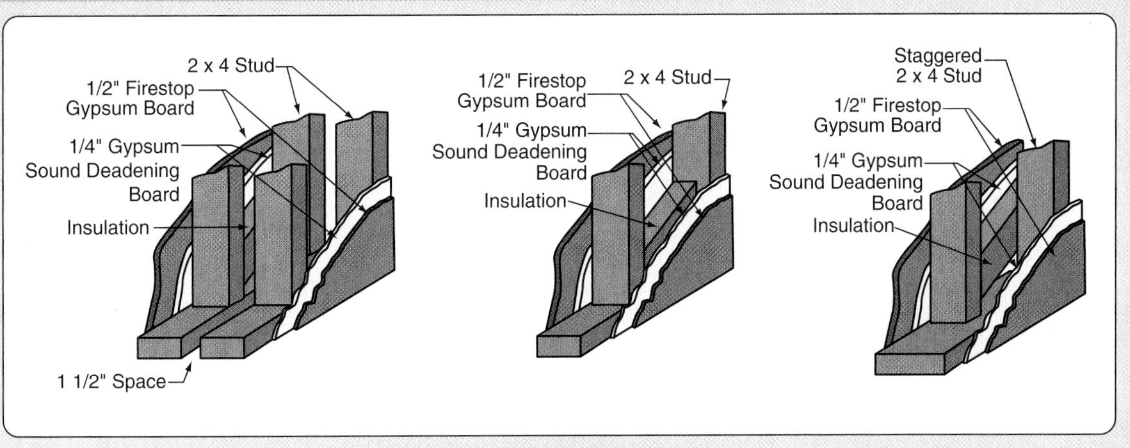

Figure 5D-1

Materials

- Scale 2 x 4 and 2 x 6 stock (scale may be 1/2 or 1/4)
- Scale 1/4" and 1/2" gypsum board
- Scale 2 1/2" fiberglass insulation
- Pencils
- Rulers
- Backsaw
- Brads
- Hammer

Procedure

1. Select the type of wall you will construct.
2. Build the model according to the design.
3. Test the sound transmission by placing a speaker on one side of the wall and a tape recorder and microphone on the other. Create a sound with the speaker. Observe the level of sound on the microphone side by using the level meter on the recorder.

Section 6

Applying Technology: Communicating Information and Ideas

Tomorrow's Technology Today

Fiber Optic Systems

Scientists and other researchers are currently working on making the information we transmit through fiber optics travel even faster than it does today. Fiber optics are extremely thin transparent fibers of glass or plastic that transmit data through light signals. They work individually and in bundles.

The fiber optic connections employed in networks today use a single ray of light to send information. The single ray is powerful, though—these systems can transmit millions of bits of data per second. The newer systems, however, divide the light ray into a spectrum of more than one hundred colors, and each channel (color) of light can move data—thus allowing the system to transmit up to *ten billion* bits of data per second. These systems will be used for advanced research and business applications and will change the way people use the Internet.

One benefit will be that information will be able to be transmitted at different speeds. This means that video, which needs faster speeds during transmission to keep from breaking up, will be able to be moved more quickly than text, eliminating some of the current break-up problems.

Through this new networking, researchers will be able to gain access quickly to vast amounts of statistics. This is especially important in today's Information Age, in which scientists continue to gather information at an ever-increasing rate. The mapping of the entire human genome is a good example of the immensity and complexity of some of the research projects scientists have undertaken in recent years.

This emerging networking technology holds the promise of helping us understand the world around us, especially in an age in which everything is interconnected. Researchers hope to have quicker access to more accurate data about everything from weather patterns to health statistics. With so much data at their fingertips, they are expected to be able to find connections and relationships that would not otherwise be seen.

These ultrafast superbroadband connections will transform the ways research and business are done in many settings. In the coming years, these new fiber optic systems will be used to connect schools, hospitals, libraries, research centers, and police and fire stations. Eventually, links will be made available to private residences as well.

The new fiber optic systems and other networking technologies are projected to continue developing at an exponential rate. The general guideline is that computer chips double their power every year and a half, but it is becoming clear that the fiber optic systems will soon surpass this rate. People are only beginning to feel the impacts of networking possibilities in their daily lives, and we can expect to see many advances in the coming years.

Chapter 19
Using Technology to Communicate

Learning Objectives

After studying this chapter, you will be able to:

- Define communication technology.
- Describe items that are communicated.
- Name the three goals of communication.
- Identify features of the communication model.
- List the types of communication.
- Identify the five types of communication technology systems.

Earlier in the text, we looked at data, information, and knowledge. You discovered that ***data*** are unorganized facts; ***information*** is organized data; and ***knowledge*** is information applied to a task. How do you obtain information? Where does information come from? These are important questions.

Think back over the day. Did you read a newspaper or listen to a radio newscast? Did you see a traffic signal or sign? Did you look at a speedometer? Did you listen to music or watch television? If you did any of these, you were using ***communication technology.*** Communication technology is a system that uses technical means to transmit information or data from one place to another or from one person to another.

Communication has always been an essential part of human life. Humans first communicated with gestures and grunts. Later they developed language, increasing their ability to communicate. These forms of communication, however, do not involve technology. There is no technical means between the sender of the message and the receiver.

Possibly the first use of communication technology was cave paintings. The "artists" used sticks, grass, or their fingers to apply paint to cave walls, **Figure 19-1.** The result was a message that could be stored. At some later date, a person could retrieve the message.

Over time, communication technology has expanded in range and complexity. To understand communication technology better, in this chapter we will discuss five aspects of communication: the major kinds of items that are communicated, the goals of communication, the communication model, types of communication, and communication systems.

Items That Are Communicated

Communication is used for a number of purposes. Most often it is used to convey ideas, exchange information, and express emotions, **Figure 19-2.** An ***idea*** is a mental image of what a person thinks something should be. You probably have ideas about what kind of music is good, how people should behave, or what activity is fun. You probably also have opinions on how to perform various tasks, such as washing an automobile, riding a bicycle, or mowing a lawn. People also have ideas on such issues as how to protect the environment and whether to allow capital punishment. Communication media can be used to share these ideas.

Information is vital to taking an active part in society. It provides a concrete foundation for decision making and action. Information can be as simple as the serving time for lunch. It can be as complex as the moon's effect on the tides. Communication media can be used to exchange information among people.

Finally, you may want to communicate ***emotions.*** Ideas and information are important. Feelings, however, are just as vital to many people. Communication media can

Figure 19-1. This is a reproduction of a painting found in an ancient Kiva (a Pueblo Indian structure) in the southwest part of the United States.

Idea

Information

Emotion

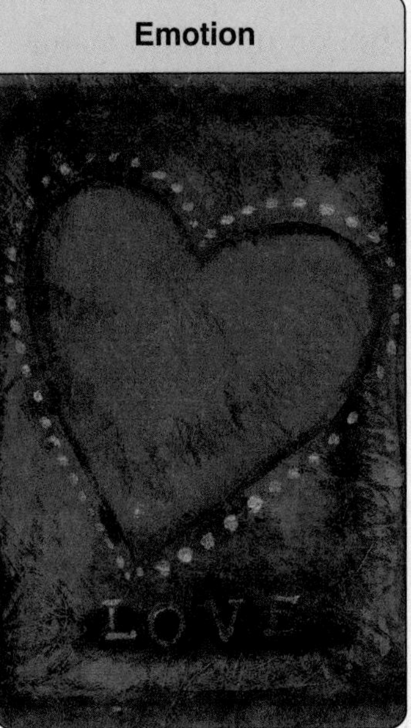

Figure 19-2. Communication technology can be used to convey ideas, information, or emotions.

convey these feelings. For example, a photograph can communicate the excitement of a sporting event. People can communicate affection for each other through greeting cards. Communication media can make us laugh or cry, make us excited or calm, and make us feel good or bad.

Goals of Communication

Each communication message is designed to impact someone. The communication can meet one or more of three basic goals:

- ***Inform*** by providing information about people, events, or relationships. We read books, magazines, and newspapers to obtain information. Radio news programs, television news programs, and documentaries are designed to provide information.
- ***Persuade*** people to act in a certain way. Examples include the "Say No to Drugs," "Buckle Up" and "Give a Hoot, Don't Pollute" campaigns. Print and electronic advertisements, billboards, and signs are typical persuasive communication media.
- ***Entertain*** people as they participate in or observe events and performances. Television programs, movies, and novels are common entertainment-type communication.

These three goals may be merged. Two new words in our language arise from this merging of goals. The first is ***infotainment,*** which means providing information in an entertaining way. You may learn as you watch a quiz show on television or play with computer simulations. Both of these are enjoyable ways of gaining new and useful information.

The second term, ***edutainment,*** takes communication one step beyond infotainment. Edutainment is more than allowing the information to be available in an entertaining way. It creates a situation in which people want to gain the information. The television program *Sesame Street* is a good example of edutainment.

The Communication Model

Communication can be thought of as a simple process, **Figure 19-3.** The action starts with the ***encoding*** of a message. Encoding means that the message is changed into a form that can be transmitted. The encoding may be electronic impulses on a magnetic tape, an arrangement of letters on a printing plate, or exposed chemicals on photographic film.

The message is then ***transmitted*** to the receiver. Transmission involves a communication channel or carrier. The ***channel*** might be electromagnetic waves broadcast through the air, electrical signals carried by a wire, pulses of light on fiber-optic cable, or printed text on paper.

The other end of the communication channel is the ***receiver.*** The receiver gathers and ***decodes*** the message. Examples of receivers are radios, which change electrical impulses into sound, and television sets, which change electrical impulses into a series of images. The human mind decodes the written and graphic messages contained in photographs and printed media.

Communication models can be more complex than the one shown in **Figure 19-3.** These models show ***interference*** in the communication channel. Interference is anything that impairs the accurate communication of a message. Static on a radio is interference. Noise in a movie theater is interference. Smudged type would be interference in printed messages.

It is important to be aware of the difference between information and noise. ***Noise*** is unwanted sounds or signals that become mixed in with the desired information. When you listen to the radio, both information and noise are present. In this example, they are both in the form of sound. The information is the sound you want to hear, and the noise is the sound that you do not want to hear. Noise on a television transmission turns into static on your television screen. Noise, then, is a type of interference. Also, noise can involve personal taste. Some people will call the music you listen to "noise." You may feel the same way about the music they prefer. You are both correct because unwanted sound is noise.

Types of Communication

One way to look at communication is in terms of the sender and receiver, **Figure 19-4.** We are all familiar with people communicating with people. This type of communicating is the most common. It works through our electronic media and our printed products. This type of communication is called ***human-to-human communication.*** It is used to inform, persuade, and entertain other people.

Other types of communication also exist, however. Have you ever reacted to a traffic light, a warning light on the dashboard of your car, or the bell that indicates the end of a class period? If so, you have participated in ***machine-to-human communication.*** This type of communication system is widely used to display machine operating conditions.

Have you keyed material into a computer or set the temperature on a thermostat? If so, you have engaged in ***human-to-machine communication.*** This type of communication system starts, changes, or ends a machine's operations.

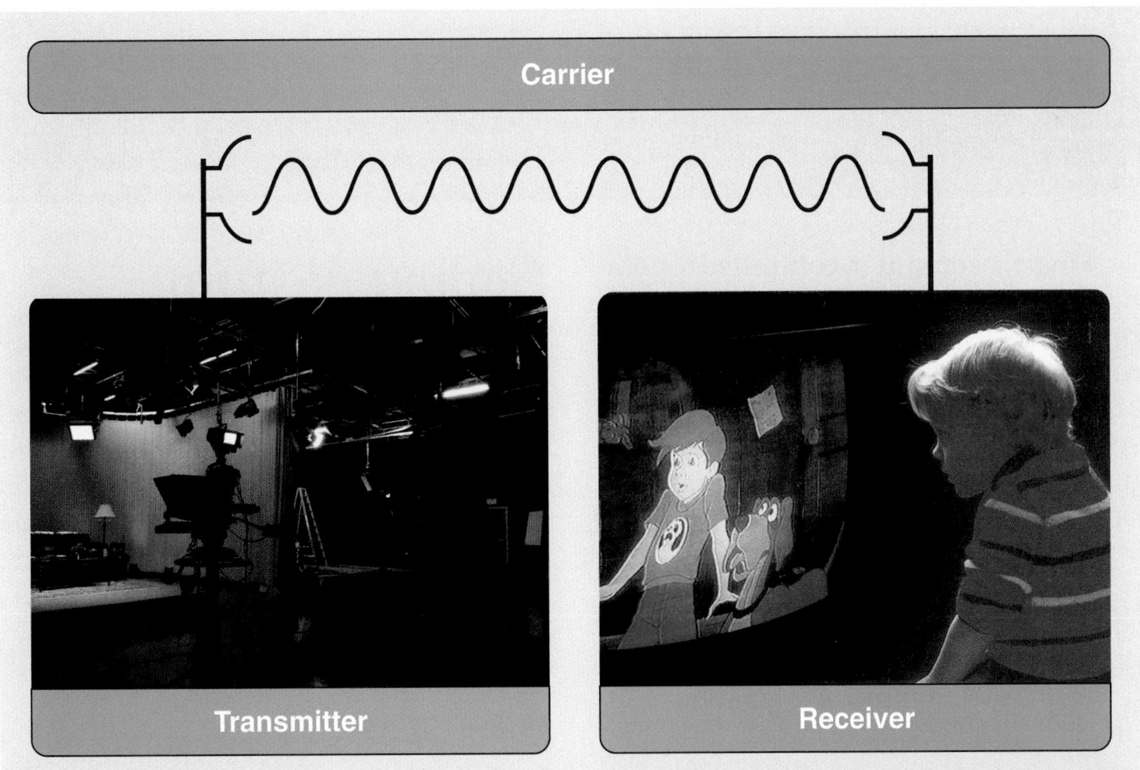

Figure 19-3. All communication systems have transmitters, channels, and receivers. This shows the television communication system. (Westinghouse Electric Co.)

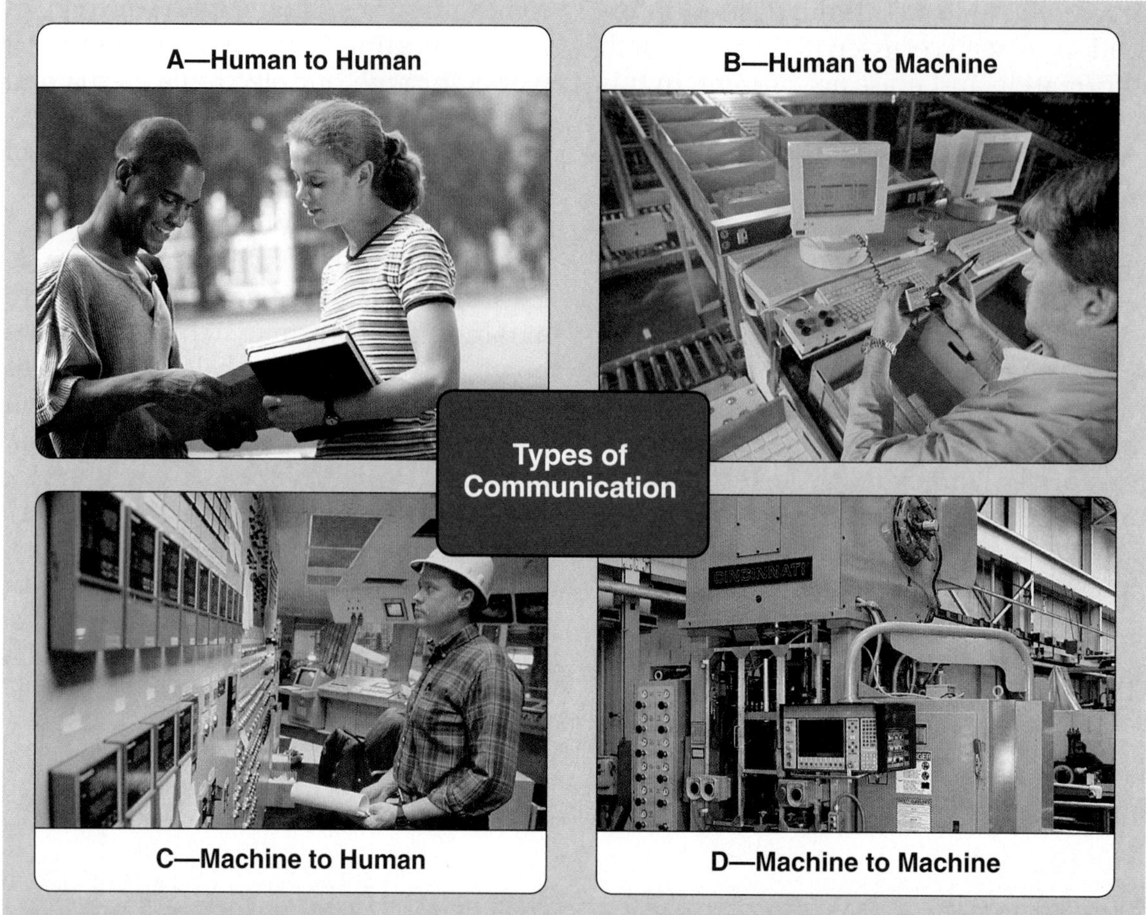

Figure 19-4. The type of communication is based on the sender and the receiver. A—Relaying information to others through printed media. B—Relaying information to a computer. C—Checking a readout from a plant system. D—Controlling a manufacturing system through a computer. (United Parcel Service, Siemens, Inland Steel Co., Cincinnati, Inc.)

Finally, computer-controlled operations use ***machine-to-machine communication.*** Modern industry is becoming more computer based. Humans enter programs and data into the computer, and then the computer directs and controls an apparatus. Examples of this type of communication include a computer controlling a printer and a thermostat controlling a furnace. Typical examples of computer-controlled operations are computer-aided design (CAD), computer-aided manufacturing (CAM), computer-integrated manufacturing (CIM), and robotics.

Communication Systems

We are bombarded with information every day. It comes in many printed and electronic forms. Still, the communication technology used to deliver the information can be divided into five main types, **Figure 19-5.** These types are:

- Printed graphic communication
- Photographic communication
- Telecommunication
- Technical graphic communication
- Computer and Internet communication

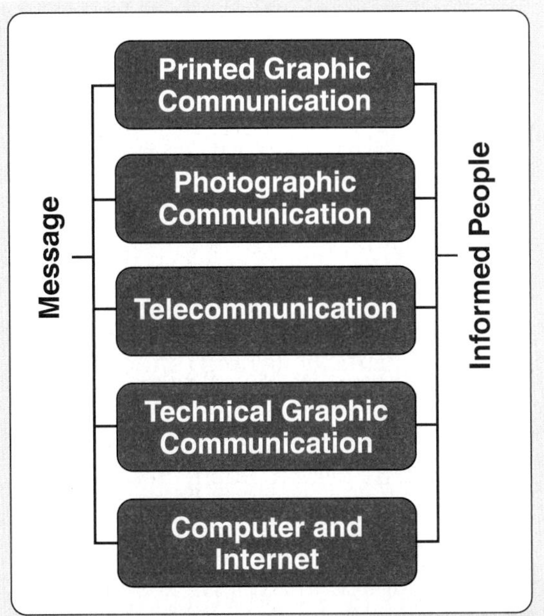

Figure 19-5. We receive information through five basic types of communication technology systems.

Printed Graphic Communication

Much of communication technology was developed to satisfy the need for mass communication. People wanted to tell their message to large numbers of other people. The first mass communication system developed was printing or ***printed graphic communication,*** **Figure 19-6.** In Chapter 7, you were introduced to movable type, the foundation of printing.

The term *printing* originally meant putting an image on paper with inked type. It now includes all of the processes used to reproduce a two-dimensional image on a material. That material may be paper, metal, plastic, cloth, or wood. These processes will be discussed further in Chapter 20.

Figure 19-6. Books are an example of printed graphic communication media.

Technology Explained

Compact Disc (CD) Systems

compact disc (CD) system: An information reproduction system using a laser to read optically encoded information, changing the optical signals into digital signals, and then recreating the information.

In 1877, Thomas Edison became the first person to record sound. The first records were wax cylinders with grooves capturing voice and music, **Figure A.** These cylinders gave way to flat disc records, which were cheaper to produce.

In these earlier systems, the information was encoded by producing wavy grooves in the surfaces of the records. A stylus, or needle, tracing the grooves read the code. As the wiggles in the grooves vibrated the stylus, it sent continuous, or analog, electrical signals to the playback unit. This unit amplified this rising and falling signal and fed it to the speaker. The electrical energy was converted to the sound you heard in the speaker.

Figure A. Here is an early phonograph. (Library of Congress)

The CD is much smaller but can hold far more music than the larger records. It can hold over one hour of music on one disc.

CD systems are different from past systems. First, the information on the CD is recorded as a digital signal, a signal in discrete steps. Digital signals can be accurately reproduced. Second, CDs use a beam of light, rather than a stylus, to pick up the coded message from the disc.

When music is recorded, its frequency is sampled 44,100 times a second. This reading is converted to a binary number. A binary code uses a series of zeros and ones to represent any number. Electricity flowing (on) or not flowing (off) can represent the ones and zeros. A CD contains billions of tiny pits in its surface. When the laser detects a change from the flat surface to a pit, or vice versa, it signals a one to the system. Flat stretches, either over the surface or over the pits in the surface, signal zeros to the system.

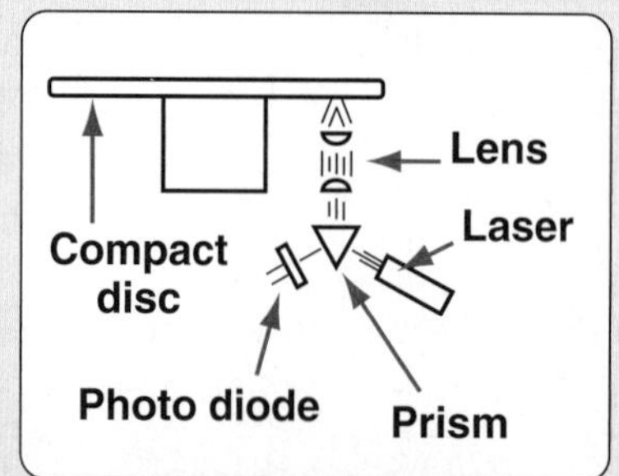

Figure B. This is the structure of a typical CD player.

A laser reads this code. When the laser hits the disc, the beam is reflected through a prism to a light-sensitive diode, **Figure B**. This diode allows a voltage flow relative to the amount of light striking it. Little light is reflected from pitted areas. Nearly all the light is reflected from flat surfaces. The diode translates the light signals into electrical signals. Electrical signals are converted from digital to analog and then fed to the speaker.

CD technology is not limited to simply reproducing sound. The CD has become a common storage medium for large amounts of read-only memory (ROM) for computers.

Photographic Communication

Some communication technologies use light to convey their message. The most common of these is photography. Photography captures light as an image on a recording medium that we call film. When the film is developed, the image becomes permanent. The film can be used in making prints of the image or in projecting the image onto a screen.

Figure 19-7. This photograph communicates specific emotions.

Figure 19-8. This customer service representative is using telecommunication to help customers.

Photography is the process of producing the image. ***Photographic communication*** is the process of using photographs to communicate a message, **Figure 19-7.** These processes will be discussed in Chapter 21.

Telecommunication

One type of communication system is playing an increasing role in modern life. This system is ***telecommunication,*** communication at a distance, **Figure 19-8.** Telecommunication includes a number of specific types of communication. Probably the most widely used are radio, television, and the telephone. Telecommunication is the subject of Chapter 22.

Technical Graphic Communication

People and companies often want to communicate specific information about a product or its parts. This information may convey the size and shape of a part. It may suggest how the parts are assembled to make a product or structure. The information might indicate how to install and operate a product. It might also tell how to adjust and maintain the product. This type of information is often communicated

Connections to Technology: **Communication**

The Power of Radio

The author H. G. Wells wrote one of his most popular science fiction books, *The War of the Worlds,* in 1898. The book was about an invasion from Mars. Little did he know the uproar the book would cause forty years later as a result of the power of the telecommunication system known as the radio.

In the 1930s one of the more popular forms of entertainment was the broadcast of various programs on the radio. Every night millions of Americans would tune in to hear their favorite comedies, dramas, and news programs. One of the more popular dramas featured plays directed by Orson Welles. For the broadcast of October 30, 1938, he decided to feature an adaptation of *The War of the Worlds.* In adapting the book for the radio, however, he apparently wanted to add more drama, so he made the play sound more like a news broadcast. He had the play's music interrupted by fake news bulletins read in dramatic tones by various actors. The "bulletins" warned that Martians had invaded New Jersey and were intending to destroy the United States. Even though Welles had noted before the program that this was just a play, many had not heard the disclaimer. The program sounded so realistic that some people thought an actual invasion had occurred, and they panicked. Some jumped into their cars to run, and others hid in their basements. Still others put wet towels on their heads to avoid the Martians' poison gas.

The disturbance caused a scandal, with some calling for more government regulation. The commotion eventually died down, but to this day the incident shows the tremendous power involved in using technology to communicate. Can you give another example of this power?

through engineering drawings or technical illustrations, **Figure 19-9.** The methods that prepare and reproduce this media are called ***technical graphic communication*** systems. Technical graphic communication was discussed at length in Chapter 12.

Computer and Internet Communication

A new system of communication is rapidly changing how people gather and use information. It is based on a key invention, the personal computer. This new type of communication uses a network of computers to share information. This network is called the ***Internet.*** It allows people to gather information using the World Wide Web and exchange messages using electronic mail. Using the Internet for communication is discussed in Chapter 23.

Figure 19-9. This designer is preparing technical illustrations for a new product. (Ohio Art Co.)

Career Corner

Advertising

Careers in advertising include advertising specialists, copywriters, media planners, media buyers, creative directors, and public opinion researchers. Jobs in advertising require good oral and written communication skills. Many people in the advertising field have college degrees in business, journalism, or mass communication. In college courses in advertising, people learn about the research used in developing advertising programs, how advertising campaigns are produced, about the role of advertising in marketing, and related computer skills.

Summary

We are rapidly moving from the industrial age into the information age. This new age is characterized by the wide-scale availability of communication media. People receive and send a constant barrage of data and information every day. Each communication message is designed to inform, persuade, entertain, or achieve some combination of these goals.

Transfer of communication is done in terms of human-to-human, human-to-machine, machine-to-human, and machine-to-machine communication. Each of these systems has a transmitter that encodes and sends the message. The message then travels over a carrier to a receiver. The receiver collects and decodes the message.

The most common forms of communication systems are printed graphic, photographic, telecommunication, technical graphic, and computer and Internet communications. Our lives are affected daily through these systems.

Key Terms

channel
communication technology
data
decode
edutainment
emotions
encoding
entertain
human-to-human communication
human-to-machine communication
idea
inform
information
infotainment
interference
Internet
knowledge
machine-to-human communication
machine-to-machine communication
noise
persuade
photographic communication
printed graphic communication
receiver
technical graphic communication
telecommunication
transmit

Test Your Knowledge

Write your answers on a separate sheet of paper. Please do not write in this book.

1. What is communication technology?
2. What are the three major items that we try to communicate?
3. List the three goals of communication.
4. *True or false?* All communication systems have transmitters, channels, and receivers.
5. An unwanted sound or signal is called _____.
6. List the four types of communication.

For Questions 7–15, match the communication system on the right to the correct example on the left. (Note: Some letters will be used more than once.)

Example	Communication System
7. _____Newspaper	a. Printed graphic
8. _____Radio	b. Photographic
9. _____Motion picture	c. Telecommunication
10. _____Electronic mail	d. Technical graphic
11. _____Blueprint of a house	e. Internet
12. _____Vacation photo	
13. _____Parts drawing in an owner's manual	
14. _____Web page	
15. _____Television	

Applying Your Knowledge

1. Select a piece of information or an emotion you want to communicate. List four ways you could communicate it and the communication system you would use (printed graphic, photographic, telecommunications, technical graphic, or Internet).
2. Design a communication message that will persuade a person to your point of view on an issue.

Chapter 20
Printed Graphic Communication

Learning Objectives

After studying this chapter, you will be able to:

- Describe printed graphic communication.
- Name products produced by printed graphic communication processes.
- Identify the six major types of printing processes.
- List the steps involved in designing a printed graphic message.
- List principles of design.
- Identify how printed messages are prepared for production.
- Explain how printed graphic messages are produced.
- List activities involved with finishing printed graphic communication products.
- Describe computer-based publishing.

Communication means many things to many people. We talk about communication messages, communication media, communication networks, and communication systems. Central to all of these is the word *communication.* Simply stated, communication is the passing of data and information from one location to another or from one person to another. When the spoken word is used to communicate, we call it language or verbal communication. When a technical means is used to convey information, it is called communication technology.

As you learned in the last chapter, a number of communication systems use technology. The oldest of these systems is ***printed graphic communication,*** or ***printing.***

Most printed communications use an alphabet to convey the message. An alphabet is a series of symbols developed to represent sounds. The first alphabet was developed in Syria around 1200 B.C. Today, various alphabets exist, including the twenty-six-letter one used for the English language. These letters can be arranged to convey a message through writing.

The process of printing takes written words and places their image on a material. The printing is applied to a material called the ***substrate.*** Originally, almost all printing was done on paper. Today, printing is done on a variety of substrates including paper, glass, plastic, cloth, ceramics, metal, and wood. The result of this mixture is a broad range of printed products. These products include newspapers, magazines, books, brochures, pamphlets, labels, stickers, clothing designs, and signs. Each of these is carefully designed and produced by a specific printing process.

In this chapter we will look at three main topics in printed graphic communication today. These topics are printing methods, steps in producing printed graphic messages, and computer-based publishing.

Printing Methods

Human beings first printed by carving an image into a block of some material. Next, they covered the block with ink. They then pressed the inked block on paper to produce the "printed word." This technique, shown in **Figure 20-1,** was the forerunner for all printing processes. Today, six major printing processes exist. These processes are:

- Relief
- Lithography
- Gravure
- Screen
- Electrostatic
- Ink-jet

Relief Printing

Relief printing is the oldest of all printing processes. It uses an image that is on a raised surface, **Figure 20-2.** Ink is applied to this raised surface, and it is pressed against the substrate. The pressure forces the ink to adhere to the substrate, producing the printed message.

Figure 20-1. This museum exhibit shows early block printing.

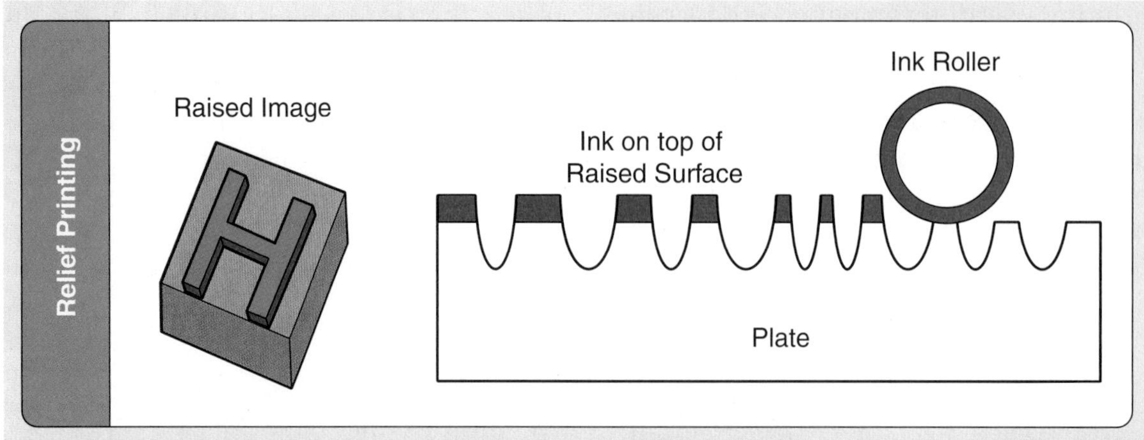

Figure 20-2. This is a diagram of relief printing. Note the raised printing surface.

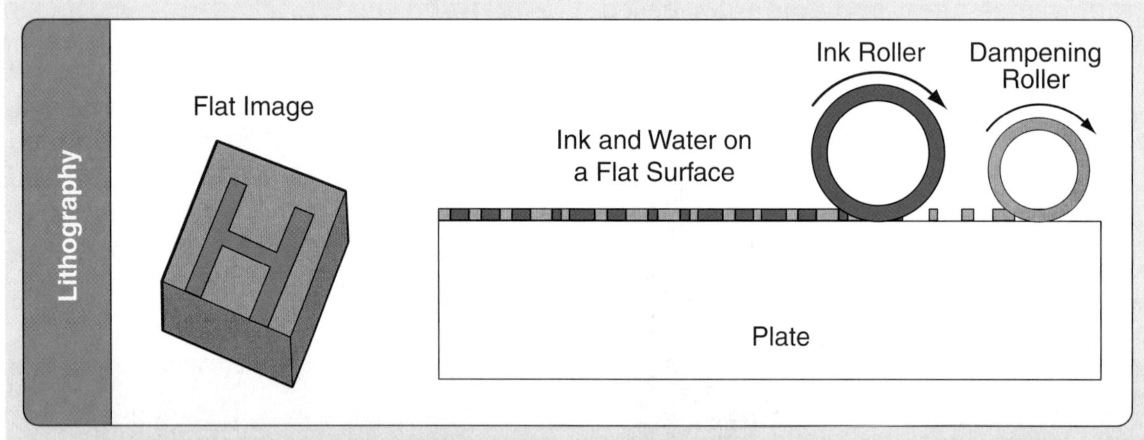

Figure 20-3. This is a diagram of lithographic printing. Note the flat printing surface of this method.

Relief printing requires that the image be reversed, or "wrong reading," on the printing block, or ***image carrier***. Once the inked block has been pressed against the substrate, the message will be forward, or "right reading." Look at a rubber stamp, and you will see an example of a reversed or "wrong-reading" relief image.

The two main types of relief printing are letterpress and flexography. ***Letterpress*** uses metal plates or metal type as the image carrier. Until the last thirty to forty years, letterpress was the most common printing process. It now accounts for less than 5 percent of all printing.

Flexography is an adaptation of letterpress. It uses a rubber or plastic image carrier. Flexography is widely used for printing packaging materials. Today, this process accounts for about 2 percent of all printing.

Lithographic Printing

Lithographic printing, or ***offset lithography,*** is the most widely used method of printing today. This process prints from a flat surface, **Figure 20-3.** Offset lithography is based on the principle that oil and water do not mix. This printing method was discovered by Alois Senefelder. Senefelder used a limestone slab as his image carrier. He first wrote a reverse, "wrong-reading," image on the stone with a grease pencil. Next, he wet the stone, which dampened the stone surface except for the areas with grease marks. Then, he rolled ink over the

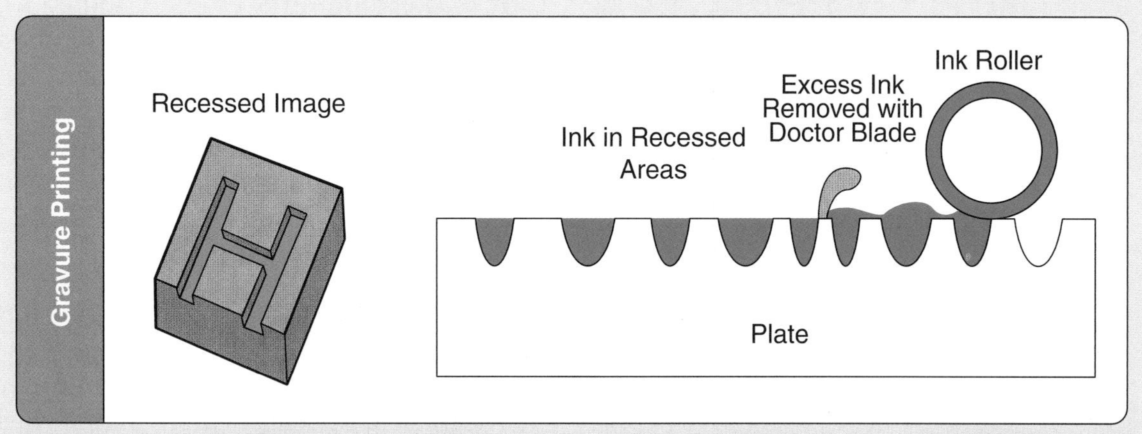

Figure 20-4. This diagram is of the gravure printing process. Note that the inked image is inset into the image carrier.

stone surface. The ink adhered to the grease-marked areas but not to the wet stone. An inked image remained, which could be transferred to paper. Finally, he placed paper over the stone and applied pressure. This caused a direct transfer of the image from the carrier to the substrate. Today, some artists still use Senefelder's basic process to produce prints of their work.

The basic lithographic process has been refined. A "right-reading" image is produced on an image carrier called an offset plate. The inked image on the plate is first transferred to a special rubber roller called an offset blanket. This reverses the image. The image becomes "right reading" again when it is transferred to paper as it is fed through the press. Because the original image is first offset to a blanket, the process is called offset lithography. Modern offset presses can produce more than 500,000 copies from a single offset printing plate.

Gravure Printing

Finely detailed items, such as paper money and postage stamps, are usually printed using the ***gravure,*** or *intaglio,* process, **Figure 20-4.** This process is the opposite of relief printing. In gravure printing, the message is chemically etched or scribed into the surface of the image carrier. The carrier is then coated with ink. Next, the surface ink is scraped off with a doctor blade. This leaves ink only in the recessed areas of the carrier. When paper is pressed very tightly against the carrier, it picks up the ink in the cavities.

Gravure is an expensive process. It is economical for very long production runs, however, because one gravure plate can withstand several million impressions. In addition to money and stamps, some magazines are printed with this process.

Screen Printing

Screen printing is a very old printing process, dating back more than one thousand years. Screen printing uses a stencil with openings that are the shape of the message. The stencil is mounted on a synthetic fabric screen. Paper is then placed beneath the screen, and ink is applied to the screen's upper surface. Finally, a squeegee is pulled across the stencil. This forces ink through the openings in the stencil to produce the printed product, **Figure 20-5.**

Screen printing is used to print on fabrics, T-shirts, drinking glasses, PC (printed circuit) boards, and many other products. It is also used to print small quantities of very large products, such as posters or billboards.

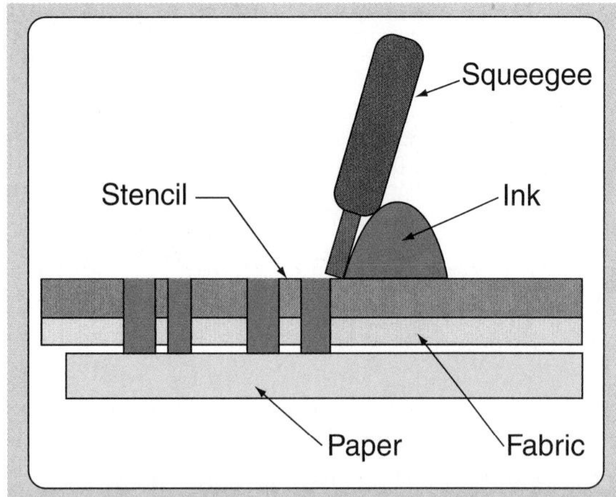

Figure 20-5. This is a diagram of screen printing, or silk screening.

Screen printing is often called ***silk screening.*** Originally, the stencils were mounted on silk fabric. Currently, however, stronger and more durable fabrics have almost completely replaced silk as the screen fabric used in the process.

Electrostatic Printing

A commonly used process today is ***electrostatic printing.*** Electrostatic printing is often called "Xerox copying." This is not the proper term because Xerox® is the trade name for one brand of electrostatic printers.

The electrostatic printing process uses a machine with a special drum. The image to be copied is placed on this machine. Then a strong light is reflected off the image and onto the drum. The reflected light creates an electrically charged likeness of the image on the drum's surface. This charge is transferred to a sheet of paper that is then passed over fine particles called toner. The toner is attracted to the charged image on the paper. Finally, the paper passes through a heating unit that fuses the toner onto it.

The early office copiers were fairly slow. Today, high-speed electrostatic copiers can produce several thousand copies an hour. Some of the newest electrostatic copiers can be directly interfaced with (joined to) microcomputers. The computer sends a chosen image directly to the copier. The paper original is eliminated, and even the most complex page layouts are easily and quickly reproduced.

Ink-Jet Printing

The newest printing process is ***ink-jet printing.*** In this process a computer generates the printed message. The computer then directs a special printer that sprays very fine drops of ink onto the paper. Because the printing head never touches the substrate, a wide range of materials can be printed by this process.

Ink-jet printing is used for coding packages and producing mailing labels. The process can use multiple printing heads, allowing the printer to produce several thousand characters per second.

Steps in Producing Printed Graphic Messages

Printed messages and products are the result of a series of planning and production activities. These activities can be grouped into three main steps, **Figure 20-6:**

1. Designing the message
2. Preparing to produce the message
3. Producing the message

Designing the Message

Communication is successful only when the intended audience receives the given data or information. We have various ways to communicate an idea or concept. Think of the concept *church,* **Figure 20-7.** We can say the word and cause people to form a mental image of a church. This process, however, requires us to be near the audience. A larger audience over a greater distance can be reached if we use printed graphic communication. Printed graphic communication can include additional words, which we call text, or ***copy.*** We can also illustrate the message. ***Illustrations*** are pictures and symbols that add interest and clarity to the printed communication.

The design of each communication follows a set procedure. This process starts with an ***audience assessment.*** For an audience assessment, the designer must determine, among other factors, who the audience is, what they like and dislike, and what their interests are, **Figure 20-8.** In short, graphic designers must get to know their audience. This background information will help the designer decide what information to tell and how to tell it. The designer generally has a great deal of information available. The designer's challenge is to select the useful information to include in the message. For example, suppose you are developing a promotional flyer for recycling. You would first have to decide who your target audience is. Will it be private citizens, business, industry, or government? Each of these groups has its own set of interests and behaviors. Your message will have to appeal to their individual interests in order to help change their personal behavior. It is difficult to

Figure 20-6. The three main steps in creating printed communication are designing the message, preparing the message, and producing the message.

Figure 20-8. An audience assessment allowed this designer to tailor his message to the audience. (AC-Rochester)

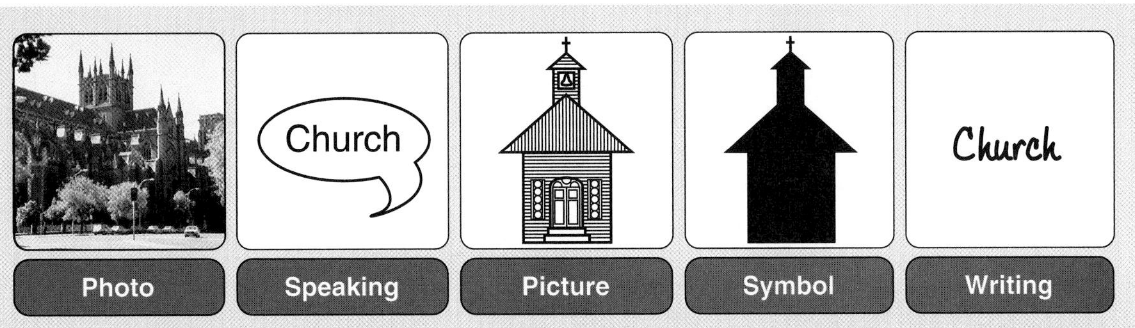

Figure 20-7. We can communicate the idea "church" in several ways.

write one message for all people. Most communication messages are targeted to a specific audience.

After getting to know your audience, you must next get them to receive the message. This prompts the "how" part of the design. To promote your recycling program, should you use a humorous appeal or an environmental concerns appeal? A message needs an attention-getting aspect, or a "hook," to have a strong chance of making an impact.

Designing the message involves two main steps, design and layout. In the ***design*** stage the message is developed. In the ***layout*** stage the message is put together.

Design

Graphic design deals with the appearance or "look" of the page. The design should attract readers and hold their attention. An important message that is never read is useless.

To develop a message, you must consider a number of design principles, **Figure 20-9.** These principles include proportion, balance, contrast, rhythm, variety, and harmony.

Proportion

The principle of ***proportion*** deals with the relative size of the parts of the design. It is concerned with the height-width relationships of the parts within the design. Good proportion will have an eye-pleasing relationship between large and small elements within the message.

Balance

People seem to enjoy a visual ***balance,*** so strong media messages will be designed with this principle in mind. Balance is accomplished by having the information on both sides of a centerline appear equal in visual weight.

You can achieve balance in two ways, **Figure 20-10.** The first way is *formal balance.* With formal balance both sides of the centerline are close to mirror images of each other. The second way is *informal balance,* in which there are equal amounts of copy and illustrations on each side of the layout, yet both sides of the centerline do not mirror each other. This layout still produces a feel of equal weight or balance when viewed.

Contrast

Contrast is used to emphasize portions of the message. You can achieve contrast by changing the color of the important elements or by printing the important elements in bold, italic, or enlarged type.

Design Principles

Proportion
Balance
Contrast
Rhythm
Variety
Harmony

Figure 20-9. Publishers of printed graphic communications use design principles in their publications.

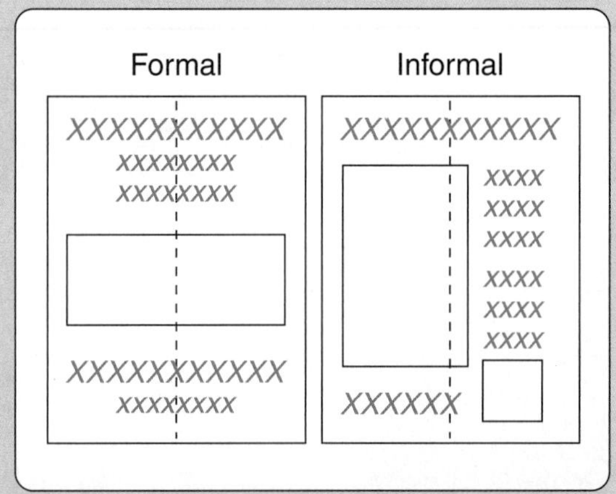

Figure 20-10. These are representations of formal and informal balance. Notice how there is a similar visual weight on both sides of the centerline.

Rhythm

The principle of ***rhythm*** deals with the flow of the communication. Rhythm in an advertisement might repeat certain elements in the design. It produces a sense of motion and guides the eye to an important feature.

Variety

People receive thousands of messages each day. ***Variety*** is the technique that makes the message unique and interesting. For example, in a full-color publication, a black-and-white message stands out. Unique layouts will also catch the reader's eye.

Harmony

We enjoy hearing harmony in music. ***Harmony*** refers to the notes blending together to form a pleasing sound. Likewise, harmony in graphic design is achieved by blending the parts of the design to create a pleasing message. The parts of a harmonized message will fit and flow together.

Layout

Layout is the physical act of designing the message. It involves positioning the copy and illustrations to form a message that communicates effectively. This task starts with sketching. First, the designer prepares a series of ***thumbnail sketches,*** **Figure 20-11.** These allow the graphic designer to experiment with various arrangements of copy and illustrations. The designer then uses these thumbnails in preparing ***rough sketches,*** which integrate and refine the ideas generated in the thumbnails. Rough sketches are drawn in the final proportion.

Finally, the designer makes a ***comprehensive layout*** of the best rough sketch, **Figure 20-12.** The comprehensive, or "comp," is the layout for the final design. The comp deals with several elements. One is typography. Typography is the typeface, size, and style of type that will be used in the layout. Also, the designer must size illustrations and crop them to fit into the layout. Cropping is selecting and separating the important information in a photograph. Finally, the designer must specify the color for the various parts.

Preparing to Produce the Message

After the design has been developed, it must be made ready for printing. This involves three basic steps: composition, pasteup, and image carrier preparation.

Figure 20-11. These are typical thumbnail sketches used in designing a graphic communication. (Ohio Art Co.)

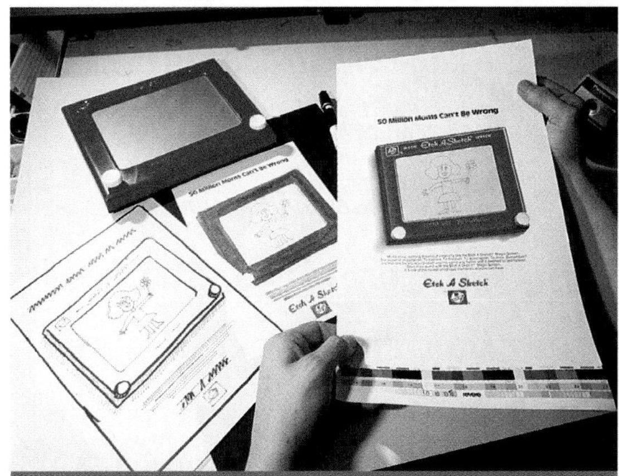

Figure 20-12. Shown here are comprehensive layouts for a communication product. (Ohio Art Co.)

Composition

Composition consists of the activities that change written words and illustrations into forms that can be used in various printing processes. In most cases, this involves typesetting and illustration preparation.

Typesetting produces the words of the message. This includes generating, justifying, and producing the type. A typeface is specified in the design. It is selected from hundreds of different available styles, **Figure 20-13.** Today, the written words are usually developed with word processing and graphic arts software. Whether to use letters with serifs or letters without serifs (sans serif) is one option in choosing a typeface. Serifs are the small flares at the ends of printed letters. **Figure 20-14** shows words printed with and without serifs. Additional groupings include traditional, contemporary, decorative, and script.

Justification (alignment of the copy) on the left, right, or both edges (also called full justification, Figure 20-14) of the paper is another part of typesetting. Letters and informal messages are generally left-justified, leaving an uneven or "ragged" right margin. Books, magazines, and newspapers justify both edges of the copy. Artistic messages are sometimes right-justified to attract attention.

Finally, the type font must be selected. Fonts include plain, bold, italics, condensed, and expanded, among other styles, Figure 20-14.

Illustration preparation is often required for useful communication. This preparation may include sizing and converting the line art or photographs. These pieces might need to be enlarged or reduced to fit the space in the layout. Other than sizing, line art often needs no preparation to be placed in a layout. Photographs, however, must be changed into different forms. They cannot be directly used in printing processes. All photographs are first changed into halftones.

To produce a halftone in a black-and-white photograph, one creates a negative of the picture with the photograph being

The large brown dog ran fast.

The large brown dog ran fast.

The large brown dog ran fast.

The large brown dog ran fast.

The large brown dog ran fast.

THE LARGE BROWN DOG RAN FAST.

Figure 20-13. The same words can be shown in many different typefaces. Designers will choose a typeface on the basis of the message they are trying to convey.

Type Style

Serif	San Serif
Johannes Gutenberg began the printing revolution more than 500 years ago. Gutenberg's method of printing is called letterpress.	Johannes Gutenberg began the printing revolution more than 500 years ago. Gutenberg's method of printing is called letterpress.
Left Justification	**Full Justification**
Johannes Gutenberg began the printing revolution more than 500 years ago. Gutenberg's method of printing is called letterpress.	Johannes Gutenberg began the printing revolution more than 500 years ago. Gutenberg's method of printing is called letterpress.
Plain	**Bold**
Johannes Gutenberg began the printing revolution more than 500 years ago. Gutenberg's method of printing is called letterpress.	**Johannes Gutenberg began the printing revolution more than 500 years ago. Gutenberg's method of printing is called letterpress.**

Figure 20-14. Notice the contrasts in these three typesetting options.

exposed through a halftone screen. When the photograph is printed, the various tones in the photograph are changed into a series of dots. The light areas are made up of small dots, and dark areas are made up of large dots. Using a magnifying glass, look at a photograph printed in a newspaper. You will see that it is actually a series of dots.

Color photographs are separated into four printers, or masters. These, as shown in **Figure 20-15,** are yellow, magenta (red), cyan (blue), and black. During the separation, the four are run through a halftone screen. Later, each color will be printed independently to re-create the color photograph.

Line drawings for multicolor screen prints are produced in several layers. One layer is necessary for each different color that will be printed, **Figure 20-16.**

Pasteup

The second step in preparing the message produces a ***pasteup,*** or ***mechanical.*** In this step, all the type and illustrations are assembled onto a sheet that looks exactly like the finished message, **Figure 20-17.** This assembled layout is called a pasteup.

Black-and-white (or any single-color) printing uses one pasteup for each image carrier (printing plate or screen). Color printing requires one pasteup for each color. The pasteup is the master copy used to prepare the image carriers, which will, in turn, produce the final printed product.

New computer-driven page composition systems generate the layouts and "pasteup" in a single step. The computer operator edits the text, manipulates the copy, and locates the spaces for illustrations right on the computer screen. The layout is sent electronically to the printer, which then produces a finished page layout. With these computer systems, the need for manual locating and pasting down of type and illustrations is eliminated. Also, the layouts are stored in the computer's memory, so changes are quick and easy. The use of these systems, often called electronic publishing, will be further discussed later in this chapter.

Figure 20-15. A color photograph such as this one of a flower is separated into four single-color printers.

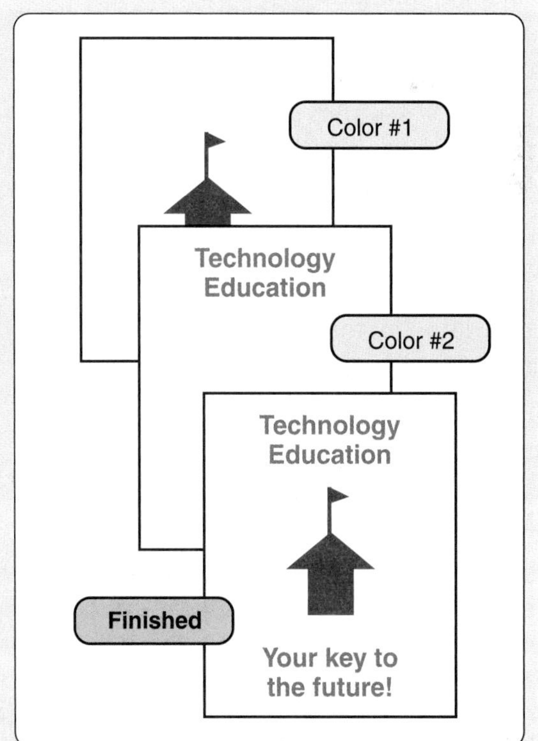

Figure 20-16. Each color on a line drawing is printed separately.

Technology Explained:

Fax Machines

fax machine: A device that sends copies of documents over telephone lines or radio waves, using digital signals.

A new communication system to the general public is the facsimile machine, more commonly known as the fax machine. A fax machine is like a copier, except it has two parts. The two parts, the scanner and the printing unit, work independently. The scanner on one fax machine sends the image of a page to the printing unit of another fax machine. The two fax machines can be many miles apart.

A fax machine changes the black-and-white images on a page into digital information. Digital information has two states, represented by *1*s and *0*s. The two states can be "on" and "off" or a higher voltage and a lower voltage. A fax machine uses an optical scanner that looks at the image on a page, **Figure A.** The scanner breaks the page into small lines. Each line is broken into small dots called *pixels.* The scanner examines each pixel to see if it is black or white. The black-and-white pixels generate a digital signal. A *1* might represent a white pixel, and a *0* might represent a black pixel (or vice versa).

A modem is used to convert the digital signals into audible tones that can be sent over telephone lines or radio waves. Newspapers and police departments have sent wirephotos over the phone lines for years. Amateur radio hobbyists use fax machines to send documents over the airwaves. Weather maps are also sent via radio. Instead of using telephone lines, the tones from the modem are fed into a transmitter and broadcast. Another person with a receiver feeds the tones back into a fax machine for printing. Today, anyone with a fax machine can send letters, pictures, and other printed information across town or around the world.

The receiving fax machine has a modem converting the tones back into digital signals. A printing system changes the digital signals back into black-and-white pixels. The pixels are grouped back into lines, and the lines are printed onto a page. The printing system in a fax machine can be a thermal printer, or it may use a photosensitive drum and toner like an office copier, **Figure B.**

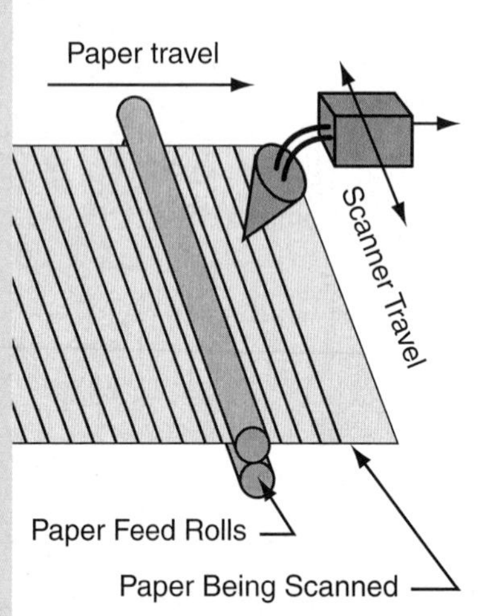

Figure A. This is a schematic diagram of how a fax scanner works. Some fax machines use moving scan heads; others move the paper over a row of sensors.

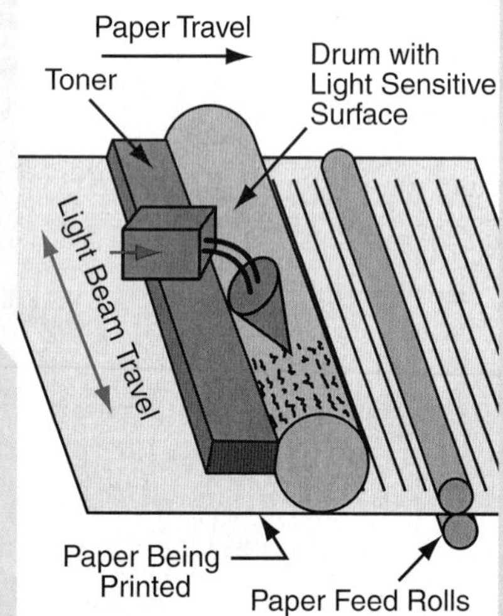

Figure B. The printing unit of a fax turns the digital signals from the modem into black-and-white pixels on paper. This diagram shows a photosensitive drum printing unit.

Figure 20-17. A graphic designer is working on this pasteup.

Image Carrier Preparation

Most printing processes use an image carrier. The image carrier is the feature that makes each process unique. Let us look at how relief, offset lithography, gravure, and screen image carriers are prepared.

Preparing Relief Image Carriers

A relief image carrier has a raised printing surface. This surface can be produced by setting type, as described in Chapter 7. The letterpress (metal type) printing process, however, is becoming less common each year.

As you learned earlier in the chapter, flexography is an emerging relief printing process. Flexography uses a flexible plastic or rubber printing plate. This image carrier is produced in one of two basic ways, casting or etching. Casting uses a mold that is formed and cured around metal type. The mold is then removed from the type and filled with a plastic material. The plastic cures (hardens) into a sheet that is removed from the mold. This becomes the printing plate.

Etching is a photochemical process. A negative of the desired image is placed over a light-sensitive material. The material and negative are held together tightly and placed under a strong light. The light-sensitive material is exposed through the open areas on the negative. A chemical developer hardens the exposed material. A detergent is then applied to wash away the unexposed material. A raised image of the message is left.

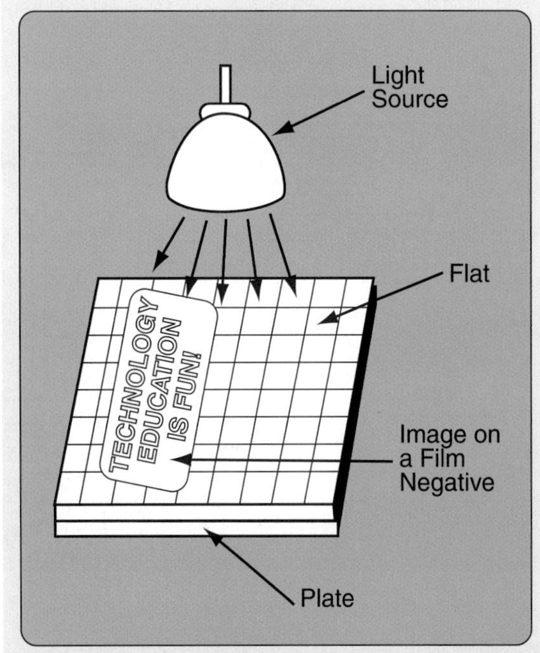

Figure 20-18. Offset plates are produced by exposing the plate through a photographic negative and then developing the resulting image.

Preparing Offset Lithographic Image Carriers

Offset lithography uses a flat printing plate, produced through a photochemical process, **Figure 20-18.** The pasteup is converted into a negative and mounted on a sheet of paper. The paper over the type and illustration area is then cut away. The resulting form is called a flat. The flat is placed on top of a sheet of metal (offset plate) that has a light-sensitive coating. The flat-and-plate assembly is placed in a vacuum table that holds them together tightly. This assembly is exposed to a very bright light. The image area on the negative allows light to pass through and expose sections of the plate. The plate is then chemically developed, producing the image carrier.

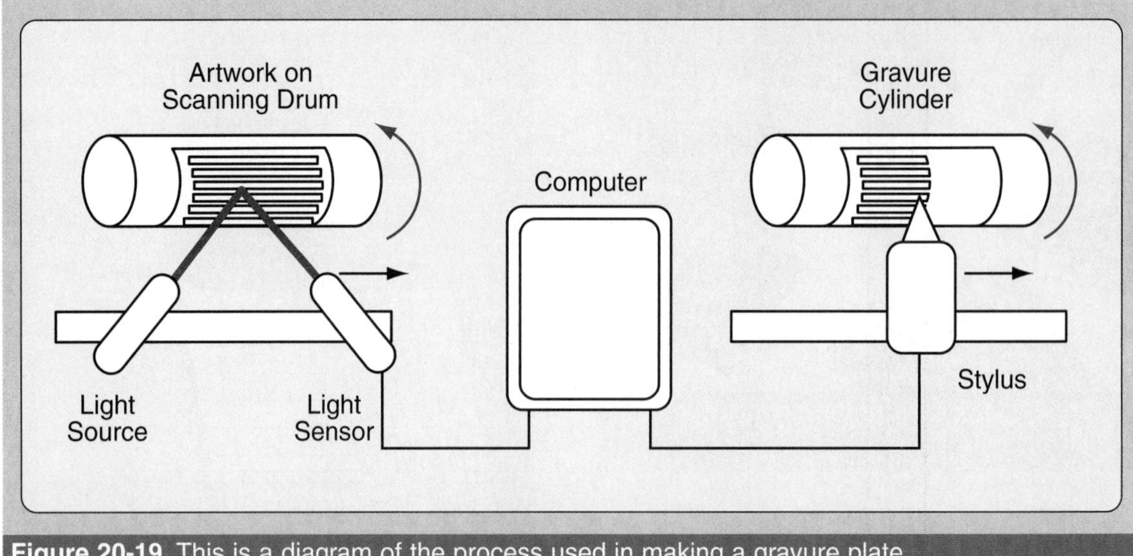

Figure 20-19. This is a diagram of the process used in making a gravure plate.

Preparing Gravure Image Carriers

The gravure image carrier, or printing cylinder, has the image engraved into it. This is generally done with an electromechanical scanner, **Figure 20-19.** This computer-controlled machine reflects light off the copy. The light intensity is then picked up by a sensor. Black areas reflect little or no light, whereas white areas reflect a high percentage of light. The gray areas reflect varying amounts of light in proportion to their shade. The amount of light reflected is communicated to the computer through the sensor. The computer processes the information and then uses the information to drive a stylus touching the gravure cylinder. The stylus carves the desired pattern into the cylinder, producing the image carrier.

Preparing Screen Image Carriers

Most screen carriers are produced using a photochemical process. First, a positive transparency of the image is produced. This transparency is placed on top of photosensitive film with a clear plastic backing. Light is passed through the positive, exposing the film. When the film is developed, the portion that was exposed to light will harden. The remaining emulsion is then rinsed away. The wet film is placed under the screen fabric. The fabric is pressed into the film, and it is allowed to dry. The clear plastic film backing is then removed, leaving a negative of the image on the screen.

Producing the Message

The third activity in printed graphic communication is producing the message. Production involves printing the message (message transfer) and finishing the product (product conversion). Two basic forms of substrate are used in the production of printed graphic messages, **Figure 20-20.** The first form is sheet. This form includes all the flat, rectangular pieces of substrate. They can be sheets of paper, glass, plastic, or metal, for example. The other form is web. The web form uses a roll of substrate that is continuously fed into the printing process.

Both of these forms require specific printing press equipment. A typical press has a feeder unit, which moves the substrate into the press; a registration unit, which positions the material for correct placement (register) of the message; the printing unit, which transfers the message from the carrier onto the substrate; and the delivery unit, which stacks the printed product.

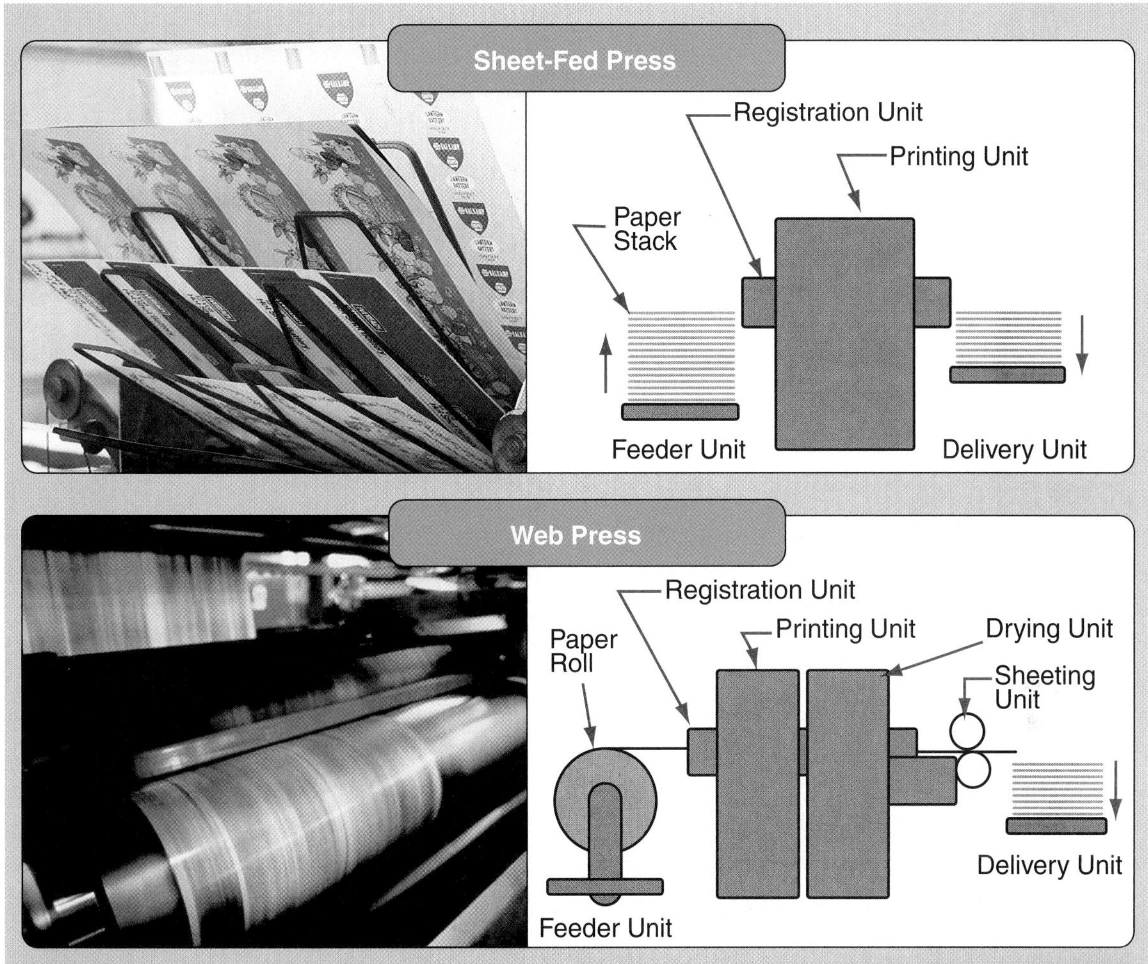

Figure 20-20. Shown are two printing press systems. The top system is printing on a sheet substrate. The bottom system is printing on a web (roll) substrate. (Ohio Art Co., Tribune Co.)

Web presses have two added features, drying units and sheeting units. Drying units set the ink so that it does not smear, and then the sheeting, or cutting, units cut the continuous ribbon of material into sheets. Some printing equipment also includes automatic folding machines, which fold and assemble the sheets into products such as newspapers or magazines.

Message Transfer

The action of placing the message on the substrate is called message transfer. The transfer process varies with each printing process. We will look at the message transfer process for four printing processes: relief, offset lithography, gravure, and screen printing.

Relief Message Transfer

Most relief printing is done on flexographic presses. This press transfers very thin ink from a reservoir to an application, or anilox, roll, **Figure 20-21.** This roll distributes a thin coating of ink on the raised surface of the rotating plate. As the plate revolves, it contacts the web of paper that is passing over the impression roll. The pressure between the plate and impression roll transfers the image to the substrate.

A special flexography press, called the Cameron press, is sometimes used to print books. It prints with a flexography "belt," which contains all the pages of the book. The paper and the flexography image carrier are moved through two press units. The first unit prints on one side of the paper while the second unit prints on the reverse side. A dryer is at the end of each printing unit to set the ink.

Each cycle of the image carrier prints one complete book. The paper is then sent through assembly and binding units at the end of the press to complete the book. The Cameron press is used to print lower-quality publications such as paperback novels.

Offset Lithography Message Transfer

As you learned earlier in the chapter, offset lithography prints from a flat surface, using the principle that oil and water do not mix. The offset press has several systems in its printing unit, **Figure 20-22.** A plate cylinder holds the image carrier. The dampening system applies a special water-based solution, called the fountain solution, wetting the plate. Next, the inking system applies ink to the plate's surface. Finally, the impression system accepts the substrate and then moves it through the printing unit and onto the delivery unit. The impression system works in two steps. First, it transfers the inked image from the plate to an offset (blanket) roll. The image is then transferred from the blanket to the substrate.

Two or four offset units may be arranged in a single press. This allows color printing to be done efficiently. A four-color press can print full-color materials in a single pass. A two-color press can print the base color (usually black) and a highlight color in one pass. This press can also be used for four-color work, but the press must print the first two colors (yellow and cyan) on one pass. Then the press must be cleaned and the ink colors changed. The second pass will print the other two colors (magenta and black) needed for a four-color run.

Gravure Message Transfer

The gravure image area is a series of very small ink reservoirs engraved in the surface of the carrier. This carrier is a metal drum that is partly suspended in a tray of ink, **Figure 20-23.** As the roll revolves, it picks up a layer of ink. A doctor blade wipes off the excess, leaving ink only in the cavities in the roll. The roll then contacts the substrate under high pressure. This action transfers the ink onto the substrate.

Often several gravure units are assembled into a single press. Each unit is used to print a different color of a multiple-color run.

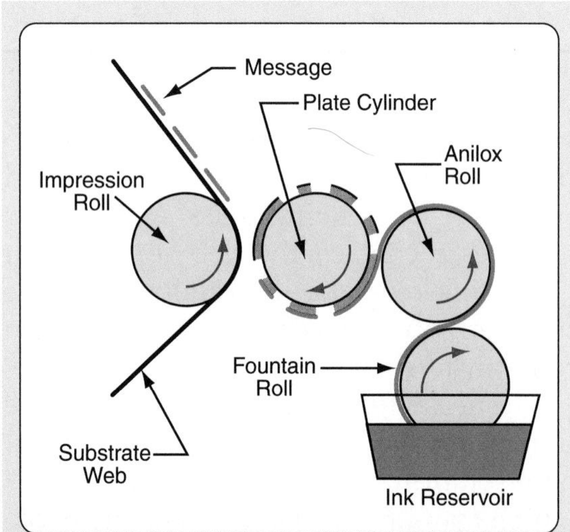

Figure 20-21. This is a diagram of a flexography press.

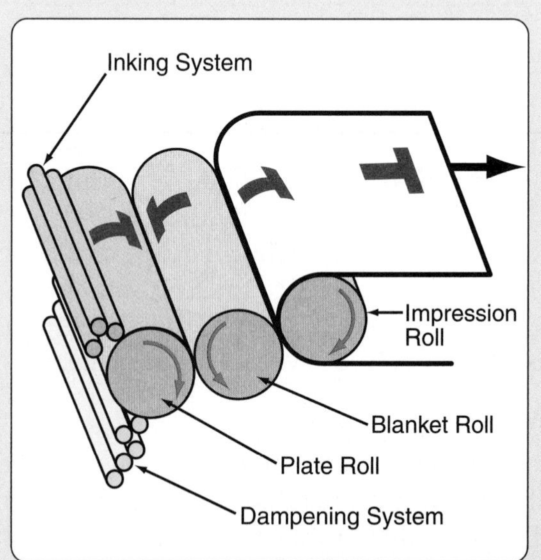

Figure 20-22. Shown here is the operation of an offset lithography press. Notice how the image is first offset before it is printed.

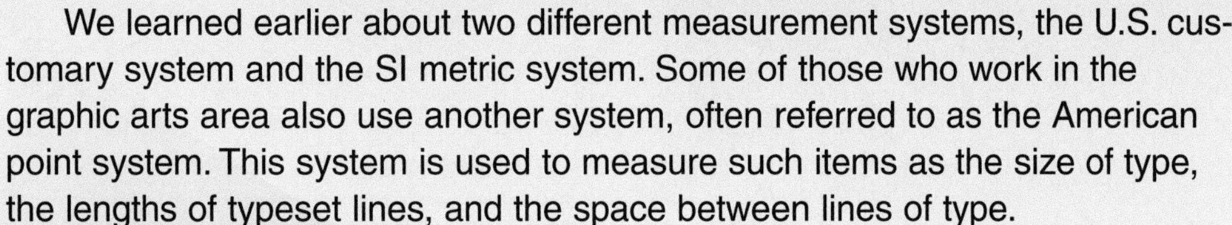

Connections to Technology: **Mathematics**

Measuring Type

We learned earlier about two different measurement systems, the U.S. customary system and the SI metric system. Some of those who work in the graphic arts area also use another system, often referred to as the American point system. This system is used to measure such items as the size of type, the lengths of typeset lines, and the space between lines of type.

The two main units of measurement in this system are the point and the pica. A point measures approximately 1/72 of an inch. A pica is approximately 1/6 of an inch. Twelve points equal one pica, and six picas equal approximately one inch.

Generally, points are used to measure the vertical height of type sizes, and picas are used to measure the lengths of lines. The vertical size of a line of type is the measurement from the top of the letters ("ascenders") to the bottoms ("descenders"). For example, for the word "deep," one would measure from the top of the *d* to the bottom of the *p*. Line length is measured from the beginning of the first word of the line to the end of the last word of the line.

Those who work with type use a typographer's rule to measure points and picas. After locating one of these rules, use it to measure a line of type in this paragraph. How many picas long is the line? What size type is being used?

The image on a gravure roll can be a continuous, repeated pattern. Therefore, this process is often used to print wallpaper, hardboard wall paneling, vinyl flooring, and vinyl shelf covering.

Screen Message Transfer

Many people have done screen printing in art classes or as a hobby. The process, as described earlier in this chapter, is quite simple. Ink is pressed through a stencil onto the substrate. Production screen printing presses use one or more stations. The substrate is automatically fed into position under the screen. A screen is lowered onto the substrate, and a squeegee automatically moves across the image. The screen is then lifted and the substrate removed, or, if more than one color is being used, it is moved to another printing station. Each station lays down a different color until the final image is produced on the product.

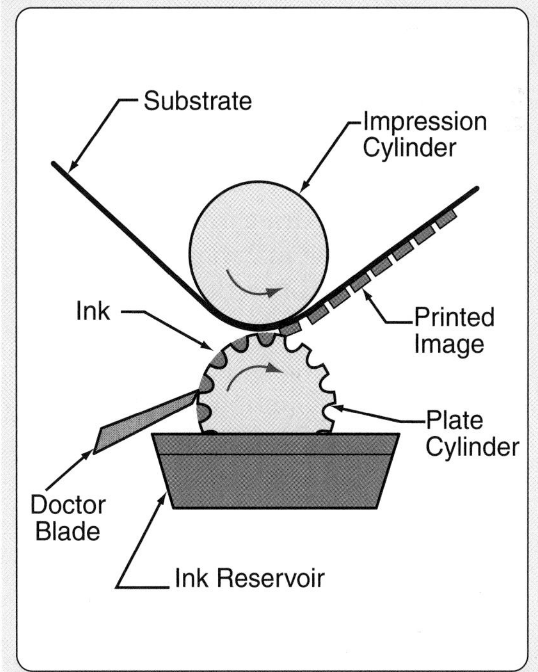

Figure 20-23. Shown here is how a gravure press operates.

Figure 20-24. The irregular shape of this package is die cut.

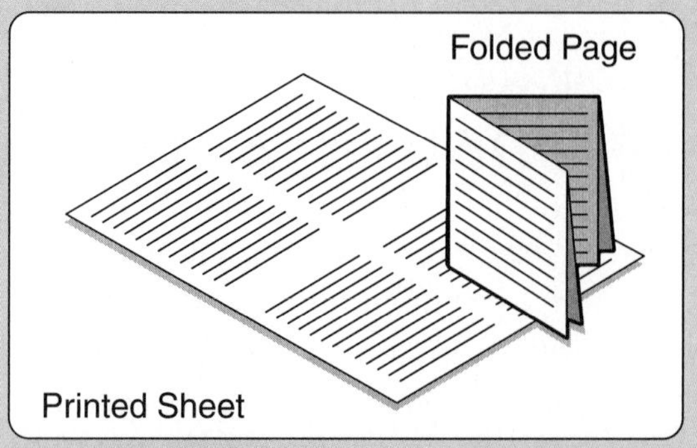

Figure 20-25. A single sheet containing multiple pages is folded to make a signature.

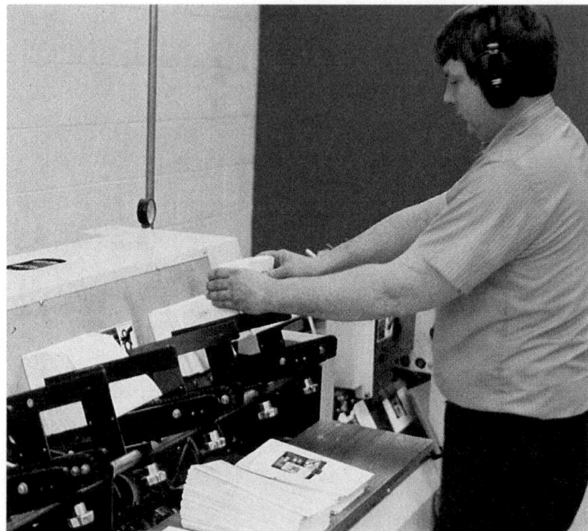

Figure 20-26. This worker is using an automatic machine to assemble signatures into a book. Note his ear protection to prevent hearing loss.

Finishing the Product

Most printed graphic communication products are not finished when they leave the printing press. They require additional work. This work may include:

- Cutting: Trimming the substrate into rectangular sheets, or cutting assembled products to their final size.
- Die cutting: Shearing irregular shapes or openings in the material, **Figure 20-24.** This process is often done with cutting dies on a relief press.
- Folding: Creasing and folding the product to form pages. Often each printed sheet will contain a number of pages, **Figure 20-25.** Properly folded sheets are called signatures.
- Drilling or punching: Creating holes in the substrate for insertion in binders.
- Assembling: Gathering and placing sheets or signatures into the final order, **Figure 20-26.** This may be done manually or with automated machines.
- Binding: Securing sheets and signatures into one unit. This may be done by gluing, stapling, or sewing the units to form a book or magazine.
- Stamping: Transferring an image to a book or album cover with a hot-stamping technique. This is often done using heated type and colored foil. The heat melts the foil and transfers the type image onto the product.
- Embossing: Pressing a raised pattern into the substrate. Paper or other material is pressed between mated dies to produce a pattern.

Computer-Based Publishing

Computers have given rise to several new graphic communication systems. These systems include desktop publishing, electronic publishing, and pagination systems.

The term ***desktop publishing*** is most often used to describe a fairly simple computer system that produces type and line illustration layouts for printed messages. The type and line illustrations are generated using two separate software packages.

This system includes a computer, a laser printer, and the desktop publishing software. Simply adding this software to the computer systems found in many businesses makes desktop publishing available to most companies. They use desktop publishing to create newsletters, announcements, flyers, and other simple communication products.

More complex systems can function as typesetting and layout systems. These systems are called ***electronic publishing.*** They produce and combine text and illustrations into one layout. Electronic publishing produces a higher quality product than desktop publishing produces.

A common electronic publishing system, as shown in **Figure 20-27,** will include:

- Computer keypad to input text copy and to provide commands to the system
- Processing unit (microcomputer) to store and process data
- Monitor to display the text and illustrations that are being manipulated
- Page layout software to direct and control computer and printer actions
- Mouse to send commands to the computer easily
- Ink-jet printer to run proofs and sample layouts
- Laser printer to produce high-quality single-page layouts

Career Corner
News Reporters

News reporters gather information, prepare stories, and make broadcasts that inform an audience about recent events. Reporters investigate leads, read documents, observe events, and interview people. They take notes, organize the material, determine the story's focus, write the story, and edit accompanying video material.

Most employers prefer individuals with a bachelor's degree in journalism or mass communications. Many larger television stations prefer candidates with a degree in a subject-matter specialty, such as economics, political science, or business. Most large broadcasters hire only experienced reporters.

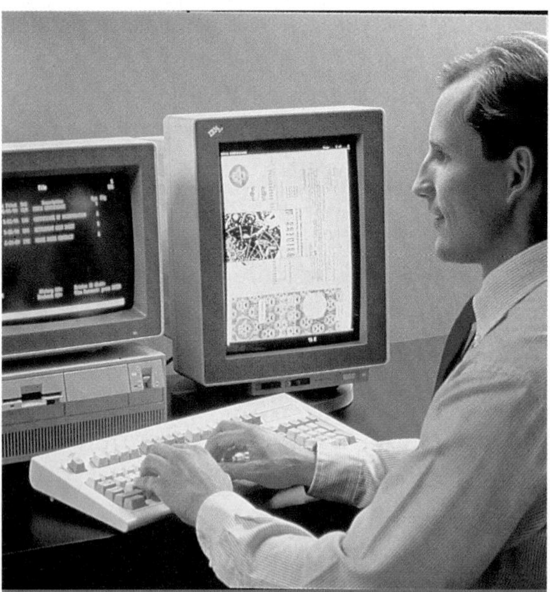

Figure 20-27. Shown is an electronic publishing system. The system is used to combine text and illustrations into an attractive final product. (IBM)

- Scanner to digitize and input illustrations into the computer

The system may also use several additional software programs. Many systems include illustration software to create line illustrations. Also, computer clip art files may be used as a source of additional line illustrations.

The most complex and expensive computer systems are ***pagination systems.*** These systems allow the operator to merge text and illustrations very accurately, **Figure 20-28.** The abilities of a pagination system may include cropping or size adjustments to photographs, color enhancement, image grade enhancement, wrapping text around images, and a complete layout control. Often the output of the system can be directly converted to film or printing plates.

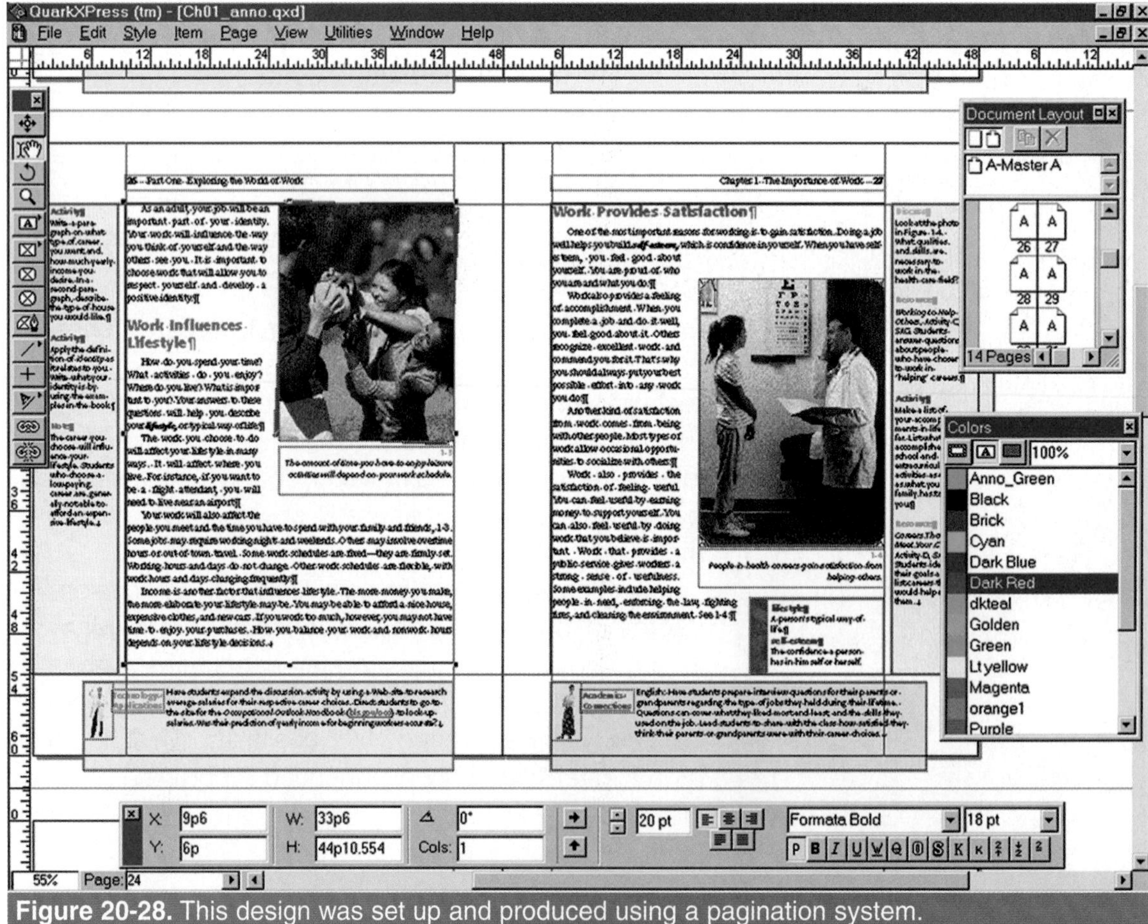

Figure 20-28. This design was set up and produced using a pagination system.

Summary

Each of us is constantly served and impacted by printed graphic communication products. These products may be books and magazines, flyers and pamphlets, or signs and posters.

Each printed product is produced using one of six common printing methods. These methods are relief, offset lithography, gravure, screen, electrostatic, and ink-jet printing. Each product follows a three-step production process. The product is designed. The design is prepared for printing. The product is printed and finished.

Computers have led to new graphic communication systems. These systems are desktop publishing, electronic publishing, and pagination systems.

Key Terms

audience assessment
balance
comprehensive layout
contrast
copy
design
desktop publishing
electronic publishing
electrostatic printing
flexography
gravure
harmony
illustration preparation
illustrations
image carrier
ink-jet printing
layout
letterpress
lithographic printing
mechanical
offset lithography
pagination systems
pasteup
printed graphic communication
printing
proportion
relief printing
rhythm
rough sketches
screen printing
silk screening
substrate
thumbnail sketches
typesetting
variety

Test Your Knowledge

Write your answers on a separate piece of paper. Please do not write in this book.

1. Define printed graphic communication.
2. Name three products produced by printed graphic communication processes.

For Questions 3–12, match the printing method on the right to the correct description on the left. (Note: Some letters will be used more than once.)

Description	**Printing Method**
3. _____Uses a stencil	a. Relief
4. _____Office copier is one type	b. Lithographic
5. _____Requires the image be reversed on printing block	c. Gravure
6. _____Computer generates the printed message	d. Screen
7. _____Based on the principle that oil and water do not mix	e. Electrostatic
8. _____Also known as intaglio process	f. Ink-jet
9. _____Is most widely used today	
10. _____Used to print on T-shirts	
11. _____Used for coding packages	
12. _____Used to print paper money	

13. What are the three main steps in producing printed products?
14. Name two principles of design.
15. The physical act of designing a message is called _____.
16. List the three basic steps in preparing a design for production.
17. *True or false?* Other than sizing, line art usually needs no other preparation for printing.
18. Briefly describe how an offset lithography message is produced.
19. List three activities involved with finishing a printed graphic communication product.
20. What is the difference between desktop publishing and electronic publishing?

Applying Your Knowledge

1. Choose or invent a school event to promote. Complete a chart of the event similar to the one below. Then prepare rough and refined sketches for a poster to promote the school event.

What is the name of your event?
Who will be your audience?
What is the theme of your event?
What is your design approach (humorous, for example)?
What type of promotional material (flyer or poster, for example) will you use?

2. Design a logo or symbol for an organization or a cause. Make a block-cut relief image carrier to print your image.

This activity develops the skills used in TSA's Promotional Graphics event.

Promotional Graphics

Activity Overview

In this activity, you will create a color graphic design appropriate for posters and T-shirts. The theme of the design can be either your school or your community.

Materials

- Paper and pencil
- Computer with graphic design software and clip art
- Color printer

Background Information

General. Consider the principles of design as you develop your project. Locate and size elements to achieve balance and proportion. Use contrast to emphasize key elements. Select elements to provide rhythm.

Elements. The design may include graphic elements, such as photographic images, clip art, or original design elements. When selecting illustrations, consider the content of the illustration (is it appropriate for the cover?) and its final printed size (will it be printed so small that details are lost?). Try to incorporate elements that are generally connected with and related to your theme.

Guidelines

- Your design cannot exceed 8" x 10" (20 cm x 25 cm).
- Identify good design examples from other similar graphics.
- Develop sketches and a rough layout using paper and pencil.
- After you have developed the rough layout, create the design using the graphic design software.
- Print the final design on a color printer.

Evaluation Criteria

Your project will be evaluated using the following criteria:

- Design elements
- Attractiveness/impact
- Relation of design to theme

This activity develops the skills used in TSA's Desktop Publishing event.

Desktop Publishing

Activity Overview

In this activity, you will create one of the following items:

- Tri-fold pamphlet designed as a recruitment tool for potential new students
- Three-column newsletter providing information about a current event or issue in your community.
- Poster advertising an upcoming school function or activity

Materials

- Paper
- Pencil
- Computer with desktop publishing software and clip art
- Color printer

Background Information

General. Consider the principles of design as you develop your project. Locate and size elements to achieve balance and proportion. Use contrast to emphasize key elements. Select elements to provide rhythm.

Elements. Do not use too many variations of font type and size. When selecting illustrations, consider the content of the illustration (is it appropriate for the piece?) and its final printed size (will it be printed so small that details are lost?). Finally, consider the use of white space.

Type. Too many fonts and type sizes can make a piece unattractive. Vary the size and font based on the function of the text. Titles are meant to draw attention. Body type should blend in to the overall design and use a typeface, size, spacing, and justification comfortable for reading. If using a justified layout, it may be necessary to break words (use a hyphen to divide onto two separate lines) to avoid excessive spacing between words.

Tri-fold pamphlet. Begin with the content of the tri-fold—what information should you provide in it? What would a potential student like to know about your school? The pamphlet cover should be designed to be visually appealing. Inside the pamphlet, use several sections of body type, each introduced with a heading. Add illustrations where appropriate.

Three-column newsletter. The title of the newsletter should run the entire width of the page. You may want to use headings that run across all three columns, or perhaps two columns. Illustrations should be sized to fit in either one or two columns. Illustrations should be selected or created to both communicate information and provide visual interest. When placing illustrations, do not disrupt the flow of text for the reader.

Poster. Include all the critical information on the poster: *what, where,* and *when.* The *what* is the attention grabber. The *where* and *when* are also critical elements. Include some advertising copy to persuade the reader to attend the event. Why should the reader attend? Will the reader be entertained, educated, or involved? What will the reader miss by *not* attending? Pay close attention to rhythm when designing the poster—your selection of elements will determine how the reader's eye "moves" when viewing.

Guidelines

- Identify good design examples from samples of tri-fold pamphlets, newsletters, and posters before beginning your design.
- Develop sketches and a rough layout using paper and pencil.
- After you have developed the rough layout, create the design using the desktop publishing software.
- Your project must be designed to be printed on 8 1/2" x 11" paper.
- Your project must include a sample of text wrapping (text flowing around an illustration).
- Your project must use at least one clip art illustration and one original illustration or graphic created using a graphics program.
- Print your project in color.

Evaluation Criteria

Your project will be evaluated using the following criteria:

- Use of design principles
- Page layout
- Selection and placement of clip art
- Font selection and usage

Chapter 21
Photographic Communication

Learning Objectives

After studying this chapter, you will be able to:

- Define photographic communication.
- List the three characteristics of light waves.
- List the three steps in developing a photographic communication message.
- Describe how photographic communication messages are designed.
- Differentiate between types of film.
- Explain the principles of a camera.
- Explain how photographic messages are reproduced.
- Differentiate between digital and traditional film photography.
- List common uses of transparencies.

"A picture is worth a thousand words" is an old saying, **Figure 21-1.** This statement suggests that it is often more effective to convey your message visually than to describe it. Pictures are an efficient method of communication. In some cases, a picture alone is used to communicate an idea or feeling. In other cases, the picture is used to supplement the written word.

The word *picture* once was used to refer only to paintings. Now, just about any two-dimensional visual representation is called a picture. A ***photograph*** is a common type of picture. The act of producing a photograph is called ***photography.*** Using these photographs to convey an idea or information is called ***photographic communication.***

To distinguish photography from photographic communication, consider the following. Your family might take snapshots at family events and during vacation travels. The main goal of the snapshots is to capture a moment in time. Later, you can look at these pictures and remember those moments. This type of photograph is designed to provide a historical record, rather than to communicate information.

On the other hand, if you have flown on a commercial jet, you have seen a safety information card. This card contains a series of photographs showing how to fasten the seat belt and exit the plane during an emergency. These photographs are designed to communicate a specific procedure. They are used as photographic communication.

Figure 21-1. This photograph shows a replica of an early Egyptian hieroglyph.

To understand photographic communication in more depth, in this chapter, we will first briefly discuss the role of light and other electromagnetic waves in the creation of photography. We will then explore the fundamentals of photographic communication, with attention focused on film photography and prints. Next, we will look at digital photography, and finally, we will discuss other types of photographic communication.

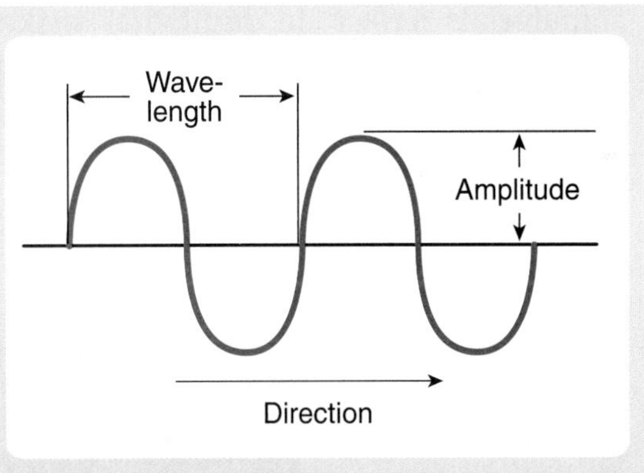

Figure 21-2. The wavelength, the amplitude, and the direction are three descriptive characteristics of a light wave.

Light and Photography

Photography is based on the understanding and appropriate use of light. Light is a wave of energy. We define light as electromagnetic radiation with wavelengths falling in a range including infrared, visible, ultraviolet, and X ray. Light waves have three central characteristics, **Figure 21-2:**

- ***Amplitude*** is the height of the wave. This indicates its strength or intensity. The height is measured from the center of one wave to the peak of that wave.
- ***Wavelength*** is the distance from the beginning to the end of one wave cycle. The wavelength of the light determines

whether the wave falls into the visible range, and if it does, what color it is. Color is more properly referred to as hue. Each hue, or color, has a set wavelength. For example, red waves are 650 nanometers (nm), or 650 billionths of a meter, in length. Closely associated with wavelength is frequency. Frequency is the number of waves that pass a point in one second.

- ***Direction*** refers to the path the light wave travels. Light waves travel in straight lines in all directions from their source. The direction might change when the wave encounters another material or force, **Figure 21-3.** If the wave passes through the material, the wave is said to be ***transmitted.*** Smooth, clear glass allows waves to pass directly through it without bending. This is called specular transmission. Etched glass allows the light to pass, but it randomly bends the light's path. This is called diffuse transmission because the light is diffused, or spread out. Some materials ***reflect*** the light that strikes them. If the angle of reflection is equal

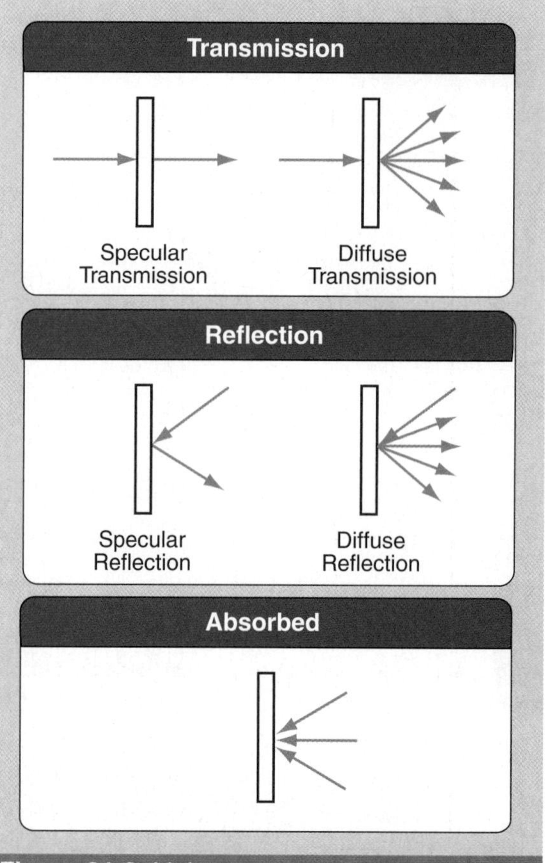

Figure 21-3. Light waves can be transmitted, reflected, or absorbed when they strike an object.

Career Corner
Commercial and Industrial Photographers

Photographers produce and preserve images that capture a scene, tell a story, or record an event. To create commercial-quality photographs, photographers must be creative and technically competent. A photographer must be able to choose appropriate subjects and present them in interesting ways.

Commercial and industrial photographers take pictures of buildings, machinery, products, and workers. Generally, they must be able to take and process standard film images and produce, manipulate, and transmit digital images. Most positions in industrial and scientific photography require a college degree in journalism or photography.

to the angle at which the wave struck the object, it is called specular reflection. Good mirrors exhibit specular reflection. Other substances reflect and diffuse the light. Rough surfaces exhibit this property, called diffuse reflection.

Finally, many materials ***absorb*** light waves. For example, black objects absorb all light that strikes them. Colored objects absorb only certain wavelengths. The other wavelengths are reflected, and your eyes receive them. This property allows your eyes to see the object's "color."

We should note that light is but one of a number of different types of electromagnetic waves. Electromagnetic waves include gamma rays, X rays, ultraviolet light, visible light, infrared rays, and radio waves. Each of these has a specific frequency range and can be divided into even more specific groups, **Figure 21-4.**

When all the wavelengths of visible light are present, we have white light. Sunlight is an example of white light. This light, however, can be divided into a group of wavelengths, each a different color. Have you ever seen a rainbow? If so, you have seen white light divided into its six basic color components: violet, blue, green, yellow, orange, and red. The raindrops in the air cause this separation. As light passes through them, they reflect and separate the various wavelengths. The shorter violet rays appear on the bottom of the rainbow, and the longer red waves are on top.

Fundamentals of Photographic Communication

Photographic communication media are produced using a common set of steps. These steps are the following:

- Designing the message
- Recording the image
- Reproducing the message

Designing a Photographic Message

Photographic messages must be designed with the same considerations used in other communication media. This approach includes answering such questions as the following:

- Who is the intended audience?
- What is the message to be communicated?

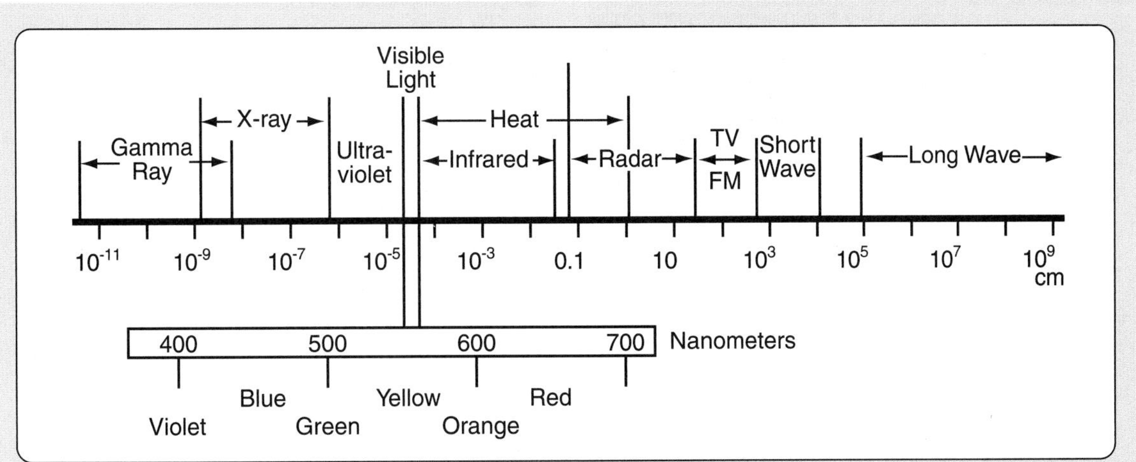

Figure 21-4. This chart shows electromagnetic waves grouped by their wavelength. Notice that visible light is only a small set of electromagnetic radiation.

- What type of photograph will best communicate the information or idea?

A photograph shows a record of light. It captures the light reflected from a series of objects the instant before the film is exposed. Some people see this as a window to reality, whereas others see it as a flat representation of a scene. The goal in designing a photographic communication message is to make the message a realistic window, rather than the flat representation.

To be photographic communication, the photograph must describe something. This can be a process (a way of doing something), an idea (a mental image), or an event. Several techniques are used in helping photographs communicate. First, photographs are ***composed,*** or designed, with a ***point of interest.*** The point of interest is the place where your eye is drawn. This is where the central message is communicated. Look at the two photographs in **Figure 21-5.** Where is your eye drawn in each of them? In one photograph, the sky is the central focus. The other uses the sky simply as a background to the central focus, the horseback riders. What message does each point of interest communicate to you?

Another technique uses ***distance*** and ***scope*** (panorama) to add interest. Look at the photographs in **Figure 21-6.** They both show the same mountain from approximately the same vantage spot. The mountain is the point of interest for both pictures. In one picture, however, the mountain is the dominant feature, whereas the other

Figure 21-5. In these photographs, we see two different points of interest. In the first photograph, the eye is drawn to the sky, whereas in the second photograph, the riders are the central point of interest.

Figure 21-6. In these photographs, the mountain is the main point of interest, but the change in distance and scope creates two different images.

Technology Explained:

Digital Theaters

digital theater: A movie theater receiving and displaying digitally formatted movies.

Most people have gone to a movie theater and are familiar with movie film, **Figure A.** This film is a series of pictures on a long strip of plastic. As the film is fed through a movie projector, the series of still pictures creates the illusion of motion. Providing each theater with a copy of the film, however, is expensive. Each copy can cost more than a thousand dollars to produce. If a movie is showing in several thousand theaters on any day, this cost is substantial.

A new system that can replace film movies is called digital theater. This process uses technology similar to the home digital videodisc (DVD) players. A special device—a TeleCine™ scanner—converts the films to digital video for theaters. At this stage, the operators can correct the film for color. They can add sound and captions to the film, using computerized equipment. These operators can also produce alternate sound tracks for foreign languages.

The finished digital film is sent to a master server. This server allows the company to distribute the film by satellite or broadcast lines. The film can also be distributed on disc, **Figure B**.

At the local theater, the digital film is received and loaded on the local server. The server can play several movies into different auditoriums at the same time. A single server can service about four different auditoriums at once. Therefore, a 12-screen theater would need three servers.

The theater server directs the video part of the film to a special digital projector. The audio portion is sent to a cinema audio processor. Advanced systems automatically dim the lights, open the curtain, and start the film without human operators.

Figure A. Here is a typical theater.

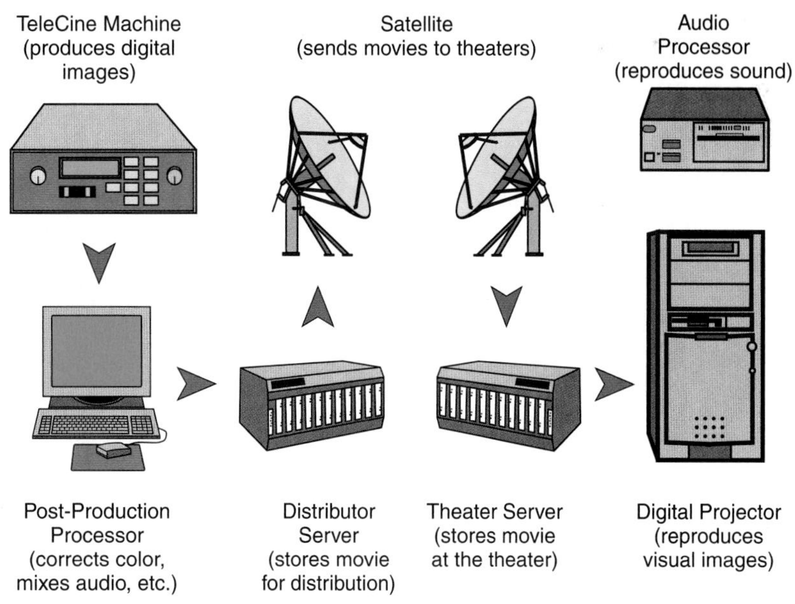

Figure B. This is how a digital movie is delivered to a theater.

picture shows it as part of a larger world. What diverse ideas do these pictures inspire?

A third technique uses color to help communicate. Look at the photographs in **Figure 21-7.** Both are pictures of a concept car. Which one communicates the idea more clearly to you? Why?

A photographic message must meet all the design criteria discussed in the last chapter. Its parts must be in balance and in harmony with one another. In addition, its size and proportion should be pleasing.

Recording the Message

Once the shot is composed, most photographic messages are produced using a two-step process. The image or message is captured on film. The film is then used to reproduce the message so the audience can receive it, **Figure 21-8.** Recording the message requires two separate procedures. These procedures are selecting the medium and exposing the film.

Figure 21-7. Notice how color adds interest to this photograph of a concept car. (Pontiac)

Selecting the Medium

The film used in producing photographs is grouped in several ways. These groupings include type, format, size, and speed.

Types of Film

The two major types of film are black-and-white film and color film. Panchromatic film is the most common form of black-and-white film. ***Panchromatic film*** reacts to all colors of visible light. It records those colors as shades of gray.

Black-and-white film is available only in negative form. ***Negative film*** produces a reverse image of the photographed scene. This means the negative will be dark where the subject is light, and the negative will be light where the subject is dark. Later, it is used to make photographic prints, **Figure 21-9.**

Color film is available in a positive (transparency) or a negative form. ***Positive transparencies,*** or slides (as small transparencies are called), produce the actual view, such that it can be viewed directly through a slide or movie projector. Color negatives are used to make color photographic prints.

Figure 21-8. A careful technical process is used to turn exposed film into photographic communication. (American Greetings Co.)

Film Formats and Sizes

Each type of film comes in several formats and sizes. Film is available in rolls and sheets. Roll film is the most commonly used format. It is loaded in a light-tight canister or cassette. The film is drawn from its container as it is exposed in a camera. Later, it is drawn back into the container and removed from the camera for developing. Both formats come in a variety of sizes. You can buy 4″ × 5″ sheet film or 35 mm (width) roll film, to name just two options.

Figure 21-9. The negative (top) and positive (bottom) are of the same photograph. (Jack Klasey)

Film Speed

Both panchromatic and color films are available in a number of speeds. An ASA/ISO number gives this speed. The speed of the film indicates its sensitivity to light. Fast film is very light sensitive, whereas a slow-speed film takes a considerable amount of light to generate an image. Fast film is used to photograph objects and events in poorly lighted situations. It is also used in photographing fast-moving objects. Slow film is used in well-lighted areas, as well as for photographing outdoors.

The classification of fast film and slow film is somewhat subjective. Note, however, that films with ASA ratings of 25 to 100 are considered slow films. Fast films have ASA ratings of 200 and above.

Exposing Film

Film is almost always exposed in a camera. This technological device is, basically, a light-tight box used to expose a light-sensitive material. A camera has three main systems:

- Film exposing system
- Viewing system
- Film feeding system

Film Exposing System

The ***shutter*** is a device that opens and closes to permit or prevent light from entering the camera. The length of time the shutter is open controls the length of time light can strike the film. Many cameras have adjustable shutters. Speeds can range from 1 second or more, down to 1/1000 of a second.

The ***diaphragm,*** or aperture control, regulates the amount of light that can enter the camera at any given moment. Generally, it is made of a series of very thin metal leaves or plates. The leaves can be adjusted to change the size of the hole (aperture) behind the camera lens. An ***f-stop number*** identifies the size of the opening. The f-stop numbers are the reverse of the apertures—as the f-stop number increases, the size of the aperture decreases. As the f-stop number decreases, the size of the aperture increases. The amount of change is specific. It doubles or halves the amount of light from stop to stop. Thus, going from f/11 to f/16 will decrease the light by one-half. Moving from f/11 to f/8 will double the amount of light.

The purpose of the ***lens*** is to focus the light on the film. The lens also inverts the image. The focal length of the lens determines the size of the image. A typical 35-mm camera has a lens with a 50-mm focal length. This means the front of the lens is 50 mm away from the film. Different lenses are used for varying effects. A 28-mm wide-angle lens produces a wide view for close objects. A 300-mm telephoto lens has a narrow angle of view and enlarges distant objects.

Viewing System

A camera captures only the light reflected from the scene in front of the lens. Therefore, it is essential that the operator points the camera precisely at the desired scene. Also, it is important that the object is in focus. Both of these conditions can be achieved using the viewing system.

Cameras use one of two systems to select and adjust the view: the viewfinder system and the single-lens reflex system, **Figure 21-10.** In the viewfinder system, the operator looks through a viewing port on the back or top of the camera. The scene is viewed through a lens on the camera's front. The camera lens and the viewing lens are separate. The view the operator sees is approximately what the camera views. Because the viewfinder lens and camera lens are separate, however, the view is not exactly what will be captured on film.

More accurate viewing is achieved using a single-lens reflex system. This system uses a series of prisms and mirrors that allow the operator to look directly through the camera lens. The view seen through the single-lens reflex is exactly the view captured on film.

Film Feeding System

Most cameras capture a series of photographs on a strip of film. This film is held in position behind the shutter. After each shot, the film is advanced so the next area of unexposed film moves into position behind the shutter.

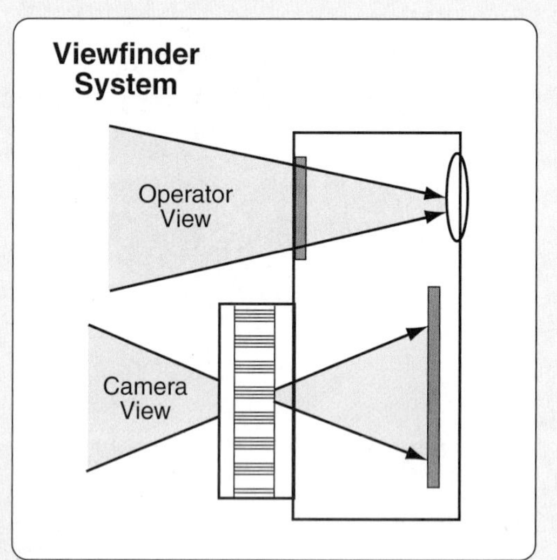

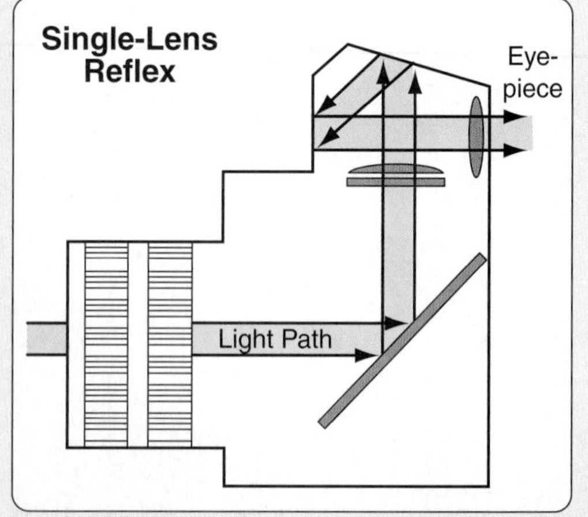

Figure 21-10. Cameras use two viewing systems. The single-lens reflex system gives a more precise view of the image to be captured.

Typically, film is available in rolls or cassettes, **Figure 21-11.** Many of the more expensive cameras use a 35-mm roll film. This film comes in lengths that allow 12, 20, 24, or 36 exposures (photographs). "Snapshot" cameras generally use film cassettes in the smaller 110 size. These hold 12 or 20 shots.

Reproducing the Message

Communication technology is often used to produce messages for a large, diverse audience. Therefore, the message must be produced in quantity. Reproducing photographic communication media requires developing film and making prints.

Developing Film

Photographic communication is not possible until the captured image is retrieved. This retrieval process is called developing the film. Developing chemically treats the light-sensitive materials in the film emulsion to bring out the image. Black-and-white film is developed with a fairly simple process. Let us explore this activity, **Figure 21-12.**

Developing film involves a series of chemical reactions. These reactions are the following:

1. Developing
2. Stopping
3. Fixing

The first step, ***developing,*** uses chemicals (developers) to alter the light-sensitive crystals in the emulsion. The crystals are changed from silver halide, which cannot be seen, to black metallic silver, which can be seen. The next step, ***stopping,*** stops the chemical action of the developer. It uses an acid solution, called ***stop bath,*** to neutralize the developer. ***Fixing*** then removes the unexposed silver halide crystals remaining in the film. This makes the image permanent. After these three steps are complete, the film is washed to remove any chemical residue. It is then dried so it can be stored or used to make prints.

Safety with Darkroom Chemicals

Use care when mixing darkroom chemicals. They can cause skin and eye irritation. Also, use tongs and an apron when transferring prints from one solution to another. Wash your hands if they come in contact with the chemicals.

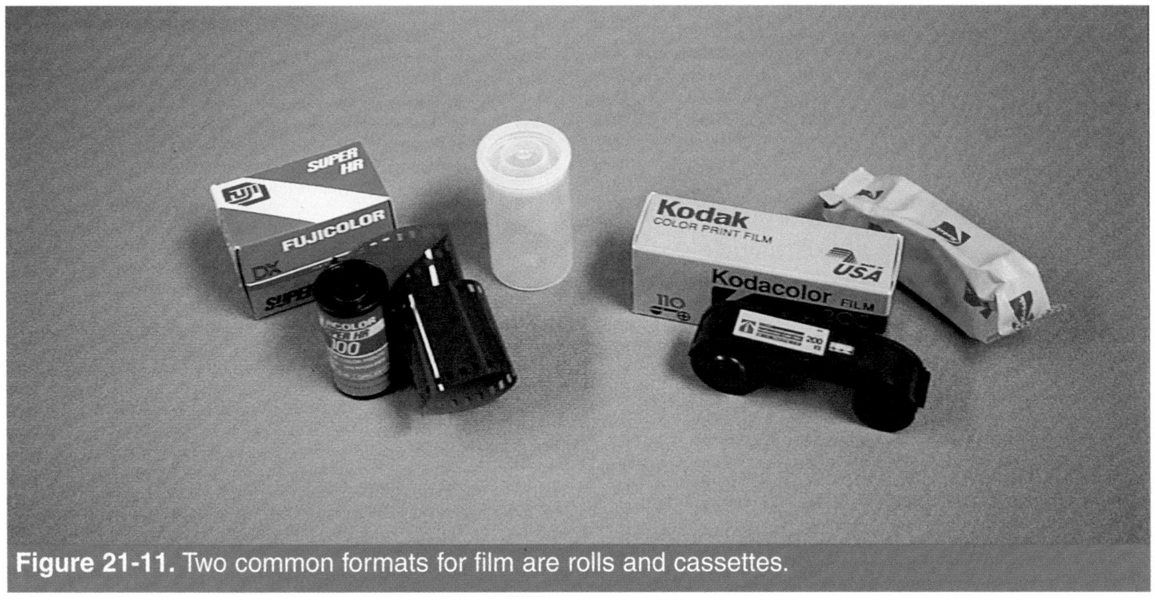

Figure 21-11. Two common formats for film are rolls and cassettes.

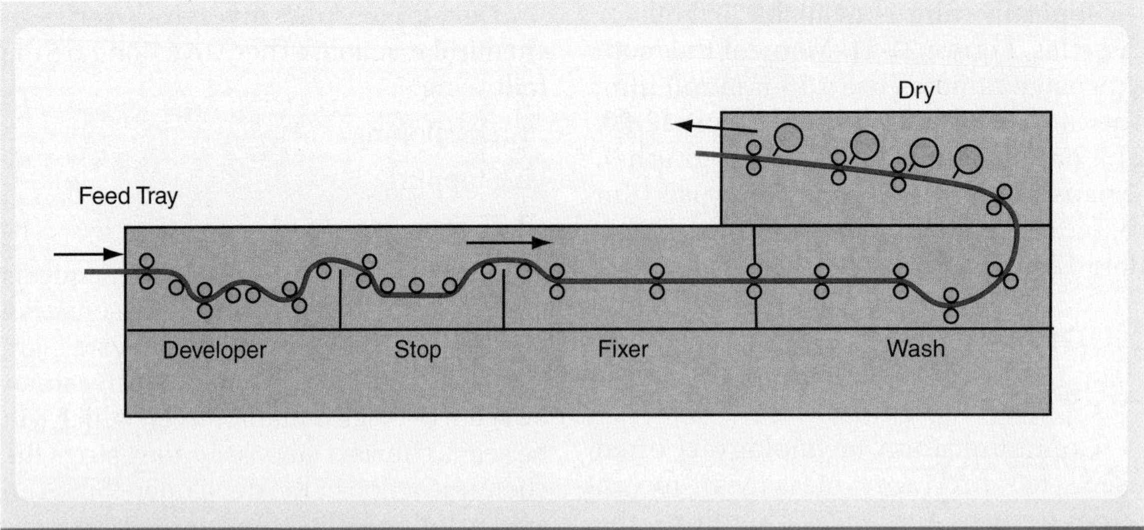

Figure 21-12. The dark line shows the path of film in an automatic film-developing machine.

Making Prints

Most types of black-and-white film produce negative (reverse) images. This negative is used to make photographic prints. Two basic techniques, contact printing and projection printing, are used to produce prints.

Contact printing is the simpler procedure. The negative is placed directly on top of a piece of light-sensitive photographic paper. The film and paper are held tightly under a sheet of glass. This assembly is exposed to a bright light. The paper is then removed and developed, with the photographer following the same steps used for developing film: developing, stopping, fixing, and washing. Note, however, that a different chemical is used for developing prints than is used for film.

It is hard to see how each negative will print out, so contact prints are widely used as proof prints. A contact print will show each picture as a small print, or photograph. This allows the photographer to select only the best negatives to make into larger prints.

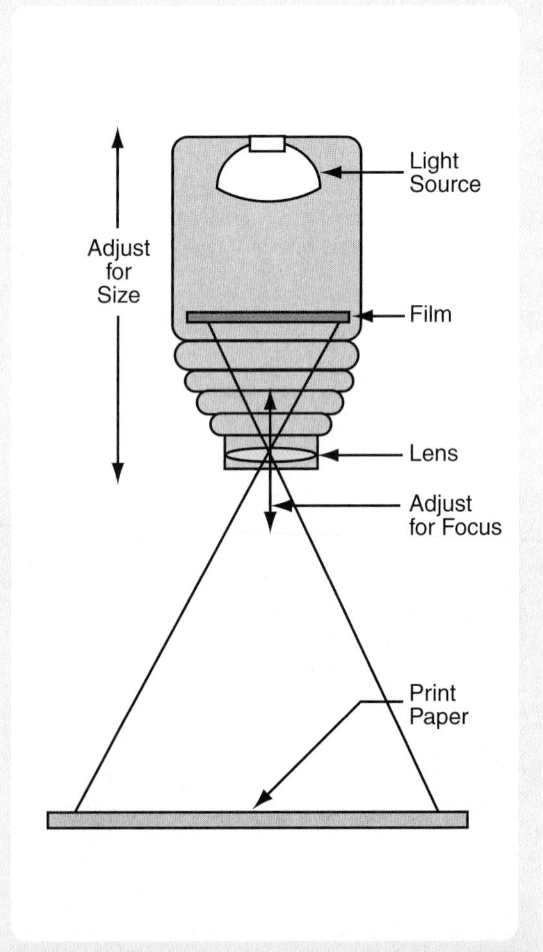

Figure 21-13. In projection printing, the operator shines light through a negative onto a piece of photographic paper.

Projection printing is done by projecting (shining) light through the negative onto a piece of photographic paper, **Figure 21-13.** Projection printing requires an enlarger to adjust and control the print size and focus. The farther the enlarger lens is away from the paper, the larger the print will be. Producing a projection print uses the following steps:

1. Cleaning and inserting the negative in the enlarger
2. Adjusting the enlarger for print size and focus
3. Exposing the photographic paper
4. Developing the print
5. Drying the print

Digital Photography

Today, many people are turning away from film cameras and toward digital cameras. What does *digital* mean? A digital device is any device that stores information in electronic form. Generally, this form uses the binary language of zeros and ones. Digital technology is everywhere you look. It is in the watches we wear, the television sets we watch, the computers we use, and now, in the cameras with which we take photographs.

Most digital cameras operate much like simple, automatic point-and-shoot film cameras. See **Figure 21-14.** Digital cameras generally have an autofocus feature and automatically adjust the shutter speed and aperture. The image to be captured is shown on a small screen on the back of the camera. When the photographer is satisfied with the image shown, a button is pressed to capture the shot. Newer, more expensive digital cameras operate like the single-lens reflex cameras described earlier. They allow the operator to view the image to be captured through the camera lens.

Figure 21-14. Digital cameras are becoming the most popular way for people to take photographs.

Even though digital cameras look and operate much like traditional cameras, there are major differences. Traditional cameras use light passing through a lens to expose film. This process of light passing through the lens makes a photochemical record of an image on film. Digital cameras also use a series of lenses to focus light and create an image. The image is recorded, however, on a digital sensor array, which is a specialized computer chip. The digital sensor has thousands of tiny square points that can sense light. These points are called picture elements, or pixels. A pixel is defined as the information stored for a single grid point in the image. Essentially, a pixel is a square dot of light that has a given color and brightness value.

Digital images are made up of thousands of small squares, just like a mosaic is made up of many colored pieces of glass. See **Figure 21-15.** The complete image is developed by an arrangement of pixels in rows and columns. The varying colors and brightness of the many pixels presents the image. Digital cameras are available with differing numbers of pixels. The following are some common camera resolutions:

- 1216 × 912 pixels, or 1 megapixel (1 million pixels), which provides good family photos.
- 1600 × 1200 pixels, or 2 megapixels, which produce 4″ × 5″ photo-lab quality prints.
- 2240 × 1680 pixels, or 4 megapixels, which provide the ability to print larger high-quality prints.

Figure 21-15. Note that the picture on the left is made up of colored square pixels as seen on the right.

- 4064 × 2704 pixels, or 11 megapixels, which produce large, high-quality prints.

The camera's circuitry processes the information the digital sensor has gathered and stores the information on a memory device. The memory device is generally a card or strip that can be used to download the information directly into a computer system. There, the information can be manipulated (color balanced, cropped, and sized, for example) using various computer software programs. The images can then be printed on an ink-jet printer or professionally printed, through a camera store or other retail business. They can also be viewed on a number of electronic devices, including computer monitors and TV sets.

Other Types of Photographic Communication

Prints are one common use for photography. They can be used to communicate directly, or they may be integrated into a printed graphic message. This was discussed in Chapter 20.

Photography is also used to produce transparencies. As you learned earlier, these are positives of the recorded scene. They show the image as the eye sees it. Three common uses of transparencies in communication are shown in **Figure 21-16.**

- ***Slides*** are single transparencies designed to be viewed independently.

Figure 21-16. Color transparencies, such as slides, filmstrips, and motion pictures, are used frequently in classrooms for graphic communication.

Connections to Technology: **History**

The Beginnings of Photojournalism

The creation of cameras and experiments with photography date back to the 1500s. Not until the mid-1800s, however, did the use of cameras begin to play a major role in how society viewed events. Perhaps the best example of the new power of photographic communication can be seen in the work of Matthew Brady, America's first photojournalist.

When the Civil War broke out in 1861, Brady was well known as a portrait artist. He asked for and received permission to create a photographic record of the war. He hired a crew and placed them at various sites where soldiers were camped or battles were expected to break out. The resulting photographs were the first visual images ordinary American citizens had ever seen of war as it was occurring.

The images put a human face on the war. People saw pictures of soldiers at their daily chores, as well as pictures of dead combatants lying on the field after a battle. The photographs were dramatic and changed Americans' perceptions. Previous artwork had glamorized and romanticized war; the new images awakened Americans to war's other side, the side that includes both acute boredom and terrible suffering. Can you think of another example when new technology affected people's perceptions of reality?

They are cut from the film and mounted in frames. Slides can be arranged into sets called slide series. Later, they can be rearranged or supplemented with new slides to communicate a different message.

- ***Filmstrips*** are a series of transparencies designed to be viewed one at a time. Their sequence is rigid, and they communicate a specific message. The film is left in strip form, as produced by a camera or filmstrip duplicator.
- ***Motion pictures*** are a series of transparencies shot over a span of time. They are, generally, shot at a rate of twenty-four frames (single transparencies) per second. This rate is fast enough that the human eye cannot distinguish each individual picture. When the transparencies are projected at the rate they were recorded, the series of still shots will give the illusion of movement.

In addition, color transparencies are used to reproduce color illustrations in books, magazines, and newspapers. Often, these transparencies are 35-mm slides. Cameras called view cameras, however, produce larger transparencies on sheet film. These transparencies, usually 4″ × 5″ or 8″ × 10″, are used in advertising and publishing. Most of the illustrations in this book were reproduced from transparencies.

Summary

Communication through photographic images is another method human beings have to express ideas, information, and feelings. To communicate effectively, however, we must know how to design, record, and reproduce the image or images correctly. Design involves knowing the audience and what message needs to be communicated. Recording involves choosing the proper film and exposing that film, based on the type of camera used. Reproducing involves developing film and making prints or transparencies.

Key Terms

absorb
amplitude
composed
contact printing
developing
diaphragm
direction
distance
filmstrip
fixing
f-stop number
lens
motion picture
negative film
panchromatic film
photograph
photographic communication
photography
point of interest
positive transparency
projection printing
reflect
scope
shutter
slide
stop bath
stopping
transmit
wavelength

Test Your Knowledge

Write your answers on a separate piece of paper. Please do not write in this book.

1. What is photographic communication?
2. List the three characteristics of light waves.
3. List the three steps involved in developing a photographic communication message.
4. Give one example of a question that must be addressed when designing a photographic message.
5. The spot to which your eye is drawn in a photograph is called the _____.
6. Name one design technique used to add interest to a photographic message.
7. What is the difference between negative film and positive transparencies?
8. What are the three main systems in a camera?

For Questions 9–14, match the parts of a camera on the right to the correct description on the left. (Note: Some letters will be used more than once.)

Description	Parts of a Camera
9. _____Opens and closes to control light	a. Shutter
10. _____Its setting is measured by f-stops	b. Diaphragm
11. _____Inverts the image	c. Lens
12. _____Its speed ranges from 1 second to 1/1000 of a second	
13. _____Its purpose is to focus light on the film	
14. _____Also called aperture control	

15. List the three steps in developing film.
16. Name the two basic techniques used to produce prints.
17. What is one of the main differences between digital and film photography?
18. Name three common uses of transparencies in photographic communication.

Applying Your Knowledge

1. Design and shoot a six-photograph series. Have the series describe how to do something, or have the series communicate an idea about pollution, energy conservation, or another important issue. Develop a layout, similar to the one below, to describe the theme and each shot you would use. Sketch the location of the major elements as you compose each photograph.

Theme or topic:		
Slide 1 (Description)	Slide 2 (Description)	Slide 3 (Description)
Sketch of the shot	Sketch of the shot	Sketch of the shot
Slide 4 (Description)	Slide 5 (Description)	Slide 6 (Description)
Sketch of the shot	Sketch of the shot	Sketch of the shot

2. Shoot and develop photographs that show (a) happiness, (b) sadness, (c) quietness, and (d) harmony with nature.

This activity develops the skills used in TSA's Imaging Technology event.

Imaging Technology

Activity Overview

In this activity, you will create a photographic display. The theme of the display can be provided by your teacher or chosen from one of the following:

- Your school
- Your local community
- Nature
- Technology

Materials

- 35 mm camera or digital camera
- Computer with photo editing software and color printer (for digital camera)
- Photo developing lab (for 35 mm camera)

Background Information

General. Check your camera's settings before you take pictures. Most often, using the camera's automatic setting will allow you to "point and shoot." This means you can concentrate on your subject, instead of worrying about camera controls. To avoid blurry photos, hold the camera firmly with both hands, and press the shutter button with a slow, steady motion. Don't "jab" the button with your finger—that will cause the camera to shake.

Portraits. Get close to your subject. Fill most of the viewfinder with your subject, instead of the background. Press the shutter button halfway to bring the subject into focus, then watch for a good expression and take the photo. Try both people and pet portraits, and try them with and without using the camera's flash. If your subject is against a bright background (such as a window), you'll probably need to use a flash unless you want to do a silhouette.

Landscapes. If the camera has a zoom lens, use the wide-angle setting to show a broad area. Try to get an interesting object (for example, a rock, bush, or fence) in the foreground to add depth to your picture. The wide-angle lens will make it easier to keep both near and distant subjects in focus.

Still life. Try for a pleasing arrangement of objects that fit the theme. Don't use your flash, since that would cause harsh shadows. Instead, use room lighting, possibly with a table or desk lamp placed just outside the picture to provide additional light. You can move the lamp around to place shadows where you want them.

Action. Try to shoot action pictures on a brightly lit day, so the camera's automatic mode will choose a fast shutter speed. This will prevent blurring. Try to arrange the situation so your subject is moving toward you, or at an angle, rather than "crossing" in front of you. This will help to stop motion. Practice with the camera so you can adjust for *shutter lag*—the time between pressing the button and the actual shutter operation.

Photo editing and enhancement. Use your computer's photo-editing software to adjust the brightness of your picture (darken or lighten as necessary). If the picture has a *color cast,* such as a very blue or very yellow appearance, your software should have the ability to remove the cast. Don't be afraid to crop away part of the picture. Many pictures can be greatly improved by eliminating extra background and focusing attention on the subject.

Guidelines

- Your display must include 15 prints. At least five prints must be color prints, and at least five prints must be black-and-white prints.
- None of the prints can be larger than 8" x 10".
- Your display should include a variety of prints, such as still life, action, product, portrait, landscape, and special effects.
- Prints may be single matted, and matting and display board must be a single color—either white, black, or gray.

Evaluation Criteria

Your project will be evaluated using the following criteria:

- Composition
- Lighting
- Relation of images to theme
- Processing and finishing
- Creativity
- Display quality

Chapter 22
Telecommunication

Learning Objectives

After studying this chapter, you will be able to:

- Define telecommunication.
- List the major types of telecommunication systems.
- Explain the difference between frequency modulation and amplitude modulation.
- Differentiate between the two types of television stations.
- List the major activities involved in designing a broadcast message.
- Name activities involved in producing a broadcast message.
- Identify how broadcast messages are delivered.
- Describe a remote-link system.
- List the three mobile communication systems.

Humans have always communicated their ideas and feelings. Early communication included people speaking to one another. Later, humans developed writing to record and transmit information. Writing allowed us to express and *store* information, opinions, and concepts. These types of communication, however, did not meet all human needs. People wanted to hear the human voice and to do so beyond the limits of face-to-face communication. They wanted to communicate their thoughts and knowledge over great distances. Out of this desire came ***telecommunication,*** **Figure 22-1,** meaning communication over distance. Telecommunication implies that a message exists and that hardware (technology) is available to deliver it.

In this chapter, we will study various aspects of telecommunication today. Included are discussions of the physics behind telecommunication and the different types of major telecommunication systems that exist. We will also examine how we communicate with major telecommunication systems and look at other telecommunication technologies.

The Physics of Telecommunication

Important in the discussion of telecommunication are the principles of electricity and electromagnetic waves. These principles can help you understand the technology you use daily.

Figure 22-1. Telecommunication means communicating over distance. This satellite dish receiver is a telecommunication system that aids in capturing messages sent from great distances.

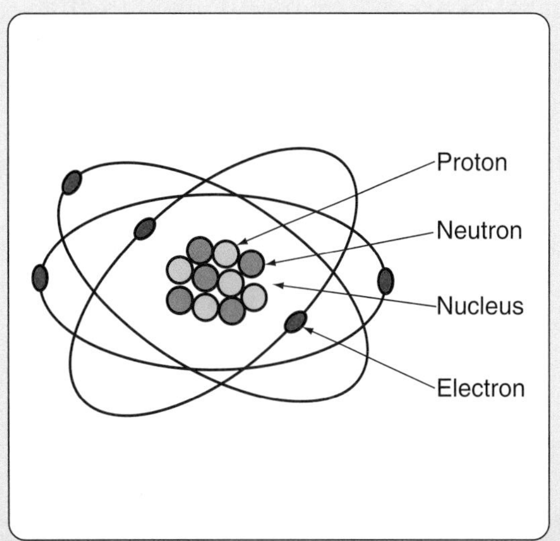

Figure 22-2. Atoms are made up of protons and neutrons, which form the nucleus, and electrons, which orbit the nucleus.

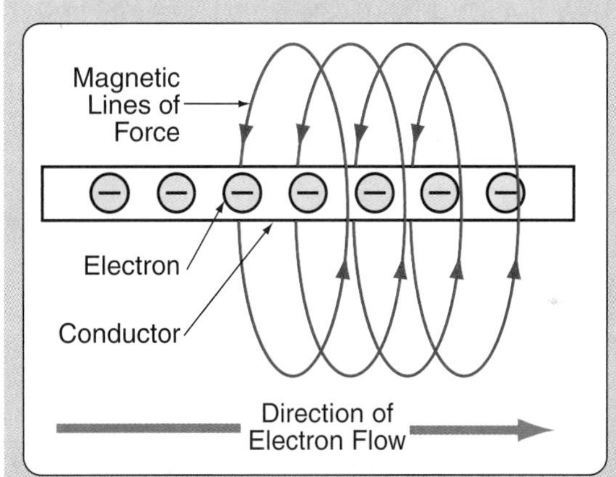

Figure 22-3. Electron movement through a conductor is called electricity. This movement creates magnetic lines of force around the conductor.

Electrical Principles

You have probably learned about the atom in your science classes. All matter is made up of atoms. Each atom has a nucleus, or center. This center is made up of positively charged particles, called protons, and neutral particles, called neutrons. A group of negatively charged particles orbit the nucleus. These particles, or bits of energy, are called electrons. The electrons are held in orbit by their attraction to the positively charged protons, **Figure 22-2.**

In certain situations, electrons will travel from one atom to another. This movement is called electricity. It most often takes place in a metal called a ***conductor.*** When electrons are flowing in one direction along the conductor, it is called ***direct current.*** If the electrons flow in both directions along the conductor, reversing at regular intervals, it is called ***alternating current.***

Movement of electrons in a conductor creates magnetic lines of force, **Figure 22-3.** This force is known as an electromagnetic force. As these lines of force increase and decrease in strength, they can cause electrons to flow in an adjacent wire. This process is called ***induction.*** The principle of induction is commonly used to change sounds into electrical signals or to change electrical signals into sound. This process is used in microphones and speakers. These technological devices are examples of ***transducers.*** Transducers change energy of one form into energy of another form.

Electromagnetic Waves

Earlier in this book, you were introduced to basic electromagnetic wave theory. You will remember that two important characteristics of waves are frequency and amplitude, **Figure 22-4.** ***Frequency*** is the number of cycles (complete wavelengths) that pass some point in one second. The number of cycles per second is measured in ***hertz.*** The basic units of measurement in telecommunication are ***kilohertz,*** or kHz (1000 cycles per second), and ***megahertz,*** or MHz (1 million cycles per second).

Amplitude measures the strength of the wave. The higher the amplitude, the stronger the signal. Telecommunication uses changes in the amplitude of the waves, the frequency of the waves, or both to carry a communication message.

Within the electromagnetic spectrum is a series of frequencies we call ***radio waves.*** These frequencies extend from around 30 hertz to 300 gigahertz (300 billion cycles per second). Below this range of frequencies is a series we call extremely low frequency (ELF). The series extends from about 10 Hz to 13.6 Hz. It is used for underwater communication. Naval commanders use these frequencies to communicate with submarines.

Radio waves are part of a series of frequencies known as ***broadcast frequencies.*** These frequencies are used for a wide range of communication systems. These systems include police and fire department radio, broadcast radio, cellular telephone, and television communication.

The Federal Communication Commission (FCC) assigns each type of communication to a range of frequencies. For example, 160.215 to 161.565 kHz is assigned for railroad communication, and 50.0 to 54.0 MHz is assigned for 6-meter amateur radio. Frequency assignments are exclusive for each particular use nationwide.

Types of Signals

Signals are fundamental to communication technology. They are used to carry information from a source to a receiver. All signals have two basic properties:

- They contain energy and power.
- They have a pattern that carries information.

The basic types of signals used to communicate information are analog and digital signals.

Analog Signals

Thomas Edison developed the first workable device that could record sound in 1877. His machine used a diaphragm attached to a needle to record sound on a foil cylinder. When a person spoke in the diaphragm, it vibrated, causing the needle to scribe a groove into the foil. In the playback mode, the opposite happened. The needle traveled along the groove of the

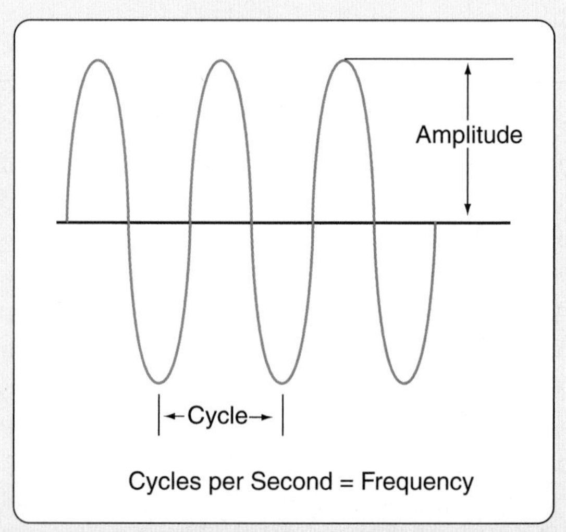

Figure 22-4. The information in radio waves is coded through varying the frequency and amplitude of the wave.

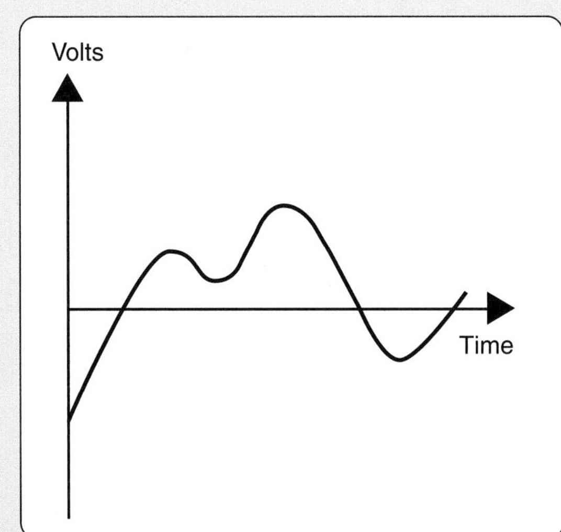

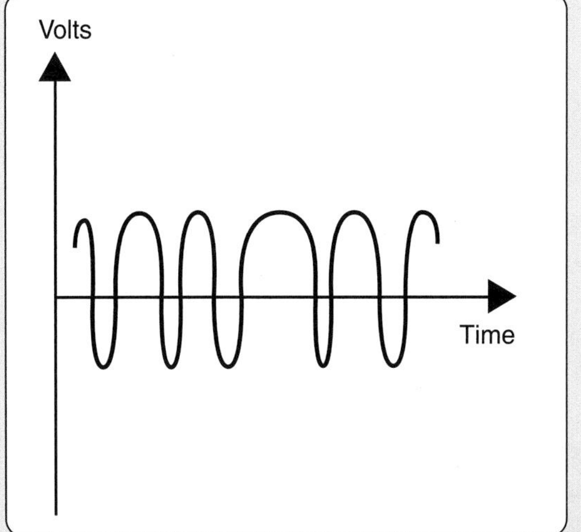

Figure 22-5. An analog signal is a continuous signal that varies in strength (amplitude), as shown on the left, or in frequency, as shown on the right.

rotating cylinder. The movement of the needle vibrated the diaphragm and reproduced the original sound. Edison's invention was improved by a number of people over the years and became the common way to record speech and music. This type of recording equipment stored the sound using analog signals.

An analog signal is a continuous electronic signal that carries information in the form of variable physical values. Information is added to the base signal by amplifying its strength (amplitude modulation, or AM) or varying its frequency (frequency modulation, or FM). See **Figure 22-5**. AM and FM will be discussed later in this chapter.

Until recently, nearly all telephone, television, and radio signals used analog signals because analog signals are fairly easy to create and transmit. Analog signals have, however, one major disadvantage. Every detail in the signal pattern is important, and outside forces can alter the pattern. These alternations create noise and distortion that make the output different from the input. This major problem has led to the development of a newer type of signal: digital signals.

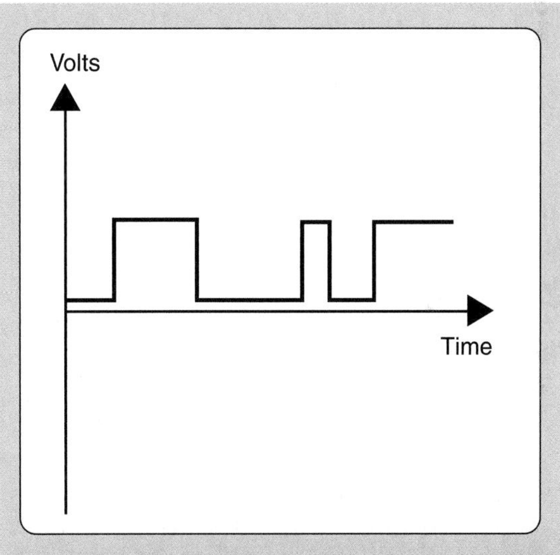

Figure 22-6. A digital signal is a series of *1*s (on) and *0*s (off).

Digital Signals

Digital technology generates, stores, processes, and transmits data using positive and nonpositive electrical states. The positive state (on) is expressed by a *1*, while the nonpositive (off) is expressed by a *0*. All digital data is, therefore, expressed as a string of *1*s and *0*s. See **Figure 22-6**.

This type of signal has several advantages over analog signals. Digital signals can be transmitted faster, therefore allowing more information to be moved in a given period of time. Also, they are more accurate and less prone to outside interference (noise and distortion). This makes digital music recordings (such as those found on compact discs) clearer than older analog long-playing (LP) recordings and cassettes. Digital recordings produce sound that is very close to the reproduced signal. They also sound the same no matter how many times they are played. Likewise, digital TV programs have clearer pictures and better sound than analog programs.

Types of Telecommunication Systems

All these frequencies are used to communicate data, information, or ideas. They are used in two major types of telecommunication systems. These types are hard-wired systems and broadcast systems, **Figure 22-7.** Each will be discussed in turn.

Hard-Wired Systems

You will remember that communication systems have three major parts: a sender, a communication channel, and a receiver. A typical ***hard-wired system*** is the telephone system, **Figure 22-8,** which was described briefly in Chapter 7. In this system, the microphone in the mouthpiece changes sound waves into electrical impulses. The frequency and duration of the electrical impulses are the coded message.

These electrical codes are usually conducted over a permanent waveguide that connects the sender and the receiver. This guide may be a copper wire or fiber-optic (glass fiber) conductor. In some cases, microwave radio signals take the place of a waveguide for a portion of the circuit. Microwaves are often used to send a message between major cities. The signals are then transferred back onto wire or cable.

A special system called ***multiplexing*** is frequently used to increase the capacity of the waveguide. This system allows several unrelated messages to travel down a single conductor at the same time. This can be done through time division multiplexing (TDM) or frequency division multiplexing (FDM).

Figure 22-7. Telecommunication systems can be grouped as hard-wired or broadcast systems. Shown on the left is a telephone (hard-wired) unit. On the right is a remote unit for a television (broadcast) station. (Harper-Fritsh Studios, Gannett)

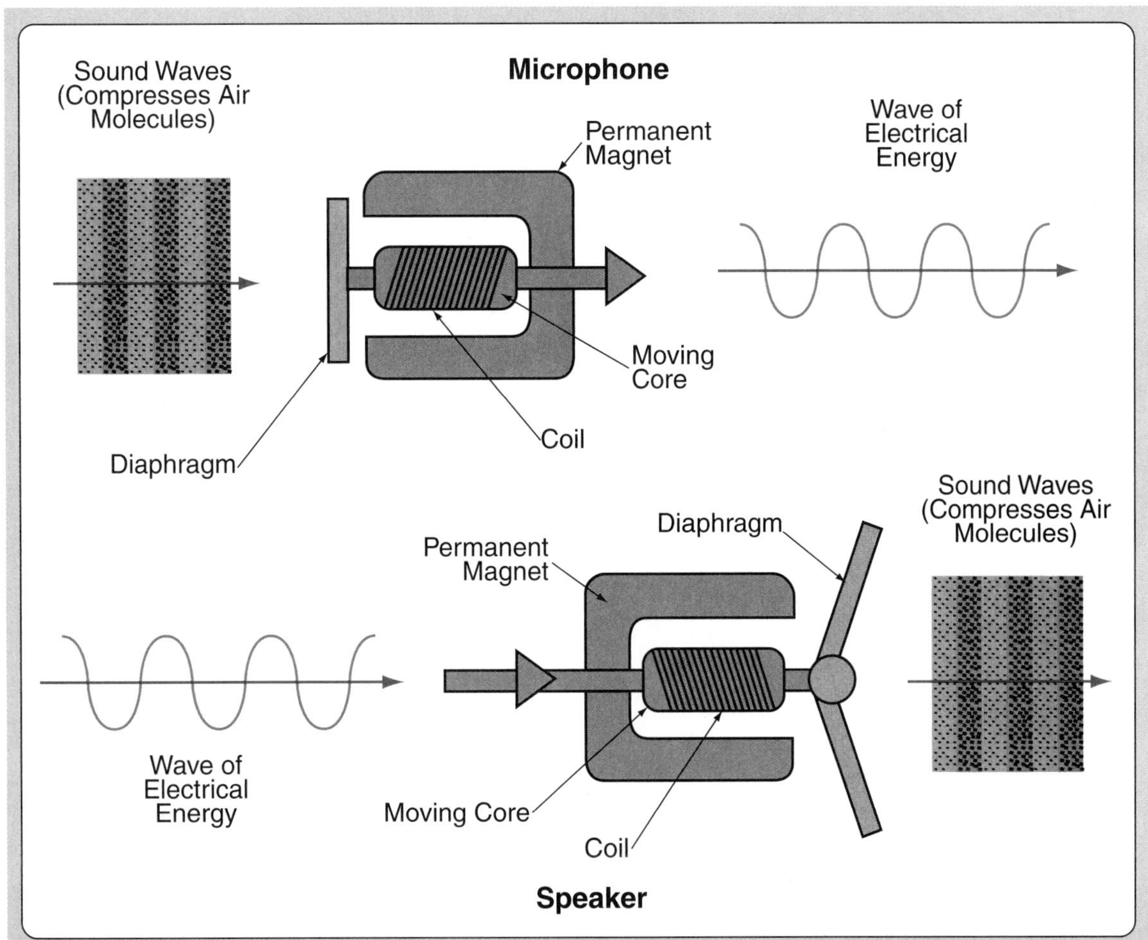

Figure 22-8. Telephone systems use coded electrical messages to communicate the spoken word. Shown here are sound waves being transferred into electrical waves by a microphone. The electrical wave is then channeled to a receiver, where a speaker decodes the signal.

Time division multiplexing (TDM) divides time into very brief segments. Several messages are also divided into small discrete bits. Electronic equipment then assigns each bit separate time slots. Next, the message bits are transmitted. The receiver collects all the bits. It sorts the bits and assembles each message back into its original form.

Frequency division multiplexing (FDM) uses a separate frequency to transmit each message. Several messages are transmitted at the same time over the same waveguide. They are blended together and channeled down the carrier. At the receiving end, the various frequencies are separated, and each message is delivered to a separate receiver. The speaker in the earpiece decodes telephone messages. There, the electrical impulses are changed back into an audible sound.

Broadcast Systems

Broadcast systems send radio waves through the air, carrying the signal from the sender to the receiver. The transmitter (sender) changes sound into a signal containing the message. This signal radiates into the atmosphere from an antenna. Another antenna attached to a receiver gathers the signal. The receiver separates the desired signal from other signals and changes it back into audible sound, **Figure 22-9.**

Generally speaking, radio signals project in all directions from an antenna.

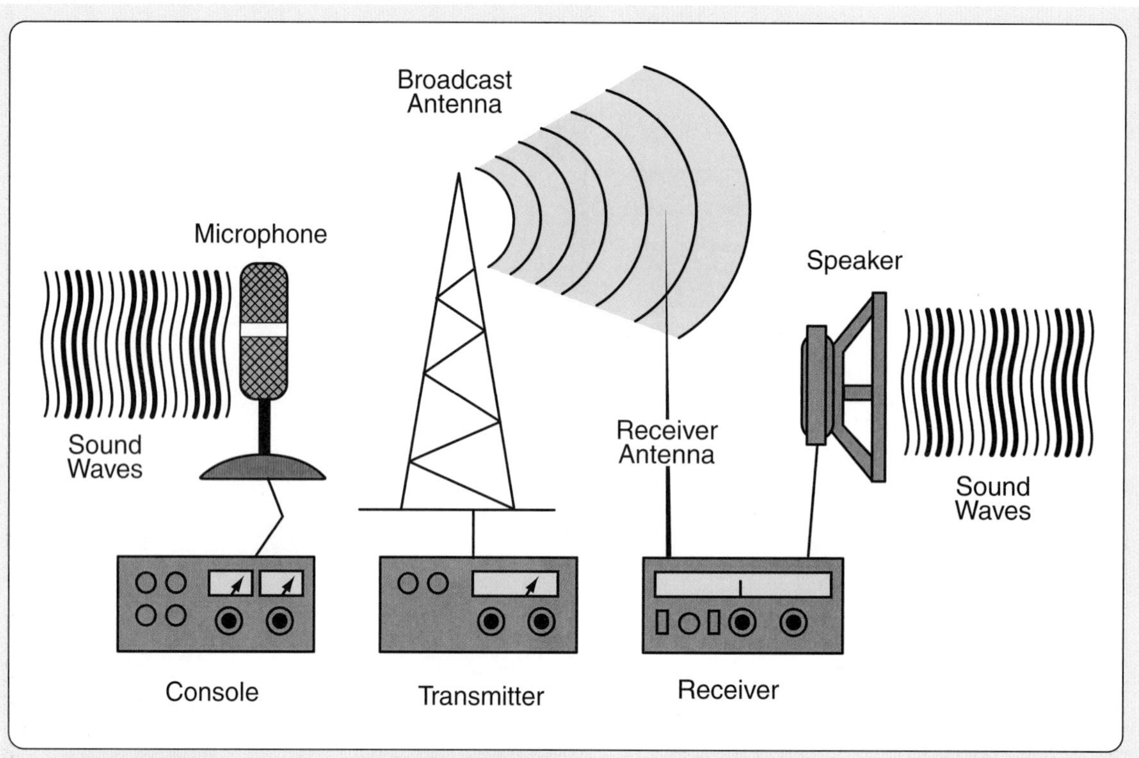

Figure 22-9. Shown here is a typical radio broadcast system.

Telephone microwave communication systems use directional antennas to focus the signal to receiving antennas. Two common broadcast systems are radio and television. Each will be discussed in turn.

Radio Broadcasting

Radio communication was the first widespread broadcast medium. Originally, it was called the *wireless* because no hard-wired connection existed between the sender (transmitter) and the receiver (radio set). All radio broadcast systems use a ***carrier frequency*** that radiates from the transmitter. It is to this carrier frequency that you tune your radio to receive a station. The code for the audible sound is imposed onto this frequency, **Figure 22-10.** The earliest radios used ***amplitude modulation (AM)*** to code the carrier frequency. These systems merged the message onto the carrier wave by changing the strength (amplitude) of the carrier signal. This type of broadcast radio

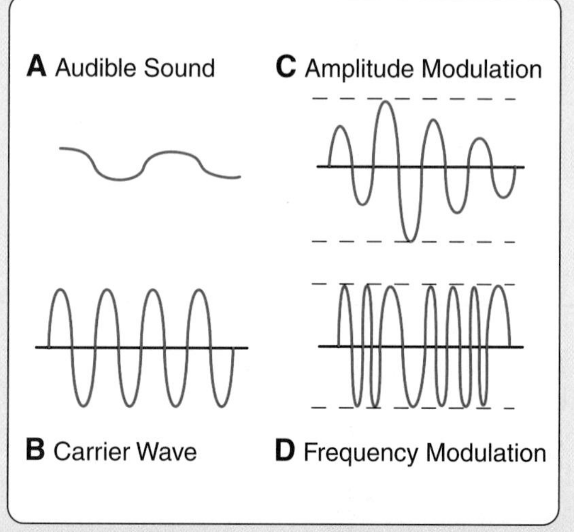

Figure 22-10. Radio waves are changed to carry the message through amplitude modulation (AM) or frequency modulation (FM). Notice how the amplitudes of the sound wave (A) and the carrier wave (B) have been blended in AM. The amplitude of the combined waveform (C) oscillates in a pattern similar to the sound wave. Notice, in the frequency modulated waveform (D), how the frequency varies in a pattern with the initial sound wave.

is assigned the frequencies between 540 and 1600 kHz.

Later, radio broadcast systems using ***frequency modulation (FM)*** were developed. These systems encode the message on the carrier wave by changing its frequency. The 200 separate FM radio broadcast frequencies range from 88.1 to 107.9 MHz. Look back to Chapter 7 for an explanation of the operation of radio communication systems. Included in that discussion is an explanation of the operation of radio transmitters and receivers.

Television Systems

Television broadcast systems are really two systems in one. These systems usea fairly broad range of frequencies within their broadcast band. Each channel is assigned a band 6 MHz in width, **Figure 22-11.** A radio-like system uses one portion of the band to send and receive the ***audio*** (sound) portion of the message. This portion uses FM to impose the audio message onto the carrier wave.

The larger portion of the band is assigned to a second system that communicates the ***video*** (visual) part of the program. This system uses AM to send the picture portion of the message. At the ends of each channel are unused frequencies. This buffer zone keeps the signals from one channel from disturbing adjoining channels.

Television systems use a microphone to capture the sound, while a camera generates the picture. The television receiver reproduces the program using a speaker for the sound and, usually, a cathode ray (TV) tube to display the picture. The operation of these devices is discussed later in this chapter.

Two basic types of television stations exist. ***Very high frequency (VHF)*** stations broadcast on channels 2–13. Each channel is assigned a specific 6-MHz range of frequencies. Channels 2–6 broadcast on frequencies between 54.0 and 88.0 MHz.

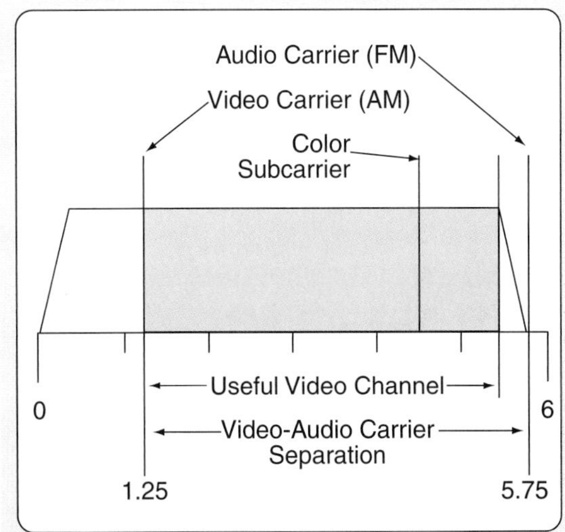

Figure 22-11. This diagram of a television broadcast channel shows the audio and visual broadcast frequencies.

Channels 7–13 use the range between 174 and 216 MHz. The frequencies between these two broadcast ranges are used for FM radio, aircraft navigation and aircraft communication operations, weather satellites, and amateur (ham) radio.

The ***ultrahigh frequency (UHF)*** stations broadcast on channels 14–83. They use the frequency range from 470 to 890 MHz. Because VHF channels can broadcast a greater distance than UHF channels, they are generally assigned to the major television network outlets and large local stations. Public broadcasting and smaller independent stations are often found on the UHF channels.

Communicating with Telecommunication Systems

Earlier, you learned that telecommunication includes a message and a delivery system. Communicating with telecommunication systems requires three distinct actions, **Figure 22-12:**

Communicating with Telecommunications Systems

Designing the Message

Producing the Message

Broadcasting the Message

Figure 22-12. Telecommunication messages move through three stages. They are first designed, then produced, and finally broadcast. (Ohio Art Co., Harris Corp., Ball State University)

- Designing the message
- Producing the message
- Delivering (broadcasting) the message

Designing the Message

Many communication messages are not carefully designed. When you talk on the telephone, you often have to adjust to circumstances. You have a basic message, or reason for the call. Yet, spontaneous interaction generally occurs between the caller and the receiver. The caller adjusts the message as he or she receives feedback from the receiver. The format and sequence of the message are not fixed.

This scenario is not typical for broadcast telecommunications. Most broadcast messages are the result of focused design activities. These activities are the following:

- Identifying the audience
- Selecting the approach
- Developing the message

To understand these steps, let us explore the development of a television program. A radio program follows a similar procedure, except for the use of visual impressions.

Identifying the Audience

The first step in designing a television program or advertisement is to select the target audience. This includes determining who the audience is and what they like. Different people find different things important. Many teenagers want to be part of "the group." Therefore, advertisements and programming directed toward this age group will emphasize this factor. Advertisements often try to convince teenagers that wearing the "right" clothes or driving the "right" car will make them more popular.

The aims of the communication are chosen through ample research. Actions that study audience needs and wants are called ***audience assessment.*** Audience assessments were discussed in Chapter 20.

Selecting the Approach

Once we have identified and described the audience, we must, as in other communication media, select the approach that will most effectively communicate our message. In advertising, this may involve a catchphrase such as "All news, all the time" or "Quality people, quality products." Included in the approach is the tone of the message. We might decide to express our message in a serious or a humorous way. A regal (dignified) technique is another option. We might also select a contemporary (present day), historical, or futuristic theme. Whatever the style, the tone must appeal to the audience.

Note that the tone must fit the situation. A program covering the effects of air pollution would probably not use a humorous approach. On the other hand, a program on circus life might have humor as an important component.

Developing the Message

The number of people who receive and act on a broadcast message determine the ultimate success of the message. Therefore, the message must be carefully crafted. This most often requires a ***script.*** A script identifies the characters, develops a situation, and communicates a story. Also, it provides the dialog for the characters and describes their movements on the stage.

The previous knowledge level of the audience must be considered. Those involved in local programming assume the audience has an understanding of local politics and geographical features (such as towns or rivers). Those who deal with national programming might need to explain these details. Four common types of scripts are used in television:

- *Full* or *detailed scripts* are complete scripts containing (a) every word to be spoken, (b) sound effects and music information, (c) all major visual effects, and (d) production notes (timing and camera angles, for example), **Figure 22-13.**

Video	Audio
Show Title	Music theme up then fade.
Host seen–head shoulders	Art has long been an important communication medium. In recent years it has grown in popularity as the availability of reasonably priced, quality paintings has grown.
Background light as the camera dollies past host to pan art on back wall	In our city, this trend is easily seen with a visit to the Artists' Guild show at the convention center.
Zoom in on Watercolor #1	Artists using many media are exhibiting the results of their talent.

Figure 22-13. Shown here is an example of a full-script format for a local public interest program.

Generally, this script is written in a two-column format. The left column contains video and production information. The right column contains audio information, including the dialog (words to be spoken) and sound effects (background sounds).

Full scripts for commercials often develop from a storyboard, **Figure 22-14.** A storyboard is a series of sketches or pictures showing all the scenes in the commercial. Its dialog and production notes are under each sketch.

- *Partial format scripts* contain complete scripts for the introduction and conclusion. The remaining parts of the show are simply outlined. This type of script is used for talk shows and sports events, where actions or responses give direction to the show. The script simply provides a skeleton around which the show develops. Often, it will contain time cues and other production notes.
- *Show format scripts* develop a list of the various film or show segments. Production notes and timing cues are included. This type of script is used for programs that follow a specific format for each show. Programs using this type of script would include morning and nighttime news programs, **Figure 22-15.**
- *Fact* or *rundown scripts* develop a list of facts or characteristics. This type of script is used when the performer ad-libs a part in an advertisement or a sportscaster uses statistics for broadcasting color commentary during a sportscast.

These scripts are designed to do three things for each program. They *establish the format, contain production direction,* and *communicate the content.*

Producing the Message

Once the design work is done, the stage is set for producing the message. Production steps include casting, rehearsing, performing, and recording.

Casting and Rehearsing

Developing a television program is a controlled activity. Many different people are involved. These people fall into three major groups: creative personnel, production personnel, and performers.

These people are all employed to deliver a service. The behind-the-camera personnel are hired like most other workers. They seek a job, and the production company interviews them. On-camera performers are employed through a process called ***casting.*** They are selected for their appearance and talent. Performers have

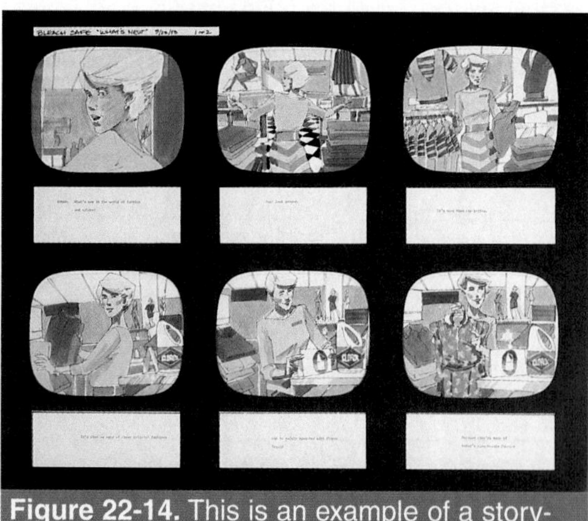

Figure 22-14. This is an example of a storyboard for a commercial. (Clorox Co.)

Figure 22-15. News programs often use show format scripts. (Gannett)

Technology Explained:

Fiber Optics

fiber optics: Channeling messages, in the form of light, through glass fibers.

Information has been transmitted with light for many years. One of the first methods used was smoke signals. More recently, flags and flashing lights have been used to convey information. All these techniques, though, are limited to the line of sight between the sender and receiver. Practical use of long-distance light communication came with the invention of optical cables.

Optical cables are channels guiding light waves, through internal reflection, over some distance. Internal reflection means that, when the light waves strike the outer edge of the fiber, they are reflected back toward the center. Optical communication of this type is called guided optical transmission. The development of glass fibers for guided optical transmission began in the 1960s.

A typical fiber-optic cable, as shown in **Figure A,** has three layers. The outside layer is a protective plastic coating. The middle layer is called cladding. It reflects the light waves back into the glass fiber. The inner layer is a strand of glass called the core. The individual core strands are as thin as human hairs. Typically they are about 0.0005" in diameter.

Several hundred of these strands are bundled into larger cables. Each of these fiber-optic waveguides can carry numerous messages. Billions of bits of information per second can move down an optical fiber, **Figure B.** This is the same amount of information contained in thousands of independent telephone conversations. As you can see, optical fibers are capable of carrying encyclopedias of information per second.

Fiber-optic cables are rapidly replacing copper circuits in telephone systems. They are smaller in diameter, use less power, and are much more resistant to interference. During thunderstorms, you will hear a great deal of static on phone lines using copper cable. Phones linked with fiber-optic cable will remain unaffected by the flashes of lightning.

Fiber optics is finding uses in areas other than communication. In the medical field, fiber optics are being used in endoscopes. These devices allow physicians to look inside the human body using fiber-optic cables with viewing lenses at one end. In many cases, this can prevent unnecessary surgery.

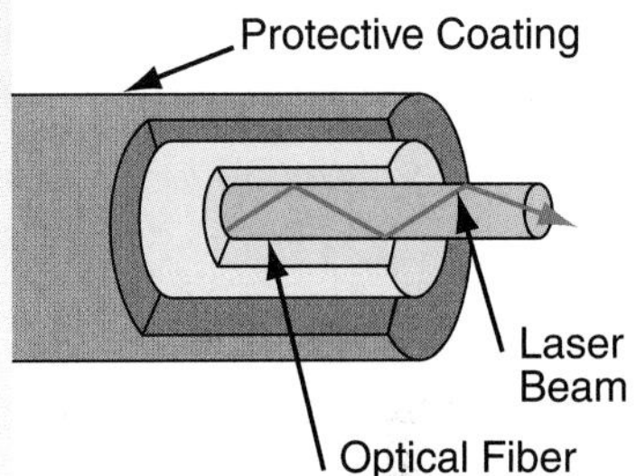

Figure A. Here is a diagram of a fiber-optic cable.

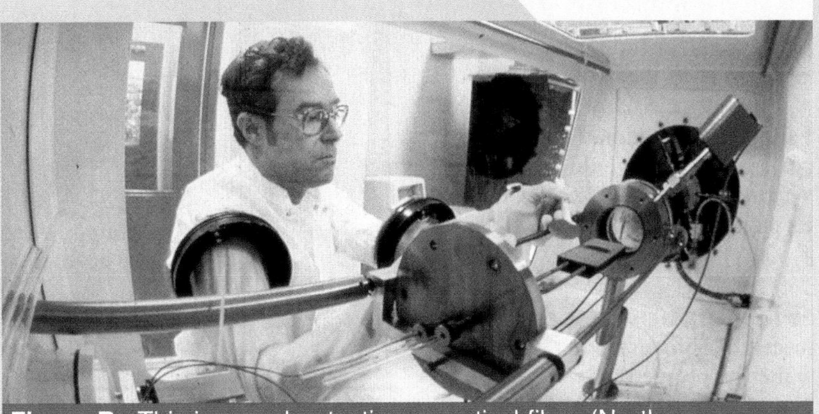

Figure B. This is a worker testing an optical fiber. (Northern Telephone Co.)

Connections to Technology: **Communication**

Advertising

Many commercial telecommunication broadcasts are supported through advertising. An important part of this advertising is a slogan. These slogans are designed to keep the product or company in mind. We probably have all listened to the radio and hours later recalled an advertising jingle. For example, many of us will remember such slogans as "When you care enough to send the very best" (Hallmark cards) and "We try harder" (Avis rental cars).

The purpose of an advertising slogan (sometimes called a tag or tagline) is to leave the brand message in the mind of a potential customer. Many slogans are also protected as trademarks. They are registered with the government trademark office, which allows the company to use the registered symbol (®).

A good slogan must meet several criteria. First, it must have the ability to be recalled. This is called a memorable slogan. Second, a good slogan should present a key benefit of the product or service. For example, Holiday Inn has used the slogan "Pleasing people the world over." Third, the slogan should differentiate the product or service. The Hallmark slogan above suggests that Hallmark cards are better than any others.

An effective slogan should generate positive feelings about the product or service. It should cause customers to purchase the item and feel good about owning it. Can you think of three examples of effective slogans?

to look the parts of the characters they will be playing. They must be able to deliver believable performances.

The production team and performers must then rehearse. This involves the actors and musicians practicing their performance. Rehearsing also involves the camera and sound technicians running through their work, moving the camera and sound booms (microphones on long shafts) through their various stations.

Performing and Recording

The goal of a television project is to capture sound and sight on tape. This process uses two important transducers. A microphone changes audible sound into electrical signals. Likewise, the television camera changes light into electrical signals. Together, these coded messages are captured on videotape. Let us look at how these technological devices work.

Audio Recording

Recording sound is essentially the same for radio, audio recordings, or television. It usually involves a microphone, a mixer, and a recording device. The microphone changes sound waves into electrical waves of a similar pattern. Each increase in the sound wave corresponds to an increase in the electrical wave. These waves form a continuous, flowing pattern we call analog. The waveform moves upward and downward as the amplitude changes.

Video Recording

Video recording uses a camera to capture the light reflecting from a scene. The lens in the camera focuses the light on the image plate in a vidicon tube. Color television cameras use a three-tube system, **Figure 22-16.** A prism breaks the beam of light into red, green, and blue segments. Each beam is focused on a different tube. This action produces separate red, green, and blue signals. The signals are then transmitted to the television receiver, which uses three guns to project the three colors onto the tube screen, producing a picture.

Storing Signals

Audio and video signals are often recorded on tape or disc. Magnetic tape recording uses a specially coated plastic tape. The coating is a metal oxide material made up of needlelike crystals. The signal from a receptor (a microphone or video recorder, for example) is fed into an amplifier and then to a recording head. The electrical impulses from the signal generate magnetic forces in the recording head. These magnetic forces realign the metallic crystals in the tape. This new arrangement forms a code corresponding with the sound or picture captured by the receptor. Later, this code can be used to reproduce the information.

Delivering (Broadcasting) the Message

All broadcast media (radio and television) use a carrier wave with a code imposed on the wave. This coded wave is sent through the air from a transmitter, or broadcast site, to receivers at distant locations. Radio waves travel in straight lines. Two types of broadcast systems have been developed that allow these waves to travel great distances, **Figure 22-17.** The first system uses direct waves. Television and FM stations use direct waves. Direct-wave transmitting and receiving antennas must have an open line of sight. Because of the curvature of the earth, these antennas must be placed within approximately 60 miles of each other. The earth will block the signal from reaching an antenna beyond this distance. This is why you have trouble receiving distant television and FM stations.

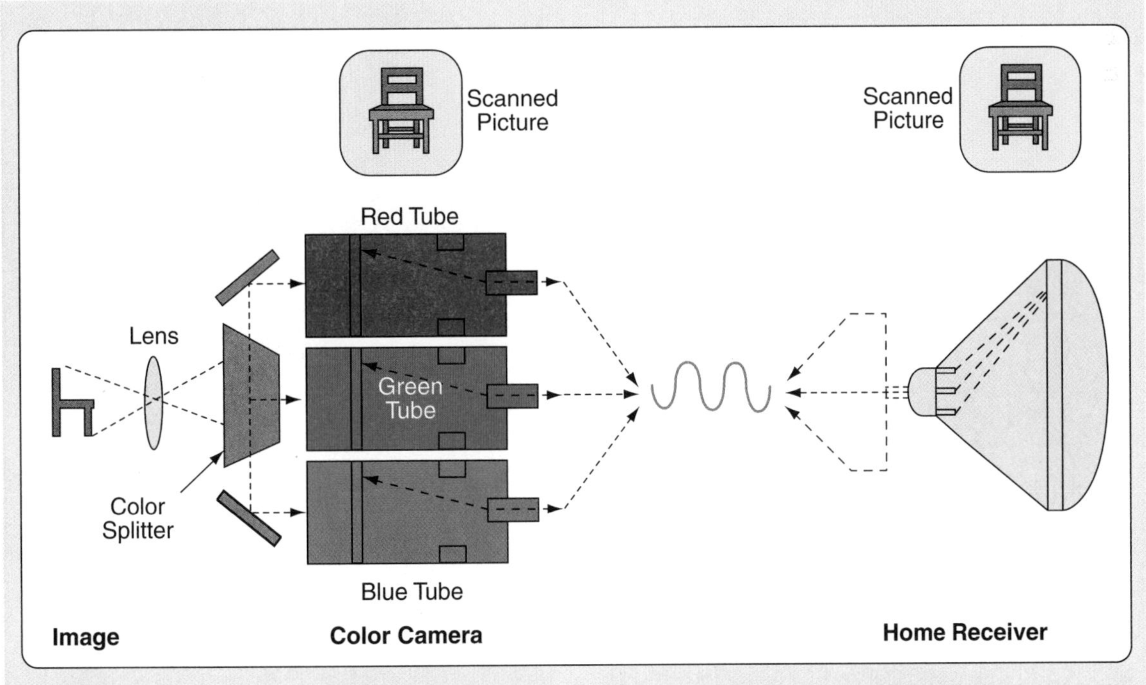

Figure 22-16. This is a simplified drawing of a color television camera system.

One broadcast station and antenna can be used to cover a specific area. This type of system is used for some television and radio broadcasts. In other cases, a series of receivers and transmitters (relays) are used to transfer the waves between broadcast towers. This type of system is used for microwave telephone communication, **Figure 22-18.**

Another way to overcome the problem caused by the curvature of the earth is to use satellite communication systems, **Figure 22-19.** The signal is transmitted to a satellite orbiting in space. One part of the satellite is a receiver; another part is a transmitter. The transmitter portion takes the message from the receiver and transmits it back to earth. This system allows a very large area of the earth to receive a message from one transmitter site. The area a satellite covers is known as the satellite's footprint. Television "superstations" and networks such as CNN use satellite communication systems. These systems have opened rural areas to more complete television programming, **Figure 22-20.**

Figure 22-18. Shown here is a microwave tower, which transmits telephone messages in straight lines.

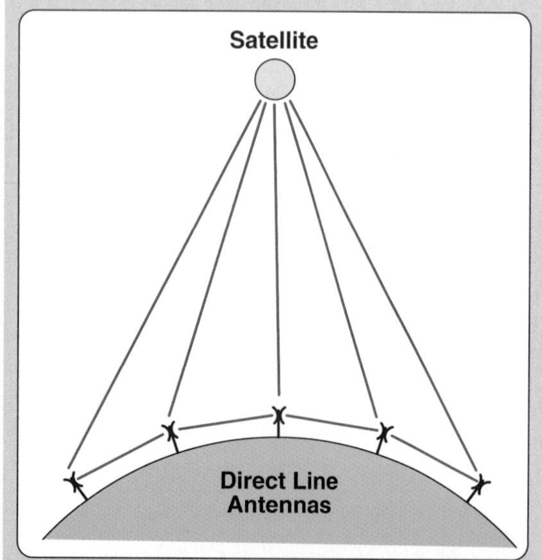

Figure 22-17. The curvature of the earth requires that direct line antennas be placed within 60 miles of one another. Satellites present another way to allow direct waves to cross long distances.

Figure 22-19. Communication satellites can send information in straight line waves over a large part of the earth. This photo shows a communication satellite being placed in orbit from the space shuttle. (NASA)

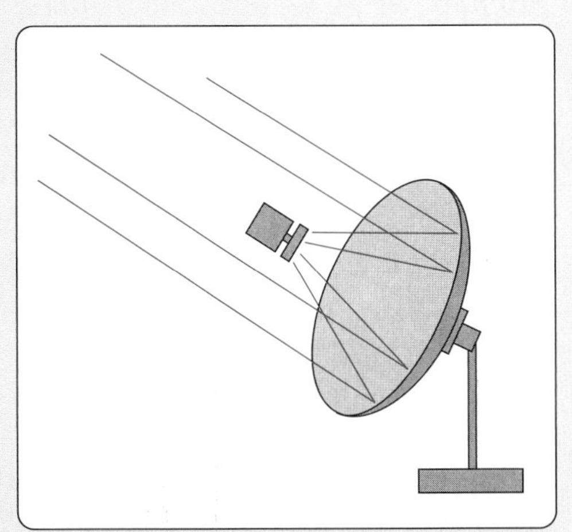

Figure 22-20. Satellite dishes, such as the dish pictured, allow many rural areas access to better communication.

The second broadcast system uses reflected waves. In this system, the signal is bounced off the ionosphere (upper part of the atmosphere). A distant antenna then receives the reflected wave. AM radio signals are transmitted through reflection. When atmospheric conditions are just right, you can receive radio stations from thousands of miles away.

Broadcast systems, as you have learned, also have several major parts. The first is a transmitter. The transmitter imposes the message on the carrier. A broadcast antenna radiates the resulting signal into the atmosphere. A receiving antenna captures this signal. The signal is fed into a receiver, where the coded message is separated from the carrier wave. A transducer then converts the electrical impulses of the message. Speakers convert audio messages to the sound you hear. Video messages are usually changed into pictures with cathode-ray tubes.

A fairly new broadcast system is the remote-link-system, **Figure 22-21.** This system allows the signal to be produced at a remote location, such as a distant sporting event, the site of a natural disaster, or the place of some other newsworthy incident. The on-site camera and transmitter generate

Figure 22-21. Remote relay communication allows us to see live television pictures from almost any spot on the globe.

the signal. The signal is transmitted from an antenna on a truck to a relay antenna. The relay antenna might be on a building or a satellite. Finally, the relay antenna relays the signal to the receiving antenna.

Other Communication Technologies

Broadcast radio and television are not the only telecommunication technology available. Mobile radio and cellular communication systems have become a vital part of personal and business communications. So, too, have satellite communication systems and sound recordings.

Mobile Communication Systems

Three basic systems are used in mobile communication. These are shown in **Figure 22-22.** They are the following:

- ***Simplex systems*** use the same frequency, or channel, for both the base and mobile transmissions. The systems allow only one-way transmission at any given time. The mobile unit cannot break into a base transmission. It must wait until the base transmission is completed.
- ***Duplex systems*** use two frequencies, or channels. Generally, the base unit transmits on one frequency, and the mobile unit replies on another. The systems allow one mobile unit to transmit information to the base station, while another unit is receiving a signal from the home base. Most of these systems, however, do not allow the mobile units to talk to one another. Their messages must be relayed through the base station.
- ***Multiplex systems*** use multiple frequencies to accommodate different types of units. The base might broadcast on one frequency. Vehicle units might use a second frequency, and

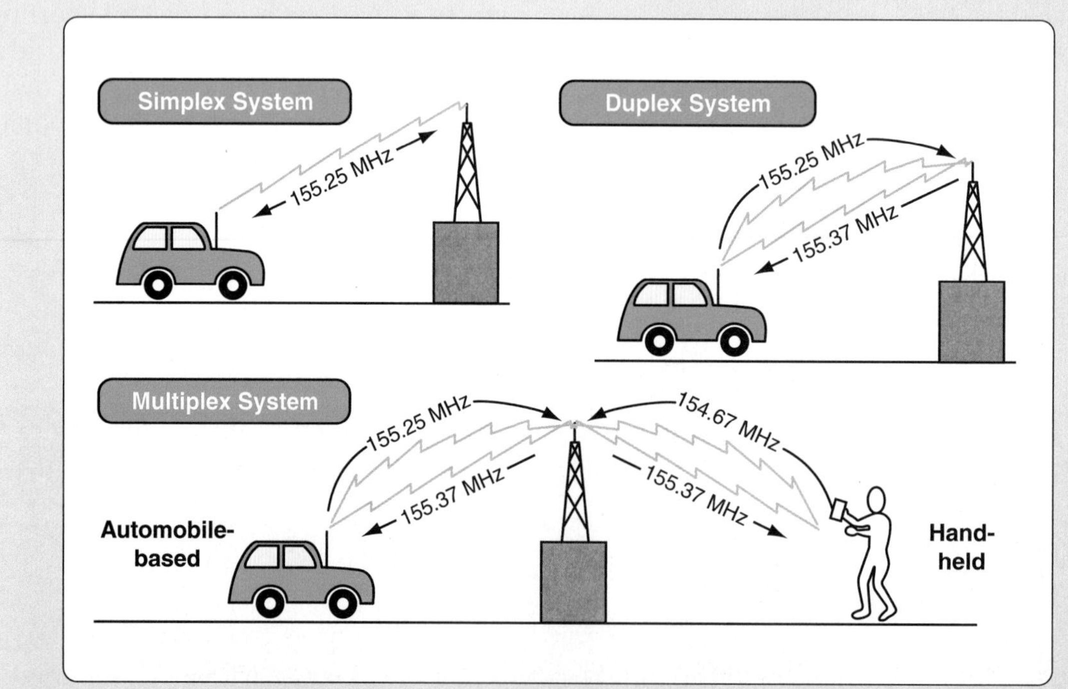

Figure 22-22. The three basic mobile communication systems vary according to the number of frequencies used.

handheld units might use a third frequency. These systems ensure that the low-power handheld units can get through to the base station.

Satellite Communication Systems

Satellite communication systems also link business areas together. Data, product orders, and other information are transmitted from one location to another. **Figure 22-23** shows a dish antenna used to communicate parts orders from an automobile dealership to a factory warehouse.

Figure 22-23. The communication antenna on top of this automobile dealership connects the office with the parts warehouse.

Sound Recording Systems

Still another communication system affecting almost all of us is sound recording, **Figure 22-24.** The fundamentals of this system were explained in the earlier parts of this chapter dealing with audio recording. Compact disc (CD) recordings are generally the system of choice today.

Figure 22-24. This audio technician is mixing (blending) sound from several microphones to make an audio recording. (Ball State University)

Summary

Telecommunication is an important part of modern life. The two major types of telecommunication are hard-wired systems, such as the telephone, and broadcast systems, which are primarily radio and television broadcasting. Telecommunication usually involves three steps. First, we design the message through audience assessment, tone selection, and script writing. Second, we produce the message through casting, rehearsing, performing, and recording. Third, we broadcast the message using direct or reflected waves. Some newer telecommunication technologies are mobile radios, cellular telephones, and satellite communication systems.

Key Terms

alternating current
amplitude
amplitude modulation (AM)
audience assessment
audio
broadcast frequency
broadcast system
carrier frequency
casting
conductor
direct current
duplex system
frequency
frequency division multiplexing (FDM)
frequency modulation (FM)
hard-wired system
hertz
induction
kilohertz
megahertz
multiplex system
multiplexing
radio wave
script
simplex system
telecommunication
time division multiplexing (TDM)
transducer
ultrahigh frequency (UHF)
very high frequency (VHF)
video

Test Your Knowledge

Write your answers on a separate piece of paper. Please do not write in this book.

1. What does telecommunication mean?
2. The term used to describe electrons flowing in one direction along a conductor is _____.
3. List the two major types of telecommunication systems.
4. What is one difference between AM and FM radio stations?
5. Television stations that broadcast on channels 14–83 are called _______ stations.
6. List the major activities involved in designing a broadcast message.
7. The two types of wave communication used in broadcasting are direct waves and _____.

8. What is a remote-link system?
9. List the three basic systems used in mobile communication.

Applying Your Knowledge

1. Select a product or an event and conduct an audience assessment for it. Record your data on a chart similar to the one below.

Product or event:
Intended audience:
Size of the audience:
What has high value to the audience?
What approaches are most likely to attract the audience's attention?

2. For the product or event selected above, (a) select a theme or approach and (b) develop a full script or a storyboard for an advertisement to promote it.
3. Produce an advertisement using the script developed above by (a) selecting production personnel and performers, (b) rehearsing the production, and (c) recording the advertisement.

This activity develops the skills used in TSA's Film Technology event.

Film Technology

Activity Overview

In this activity, you will develop a storyboard, script, and finished film or video for one of the following areas:

- The arts
- Social studies
- Science
- Technology

Possible subjects include social study documentaries, nature films, advertisements, comedies, or dramas.

Materials

- Paper and pencil
- 3" x 5" note cards
- Three-ring binder
- Video camera
- Computer with video production software

Background Information

Planning the video. Plan your film as a story, making sure it has a beginning, middle, and ending. List the important points you want to cover, and then use those points to create a simple outline. Think visually—remember that your story will be told mostly with pictures, instead of words.

Storyboard. To help yourself and the others involved in the project visualize the video, make a storyboard. This is a series of simple pictures, almost like a comic book, which will show each change in what the viewer will see. Make each individual storyboard sketch on a separate note card. This will allow you to try different combinations to get the most effective sequence for the video. When you have a final sequence, number your cards.

Scripting. Even though the pictures (video) will tell most of the story, the spoken words (audio) will tie the pictures together. Your script will follow the sequence of the storyboard, with a brief description of each numbered shot and the actual words that will be recorded to accompany that shot. Sometimes, audio is not fully scripted.

Production planning. Film is almost never shot in the final program sequence, so careful planning is needed to make the most effective use of time and resources. For each numbered shot in your storyboard, your production plan should list the location of the shot, any on-camera people involved, any props or special materials needed, and (where relevant) the time of day the shot must be made. For initial planning, note cards with the information for each shot are useful. They can be sequenced to make efficient use of your resources—for example, grouping together all the shots at one location or involving the same people.

Shooting the film. To provide visual interest in the finished video, be sure to vary your types of shots—don't shoot all close-ups or all wide shots. Suit the shot type to the subject and to the flow of the video as shown on your storyboard. To help identify each shot, make a large card with the scene number written in bold marker. Shoot a few seconds of the card at the beginning of the shot. To provide editing flexibility, shoot a variety of cutaway shots. These are used as transitions between other shots. For example, if you are showing a school assembly program, include a few close-ups of members of the audience listening to the speaker. When shooting interviews, you would normally concentrate on the person being interviewed. When the interview is finished, shoot some close-up shots of your reporter asking questions and some where she is just looking interested and nodding.

Camera and microphones. If you are shooting in a location where you depend on battery power for the camera, be sure to have one or more fully charged "backup" batteries available. When lighting conditions change (such as moving from indoor to outdoor settings), be sure to check the camera's white balance and adjust it if necessary, to avoid adding a color cast to your video. When shooting general scenes, such as a cafeteria or sporting event, record the "live" audio, which can later be used in post-production. For recorded narration or interviews, avoid using the camera's built-in microphone, if possible. Much better audio quality will result from using a microphone (wired or wireless) placed close to the subject being recorded.

Post-production. When editing your scenes, use various visual effects, such as dissolves, fades, and zooms. Music, sound effects, and additional voice-over narration can be added during post-production. Use cutaway shots to avoid disturbing jump cuts in the video. These occur, for example, when an interview must be edited to eliminate unwanted material. If the camera is focused on the speaker, changes in expression or head position between the adjoining shots will make the cut obvious. By inserting a second or so of video showing the interviewer looking interested, the change will not be noticed.

Guidelines

- Final video must be three to five minutes in length.

The following items must be included in your final printed report:

- Cover page
- Table of contents
- Storyboard
- Script
- Description of editing techniques

Evaluation Criteria

Your project will be evaluated using the following criteria:

- Creativity and originality
- Correlation of storyboard and script to finished video
- Technical quality and skill

Career Corner
Radio and Television Broadcasting

Broadcasting is experiencing rapid change brought about by expanded use of computers and interactive media. Although the on-air positions are most recognized, there are more off-camera and off-microphone jobs than on-air positions. Many of these technical, sales, and administrative positions pay as well or better than the ones on-air performers hold. Nonbroadcast careers in electronic media, radio-television, and broadcasting include broadcasting station manager, film or tape librarian, community relations director, unit manager, film editor, news director, news writer, transmitter engineer, technical director, advertising sales coordinator, producer, account executive, and floor manager.

Chapter 23
Computer and Internet Communication

Learning Objectives

After studying this chapter, you will be able to:

- List the parts of a computer system.
- Name the types of computer networks.
- List common uses of the Internet.
- Explain how computers are connected using the Internet.
- Define terms used in accessing the Internet.
- Identify the parts of domain names.
- Explain how Web browsers and Web servers work.
- Explain what hypertext markup language is.
- Explain what a hyperlink is.
- Define how search engines work.
- Explain how electronic mail works.

The world entered a new age with the advent of the widespread use of computers. This age has been called by different names, including the Computer Age and the Information Age. Unlike other ages, this age was launched by a single technological advancement: the invention and development of the personal computer.

A computer is a machine that performs a number of tasks controlled by a set of instructions. Computers are used in all areas of personal and business life, **Figure 23-1.** For example, computers are used to scan the price of items at the grocery checkout area, maintain financial records, and operate banking machines. They select channels on the television, create images from DVD (digital video disc) players, and operate game players. Computers maintain temperatures in homes, control microwave ovens, and sequence the cycles in washing machines. They control automobile systems, operate home security systems, and manage sports scoreboards.

In this chapter, we will explore the use of computers as a primary communication tool. First, we will look at computer systems in general. We will then examine

Figure 23-1. Computers are used in all areas of life. (Apple Computer)

the expansion of communication resulting from the networking of computer systems, with particular attention paid to the types of communication allowed by way of the Internet.

Computer Systems: An Overview

You learned in Chapter 7 that the computer includes physical equipment (hardware) and operating instructions (software). Hardware includes the computer itself and devices attached to it, **Figure 23-2.** The four main units of hardware are the central processing unit, the memory, the input devices, and the output devices.

The ***central processing unit*** (CPU) is the heart of a computer. This is the working part of the computer that carries out instructions. The CPU is a microprocessor chip. It is a piece of silicon containing millions of electrical components.

Figure 23-2. Computer systems have input devices, a central processing unit, memory, and output devices. (Apple Computer)

The other major part of a computer itself is called ***memory.*** This is where the computer stores its data and operating instructions. The computer has two types of memory:

- ***Read-only memory*** (ROM): Memory that can be read by the computer but cannot be changed

- ***Random-access memory*** (RAM): Memory that can be read or changed by both the computer and the user

In addition to the internal memory, data can be stored outside the basic computer circuitry on devices called ***external storage devices.*** These devices include hard drives, floppy discs, CD-ROMs (compact disc—read-only memory), and DVDs.

Attached to the computer are input and output devices. ***Input devices*** allow the operator to enter data into the computer's operating system. Typical input devices are the keyboard and mouse. Other input devices are:

- A joystick that is used in video and computer games
- A scanner that converts images, such as drawings and photographs
- A microphone that can be used to gather sound information
- A touch panel that senses where a person places his or her finger, **Figure 23-3**
- A pen with a light that can be used to draw images or select objects from a video display

The data generated by the computer operations are called output. These data can be viewed on a number of ***output devices.*** The most common is the video display monitor, or the flat screen liquid crystal display. Other output devices include:

- Dot matrix, ink-jet, and laser printers, **Figure 23-4**
- Speakers
- Overhead projectors

The other major part of a computer system, as we noted earlier, is called software. This term is used to describe the instructions that direct the computer to perform specific tasks. For example, software programs allow people to use computers to write text, prepare drawings, maintain financial records, and perform hundreds of other functions, **Figure 23-5.**

Networks

Computers can be used alone to do specific tasks. Also, they can be connected into ***networks.*** Through a network connection, a computer can communicate with other computers. Networking allows people to exchange data rapidly and to share output devices and hard disc storage. Networks are essential for computer communication techniques.

Two basic types of networks exist. The first is called a ***local area network*** (LAN). This system is generally used in a single

Figure 23-3. This farmer is using a touch screen to enter data. (Deere and Co.)

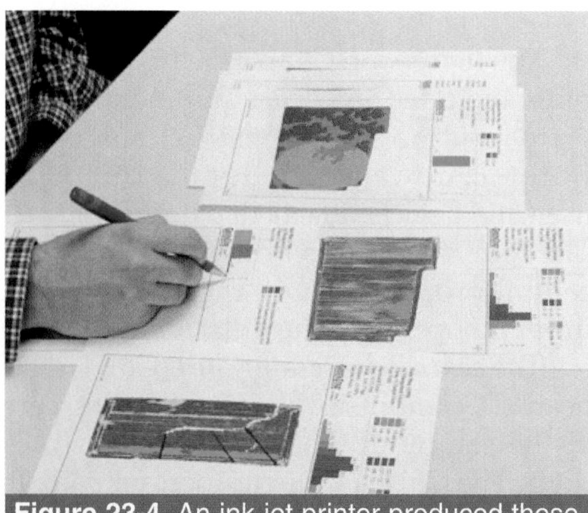

Figure 23-4. An ink-jet printer produced these pages. (Deere and Co.)

building or site. It connects several personal computers (PCs) or workstations to a special computer called the ***server,*** **Figure 23-6.** The server is used to store programs and data for the network.

The second type of network is called a ***wide area network*** (WAN). These networks cover large geographical areas. They are used to connect computers in distant cities and countries. The largest WAN is the Internet, which we will now examine in more depth.

The Internet

The term ***Internet*** means *inter*connected *net*works. The Internet is a computer-based global information system. No one really owns it because much of its creation and initial workings emerged from research that was federally funded. The Internet is a collection of large and small independent networks called *Intranets.* Each of these Intranets can link hundreds of computers in a company or other institution. The Internet connects all these Intranets so they can share selected information. The Internet has become a system that provides inexpensive and efficient communication for people all over the world.

People use the Internet for various activities, **Figure 23-7.** Four common uses are:

- The World Wide Web (WWW): A way to find and display multi-media documents
- Electronic mail (e-mail): A way to send and receive written messages
- Newsgroups or chat rooms: A way to carry on interactive discussions using written text
- Electronic commerce (e-commerce): A way to buy and sell goods

Every computer connected to the Internet is part of a network. Companies have their own LAN. A home computer is connected into a network formed by the ***Internet service provider*** (ISP). The Internet can be described as a network of networks. This complex network is a collection of backbones, access points, and routers.

Figure 23-5. Software enabled the computer to produce this graphic layout. (Deere and Co.)

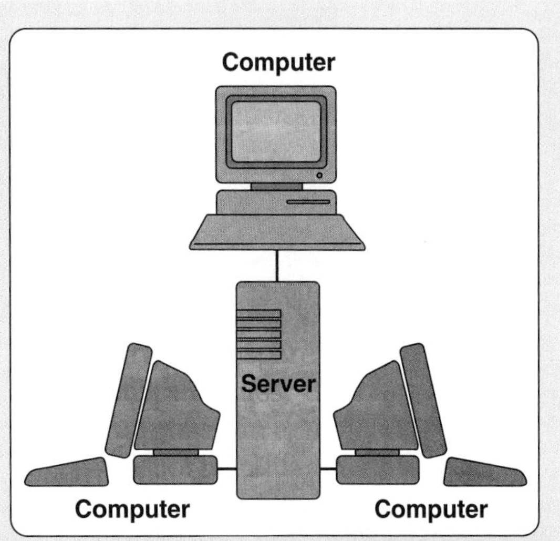

Figure 23-6. A local-area network connects personal computers to a server.

Large communications companies have built their own communication lines called ***backbones.*** These are typically fiber-optic lines that connect regions in their systems.

These companies have also established connection points in each region called a ***point of presence*** (POP). These connection points let local users access the company's network. The POP is often a local or toll-free telephone phone number or a dedicated line. The various communication companies connect their systems at ***network access points*** (NAPs). NAPs allow a customer of one company to connect with a customer of another company. Many large Internet service providers interconnect at NAPs in various cities. Therefore, the Internet is a collection of large corporate networks that agree to intercommunicate at the NAPs.

The Internet also contains many ***routers.*** These devices are specialized computers that determine how to send information from one computer to another.

To use the Internet, an individual must first have access and a domain name. These features are discussed in more detail as follows.

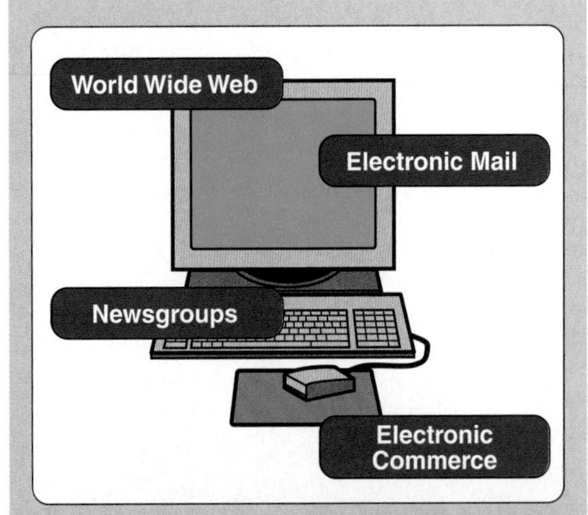

Figure 23-7. Through the Internet, an individual can access such features as the World Wide Web, electronic mail, newsgroups, and electronic commerce.

Internet Access

Internet access is the term that refers to the way a computer is connected to the Internet. This access is provided in two basic ways. The first is ***dial-up access.*** In this system, computers are connected to the Internet through a modem. A modem is a device that can convert data into signals that a telephone system can recognize. It also converts these signals back into data. The Internet is accessed by special software. It places a telephone call to a company that provides Internet service. This company is the Internet service provider, **Figure 23-8.** The ISP modem answers calls from the user's modem. It receives and transmits the signals through the telephone lines.

The second system is called ***dedicated access.*** In this system, the subscriber's computer is directly connected to the Internet at all times. Many companies have their own dedicated Internet connections. Digital subscriber lines (DSL) or cable modems often connect smaller business and home computers to the Internet. A DSL connection uses telephone lines for the connection. Cable modems use cable television lines to connect to the Internet. Both DSL and cable modems send data over the wires using unique frequencies or channels. This keeps the Internet communication from interfering with other signals on the wires or cables.

Every computer that is connected to the Internet has its own identifying number. This number is called an ***Internet protocol address.*** Internet protocol (IP) is the computer language used to communicate over the Internet. A typical IP address is a series of numbers, such as 12.225.103.96. The numbers are used to create groups of IP addresses that can be assigned to specific businesses, government agencies, or other entities, including individuals. This number contains two sections. The first is the *net section,* which identifies the network to

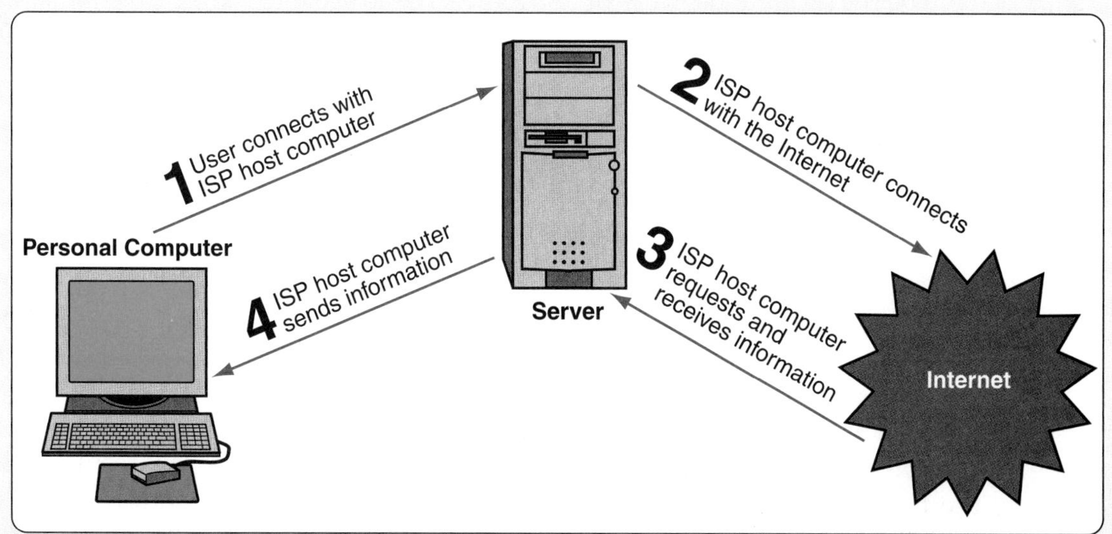

Figure 23-8. Through an Internet service provider (ISP), an individual can use a personal computer to access information from Internet sources.

which the computer is connected. The second is the *host section,* which identifies the actual computer on the network.

Internet Domains

Every time a person uses the Internet, he or she uses a ***domain name,*** **Figure 23-9.** Each site on the World Wide Web has a ***uniform resource locator*** (URL). This code is a method of naming documents or sites on the World Wide Web. A URL is a series of characters that identifies the type and name of a document. This locator also includes the domain name of the computer on which the page is located. For example, the URL of the International Technology Education Association (ITEA) home page is *http://www.iteaconnect.org.*

Note, too, that if you use e-mail, you will have an address. For example, the e-mail address for the ITEA is *itea@iris.org.*

The domain name server (DNS) translates these names. The server changes the name that humans can read into machine language that a computer can read. A domain name may have three or four segments. The first series of letters (working backward) are called top-level domains. In four segment codes, such as "http://www.technology.org.uk," the group of letters on the far right identifies the country (i.e., UK = United Kingdom, AU = Australia). The next set of letters to the right identifies the type of site. There were seven original types:

- Com — Commercial organizations
- Edu — Four-year colleges and universities
- Gov — Government institutions
- Int — International organizations
- Mil — Military groups
- Net — Organizations directly involved in Internet operations
- Org — Organizations other than those above

The group called the Internet Corporation for Assigned Names and Numbers is in charge of assigning domain names. Because of the need to increase the range of top-level domains available, the group has added more codes, including aero (air transport industry), biz (businesses), and info (unrestricted use). Most likely, even more codes will be developed to allow for new domain names.

In three-segment names, such as http://www.dtonline.org, the country code is not used. Instead, the name ends with the three-letter type code.

Following from right to left, after the top-level domain code is a second-level domain code. This code identifies the specific organization, agency, or business. In the ITEA Web address the "iteaconnect" is the organization identifier.

The left segment of the address contains the host name. In the ITEA example, the "www" is the host. It specifies the computer at the site that will receive the message.

Finally, the prefix to the domain name is important. The part of the URL before the colon indicates the format used to retrieve the document. The prefix "http" means the document is on the World Wide Web. If instead you see the prefix "ftp," it means the document can be accessed through the file transfer protocol. This protocol allows the user to retrieve and modify files that are on another computer connected to the Internet.

As we said earlier, once a person has access and a domain name, he or she can use the Internet. The first use we will look at is the World Wide Web, and then we will explore other popular uses, including electronic mail, newsgroups and chat rooms, and electronic commerce.

The World Wide Web

The ***World Wide Web*** (WWW) is a computer-based network of information resources. It is often called simply the Web.

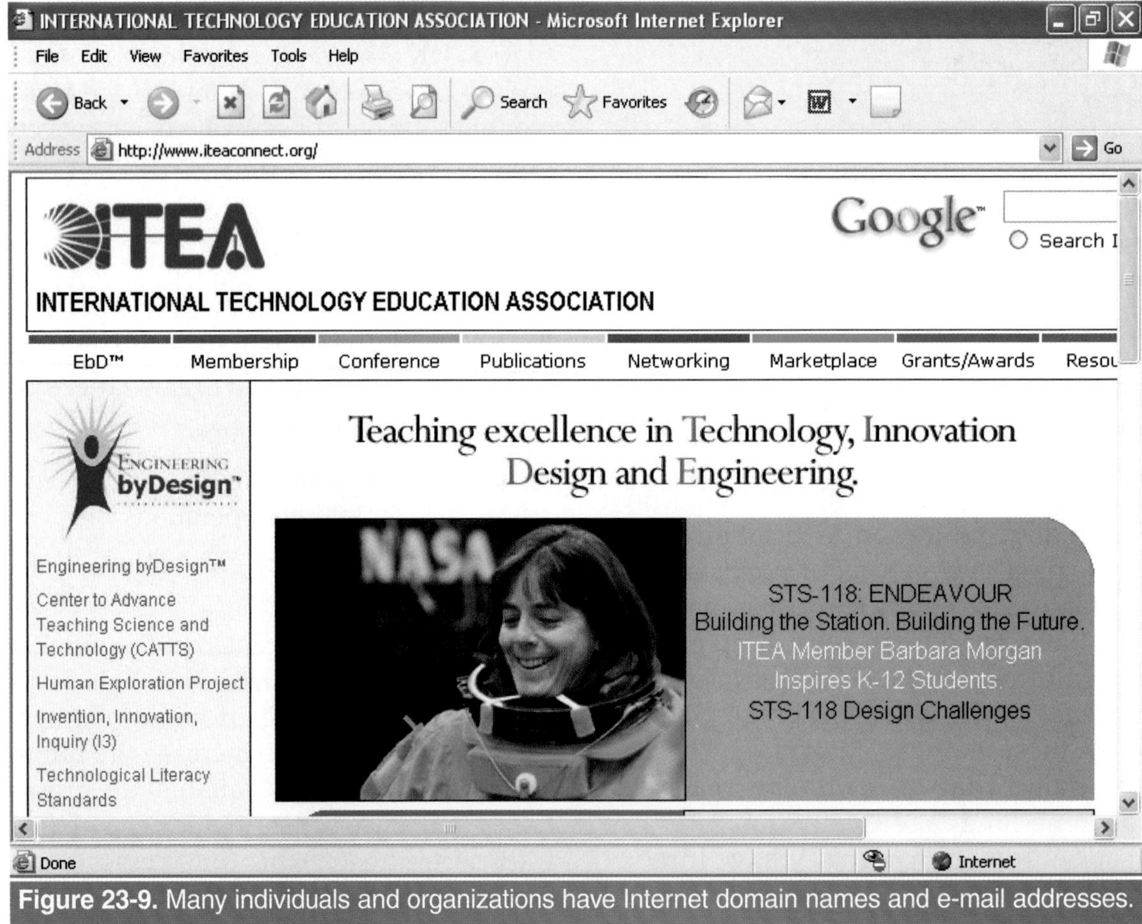

Figure 23-9. Many individuals and organizations have Internet domain names and e-mail addresses.

It was developed in 1993 to allow people to view information and images on the Internet. The Web provides companies, universities, government agencies, and other organizations and individuals a way to present information. Most information on the Web is free; however, some sites charge a subscription fee for user access.

The Web has rapidly become a source of information for millions of people. It allows individuals to read text, view digital images (photos and drawings, for example), listen to sounds (music and speeches, for example), and access multimedia presentations (images with integrated sound).

The Web has numerous features. Here we will look at browsers, servers, pages, links, and search engines.

Web Browsers

Individual computers are connected to the Web through one of the Internet connections already discussed. Sites on the WWW are accessed using a Web browser, **Figure 23-10.** A *browser* is a software program such as Netscape or Internet Explorer. The program acts as the interface between the user and the World Wide Web. Browsers know how to find a Web server on the Internet. They can request a page and deliver it to a personal computer. Finally, they can format the page so that it is correctly displayed on the computer monitor.

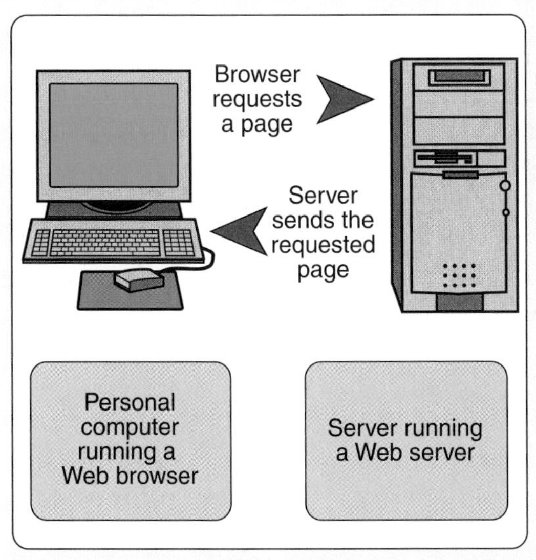

Figure 23-10. People access the World Wide Web through a Web browser.

Most browsers are graphic media, **Figure 23-11.** They allow the user to view images on their computer. "Point-and-click" or button functions allow the person to access resources on the Internet.

Web Servers

As noted earlier, servers are special computers used to store programs and data for the network. Web servers are operated by the organizations that develop and share information. The servers hold Web documents and related media. Servers contain computer software that can respond to a browser's request for a page. They can deliver the selected page to the Web browser through the Internet.

Each document on the server has an address. This address is also a URL. If the site has one hundred Web pages, it will have one hundred URLs. For example, the URL of ITEA's home page is http://www.iteaconnect.org/index.html. The URL for the page that appears when the "About ITEA" link is clicked is http://www.iteaconnect.org/AboutITEA/about.htm.

Web Pages

A Web *page* is a text file that someone creates to share information or ideas. It contains the text of the message and a set of *hypertext markup language* (HTML) tags. These tags or codes tell the receiving computer how a page should look. Look at **Figure 23-11** again to see a format created by HTML tags. It allows the developer to specify fonts and colors, create headlines, format text, and present graphics on a page.

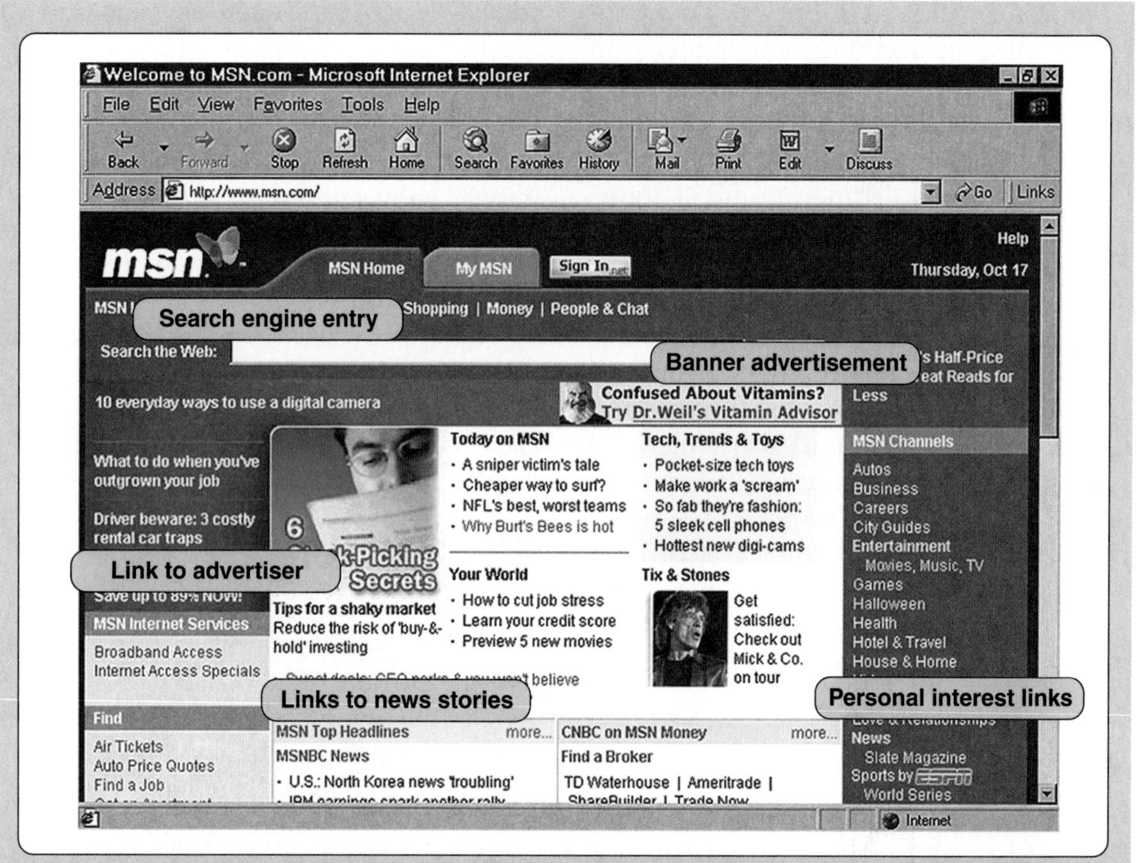

Figure 23-11. Shown here is the browser page for an Internet service provider. (Microsoft Corporation)

Links

The URLs and HTMLs allow Web sites to connect a Web page with other pages and Web sites. These connections are called ***hyperlinks.*** They are generally referred to as simply ***links.*** They can be underlined phrases, buttons, or other means that can be selected, **Figure 23-12.** They allow the operator to point and click on the link and be connected to the selected site or page. Hyperlinks allow users to move between Web pages in no particular order.

Search Engines

Often people want to find information about a specific subject, but they do not know where this information is on the Web. ***Search engines*** allow individuals to search the Web by topic. Search engines are special sites on the Internet that operate on the principle of key words. The search engine operators prepare and maintain an index of the major Internet sites using these key words. When a person accesses the search site, he or she can search the site's index for these words. For example, you could search for sites related to "design and technology." The search would produce all sites that have design and technology in their titles and main descriptions. The search results would contain links that would allow you to point and click on any entry to access the site.

Many different search engines exist. Examples include Alta Vista, Excite, Google, and Yahoo. One single search engine cannot cover every available Web resource. Their operators select what they

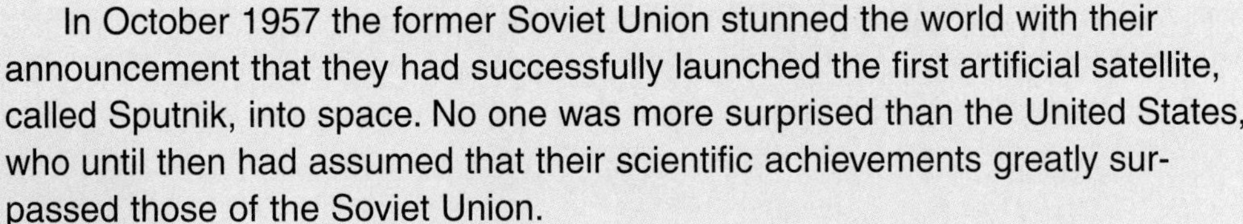

In October 1957 the former Soviet Union stunned the world with their announcement that they had successfully launched the first artificial satellite, called Sputnik, into space. No one was more surprised than the United States, who until then had assumed that their scientific achievements greatly surpassed those of the Soviet Union.

The United States was also very concerned because of the military implications behind Sputnik. At that time the United States and the Soviet Union were engaged in what has been called the cold war. Whoever achieved technological superiority might be able to do great damage to the other country's military capabilities. One immediate response by the United States, therefore, was to put more funds into the research and development efforts of the Department of Defense.

One result was the department's creation of the Advanced Research Projects Agency (ARPA). Among other goals, the agency was interested in helping to create technology that would enable computers to communicate with one another and to do so in such a way that some computers could still be in contact even if others had been shut down because of a military attack. This agency funded the research efforts of scientists at some major corporations and at four universities who were also working on the idea of computer networking.

After years of testing and retesting various methods, in 1972 the scientists unveiled ARPANET, a computer networking system based on the idea of breaking down data into labeled packets that could be forwarded from computer to computer. This effort led to the development of the network control protocol (NCP) to transfer data, which allowed communication between hosts on the same network. Development continued, eventually resulting in the transmission control protocol/Internet protocol (TCP/IP) technology. Now computer networks could interconnect and communicate with one another.

Thus what started as a military concern became a new way of communication for millions of people. ARPANET became the Internet, and the Internet today continues to expand.

In 1972 ARPA became the Defense Advanced Research Projects Agency (DARPA), which is still in operation. Can you identify some projects in which it is involved today?

think are the most appropriate sites for each key word in their index. Therefore, the results from using the same key word will vary from one search engine to the next.

To use a search engine, the computer operator must first open the Internet software that accesses the Internet Service Provider's service. Then he or she should:

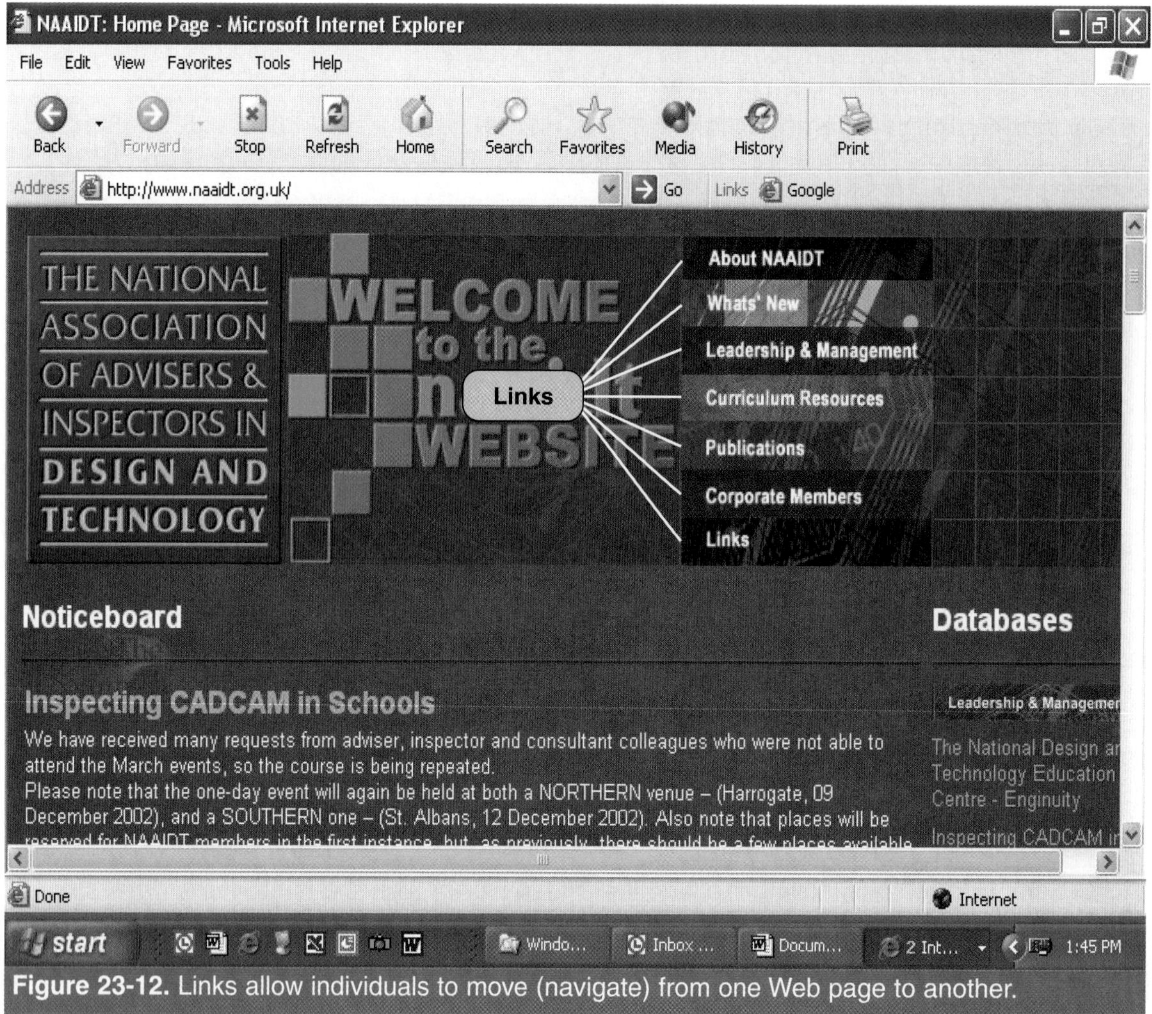

Figure 23-12. Links allow individuals to move (navigate) from one Web page to another.

- Determine the key words that will be used to find appropriate information.
- Review the names and descriptions of the sites displayed by the search engine.
- Select and click on sites that seem promising, **Figure 23-13.**

Electronic Mail

Every day, millions of personal messages travel over the Internet. These messages can move over the Internet because of a major computer-based communication tool called electronic mail or e-mail. The system allows an individual to send a message to another computer or a number of different computers, **Figure 23-14.** E-mail operates on a set of Internet rules called the simple mail transfer protocol (SMTP).

This protocol establishes a client/server relationship. A program such as Outlook Express on an individual's computer allows the person (client) to interact with a server computer. It lets the client compose, send, receive, read, and reply to messages.

E-Mail System

The e-mail systems that most people use have two servers. The first one handles outgoing mail. It is called the SMTP server because it operates on the simple mail transfer protocol. The second server handles incoming mail and is called the POP3

Figure 23-13. Shown here is the image of a typical search engine page. (Ask.com)

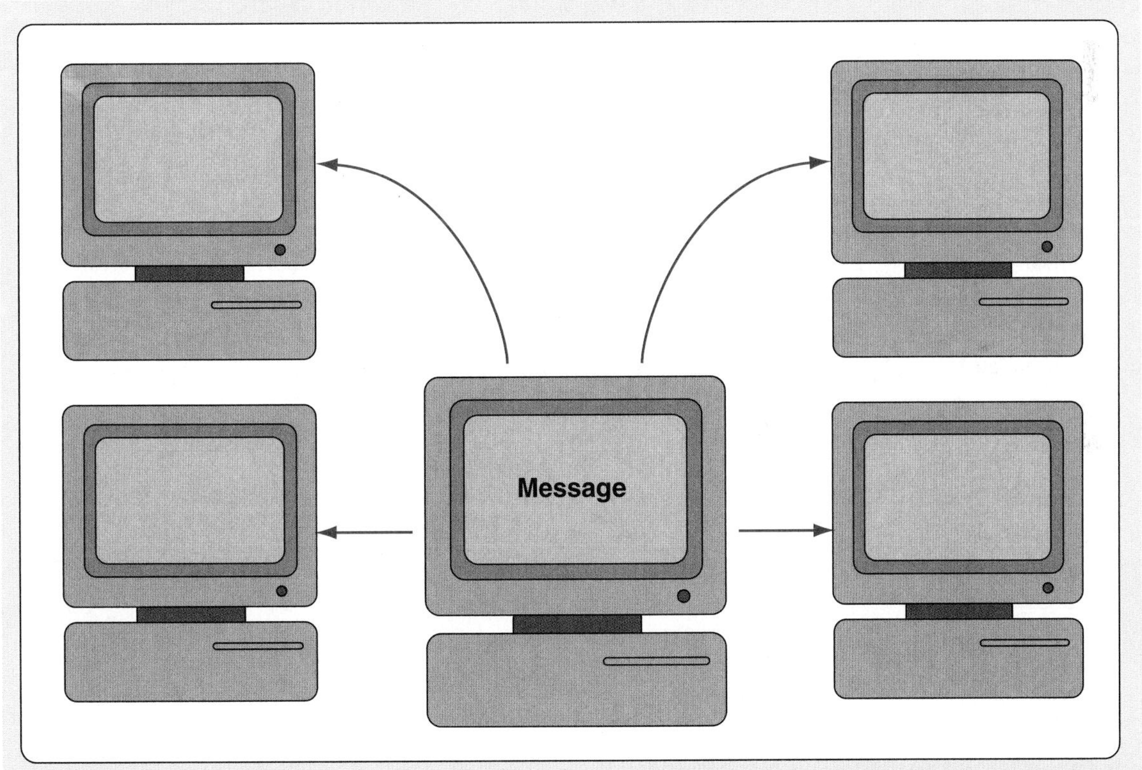

Figure 23-14. E-mail systems can deliver messages between two or more computers.

server. It operates on the post office protocol. The two protocols are the rules that allow the computers that handle incoming and outgoing mail to "talk" to one another. They allow the computers to become a communication system.

E-Mail Messages

E-mail messages are simple text messages. They are a "letter" in electronic form. They are prepared and processed by special computer software that has several

Technology Explained:

Virtual Reality

virtual reality: A computer interface allowing a user to interact with three-dimensional, computer-generated images.

Computer technology has taken one giant step after another over the past 30 years. As the computer has become more complex, it has branched off into many different areas. Virtual reality is one of the more spectacular areas of computer technology, **Figure A.**

Virtual reality was born in the 1960s, though in a very crude form. In 1981, the first practical application of a virtual reality system was exhibited. The Visually Coupled Airborne Systems Simulator trained pilots to fly complex, high-speed aircraft. The term *virtual reality* did not, however, appear until the mid-1980s. Jaron Lanier, the founder of VPL Research, coined the term.

Virtual reality allows a user to interact with an environment a computer creates. Therefore, complex equipment is needed, in addition to the computer. Computer software, a headset, and some tool relaying the user's movements are required. This tool is often a glove transmitting its relative position and many finger movements to the computer.

Uses for virtual reality are many, with new uses being developed everywhere. The military has used virtual reality to train fighter pilots for some time. Companies use virtual reality to train new employees. Architects allow clients to tour new homes during the design stage, **Figure B.** Clients can make changes based on what they see *before construction begins.* Of course, virtual reality has recreational uses as well. Complex systems can be found in expensive arcades. Simpler systems are available for home video games.

There are impressive prospects for the future of this technology. Soon, surgeons may have virtual reality training facilities. They could train on virtual patients with virtual scalpels. Scientists, as well, could do research with virtual reality. Areas such as molecular modeling and engineering could see significant benefits.

Figure A. Employees manipulate a virtual front-end loader. (Caterpillar Inc.)

Figure B. Virtual objects can be moved about in this virtual office space.

features. The most important are the message-generation window, the mailbox, and the address book.

The message-generation window allows the user to compose, edit, and spell-check messages, **Figure 23-15.** The mailbox is a feature that contains the messages that have been sent and received. An address book contains the e-mail addresses of people that the client communicates with regularly.

In addition, most software allows the user to make attachments to the message. That is, the person can attach photos, documents, sound files, and other electronic media to the e-mail message.

To send a message, the user must complete several tasks, including:

- Open the e-mail software
- Select "New Message" on a menu or dialog box
- Enter an e-mail address or select one from the address book
- Enter a title for the message
- Prepare, edit, and spell-check the message
- Select any appropriate attachments
- Send the message

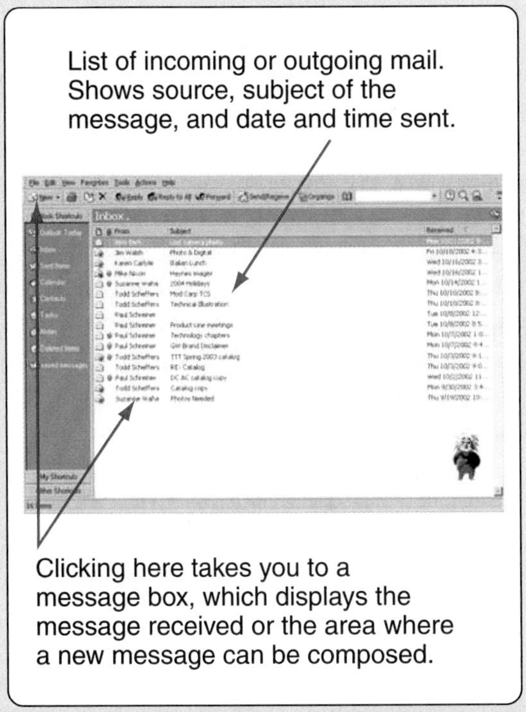

To receive a message, the user can point-and-click on the message in the in box. The message will open in the message window. Attachments may be opened in the same manner or by way of other computer software such as graphics or word processing programs.

List of incoming or outgoing mail. Shows source, subject of the message, and date and time sent.

Clicking here takes you to a message box, which displays the message received or the area where a new message can be composed.

Figure 23-15. Shown here is a typical e-mail software page.

E-Mail Mailing Lists

Sometimes a single message needs to be sent to a number of people. In this case the sender uses a mailing list. The sender may develop mailing lists or people may join a list so that they receive regular news or product updates.

A mailing list works fairly simply. First, a message is sent to a single e-mail address. This address is called the mailing list address. Then, the same message is re-sent or reflected to the other people on the list.

Newsgroups and Chat Rooms

E-mail is the most popular means of communicating over the Internet. It allows people to send messages directly to an individual or group. E-mail works when the recipient of the message is identified and an immediate answer is not needed.

Several other communication approaches are available on the Internet, however. They include:

- *Chat:* An electronic conversation between two or more people. Chat room software allows a group of people to keyboard messages that can be seen by everyone in the "chat room."
- *Newsgroups and forums:* An ongoing public discussion about a particular topic. All people who subscribe to (join) the group have access to all messages and can reply to any or all of them.

Career Corner
Computer Programmers

Computer programmers write, test, and maintain the programs that computers follow to perform their functions. These programs are written to specifications that computer software engineers and systems analysts establish. A program is designed, and then the programmer converts the design into a logical series of instructions the computer can follow. These instructions are coded in a conventional programming language. Programmers update, repair, modify, and expand existing programs. They test a program and make changes until it produces the correct results.

Programmers often are grouped into two broad types: applications programmers and systems programmers. Applications programmers write programs to handle a specific job. Systems programmers write programs to maintain and control computer systems' software. A bachelor's degree is commonly required for programmers, but some positions require only two-year degrees or certificates.

- *Instant messaging:* An approach that allows people to develop a list of individuals with whom they wish to interact. They can send and receive messages to any of the people in the list who are online at the time.
- *IP telephony:* A system that uses the Internet for voice messages. All the people who are online must have a computer and special software. This system requires a special phone that digitizes the voice (sound message) and sends it over the Internet to a second IP telephone.

Electronic Commerce

People have participated in commerce for hundreds of years. Simply put, commerce is the exchange of goods and services for money. In recent years, a new type of commerce has developed. It is called ***electronic commerce*** or e-commerce. It involves selling products and services over the Internet.

This system connects the two major players in commerce. These players are the buyers who want to purchase a good or service and the sellers who have goods and services to sell.

An e-commerce site on the Internet must provide the essential elements of commerce. The first element is merchandising. The product must be displayed so that the potential customer can view it. Pictures and text display describe the features of the product. Often a series of links allow the customer to narrow the choices until a single product is selected, **Figure 23-16.**

Second, the customer must be able to select the items he or she wants to buy. The site generally provides an order form so that the customer can purchase the product.

Third, the site must have a way to close the sale. It must have a way to collect money and issue a receipt. Generally the site provides a way for the customer to use a credit card to pay for the product. An electronic receipt is usually sent to the customer confirming the order.

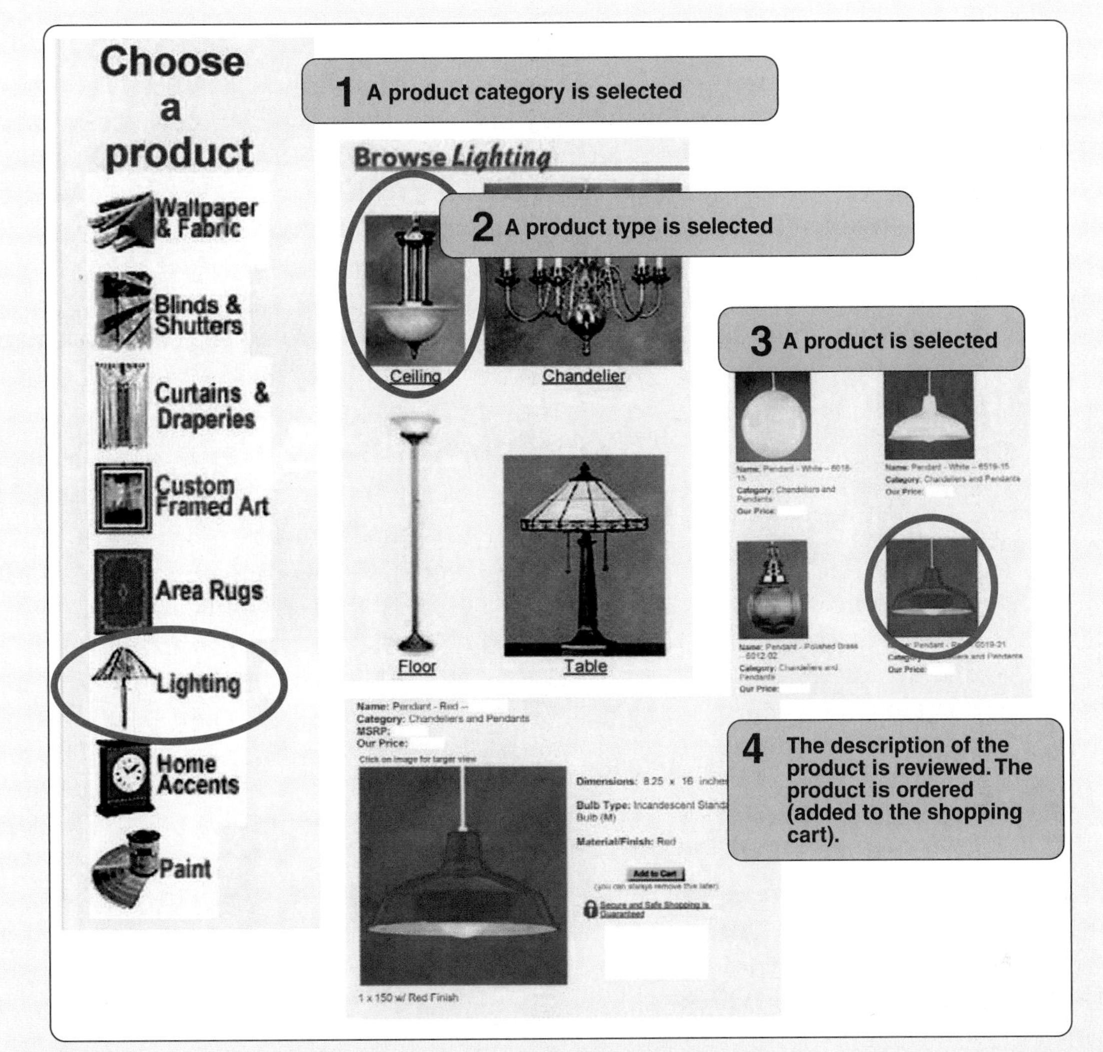

Figure 23-16. An e-commerce site uses a series of links to guide the customer to a product to buy. (American Blind and Wallpaper Factory)

The e-commerce company must then fulfill the order. The merchandise must be sent to the customer. The company might send packages through the U.S. Postal Service or by some private package service. They might deliver printed matter and software over the Internet.

A way to return unwanted, damaged, or defective merchandise must also be in place. Finally, the company should provide technical support for complex products.

We should also note that like all businesses, e-commerce companies must attract customers. They must advertise their site and its products. This can be done by traditional ways such as print, radio, and television advertising.

Another avenue is called banner advertising. These advertisements are small rectangular messages that appear on Web pages. A banner ad is a special hypertext link that instructs a Web server to display a particular Web page containing graphics and sometimes animation.

The banner ad encourages the Internet user to click on the banner ad and go to the advertiser's Web site. The ad also builds name identification for a product, company, or Web site.

Summary

The introduction and use of the personal computer dramatically changed our lives, especially in the ways we communicate. By connecting computers together through networks, we have changed our methods of communication even more. The two basic types of networks are local area networks and wide area networks. The largest wide area network is the Internet, a collection of both small and large independent networks. We can use the Internet by accessing it correctly. Once on the Internet we can engage in such activities as obtaining information from the World Wide Web, sending and receiving electronic mail, joining newsgroups, and buying and selling commercial products.

Key Terms

backbones
browser
central processing unit
dedicated access
dial-up access
domain name
electronic commerce
external storage devices
hyperlinks
hypertext markup language
input devices
Internet
Internet access
Internet protocol address
Internet service provider
links
local area network
memory
network access points
networks
output devices
page
point of presence
random-access memory
read-only memory
routers
search engines
server
uniform resource locator
wide area network
World Wide Web

Test Your Knowledge

Write your answers on a separate piece of paper. Please do not write in this book.

1. The Information Age resulted from the invention of the _____.
2. Give one example of an external storage device for a computer system.
3. Name the two common types of computer networks.
4. List two common uses of the Internet.
5. *True or false?* A connection point established by a company in a particular region is called a point of presence.
6. The two basic ways computers are connected to the Internet are dedicated access and _____ access.
7. What is an Internet protocol address?

8. Define the term *uniform resource locator.*
9. *True or false?* The prefix "http" in a domain name means the document is on the World Wide Web.
10. Software that allows a user to connect with the World Wide Web is called a(n) _____.
11. Define the term *hypertext markup language.*
12. Connections that allow users to move from one Web page to another are called _____.
13. What is a search engine?
14. The server that handles incoming electronic mail is called the _____ server.
15. The system that allows computers to carry voice messages is called _____.

Applying Your Knowledge

1. Use a search engine such as Alta Vista, Excite, or Yahoo to locate Internet sites that explain how things work. Select one device and print out the information the site supplies. Use the same search of key words on a second search engine and compare the results you get.
2. Design a layout for an e-commerce page to sell school supplies. Sketch what each page will contain and the links you will use so people can navigate your site.

This activity develops the skills used in TSA's Cyberspace Pursuit event.

Cyberspace Pursuit

Activity Overview

In this activity, you will create a Web site composed of four components:

- An overview of your school's technology education program
- General information about your school
- Historical information about your school
- A page of links to related/interesting Web sites

Materials

- Paper
- Pencil
- Computer with Internet access and Web page development software

Background Information

Design. Careful planning is critical when developing a Web site. Create a sketch for each Web page. List the elements to be included on each page. Include lines showing the links between Web pages. Consider how a user will navigate within the Web site to be sure you have the necessary links.

Navigation and functionality. The visual appearance of your Web page is important, but not as important as easy navigation and functionality. Design your Web page so that links can be easily located. Select type for links that is easy to read.

Type. Too many fonts and type sizes can make your Web page unattractive. Vary the size and font based on the function of the text. Titles are meant to draw attention. Body type should blend in to the overall design and use a typeface, size, spacing, and justification comfortable for reading. If you use an unusual type font, a person viewing your Web page may not have that font on their computer. In this situation, another font is substituted. This will cause your Web page to have an unintended (and, most likely, less attractive) appearance. Use common type fonts such as Arial, Times New Roman, Tahoma, or Courier. If you want to use an unusual font, create the text as an image file and insert it into the Web page.

Images. Images can add to the visual appeal of your Web pages, but using too many images can clutter a page and cause it to load slowly. Use a compressed image file format (such as jpeg) for faster loading, and use the lowest acceptable resolution.

Guidelines

- Review at least five Web sites, evaluating the design of the sites in terms of attractiveness and usability.
- Your Web site must have a home page containing separate links to each of the four components.
- There is no minimum or maximum number of pages for the individual components.
- The home page and each of the four components must contain both text and graphics.
- All pages must include a link to the home page.
- Use pencil and paper to prepare a rough sketch for each page and an organization chart showing how pages are linked.
- After you've developed the rough sketches, create the Web pages.
- Test the completed design to make sure all links work properly.

Evaluation Criteria

Your Web site will be evaluated using the following criteria:

- Web page design
- Originality
- Content of Web pages
- Functionality

Section 6 Activities

Activity 6A

Design Problem

Background

Communication technology is used to deliver information, project ideas, and generate feelings. Often, these goals are incorporated in advertising. Communication can promote a product or idea by delivering information and persuading people to act in a certain way.

Situation

You are employed as an advertising designer for Breckinridge and Rice Agency. One of your clients wants to promote technological literacy as a public service effort.

Challenge

Design a full page (6 3/4" x 9 1/2") magazine advertisement promoting technological literacy. The ad should encourage students to select at least one class in technology education to help them understand technology as it impacts their lives. In this process, develop a theme, prepare a layout, and specify the type size and style, ink colors, and type and location of the photograph for your design.

Optional

1. Print the advertisement as a flyer that can be sent home with students in your school.
2. Convert the layout to a poster, print copies, and post them on bulletin boards in your school.

Activity 6B

Production Problem

Background

Photographs are used either to capture an event for historical purposes or to communicate information and feelings. The first use allows people to record family experiences, chronicle trips, or capture specific happenings. The second use is communication. This use attempts to give directions, develop attitudes, or communicate feelings.

Challenge

You are a communication designer for a publisher of children's instructional books. The publisher is developing a series of simple how-to booklets for children four to eight years old. Select a task a child of this age must master. Design and produce a six picture set showing the child how to systematically complete the task.

Equipment

- Layout sheets (photocopy the form included with this activity)
- Pencils and felt-tip pens
- Camera and film (black-and-white print or color slides) or digital camera
- Darkroom equipment and chemicals (optional)

Procedure

Designing the Product

1. Select a task.
2. List the steps needed to complete the task.

3. Group steps that can be shown in a single photograph (shot).
4. Develop a layout for each shot, using the layout sheets. **See Figure 6B-1.**

Producing the Product

1. Gather the items (props) needed for each shot.
2. Set up shot #l.
3. Shoot at least three photographs of the shot, using different f-stops to bracket the exposures. (Generally, one photograph is shot using the f-stop the light meter indicates. Then, a second shot is exposed using one f-stop above, and a third is exposed using one f-stop below the first setting.)
4. Repeat steps 2 and 3 until all six shots are taken.
5. Process the film. If black-and-white print film is used, do the following:
   A. Make a contact print of the strip of negatives.
   B. Select the best negative for each shot.
   C. Make a print of each of the six selected negatives.
   D. Mount the prints for display.

Optional

1. For black-and-white prints, make a storybook by producing a printed narrative for each picture. Print the story for each photograph on a separate page, and then mount the picture above the narrative. Design and produce a cover for your book.
2. For slides, prepare a script and record a narrative for the series. Produce a title slide and an end slide for the series.

Series Title: ____________________

Shot #: ____________________

Description: ____________________

Figure 6B-1

Section 7

Applying Technology: Transporting People and Cargo

Tomorrow's Technology Today
Human Transporters

If you have ever groaned at the thought of having to walk to school or wished there was a quicker way to get across town to accomplish all of the errands on your to-do list, you may be in luck. New technological products, such as the Segway™ Human Transporter, are expected to become common personal vehicles in the coming years, and they may be the perfect solution for local transportation. These devices are both effective and efficient, and they hold the promise of many benefits to industry and to the general public.

The Segway Human Transporter is the first vehicle of this kind that has been put on the market. It is a transportation device for a single person to use in any pedestrian environment. These vehicles are battery-operated, and a single charge can usually last up to 17 miles. The transporters are also self-balancing and have no turning radius, making them extremely similar to natural human movement. They change direction when the riders lean in any direction, and because of this, riding on a transporter is like gliding, or almost flying.

Human transporters allow people to travel quicker and accomplish more in shorter amounts of time. They can travel over three times faster than people can walk, and they can carry up to 300 lbs. These vehicles are less expensive than regular vehicles, and since they run on batteries, there is no cost for fuel. With additional cost savings in insurance, maintenance, and parking fees, owning this type of vehicle could save you thousands of dollars a year. In addition to these benefits, if using these transporters catches on, environmental problems and traffic congestion may be greatly reduced.

Besides these benefits to the public, human transporters may also prove to be very helpful in certain industries. They were used in a study with postal workers, who found them easy to use and very effective in reducing the time it took to deliver mail along a regular route. The workers also appreciated the relief of physical strain from carrying packages. Experimental studies have also been done with these vehicles in manufacturing, distribution, and other corporate warehouses. The results of these studies show that these transporters could result in production gains as high as 12%.

Other potential users of human transporters include law enforcement and emergency workers, security guards, maintenance personnel, and airport workers. These people could do their jobs more effectively if they could easily travel down busy streets and through crowded areas. Human transporters might also soon be used at places such as entertainment events and conventions, parks and recreation areas, hotels, restaurants, and amusement parks. The possibilities are nearly endless!

Chapter 24
Using Technology to Transport

Learning Objectives

After studying this chapter, you will be able to:

- Define transportation.
- List types of transportation systems.
- Identify components of a transportation system.
- Define what is meant by place utility.
- Describe transportation pathways.
- Name types of transportation vehicles.
- Identify types of transportation support structures.

The development of transportation and civilization are closely related, **Figure 24-1.** Without transportation, humans are restricted to a very small area. In prehistoric times, for example, a traveler had to make his or her journey out from home and back in one day. This distance was less than 25 miles (40 km). The domestication of animals and their use as beasts of burden and for riding enlarged this travel area. Now people could travel 40 to 50 miles (64 to 80 km) in a day. More travel led people to design and build roads. As early as 30,000 B.C., established transportation routes existed. Only the very brave ventured far from home, however. In fact, human beings tended to stay within a restricted area of travel until recent times.

The development of sailing ships, followed much later by the railroad, enlarged the area of travel. Still, worldwide travel was available only to wealthy or adventurous people. Travel as we know it today is a recent development. The advent of jet-powered aircraft after World War II made the far reaches of the globe available to many people. In one day, we can eat breakfast in New York, lunch in Los Angeles, and be back in New York for a late dinner.

Transportation: A Definition

Transportation is one of the basic areas of technological activity. We use transportation systems today, and we will continue to use them in the future. Simply stated, the word ***transportation*** can be defined as *all acts that relocate humans or their possessions.* ***Transportation technology*** provides for this movement, *using technical means to extend human ability.* Transportation technology extends our ability beyond our own muscle power and our ability to walk.

Transportation is also an interaction among physical elements, people, and the environment, **Figure 24-2.** Transportation provides mobility for people and goods. Transportation uses such resources as materials, energy, money, and time. It has a level of risk to the cargo and passengers that can result in damage, injury, or death. It has a societal impact in that it affects such areas as employment and pollution.

The Importance of Transportation

Transportation is so important that it has become a part of human culture. Try to imagine life without well-developed transportation systems. We think of transportation in the same light as food, clothing, and shelter. It has become a basic need. For example, transportation takes us to work, opens up areas for recreation, allows for easy shopping, and helps keep families in touch.

To help understand how important transportation is, consider the following:

- About 20 percent of the gross national product (the total of all goods and services sold in the United States) is related to transportation.

Figure 24-1. Transportation systems have evolved to meet our changing needs. All the transportation vehicles shown above—trains, automobiles, airplanes, and spacecraft—were new transportation vehicles at one time.

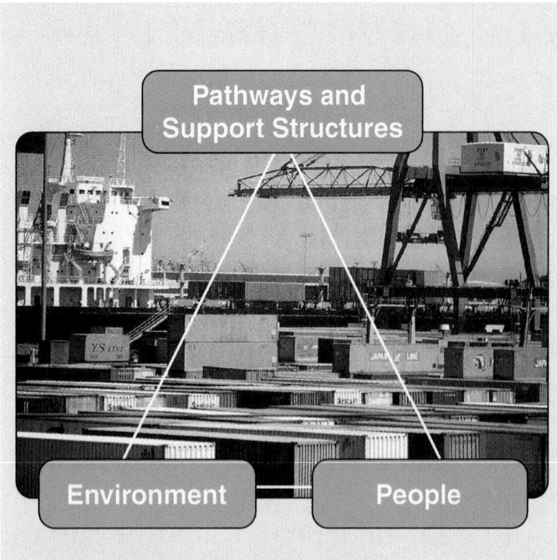

Figure 24-2. Transportation systems are an interactive system of physical elements (vehicles and pathways, for example), people, and the environment.

- We have more than 110 million passenger cars in the country—nearly one for every two people.
- Nearly half the citizens of the country are licensed to drive an automobile.
- Almost 15 percent of personal spending goes toward transportation services.
- Nearly 64 percent of all petroleum is used for transportation, and 25 percent of all petroleum is consumed by personal cars.

Transportation As a System

Transportation, like all other technologies, can be viewed as a system. It is a series of parts that are interrelated. The parts work together to meet a goal. Transportation uses people, artifacts, vehicles, pathways, energy, information, materials, finances, and time. These parts work together to relocate people and goods.

Types of Transportation Systems

People use four environments, or "modes," for transportation. Transportation systems have been developed for land, water, air, and space, **Figure 24-3.**

Land

Humans can move over land with ease. The earliest transportation systems were designed to move people over the land. ***Land transportation systems*** move people and goods *on the surface of the earth* from place to place. Land transportation systems have developed into three major types. These are:

- Highway systems: Automobiles, buses, and trucks
- Rail systems: Freight, passenger, and mass transit systems
- Continuous flow systems: Pipelines, conveyors, and cables

Water

The next system developed was water transportation. From the humble hollow-log canoe, water transportation has grown to be an important mode for moving people and cargo. ***Water transportation systems*** use water to support the vehicle. Water transportation includes inland waterways (rivers and lakes) and ocean-going systems.

Air

Air transportation became practical in the twentieth century. Orville and Wilbur Wright made the first successful flight in a power-driven airplane in 1903. This flight took place at Kitty Hawk, North Carolina. During the succeeding years, air travel

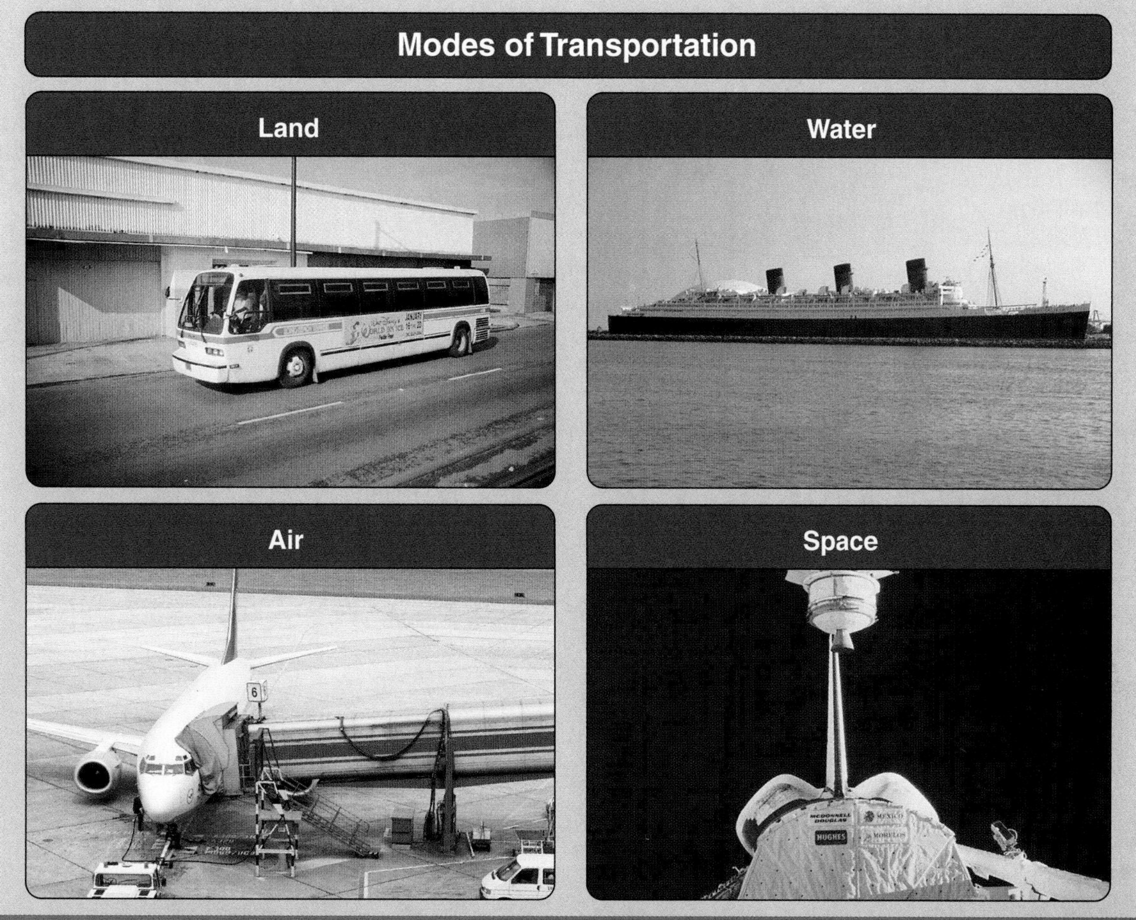

Figure 24-3. Transportation systems operate on land, in water, and through air and space. (MITS, BEA Systems, NASA)

has become a large industry. ***Air transportation systems*** use airplanes and helicopters to lift passengers and cargo into the air so they can be moved from place to place. Today, air transportation includes commercial aviation (passengers and freight) and general aviation (private and corporate aircraft).

Space

Space transportation systems are the fourth transportation mode we use. Space transportation can best be described as *emerging*. It started in 1957 with the launch of Sputnik, a Soviet satellite. Since 1957, space exploration has expanded our knowledge of the universe. Humans have traveled to the moon. The space shuttle now lifts satellites and scientific experiments into space.

On the horizon is personal space travel. Hypersonic aircraft will merge air travel and space travel technologies. People will then be able to travel anywhere on the globe in a matter of hours.

Transportation System Components

Within each of the transportation systems are three major components. These, as shown in **Figure 24-4,** are:

- Pathways
- Vehicles
- Support structures

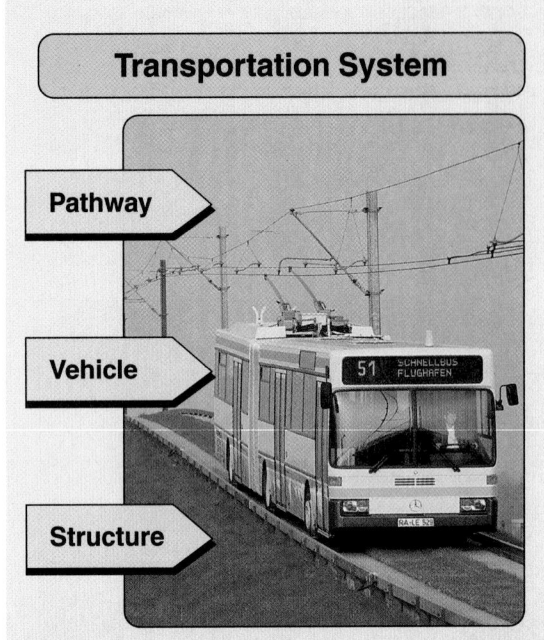

Figure 24-4. Transportation systems use pathways, vehicles, and support structures. This photo shows a guided bus system. The pathway is the concrete guideway, and the vehicle is the bus. Support structures are the stations where passengers board the bus. (DaimlerChrysler)

Pathways

Transportation systems are designed to move people and cargo from one place to another. This movement provides ***place utility.*** People value being able to move things from one place to another. The things that are moved can be food, products, or people. We are willing to pay someone else to transport us or our goods from place to place.

To deliver place utility, a transportation system links distant locations with a network of ***pathways,*** **Figure 24-5.** Pathways on land are readily visible. Land pathways are the streets, roads, highways, rail lines, and pipelines that people have built. These pathways form a network that connects most areas of the United States. Land pathways are the result of construction activities used to fulfill the needs of transportation.

Figure 24-5. Pathways include roads and rail lines (top); rivers, canals, and oceans (middle); and air and space routes (bottom).

Many water pathways are less visible. They include all navigable bodies of water. *Navigable* means that a body of water is deep enough for ships to use. Water pathways can be grouped into two categories: inland waterways and oceans. Inland waterways are rivers, lakes, and bays. Inland waterways also include human-made canals that allow water transportation across areas that lack navigable rivers or lakes. Often, the canals use aqueducts to cross streams. Canals use locks to raise or lower watercraft as the terrain changes, **Figure 24-6.**

Connections to Technology: Science

Newton's First Law of Motion

Many scientific principles come into play when we use the various kinds of transportation systems now available. Among those principles are Newton's laws of motion.

For example, we can use Newton's first law of motion to explain why we need seat belts in our automobiles. Newton's first law of motion states that a moving object will continue moving at the same speed in the same direction unless some force acts on it. This tendency of an object to resist change is also called *inertia.*

Thus, when we are traveling in an automobile at, say, 50 miles per hour, we continue moving at this speed even when we forcefully apply the brakes of the car. We have applied force (the brakes) to stop the car, but such force works only on the car, not us. Therefore, we need the seat belt because it provides the force needed to stop us from continuing to move and perhaps hitting the windshield.

Newton's first law of motion also involves the concept of *friction.* Using another method of transportation as an example, can you think of a way friction affects this object when it is moving?

Figure 24-6. Canals are constructed waterways. They include a waterway, locks (left) to raise or lower the ship, and aqueducts (right) to cross streams and gullies.

The oceans and seas provide a vast water pathway. The major land masses on earth are touched by an ocean or sea. Therefore, the continents can be easily served by an ocean transportation network.

Some pathways are hard to see. Air and space routes are totally invisible. They are defined by humans and appear only on maps. Airplanes and spacecraft can travel in almost any direction. Only human decisions establish the correct pathway. The pathway is chosen to ensure efficient and safe travel. Airplanes use pathways to take us easily over land and water barriers. Airplanes allow us to travel thousands of miles in a few hours. Covering these distances before airplanes or railroads took *months* of travel by wagon.

Vehicles

All transportation systems, except for continuous flow systems, use vehicles. ***Vehicles*** are technological artifacts designed to carry people and cargo on a pathway. Vehicles are designed to contain and protect the cargo as it is moved from place to place. The demands on the vehicle change with the type of cargo, **Figure 24-7.**

People need to be protected from the environment through which the vehicle travels, and they desire to travel in comfort. Passenger vehicles should provide such features as proper seating and good lighting. Vehicles that carry people long distances should provide food and rest rooms. Also, passenger vehicles should include safety features because of the possibility of accidents. Impact-absorbing construction, seat belts, fire protection, and other safety measures should be designed and built into the vehicle.

Cargo vehicles are designed to protect the cargo from damage by motion and the outside environment. Cargo should be cushioned so that is does not become broken, dented, or scratched. Fumes, gases, or liquids that could affect the cargo must be kept at a distance. Finally, the cargo should be protected from theft and vandalism.

Support Structures

All transportation systems use structures. Many transportation pathways use human-built structures. Rail lines, canals, and pipelines are examples of constructed works.

Figure 24-7. Vehicles such as these aircraft are designed differently for cargo (left) and people (right). (Alaska Airlines, United Airlines)

Career Corner
Automotive Mechanics

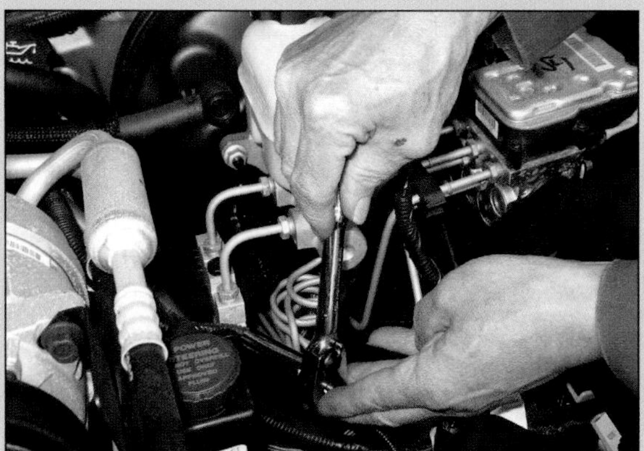

Automotive mechanics use skill and knowledge to inspect, maintain, and repair automobiles and light trucks. They review the description of a problem, use a diagnostic approach to locate the problem, repair the vehicle, and test the repair. Automotive mechanics must be able to use a variety of testing equipment, including onboard and stationary diagnostic equipment and computers. Most automotive mechanic jobs require formal training in high school or in a postsecondary vocational school. Some mechanics learn the trade by assisting and learning from experienced workers.

Note, however, that transportation systems need structures other than just pathways. They also need structures called terminals. ***Terminals*** are where transportation activities begin and end. They are used to gather and then load and unload passengers and goods, **Figure 24-8.**

Terminals provide for passenger comfort and cargo protection before the people and cargo are loaded into transportation vehicles. Terminals also provide connections for various transportation systems. Some allow truck shipping to connect with air, rail, or ocean shipping. Other terminals connect automobile, rail, and bus systems with air transportation systems. These types of terminals allow for ***intermodal shipping.*** Intermodal shipping means that people or cargo travel on two or more modes of transport before they reach their destination. An example of intermodal shipping is semi-truck trailers that are hauled on railcars.

Figure 24-8. Transportation terminals are at the start and the end of transportation pathways. This photo shows an air terminal in Chicago.

Other structures are used to control transportation systems. These structures are communication towers, radar antenna, traffic signals, and signs. These structures help vehicle operators to stay on course and observe rules and regulations.

Summary

Transportation is essential to modern-day life. We all use it and expect it to make our lives better. Each of us can choose from a variety of transportation systems to get to a place. These systems include land, water, and air transport. In the future we may also be using space as a transportation link.

All transportation systems have pathways and support structures. Most of them use vehicles to carry people and cargo from one point to another.

Key Terms

air transportation systems
intermodal shipping
land transportation systems
pathways
place utility
space transportation systems
terminals
transportation
transportation technology
vehicles
water transportation systems

Test Your Knowledge

Write your answers on a separate piece of paper. Please do not write in this book.

1. Define the term *transportation* as used in this chapter.
2. *True or false?* Walking to school is an example of transportation technology.
3. List the four types of transportation systems.
4. Define the term *place utility.*
5. *True or false?* Pathways can be visible or invisible.
6. Water pathways can be grouped into the two categories of inland waterways and _____.
7. Name one type of transportation vehicle.
8. What is a terminal?
9. Using more than one mode of transportation to ship cargo is called _____.
10. *True or false?* A traffic signal is an example of a transportation support structure.

Applying Your Knowledge

1. Select a transportation system you use and describe it in terms of _____.
   a. The pathway
   b. The vehicle
   c. The structures used

Enter your data on a chart similar to the one below.

Type of Transportation System:		
	Name	**Description**
Pathway:		
Vehicle:		
Structure:		

2. Obtain a number of bus, rail, ship, or airline route maps and road maps. Choose a place that you would like to visit. Plan a trip that will use intermodal transportation. Be sure your route uses at least two different transportation modes.

Chapter 25
Transportation Vehicles

Learning Objectives

After studying this chapter, you will be able to:

- Identify the five systems present in a transportation vehicle.
- Identify the types of structural systems used in transportation vehicles.
- Name common propulsion systems used in transportation vehicles.
- Describe common suspension systems used in transportation vehicles.
- Identify common guidance and control systems used in transportation vehicles.

As we noted in the last chapter, transportation can be defined as *all acts that relocate humans or their possessions.* Most transportation activities use technology to make them more efficient.

This technology often takes the form of a vehicle, **Figure 25-1.** A simple definition of a ***vehicle*** is a powered carrier that supports, protects, and moves cargo or people within a transportation system. Vehicles are used in all four environments of travel. They are used on land and in water, air, and space. Each of these areas will be discussed in turn after we examine vehicular systems in general.

Vehicular Systems: An Overview

All vehicles share some common systems. These are the structure, propulsion, suspension, guidance, and control systems, **Figure 25-2.**

The Structural System

All vehicles are designed to meet a common goal: to contain and move people and goods. People want to arrive safely and in comfort. Cargo must be protected from the weather, damage, and theft. The ***structural system*** of the vehicle helps to do these things. The structural system is composed of the physical frame and covering. This system provides spaces for people and cargo, power and control systems, and other devices.

The Propulsion System

Transportation vehicles are designed to move along a pathway. This pathway may be a highway, rail line, river, ocean, or air route. A vehicle must have a force to propel it from its starting point to its destination. This force is produced by the ***propulsion system.*** It uses energy to produce power for motion.

Propulsion systems range from the simple pedal, chain, and wheel system of a bicycle to complex heat engines such as gasoline, diesel, and rocket engines, **Figure 25-3.** The type of engine to be used in a vehicle is determined by several factors. These factors include:

- The environment in which the vehicle travels
- Fuel availability and cost
- The forces that must be overcome, such as vehicle and cargo weight, rolling friction, and water or air resistance

The engine must match the job. For example, using a jet engine to propel an automobile is overkill. The capabilities of a jet engine do not match the job of moving a car. Likewise, using a large diesel engine to power an airplane is not the best choice. The diesel engine would be too heavy. Therefore, we use many different sizes and many different types of engines in transportation vehicles.

Figure 25-1. Most transportation systems require a vehicle.

The Suspension System

All vehicles and cargo have weight. This weight must be supported as the vehicle moves along the pathway. Proper support is produced by a ***suspension system.*** Suspension systems include:

- Wheels, axles, and springs on land vehicles
- The wings on an airplane
- The hull of a ship

The Guidance System

Handling any vehicle requires information. The operator must know his or her location, speed, and direction of travel. Information about traffic conditions and rules is also required. This information is provided by the ***guidance system.***

Figure 25-2. Transportation vehicles have five basic systems.

Figure 25-3. Propulsion systems range from simple to complex. (Airbus)

Figure 25-4. Transportation vehicles require control in three degrees of freedom. The degrees are (1) forward-backward, (2) left-right, and (3) up-down. The steam locomotive shown in the photo has one degree of freedom.

Guidance information can be as simple as a speedometer reading or a traffic light. Guidance systems can also be quite complex. Instrument landing systems (ILS) and land-based satellite tracking stations are examples of complex guidance systems.

The Control System

A vehicle moves from its origin to its destination to relocate people and cargo. As a vehicle moves, it requires control, **Figure 25-4.** The ***control system*** consists of two types: ***speed control*** and ***direction control.*** The vehicle can be made to go faster through acceleration or slowed down by braking or coasting. How a vehicle changes its direction depends on its environment of travel. Land vehicles turn their wheels or follow a track. Ships move their rudder. Airplanes adjust ailerons, rudders, and flaps. Spacecraft use rocket thrusters.

The system of controls used will depend on the ***degrees of freedom*** a vehicles has. For example, rail vehicles have one degree of freedom. They move forward or backward by using speed control. Gravity and the rail eliminate the possibility of up-down and left-right movement.

Automobiles and ships have more freedom. They can move forward and backward and can change their direction by turning left or right. They have two degrees of freedom.

Aircraft, spacecraft, and submarines have the most freedom: three degrees. They can move forward and backward, left and right, and change their altitude by moving up and down.

Land Transportation Vehicles

Land transportation includes all movement of people and goods on or under the surface of the earth. It includes highway, rail, material, and on-site transportation systems.

Highway systems use automobiles, trucks, and buses to move people and cargo. Rail systems move people and cargo from place to place. Material transportation systems include pipelines and conveyors. On-site transportation systems include systems commonly found in such places as factories, stores, and hospitals. These systems might be forklifts, tractors and carts, chutes, and conveyors.

Structure

Each land transportation system requires special vehicles. The structure a vehicle has is based on its use. Passenger vehicles are different from cargo-carrying vehicles. All the vehicles, however, have three basic structural units, as shown in **Figure 25-5:**

- Passenger or operator unit
- Cargo unit
- Power unit

The size and location of these units or compartments will vary with the type of vehicle. Freight and some passenger rail systems place the power and operator units in one vehicle called a ***locomotive.*** The locomotive pulls the cargo and passenger units. Mass transit systems have power units (electric motors) and passenger units in each car. The operator unit is located in the front car, **Figure 25-6.**

Figure 25-5. All vehicles have operator, power, and cargo or passenger units. (Freightliner Corp.)

Figure 25-6. The light rail vehicles (top) integrate operator and passenger areas. The freight rail system (bottom) uses a locomotive that combines operator and power units. The locomotive pulls railcars, the cargo units. (Long Beach Light Rail, Norfolk Southern)

Standard automobiles and delivery trucks have all three units in one vehicle. Most long-distance trucks, however, place the power and operator units in the ***tractor.*** The cargo unit is the one or more ***trailers*** attached to the tractor.

The passenger and operator unit must be designed for comfort and ease of operation. This requires the use of ergonomic principles. Ergonomics is the study of how people interact with the things they use. Ergonomic vehicle design requires that the seats adjust for people of different sizes. The instruments that provide guidance data must be easily seen, **Figure 25-7.** The operating controls, such as for steering and braking, must be easy to reach and operate.

The vehicle must also have an appropriate structural design. The vehicle must provide for safety and for operator, passenger, and cargo protection. Most vehicles have a reinforced frame with a skin. The skin protects the vehicle interior from the outside environment. The frame supports the skin and carries the weight of passengers and cargo. The frame also provides crash protection by absorbing the impact when vehicles are involved in accidents, **Figure 25-8.**

Figure 25-7. This aircraft flight deck was designed using ergonomic principles for easy operation. Ergonomic principles are also used in land, water, and space vehicles. (Boeing)

Propulsion

Most land transportation vehicles move along their pathway on rolling wheels. The rotation is produced by two systems: power generation and power transmission. A few land transportation vehicles use a diesel-electric system or an electrical system only.

The Power Generation System

The ***power generation system*** uses an engine as an energy converter. The engine produces the power needed to propel the vehicle. The most common engine in land vehicles is the ***internal combustion engine.*** This name means that fuel is burned *inside* the engine to convert energy from one form to another. The chemical energy in the fuel is first changed to heat energy. The heat energy is then converted to mechanical (rotating) energy by the piston and crankshaft to move the vehicle.

Land vehicles normally use either a four-stroke cycle gasoline engine (described in Chapter 7) or a diesel engine. Some rail vehicles use electric motors for propulsion. Electrically powered vehicles for highway use are on the horizon.

Figure 25-8. This vehicle is being crash tested to evaluate its structural design. (Arvin Industries)

As noted in Chapter 7, the internal combustion gasoline engine uses four strokes to make a complete cycle. The first stroke is called the intake stroke. The intake valve opens, and the piston moves down. A fuel-air mixture is drawn into the cylinder. When the piston is at the bottom of the intake stroke, the valve closes. The piston moves upward, beginning the compression stroke. The piston compresses the fuel-air mixture. When the piston is at the top of its travel, a spark plug ignites the fuel-air mixture. The burning fuel expands rapidly and drives the piston down. This is called the power stroke. At the end of the power stroke the exhaust valve opens. The piston moves upward to force burnt gases from the cylinder. The last stroke is called the exhaust stroke. The exhaust valve closes, and the engine is ready to repeat the four strokes—intake, compression, power, and exhaust. You may want to look back to **Figure 7-22** to see a drawing of these four strokes.

Many large vehicles, such as buses, heavy trucks, and locomotives, are powered by diesel engines. The most common diesel engine uses a four-stroke cycle similar to a gasoline engine, **Figure 25-9.** During the intake stroke, however, only air is drawn in. Thus, only air is compressed. As the air is compressed it becomes very hot. When the piston is at the top of the compression stroke, fuel is injected directly into the cylinder. The fuel touches the hot air and burns rapidly. This causes the power stroke, which is followed by the exhaust stroke. These last two strokes are exactly like those of a gasoline engine.

The Power Transmission System

The power of an engine is worthless unless it can be controlled and directed to do work. This is the task of the ***power transmission system.***

The transmission connects the engine with the drive wheels, **Figure 25-10.** This is done in automobiles and trucks by mechanical or fluid devices. Often a transmission has several input/output ratios, commonly called "speeds." A five-speed transmission has five ratios. The different ratios allow the power of the engine to be used efficiently. The transmission provides high torque (rotating force) at low speeds as the vehicle starts to move. Later, as gravity, rolling resistance, and air resistance are overcome, low-torque—high-speed outputs are used.

Mechanical transmissions (also called manual transmissions) have a clutch between the engine and the transmission. This allows the operator to disconnect the

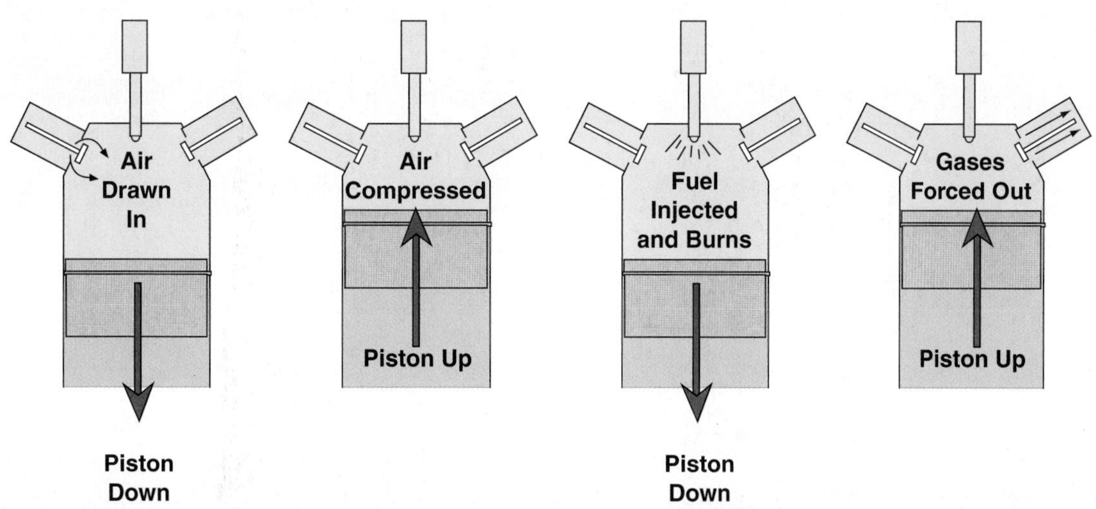

Figure 25-9. These drawings show how a diesel engine works.

engine so that the transmission ratio can be changed. This action is called shifting gears. ***Automatic transmissions*** use valves to change hydraulic pressure so that the transmission shifts its input/output ratios.

When a vehicle makes a turn the wheels on the outside must be able to turn faster than those on the inside. The outside wheels turn faster because they have farther to travel. A device called a differential allows the wheels to turn at different speeds. The differential is a set of gears that independently drive each axle. The axles are connected to the vehicle's drive wheels. In rear wheel drive automobiles and trucks, this device is separate from the transmission. Front wheel drive cars combine the transmission and differential into a single unit called a transaxle.

The Diesel-Electric System

Freight and passenger locomotives use a different type of power generation and transmission system. This type is called a diesel-electric system, **Figure 25-11.** This system uses a very large (8 to 12 cylinder) diesel engine to convert the fuel to mechanical motion. This motion is used to drive an electric generator. The electricity is transmitted to electric motors that are geared to the drive wheels of the locomotive. This motor-generator unit is relatively light. Therefore, extra weight is added to the locomotive to give it better traction. This weight is called ***ballast.*** This extra weight is created by making the frame of the locomotive out of heavier steel plates than needed.

The Electric System

Some rail vehicles are powered only by an electrical system. The vehicle has a moving connection that remains in contact with the electric conductor. Overhead wires or a third rail are used as the conductor. The electricity is conducted to motors that turn the vehicle's wheels.

Suspension

Suspension systems keep the vehicle in contact with the road or rail. They also separate the passenger compartment from the drive system to increase passenger and operator comfort.

The suspension system of a vehicle has three major parts:

- Wheels
- Axles
- Springs and shocks

The wheels provide traction and roll along the road or rail. The wheel also spreads the weight of the vehicle onto the

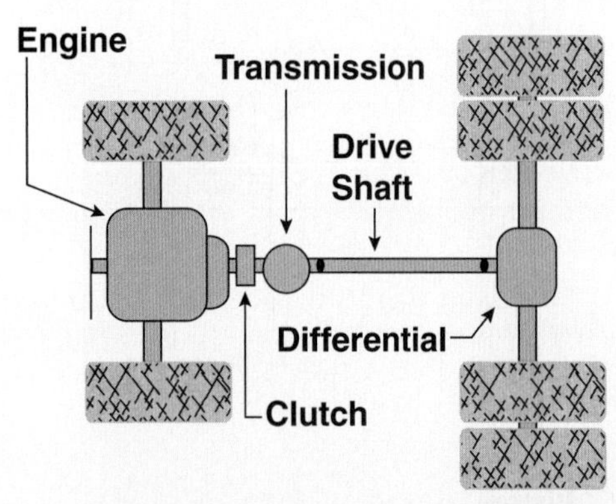

Figure 25-10. The transmission joins the engine to the rest of the drive train.

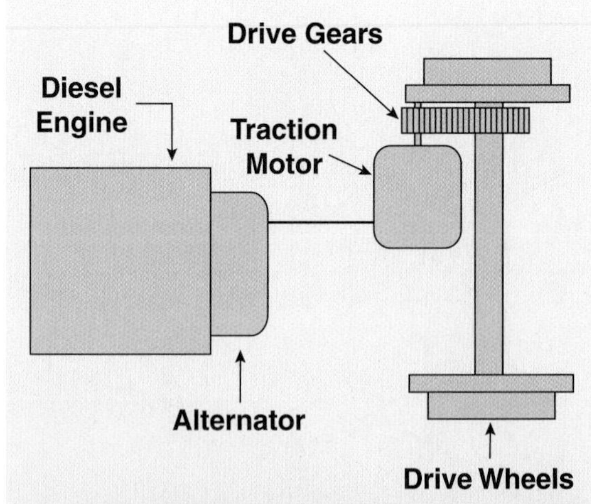

Figure 25-11. This drawing shows a diesel-electric traction system used to power railroad locomotives.

road or rail. A wheel can be used to absorb shock from bumps in a road or rail. Often rubber tires are attached to the wheels to absorb shock and increase traction. Most rail systems use steel wheels and depend on friction between the wheels and rails for traction.

Axles carry the load of the vehicle to the wheels. The axles support a set of springs. The springs absorb movements of the axles and wheels. Springs, however, give a very "bouncy" ride. This motion is prevented by the use of shock absorbers. The shock absorbers are attached between the axle and the frame to dampen the spring action, **Figure 25-12.** The result is a controlled, smooth ride.

A variation of the suspension system uses torsion bars. Instead of a spring, a steel bar absorbs the movement of the wheels and axles by twisting. The torsion bar untwists to release the absorbed energy. Shock absorbers are used to dampen the motion of the torsion bar.

Guidance and Control

Most land transportation vehicles are controlled manually by an operator. The speed and direction of a vehicle can be changed by the operator. Throttles and accelerator pedals are used to control engine speed. Transmissions are shifted to increase the power being delivered to the drive wheels. Brakes can be applied to slow the vehicle. The wheels can be turned to take the vehicle in a new direction.

These actions require that the operator make decisions. These decisions are made on the basis of visual information and judgment. The operator receives information from signs and signals, **Figure 25-13.** The operator also observes traffic conditions. Gages and instruments provide information on vehicle speed and operating conditions. Because all this information must be considered and acted on, the operator's judgment also becomes important. Inexperienced operators and those influenced by drugs and alcohol use may make poor judgments. Their actions may cause accidents and human injury or death.

Water Transportation Vehicles

More than 70 percent of the earth's surface is covered by water. Thus, water transportation is an important form of travel. Water transportation includes all vehicles that carry passengers or cargo over or under water. Water transportation on

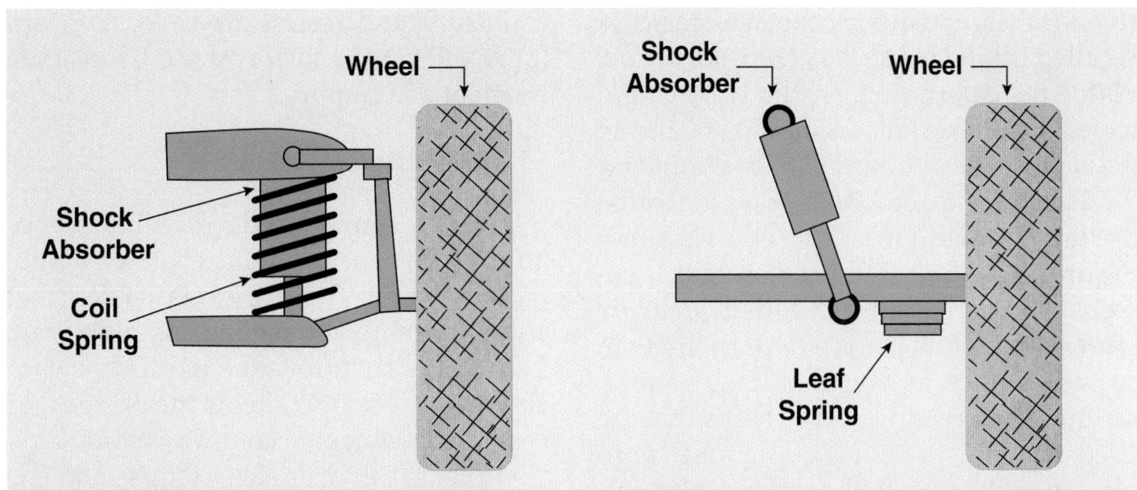

Figure 25-12. This drawing shows two types of suspension systems used on automobiles.

Figure 25-13. Signs and signals, such as highway signs (left) and rail signals (right), give operators guidance.

Figure 25-14. The three types of boats and ships are pleasure, commercial, and military.

rivers, lakes, and along coastal waterways is called ***inland waterway transportation.*** Water transportation on the oceans and large inland lakes such as the Great Lakes in the United States is called ***maritime shipping.***

There are three kinds of water transportation vehicles (called vessels), **Figure 25-14.** Vessels owned by private citizens for recreation are called ***pleasure craft.*** They allow people to participate in activities such as water skiing, fishing, sailing, and cruising. Large ships that are used for transporting people and cargo for a profit are called ***commercial ships.*** Some ships provide for the defense of a country. These vessels are called ***military*** (or naval) ***ships*** and are owned by the government of a country.

Structure

What a ship is used for will influence how it is designed. All ships have two basic parts, however. The hull forms the shell that allows the ship to float and contain a load. The superstructure is the part of the ship above the deck. It contains the bridge and crew or passenger accommodations.

Like land vehicles, ships can be designed to carry either passengers or

Figure 25-15. Commercial ships can be either passenger or merchant ships.

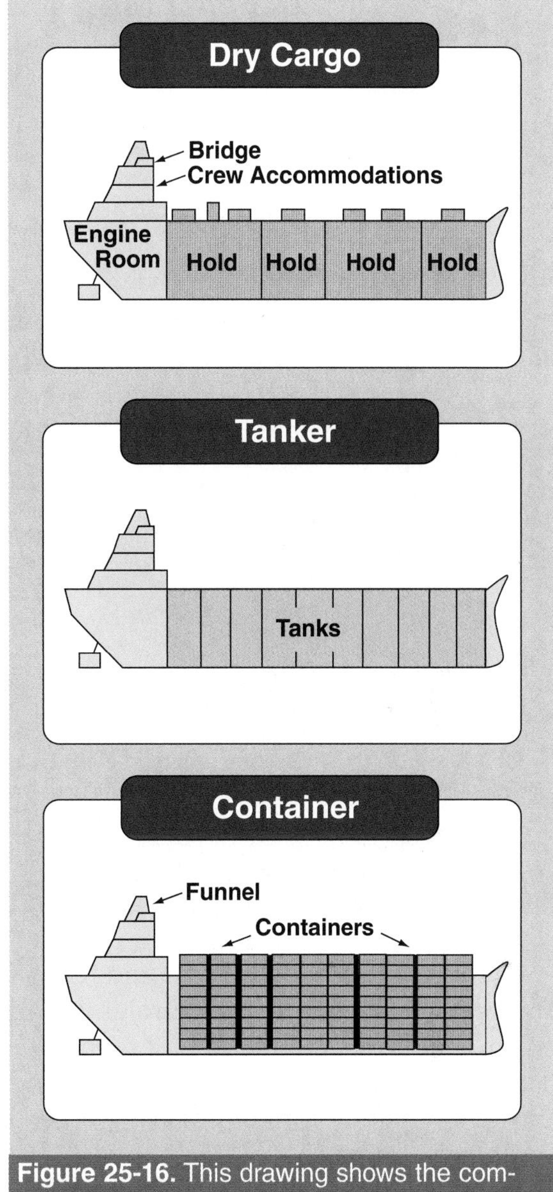

Figure 25-16. This drawing shows the common types of merchant marine vessels.

cargo, **Figure 25-15.** Vessels that carry people are called ***passenger ships*** or passenger liners. Cargo-carrying ships are called ***merchant ships.***

Passenger ships include oceangoing liners, cruise ships and, for short distances, ferries. Ocean liners and cruise ships are miniature cities. They must supply all the needs of the passengers. They must thus include such areas as sleeping quarters, kitchens, dining space, recreational areas, and retail shops, such as haircutting and clothes-cleaning establishments.

Merchant ships are usually designed for one type of cargo. The types of cargo can include dry cargo, liquids, gases, and cargo containers, **Figure 25-16.** ***Dry-cargo ships*** are used to haul both crated and bulk cargo. The term *bulk cargo* describes the loose commodities that are loaded into the holds (compartments) of ships. Grain, iron ore, and coal are typical kinds of bulk cargo. Other dry cargo is contained in crates that are loaded into various holds of a ship. Land vehicles, such as cars and trucks, are another type of bulk cargo. They are often transported on special vessels called RORO ships (for *R*oll *O*n-*R*oll *O*ff). The vehicles are driven onto the ship and off the ship through large hatches (doors) in the hull.

Figure 25-17. Containers are often shipped from inland points (left) and then unloaded at the dock (center). Later, the containers are loaded onto container ships (right).

Figure 25-18. Tugboats pull or push barges to carry large loads on inland waterways. (American Petroleum Institute)

Tankers are used to move liquids across the oceans. Tankers carry petroleum and other chemicals in a series of large tanks. The liquid cargo can be pumped into and out of the tanks. Tankers have also been built to move gases.

Container ships are a newer and faster way to ship large quantities of goods, **Figure 25-17.** The shipments are loaded into large steel containers that resemble semi-truck trailers without wheels. The containers are sealed and then loaded into the hold and stacked on the deck of the ship. The loading process that once took days is now done in hours.

Large ships sail the oceans, whereas smaller ships are used on such lakes like the Great Lakes in North America. Most shipping on inland waterways, however, is done with tugboats and barges, **Figure 25-18.** These vessels are "water trains." The tugboat acts like the locomotive. It contains the power generation and transmission system and the operation controls. The barges are the waterway "rail cars." Various types of cargo have specifically designed containers. The tug pushes or pulls a group of barges that are lashed together.

Propulsion

Most commercial ships are propelled through the water by a propeller that is driven by an engine. This engine can be a steam turbine or diesel engine. Often the engine is located inside the ship or boat. Small boats, however, might be powered by a motor (internal combustion gasoline engine) attached to the stern (the back) of the boat. This type of power source is called an ***outboard motor,*** **Figure 25-19.**

Some military ships and submarines are powered by nuclear reactors. Heat from the nuclear reactor is used to turn water into steam. This steam is used to turn a turbine. The turbine turns the propeller using a shaft.

The propeller (called a "prop") is a device with a group of blades that radiate out from the center. Propellers can have from two to six blades. The propeller attaches to a shaft at the center. Propellers range in size from 2 to 3" (51 to 76 mm) to more than 30′ (9 m) in diameter. Each blade of the propeller is shaped to "bite" into the water much like a window fan "bites" into the air. The rotation of the propeller forces

Figure 25-19. Small boats can be powered by outboard engines. (OMC Power Products)

the water past it. As we learned earlier, one law of physics tells us that for every action there is an equal and opposite reaction (Newton's third law of motion). Therefore, the action of forcing the water through the spinning propeller causes an opposite reaction that pushes the boat forward.

Not all boats are powered by engines, however. Some use wind power. One of the earliest sources of propulsion was the sail. A sail catches the wind and pushes the boat through the water, **Figure 25-20.**

Figure 25-20. Sails are widely used on pleasure craft. Wind power was once the leading way to propel ships.

Suspension

For water transportation vehicles, the hull is the primary component of the suspension system. The hull must be carefully designed to ensure that the boat remains stable and afloat. If the hull is not properly designed, a boat could roll onto its side and sink.

The shape of the hull has a great effect on the stability of the boat. Boats with round-bottom hulls are relatively unstable. V-bottom hulls, which are shaped like the letter *V*, provide greater stability than round-bottom hulls. Flat-bottom hulls are the most stable. See **Figure 25-21.**

The weight distribution in the boat also affects stability. If the boat's load is evenly spread over the hull, the boat should be very stable. If the boat is unevenly loaded, it may lean in the water. When the boat is unevenly loaded, the propulsion system and the control system will not work as efficiently as when the load is evenly distributed. In addition, an unevenly loaded boat is more likely to sink.

When an object is immersed in a fluid, the fluid tends to push the object up. If you lift your arms away from your body while swimming, you will notice that the water tends to push your arms toward the surface. This tendency is called buoyancy. ***Buoyancy*** is the upward force exerted on an object immersed in a fluid.

Technology Explained
Maglev Train

maglev train: A transportation vehicle using magnetism to suspend and propel itself along a guideway.

The rough ride and noise traditional rail systems produce have been a drawback to their use for passenger travel. The steel wheels and rails used in most rail systems cause these drawbacks. Improvements in rails, wheels, and maintenance have made trains quieter. Some trains, such as monorails, use rubber tires or inserts in steel wheels. Because of the contact between the wheels and rails, however, some noise is still produced. In the 1960s a new land vehicle appeared that did not touch the rails. This type of vehicle is called a *magnetically levitated train,* or *maglev.* Maglevs use magnetic forces to support and move along special pathways called *guideways.*

Figure A. This is the Transrapid™ 06 car. It uses magnetic *attraction* (magnets pulling toward each other) to suspend the vehicle. (Transrapid International)

There are two types of maglevs: attraction and repulsion. An attraction maglev uses magnets pulling *toward* each other to support the train, **Figure A.** A repulsion maglev uses magnets pushing away from each other to support the train, **Figure B.** The guideways differ between the two types of maglevs. Attraction maglevs wrap around the guideways. Repulsion maglevs sit in trough-like guideways.

The maglev train includes a passenger compartment providing a quiet and comfortable environment for the passengers. Beneath the passenger compartment is the suspension system. Attraction maglevs use electromagnets to support the train. The electromagnets pull toward a rail in the guideway, lifting the train. The amount of electricity is controlled so the electromagnets and rail never touch. Repulsion maglevs use superconducting magnets to induce magnetic fields in coils in the guideway. The maglev floats on these magnetic fields.

Figure B. Magnet *repulsion* can also be used to suspend a maglev. The MLU-002 maglev uses magnets pushing away from each other for suspension. (Japan Railways Group)

A device called a *linear induction motor,* or LIM, propels maglevs. A LIM works like a standard electric motor, using the principle that like poles of a magnet repel each other, while unlike poles attract. A standard electric motor produces rotary motion. A LIM is laid out flat in a straight line. Thus, a LIM produces *linear,* or straight-line, motion.

A LIM uses alternating electric current to create magnetic fields. The direction of the electric current changes many times a second. When this happens, the lines of magnetic force collapse and change direction. This causes the rail to be attracted to the magnet, and the vehicle moves along the rail. Then, the current switches direction, causing the magnet to repel the rail. This cycle repeats over and over and pushes the maglev along the guideway.

You can think of the buoyant force as a force trying to hold an object "up" in the water. In order for the object to float, the buoyant force must be equal to the weight of the object. If the weight of the object is greater than the buoyant force, the object sinks.

When an object is placed in a fluid, the fluid must move to make room for the object. Displaced fluid is the fluid that had previously occupied the space now occupied by the immersed object. The volume of displaced fluid is identical to the volume of the object that is submerged. The weight of the displaced fluid can be calculated by multiplying the volume by the density of the fluid. The buoyant force is equal to the weight of the displaced fluid.

To determine whether an object will float, you simply compare the weight of the object to the weight of the displaced water (buoyant force). If the weight of the object is greater, the object sinks. If the buoyant force is greater, the object floats.

We should note that three special types of boats use unique suspension principles. The first is the hovercraft, **Figure 25-22.** A ***hovercraft*** is suspended on a cushion of air.

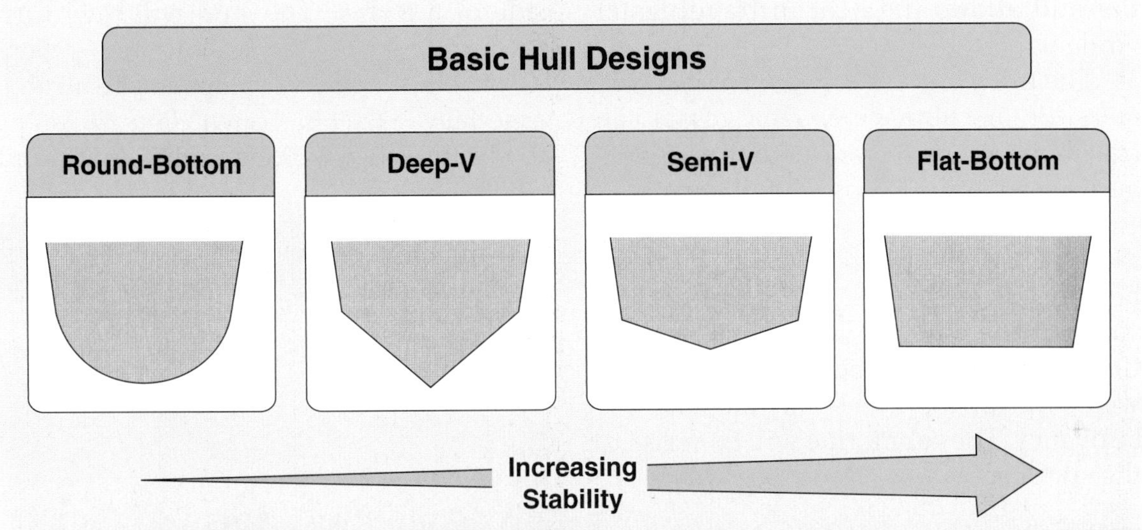

Figure 25-21. Basic hull designs range from the round-bottom kind to the flat-bottom kind. The flat-bottom hull is the most stable.

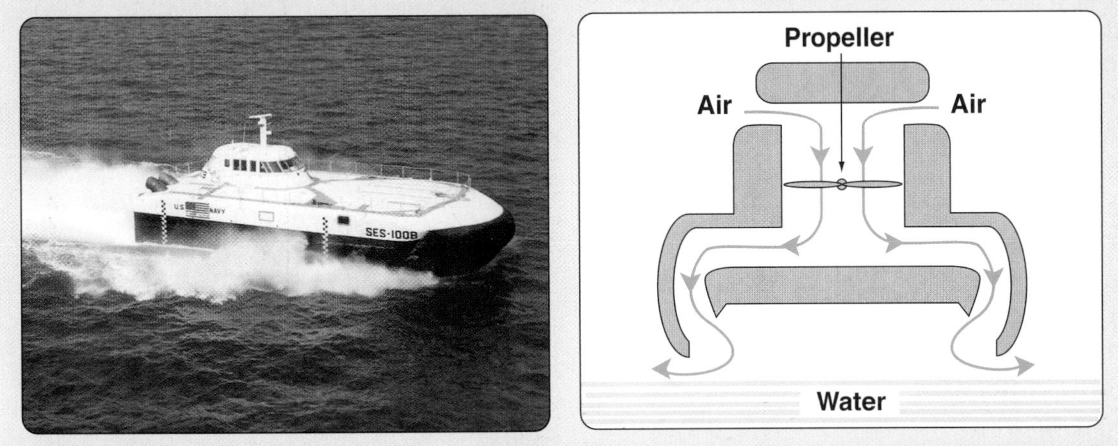

Figure 25-22. A hovercraft is suspended above the water on a cushion of air.

The air is forced by large fans into a cavity under the boat. As the air escapes this pocket, the boat is lifted above the water. Hovercrafts are used over shallow water, swamps, and marshy land and where speed is important.

The second type of special boat is the hydrofoil. A ***hydrofoil*** has a normal hull and set of underwater "wings." These wings are called hydrofoils. A jet engine provides power for the boat. As the boat's speed increases, the water passing over the hydrofoils produces lift. The lift causes the hull to rise out of the water. The reduced friction between the water and the hull allows the boat to travel faster while using less fuel.

The third type of special boat is a ***submersible*** or submarine. This vessel can travel on the surface and under water. Its buoyancy is adjusted by allowing water to enter or forcing it out of special tanks. As the tanks fill with water the vessel becomes heavier, or less buoyant. Therefore, the submarine sinks into the water. Compressed air can be used to force the water out of the tanks and increase the buoyancy. The submarine will then rise to the surface.

Guidance and Control

The operator (called a "skipper" or "captain") of a ship is responsible for its course (path) and safety. He or she obtains guidance from a number of sources. A compass and the charts of rivers, harbors, and oceans are the bases for navigating ships of any kind. In or near a harbor, lighthouses and lighted buoys identify safe channels for navigation or mark hazards, **Figure 25-23.** Flags and radio broadcasts communicate weather conditions. Special electronic systems help to pinpoint the ship's location and indicate the depth of the waterway.

This information is used as the operator makes judgments about the best route. The path a ship follows is controlled in various ways, **Figure 25-24.** Ships that have their own power source are generally guided with a ***rudder.*** This is a large flat plate at the stern of the ship. When it is turned away from the ship's course, it will deflect the water passing under the hull. This deflection forces the ship into a turn. Unlike most land vehicles, where the front of the vehicle contains the guidance and control systems, the back of the ship changes its path and causes the vehicle to turn.

Some large ships have two propellers or "twin screws." Increasing the speed of one propeller will create additional force on that side of the ship. The ship will then turn toward the side with the slower propeller.

Another way large ships can be turned, especially in docking, is with ***bow thrusters.*** A bow thruster is a propeller that is mounted

Figure 25-23. Safe channels are marked by lighted buoys and lighthouses.

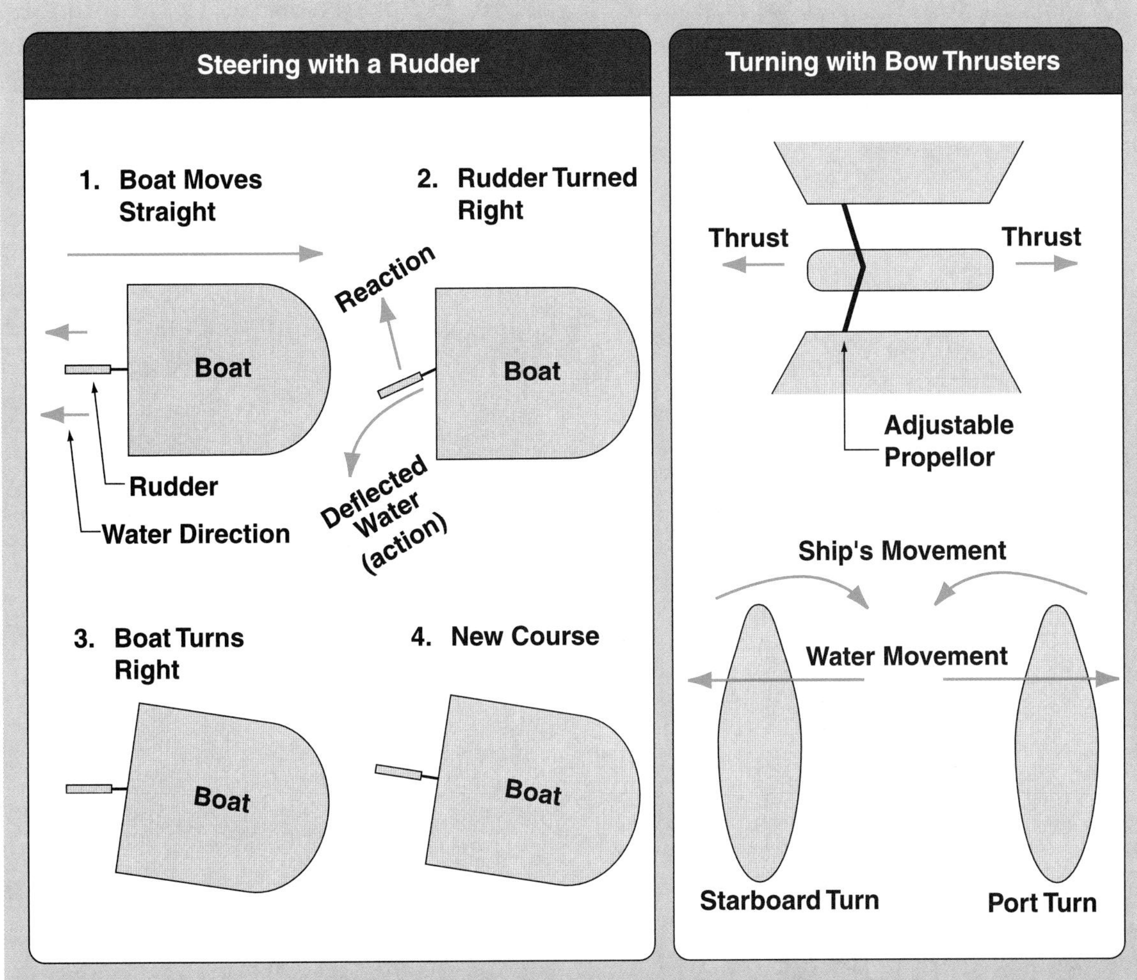

Figure 25-24. Boats and ships use rudders to change their direction. Some large ships have bow thrusters to make small changes while docking.

at a right angle to the keel. The blades of the propeller can be adjusted to provide thrust in either direction. The ship will turn opposite the thrust.

Sailboats are controlled by a combination of rudder and sail. The rudder uses the force of the water to turn the boat. Sails use forces produced by the wind to aid the rudder in guiding the boat.

Air Transportation Vehicles

The newest transportation vehicle in widespread use is the airplane. In the past sixty years the airplane has changed from a vehicle for the wealthy to an everyday transportation device.

Airplanes can be divided into two groups: general and commercial. ***General aviation*** is travel for pleasure or business on an aircraft owned by a person or a business. The airplane is not available for the use of the general public.

Commercial aviation includes the airplanes used by businesses that make money by transporting people and cargo. These companies include scheduled airlines, commuter airlines, air-freight carriers, and overnight package companies.

Structure

Three major types of air transportation vehicles exist, **Figure 25-25.** They are lighter-than-air, fixed-wing, and rotary-wing aircraft.

Lighter-than-Air Vehicles

Lighter-than-air vehicles use either a light gas (such as helium) or hot air to produce lift. These were the first air vehicles to be used by people. The earliest was a hot-air balloon developed in France in 1783. This balloon, like present-day hot-air balloons, was basically a fabric bag with an opening in the bottom. This opening allows warm air to enter the balloon and displace heavier, cold air. Because warm air is less dense than cold air, it rises to the top of the bag. When enough warm air has built up in the balloon, it rises off the ground.

Blimps and dirigibles are more advanced lighter-than-air vehicles. A ***blimp*** is a nonrigid aircraft, meaning it has no frame. The envelope (the bag that is filled with a light gas, usually helium) determines its shape. Slung beneath the envelope is an operator and passenger compartment. One or more engines are attached to the compartment to give the blimp forward motion.

Dirigibles are rigid airships with a metal frame that is covered with a skin of fabric. They used hydrogen gas to give maximum lift. Dirigibles were used for trans-Atlantic passenger service in the 1930s. Hydrogen is highly flammable, however, and several disastrous accidents occurred because of its use. As a result, dirigibles ceased to be used.

Fixed-Wing Aircraft

Today most passenger and cargo aircraft are ***fixed-wing aircraft.*** They all use similar structures. The flight crew, passenger, and cargo units are contained in a body called the ***fuselage.*** One or more wings attached to the fuselage provide the lift necessary to fly. A ***tail assembly*** provides steering capability for the aircraft.

As we noted in the last chapter, the Wright brothers developed the first successful fixed-wing aircraft. Their *Flyer* was first flown in 1903. Later, Charles Lindberg made his historic trans-Atlantic flight in a fixed-wing aircraft called the *Spirit of St. Louis,* **Figure 25-26.**

Rotary-Wing Aircraft

The helicopter is the most common ***rotary-wing aircraft*** in use today. The body of the craft contains the operating and cargo unit and encloses the engine. Above the engine is a set of blades. The blades are adjustable to provide both lift and forward and backward motion. The tail has a second, smaller rotor that keeps the body from spinning in response to the motion of the main rotor.

Helicopters are widely used by the military. In civilian life they are used for emergency transportation, law enforcement, and specialized applications. These special

Figure 25-25. The three types of aircraft are: lighter-than-air (left), fixed wing (center), and rotary wing (right). (Goodyear, Grumman Corp., Bell Helicopter)

jobs include communication tower erection, high-rise building construction, flying to remote areas, and aerial logging.

Propulsion

Aircraft use two major types of propulsion systems: propellers or jet engines. Smaller aircraft use propellers attached to internal combustion engines. The engine operates on the same principle as the automotive internal combustion engine. Some large aircraft use a variation of the jet engine to turn the propeller.

Most business and commercial aircraft are powered by one of three types of ***jet engines:*** turbojet, turbofan, and turboprop, **Figure 25-27.** The first type to be used was the ***turbojet engine.*** This engine was developed during World War II. It operates through the following steps:

Figure 25-26. These historic aircraft, the *Wright Flyer* (top) and the *Spirit of St. Louis* (bottom), are on display at the Smithsonian Institution.

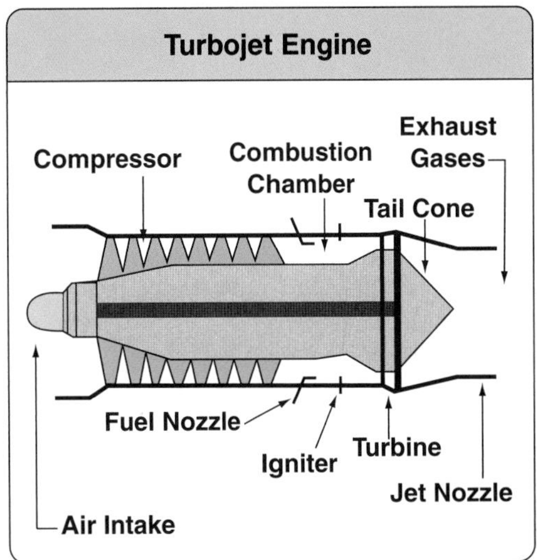

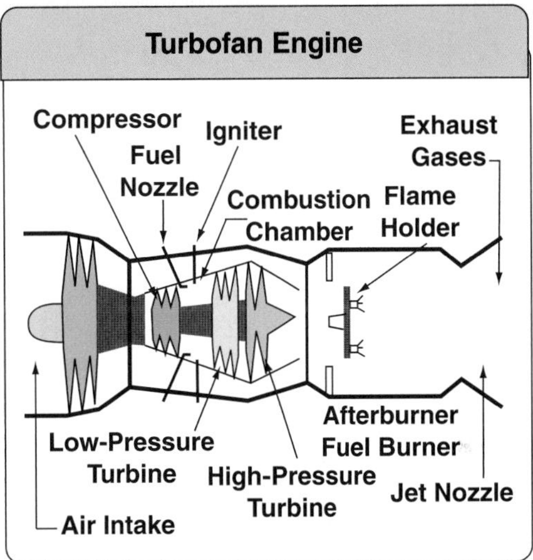

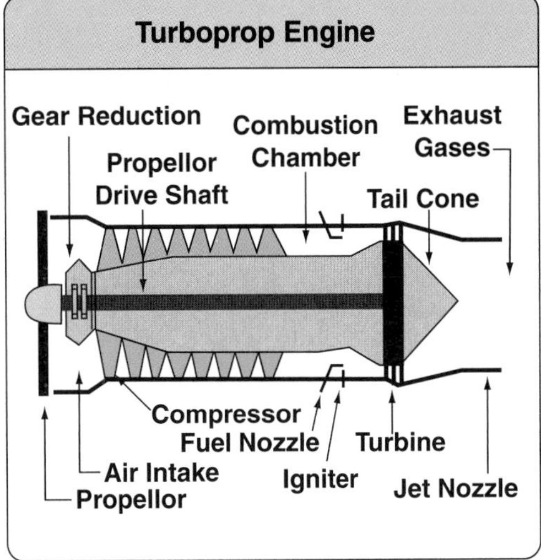

Figure 25-27. Three types of jet engines are in use today.

1. Air is drawn into the front of the engine.
2. Air is compressed at the front section of the engine.
3. The compressed air is fed into a combustion chamber.
4. The fuel is mixed with the compressed air.
5. The fuel-air mixture is allowed to ignite and burn rapidly.
6. As a result, the rapidly expanding hot gases exit the rear of the engine.

The exiting hot gases serve two functions. The gases turn a turbine that operates the engine compression section. The gases also produce the thrust to move the aircraft.

The turbojet engine operates at high speeds and is used in military aircraft. Early commercial jet airliners also used these engines.

The ***turbofan engine*** is the engine of choice for most commercial aircraft in use today. The turbofan operates at lower speeds than a turbojet engine. Also, a turbofan uses less fuel to produce the same power. In this engine the turbine drives a fan at the front of the engine. The fan compresses the incoming air. The compressed air is then divided into two streams. One stream of air enters the compressor section. In this section the air is compressed further, fuel is injected, and the fuel-air mixture is ignited. The other stream of air flows around the combustion chamber. This stream is used to cool the engine and reduce noise. In the rear section of the engine, the exhaust gases and cool air mix. If necessary, additional fuel can be injected and ignited to provide additional thrust. This arrangement is called an afterburner.

Another variation of the jet engine is the ***turboprop engine.*** It operates in the same manner as a turbojet engine. The turbine, however, also drives a propeller that provides the thrust to move the aircraft. Turboprop engines operate more efficiently at low speeds than turbojet or turbofan engines. Therefore, they are widely used on commuter aircraft.

A new type of jet engine has been developed also. It is called a propfan. The propfan differs from a turboprop in that two propellers are driven. The propellers rotate in opposite directions. Propfan engines promise to be fuel efficient while operating at high speeds.

Suspension

Air transportation vehicles are suspended in the atmosphere. Such a state requires knowledge of the principles of physics. All air transportation vehicles depend on the fact that an area with less dense air will allow heavier air to force the vehicle up. This force must be greater than the force of gravity that draws the craft toward the earth. As with structure, suspension varies according to the type of vehicle.

Lighter-than-Air Vehicles

Lighter-than-air vehicles use air weight differences to cause the vehicle to rise and be suspended in the atmosphere. Two ways are used to cause this difference in weight.

The first method uses a closed envelope filled with a light gas. As you learned earlier, blimps use helium and dirigibles used hydrogen to generate lift. The vehicle is made to rise and lower much like a submarine. Inside the vehicle are air tanks or bags. The tanks are filled with outside air to make the aircraft heavier. Therefore, the aircraft will descend. Removing the air from the tanks or bags will make the craft light, and it will rise.

The second system uses circulating, warm air in an open-ended envelope to cause the craft (the hot-air balloon, **Figure 25-28**) to ascend and descend. The balloon has a burner suspended below its opening.

To cause the balloon to rise the burner is ignited. The flame heats air, which rises into the balloon. The warm, rising air displaces colder, heavier air. The warmer air is less dense and makes the balloon lighter than the air it displaces in the atmosphere. Therefore, it rises until its displaced air is equivalent to its weight. As the balloon rises, it enters air that is less dense (lighter).

Also, the air in the balloon cools and becomes heavier. Therefore, the burner must continue to operate to keep the balloon flying. To descend to earth the operator turns the burner off. During flight the operator uses the burner intermittently (turns it on and off) to keep the balloon at a constant height. Running the burner all the time might cause the balloon to rise too high. Not running the burner often enough will allow the balloon to descend too low over the earth.

Fixed-Wing Aircraft

As noted earlier, the most common aircraft is the fixed-wing airplane. An airplane has four major forces that affect its ability to fly. These forces, as shown in **Figure 25-29,** are:

- Thrust: The force that causes the aircraft to move forward
- Lift: The force that holds or lifts the craft in the air
- Drag: The air resistance force that opposes the vehicle's forward motion
- Weight: The pull of gravity that causes the craft to descend

Critical for all flight is ***lift.*** It is generated by air flowing over the wing of the aircraft. The *wing* is shaped to form an ***airfoil,*** **Figure 25-30.** The wing separates the air into two streams. The airfoil shape causes the upper stream to move farther than the

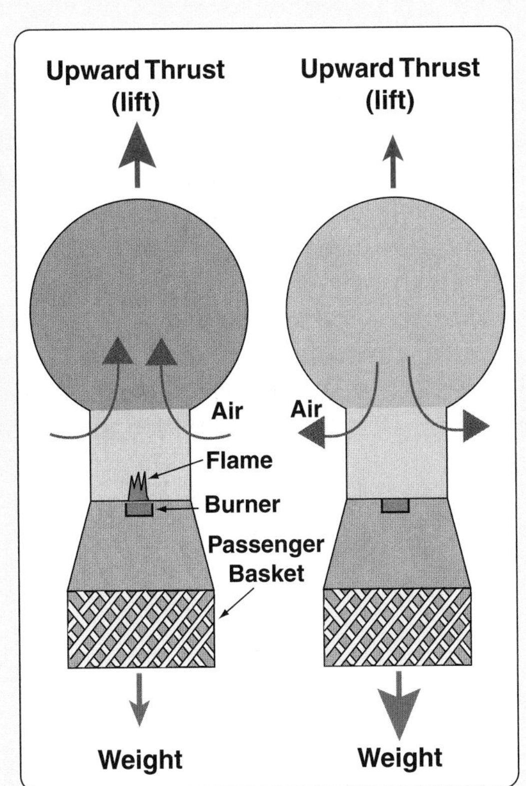

Figure 25-28. A hot-air balloon uses warm air to produce lift.

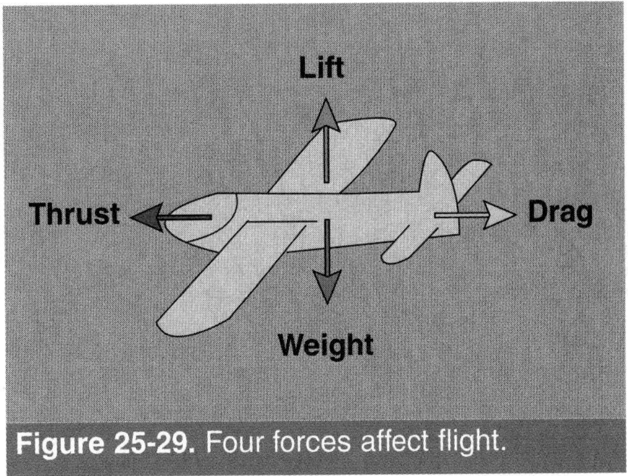

Figure 25-29. Four forces affect flight.

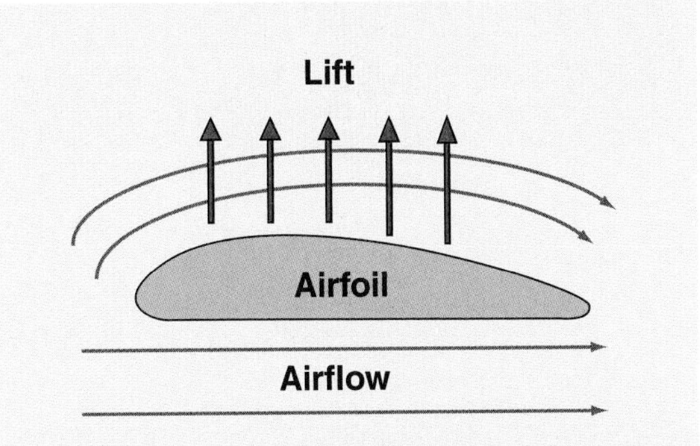

Figure 25-30. Air flows a greater distance over the top of an airfoil and produces lift.

lower stream. Therefore, the upper stream speeds up. This increased speed causes a decrease in pressure. The high-pressure air below the wing forces the wing up. This gives the plane the required lift.

The greater the slope of the upper surface of the wing, the greater the lift. Note, however, that drag is also increased because of the larger front profile of the wing. An airplane needs greater lift during takeoff and landing. Devices called flaps on the leading and trailing edges of the wings are extended, **Figure 25-31.** The flaps increase the lift. They are also used to slow the aircraft during landings.

Rotary-Wing Aircraft

The blades on a helicopter operate like any other airfoil. As they rotate, they generate lift. The angle of the blades is increased to generate lift. This will cause the helicopter to rise vertically. When the proper altitude is reached the pitch (angle) of the blades is changed. The pitch is adjusted so that the lift equals the weight. When the weight equals the lift, the helicopter will *hover.* Reducing the pitch will allow the helicopter to descend.

Forward flight is accomplished using a complex mechanism. This mechanism changes the pitch of the blades and tilts the blades as they rotate. In forward flight, the blade is tilted slightly forward. Also, the pitch of the blade is increased in the rear part of their travel. The pitch is reduced as they rotate in the front part of their travel. This causes the craft to have more lift in the back than in the front. Therefore, the helicopter leans slightly forward. The combination of the lift and thrust of the rotors causes the helicopter to move forward.

Backward flight is accomplished by reversing the tilt. The rotors are pitched the most in the front part of their travel. This causes the craft to tilt slightly backward and travel in reverse. Helicopters can move left and right in the same way, by tilting the rotor and adjusting the pitch.

Guidance and Control

Guidance for the pilot of a commercial aircraft comes in many forms. Ground personnel help the pilot bring the aircraft into the terminal safely, **Figure 25-32.** Control tower personnel are in radio contact as the airplane taxies from the terminal and is cleared for takeoff. Once in the air, regional air traffic controllers take over. They monitor and direct the aircraft's progress across their region and hand the craft off to controllers in adjoining regional centers as the vehicle leaves their region. Operators of general aviation aircraft do not normally use regional air traffic control centers. They use visual flight rules to govern their movements.

Various instruments help pilots monitor and control the aircraft properly, **Figure 25-33.** These instruments are called ***avionics.*** This term is derived from the

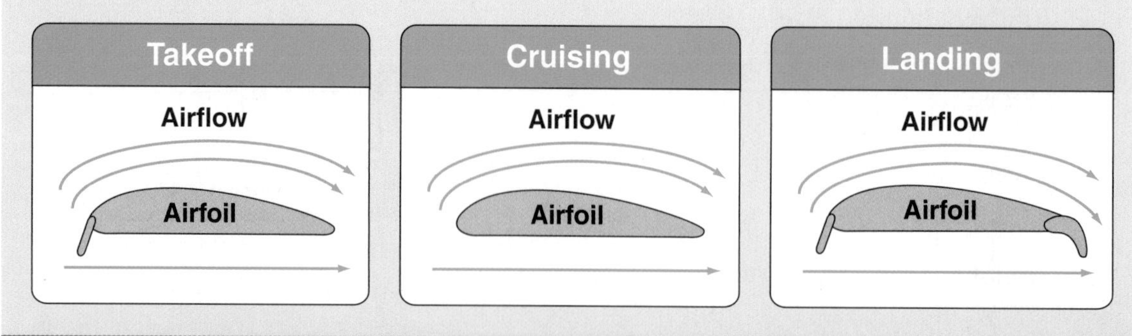

Figure 25-31. These drawings show the airflow over the wing during takeoff, cruising, and landing. Devices called flaps are extended during takeoff and landing. The flaps increase the lift.

Figure 25-32. This ramp attendant is helping a pilot guide an aircraft to the jetway, a passenger walkway.

Figure 25-33. This passenger jet cockpit contains a number of instruments. (Alaska Airlines)

words *avi*ation electr*onics*. It includes all electronic instruments and systems that provide navigation and operating data to the pilot. The amount of avionics an airplane carries depends on its size and cost. Large passenger airliners have a wide variety of avionics. A small general aviation airplane will have just basic equipment.

Avionics include various functions and systems. These functions and systems are:

- Communication systems: Short-distance radio systems operate on frequencies between 118 and 135.975 MHz. These radios allow pilots to communicate with air-traffic controllers and other people on the ground.
- Automatic pilot: The automatic pilot is an electronic and computer-based system that monitors an aircraft's position and adjusts the control surfaces to keep it on course.
- Instrument landing system: This system (called ILS) helps airplanes to land in bad weather. ILS uses a series of radio beams to help the pilot guide the airplane to the runway. A wide, vertical beam helps the pilot stay on the center of the runway. A wide horizontal radio beam tells the pilot if the plane is above or below the correct approach path. Vertical beams serve as distance markers to tell the pilot how far away he is from the runway. Pilots in aircraft without ILS equipment use approach lighting to guide the plane in for landing, **Figure 25-34.**
- Weather radar: This system provides pilots with up-to-date weather conditions in front of the aircraft.
- Navigation systems: These systems guide the plane along its course and indicate air speed.
- Engine and flight instruments: These instruments give pilots information about how the engine or engines are operating. Flight instruments tell the direction (in three degrees of freedom) the airplane is flying.

Space Transportation Vehicles

Space travel may be the transportation system of the future. Today, however, its role is limited. Space travel is restricted to conducting scientific experiments and placing communication, weather, and surveillance satellites into orbit.

Figure 25-34. This photo shows a jet landing over runway approach lighting. (United Airlines)

Types of Space Travel

Space travel can be classified in two ways:

- Unmanned or manned flight
- Earth orbit or outer space travel

Unmanned or Manned Flight

The first space flights were ***unmanned space flights.*** These flights used rockets to place a payload into orbit. The term *payload* is usually used to describe the items for which a flight is made. Typically, the payload was either a scientific experiment or a communication satellite, **Figure 25-35.** Today, unmanned space vehicles continue to launch these and other payloads.

A ***manned space flight*** carries human beings into space and returns them safely to the earth. The first manned space flight was flown on April 12, 1961, when a Soviet cosmonaut named Yuri Gagarin made one orbit of the earth. Manned space flight in the United States also started in 1961. Astronaut Alan Shepard completed a suborbital flight less than one month after Gagarin's flight. This was followed by John Glenn's first orbital flight in the Mercury program. The Mercury program was followed by the Gemini and Apollo programs, which ended with the first man on the moon, **Figure 25-36.** All these programs used capsules attached to the nose of a rocket to place humans into orbit or into outer space.

The latest manned flights are carried out by space shuttles. A shuttle rockets into orbit, where its crew carries out experiments and places satellites into orbit, **Figure 25-37.** At the end of the mission the shuttle returns to earth. The shuttle glides in for a landing like an airplane. After landing, the shuttle is prepared for its next flight.

Figure 25-35. Satellites are a typical payload for unmanned space flights.

Figure 25-36. This model of the moon lander is on display at the Smithsonian Institution.

Earth Orbit and Outer Space Travel

The two types of space travel are earth orbit and outer space travel. ***Earth orbit travel*** is fulfilled by communication satellites and space shuttles. They travel at about 18,000 miles per hour in an elliptical orbit around the earth. This orbit comes *closest* to earth at a point called the ***perigee.*** The *farthest* distance away from earth is called the ***apogee.***

Satellites can be placed into two types of orbits. One type allows the satellite to circle the earth, viewing a complete path around the globe. This type of orbit can be used for such objectives as weather monitoring, geological and agricultural surveys, and military surveillance.

The second type of orbit is called ***geosynchronous orbit.*** In this type of orbit the satellite travels the same speed the earth is turning. Therefore, the satellite stays over a single point on the globe. This type of orbit is used for communication and weather satellites.

Outer space travel places the spacecraft in earth orbit first. Then the speed of the spacecraft is increased to more than 25,000 mph. At this speed the spacecraft is "thrown" into a path that will cause it to move out of the earth's gravitational field. The spacecraft can then travel to distant planets and out of the solar system. *Voyager 1* and *Voyager 2* are examples of this type of spacecraft. They moved out of earth orbit and traveled past a number of planets. They are now traveling out of the solar system into what is called deep space.

Structure

A space vehicle and its launch systems have three major parts. These parts are a rocket engine to place the vehicle or payload into space, an operating section, and a cargo and passenger compartment. As you have learned, early space programs integrated these components into a single launch vehicle or rocket. The engine was at the rear. The operating controls were placed above and around the engine. The cargo or passenger capsule formed the nose of the rocket.

The space shuttle is a newer type of space vehicle. It uses two distinct systems:

- Two solid fuel rockets strapped to the external fuel tank
- The shuttle orbiter

The solid fuel rockets lift the orbiter off the launch pad and give it initial acceleration. After they exhaust their fuel, they fall away. The solid fuel rockets parachute into the ocean and are recovered to be used again. At this point the three engines in the orbiter take over. They use the fuel in the external tank to place the shuttle into orbit. As it enters orbit, the external fuel tank falls away and burns up as it falls toward the earth.

The shuttle is about the size of a small airliner (such as a Boeing 717)—122′ (37 m) long with a 78′ (24 m) wingspan. Its crew operates it from the flight deck at the front of the orbiter. The middle of the shuttle is a large cargo bay. It can carry nearly 30 tons (27 metric tons) into orbit and almost 15 tons (13.5 metric tons) on re-entry.

Propulsion

Space transportation depends on rocket propulsion systems. The rocket is based on

Figure 25-37. This illustration is an artist's representation of a space shuttle experiment. (NASA)

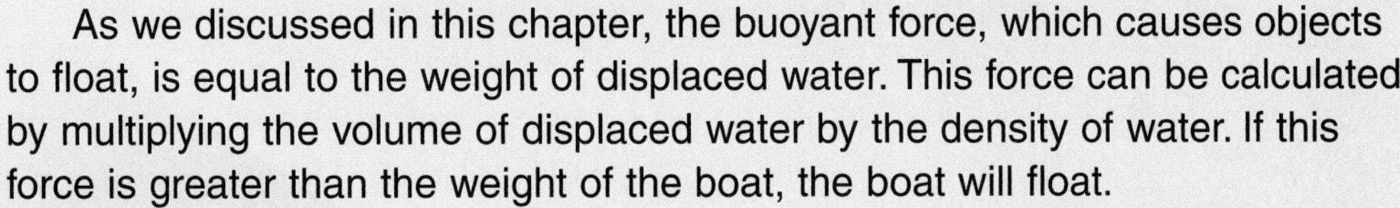

As we discussed in this chapter, the buoyant force, which causes objects to float, is equal to the weight of displaced water. This force can be calculated by multiplying the volume of displaced water by the density of water. If this force is greater than the weight of the boat, the boat will float.

For example, a fully loaded flat-bottom boat weighs 50,000 lbs. The boat is 20′ long, 8′ wide, and 6′ deep. To determine whether the boat will float, we need to calculate the weight of the displaced water and compare it to the boat's weight. First, we calculate the volume of the boat:

Volume = length x width x depth
= 2′ x 8′ x 6′
= 960 ft^3

To determine the weight of displaced water, we would then multiply the volume of the boat by the density of water. The density of water is 62.4 lbs/ft^3 (pounds per cubic foot):

Weight = volume x density
= 960 ft^3 x 62.4 lbs/ft^3
= 59,904 lbs.

The buoyant force (59,904 lbs.) is greater than the weight of the boat (50,000 lbs.). Therefore, the boat will float.

If a fully loaded flat-bottom boat is 25′ long, 10′ wide, and 6′ deep and weighs 75,000 lbs., will it float? Why or why not?

Newton's third law of motion that we discussed earlier (for every action there is an equal and opposite reaction). Thus, rocket engines can be called *reaction* engines. The rocket applies this principle by:

- Burning fuel inside the engine
- Allowing pressure from the exhaust gases to build up
- Directing the pressurized gases out of an opening at one end of the engine

This action produces motion that is in the opposite direction from the exiting gases. If you have ever blown up a balloon and allowed it to fly around as the air escapes, you have seen the principle of a rocket engine.

The fuel used in rockets is either liquid or solid. Early rockets were developed in China and India before 1800. The developers used gunpowder as a solid fuel. The first modern liquid fuel rocket was designed and built by Robert Goddard in 1926.

Today's ***liquid fuel rockets*** have two tanks. One contains the fuel or propellant. The other contains oxygen—the oxidizer. The fuel and oxidizer are fed into the combustion chamber. There they combine and burn to generate the thrust needed to propel the rocket and its payload into orbit.

Liquid fuel rockets have several distinct advantages. First, the amount of thrust can be controlled by the amount of fuel and oxidizer fed into the engine. Second, a

Career Corner
Railroad Conductors

Railroad conductors coordinate the activities of train crews. Freight train conductors review schedules, switching orders, waybills, and shipping records to obtain cargo loading and unloading information. Passenger train conductors ensure passenger safety and comfort as they go about collecting tickets, making announcements, and coordinating passenger service activities.

Conductors might be expected to work nights, weekends, and holidays because trains operate around the clock, seven days a week. The more senior conductors receive the more desirable shifts. Railroads require that applicants have a minimum of a high school diploma. Almost all railroad workers begin working as yard laborers and, later, might have the opportunity to train to become conductors.

liquid fuel rocket engine can be used intermittently. Finally, the rocket engine can be recovered and reused after the space flight.

Solid fuel rockets use a powder or a spongelike mixture of fuel and oxidizer. Once the mixture is ignited, it burns without outside control. Therefore, the thrust cannot be changed. Also, the mixture will burn completely; it cannot be stopped once it is started.

Areas of Operation

Spacecraft and satellites operate in several regions. The lowest is called the ***troposphere,*** which includes the first 6 miles (9.7 km) of space above the earth. General aviation and commuter aircraft operate in this region.

Above this region is the ***stratosphere.*** This region extends from 7 to 22 miles (11 to 35 km). Commercial and military jet aircraft operate in the lower part of this region. The upper part of the stratosphere is called the ***ozone layer.*** This layer absorbs much of the sun's ultraviolet radiation. Evidence of damage to this layer has caused great concern about global warming and the health of the planet.

The next layer is the ***mesosphere.*** This region extends from 22 to 50 miles (35 to 80 km) above the earth. The thermosphere or ***ionosphere*** lies just above the mesosphere. It ranges from 50 to 62 miles (80 to 99 km) above the earth. Many satellites operate in this layer of the atmosphere. The last layer is called the ***exosphere*** and blends directly into outer space.

Summary

Transportation involves all actions that move people and goods from one place to another. Today four types of transportation exist: land, water, air, and space. All these systems use vehicles to move the people or cargo. Each of these vehicles has structural, propulsion, suspension, guidance, and control systems. The structure is the physical frame and covering. Structural systems provide spaces and protection for people, cargo, and other devices. The propulsion system uses energy to create power for motion. Suspension systems support the weight of the vehicles and cargo. Control systems affect the speed and direction of vehicles. Guidance systems provide information for control.

Key Terms

airfoil
apogee
automatic transmissions
avionics
ballast
blimp
bow thrusters
buoyancy
commercial aviation
commercial ships
container ships
control system
degrees of freedom
direction control
dirigibles
dry-cargo ships
earth orbit travel
exosphere
fixed-wing aircraft
fuselage
general aviation
geosynchronous orbit
guidance system
hovercraft
hydrofoil
inland waterway transportation
internal combustion engine
ionosphere
jet engines
lift
lighter-than-air vehicles
liquid fuel rockets
locomotive
manned space flight
maritime shipping
mechanical transmissions
merchant ships
mesosphere
military ships
outboard motor
ozone layer
passenger ships
perigee
pleasure craft
power generation system
power transmission system
propulsion system
rotary-wing aircraft
rudder
solid fuel rockets
speed control
stratosphere
structural system
submersible
suspension system
tail assembly
tankers
tractor
trailers
troposphere
turbofan engine
turbojet engine
turboprop engine
unmanned space flights
vehicle
wing

Test Your Knowledge

Write your answers on a separate piece of paper. Please do not write in this book. For Questions 1–8, match the vehicular system on the right to the correct description on the left. (Note: Some letters will be used more than once.)

Description

1. _____Involves degrees of freedom
2. _____A speedometer is one example
3. _____Includes the physical frame
4. _____The wing on an airplane is an example
5. _____A bicycle pedal is one example
6. _____Provides spaces for cargo
7. _____The hull of a ship is an example
8. _____Involves direction the vehicle moves

Vehicular System

a. Structure
b. Propulsion
c. Suspension
d. Guidance
e. Control

9. What are the three structural units common to all land transportation vehicles?
10. The most common engine in land transportation vehicles is the _____ engine.
11. *True or false?* Most land transportation vehicles are controlled manually by an operator.
12. Ships that carry cargo in sealed steel boxes are called _____.
13. Ships and barges float because of the scientific principle of _____.
14. *True or false?* Friction plays a role in the suspension system of a hydrofoil.
15. What is a fuselage?
16. List the three major types of jet engines.
17. What are the four main forces that affect a fixed-wing airplane's ability to fly?
18. A satellite that travels at the same speed that the earth is turning is said to be in _____ orbit.
19. What are the two major systems that comprise a space shuttle?
20. *True or false?* Rocket propulsion systems are based on Newton's first law of motion.

Applying Your Knowledge

1. Using a rubber band for power, design a land vehicle that has all five vehicle systems. Analyze the vehicle and explain the features of each system.
2. Build and fly either a model airplane or a model rocket.
3. Design a cargo container for a raw egg that can withstand the impact of a 15′ (4.6 m) fall.

Chapter 26
Operating Transportation Systems

Learning Objectives

After studying this chapter, you will be able to:

- Describe how the speed of transportation has increased over time.
- Explain the differences between personal and commercial transportation systems.
- Identify the common components of all transportation systems.
- Name the major actions of transportation systems.
- Describe maintenance and repair activities involved in transportation systems.
- Identify various regulatory agencies that have an effect on transportation systems.

For each of us transportation is an important part of everyday life. When we walk we are using a common form of transportation. We might also use many different kinds of transportation technology. We use vehicles and systems to improve our ability to move from one place to another.

Technology has increased the speed of moving cargo and relocating people. Over time a great change has occurred in the speed of travel, **Figure 26-1.** In early times, for example, we depended on our ability to walk. A human can walk at an average of one mile per hour (1.6 km/h) over a great distance. Later, animals pulled wagons, and sails captured wind to push boats. These advancements increased the travel speed by nearly three times. The development of canals and steamships also made travel faster. The advent of the railroad multiplied this speed by twentyfold to 50 mph (80 km/h). Today, jet aircraft move people at about 600 mph (966 km/h), and satellites move at 18,000 mph (28,968 km/h).

The advancements in the forms of travel and in the rates of speed have increased the kinds of operations involved in different transportation systems. In this chapter we will look at the types of transportation and various aspects of their successful operation.

Types of Transportation

Through the course of their lives, most people use two types of transportation. These, as shown in **Figure 26-2,** are:

- Personal transportation
- Commercial transportation

Personal transportation is travel using a vehicle owned by one person. The most common personal transportation vehicle is the automobile. Other typical personal vehicles are bicycles, motorcycles, mopeds, and skateboards.

Personal transportation vehicles are very flexible. They allows us to travel when we want and can go almost anyplace. Personal transportation (except for bicycles) is not always energy efficient, however. A car burns the same amount of fuel whether one or five people ride in it. We often drive around alone in our cars. The passenger seats are empty and, therefore, are underutilized.

Most personal transportation systems use public routes. These pathways are built using public funds generated by taxes and fees. Most often these pathways are the streets, roads, and highways that crisscross the nation.

Commercial transportation includes all the enterprises that move people and goods for money. These are the land, water,

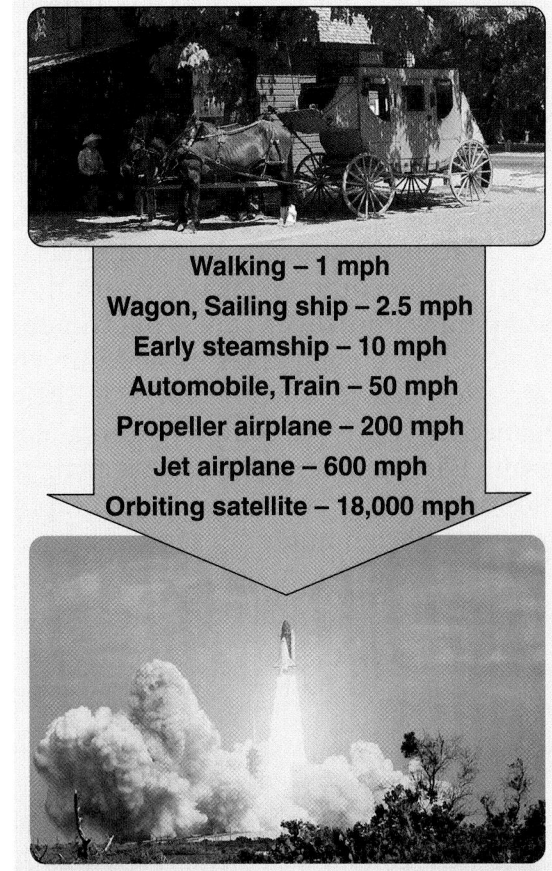

Figure 26-1. The speed of transportation has increased throughout history. (NASA)

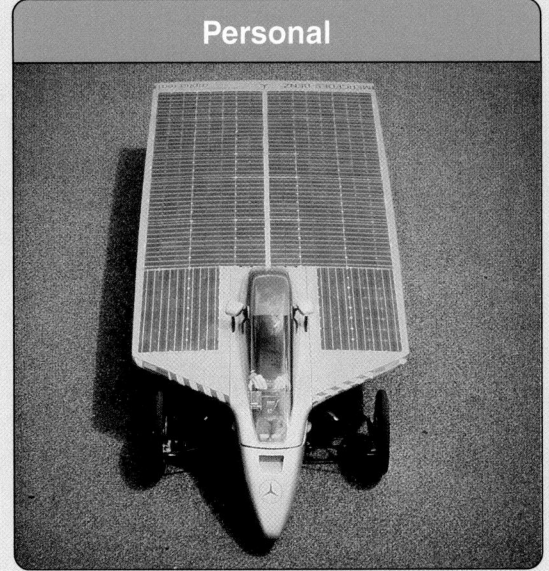

Figure 26-2. The solar-powered automobile may be the personal transportation vehicle of the future. The cruise ship is an example of a commercial vehicle. (DaimlerChrysler)

and air carriers that operate locally, nationally, and internationally. These carriers include taxi, bus, truck, rail, ship, and air transport companies. Most are profit-centered enterprises that use private funds to finance their initial operation. These companies are owned by people who expect to make a profit on their investment. Some commercial transportation companies, however, are government owned. These companies provide transportation services where there is little chance for a private company to make a profit. For example, public transit companies operate city and regional bus, rapid transit, and commuter rail services.

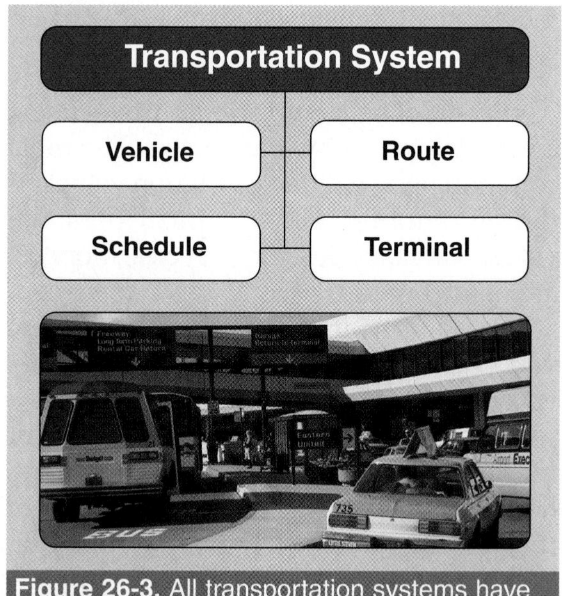

Figure 26-3. All transportation systems have vehicles, routes, schedules, and terminals.

Components of a Transportation System

All commercial transportation systems share some common elements, **Figure 26-3.** They all have:

- A vehicle
- A route the vehicle travels from the origin to the destination
- An established schedule for the movement of people and goods
- Terminals at the origin and destination points of the system

Vehicles

You studied vehicles in Chapter 25. You learned that vehicles are powered carriers that support, move, and protect people and cargo. You also learned that all vehicles have structure, propulsion, suspension, guidance, and control systems.

Transportation Routes

All transportation vehicles move from a point of origin to a destination. The path a vehicle follows is called a ***route.*** Personal transportation vehicles follow individual routes. These routes depend on the destination and the purpose of the trip. For example, a person hurrying to get to work might follow the most direct route or the one having the shortest travel time. On the other hand, a person on vacation may take a less direct, but more scenic route.

Commercial transportation vehicles often travel on *specific routes.* This means a vehicle follows one path from its origin to its destination, **Figure 26-4.**

Typically, transportation systems are made up of many individual routes. These routes are designed to collect vehicle traffic. To understand this concept, consider the road system shown in **Figure 26-5.** These roads are typical of those found in the midwestern United States. In this part of the country there are often roads on each section line. A section is a one-mile square piece of land. The smallest of these paths gives access to individual parcels of land, homes, and farms. These paths may be dirt, gravel or paved roads. This series of local access roads feeds into collector roads. Collector roads are larger pathways designed to carry more traffic. The collector roads feed into arterial roads. These are

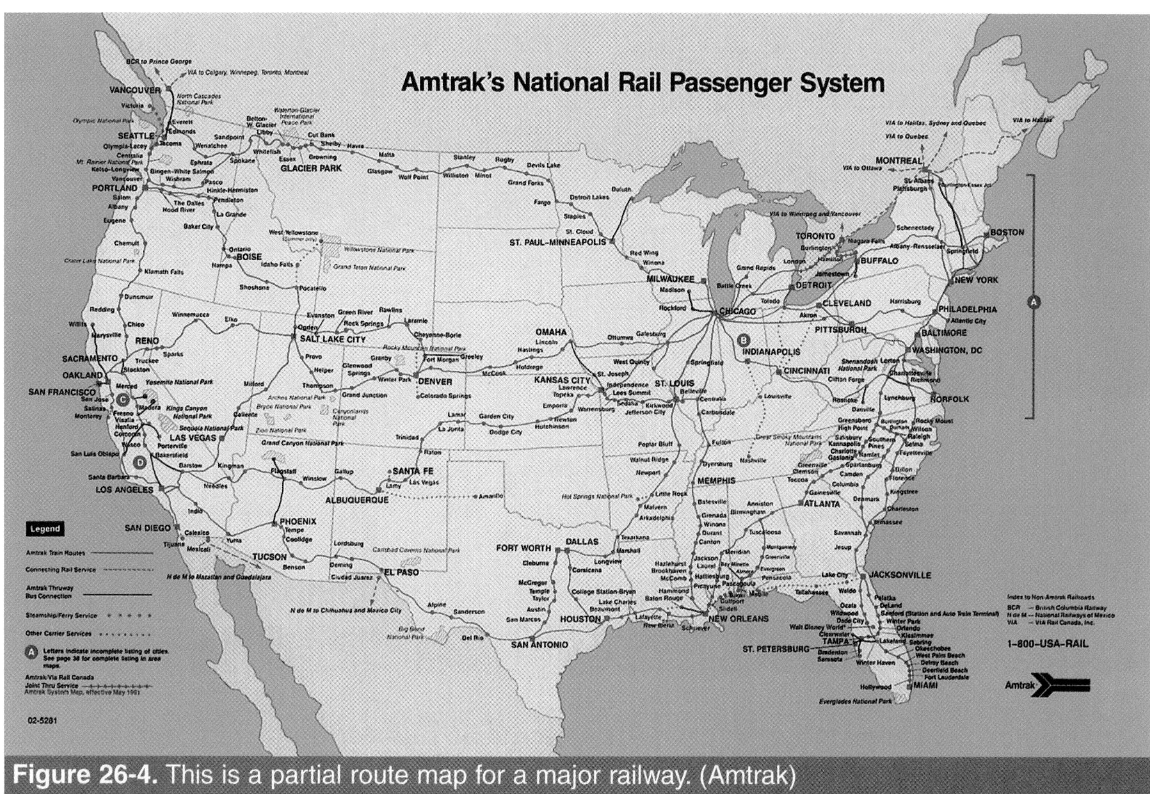

Figure 26-4. This is a partial route map for a major railway. (Amtrak)

high-speed highways that connect cities and towns. Some highways are *limited access highways.* On this type of highway, vehicles can enter and leave only at interchange points. Thus, the access to some land along these highways is limited. Larger U.S. highways and the interstate highway system are good examples of arterial highways.

Airlines use a route pattern called a ***hub-and-spoke system,*** **Figure 26-6.** Note that these routes are less apparent because they are pathways through the sky. A hub-and-spoke system is made up of small local airports and large regional airports. The local airports are on the end of two-way routes that radiate from the regional airport. Their routes appear to be spokes attached to a hub of a wheel.

The local airport serves as collecting and dispersing points for the system. The passengers normally arrive at a local airport using one or more transportation modes. They may arrive by personal car, taxi, shuttle van, bus, or train.

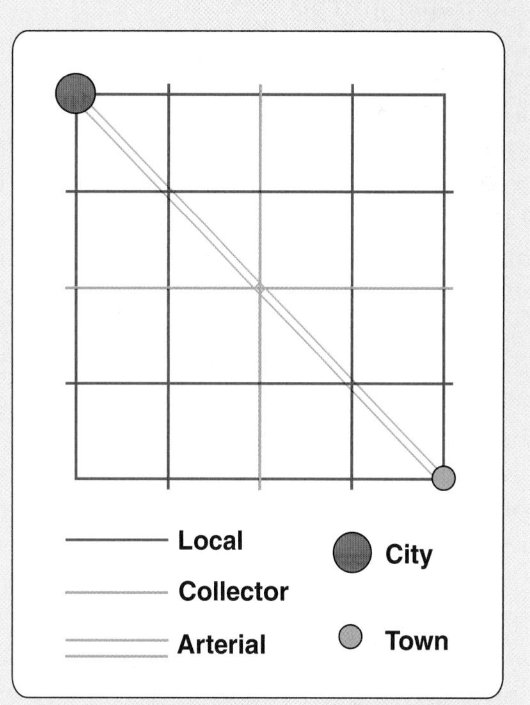

Figure 26-5. This map shows how the various parts of a highway transportation system feed into each other to provide local access and connections.

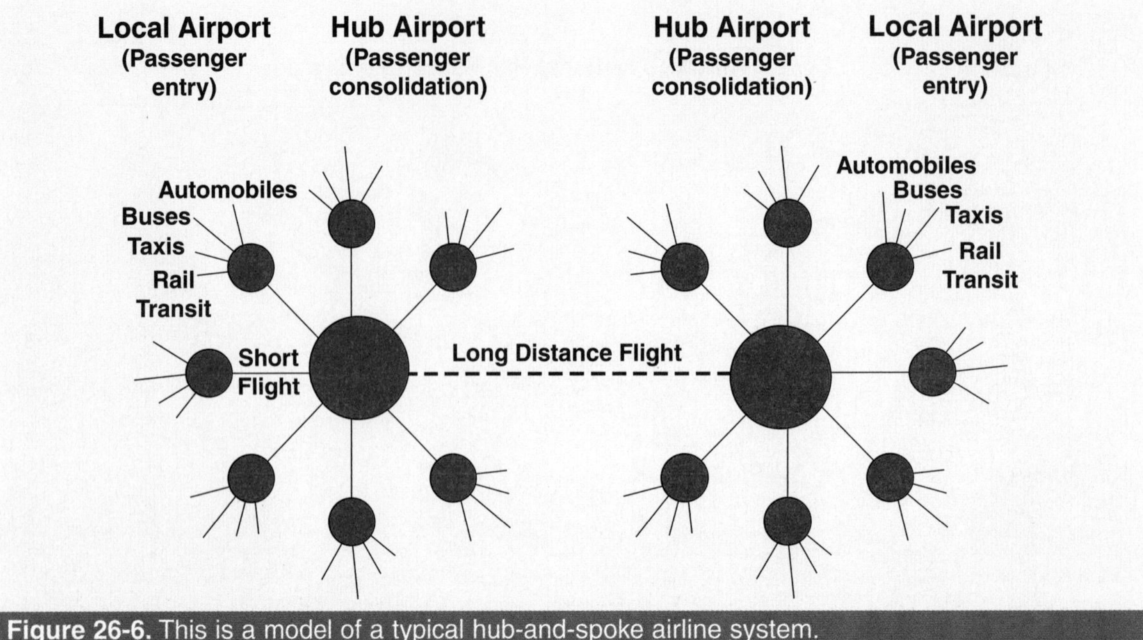

Figure 26-6. This is a model of a typical hub-and-spoke airline system.

At scheduled times, flights leave the local airport and travel to the hub airport. Small- and medium-sized jet aircraft are used, or these routes are flown by small, commuter airlines.

The passengers arriving at the hub airport can transfer to other flights that travel to their destination. The simplest hub-and-spoke systems use one hub. For example, a wave of aircraft may arrive from points to the east and south of the hub. The same airplanes will then be used for flights to the west and north. Later, the flow is reversed with arrivals from the west and north and departures that travel to the east and south.

More complex systems tie two or more hubs together. Some passengers use one of the hub airports to travel within the region of the country it serves. Other passengers use both hubs. They fly into one hub and then take a flight that connects it to a second hub across the country. The final flight segment will take them from the second hub to their destination. This system also allows the hub city to be served from a number of directions.

Often the passenger who transfers at a hub never leaves the airport. For example, O'Hare International Airport in Chicago is a major hub terminal. Chicago is also a major commercial and industrial city. Some passengers arriving at O'Hare will stay in Chicago for personal or business reasons. More than half of the 60-million-plus passengers who use the airport each year, however, never step foot outside of the airport. They simply arrive on one flight and leave on another.

Airlines are not the only transportation systems using hub-and-spoke type systems. Trucking lines use small trucks to pick up and deliver freight. These vehicles work from a local terminal that collects and dispenses cargo for a city or local region. The freight might then move to a regional terminal where it is sorted for long-distance transfer on large cross-country trucks. At the other end the freight is separated into smaller shipments. These shipments are delivered to local terminals and on to their final destination.

Parcel services such as Federal Express and United Parcel Service also gather and group shipments using the hub-and-spoke principle. Likewise, many railroads and interstate bus systems use this principle.

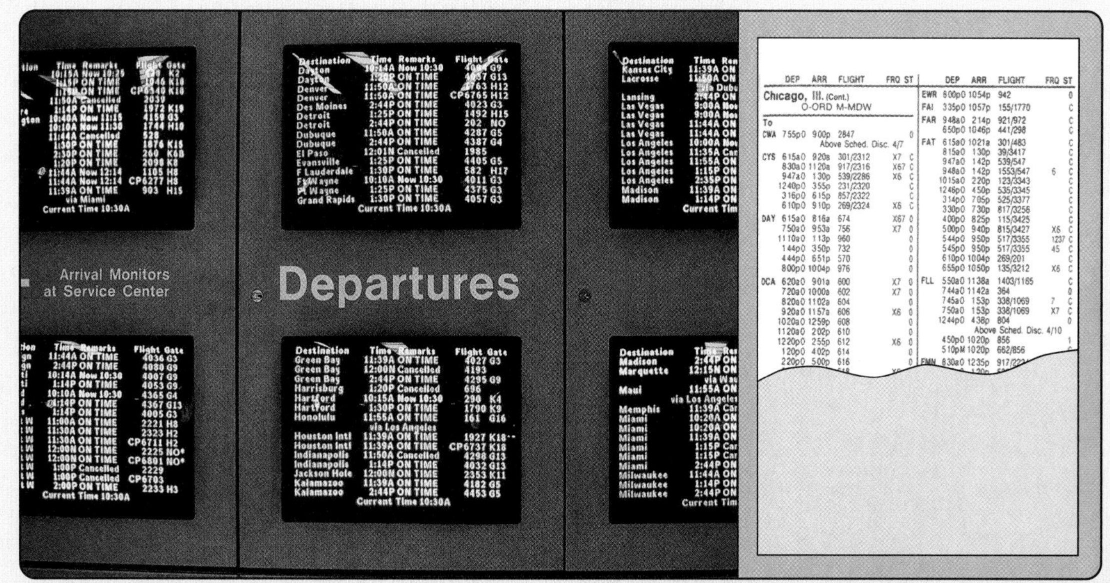

Figure 26-7. A schedule might be displayed on television monitors in a terminal (left) and contained in a printed schedule (right). (United Airlines, Bruce Kincheloe)

Transportation Schedules

People travel for various reasons. They may be going to attend a conference, to visit other people, to participate in a business meeting, or to watch a sports event. Therefore, the arrival time is very important. Travelers will select a transportation mode by how long the trip takes and the expected arrival time.

This information is presented in transportation schedules. ***Schedules*** list the departure and arrival times for each trip. These schedules are available for all commercial transportation systems. These include freight hauling and passenger travel on bus, train, or airline systems.

The information is often presented in printed form. Schedules for passenger travel are often displayed on television monitors or on schedule boards at the terminal, **Figure 26-7.**

Transportation Terminals

Terminals are important parts of commercial transportation systems. ***Terminals*** are places where passengers and cargo are loaded onto and unloaded from vehicles. These structures are most often located at the origin and the destination of a transportation route.

Cargo systems have terminals that provide storage for goods awaiting shipment. They also have methods to load cargo onto transportation vehicles, **Figure 26-8.** Finally, they have spaces for vehicles to stop as they are loaded and unloaded. These places might be a dock for ships, an apron for aircraft, or a loading dock on a warehouse for trucks and railcars.

Passenger terminals are more complex than cargo terminals. People need and demand more attention and comfort. Many terminals resemble small cities. They include at least six different areas, **Figure 26-9:**

- Passenger arrival areas that allow vehicles from other modes to unload passengers. Also, parking lots and bus shuttle services are provided at many large terminals.
- Processing areas where passengers may purchase or validate tickets and check in luggage.

Figure 26-8. The water transportation terminal includes warehouses, storage yards, and a dock for berthing and loading/unloading ships. The rail terminal includes a rail yard for sorting and storing rail cars, warehouses for product storage, and a loading/unloading area. (Norfolk Southern)

- Passenger movement systems to allow passengers to move from ticket counters to loading areas. Such systems include moving sidewalks, escalators, and automatic people movers.
- Passenger service areas that provide food, gifts, and reading materials. Also, rest rooms and VIP (very important person) or frequent-traveler lounges are available.
- Passenger waiting areas where people can sit while waiting to board their vehicle.
- Passenger boarding areas and mechanisms that allow people to board the vehicle.

Many functions in passenger terminals occur behind the scenes. Some workers sort baggage for departing travel. Others return baggage to arriving passengers. System managers direct the arrival, loading, unloading, and departure of vehicles.

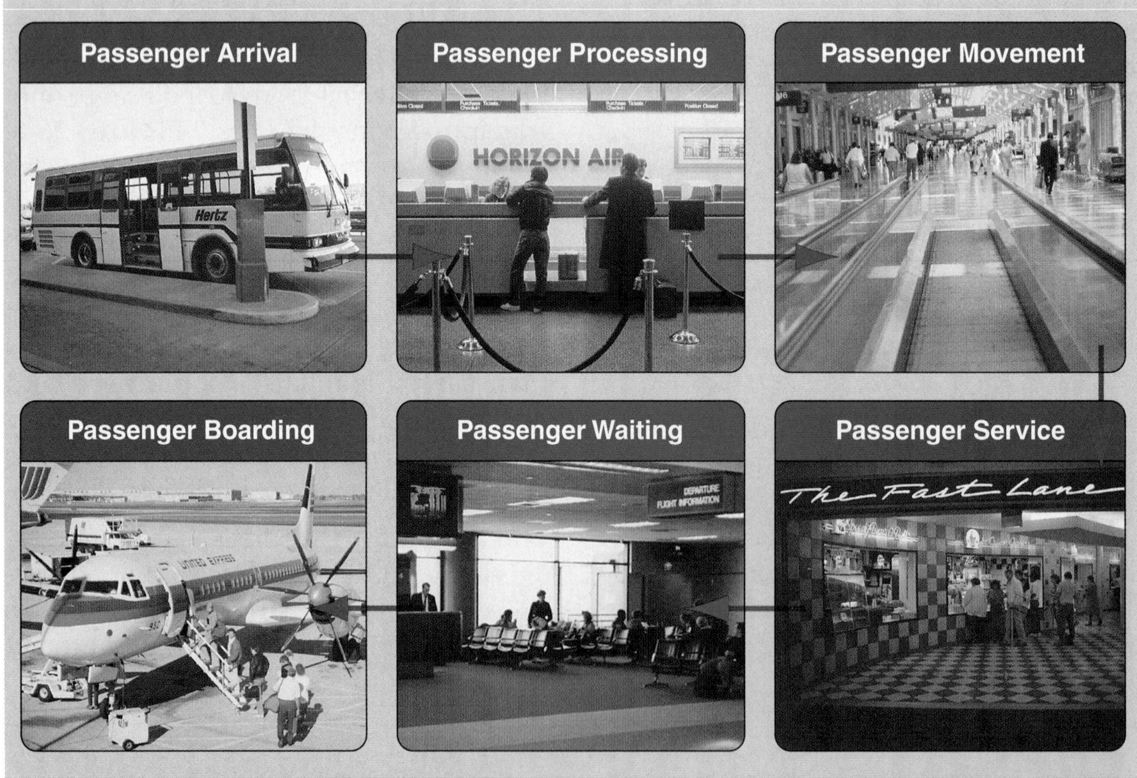

Figure 26-9. Passenger terminals serve a number of purposes.

Operating crews have areas to prepare for trips and to rest between assignments.

Transporting People and Cargo

You have been introduced to transportation vehicles, routes, schedules, and terminals. These things exist for one reason only: to move people and cargo from one point to another. This task requires three major actions, as shown in **Figure 26-10.** These actions are loading, moving, and unloading.

Loading

Loading involves placing cargo onto or allowing people to board a vehicle. It entails checking bills of lading (shipping tickets) for cargo or passenger tickets. Then, the cargo must be loaded or the people allowed to board the vehicle. During loading, cargo must be secured so that it does not shift during the trip. Unwanted movement can cause damage to the freight.

People must be securely seated before a land or air trip can begin. Air transportation regulations, especially, require that all passengers be properly seated. They must have their seat belts securely fastened before the plane can leave the terminal. Likewise, many states require seat belt use in automobiles. Bus and rail transportation systems do not normally have this requirement.

At the same time that the people and cargo are loaded, workers complete a number of other tasks, **Figure 26-11.** They check to be sure that vehicles receive scheduled service. They add fuel, check oil levels, deliver drinking water and food, and remove wastes. Also, they complete routine maintenance and minor repairs.

Moving

Moving people and cargo involves driving a vehicle, piloting a ship, or flying an airplane. Still, this is only part of the task. It also involves monitoring and controlling the progress of the vehicle. In air transportation, air traffic controllers talk to the pilot of the aircraft. In railroad systems, employees monitor the progress of a train and switch it onto appropriate tracks, **Figure 26-12.**

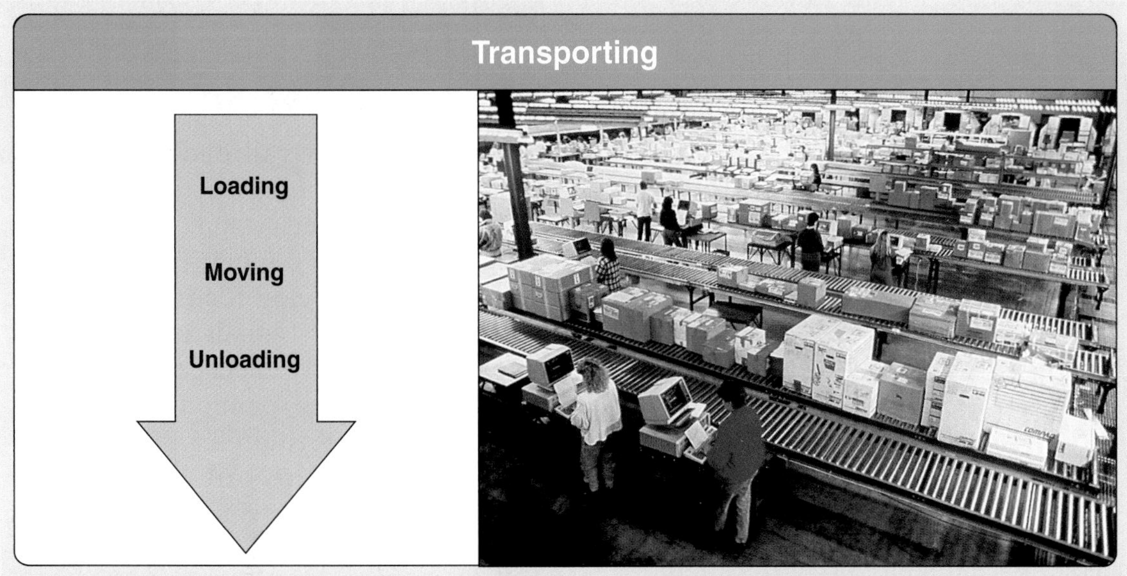

Figure 26-10. Transportation involves loading, moving, and unloading people and goods.

Figure 26-11. During the loading phase of a trip, passengers and cargo are loaded, the vehicle is serviced, and routine maintenance is performed. (United Airlines)

Unloading

Unloading passengers and cargo is the opposite of loading them. The passengers are allowed to get off the vehicle. Cargo is untied and lifted out of the vehicle.

Both people and cargo may enter the transportation system again by being loaded onto or allowed to board another vehicle. This might be a vehicle of the same or a different mode. For example, airline passengers might get off one plane and board another one at a hub airport. This is called "making a connection." Passengers might also leave the airport and board a bus or train or rent an automobile.

Figure 26-12. This railroad employee is using computer controls to monitor part of a rail system. The computer allows him to signal engineers and conductors on trains. He can also open and close switches to reroute the train onto specific tracks. (Norfolk Southern)

Maintaining Transportation Systems

Transportation systems are made up of a group of technological devices and structures. Each of these requires maintenance. Typically, the roadways, rail lines, navigational structures, and terminals are maintained using construction knowledge. It was this knowledge that was used to build the components and, therefore, is needed to maintain and repair them.

Structure Repairs

Roadway and railroad track repair ensures that support structures meet the requirements of the system, **Figure 26-13.** This kind of repair might involve patching holes and resurfacing roads or replacing rails and cross ties. Signals and signs must also be periodically checked, repaired, or replaced. Likewise, painted surfaces and stripes must be repainted.

Terminals also require repair of their functional parts. Their roofs will need to be periodically repaired. Exterior surfaces will need painting or re-siding. Entrances might need new steps and sidewalks. All these repairs ensure that the building can fulfill its function: to protect people and cargo.

Appearance is also important to passengers. They want and expect a clean, bright, and well-maintained terminal. Therefore, walls might be repainted or wall coverings replaced. New carpeting and lighting fixtures might be installed. Seating might be repaired or replaced.

Vehicle Repairs

Everyone wants to travel in comfort and in safety. This requires service and repair of vehicles and other equipment, **Figure 26-14.** You might remember from Chapter 18 that ***maintenance*** is the routine tasks performed to keep a vehicle operating properly. Generally, when a vehicle (bus, airplane, or train) reaches a terminal it receives daily service. Workers check lubricants, adjust controls and gages, and perform other simple tasks. They also carry out more extensive servicing, such as engine overhauls, on a fixed schedule. Look at an automobile owner's manual. It details the servicing required for the vehicle on the basis of fixed times or mileage.

Repair involves diagnosing technical problems, replacing worn or damaged parts, and testing the repaired vehicle. Minor repairs and parts replacement might take place in the field (at the terminal). Extensive repairs are generally completed at a vehicle repair and servicing center.

Figure 26-13. Transportation support structures must be kept in good repair. The first photo shows a railroad track repair device, and the second photo shows a waterway being dredged.

Regulating Transportation Systems

Transportation is considered essential for the general welfare of a country and its people. Therefore, it is subject to ***regulation*** so that all people will be served. This regulation takes place on two levels. The first level is ***domestic transportation.*** This type of transportation occurs within the geographic boundaries of one country. The other level is ***international transportation,***

Figure 26-14. Transportation vehicles must be maintained continuously and repaired when needed. (Alaska Airlines)

Career Corner
Bus Drivers

Bus drivers pick up and drop off passengers at bus stops, stations, and other locations on an established route. Drivers must maintain a set time schedule and operate vehicles safely. Local-transit drivers collect fares, answer riders' questions about schedules and routes, submit daily trip reports, and report mechanical problems. They might make several trips each day around the same city or suburban streets. The drivers usually have a five-day workweek. Saturdays and Sundays are considered regular workdays. Bus drivers must meet state and federal qualifications and hold a commercial driver's license (CDL). A CDL requires applicants to pass a written test on driving rules and regulations and demonstrate safe bus-driving skills.

which moves passengers and cargo between nations.

In the United States, domestic transportation is regulated by local, state, and federal agencies. For example, cities often control the number of taxicabs and the area they serve. Buses and rapid transit lines are often owned and operated by city or county governments. Therefore, these governments regulate routes and the level of bus service.

State agencies are responsible for building and maintaining the highway system. Their actions regulate their use, set speed limits, and control other operating conditions.

Much of the traffic in the United States crosses state lines. This type of traffic involves ***interstate commerce.*** This means that *business dealings* extend across state lines.

Therefore, the federal government plays an active regulatory role over much of the transportation industry. This is done by a number of agencies, including the following:

- The Department of Transportation: Coordinates federal programs and policies to promote fast and safe transportation. Two of the most important parts of the department are:
  - The Federal Aviation Administration: Regulates airspace and air safety and provides navigational aids. The FAA licenses pilots and determines the airworthiness of aircraft. The FAA also regulates the economics of air transportation.
  - The Federal Highway Administration: Sets highway construction standards for federally financed roads and regulates the safety of interstate trucking.
- The Federal Maritime Commission: Regulates and promotes shipping on the oceans.

Connections to Technology: **Mathematics**

Relating Speed, Time, and Distance

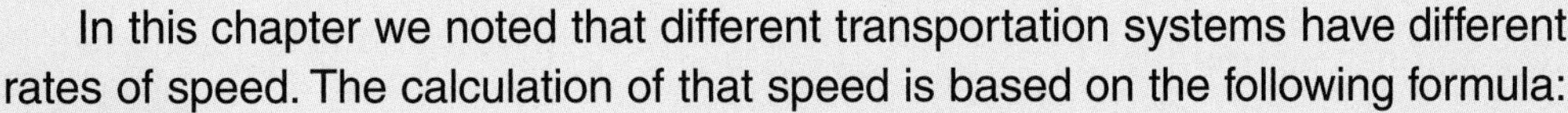

In this chapter we noted that different transportation systems have different rates of speed. The calculation of that speed is based on the following formula:

$$\text{Speed} = \frac{\text{Distance}}{\text{Time}}$$

We thus know that if we have traveled 10 miles (distance) in 1 hour (time), our rate of speed was 10 miles per hour:

$$\text{10 miles} = \frac{\text{10 mph}}{\text{1 hour}}$$

As we have learned in our mathematics courses, we can also determine the other parts of the formula by changing the equation. For example:

$$\text{Time} = \frac{\text{Distance}}{\text{Speed}} \quad \text{and} \quad \text{Distance} = \text{Speed} \times \text{Time}$$

Therefore, if you lived 1 mile from school and you wanted to know how long it would take to get to school if you walked at a rate of 2 miles per hour, you could calculate the time as follows:

$$\text{Time} = \frac{\text{1 mile (distance)}}{\text{2 mph (speed)}} = \text{1/2 hour, or 30 minutes}$$

If you wanted to know how far you had traveled and knew you had gone at a rate of 2 miles per hour for 1/2 hour, you would calculate as follows:

$$\text{Distance} = \text{2 mph (speed)} \times \text{1/2 hour (time)} = \text{1 mile}$$

How long would it take you to travel the 1 mile to school if you rode a bicycle and traveled at 10 miles per hour?

In addition, the National Transportation Safety Board promotes safe transportation and investigates serious accidents. The Environmental Protection Agency sets controls on auto emissions, traffic control plans, and other transportation issues.

International transportation involves diplomacy. *Diplomacy* entails conducting negotiations between countries that end up with agreements. These agreements may establish transportation routes between two countries, set ***fares*** (the cost of a ticket), and approve schedules. These actions often consider the needs of the passengers but might also attempt to protect one country's transportation companies.

Summary

Moving people and cargo may be completed by personal, privately owned vehicles. They serve only the owner, however, and cannot move large quantities of goods or large numbers of people. Therefore, commercial transportation systems have developed. These systems include vehicles, routes, schedules, and terminals. Through these systems people and cargo are loaded, moved, and unloaded. Because of the wear and tear involved in such movement, the structures and vehicles involved in the system must be maintained and repaired.

Each transportation system is regulated by various local, state, and federal agencies. Their goal is to encourage the development of an integrated transportation system that serves the individuals and society as a whole.

Key Terms

commercial transportation
domestic transportation
fares
hub-and-spoke system
international transportation
interstate commerce
maintenance
personal transportation
regulation
repair
route
schedules
terminals

Test Your Knowledge

Write your answers on a separate piece of paper. Please do not write in this book.

1. Name one technological development and how it affected the speed of travel.
2. Name one difference between personal and commercial transportation.
3. List the four components of a transportation system.
4. What is a hub-and-spoke system?
5. *True or false?* Cargo terminals are more complex than passenger terminals.
6. What are the three major actions performed by transportation systems?
7. Name one maintenance activity involved in transportation systems.
8. Give one example of how a city might regulate a taxicab company.
9. What is a major purpose of the Federal Highway Administration?
10. *True or false?* The Federal Maritime Commission regulates airspace.

Applying Your Knowledge

1. Obtain a rail, bus, or airline schedule. Use the schedule to plan a trip that will take you from your city to a major destination. List the departure time, arrival time, and travel time for each segment of the trip and for the entire trip.
2. Obtain a map and plot the routes for the trip you developed above.
3. Visit a local bus terminal or airport. Make a rough floor plan of the terminal. Indicate the various passenger areas.
4. Develop a drawing for a new terminal to replace the one you visited.

This activity develops the skills used in TSA's System Control Technology event.

System Control Technology

Activity Overview

In this activity, you will build a computer-controlled mechanical model representing controls for an elevator serving two levels. You will also prepare a report explaining your design and listing directions for operation.

Materials

- Pencil and paper
- Two (2) touch sensors
- Two (2) lights
- Two (2) motors
- Computer hardware and software control system

Background Information

General. You will use the sensors, lights, and motors to model an elevator serving two floors. The sensors represent the elevator call button. The motors represent the elevator doors—one representing the first-floor elevator doors, the other representing the second-floor elevator doors. A running motor represents an open elevator door. The lights represent the location of the elevator car. When the first-floor light is on, the elevator car is at that floor. When the second-floor light is on, the elevator is at that floor.

Control conditions. Your control system must adhere to the following guidelines:

- The elevator car can be at only one floor at any given time.
- The elevator car must be at the floor before the doors can open.
- The elevator car cannot leave the floor while the doors are open.
- When a button is pushed, the elevator car is called to the floor, the doors open and close, the car moves to the other floor, and the doors open and close.
- The car remains at the current floor until called.

- There must be a two-second delay between when the car arrives at the floor and when the doors open.
- The doors remain open for five seconds and then close automatically.
- If the elevator call button is pressed while the doors are open, the doors remain open for five seconds from the time when the button is pushed.
- The car must wait two seconds after the door closes before going to the other floor.

Report. Your report must include the following:

- Description of solution
- Instructions for operation
- Printout of control program

Guidelines

- Create rough schematics of design solution and control logic.
- Sketch the final design before constructing it.
- After constructing the model and writing the control program, test the model. Before you begin testing, create a list of the various conditions and situations that need to be tested.

Evaluation Criteria

Your project will be evaluated using the following criteria:

- Report
- Model functionality, dependability, and ingenuity
- Computer program logic and functionality

Section 7 Activities

Activity 7A

Design Problem

Background

Most transportation systems use a vehicle to move people and cargo to their destinations. The vehicle must have a propulsion system to cause it to move along its pathway.

Situation

You are employed as a designer for the Technology Kits Company. This company markets simple kits that are fun to assemble and teach basic principles of technology.

Challenge

You have been chosen to design a transportation vehicle kit. You may do any or all of the following:

1. Use air, land, or water as the medium of travel for the vehicle.
2. Utilize a rubber band, mousetrap, or spring as the power source.
3. Use any material commonly available in a craft store, such as tongue depressors, balsa strips, wheels, and dowels.

Your supervisor wants a working model, a parts list for the kit, and a set of assembly directions for the purchaser.

Optional

1. Design a package for the product.
2. Create a catalog advertisement for the kit.

Activity 7B

Fabrication Problem

Background

The internal combustion engine is the most popular source of power for land vehicles. These engines use petroleum-based fuels, which, in the near future, will be in short supply. Internal combustion engines also produce emissions that pollute the environment. Developing alternative fuel systems for land vehicles is a major challenge facing designers and engineers.

Challenge

Build a land vehicle a propeller can power. See **Figure 7B-1.** Develop an electric power system for the vehicle. Then, test the performance of the vehicle. Finally, change the number of blades on the propeller.

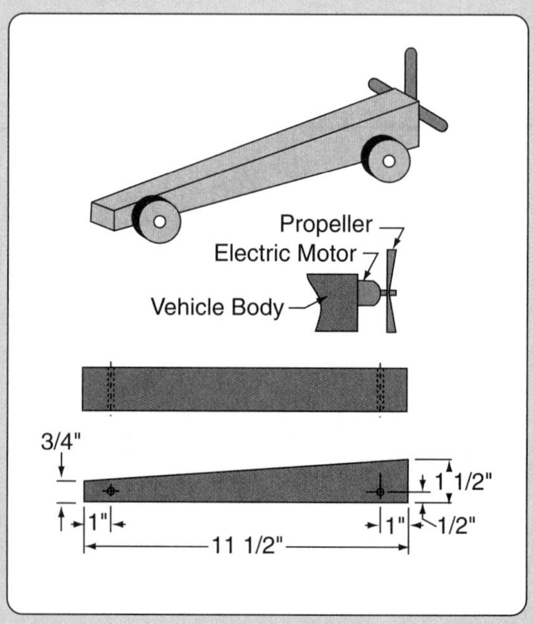

Figure 7B-1

Materials

- One piece 1 1/2" x 1 1/2" x 11 1/2" wood (pine, fir, or spruce)
- Four wheels (available from many hobby stores)
- Two pieces 1/8" diameter x 3" welding rod (for use as the axles)
- Two pieces, 1 1/2" long, of soda straw
- One low-voltage electric motor (available from a hobby or an electronics store)
- One battery that matches the motor voltage
- 12" electrical hookup wire
- One 4-blade and one 3-blade propeller (available from a hobby store)

Procedure

Building the Prototype

1. Obtain a block of wood 1 1/2" x 1 1/2" x 11 1/2" long.
2. Lay out the block, including the following:
   A. A taper from 3/4" at one end to 1 1/2" at the other end.
   B. Axle holes 1" from each end and 1/2" above the bottom of the vehicle block.
3. Cut the block along the taper.

Safety

Follow all safety rules your teacher demonstrates!

4. Drill two holes for the axles. NOTE: The diameter of the holes should be the same as the outside diameter of the soda straw axle bearings.
5. Measure the diameter of the electric motor.
6. Drill a hole the same diameter of the electric motor into the tall end of the body. NOTE: The depth of the hole should allow at least 1/2" of the motor to extend out of the block. The diameter of the propeller and the depth of the hole determine the vertical location of the hole. The motor should be as high in the vehicle as possible.
7. Drill a hole from the top of the vehicle into the motor hole. NOTE: This hole will allow the motor wires to be drawn out of the mounting hole.
8. Cut the axles to length.
9. Mount the axles and wheels onto the body.

Installing the Propulsion System

1. Feed the motor wires into the motor hole and out the wire escape hole.
2. Press the motor into the body.
3. Attach one motor wire to the battery.
4. Mount the battery onto the vehicle.
5. Attach the 3-blade propeller to the motor.

Safety

The propeller will now rotate when the motor spins. Be sure to keep your hands out of the way.

Testing the Vehicle

1. Attach the second wire to the motor.
2. Place the vehicle on the floor.
3. Test the distance it will travel in a given period of time.
4. Disconnect one wire from the motor.
5. Change the propeller to the 4-blade model.
6. Repeat steps 1 through 4.

Optional

If the propellers are made of stamped metal, change the pitch (angle) of the blades. Test the propeller again to check the performance.

Section 8

Applying Technology: Using Energy

Tomorrow's Technology Today
Fuel Cells in Automobiles

The automobiles of the future will probably look much different, both inside and out, from what we see today. Engineers and others have developed a new automobile platform designed for use with a fuel-cell engine and a hydrogen fuel tank. Some reports say the new platform looks like a giant skateboard.

All mechanical controls will be eliminated, including the internal combustion engine, exhaust system, and brake and accelerator pedals. Electric motors in the hub of each wheel will propel the new chassis, making each vehicle all-wheel-drive. All the controls will be located in the steering mechanism.

The fuel cell is a key concept for these cars of the future. A fuel cell is a system that uses stored hydrogen and oxygen from the air to create electrical energy. Fuel cells will most likely be widely seen in homes in the near future in products such as laptop computers, cellular telephones, vacuum cleaners, and hearing aids. Used in these ways, a fuel cell provides much longer life than a battery does, and it can be recharged quickly. Using fuel cells to power automobiles, however, is a much more difficult application. Recently, researchers have made major breakthroughs in creating a fuel cell with commercial prospects, and several automobile companies are now working on new vehicles with fuel-cell engines.

We probably will not see these vehicles mass-produced for a few years, but it will not be too long before the internal combustion engine is a thing of the past. While researchers and automobile manufacturers work on developing fuel-cell vehicles that run on pure hydrogen, gasoline-fed fuel cells will serve as a transitional technology. Some of these, already on the market, cut carbon dioxide emissions by 50 percent and can get up to 40 miles per gallon of fuel.

Car manufacturers hope to have vehicles powered entirely by fuel cells available to consumers in just a few years. The advantages of the fuel-cell engine are numerous: it has no moving parts; it is a quiet, reliable source of power; it will lead to a cleaner environment (the only emission is water vapor); and it will create less dependence on foreign oil. Another advantage is that the vehicle, while sitting in the garage, can generate enough electricity to power a home.

Some obstacles must be overcome first, however. It currently costs two to three times the expense of running a traditional power plant to produce a fuel cell powerful enough to operate a car. In addition, a new fueling infrastructure (a replacement for gas stations as we know them) will have to be installed across the country to distribute a usable fuel, such as methanol. Hopes are high for the emergence of this new technology, but several years will pass before it replaces the current technology that is presently on every street in America.

Chapter 27
Energy: The Foundation of Technology

Learning Objectives

After studying this chapter, you will be able to:

- Define energy.
- Describe the difference between potential and kinetic energy.
- Identify the differences among energy, work, and power.
- Identify the six major forms of energy.
- Identify exhaustible, renewable, and inexhaustible energy sources.
- Explain how energy technology can make our lives better and how energy technology can cause damage.

We use the word *energy* in different ways. We might say, "I don't have the energy to mow the lawn." We might worry about our energy dependence on foreign petroleum. We might hear people talking about energy conservation. Not everyone using the word knows exactly what it means, however.

Energy comes from the Greek word *energeia,* which means *work.* As time passed, the word came to describe the force that makes things move. Today, ***energy*** is defined as the *ability to do work.* This ability includes a broad spectrum of acts. Energy is used in simple human tasks, such as walking. Energy can be obtained from petroleum, which is then used to power a ship across the ocean, **Figure 27-1.** Energy can be used to provide motion in vehicles or machines. It can be used to produce heat or light. Energy is fundamental to our communication technologies. It is used in manufacturing products and constructing structures. Energy is everywhere and is used by all of us.

In this chapter we will look at energy from a variety of angles. We will discuss the types of energy, the connection between energy and work and power, and the forms of energy and their interrelationships. We will also examine the various sources of energy and whether energy technology is helpful or harmful.

Types of Energy

Two types of energy exist. Energy can be associated either with a force doing the work or with a force that has the *capability* of doing work. Energy that is involved in

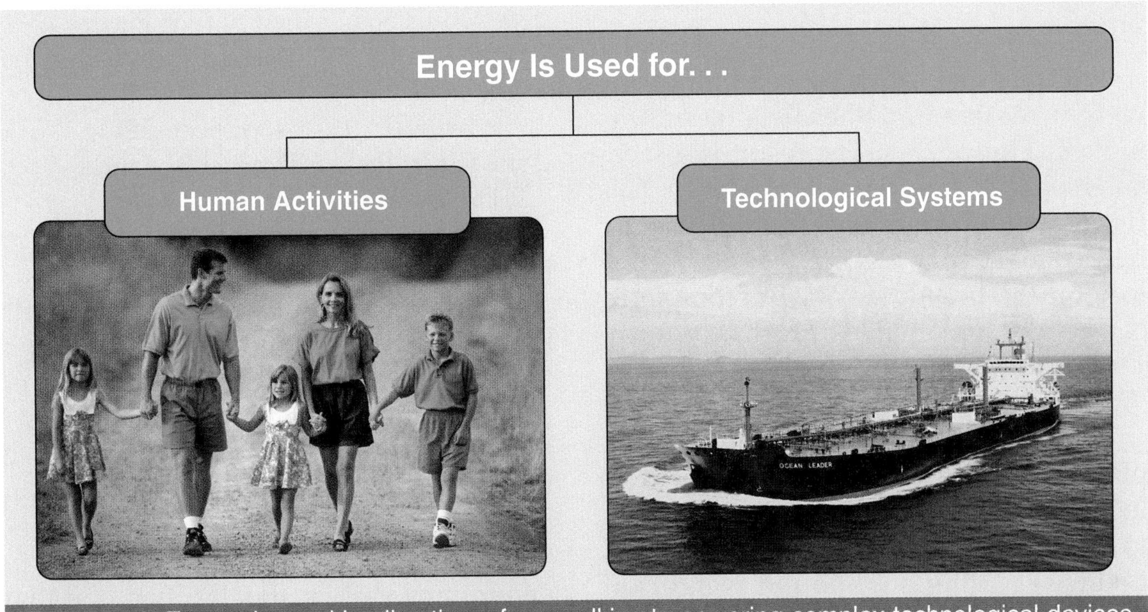

Figure 27-1. Energy is used in all actions, from walking to powering complex technological devices.

moving something is called ***kinetic energy.*** It is the energy in motion. A hammer striking a nail is an example of a technological act that uses kinetic energy. A sail capturing the wind to power a boat uses kinetic energy. A river carrying a boat or turning a waterwheel are still other examples of kinetic energy.

Not all energy is being used at any given time, however. Some energy is stored for later use. Energy in this condition has the capability, or potential, of doing work when it is needed. For example, water stored behind a hydroelectric dam possesses energy. It will release this energy to turn a turbine when it flows through the power-generating plant. This stored energy is called ***potential energy,*** **Figure 27-2.** A flashlight battery and a gallon of gasoline are other examples of potential energy.

Figure 27-2. The coal being mined has potential energy. The energy will be released when the coal is burned. (AMAX)

Energy, Work, and Power

It takes energy to do work. For example, you must eat well if you plan to run a marathon or hoe a garden. When you do one of these things, you might say, "I really worked hard!" But what does this mean? In scientific terms, ***work*** is *applying a force that moves a mass a distance in the direction of the applied force.* You might have lifted boards off the floor and placed them on a table. That was work. The boards had mass, and they were moved some distance.

Measuring Work

We measure work by combining the weight moved and the distance it was moved, **Figure 27-3.** The result is a measurement called ***foot-pounds (ft.-lbs.).*** This figure will tell you the amount of energy needed to move an object from one location to another. The amount of work completed can be measured with the formula:

Work (in foot-pounds) = Force or Weight (in pounds) x Distance (in feet)

Suppose you weigh 140 pounds. You plan on walking across a room that is 40 feet wide. You need 5600 ft.-lbs. of energy to complete the task (140 pounds x 40 feet). Likewise, lifting a 20-pound weight off the floor and placing it on top of a 36-in.-high table requires 60 ft.-lbs. of energy (20 pounds x 3 feet [36 inches]).

In the metric system, work is measured in newtons per meter, or ***joules*** (J). The force or weight is measured in newtons, and the distance is measured in meters. The metric work formula is:

Work (in joules) = Force or Weight (in newtons) x Distance (in meters)

Measuring Power

Work is done in a context of time. Measuring the *rate at which work is done* gives you a term called ***power,*** **Figure 27-3.** Power can be calculated by dividing the work done by the time taken.

Power (in ft.-lbs. / second) = Work done (in ft.-lbs.) / Time (in seconds)

The metric version is:

Power (in watts) = Work done (in joules) / Time (in seconds)

Two common power measurements are the horsepower and the kilowatt hour. ***Horsepower*** is used to describe the power output of many mechanical systems. One horsepower is the force that is needed to move 550 pounds one foot (550 ft.-lbs.) in one second. The factor of time is important to power. A motor that lifts 550 pounds in one minute can be smaller than one that lifts 550 pounds the same distance in one second. Likewise, the engine that moves a car from 0 to 60 mph in 7 seconds must be more powerful than one that does the same job in 9 seconds.

The term horsepower is used in several different ways. The theoretical, or *indicated horsepower,* is the rated horsepower of an engine or motor. This number suggests the maximum power that can be expected from the device under ideal operating conditions. Most often, this amount of power is not available from the device.

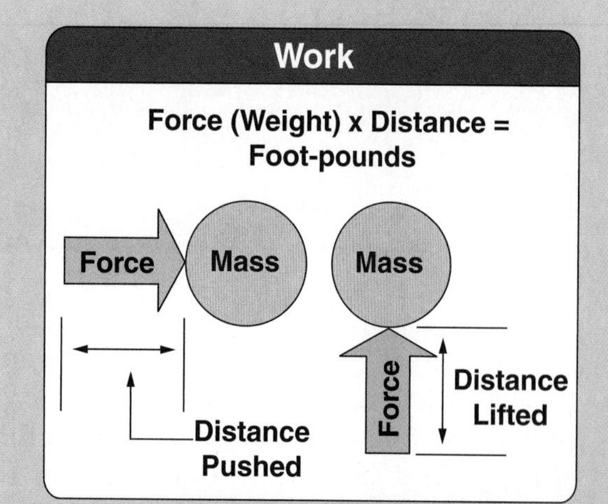

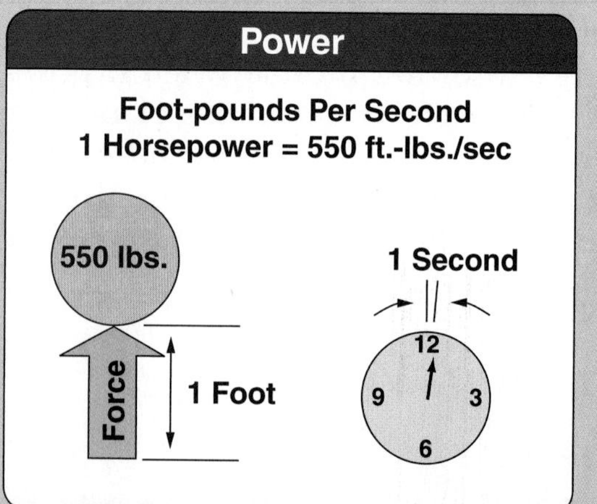

Figure 27-3. Work is done when a force moves a mass over a distance. Work done per unit time is power.

The *brake horsepower* is the power delivered at the rear of an engine operating under normal conditions. *Drawbar horsepower* is the power delivered to the hitch of tractors. *Frictional horsepower* is the power needed to overcome the internal friction of the technological device.

One ***watt*** is equal to one joule of work per second. One ***kilowatt hour*** is the work that 1000 watts will complete in one hour.

In electrical apparatus, the power consumed is determined by the resistance of the device. Its wattage rating is the product of the electrical current flowing through it and the voltage drop across it. The formula can be expressed as follows:

Power (in watts) = I (current in amperes) x E (electromotive force in volts)

Forms of Energy

Energy is everywhere we look. It is in the fires that burn coal and wood. Energy is in sunlight, wind, and moving water. In fact, we could not exist without the aid of energy.

The hundreds of examples of energy can be grouped into six major forms. These are, as seen in **Figure 27-4:**

- Mechanical
- Radiant
- Chemical
- Thermal
- Electrical
- Nuclear

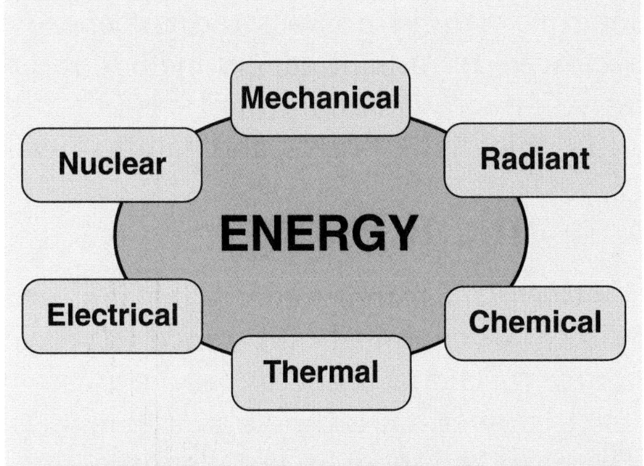

Figure 27-4. Energy can be grouped into six major forms.

These energy sources do work that ends up as motion, light, or heat. They are used to power manufacturing machines, light buildings, propel vehicles, and produce communication messages.

Mechanical Energy

Most of us are familiar with ***mechanical energy.*** Often, it is produced by motion of technological devices. We associate machines with mechanical energy. This is correct, but it does not include all types of mechanical energy. Wind and moving water have motion and thus are also sources of mechanical energy, **Figure 27-5.**

Figure 27-5. This waterfall is a natural source of mechanical energy.

Connections to Technology: **History**

Origin of Horsepower

After developing the first truly practical steam engine, James Watt needed to describe its output in a way that potential customers could understand. In those days, horses drove much of the machinery. Watt decided the best way to describe the output of an engine was in terms of the number of horses that it could replace.

In 1782 Watt began working on the formula that would eventually lead to the unit of power known as *horsepower.* After consulting with experts who designed horse-driven machinery, Watt determined that an average horse could perform 22,000 foot-pounds of work per minute.

Since the rotary-motion steam engine was new technology, potential customers were skeptical. Watt was afraid that if factory or mine machinery failed because the engine used to power it was too weak, the public would blame the steam engine technology and not the miscalculation of the purchaser. Watt decided that it would be better to understate the power of his engines.

If a customer bought a two-horsepower engine, Watt wanted that engine to be able to do the same amount of work as three actual horses. In order to accomplish this, Watt added 50 percent to his estimation of the amount of work that a horse could do over a given time. This caused one horsepower to be equivalent to 33,000 foot-pounds of work per minute, which is the value still used today.

As technology improves, the amount of horsepower available to the average person increases. Research the horsepower rating for several types of automobiles. How many actual horses would be required to produce the same amount of power? (Remember, a horse can only produce 0.67 horsepower.)

Radiant Energy

Radiant energy is energy in the form of electromagnetic waves. You learned about these waves in Chapter 22. They extend from ELF radio waves (long waves) to gamma rays (short waves). Cool objects give off longer waves than hot objects. Low-frequency waves contain less energy than high-frequency waves of the same amplitude.

The main source of radiant energy is the sun. Radiant energy is also emitted by objects heated with a flame or from a lightbulb in a lamp. Sometimes radiant energy is called light energy. This is not completely correct because many waves with wavelengths longer or shorter than the wavelengths of light possess radiant energy. Examples of radiant energy include radio waves; microwaves, infrared, visible, and ultraviolet light; X-rays, and gamma rays.

Chemical Energy

Chemical energy is energy that is stored within a chemical substance. Typical sources of chemical energy are the fuels we use to power our technological machines. The most common are petroleum, natural gas, and coal. Wood, grains (such as corn), and biomass (organic garbage) are less-frequently used sources of chemical energy.

Chemical energy is released when a substance is put through a chemical reaction. This may be done by rapid oxidation (burning) or other chemical actions, such as digestion and reduction.

Thermal Energy

Thermal energy is another name for heat energy. Thermal energy cannot be seen directly. You can see its effects, however, by watching the heated airwaves above a road on a very hot day. Thermal energy is usually felt. The energy strikes a surface, such as your skin, and elevates its temperature.

Thermal energy is created by the internal movement of atoms in a substance. These particles are always in motion. If the atoms move or vibrate rapidly, they give off heat, or thermal energy. The faster they move, the more heat they give off.

Heat energy is widely used in technological devices. It provides the energy for our heating systems and some electrical generating plants.

Electrical Energy

Electrical energy is associated with electrons moving along a conductor. This conductor may be a wire in a human-developed electrical system. The conductor could also be the air, as with lightning. Lightning is a natural source of electrical energy. Electrical energy is used as a basic source for other forms of energy. It is often converted into heat energy (for example, to warm buildings) and into light energy (for example, to illuminate our homes).

Nuclear Energy

Nuclear energy is associated with the internal bonds of atoms. When atoms are split, they release vast quantities of energy. This process is called ***fission.*** Likewise, combining two atoms into a new, larger atom releases large amounts of energy. This process is called ***fusion.***

Energy Is Interrelated

All these forms of energy are related to one another. Radiant energy can be used to produce heat. If you have ever been sunburned, you have experienced this relationship. The *radiant* energy of the sun *heated* your skin until it burned.

A fire causes fuel to undergo a chemical action. For example, coal may be changed into carbon dioxide and water. In the process of this chemical action, heat is given off, **Figure 27-6.**

The mechanical motion of an electrical generator causes magnetic lines of force to cut across any nearby conductor. This process induces an electrical current in the conductor.

Sources of Energy

Energy is a basic input to all technological systems. All energy comes in one of three basic types of resources. As shown in **Figure 27-7**, these are:

- Exhaustible energy resources
- Renewable energy resources
- Inexhaustible energy resources

Figure 27-6. This person is burning fuel (wood) to generate thermal (heat) energy to cook food.

Exhaustible

Renewable

Inexhaustible

Figure 27-7. The three types of energy resources are exhaustible, renewable, and inexhaustible. Pictured here are oil (exhaustible), wood (renewable), and wind (inexhaustible). (Marathon, U.S. Department of Energy)

Exhaustible Energy Resources

Exhaustible energy resources are those materials that cannot be replaced. Once they are used up, we will no longer have that source. The most common exhaustible resources are petroleum, natural gas, and coal. These resources are called ***fossil fuels.*** They originated from living matter. Millions of years ago plant and animal matter were buried under the earth. Over time, this matter was subjected to pressure, and it decayed. This resulted in deposits of solid (coal and peat), liquid (petroleum), and gaseous (natural gas) fuels. These deposits have been found in many locations on the earth. Chapter 14 described how the deposits are located and the fuels extracted.

Uranium is another exhaustible energy source. It is an element that developed when the solar system came into being. Uranium is a radioactive mineral that is used in nuclear power plants.

Renewable Energy Resources

Renewable energy resources are biological materials that can be grown and harvested. Their supply is directly affected by human propagation, growing, and harvesting activities. These activities can be improved by practices known as bio-related technology or ***biotechnology***. These activities improve the types and the quantities of resources that are grown. One activity is gene modification, which changes the structure of the organism. The new plant or animal may have improved growth characteristics or reduced susceptibility to disease and parasites. Another activity involves careful management and land use practices to ensure a steady supply of the renewable resources.

The most common renewable energy resources are wood and grains. They can be burned directly to generate thermal energy. Corn is often converted to alcohol (ethanol), which then can be used as a fuel.

Organic matter such as garbage, sewage, straw, animal waste, and other waste can be an energy resource. They are often referred to as ***biomass*** resources. The prefix *bio-* means having a biological, or living, origin. The resources can be traced back to plant or animal matter.

These organic materials can be burned directly as ***biofuels.*** Also, these materials can be converted into methane, a highly flammable gas. This process generates a ***biogas,*** which can replace some exhaustible fuel resources.

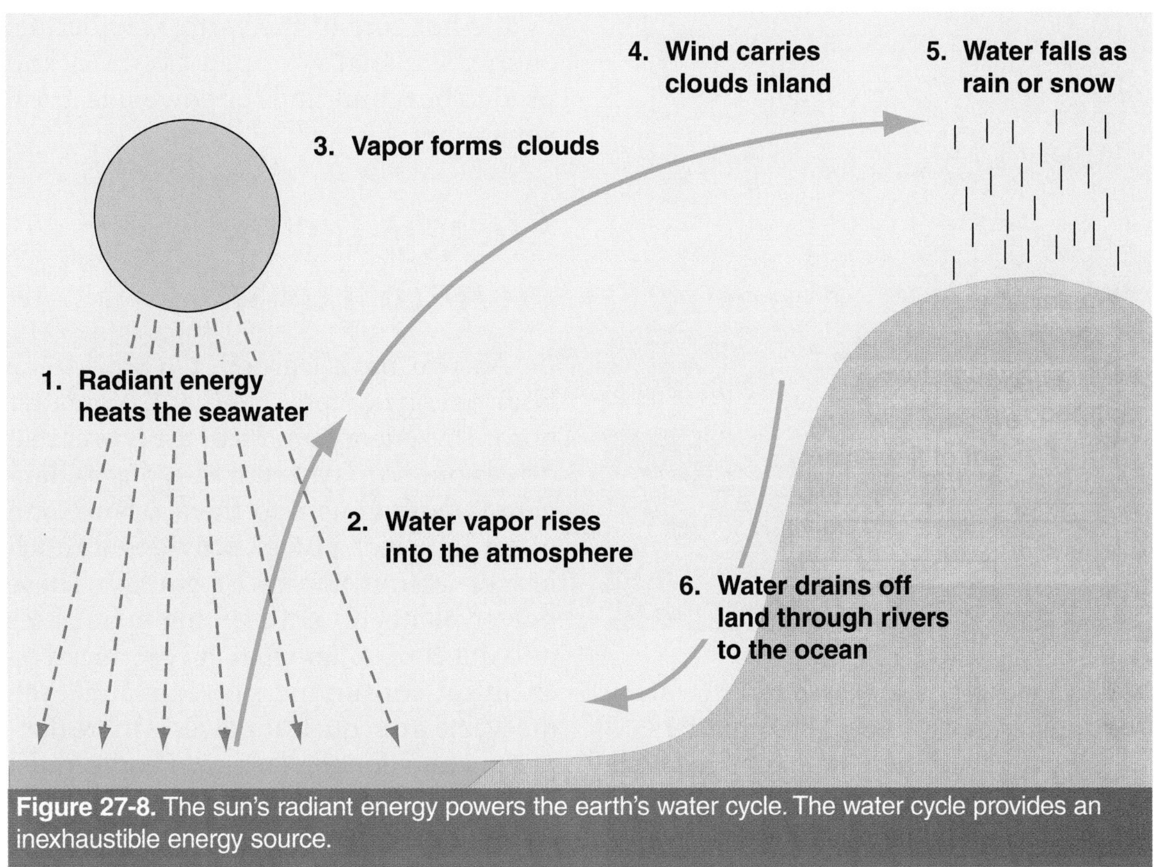

Figure 27-8. The sun's radiant energy powers the earth's water cycle. The water cycle provides an inexhaustible energy source.

Inexhaustible Energy Resources

Inexhaustible energy resources are part of the ***solar weather system*** that exists on earth. This natural cycle starts with solar energy. About one-third of the solar energy that reaches the earth's atmosphere is reflected back into space. The other two-thirds enters the atmosphere. Much of this solar energy is absorbed by the atmosphere.

About 23 percent of this energy powers what can be called a ***water cycle***, **Figure 27-8.** A small portion of the earth's water is in rivers and lakes. The majority is in the oceans that cover much of the globe. The solar energy causes the water in the oceans to heat and evaporate. The warm water vapor rises into the atmosphere and forms clouds. The clouds rise and are carried inland by the wind. As the clouds travel upward, the water vapor cools. This cooling effect condenses the water vapor into droplets. The droplets fall to the earth in the form of rain or snow. Much of the water runs off the land and collects in rivers. From there the water flows into the ocean, where it begins the cycle once again. Not all the water follows this exact pattern.

Some quantity of water is used by plants in their respiratory cycle. Some water ends up in lakes, from which it evaporates into the air to join clouds. Other portions flow underground into rivers and oceans.

Solar energy also heats the land, but the heating effects on the oceans and the land are different. Because different amounts of solar energy strike various areas of the globe, temperature differences are created. Warmer air rises and is replaced by cooler air. This air movement is called wind.

The water cycle, winds, and *direct solar energy* become inexhaustible energy resources. They produce energy through

Figure 27-9. This geyser is produced by geothermal energy.

hydroelectric power generators (water), wind generators, and solar converters.

Another inexhaustible energy resource is geothermal energy. The geysers found in Yellowstone National Park and other locations are examples of this energy resource, **Figure 27-9**. This energy source comes up from the earth (*geo*) in the form of heat (*thermal*). It uses water that has been heated by the hot core of the earth. Geothermal energy is usually tapped by wells and used to heat buildings or power electrical generators.

Energy Technology: Help or Harm

As you have learned, the world is a large energy system. Fuels and biomatter burn. This energy produces the power that drives our society. Yet, it also places dangerous pollutants into the air, land, and water. Nuclear power provides a much longer-lasting source of energy. These power plants do not leak the toxic gases into the atmosphere that are produced by chemical consuming power plants. Yet, the waste from nuclear power is more dangerous and longer lasting. People must look at both the good and bad effects of technology. Properly used, energy allows us to produce the goods needed to survive and grow. Improperly used, technology can waste energy resources and create human suffering.

Career Corner
Power Plant Operators

Power plant operators control and monitor boilers, turbines, and generators in electrical power-generating plants. They monitor instruments to maintain voltage and regulate electricity flow from the plant. These operators start or stop generators and connect or disconnect them from circuits to meet changing power requirements.

To maintain round-the-clock operations, power plant operators usually work 8- or 12-hour shifts. Employers seek high school graduates with strong mathematics and science skills for entry-level power plant operator positions. Most power plant operators receive extensive on-the-job and classroom instruction. Several years of training and experience are required to become a fully qualified operator.

Summary

Energy is a basic need for all technological activities. It is the foundation for power generation and work. Energy takes the forms of mechanical, radiant, chemical, thermal, electrical, and nuclear energy. These energy forms are derived from exhaustible, renewable, and inexhaustible resources. Each form of energy has positive and negative impacts.

Key Terms

biofuels
biogas
biomass
biotechnology
chemical energy
electrical energy
energy
exhaustible energy resources
fission
foot-pounds
fossil fuels
fusion
horsepower
inexhaustible energy resources
joules
kilowatt hour
kinetic energy
mechanical energy
nuclear energy
potential energy
power
radiant energy
renewable energy resources
solar weather system
thermal energy
water cycle
watt
work

Test Your Knowledge

Write your answers on a separate piece of paper. Please do not write in this book.

1. Define energy.
2. What is the difference between kinetic and potential energy?

For Questions 3–9, match the term on the right to the correct description on the left. (Note: Some letters will be used more than once.)

Description	Term
3. _____Can be defined as applying a force that moves a mass a distance in the direction of the applied force	a. Energy
4. _____One common measurement is the kilowatt hour	b. Work
5. _____Is measured by combining the weight moved and the distance moved	c. Power
6. _____Can be defined as the ability to do work	

Description	Term
7. _____Can be defined as the rate at which work is done	a. Energy
8. _____Is either in motion or stored	b. Work
9. _____In metric system is measured in newtons or joules	c. Power

10. *True or false?* Wind is one source of mechanical energy.
11. The main source of radiant energy is _____.
12. *True or false?* Chemical energy is also known as heat energy.
13. Two processes associated with nuclear energy are fission and _____.
14. *True or false?* All forms of energy are interrelated.

For Questions 15–23, match the type of energy resource on the right to the proper description on the left. (Note: Some letters will be used more than once.)

Description	Type of Resource
15. _____Corn	a. Exhaustible
16. _____Falling water	b. Renewable
17. _____Wood	c. Inexhaustible
18. _____Wind	
19. _____Natural gas	
20. _____Petroleum	
21. _____Sunshine	
22. _____Biomass materials	
23. _____Coal	

24. Give one example of how energy technology can improve our lives.
25. Give one negative effect of energy technology.

Applying Your Knowledge

1. Construct a simple device to change wind energy into rotating mechanical motion.
2. Identify a renewable energy resource. Describe it in terms of the factors shown in the chart below. (Prepare a similar chart for your chosen resource.)

Energy resource:
Exhaustible resources it can replace:
Present location of the resource:
Advantages of using the resource:
Problems associated with producing the resource:
Problems associated with using the resource:

Chapter 28
Energy Conversion Systems

Learning Objectives

After studying this chapter, you will be able to:

- Describe devices that convert inexhaustible energy into mechanical motion.
- Describe the main ways solar energy is converted into other forms of energy.
- Explain the differences between passive and active solar conversion systems.
- Describe a common geothermal energy conversion system.
- Explain the operation of a common biomass converter.
- Describe heat engines in terms of energy conversion.
- Explain the differences between internal and external combustion engines.
- Describe common ways to heat homes and buildings.
- Describe the major parts of an electric energy generation and conversion system.
- Describe the common energy input systems for electric generation plants.
- Explain how energy is applied to work.

Science tells us that *energy can neither be created nor destroyed.* A great deal of human action, however, is devoted to converting energy from one form into another form. For example, we burn fuels to change water into steam, which contains energy in the form of heat. The steam may then be passed through devices used to warm rooms or dry lumber.

As you can see in this example, energy converters should be viewed as part of a larger system. An energy converter is a unique device. A converter has energy as its input and its output. Mechanical energy is the input to a turbine in a hydroelectric generator. Electricity (electrical energy) is the output. This same electricity can be the *input* to several other energy converters. Incandescent lamps convert electrical energy into light (radiant energy). Motors convert electrical energy into rotary motion (mechanical energy). Resistance heaters change electrical energy into heat (thermal energy).

Your body is an energy converter. It converts food (your fuel) into energy, which moves muscles allowing you to walk, talk, and see. Likewise, an automobile engine is an energy converter. The engine converts the potential energy in

Figure 28-1. Energy is the foundation for technology. Energy powers our machines and carries us across long distances. (Apple, Jack Klasey)

gasoline into heat energy to produce mechanical motion. Energy converters power our factories, propel our transportation vehicles, heat and light our homes, and help produce our communication messages, **Figure 28-1.**

Humans have developed hundreds of energy converters to meet their needs. In this chapter we will explore four broad categories of energy conversion systems:

- Inexhaustible energy converters
- Renewable energy converters
- Thermal energy converters
- Electrical energy converters

After our discussion of energy converters, we will examine how we apply energy to do work.

Inexhaustible Energy Converters

The earliest energy conversion technologies were designed to power simple devices. These devices fall into a category that mechanical engineers call ***prime movers.*** A prime mover is any device that changes a natural source of energy into mechanical power.

Most early prime movers used inexhaustible energy sources. Specifically, they used wind and water power. Almost all societies used energy converters in transportation. Wind and flowing water helped move their boats.

Several uses for energy converters emerged, however. On land, two important technological devices were developed to harness these forces. They were windmills and waterwheels.

These two devices convert natural mechanical energy (flowing air or running water) into controlled mechanical energy (rotary motion). For example, they can produce the motion needed to power a water pump or an electric generator.

Other important converters use solar, geothermal, and ocean energy. These converters can be used to produce energy needed to heat and light our homes or power other technological devices.

Wind Energy Conversion

The sun is the original source of most of the energy on the earth. This energy is stored in growing plants and animals and in decayed organic matter—peat, coal, natural gas, and petroleum.

The sun also causes the winds to blow all over the earth. An unequal heating of the earth's surface produces these air currents. Each day the sun's rays heat the land and water masses they touch. Not all areas are touched at the same time or with equal energy, however. Polar areas receive less solar energy than do areas near the equator. Areas under cloud cover receive less solar heating than do areas in direct sunlight.

Figure 28-2. Sails were one of the earliest technological devices to harness wind power.

The heat from the land and water warms the air above it. The warm air rises, and cooler air moves in to replace it. This movement produces air currents we call wind.

The air above the hot areas near the equator is always rising. The cooler polar air moves toward the equator. In addition, air above water heats and cools more rapidly than air over land. Thus, during the morning hours, the air above the water warms quickly and rises. Cool air from the land moves in to replace it. In the evenings, the air above the land stays warmer longer than the air above the water. The cooler sea air moves inland to replace the rising land air.

Early humans designed technological devices to use these air currents. An early use of wind power was the ship's sail, **Figure 28-2.** This device was developed in Egypt around 12,000 years ago. Sails remained the primary power for ships until the development of the steam engine in the late 1700s.

Wind power, the principle of the sail, was adapted to land applications with the development of the ***windmill.*** Its first use was probably in the Middle East around 200 B.C., **Figure 28-3.** These mills were used for grinding grains into flour. This small start developed over the years, leading to today's windmills and turbines. The modern windmill is primarily used to pump water for livestock on large western cattle ranches. The wind turbine is used to power electric generators.

Windmill and wind turbine designs can be grouped into two classes: horizontal axis and vertical axis, **Figure 28-4.** The horizontal axis design has one or more blades connected to a horizontal shaft. The wind flows over the blades, causing them to turn. To see this action, place a household cooling fan in front of a blast of air. The blades will turn even though the fan's power is off.

Vertical shaft designs have blades arranged around the shaft. As the wind

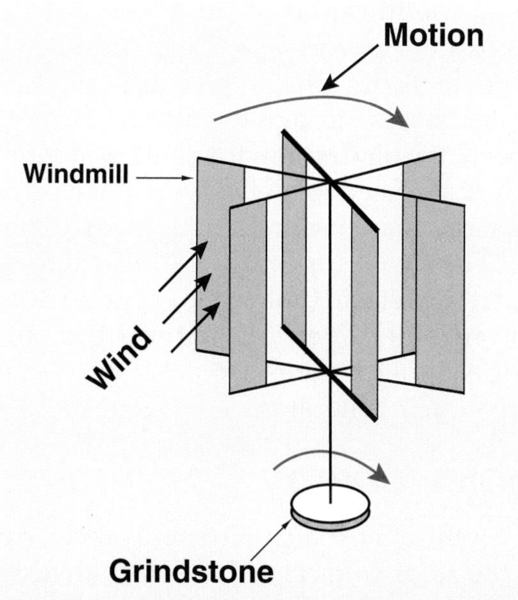

Figure 28-3. This drawing is of a design thought to be used in ancient windmills. The windmill harnessed the wind for ease in grinding.

blows past the blades, torque (turning force) is generated, rotating the shaft. The most common vertical shaft devices are the Darrieus and Savonius wind turbines.

Currently, a great deal of experimentation is being done to develop efficient wind turbines that can power electric generators. Large numbers of these devices are grouped together in "wind farms" located in various parts in the western United States, **Figure 28-5.**

Water Energy Conversion

A moving gas (air) powers windmills. Likewise, moving *liquids* can provide the energy to power technological devices. One of the earliest devices used to capture this energy was the ***waterwheel.*** This device is essentially a series of paddles that extend outward from a shaft. The

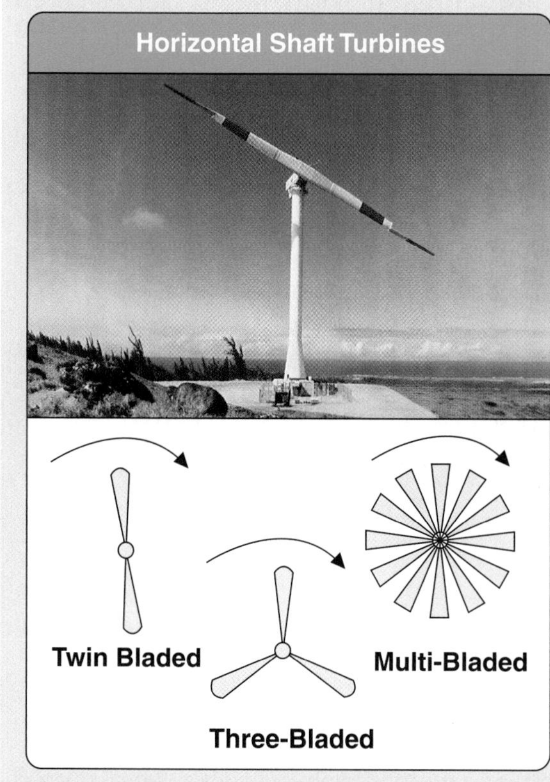

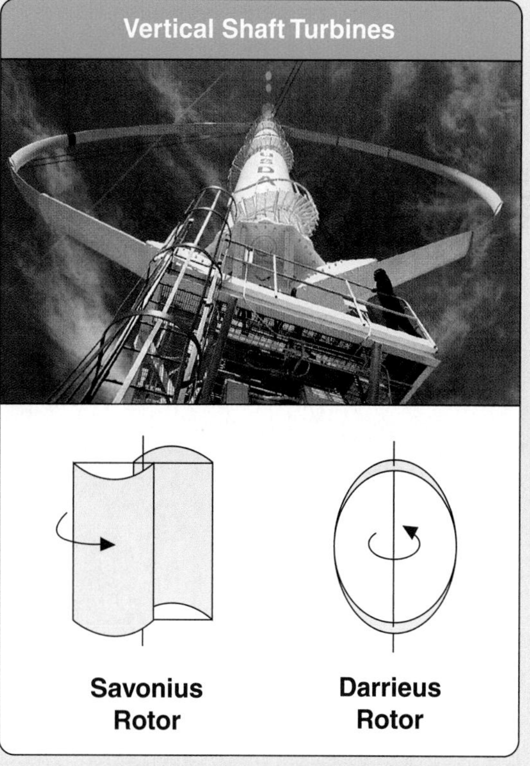

Figure 28-4. Shown are the two classes of wind turbines. On the left are examples of wind turbines with their axes parallel with the airflow (horizontal shaft). On the right are examples with their axes at right angles to the wind (vertical shaft). (U.S. Department of Energy)

Figure 28-5. Shown here is a windmill farm in California.

flowing water drives the paddles, causing the wheel to rotate.

Waterwheels powered the first factories of the Industrial Revolution. The wheels were produced in two basic designs, undershot and overshot, **Figure 28-6.** The undershot waterwheel is powered by water rushing under it. The overshot wheel is powered by water falling onto it from an overhead trough or pipe.

A modification of the waterwheel is the water turbine. The water turbine is a series of blades arranged around a shaft. As water passes through the turbine at a high speed, the blades spin the shaft. Water turbines are used widely to power electric generators in hydroelectric power plants.

The steam turbine is a similar device, but it uses steam (hot water vapor) to drive the turbine. Coal, oil, natural gas, and nuclear power plants use steam turbines to drive their generators.

Solar Energy Conversion

A third inexhaustible energy converter is the ***solar converter.*** The solar converter uses the constant energy source of the sun. The sun generates 3.8×10^{20} MW (380 million million million megawatts) of power through internal nuclear fusion.

The need to conserve exhaustible energy resources has brought solar energy into consideration as a replacement resource. Most solar energy conversion

Career Corner
Power Line Installers

Power line installers construct electrical power lines using construction equipment, such as trenchers and boring machines. They use digger derricks, which are trucks equipped with augers and cranes that dig holes and set utility poles. Line installers string cable along the poles and towers or in underground trenches or conduit. Also, line installers maintain electrical, telecommunications, and cable television lines. They identify problems and repair or replace defective cables or equipment.

Line installers often encounter serious hazards and must follow safety procedures to reduce the risks of danger. They must have a high school diploma and are trained on the job. Many employers prefer a technical knowledge of electricity and electronics obtained through vocational or technical programs, community colleges, or the armed forces.

systems have two major parts—a collection system and a storage system.

Solar energy is very intense before it reaches the earth. The energy could produce temperatures approaching 10,800° F. As it reaches the earth's protective ozone layer, however, much of this energy is absorbed and heats our atmosphere. Additional energy is absorbed by water vapor in the air. Some solar energy reaches the earth and is available as an inexhaustible energy source. The amount of this energy that is available depends on the inclination (height above the horizon) of the sun and the atmospheric conditions (cloud cover) over the earth. The term ***insolation*** is used to describe the solar energy available in a specific location at any given time. Insolation varies with the seasons and the weather. The maximum insolation on a clear, summer day is about 1000 MW per square kilometer (0.38 sq. mile).

Solar collectors depend on the principle that black surfaces absorb most of the solar energy that strikes them. This causes black surfaces to gather heat when they are exposed to sunlight.

Typical solar collectors can be grouped into two categories—passive collectors and active collectors. Each will be discussed in turn.

Passive Collectors

Passive collectors directly collect, store, and distribute the heat they convert from solar energy. Actually, an entire house is a solar collector in a passive solar collection system. The building sits quietly in position as the sun heats it. This can be done in three ways as shown in **Figure 28-7.** A ***direct gain solar system*** allows the radiant energy to enter the home through windows, heating inside surfaces.

An ***indirect gain solar system*** uses a black concrete or masonry wall (Trombe wall) that has glass panels in front of it, Figure 28-7 (B). The wall has openings at its bottom and top. The wall heats up as the sunlight strikes its surface. In turn, the air between the wall and the glass panels becomes heated by energy radiating from the wall. The warm air rises and flows into the building through the openings at the top of the wall. This creates natural convection currents that draw cooler, heavier air into the openings at the bottom. This new air, in turn, is heated and rises.

The Trombe wall also retains a great deal of heat. Consequently, after the sun sets, the wall continues to radiate the heat and warms the air between the wall and the glass panels.

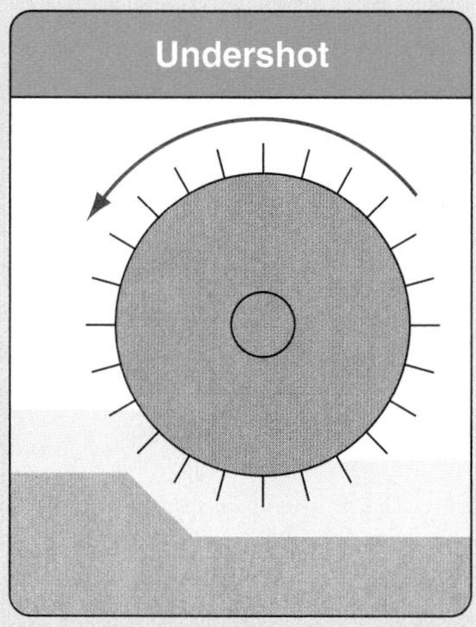

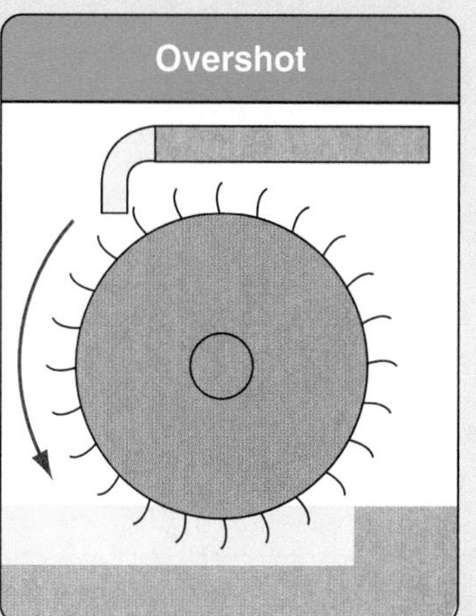

Figure 28-6. Shown here are the two common types of waterwheels.

Adobe homes in the southwestern part of the United States use a similar solar heating principle. The solar energy that strikes the adobe brick warms the surface. During the day the energy slowly penetrates the thick wall. This penetration takes about twelve hours. In the evening, the heat finally reaches the inside of the dwelling and provides warmth during the cool nights. By morning the wall is cool, and it insulates the rooms from the daytime heat. The twelve-hour lag between heating and cooling makes a very effective daytime cooling and nighttime heating system for areas that have hot days and cool nights.

An ***isolated solar system*** uses solar collectors or greenhouses that are separate from the house. These collectors are built below the level of the house. The heat generated in the collector can be channeled directly to heat the home. Additional heat is stored in a thermal mass (rock bed) for night and cloudy day heating.

Active Collectors

Active collectors use pumps to circulate the water that collects, stores, and distributes the heat that they convert from solar energy. These systems are used in many areas to provide hot water and room heat (space heating). Two major types of active collector systems exist—indirect and direct. The typical ***indirect active solar system*** has a series of collectors. Each collector has a black surface to absorb solar energy. Above or below this surface is a network of tubes or pipes. Water is circulated through these channels. As the fluid passes a warm black surface, it absorbs heat. The warm water is then pumped to a heat exchanger, where it can heat water for domestic use or provide thermal energy for a heating system, **Figure 28-8.**

A ***direct active solar system*** does not have a heat exchanger. The water that is circulated in the system is used as domestic hot water. The water flows directly to such areas as household faucets, washing machines, and showers.

A more recent use of active solar converters is in electric power generation. The converters produce steam that drives the turbines in the generation plant, **Figure 28-9.**

Another type of solar converter is the ***photovoltaic*** (solar) ***cell.*** The cell can be a small device that powers a pocket calculator or part of an array of units that

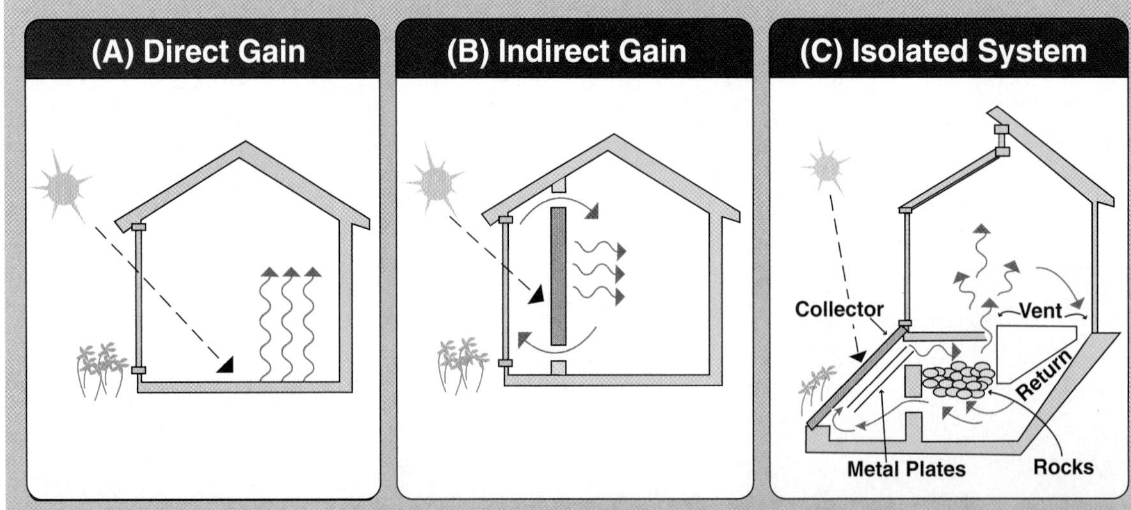

Figure 28-7. Three types of passive solar systems can be used to heat a home. A–Demonstrates the simplest system, direct gain. B–Shows how a Trombe wall, from the indirect system, works. C–Diagrams one type of isolated system. (PPG Industries)

provide electricity for such devices as satellites and solar-powered vehicles, **Figure 28-10.**

The cells are made of certain semiconductor materials such as crystalline silicon. The cell is impacted by small bundles of light energy called photons. These photons strike the cell, causing electrons to dislodge from the silicon wafer. The electrons move in one direction across the wafer. A wire attached to the wafer provides a path for the electrons to enter an electrical circuit. A second wire allows electrons from the circuit to return to the wafer. This current flow supplies the power for each device.

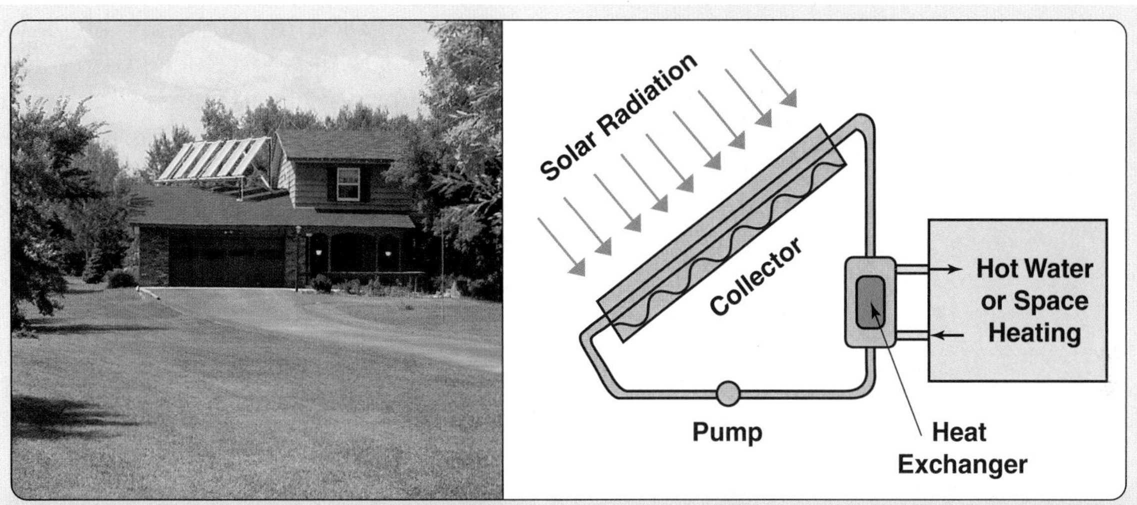

Figure 28-8. The solar panels on the building power a system similar to the one in the diagram. (U.S. Department of Energy)

Figure 28-9. This large bank of solar collectors collects solar energy to produce steam for an electrical generating plant located in the Mojave Desert in California.

Figure 28-10. This experimental electric-powered car uses photovoltaic cells to convert solar energy into electricity. (DaimlerChrysler)

Geothermal Energy Conversion

Geothermal energy is heat originating in the molten core of the earth. This energy can be found at great depths all over the planet. At certain locations, however, it reaches the surface. The energy appears as volcanoes, hot springs, and geysers, **Figure 28-11.** This energy is tapped in a number of ways. Electricity is readily produced using geothermal energy to form steam that drives a steam turbine–powered generator.

Other applications use geothermal energy for direct heating. Geothermal heat pumps use the constant 55°F temperature of ground water to heat homes. Often the water is pumped from a well that extends into an underground aquifer. The water enters the heat pump, which removes heat from the water and transfers it to the

Figure 28-11. The Old Faithful geyser in Yellowstone National Park is produced by geothermal energy. (NPS Photos)

dwelling. The cool water is returned to the aquifer through a second well. In Klamath Falls, Oregon, a number of buildings are heated and a section of highway is de-iced with hot water from geothermal wells. In Sandy, Utah, 14.6 acres of greenhouses are heated with geothermal energy.

Ocean Energy Conversion

Ocean energy is an inexhaustible source that has only recently been considered a major source of energy for the coming generations. The oceans cover more than 70 percent of the globe. They contain two important sources of energy, thermal and mechanical (wave and tide motion).

Ocean Thermal Energy Conversion Systems

Ocean thermal energy conversion (OTEC) systems use the differences in temperature between the various depths of the ocean. The basic system has three steps. First, warm ocean water is used to evaporate a *working fluid*. Second, the vapors are fed into a turbine that turns an electrical generator. Finally, cold ocean water is used to condense the vapors to complete the energy conversion cycle. The process requires water with at least a 38°F difference. This occurs only at the equator.

Ocean Mechanical Energy Conversion Systems

Ocean mechanical energy conversion systems use the mechanical energy in the oceans to generate power. Two sources of mechanical energy are tapped—wave energy and tidal energy.

Wave Energy Conversion

Presently, only two small wave generation plants exist in the world. Several devices are being developed, however, that show commercial promise. The first is a *mechanical surface follower,* which is used as a navigational aid. This device is one of the simplest designs. It is a buoy that floats in the water. Inside the device is a mechanism that uses the up-and-down movement of the buoy to ring a bell or blow a whistle.

The second system is a *pressure-activated device.* This is also a buoy that is used for navigational aid. The device uses the bobbing action created by the waves to compress air in a cylinder. As the water rises, it compresses the air. When the buoy falls, the compressed air is released, powering a small generator. The resulting electricity can power a navigational light.

Tidal Energy Conversion

Tidal energy devices use the difference between the height of the ocean at high tide and low tide to generate power. As the ocean rises, water is allowed to flow over a dam into a basin. As the tide recedes, the water flows back through turbines. This generates electricity, **Figure 28-12.**

A system in France creates power from water flow in both directions. The rising tide drives the turbine in one direction. Later, the falling tide powers it in the other direction.

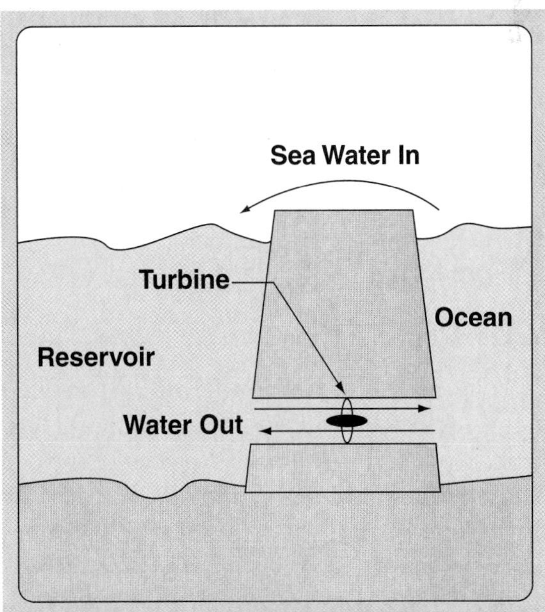

Figure 28-12. This diagram demonstrates a tidal power ocean energy conversion system.

Renewable Energy Converters

Our early ancestors depended heavily on renewable energy resources. They burned wood and cattle dung to heat their homes and cook their food. In 1850, these energy sources provided 90 percent of our energy needs. In many parts of the world, these resources are still very important. In the United States today, however, they supply just a small percentage of our energy needs.

Biomass resources are one source of renewable energy being actively considered as an alternate energy supply. Biomass resources are all the vegetable wastes and animal wastes generated through biological actions. Most biofuels come from three sources:

- Forest products industry—sawdust, bark, wood shavings, scrap lumber, and paper, for example
- Agriculture and food processing—corncobs, nutshells, fruit pits, grain hulls, sugar cane bagasse, and manure, for example
- Municipal waste—sewage and solid waste (garbage)

Bioenergy conversion is completed using one of two basic processes. These processes are thermochemical conversion and biochemical conversion.

Thermochemical Conversion

Thermochemical conversion produces a chemical reaction by applying heat. The most common method is *direct combustion.* The biofuel is burned to produce heat for buildings or to produce steam to power electric generation plants. This system is widely used in the forest products industry. Burning mill waste produces the steam that heats buildings, dries lumber, and operates the processing equipment.

A second thermochemical process is ***pyrolysis.*** In this process, the material is heated in the absence of oxygen. The heat causes the biofuel to form liquids, solids, or gases. The solids are carbon and ash. The liquids are much like petroleum and require further processing. The gases are flammable hydrocarbons. All these materials can be directly or indirectly used as fuels. Other names used for this process are ***liquidification*** and ***gasification.***

A third process is ***liquefaction.*** In this process, the biofuel is heated at moderate temperatures under high pressure. During heating, steam and carbon monoxide, or hydrogen and carbon monoxide, are present. A chemical action takes place that converts the material into an oil that has more oxygen than petroleum. This oil requires extensive refining to develop usable fuels.

Biochemical Conversion

Biochemical processes use chemical reactions caused by fungi, enzymes, or other microorganisms. The two common biochemical conversion processes are anaerobic digestion and fermentation.

Anaerobic digestion is a controlled decaying process that takes place without oxygen. The material used is agriculture waste, manure, algae, seaweed (kelp), municipal solid waste, and paper. The reaction produces methane (a flammable gas) as the biomaterials decay, **Figure 28-13.**

Fermentation is a very old process that uses yeast (a living organism) to decompose the material. The yeast changes carbohydrates into ethyl alcohol (ethanol). Grain, particularly corn, is often used for this process. The ethanol can be directly burned or can be mixed with gasoline as an automobile fuel.

Thermal Energy Converters

Heat and thermal energy have had important parts in history. The Industrial Revolution was greatly dependent on heat engines. These engines have been replaced by electrical motors in most industrial applications. Transportation is one exception.

We still depend on thermal energy, however. The comfort of the home you live in depends on heating and cooling systems. Many industrial processes use heat to cook, cure, or dry materials and products. Here we will explore two major applications of thermal energy—heat engines and space heating.

Heat Engines

Today, most of our transportation systems are based on fossil-fuel–powered engines. These technological devices burn fuel to produce heat. In turn, the heat is converted into mechanical energy. All heat engines can be classified as either internal combustion engines or external combustion engines.

These classifications are based on the location of the thermal energy source. Internal combustion engines burn the fuel within the engine; external combustion engines burn the fuel away from the engine.

Gas Storage Chamber
Biomaterial
Gas Outlet
Digester

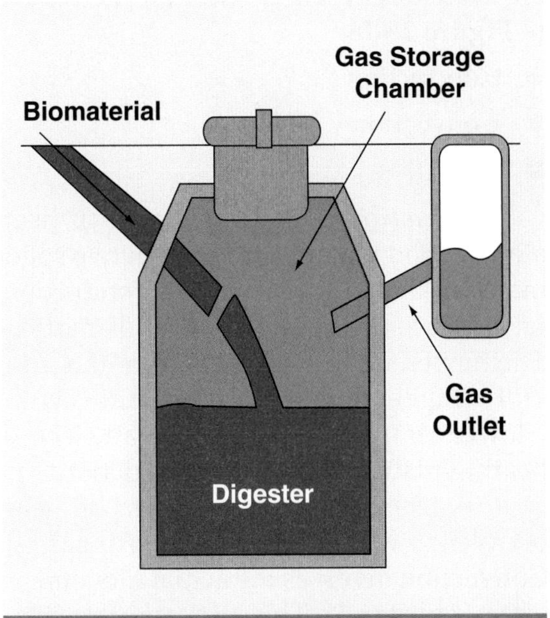

Figure 28-13. This is a diagram of a biomass converter. The product is methane, a highly flammable gas.

Internal Combustion Engines

We discussed some features of ***internal combustion engines*** in earlier chapters of this book. For example, in Chapter 7 we learned about the gasoline and diesel engines that are widely used in land and water transportation vehicles, **Figure 28-14.** Jet and rocket engines that were introduced in Chapter 25 are also internal combustion engines.

To create power, all of these engines use expanding gases produced by burning fuel. They change heat energy into mechanical motion. Let us review the common gasoline-powered internal combustion engine. You learned that the engine operates on a cycle that has four strokes: intake, compression, power, and exhaust.

During the intake stroke, the piston moves downward to create a partial vacuum. Then, a fuel and air mixture is introduced into the cylinder.

Figure 28-14. This internal combustion engine is used in Indy-class race cars.

The intake stroke is followed by the compression stroke. During this stroke, the piston moves upward, compressing the fuel-air mixture into the small cavity at the top of the cylinder.

Then, the power stroke starts with an electrical spark produced by the spark plug. This action ignites the compressed fuel-air mixture, which expands. The resulting gases force the piston downward in a powerful movement.

The final stroke is the exhaust stroke. During this stroke, the piston moves upward to force the exhaust gases and water vapor from the cylinder. At the end of this stroke, the engine is ready to repeat the four-stroke cycle.

External Combustion Engines

Most ***external combustion engines*** are steam engines. We also explored the development and operation of these engines in Chapter 7. The steam engine uses the principle that steam (hot water vapor) occupies more space than the water from which it came. In fact, one cubic centimeter (cc) of water produces 1700 cc of steam.

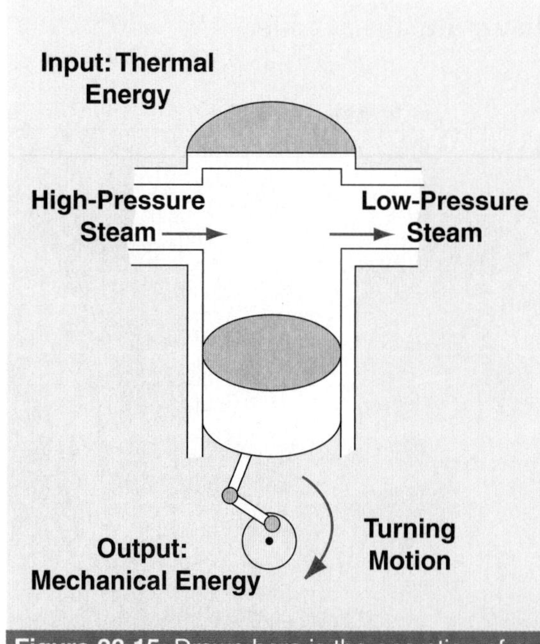

Figure 28-15. Drawn here is the operation of a simple steam engine.

The operation of the steam engine is simple, **Figure 28-15.** Water is heated in a boiler until it changes into steam. This high-pressure steam is introduced into a closed cylinder that has a free-moving piston in it. The steam forces the piston down. Next, cold water is introduced into the cylinder, condensing the steam. The resulting water takes up only 1/1700th as much space as the steam did, so a vacuum is formed in the cylinder. This causes the piston to be drawn up. At the top of the piston's stroke, a fresh supply of high-pressure steam is introduced into the cylinder. The engine repeats its cycle.

The reciprocating (up and down) motion of the engine is changed to rotary motion by a flywheel. This rotary motion can be used to power any number of technological devices. In past times, steam engines powered ships, locomotives, cars, and many machines in factories.

Space Heating

An important use of thermal energy is heating buildings and other enclosed spaces. Three basic types of heat transfer are used to heat space. These are, as shown in **Figure 28-16:**

- Conduction
- Convection
- Radiation

Conduction is the movement of heat along a solid material or between two solid materials that touch each other. Conduction takes place without any flow of matter. The movement of energy is from the area with a higher temperature to the area with a lower temperature. A pan on an electric heating plate is heated by conduction.

Convection is the transfer of heat between or within fluids (liquids or gases). Convection involves the actual movement of the substance. This process uses currents between colder areas and warmer areas within the material. Convection can occur through natural action or through the use of technological devices. The wind is an example of natural convection action. Forced hot-air heating systems are technological devices that warm homes with convection currents.

Radiation is heat transfer using electromagnetic waves. The strength of the radiation is directly related to the temperature of the radiating medium. Hot objects radiate more heat than cooler objects. The heat that people feel on a bright, sunny day is from solar radiation. Also, if you bring your hand close to a hot metal bar you can feel the heat radiate from it.

Radiation heats only the solid objects it strikes; it does not heat the air it travels through. Radiant heaters in warehouses keep workers warm in a building, while the air still feels cold.

Heat Production

A number of methods are used to produce thermal energy to heat materials and buildings. These include burning fuels, capturing heat from the surroundings, and converting electrical energy.

Fuel Conversion

Typical ***fuel converters*** include fossil fuel furnaces, wood burning stoves, and fireplaces. A furnace has a firebox, a heat exchanger, and a means of heat distribution, **Figure 28-17.** The fuel is burned in the firebox to generate thermal energy. Convection currents pass through the cells of the heat exchanger and raise its temperature. This thermal energy is transferred in the heat distribution chamber to a heating medium (water or air), which is then passed over or through the heat exchanger.

In some systems, water is heated or turned to steam. The fluid is then piped to radiators in various locations. These radiators use convection and radiation currents to heat the room. Other systems blow air through ducts to areas needing heat. Convection currents circulate the warm air within the enclosure.

Atmospheric Heat

The atmosphere has heat available no matter how cold the day seems. The standard device used to capture this heat is

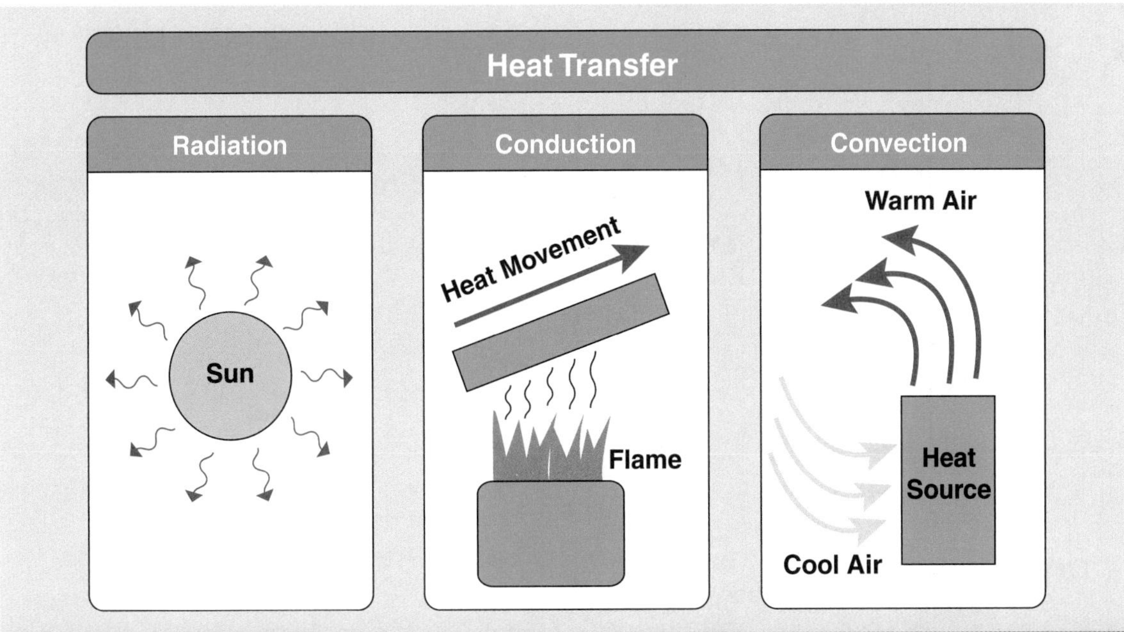

Figure 28-16. Shown are examples of radiation, conduction, and convection, the three important heating processes.

called a ***heat pump.*** It is actually a refrigeration unit that can be run in two directions. In one direction, the pump removes heat from the room and releases it into the atmosphere. This is part of what an air conditioning (cooling) unit does. When a heat pump is operated in the opposite direction, the pump takes heat from the outside air and releases the heat inside a building.

Heat pumps work on a simple principle: when a liquid vaporizes, it absorbs heat, and when a liquid is compressed, it releases heat. The system consists of a compressor, cooling or condenser coils, evaporator coils, and a refrigerant (volatile liquid), **Figure 28-18.**

In a heat pump, a heat transfer medium such as ammonia is allowed to vaporize in the evaporator coils. The heat needed to complete this task is drawn from the material around the coils. This may be air (atmospheric heat pump) or water from a well (groundwater heat pump).

The refrigerant gas is then compressed. This action causes the material to give off heat through the condenser coils. This heat may be used to warm air or water, which is then transferred to the rooms needing heat.

The system may be reversed to produce cooling for air conditioning. The heat for the evaporation is drawn from within the building. The heat from compressing the gas is expelled into the outside atmosphere.

Electric Heat

We can use electricity in a variety of ways to heat a building . The heat pump described above is one way. It is electricity that powers the compressor that draws heat from the air. Another method that uses electricity is a furnace that uses an electric resistance heater.

One common method of heating uses electric resistance heaters in each room. These heaters have special wires that have a high resistance to electrical current. The wires become very hot when electricity passes through them. The hot wires warm the air around them, and convection currents transfer the heat to all parts of the room.

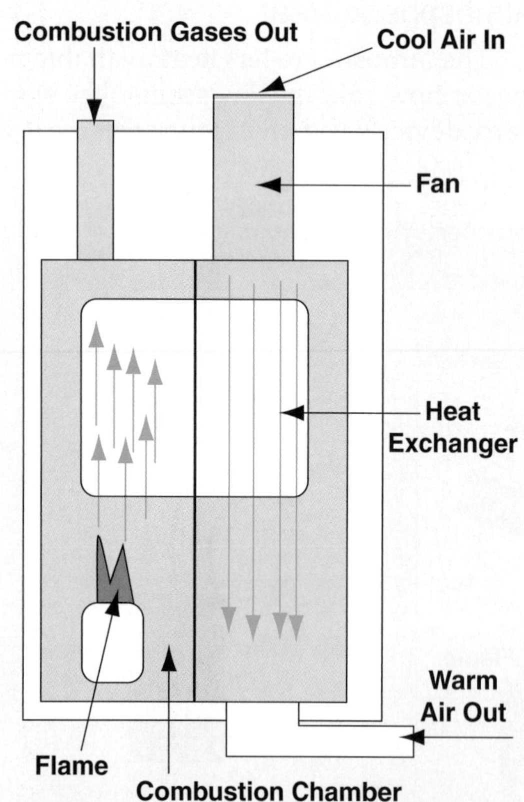

Figure 28-17. This drawing shows the airflow through a gas-fired hot-air furnace.

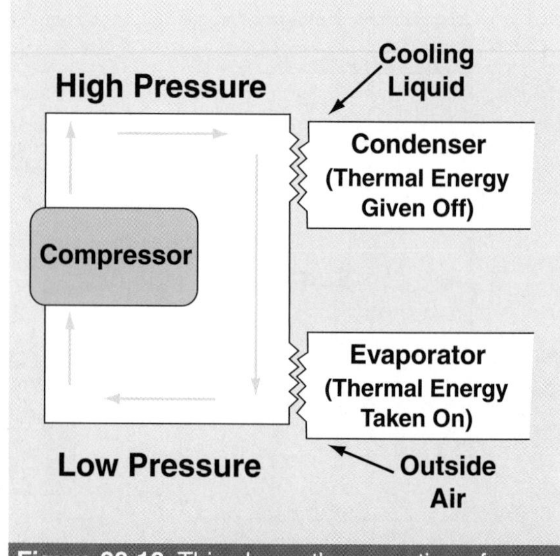

Figure 28-18. This shows the operation of a simple heat pump.

Another type of electric heating heats with radiation (radiant heating). This system uses high-resistance wires installed in the ceiling. When electricity passes through them, they become warm. This warmth radiates into the room much like heat radiated from a hot bar of steel. Objects in the room are warmed by the electromagnetic waves emitted from the system.

Electrical Energy Converters

Life in the United States is closely linked with our electrical generation and distribution systems. Electricity is very important in our society. Without electricity, stores would close, food would spoil, and many homes would become dark and cold. Let us examine the typical electrical generation and distribution system.

Generation

Electricity generation uses the principles of electromechanical energy conversion. In most commercial systems, water or steam is used to turn a turbine. A water-powered plant is called a ***hydroelectric generating plant,*** **Figure 28-19.** This plant uses a dam to develop a water reservoir. This water is channeled through large pipes into the turbines in the generating plant.

A steam-powered electrical plant uses fossil fuels or nuclear energy to produce steam to drive the generator's turbines. Fossil fuel electrical plants burn coal, natural gas, or fuel oils to produce thermal energy.

A nuclear plant uses atomic reactions to heat water in a primary system. The heated water is used to produce steam in a secondary system. The steam then drives the generator's turbine, **Figure 28-20.** Keeping the loops separate prevents the water in the reactor from entering the steam turbines. This reduces the hazards for workers in the plant and for people living near the plant.

Steam and water turbines have a series of blades attached to a shaft. As the water or steam strikes the turbine blades, the shaft turns. This shaft is attached to a generator, which changes the mechanical energy (motion) into electrical energy.

The electrical generator is the opposite of the electric motor described in Chapter 7. As we noted in that chapter, two laws of physics are directly applied. The first law states that like poles of a magnet

Figure 28-19. Hydroelectric power. A—This is Shasta Dam in northern California. Note the water delivery pipes that lead to the power plant in the lower right portion of the picture. B—This is inside a hydroelectric plant. These are the generators above the water turbines. (U.S. Department of Energy)

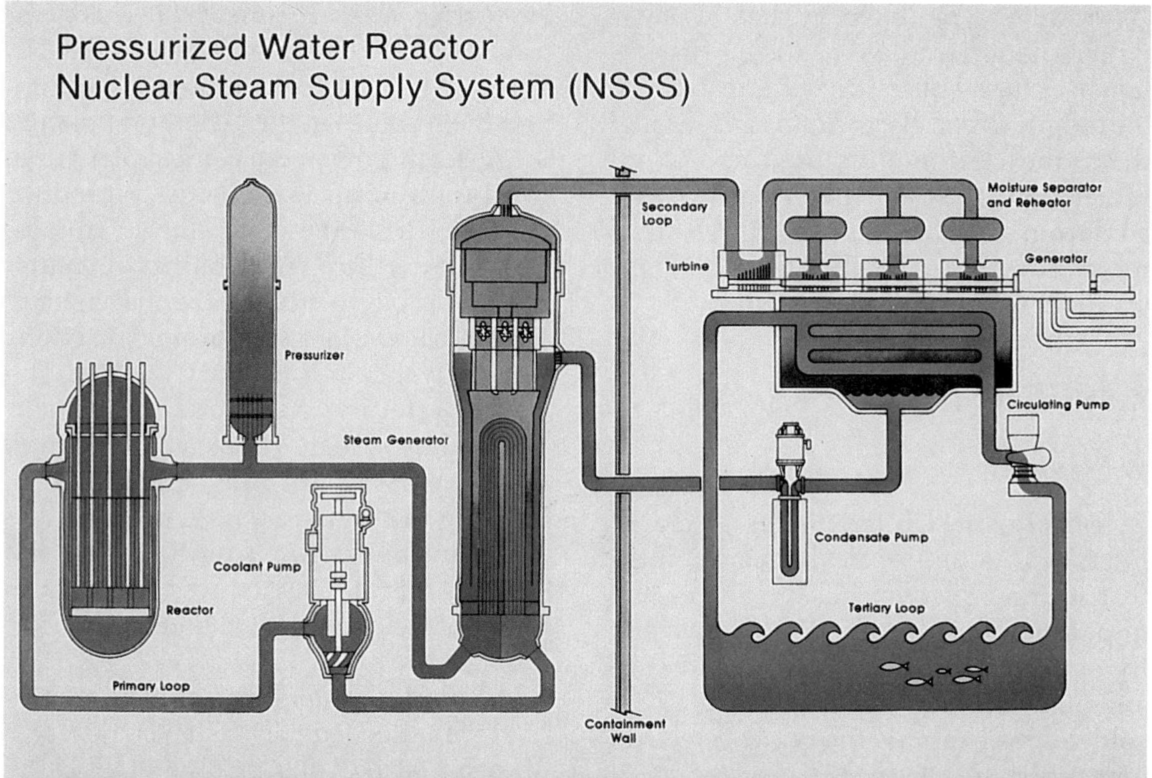

Figure 28-20. This is a drawing of a nuclear steam supply system. Notice how the water in the primary loop is used to heat the water in the secondary system. The contents of the two loops remain separate. (Westinghouse Electric Corp.)

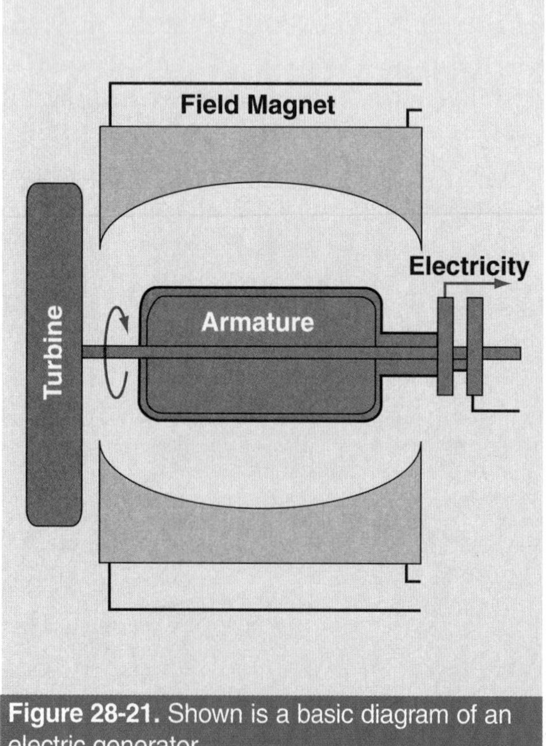

Figure 28-21. Shown is a basic diagram of an electric generator.

repel one another, and unlike poles attract one another. The second law states that current flowing in a wire creates an electromagnetic field around the conductor.

Look at **Figure 28-21.** You will note that there are two magnets. Like a motor, the outside magnet of the generator is a stationary electromagnet called the field magnet. The inside magnet is a series of wires wound on a core. This part is called the armature and is able to rotate on its axis.

An electrical current is allowed to flow in the coils of the field magnet. This action produces an electromagnetic field around the field magnet that cuts through the armature. When the armature is spun by the water or steam turbine, the wires on the armature cut through the magnetic lines of force around the field magnet. This induces a current in the armature. This current is drawn off through commutators and fed into the distribution system.

Distribution

The electricity produced in the generating plant is passed through a step-up transformer. This transformer is called a step-up transformer because it steps up, or increases, the output voltage of the electrical current. This very high voltage reduces power losses in transmission.

The high-voltage electrical current is usually carried to distant locations by large transmission lines supported on tall steel towers, **Figure 28-22.** When the current reaches the area in which it will be used, the electricity flows through another transformer, which reduces, or steps down, the voltage. This lower-voltage electrical current moves along the distribution lines. Just before it reaches its final destination, the electricity enters another step-down transformer. This transformer generally reduces the current to 110 and 220 volts for residential use. Some industrial applications use 440 or 880 volts.

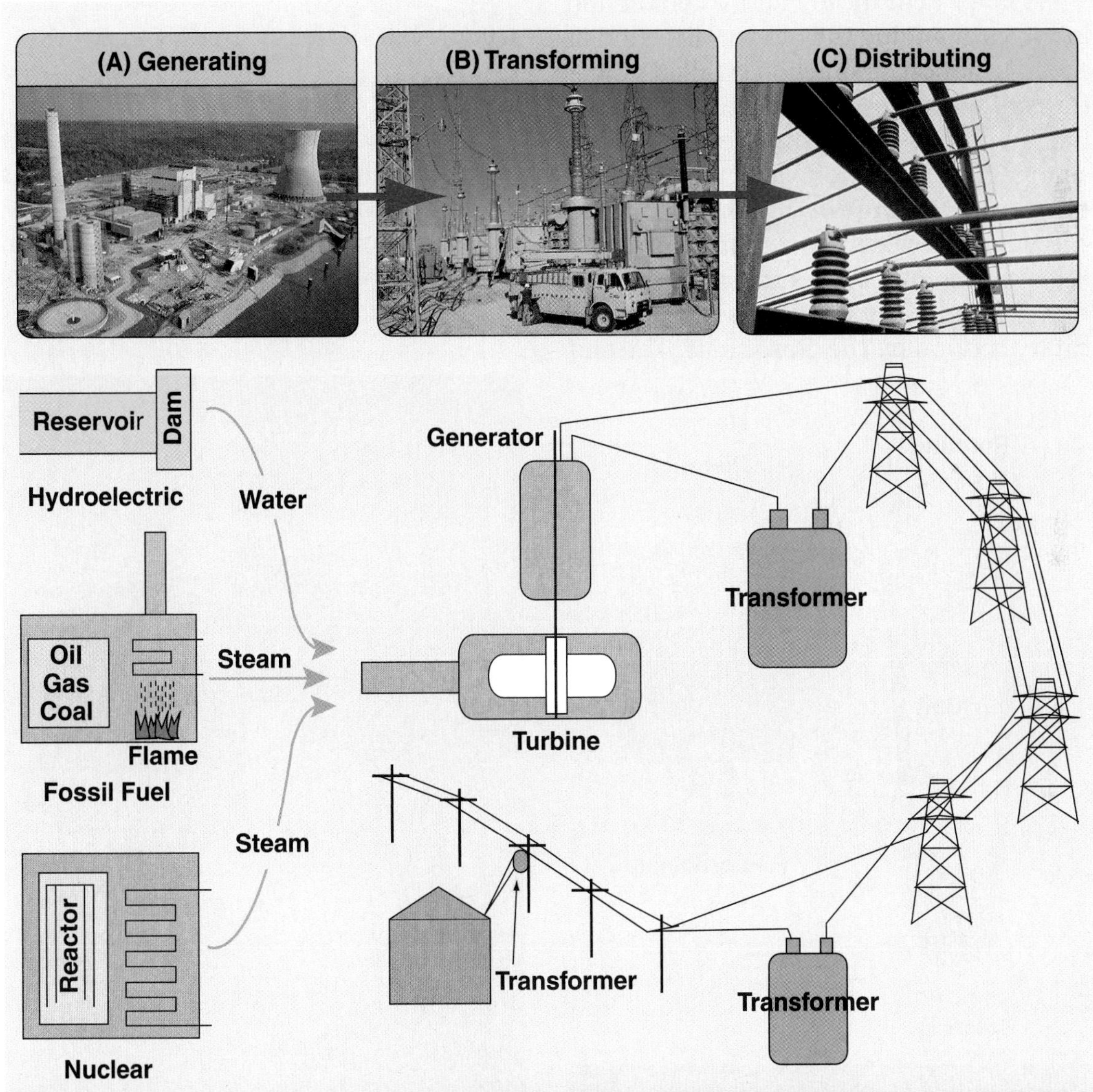

Figure 28-22. Electrical systems generate, transform, and distribute electricity. A—A nuclear power plant generating the electricity. B—A giant transformer station stepping up and distributing electricity. C—High-voltage power lines for taking electric power over long distances.

Applying Energy to Do Work

Work involves moving a load. Therefore, motion is always present while work is done. Three important types of motion need to be discussed:

- ***Rotary motion*** (spinning around an axis)
- ***Linear motion*** (moving in a straight line)
- ***Reciprocating motion*** (moving back and forth)

A main activity in energy conversion involves changing the type or direction of a load's motion. This action is called power transmission. Power transmission takes the energy generated by a converter and changes it into motion. Look at the example shown in **Figure 28-23.** The figure shows the reciprocating motion of a piston in the cylinder of an internal combustion engine. This motion is changed to rotary motion by the crankshaft. The end of the crankshaft is attached to a pulley that drives the V-belt. The belt travels in a linear motion around a second pulley. The second pulley changes the linear movement back into rotary motion. Also, note that the two pulleys rotate in opposite directions. Applying motion to perform work often requires both changing the type of motion and its direction.

Two basic types of systems are used to change the type, direction, or speed of a force. These are mechanical power and fluid power (or fluidic) systems.

Mechanical Power Systems

Mechanical systems use moving parts to transfer the motion, **Figure 28-24.** This is the oldest method of transferring energy. Various mechanical methods are used in technological devices. **Figure 28-25** shows six common techniques:

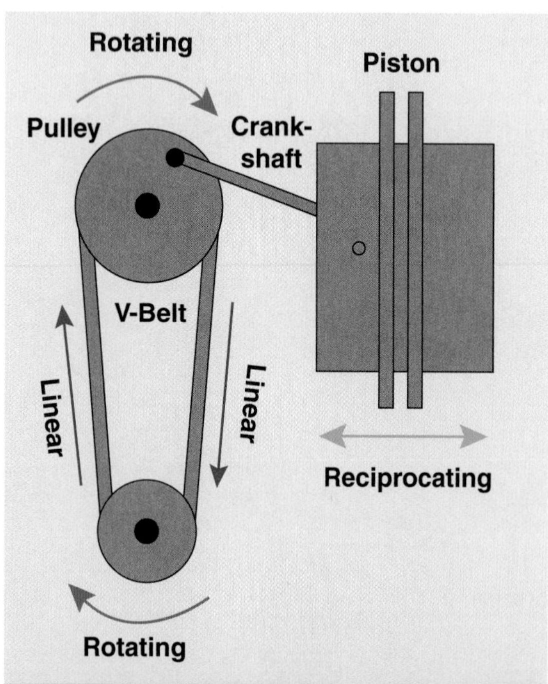

Figure 28-23. Shown here are mechanisms that change the type and direction of motion. This system can change rotating motion into reciprocating motion, or vice versa.

Figure 28-24. This crane uses a mechanical means to lift a load. (American Electric Power)

- ***Lever:*** A device that *changes the direction or intensity of a linear force.* A downward force may be applied to one end of a lever. This causes the lever arm to pivot on its fulcrum. The opposite end moves in an upward direction. You will remember from Chapter 4 that the location of the fulcrum determines whether the device multiplies the amount of the output force or the distance it moves. Many door handle mechanisms in automobiles transfer motion with levers.
- ***Crank:*** A pivot pin near the outside edge of a wheel or disk that *changes reciprocating motion into rotating motion.* The diameter of the swing of the crank will determine whether the amount of the force or distance of the force is multiplied. An internal combustion engine transfers power from the piston to the transmission using this type of drive.
- ***Gears:*** Two or more wheels with teeth on their circumference that *change the direction of a rotating force.* The relative diameters of the input and output gears determine whether the system is a force multiplier (the output gear rotates faster) or a distance multiplier (the output gear turns over a greater area). If a smaller input gear is used, the unit will increase the output force and reduce its speed. If a larger input gear is used, the unit will decrease the output force and increase its speed. Some automobile transmissions use this type of power transmission system.
- ***Cam:*** A pear-shaped disk with an off-center pivot point that is used to *change*

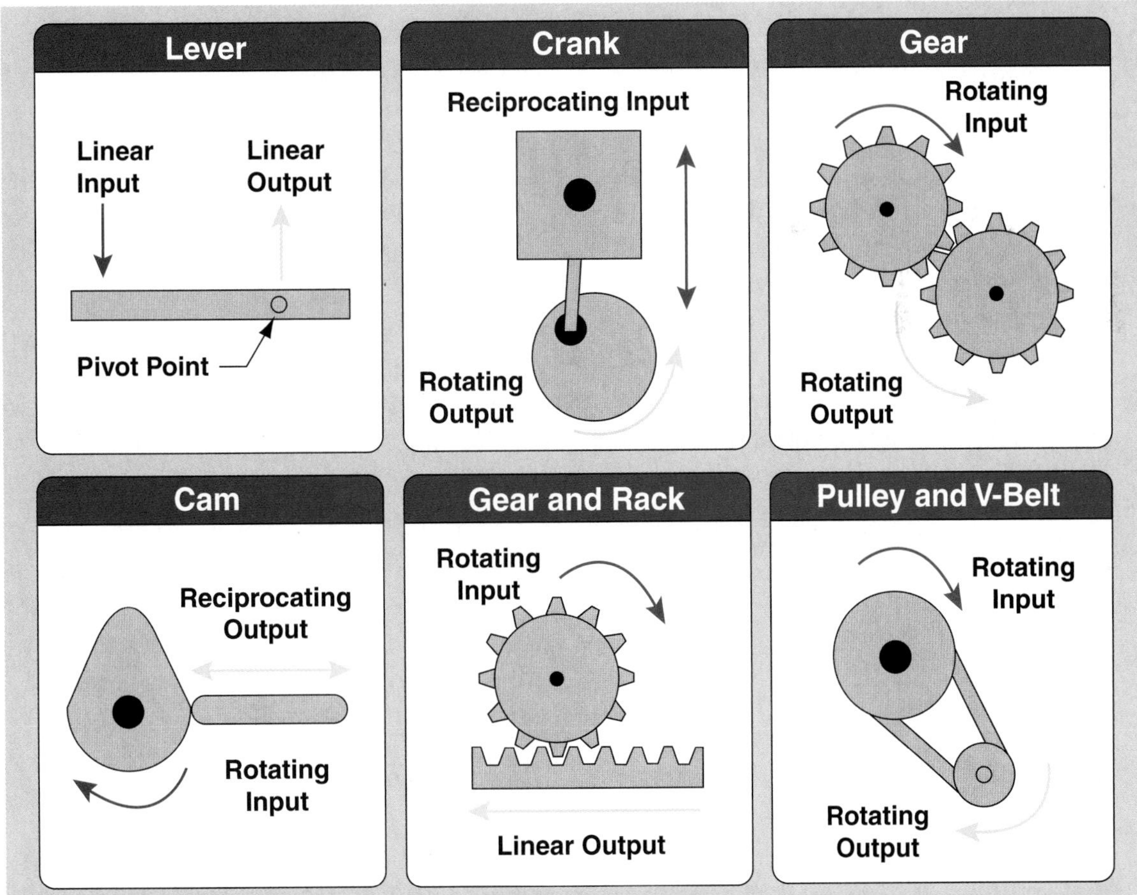

Figure 28-25. There are six important mechanical techniques that change the type, direction, or speed of a moving force.

Connections to Technology: **Science**

Laws of Gases

As we noted in this chapter, the internal combustion engine is powered by the expansion of gas produced by burning fuel. We were able to create and now use this artifact successfully because of various scientific discoveries regarding the properties of gases.

For example, the scientist Robert Boyle studied various gases and saw that a relationship exists among gases' pressure and volume. Boyle's law states that when a temperature is held constant, the volume of a fixed mass of gas varies inversely with the pressure. That is, when the pressure increases, the volume of gas decreases. The pressure doubles when the gas is compressed to half its volume. In the internal combustion engine, the pressure increases against the cylinder walls and piston.

Another scientist, Jacques Charles, also studied the behavior of gases. His law states that the volume of a fixed mass of gas at a constant pressure varies directly with absolute temperature. That is, if pressure is kept constant, volume and temperature are directly related—as the temperature increases, the volume increases. How does this law apply to the internal combustion engine?

rotating motion into reciprocating motion. A cam with a large lobe (extended portion) creates longer strokes for the reciprocating member. The force is reduced, however. The valves in an internal combustion engine are opened using a cam system.

- ***Gear and rack:*** A rotating gear meshes with a bar that has gear teeth along its length (rack) that *changes rotating motion into linear motion.* As the gear turns, it slides the rack forward or backward. Rack-and-pinion (gear) steering for automobiles uses this system.
- ***Pulleys and V-belt:*** Two pulleys with a V-belt stretched between them that *change the speed or power of a motion.* As one pulley turns, the V-belt moves, which in turn rotates the second pulley. The two pulleys rotate in the opposite direction. If the force is applied to a larger pulley, the smaller pulley turns faster but has less power. If the power is applied to the smaller diameter pulley, the larger diameter pulley turns slower, but it turns over a greater area. Many machines are driven by pulleys and V-belts.

Under heavy loads the V-belt can slip as it drives the pulleys. The problem is overcome by using gears and a chain drive similar to the one used on bicycles.

Fluid Power Systems

Fluid power systems use either liquids or gases to transfer power from one place to another. Systems that use air as the transfer medium are called ***pneumatic systems.*** Liquids (usually oil) are used in ***hydraulic systems.***

Generally, these systems contain two cylinders with movable pistons, a pump, valves to control the flow, and piping to connect the components. **Figure 28-26** shows the three basic uses for hydraulic systems.

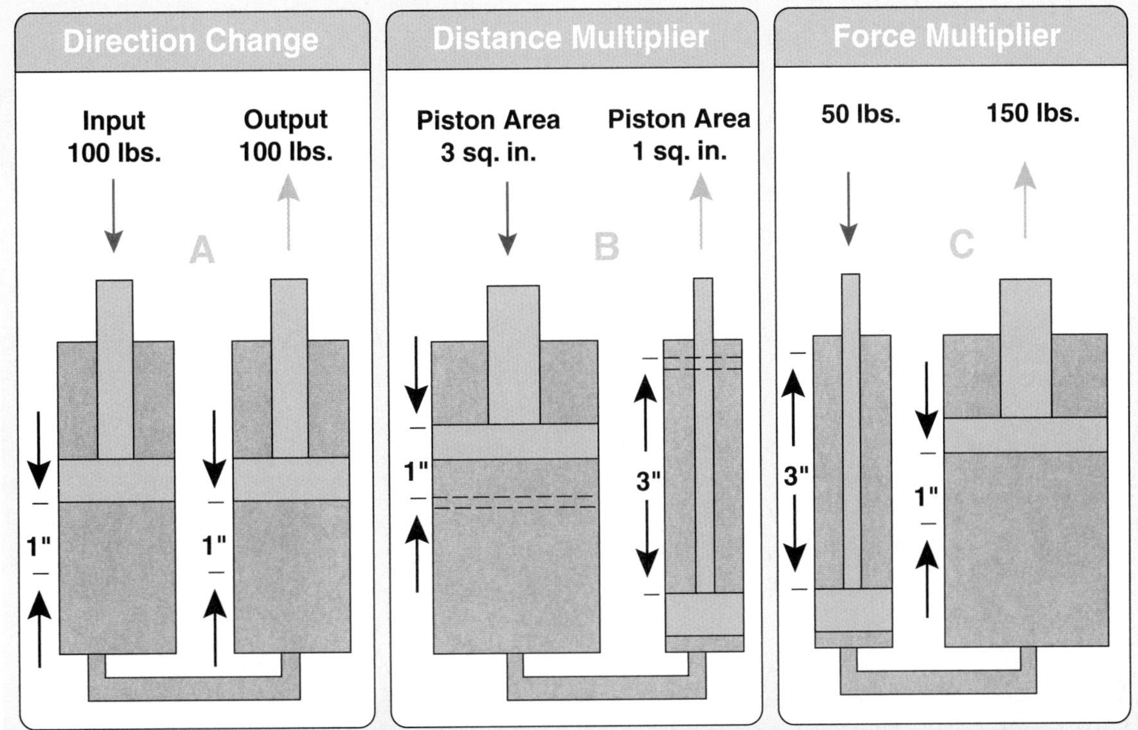

Figure 28-26. Hydraulic systems can be used in three ways. A—Change the direction of a force. B—Increase the distance of the force. C—Increase the strength of the force.

The view on the left shows a typical power transfer system. It neither increases the distance nor the intensity of the force. The system simply moves the force from one location to another, and it changes the force's direction. The fluid is forced downward by the piston in the left cylinder. This action causes the fluid to flow into the right cylinder. The fluid's movement causes the right piston to move upward.

The center drawing shows a distance multiplier. The left piston is forced downward one inch. Because the piston has an area of three square inches, the motion displaces three cubic inches of fluid. This causes three cubic inches of fluid to flow into the right cylinder. The right piston's area is only one square inch. Therefore, the piston must move upward three inches to accommodate the three cubic inches of fluid.

The drawing on the right shows a force-multiplying system. A force of 50 pounds is applied to the left piston, which has an area of one square inch. Thus, the original force is 50 pounds per square inch (psi) on the fluid beneath the piston. This force is transferred to the right piston, which has a three-square-inch area. The 50 pounds per square inch force from the piston on the left exerts a 150 pound per square inch upward force to the piston on the right.

Pneumatic systems operate in a similar way. The force and distance movement calculations are more difficult, however. Liquids do not compress, so nearly all the force is transferred from one cylinder to the other. A slight amount of the force is used to overcome the friction of the piston and the fluid in the pipes.

Air can be compressed, however. Therefore, some of the force in pneumatic systems is used in reducing the volume of the air in the system. The remainder of the force is applied to moving a load.

Summary

We spend much time and effort converting energy into different forms. The four major categories of energy conversion systems are inexhaustible energy converters, renewable energy converters, thermal energy converters, and electric energy converters.

Such devices as windmills, waterwheels, and solar collectors convert the inexhaustible energy sources of wind, water, and sun, respectively, into other forms. We use thermochemical and biochemical conversion systems to change renewable energy sources. Two major sources of thermal energy conversion are fuel converters and heat pumps. Finally, electricity produced in a generating plant is passed through transformers that increase and decrease output voltage.

We apply energy to do work. Motion is always present in work, and one activity of energy conversion involves changing motion. This action is called power transmission. The two basic systems used to change motion are mechanical power and fluid power systems.

Key Terms

active collectors
anaerobic digestion
biochemical processes
biomass resource
cam
conduction
convection
crank
direct active solar system
direct gain solar system
external combustion engines
fermentation
fuel converters
gasification
gear and rack
gears
heat pump
hydraulic systems
hydroelectric generating plant
indirect active solar system
indirect gain solar system
insolation
internal combustion engines
isolated solar system
lever
linear motion
liquefaction
liquidification
ocean mechanical energy conversion systems
ocean thermal energy conversion systems
passive collectors
photovoltaic cell
pneumatic systems
prime mover
pulleys and V-belt
pyrolysis
radiation
reciprocating motion
rotary motion
solar converter
thermochemical conversion
waterwheel
windmill

Test Your Knowledge

Write your answers on a separate piece of paper. Please do not write in this book.

1. *True or false?* An energy converter has energy as both its input and output.
2. Name one device created by earlier people that used wind energy to fulfill a need.
3. What is one difference between passive and active solar collectors?
4. Name one geothermal energy conversion system.
5. Name one of the two types of ocean energy conversion systems.
6. Biomass conversion uses two processes: thermochemical and _____.
7. Name one difference between internal and external combustion engines.
8. Name one way thermal energy is used to heat buildings and homes.
9. *True or false?* A steam-powered electrical plant might use fossil fuels to produce steam.
10. The two basic systems used to change motion are mechanical power systems and _____ systems.

Applying Your Knowledge

1. Design and construct a passive solar device that will heat the air in a shoe box. Prepare a sketch of the device and an explanation of how it works.
2. Select a major energy conversion system such as a coal-powered electric generating plant. Explain the advantages and disadvantages of the conversion system. Next, list at least two energy converters that could be used to replace the system you selected. List their advantages and disadvantages.

Section 8 Activities

Activity 8A

Design Problem

Background

All technological devices use energy. Often, this energy is converted from one form to another. For example, a motor changes electrical energy into mechanical energy. In addition, the motion an energy converter produces is often transformed. For example, a crankshaft changes the reciprocating motion of an engine's piston to rotating motion.

Situation

You are employed as a designer for a children's technology museum. The curator wants a series of models showing how humans change types of motion. She wants models demonstrating how the rotating motion a vertical waterwheel produces is changed into (1) reciprocating motion for a colonial sawmill or (2) pounding motion for a forge. See **Figure 8A-1.**

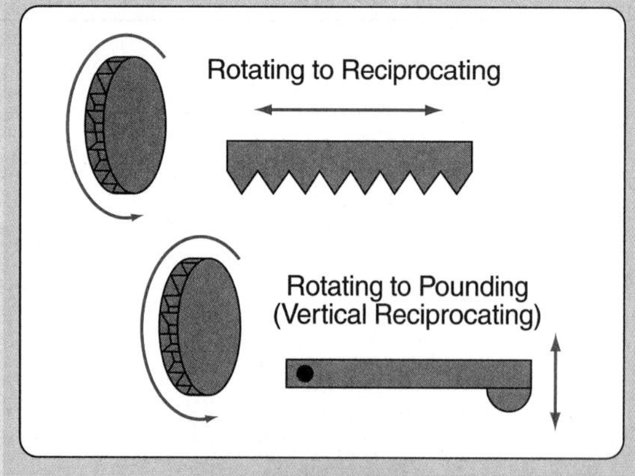

Figure 8A-1

Challenge

Select one of the two models listed above. Design and build a working model, using wood, cardboard, or other easily worked materials. See **Figure 8A-2.** This model converts vertical rotating motion into horizontal rotating motion.

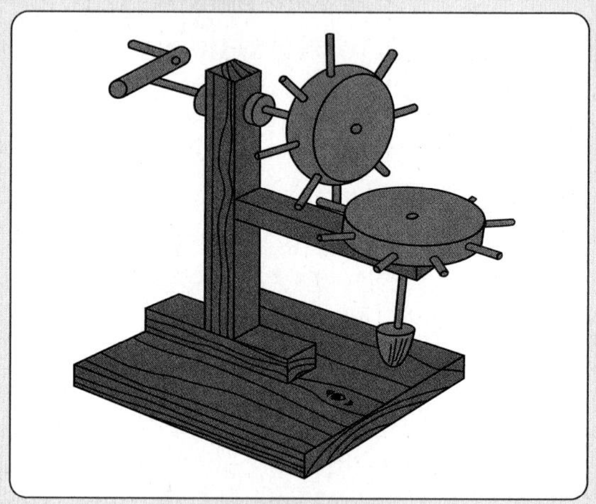

Figure 8A-2

Optional

Prepare a display, using your model, explaining your model's function and history.

Activity 8B

Fabrication Problem

Background

Using inexhaustible energy sources is a major challenge for the future. An important source is wind.

Challenge

Build and test a working model of a wind powered electric generation system.

Materials

- One piece 3/4" x 5 1/2" x 8" wood base (pine, fir, or spruce)
- One piece 3/4" x 18" strip of thin gauge sheet metal
- Two 1/2" No. 6 sheet metal screws
- One small electric motor
- One multimeter
- 24" electrical hookup wire
- One 3-blade or 4-blade propeller

Procedure

Building the Prototype

1. Obtain a piece of wood with dimensions of 3/4" x 5 1/2" x 8".
2. Cut a strip of sheet metal 3/4" wide and 18" long.

Safety

Follow all safety rules your teacher demonstrates!

3. Drill a 1/8" diameter hole 1/2" from each end of the strip of sheet metal.
4. Form the sheet metal strip around the motor. See **Figure 8B-1.** This forms the motor mount.
5. Bend the sheet metal strip to form 1" long tabs that can be mounted to the wood base.

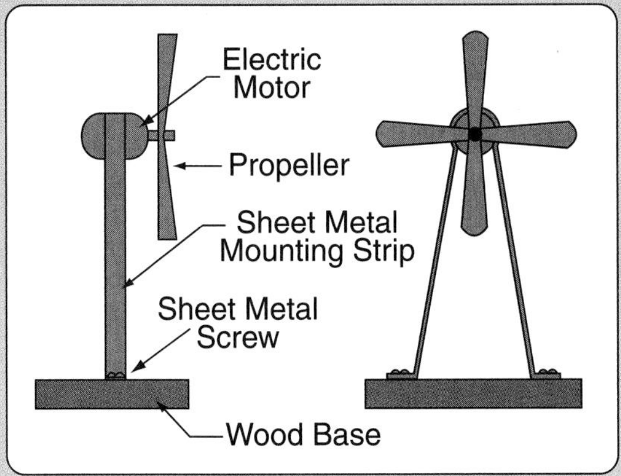

Figure 8B-1

6. Use the two sheet metal screws to attach the motor mount to the base.
7. Attach the propeller to the motor.
8. Press the motor into the loop of the motor mount.
9. Attach a piece of hookup wire to each wire coming out of the motor.
10. Solder your joints and wrap them with electrical tape.
11. Set the multimeter to read current.
12. Attach the wires to the multimeter. See **Figure 8B-2.**

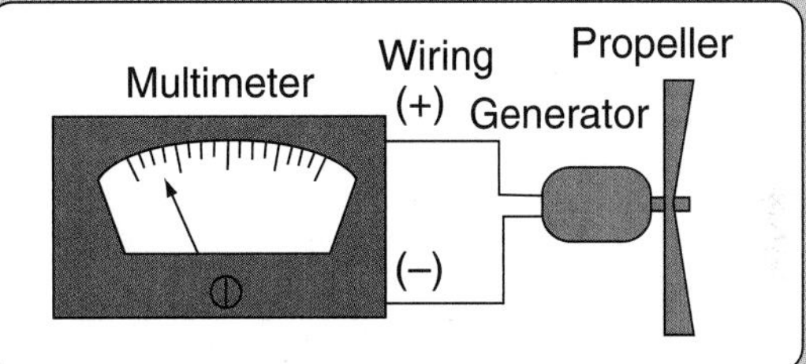

Figure 8B-2

Testing the Device

1. Place the wind powered generator in front of a window fan.
2. Turn the fan on to its low setting.
3. Observe the rotation of the propeller.
4. Note and record the meter reading on the multimeter.
5. Repeat steps 3 and 4 using the fan's medium setting.
6. Repeat steps 3 and 4 for the fan's high setting.
7. Turn off the fan.

Optional

1. Change the propeller to a model with a different number of blades or change the angle at which the wind is hitting the propeller to 30° off center.
2. Repeat steps 1 through 4.
3. Write a report explaining your observations.

Section 9

Applying Technology: Meeting Needs through Biorelated Technologies

Tomorrow's Technology Today
Genetically Modified Foods

If you have ever spoken with a farmer, you are probably aware that farmers face many obstacles in producing the world's food supply. Besides dealing with insects and other pests, weeds, the weather, and rotting food, the world's population is increasing at a higher rate than our current food production capabilities, placing additional burdens on our food producers. Fortunately, the latest research in genetic engineering suggests that we may soon have the answer to all of these problems. Genetically modified (GM) foods may be able to put these concerns in the past and offer even more exciting benefits.

GM foods can be created by altering the genetic makeup of living organisms to produce organisms with desired characteristics. Specific genes can be inserted into a plant, or the genes from two or more different species can be combined to produce a transgenic organism. Using these techniques, scientists have already been successful in creating plants that produce human insulin, crops that are resistant to herbicides, and plants that are unaffected by insects and fungus. They have also successfully engineered vegetables that can remain on the vine longer to ripen, such as the Flavr Savr® tomato. These tomatoes have less of a ripening gene, so they are firmer and tastier when they reach the grocery store.

Besides these benefits, GM foods have the potential to keep people healthier. For example, almost half of the world's population depends on eating rice daily for most of its caloric intake. This kind of diet can cause deficiencies in vitamin A, which is already a serious health problem in many countries. A new kind of rice, called Golden Rice®, has three extra genes inserted, which increase the synthesis of beta-carotene, a producer of vitamin A. This rice is being tested and evaluated for safety, but if it is deemed acceptable, it may greatly improve the health of much of the world's population.

Further down the road, there are even more exciting possible applications for GM foods. Research is currently being done on ways to introduce other vitamins into foods. If this is successful, you may never need to take a multivitamin in pill form again. There also is the possibility that vaccines may someday be inserted into foods. This would allow you to eat a GM banana instead of going to the doctor and getting a shot!

All of this sounds promising, but great debate also exists over the use of GM foods. Many scientists are concerned about the accidental creation of "superweeds," weeds that are resistant to herbicides, and inadvertent harm that may be done to other wildlife. Economists worry that GM foods would hurt traditional farmers and increase the gap between wealthy and poor countries. Many consumers are concerned about the safety of consuming artificially modified foods, especially when these foods are not labeled as such. All of these concerns are being studied, but the research and development continue.

Obviously we need to consider the benefits and risks regarding the use of GM foods. There is no doubt, however, that this emerging technology holds exciting promise.

Chapter 29
Agricultural and Related Biotechnologies

Learning Objectives

After studying this chapter, you will be able to:

- Define agriculture.
- Explain the difference between science and technology in agriculture.
- List the major types of agriculture.
- Name major crops grown on farms.
- Identify major equipment used in crop production.
- Describe how technology is used in hydroponics.
- List the types of livestock farms.
- Describe the technology used in raising livestock.
- Describe how technology is used in aquaculture.
- Define biotechnology.
- Describe ways that biotechnology can be used in agriculture.

People have a number of basic needs and wants. Two of the primary needs are food and clothing. Both of these needs are directly addressed by agriculture. This human activity involves using science and technology to grow crops and raise livestock. ***Agriculture*** is people using materials, information, and machines to produce the food and natural fibers. It takes place on farms and ranches around the world.

Modern farming uses both scientific and technological knowledge. Two important branches of the life sciences support agriculture. These areas of the life sciences are plant science and animal science. They are used in cross-pollinating plants to improve crops and in crossbreeding animals to improve livestock. Other sciences help farms manage their operations. For example, the science of weather allows farmers to plan planting and harvesting activities. The sciences that describe the seasons allow farmers to select appropriate crops and livestock to raise. Knowledge of plant nutrients allows them to select appropriate fertilizers. These and other sciences have helped farmers become more efficient.

Figure 29-1. Technological advancements have greatly changed agriculture.

Technological advancements have also caused many massive changes in farming, **Figure 29-1.** For example, they have caused changes in the size of farms and how farmers go about their work. New and modern machines and equipment allow for greater production. Fewer people are needed to grow more food on fewer acres. Technological advances have helped people work with greater ease and efficiency. These advancements can be attributed to ***agricultural technology.*** This technology uses technical means (machines and equipment) to help plant, grow, and harvest crops and raise livestock.

Types of Agriculture

Agriculture involves three major components that are managed by people: (1) land, (2) buildings and machinery, and (3) crops and livestock. This description suggests that there are two types of agriculture, **Figure 29-2.** The first is ***crop production,*** which grows plants for various uses. It provides food for humans, feed for animals, and natural fibers for many applications. Crop production produces trees and plants for ornamental use such as landscaping. It grows trees for lumber and paper production. It produces basic ingredients for medicines and health care products. Crop production also provides materials for many industrial processes, such as textile weaving, plywood manufacture, and food processing.

The second type of agriculture is called ***animal husbandry.*** This area involves breeding, raising, and training animals. These animals may be used for food and fiber for humans. In some cases they are used to do physical work. This is especially true in developing countries. Many animals are also raised as hobbies or are used for riding and racing.

The United States uses more than 40 percent of its total land for farming activities. A total of about 350,000,000 acres are used for crop production. About half of this land is used for raising wheat and corn. Another billion acres is used for pastures, ranges, and forests.

Crop Production

Many different crops are raised on North American farms and ranches. Some of these crops were originally found in other areas of the world. For example, China and central Asia gave us lettuce, onions, peas, sugar cane, and soybeans. Rice, sugar cane, bananas, and citrus fruits came from Asia. The Middle East, southern Europe, and North Africa gave us wheat, barley, and oats along with alfalfa and sugar beets. Corn, beans, tomatoes, potatoes, peanuts, tobacco, and sunflowers came from North and South America.

Crop Production

Animal Husbandry

Figure 29-2. Agriculture involves growing crops and livestock. (U.S. Department of Agriculture)

Figure 29-3. Grain such as this corn is grown widely for human and animal food. (U.S. Department of Agriculture)

A number of major crops are grown in North America. The most widely grown are forms of ***grain,*** **Figure 29-3.** Grain crops are members of the grass family that have large edible seeds. The commonly grown grains are wheat, rice, corn, barley, oats, rye, and sorghum. All of these grains can be used in food products. Corn, barley, oats, and sorghum are widely used in animal feed.

Vegetables are another important farm crop. Vegetables have edible leaves, stems, roots, and seeds that provide important vitamins and minerals for the daily diet. Vegetables include root crops such as beets, carrots, radishes, and potatoes. They also include leaf crops such as lettuce, spinach, and celery. Other vegetables provide food from their fruit and seeds. This group includes sweet corn, peas, beans, melons, squash, and tomatoes.

Fruits and ***berries*** are grown in many parts of the country, **Figure 29-4.** These plants are cultivated for their edible parts. The major fruit crops grown in temperate climates include apples, peaches, pears, plums, and cherries. Citrus fruits (oranges, lemons, limes, grapefruits, and tangerines), olives, and figs are grown in warmer climates. Tropical fruits include bananas, dates, and pineapples.

Nuts are grown in selected parts of the country. They are grown for their hard-shelled seeds. Walnuts, pecans, chestnuts, almonds, and filberts (hazelnuts) are grown in temperate climates. Palm-oil nuts and coconuts are grown in more tropical areas. Peanuts and coconuts are the most important nut crops and are significant sources of food and oil.

Forage crops are grown for animal feed, **Figure 29-5.** These plants include hay crops such as alfalfa and clover. Grasses used for pasture and hay are also included in this group.

A number of ***nonfood crops*** are grown on farms. These plants include tobacco, cotton, and rubber. Nonfood crops also include nursery stock grown for landscape use and Christmas trees.

Technology in Crop Production

Crops are no different from any other living thing. They have a set life cycle. They are "born" when a seed germinates, continue through a growing cycle, and then mature. After a time they die. To be of benefit to people and animals, the crop must be harvested before it spoils or shatters.

Figure 29-4. Berries and fruit are important crops grown on American farms. (U.S. Department of Agriculture)

Figure 29-5. Forage crops provide animal feed. (Case IH)

Farming takes advantage of a plant's life cycle through four major processes. These processes are planting, growing, harvesting, and, in some cases, storing.

In the past, farming was a very labor-intensive activity. Most of the population was involved in raising crops and animals. Technological advancements during the past two hundred years, however, have changed all this in many countries. Now just a small percentage of the population in developed countries is involved in agriculture. Farming has become very equipment intensive. To a large extent, machines and equipment have replaced human and animal labor.

Farm equipment is used at all stages of crop production. This equipment can be classified into eight major groups:

- Power or pulling equipment
- Tillage equipment
- Planting equipment
- Pest-control equipment
- Irrigation equipment
- Harvesting equipment
- Transportation equipment
- Storage equipment

Power or Pulling Equipment

People have a long history of replacing human power with other power sources. Over most of recorded history, people used animals to pull loads. The invention of the agriculture ***tractor*** in 1890 changed this, however. During the 1900s this new power source replaced animal power on most farms. Today, the farm tractor provides the power to pull all types of farm equipment.

Farm tractors can be either wheel tractors or track machines, **Figure 29-6.** Most agricultural tractors are the wheel type. Some wheel tractors have rear power wheels. This tractor usually has smaller front wheels. Other wheel tractors have power to all wheels. These tractors generally have the same size tires on all wheels. For additional traction, both types of wheel tractors may have dual drive wheels.

Track-type tractors are used for special purposes and are generally slower than wheel-type tractors. They are more suited for muddy fields. They sink less and have less slippage. Also, they compact the soil less and do not produce wheel ruts.

Figure 29-6. The two kinds of farm tractors are wheel and track tractors. (Case IH, Deere and Co.)

Most tractors are designed on a unit principle. The engine, transmission, and gearbox are a single unit. This structure provides a rigid backbone for the machine. The steering and drive wheels are attached to the basic unit. Implements are attached to it or pulled behind the tractor.

Tillage Equipment

The soil must be prepared before crops can be planted. Residue from previous crops must be handled. The seedbed must be conditioned by breaking and pulverizing the soil. This process is called ***tillage*** or tilling the soil.

The most important piece of tilling equipment is the ***plow.*** It performs the same task as the spade that is used in flower gardens. It breaks, raises, and turns the soil. This process loosens the ground and brings new soil into contact with the atmosphere.

Three major types of plows exist: the moldboard plow, the disc plow, and the chisel plow. The *moldboard plow* is made up of a frame and several plowshares, **Figure 29-7.** When the plow is pulled through the earth, it cuts and rolls the soil.

Discs plows have a frame that has several discs mounted on an axle, **Figure 29-8.** The discs and axles are set at a steep angle to the direction of travel. When pulled, the discs turn and then cut and loosen the soil.

Chisel plows have a set of shaped chisels attached to a frame. When the plow is pulled through the earth it breaks up the soil. It does not lift and turn the soil, however. Chisel plows are used in grain stubble and where the soil needs little tilling.

Planting Equipment

Once the soil is prepared, the crop can be planted. Planting involves two actions that can be done separately or together.

The first action involves *applying fertilizer.* ***Fertilizer*** is a liquid, powder, or pellet that contains important chemicals. It primarily delivers nitrogen, phosphorus, and potassium that will encourage and support plant growth. The fertilizer can be applied before, during, or after seeds are planted.

Figure 29-7. This moldboard plow is cutting and rolling the soil. (Case IH)

Figure 29-8. This disc plow is cutting and loosening the soil. Notice the close-up of the discs on the right. (Deere and Co., U.S. Department of Agriculture)

Figure 29-9. This is a specially designed grain drill that plants seeds in soil that has not been tilled. (Deere and Co.)

The fertilizer is applied with special equipment or along with a seed planter. In some cases, fertilizer is scattered (broadcast) on the ground before final tilling and planting. Liquid and gaseous (anhydrous ammonia) fertilizers can be injected into the soil by a machine with a series of knives.

The second action is *planting the crop.* This may involve putting seeds or starter plants into the fields. Grains, grasses, and many vegetables are started from seeds planted directly in soil. Some vegetables, such as tomatoes, cabbage, cauliflower, and broccoli, are started from seeds in nursery greenhouses.

The most widely used seed planter is the ***grain drill,*** **Figure 29-9.** It is pulled behind a tractor. As it moves across the field, it opens a narrow trench, drops seeds, and closes the trench.

Pest Control Equipment

In nature, not all plants that sprout grow to maturity. Diseases and insects kill some of them. Neighboring plants crowd others out. Farm crops face many of the same dangers. Farmers can take action to reduce these dangers by using chemical and nonchemical pest control measures. These techniques use special machines and devices.

Chemical control techniques generally use a liquid spray to control weeds and insects. These sprays include herbicides to control weeds and pesticides to control diseases and insects. Ground equipment and aircraft can apply both of these materials.

Nonchemical pest control uses machines and other devices to control pests. A ***cultivator*** is the machine that usually performs this task. These machines are a series of hoe-shaped blades that are pulled through the ground, **Figure 29-10.** The blades break the crust and allow rain and irrigation water to enter the soil. They also cut off and pull out weeds. (A weed is any plant that is out of place.)

Irrigation Equipment

Rainfall is sufficient to raise crops in some parts of the world but not all. Many places are too dry or the rain comes at the wrong time for successful farming. Irrigation systems can be used to support agriculture in these dry or unpredictable climates.

Irrigation is artificial watering to maintain plant growth. Irrigation systems must have:

- A constant and dependable source of water
- A series of canals and ditches to move the water from its source to various crops
- A way to control the flow of the water in the ditches and canals

The common sources of irrigation water are lakes, rivers, or underground aquifers. Various ways are used to control the water at its source. Water in a river can be restricted by a dam to form a reservoir, **Figure 29-11.** A dam at the outlet of a natural lake can control the level of the lake and divert water for irrigation. A well and pump can be used to obtain underground water, **Figure 29-12.**

The water is moved from the source to farm fields by canals and pipes. There, one of four irrigation methods can be used to water the crops.

If the land is level, ***flood irrigation*** can be used. These systems use a large quantity of water that advances across the fields. The water is brought to one end of the field by ditches or pipes. The water is released from lateral ditches or pipes with holes along their length. As additional

Figure 29-10. This field of potatoes is being cultivated to remove weeds and loosen the soil. (U.S. Department of Agriculture)

Figure 29-11. A dam such as this one can be used to form a reservoir. The water can be used for irrigation, to generate electricity, and for recreation.

Figure 29-12. This pump of a deep well supplies irrigation water for a sprinkler system.

water is released, gravity causes the water to flow across the field. In some cases, another set of ditches or pipes carries off excess water.

In row crops ***furrow irrigation*** can be used. This system uses small ditches, called furrows, that are created between the rows of plants. The furrows guide the water as it flows from one end of the field to the other, **Figure 29-13.** Pipes or siphon tubes are often used to control the flow and allocate water to various sections of the field.

Another irrigation system is called ***sprinkler irrigation,*** **Figure 29-14.** This approach produces artificial rain to water crops. It is used on uneven ground or where the amount of water applied must be closely controlled. Sprinkler irrigation uses less water than flood irrigation systems.

In sprinkler irrigation, a pump is used to force water from the source into the main distribution lines. The pressurized water flows through the main lines to a series of lateral pipes. These pipes extend at right angles from the main line. Each lateral line has a series of small-diameter

Figure 29-13. The furrows between these rows of potatoes are used to carry irrigation water. (U.S. Department of Agriculture)

Figure 29-14. This sprinkler irrigation system is operating on a farm that raises nursery plants.

pipes sticking upward at even intervals. At the end of each standpipe is a sprinkler head. The water in the lateral lines enters the sprinkler heads, which spray water onto the land. Valves between the main and lateral lines shut off or control the water flow.

Many sprinkler systems have several straight sprinkler (lateral) lines that are used to irrigate the fields. Each line applies water on a long, narrow band across the field. The water is allowed to run for a set time to irrigate the strip of land on each side of the sprinkler lines. Then the water is shut off the lateral lines with a valve, and the lines are disconnected. They are moved by hand or rolled under power to the next position (set) along the main line. Here they irrigate the next strip across the field. After a series of moves the entire field is irrigated.

In some cases, many sprinkler lines are used to cover the entire field. When the lines are turned on, the entire field is irrigated at once. This is called a solid-set sprinkler system. This approach eliminates the need to move individual lines. It is used where the crop must be watered often or where frost protection must be provided using the sprinkler system.

Another type of sprinkler system is called a ***pivot sprinkler.*** These systems use one long line. It is attached at one end to a water source. The line pivots around this point on large wheels that are powered by electric motors. The line is constantly moving very slowly in a circle. Sprinkler or mist heads apply the water as the line pivots.

Drip irrigation is the fourth type of irrigation. The system delivers water slowly to the base of the plants. Drip systems use main lines to bring water near the plants. Individual tubes or emitters extend from the main lines to each plant. These tubes apply water, which soaks into the ground around the roots. This system ensures that each plant is properly watered, and it reduces the amount of water lost to evaporation. Drip systems are used in many orchards and vineyards (grape fields).

Harvesting Equipment

To be of value, a mature crop must be harvested. Different harvesting machines exist. The most widely used is the grain ***combine,*** **Figure 29-15.** This machine can be used to harvest a wide range of grains and other seed crops.

There are several stages in the operation of a combine, **Figure 29-16.** First, a rotating reel pulls the grain into a cutter bar. There the tops of the plants that contain the grain or seeds are cut off and drawn to the center of the machine. The heads and straw move into the machine, where a revolving cylinder separates the grain from the heads. The grain and straw move onto straw walkers. These devices move the straw to the back of the machine. As the straw moves, the grain falls through holes in the walkers onto a grain pan. There blasts of air blow away chaff and other lightweight materials. The remaining grain is lifted from the grain pan into storage

Figure 29-15. This combine is harvesting a grain crop. (Deere and Co.)

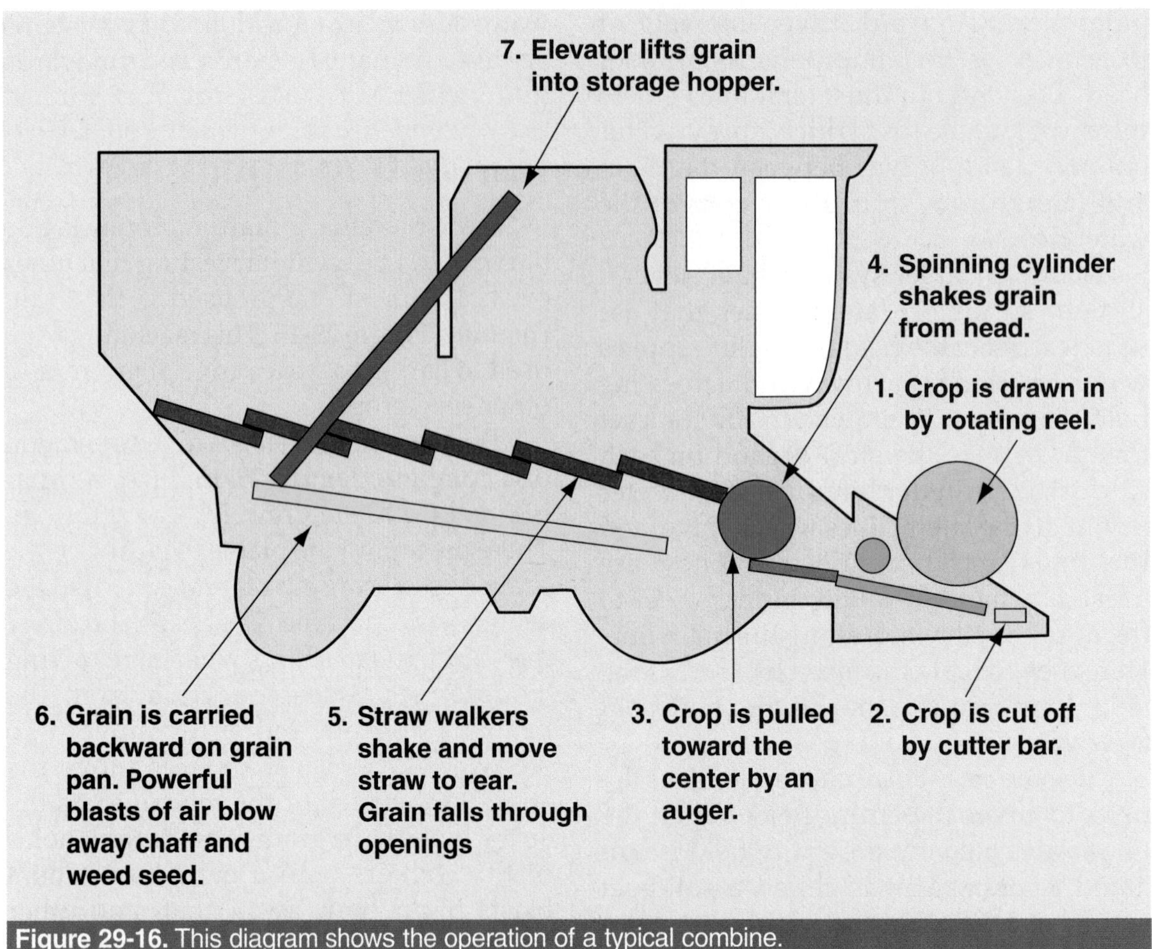

Figure 29-16. This diagram shows the operation of a typical combine.

hoppers by an auger. The unwanted straw and waste materials are conveyed out the back of the machine and drop onto the ground. When the storage hopper is full, the grain is unloaded onto a truck or wagon by another auger.

Specialized harvesting machines have been developed for other crops, **Figure 29-17.** Cotton is removed from the plants with a cotton picker. The picker strips the cotton bolls from the plant and deposits them in a bin.

Potatoes and onions are dug from the ground with mechanical diggers or harvesters. The crop moves across conveyors, where people remove rocks and dirt clods. The remaining potatoes or onions are conveyed onto trucks.

Peanuts are dug using special-purpose harvesters. Corn can be harvested with either a combine or a corn picker that strips the ears from the stalks.

Vegetables can be harvested by special-purpose machines or by hand. Special mechanical pickers are used to harvest green beans, sugar peas, and sweet corn. The vegetables are stripped from the plants in one pass. The crop is conveyed into a bin that is later dumped into trucks for delivery to canning or freezing plants.

Fruits are often picked by hand and placed in boxes. Special machines may be used for some fruits and nuts, however. These machines shake the tree, causing the fruit or nuts to fall into raised catching frames. The crop is gathered from the frames and hauled from the orchards.

Grass and alfalfa hay are harvested by a series of special machines. The crop must first be cut and laid on the ground to dry.

Figure 29-17. Shown here are some special-purpose harvesting machines. (Deere and Co., Case IH)

In some cases a mover might be used to cut the plants and let them fall on the ground. After the hay has dried for a day or more, a rake is used to gather it into ***windrows*** (bands of hay). In large-scale hay operations, however, a windrower or ***swather*** is used for these processes, **Figure 29-18.** This machine cuts and windrows the hay in one pass over the field.

After the hay in the windrows has dried, it is usually baled. A machine called a ***baler*** is used to gather, compact, and contain the hay. Balers produce two basic types of bales: square and round. The most common is the *square bale,* which can weight up to several hundred pounds. The standard bale, however, is about 4 feet long and weights 75 to 125 pounds. The bales are bound with either baling twine or wire, making the hay easy to handle and to store.

Transportation Equipment

Crops grow in fields that are spread over a farm. The harvested crops must be moved from the fields to storage or processing plants. This movement can be of two types. First, the crop may be moved from the field to a central location or storage facilities on the farm.

Second, the crop may be moved on roads and highways from the farm to processing plants or commercial storage sites. In some cases, the crops are moved directly from the harvesting equipment to processing plants or central storage, **Figure 29-19.**

Often, crops are moved short distances with wagons and medium-duty trucks, **Figure 29-20.** Longer hauls are done on heavy-duty (semi) trucks, railcars, barges, and ships.

Figure 29-19. This truck will move the sugar peas from the harvester to a vegetable freezing plant.

Figure 29-18. This windrower or swather is cutting and windrowing hay. Other swathers are self-propelled, eliminating the need for a tractor. (Deere and Co.)

Storage Equipment

Many crops are stored on farms or in commercial locations before they are sent to processing plants. Most grain crops are stored in silos or buildings at grain elevators, **Figure 29-21.** Here the grain is dumped into a pit with sloping sides. At the bottom of the pit is a grain elevator. The most common type of elevator is a bucket elevator, **Figure 29-22.** It lifts the grain using a series of buckets or pans attached to a moving belt. The moving bucket digs into the pile of grain in the pit and moves it upward. When the grain reaches the top of the elevator, it is dumped in a shoot or pipe that leads to the top of the silo.

Hay is stored in a dry location so that rain and snow will not damage it. The most common hay storage structure is a hay barn. It is simply a roof attached to long poles and does not have enclosed sides or ends, **Figure 29-23.** In very wet climates, the sides may be enclosed to give added protection.

Many vegetables and fruits are stored in climate-controlled (cold-storage) buildings. These insulated buildings use cooling equipment to maintain the crop at an appropriate temperature. The crops will be transported from these storage sites to processing plants throughout the country and world, as demand requires.

Figure 29-20. This tractor with attached wagon is being used to move grapes from the field (vineyard) to storage.

Figure 29-21. Seen here is a typical rural grain elevator.

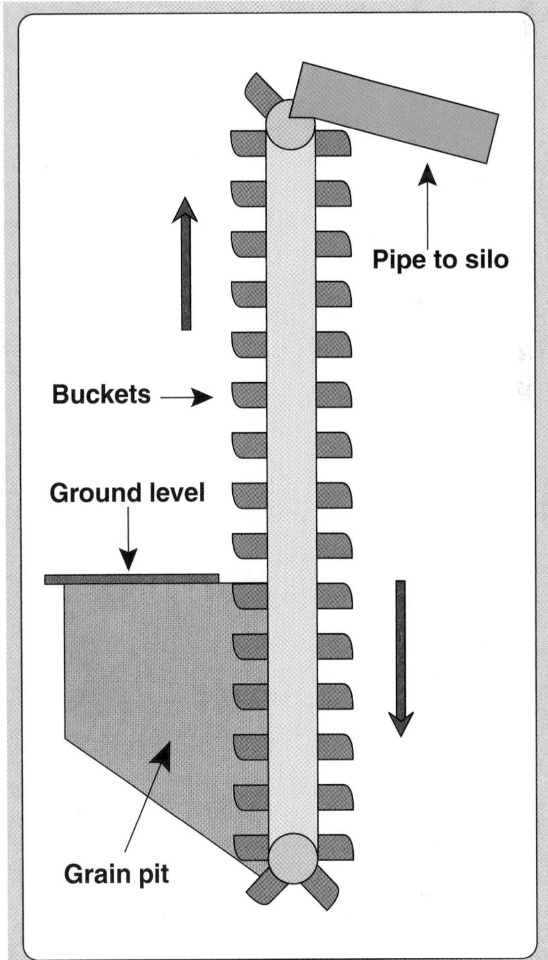

Figure 29-22. This diagram shows how a bucket elevator operates.

A Special Type of Crop Production

A unique type of crop growing is ***hydroponics.*** Hydroponics can be defined as the growing of plants in nutrient solutions without soil. Hydroponic systems have two things in common:

- The nutrients are supplied in liquid solutions.
- Porous materials (peat, sand, gravel, or glass wool) that draw the nutrient solution from its source to the roots support the plants.

Various hydroponic systems exist. The most practical commercial method is subirrigation. In this system, the plants are grown in trays filled with a coarse material such as gravel or cinders. At specific times, the materials are flooded with nutrient solution. The solution is allowed to drain off after each flooding. Each flooding feeds the plants as they grow toward maturity.

Hydroponic methods can be used to grow plants in greenhouses. They can also be used in areas where the soil or climate is not suitable for the crop. For example, hydroponics is used to grow tomatoes for winter use.

Figure 29-23. This hay shed is protecting the hay from rain and snow.

Raising Livestock

A major activity of agriculture is raising livestock, **Figure 29-24.** This includes the breeding and care of cattle, horses, swine (pigs), sheep and goats, horses, and poultry (chickens and turkeys). These animals are raised to provide meat, milk, or

Career Corner
Agricultural Workers

Agricultural workers are essential to growing and harvesting grains, fruits, vegetables, nuts, fiber, trees, shrubs, and other crops. They generally work on farms, on ranches, or in nurseries. These workers do many different jobs that require unique knowledge and skills they acquire through on-the-job training. Most farm workers receive low pay and often must perform strenuous work outdoors in all kinds of weather.

Figure 29-24. Livestock raising is a major part of agriculture.

materials for clothing and for recreational purposes.

Most livestock are raised on single-purpose farms. These farms include cattle ranches, dairies, and swine, horse, and poultry farms.

Livestock raising involves various technologies. These include:

- Constructing and maintaining livestock buildings
- Constructing and maintaining fences and fencing to establish feedlots and pastures
- Operating and maintaining buildings and machines used to feed animals
- Constructing, operating, and maintaining animal waste facilities

Livestock Buildings

Most livestock operations require specialized buildings. They are used to house animals and feed-processing equipment. The types of buildings will vary with the type of livestock being raised. Barns and sheds are built to protect animals from the weather, **Figure 29-25.** Some facilities house animals while their young are being born. Hog factory buildings contain animals in pens as they are raised for meat. Poultry houses contain chickens and turkeys raised for food. Laying or hen houses hold chickens that are raised for egg production. Stables house horses that are raised for pleasure riding and racing.

The processes for constructing agricultural buildings are the same as are used for other buildings. These processes are discussed in Chapter 17.

Figure 29-25. This barn protects dairy cattle from the weather.

Fences and Fencing

Farmers and ranchers construct barriers to separate their land from those of neighbors. Also, barriers that separate the farm into fields are needed. These barriers are called fences. Without fencing, livestock roam freely, damaging other crops.

Early barriers were made of rock, tree limbs, mud, and other natural materials. Later, smooth wire was adopted as a fencing material. It was produced by drawing hot iron through dies (block of steel with a hole in it). The smooth wire did not always keep the animals in the enclosure, however. This led to the development of barbed wire, which is a smooth wire with pointed wire added along it length, **Figure 29-26.** The technological advancement was so popular that more than 570 patents were issued for different designs for barbed wire.

Barbed wire was widely used because it was ideal for western conditions. The vast prairies and plains were divided and separated by barbed-wire fences. This led to range management, farming, and settlement.

Today, farmers and ranchers use several types of fences, including rail, barbed wire, woven wire, and electric. The type of fence selected depends on the livestock and crops being raised. Horses can run through a fence, causing injury to them. Cattle try to crawl over fences, whereas sheep try to crawl under them. Hogs attempt to root under a fence.

Rail fences are often used as border fences around farm buildings or homes. They are also popular on horse farms. Rail fences are made from posts set in the ground with boards or rails attached between the posts. The common materials used are treated wood, painted wood, vinyl-coated wooden boards, and PVC plastic.

Barbed-wire fences consist of wood or steel posts with strands of barbed wire attached to them. They usually have posts spaced 10 to 12 feet apart and use three to five strands of wire.

Woven-wire fences use posts with a special wire product, called woven wire, attached to them. The woven wire consists of a number of horizontal smooth wires

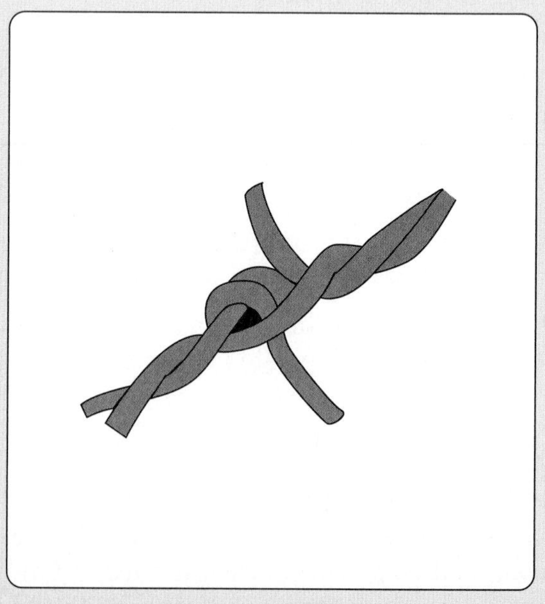

Figure 29-26. Barbed wire opened the West to ranching and farming. Left is a coil of barbed wire. Right is a reproduction of the 1874 patent drawing for barbed wire.

held apart by vertical (stay) wires. The horizontal wires are arranged with narrow spacing at the bottom and wider spacing at the top. The height of the fence will depend on the size and jumping ability of the animals that are contained. Typically the fences are 26 to 48 inches tall.

Cable-wire fences consist of 3/8-inch steel wire cables stretched from one anchor post to another. Heavy springs are attached to one end of each cable to absorb shock. The other end is rigidly attached to another anchor post. Each cable passes through holes in a number of line posts that are set between the two anchor (end or corner) posts, **Figure 29-27.**

Electric fences are temporary or permanent fences that use electrical charges (shocks) to contain animals in a field. They can be a separate fence or a strand of electric wire added to another type of fence.

Figure 29-27. This cable-wire fence has a pipe on top to add strength. (U.S. Department of Agriculture)

Buildings and Machines for Feeding

Most livestock farms have livestock-feeding equipment and buildings. The buildings contain machines that grind and mix feed for the animals. As noted earlier, hay barns may be built to protect hay from the weather. Silos are used to contain and protect grain and hay (silage), **Figure 29-28.**

Feed troughs or bunkers are required so that the animals can eat grain and hay. Water is provided using manufactured pumps and tanks.

Figure 29-28. These silos contain feed for a dairy.

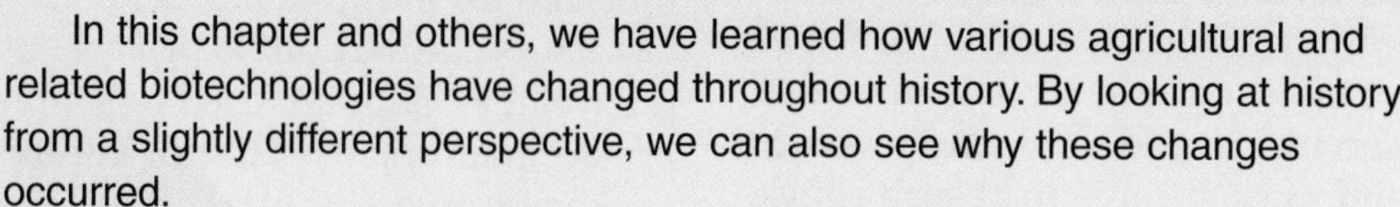

Connections to Technology: **History**

The Homestead Act and the Morrill Act

In this chapter and others, we have learned how various agricultural and related biotechnologies have changed throughout history. By looking at history from a slightly different perspective, we can also see why these changes occurred.

For example, two political events in 1862 forever changed the kinds of crop and livestock production that would be developed in the United States. In May of 1862, Congress, in an effort to promote movement to the west, passed the Homestead Act. This act gave 160 acres of land free of charge to any adult citizen who would live on the land for five years and develop it. In July of 1862, Congress passed the Morrill Act. Under this act, the federal government gave each state 30,000 acres of land for each congressional representative that state had. The land was to be sold, with the resulting monies used to establish colleges of agriculture and the mechanical arts.

Each act influenced agriculture in various ways. Because of the Homestead Act, for example, thousands of people moved west. They discovered that the land in the west was ideal for raising livestock. They also found that the dry conditions actually helped the growth of wheat. By the 1900s the plains area of the United States had become one of the world's major sources of that crop. Moreover, because of the growth of these types of farming in the west and the resulting competition, farmers in the east started turning to dairy and truck farming.

Because of the Morrill Act, many colleges and agriculture departments within universities were established. These colleges and departments had the research capabilities to help those in agriculture improve existing methods and products. Their pioneering research efforts continue to improve agricultural and related biotechnologies in the United Sates to this day.

Another event in the 1860s also greatly affected U.S. agriculture. That event was the completion of the transcontinental railroad at Provo, Utah, in 1869. Can you think of some reasons why this event made such an impact on agriculture?

Animal Waste Disposal Facilities

A major challenge for large-scale livestock production is animal waste disposal. The waste must be controlled so that it does not pollute streams, lakes, and underground water. Livestock farmers must plan to collect, store, treat, and apply animal waste to land properly. This requires that they identify sites for waste disposal and use appropriate land application procedures.

All of these actions require technological actions. They need a combination of structures and practices that serve the animal-feeding operation. Typically, this requires collection both of animal wastes and of other kinds of wastes, including feed and litter. In many cases, a lagoon is built to contain these wastes. This lagoon is a confined body of wastewater that holds animal and other waste. This waste is periodically removed from the lagoon and applied to land.

The type of equipment used to apply livestock waste to land depends on the type and consistency of the waste. Dry litter can be applied with a box (manure) spreader, **Figure 29-29.** Lagoon waste is often handled in two forms: wastewater and slurry. Wastewater is waste that contains less than 2 percent solids. A slurry mixture is agitated sludge and wastewater. Many farmers apply wastewater using their regular irrigation systems such as the ones discussed earlier in the chapter. Slurries require special pumping equipment and sprinklers that have a large nozzle.

Another technique is to inject the waste into the soil using special application equipment. This equipment uses a tank to haul the waste. When the tank arrives in the field, the waste is inserted into the soil using knife applicators. Animal manure also can be dried and sold to homeowners for lawn and garden fertilizer.

A Special Type of Animal Raising

A special kind of animal raising is called ***aquaculture.*** Aquaculture can be defined as growing and harvesting aquatic (water) organisms in controlled conditions. Aquaculture involves raising fish, shellfish, and aquatic plants. Aquaculture is considered an agricultural activity. Many differences exist between aquaculture and traditional agriculture, however. Aquaculture mainly produces protein crops. Traditional farming focuses on starchy staple crops.

Aquaculture also has waste disposal problems like livestock raising does. In traditional livestock operations, however, animal waste can be disposed of off-site. It can be spread on fields as fertilizer. In contrast, animal waste in aquaculture stays in the ponds. Therefore, careful management is required to maintain the water of the ponds.

Aquaculture is also different from traditional fishing. Traditional fishing is called capture fishing. It involves locating and capturing fish in nature (oceans, lakes, and rivers). In contrast, aquaculture requires human action in raising fish.

Figure 29-29. Lagoon waste is being applied with a special purpose applicator (left). The applicator applied the waste by knifing it into the soil (right).

Aquaculture requires machines and equipment to perform a number of tasks. An environment must be constructed. Most aquaculture is done in ponds dug in the earth. These ponds usually have water inlets and outlets. The ponds are stocked with young fish. Fertilizer is often added to the ponds. This promotes the growth of food supplies for the fish. The fish are fed a diet that promotes their growth. When they have reached market size, they are harvested. In a complete harvest, the pond is drained. All the animals are removed from the pond. In a partial harvest, some of the animals are removed using a net, **Figure 29-30.**

Figure 29-30. These workers are harvesting catfish at a fish farm.

Agriculture and Biotechnology

People have used scientific activities to improve plants and animals for hundreds of years. They have used selective breeding of livestock and cross-pollination of plants to create new or improved plants and animals.

In recent history, technology has also been used for this goal. This technology is called ***biotechnology.*** But what is biotechnology? Biotechnology can be defined as *using biological agents in processes to produce goods or services.* The biological agents are generally microorganisms (very small living things), enzymes (a special group of proteins), or animal and plant cells. They are used as a catalyst in the selected process. The word ***catalyst*** means they are used to cause a reaction. The catalyst, however, does not enter into the reaction itself.

The term *biotechnology* is fairly new. The practice, however, can be traced back into distant history. Evidence suggests that the Babylonians used biotechnology to brew beer as early as 6000 B.C. As far back as 4000 B.C., the Egyptians used biotechnology to produce bread. During World War I scientists used an additive to change the output of a yeast fermentation process. The result was glycerol instead of ethanol. The glycerol was a basic input to explosives manufacturing. During World War II, scientists used the fermentation process to produce ***antibiotics*** (anti-bodies).

Today biotechnology has a major impact on modern agriculture. Agricultural biotechnology is being used to create, improve, or modify plants, animals, and microorganisms. It is being used to produce new pest-resistant and chemical-tolerant crops. This is done through a process called ***genetic engineering.***

These new crops have helped combat diseases. For example, golden rice was developed using genetic engineering. This new rice provides infants in developing countries with beta-carotene to fight blindness.

Biotechnology is a major factor in *increasing crop yields.* It has helped produce more food on the same number of acres. For example, biotechnology has been used to produce soybeans that are resistant to certain herbicides. Also, it has been used to develop a cotton plant that is resistant to major pests.

Biotechnology can be used to *promote human health.* The nutritional value of foods can be improved using genetic engineering.

Genetic engineering is based on the fact that every living thing carries a genetic code (blueprint) that determines precisely the traits it will have. This code was linked to a major discovery called recombinant DNA. The structure of DNA is a double helix (spiral) structure. It consists of a jigsaw-like fit of biochemicals. The two strands have biochemical bonds between them.

The DNA molecule may be considered a set of plans for living organisms. It carries the genetic code that determines the traits of living organisms. Scientists can use enzymes to cut the DNA chain cleanly at any point. The enzyme selected will determine where the chain is cut. Then two desirable parts can be spliced back together. This produces an organism with a new set of traits. The process is often called ***gene splicing,*** **Figure 29-31.**

This process allows scientists to "engineer" plants that have specific characteristics. For example, resistance to specific diseases can be engineered into the plant. This could reduce the need for pesticides to control insect damage to crops.

This activity has received many headlines in newspapers and magazines. It is controversial. Some people think it will make life better. Others think that we should not change the genetic structure of living things.

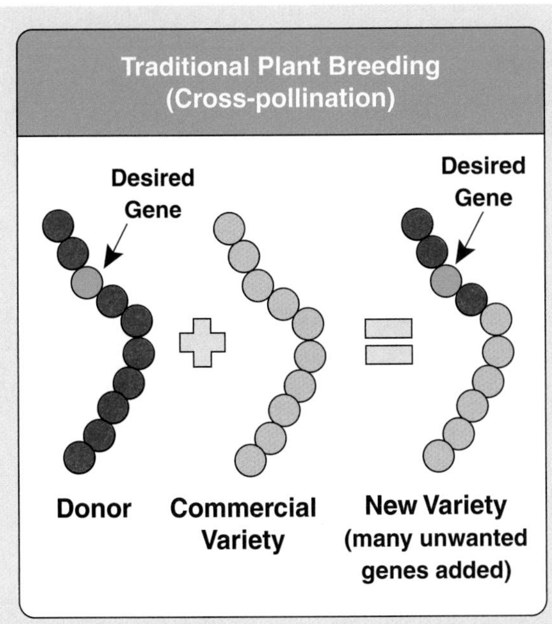

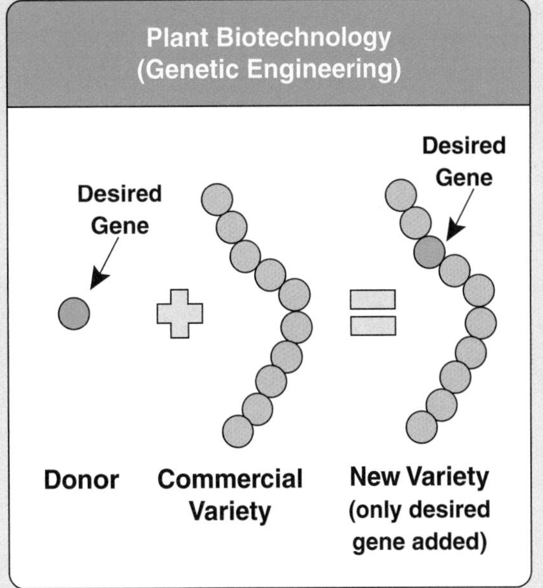

Figure 29-31. Genetic engineering can introduce a desired gene in an existing plant.

Summary

Agriculture includes two major activities that take place on farms and ranches. These activities are growing crops and raising livestock. Crop production involves using machinery and equipment in planting, growing, harvesting, and storing plants and plant products. These tasks require various kinds of equipment: power or pulling (tractors), tilling, planting, pest control, irrigation, harvesting, transportation, and storage equipment.

Livestock operations use structures and equipment to feed and care for cattle, swine, sheep and goats, poultry, and horses. Such structures and equipment include livestock buildings, fences and fencing, buildings and machines for feedings, and facilities for animal waste.

Agriculture also employs biotechnology. This type of technology applies biological organisms to production processes. We use biotechnology to produce new strains of crops and drugs.

Key Terms

agriculture
agricultural technology
animal husbandry
antibiotics
aquaculture
baler
barbed-wire fences
berries
biotechnology
cable-wire fences
catalyst
combine
crop production
cultivator
drip irrigation
electric fences
fertilizer
flood irrigation
forage crops
fruits
furrow irrigation
gene splicing
genetic engineering
grain
grain drill
hydroponics
irrigation
nonfood crops
nuts
pivot sprinkler
plow
rail fences
sprinkler irrigation
swather
tillage
tractor
vegetables
windrows
woven-wire fences

Test Your Knowledge

Write your answers on a separate piece of paper. Please do not write in this book.

1. What is agriculture?
2. Name one difference between the role of science and the role of technology in agriculture.
3. What are the two branches of agriculture?
4. Members of the grass family that have edible seeds are called _____.

5. Grass and hay crops grown for animal feed are called _____ crops.
6. The four agricultural processes that use technology are planting, growing, harvesting, and _____.
7. *True or false?* The most common machine used for pulling machinery is the track-type tractor.
8. The most important piece of tilling equipment is the _____.
9. *True or false?* A chemical applied to the soil to add nutrients is called fertilizer.
10. *True or false?* A cultivator is a piece of equipment used in nonchemical pest control.
11. The irrigation system that produces artificial rain is called _____.
12. Name one type of harvesting equipment.
13. Growing plants in nutrient solutions without soil is called _____.
14. Name one kind of livestock farm.
15. Name one kind of fencing used to contain livestock.
16. For what purpose do farmers use silos?
17. The two kinds of lagoon waste are wastewater and _____.
18. The practice of growing and harvesting fish in controlled conditions is called _____.
19. Define biotechnology.
20. Name one form of genetic engineering.

Applying Your Knowledge

1. Farmers use many types of equipment during the growing season. Some equipment is used at different times throughout the season; other equipment is used only part of the season. Using a form like the one below, list the names of some pieces of equipment ("plow" or "combine," for example) used at various times in the growing season.

Spring	Summer	Fall

2. Genetic engineering to modify different kinds of food crops has caused controversy. Research the topic of genetically modified products (such as corn) and make two lists: one of the positive effects (benefits) of genetically modified products and one of the negative effects, or risks. Compare the lists and try to form an opinion on the subject.

This activity develops the skills used in TSA's Agriculture and Biotechnology Design event.

Agriculture and Biotechnology Design

Activity Overview

In this activity, you will research a contemporary problem, issue, or technology related to agriculture or biotechnology, prepare a report, and create a display. You will prepare an oral presentation incorporating a PowerPoint presentation.

Your report must be contained in a three-ring binder and consist of the following items:

- Cover page
- Definition of problem, issue, or technology
- Report on topic (4–10 pages)
- Printout of PowerPoint slides (three slides per page)
- List of sources and references

Materials

- Three-ring binder
- Materials appropriate for a tabletop display (will vary greatly)
- Computer with PowerPoint (or similar) presentation software

Background Information

Selection. Before selecting the theme for your project, use brainstorming techniques to develop a list of possible themes. Some contemporary topics include:

- Waste management
- DNA testing
- Soil conservation techniques

- Human Genome Project
- Genetically modified food
- Cloning
- Aquaculture
- Food production techniques
- Irradiation

Research. Use a variety of sources to research your theme. Do not rely solely on information you find on the Internet. Use books and periodicals available at your local library. Research the historical developments of the topic. Did an individual develop the technology, or was the technology developed by a corporation? What were some previous technologies that allowed this to become a reality? How was the technology received by the public, and was the response expected?

PowerPoint presentation. When developing your PowerPoint presentation, consider the following design guidelines:

- Develop a general slide design and use it for all your slides.
- Keep the design simple—do not use more than two type fonts. Select type fonts that are easy to read. Be sure that the type size is large enough to be seen from the rear of the room in which you will be presenting.
- Include a title on each slide.
- Do not attempt to squeeze an abundance of information on a single slide—create multiple slides instead.

Guidelines

- Research should focus on any cultural, social, economic, or political impacts. Both opportunities and risks should be addressed.
- The display can be no larger than 18″ deep × 3′ wide × 3′ high.
- If a source of electricity is desired, only dry cells or photovoltaic cells can be used.
- The oral presentation can be up to 10 minutes in length.

Evaluation Criteria

Your project will be evaluated using the following criteria:

- Content and accuracy of report
- Attractiveness and creativeness of display
- Communication skills and presentation design of oral presentation

Chapter 30
Food-Processing Technologies

Learning Objectives

After studying this chapter, you will be able to:

- Explain the role of food in daily life.
- Define food-processing technology.
- Describe the difference between primary and secondary food processing.
- Describe ways used to process foods.
- Identify ways to preserve foods.
- Describe the steps that might be used to develop a new food product.
- Describe the activities involved in manufacturing a new food product.

Humans have always needed an adequate food supply. Food is needed to sustain life and promote growth. Try to remember what you ate yesterday, **Figure 30-1.** It may have been some fruit, vegetables, and meat. You may have eaten several slices of bread and some cereal. All of these substances are food.

Food is one of the three essentials of life. The others are clothing and shelter. Without a proper supply of food, human life is not possible. Food sustains life by providing seven basic components, **Figure 30-2:**

- Carbohydrates: Carbohydrates provide the body with its basic fuel.
- Proteins: Proteins are amino acids that provide the building material cells need to maintain their structure and growth.
- Fats: Fats provide triglycerides, which are stored in muscle cells and fat cells and are burned as fuel.
- Vitamins: Vitamins are organic substances that are essential in the life of most animals and some plants. The human body needs many different vitamins including vitamin A, vitamin B, vitamin B_1 (thiamin), vitamin B_2 (riboflavin), vitamin B_3 (niacin), vitamin B_6, vitamin B_{12}, folic acid, vitamin C, vitamin D, vitamin E, vitamin K, pantothenic acid, and biotin.
- Minerals: Minerals are elements that the body needs to create specific molecules. Common minerals that the body needs are calcium, chlorine, chromium, copper, fluorine, iodine, iron, magnesium, manganese, molybdenum, phosphorus, potassium, selenium, sodium, and zinc.

- Fiber: Fiber is the substance people eat that their bodies cannot digest. The three major fibers in food are cellulose, hemicellulose, and pectin.
- Water: The human body is about 60 percent water. About 40 ounces of it are lost each day and must be replaced. We do this through our consumption of moist foods and drinks.

To supply a reliable source of food, people have developed food-processing technology. Food-processing technology involves using knowledge, machines, and techniques to convert agricultural products into foods that have specific textures, appearances, and nutritional properties. These technological actions transform animal and vegetable materials into safe and edible food for humans. Also, food-processing technology includes processes used to make food more tasty and convenient to prepare. In addition, it involves actions that are used to extend the shelf life of perishable foods.

Food-processing technology includes two basic types of processes, **Figure 30-3.** These are:

- Primary food processing: Technological actions that change raw agricultural materials into food commodities or ingredients
- Secondary food processing: Technological actions that convert food commodities and ingredients into edible products

Primary food-processing technology produces the basic ingredients for food. For example, this type of technology changes wheat into flour. It changes animal carcasses into hamburger and transforms raw milk into pasteurized milk.

Secondary food-processing technology is used to make finished food products. For example, it converts flour and other

Figure 30-1. People eat many different types of food.

Figure 30-2. Food contains seven basic ingredients: carbohydrates, proteins, fats, vitamins, minerals, fiber, and water.

Primary Processes
Convert agricultural products into food commodities

Secondary Processes
Convert food commodities into edible foods

Figure 30-3. Agricultural products go through primary and secondary processing before becoming edible food. (Wheat Foods Council)

ingredients into bread and changes hamburger and other ingredients into lasagna.

Both of these processes are critical to our production of food. We will now look at each one in more detail.

Primary Food Processing

Most agricultural products are not eaten directly from the fields in which they are grown. They are processed using technical means (machines) to change their form, appearance, or usefulness. Primary food processes include both material conversion (processing) and food preservation.

Material Conversion

Few of us eat products that come directly from farm fields or gardens. Most of our food is processed in one way or another. This processing may be as simple as washing, grading, and packing fresh fruits and vegetables, **Figure 30-4.** In other cases, the produce may be preserved using one of several techniques. It may be canned or frozen for later use. It may be dried or cured. These and other preservation techniques will be discussed later. All of these actions constitute technology. They involve using machines to change the condition of the food.

Many food-processing techniques convert raw agricultural products into a different form. They may change the product so that it has a different nature and appearance. For example, milk may be changed into cheese or butter. Corn and soybeans may be processed to produce cooking oils. Sugar cane or beets may be refined into granulated sugar. These processes convert the original farm product into an entirely new ingredient.

Food can be processed in many ways. These ways can be grouped under the three headings of mechanical processing, thermal (heat) processing, and chemical processing. ***Mechanical processing*** uses machines to change the form of the food

Figure 30-4. Food processing can be as simple as receiving, washing, and grading produce. (Northwest Cherry Growers)

product physically. It might crush, slice, grind, or scrape the material to form a new ingredient. ***Thermal processing*** uses heat as the primary energy to convert a food. It might use this energy to melt, cook, blanch, or roast the material. ***Chemical processing*** uses energy to cause a basic chemical change in the food. It might involve pickling, fermenting, coagulating, and other similar actions. A representative example of physical, thermal, and chemical processing follows.

Mechanical Processing: Flour Milling

Flour is finely ground grain, such as wheat, rye, corn, or rice. Wheat flour is the most commonly produced flour in the Western world. The composition of this flour depends on the type of wheat used and the milling processes employed. In a common milling process, four major steps are used. These include, **Figure 30-5:**

- *Grain receiving and cleaning:* Wheat is transported from farms to the mill, where it is unloaded and stored in bins. From the bins, the wheat is moved to cleaners, which remove all impurities. Magnetic separators remove iron and steel particles; screens remove larger stones, sticks, and other materials; air blasts remove lighter impurities; and special separators remove other grains (such as oats) and weed seeds. The clean wheat is stored in hopper bins to await milling.
- *Conditioning:* The clean wheat is prepared for milling by a conditioning process. Water is added to make it easier to separate the bran (outside hull of the grain) from the endosperm (flour portion) of the wheat kernel. The wet wheat is stored in tempering bins. From the tempering bins, wheat is moved to the flour mill for ***milling*** (grinding or processing).
- *Milling:* The conditioned wheat is milled using roller mills, sifters, and auxiliary equipment. The first mill is called the first break. It uses corrugated rolls to break wheat into coarse particles. Then a sifter and purifier separate

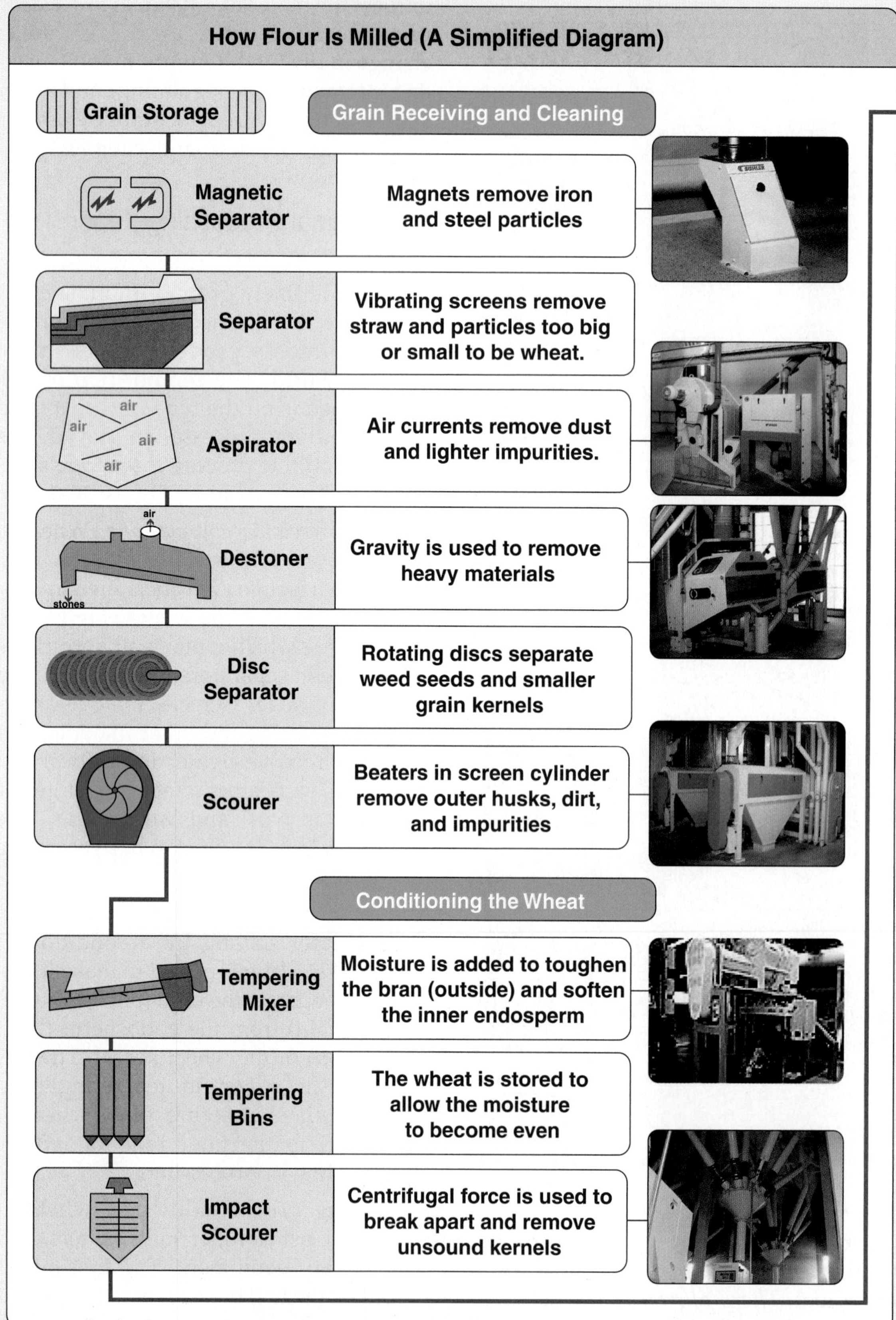

Figure 30-5. The flour milling process involves four steps. (Buhler Group, North Dakota Mill)

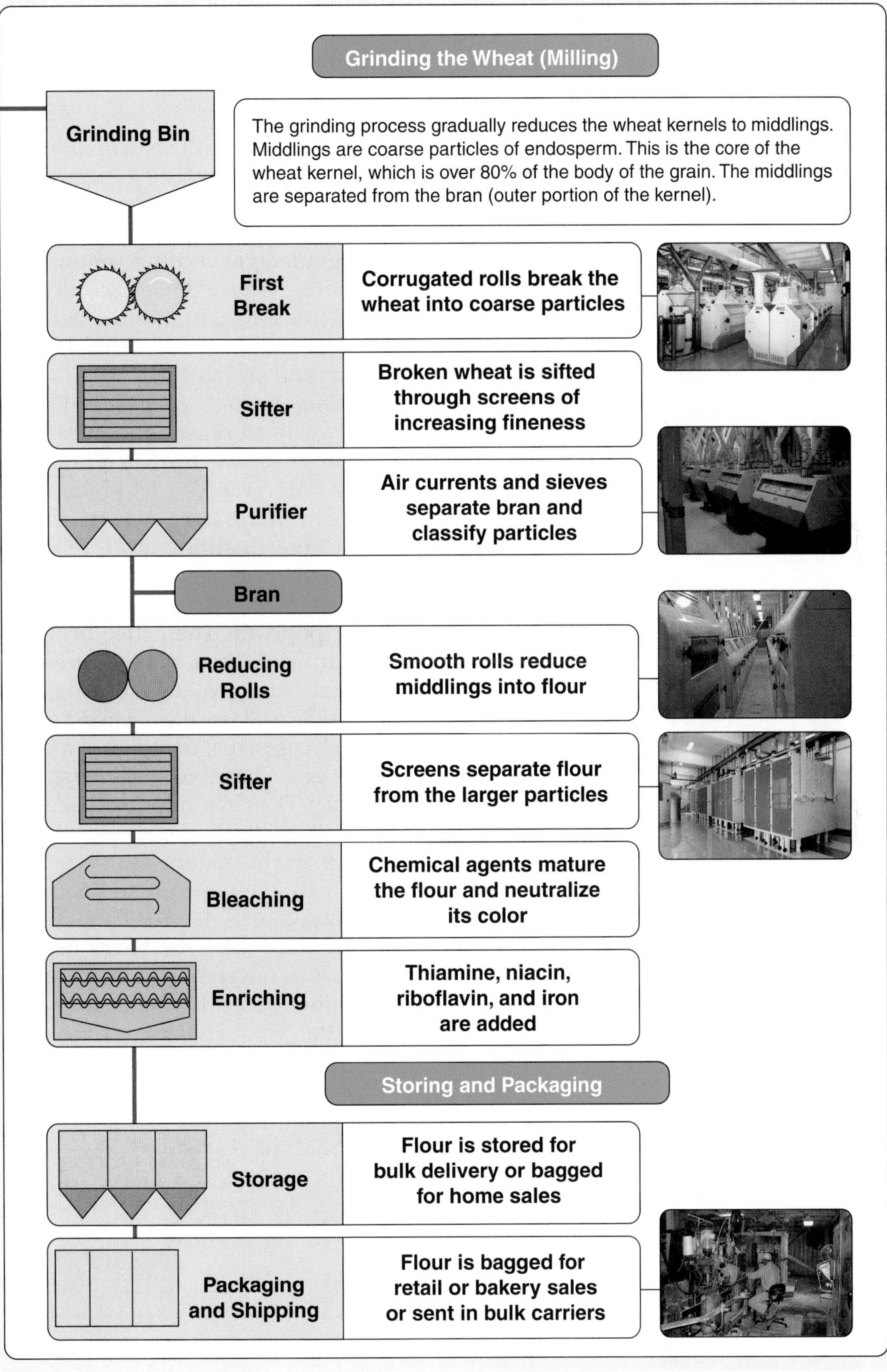
Grinding the Wheat (Milling)
Grinding Bin
The grinding process gradually reduces the wheat kernels to middlings. Middlings are coarse particles of endosperm. This is the core of the wheat kernel, which is over 80% of the body of the grain. The middlings are separated from the bran (outer portion of the kernel).
First Break
Corrugated rolls break the wheat into coarse particles
Sifter
Broken wheat is sifted through screens of increasing fineness
Purifier
Air currents and sieves separate bran and classify particles
Bran
Reducing Rolls
Smooth rolls reduce middlings into flour
Sifter
Screens separate flour from the larger particles
Bleaching
Chemical agents mature the flour and neutralize its color
Enriching
Thiamine, niacin, riboflavin, and iron are added
Storing and Packaging
Storage
Flour is stored for bulk delivery or bagged for home sales
Packaging and Shipping
Flour is bagged for retail or bakery sales or sent in bulk carriers

the bran and classify the remaining particles (called middlings) by size. The middlings move to reducing rolls that change them into flour. The flour moves through additional sifters and reducing rolls until a uniform product emerges. This product is bleached to produce even-colored flour. Vitamins and minerals may be added to produce "enriched" flour.

- *Storage and packaging:* The milled, bleached, and enriched flour is conveyed to steel tanks. From there it may be moved in tank cars to bakeries or other commercial customers. Flour for retail customers is packaged in bags and sent to grocery stores.

Thermal Processing: Coffee Roasting

Coffee is the fruit of a tree. Originally coffee was a food, not a drink. People mixed coffee beans with animal fat to produce a high-energy food. Later the beans were roasted and ground so a drink could be made.

Coffee beans are found inside the red cherry–shaped fruit of a coffee tree. Inside the fruit are two coffee beans, which are referred to as the greens.

The coffee beans are picked and shipped to the roasting factory. Upon arrival, the coffee is inspected for freshness and color. The inspected coffee beans are sent to a coffee pulper. This machine contains a metal cylinder with stripping knobs that removes the husks from the coffee beans. The beans are then placed in a fermentation tank where the film covering them is removed. Then the beans are washed and dried.

In the next stage, the parchment cover is removed from the green beans. Then the coffee is graded and classified according to size, weight, and quality.

The graded green coffee is now ready for roasting, **Figure 30-6.** Basically, two types of coffee-roasting techniques exist: the traditional drum- or barrel-roasting technique and air roasting.

The *drum-roasting technique* is the most common way to roast coffee beans. Its operation is similar to that of a clothes dryer. In this process, the green coffee beans are heated to temperatures of about 550° F in large, rotating drums. The beans tumble in the cylinder and are roasted.

The *air-roasting technique* is similar to popping popcorn. The coffee beans are roasted at the desired temperature using hot air. They are constantly moving in the air blast, which allows them to roast evenly.

In both cases, the first stage of roasting turns the green beans to a yellowish color as they dry. They begin to smell like toast or popcorn. In the second step, called the first crack, the bean doubles in size. They "pop" much like popcorn and become a light brown color.

In the third step, the beans begin to brown as their oils start to emerge. A chemical reaction caused by the heat and oils produce the flavor and aroma of coffee. A

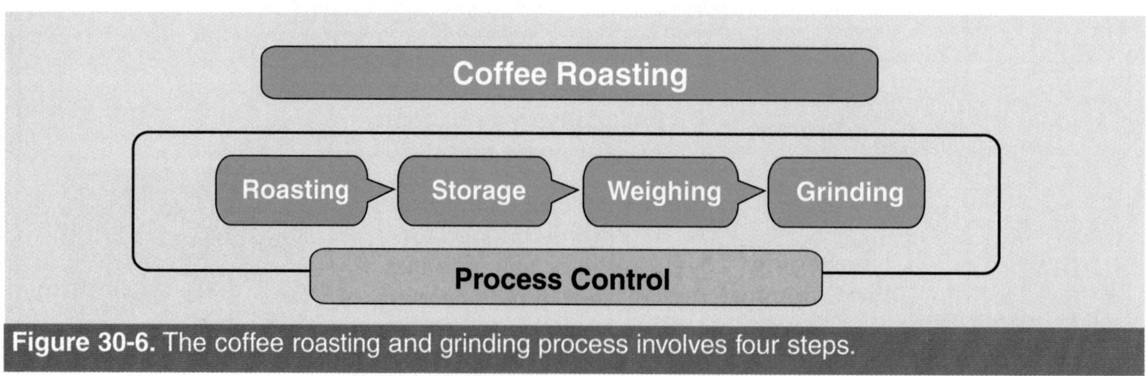

Figure 30-6. The coffee roasting and grinding process involves four steps.

second "pop" occurs several minutes later. At this point the bean is fully roasted and can be moved to storage.

The roasted coffee moves out of storage and is weighed. It then is ground using mechanical grinding machines. This allows people to get the most flavor out of the bean when coffee drinks are prepared. The ground coffee is packaged in airtight bags or cans to help maintain its freshness.

Chemical Processing: Cheddar Cheese Manufacturing

Cheese is a dairy product made from milk protein. It is a coagulated, compressed, and ripened curd of milk that has been separated from the whey (moisture in milk). It is made in hundreds of types and forms. The following describes how cheddar cheese is made through a chemical conversion process, **Figure 30-7:**

- *Receiving milk:* The cheese-making process starts with milk being delivered from local dairies. It moves from the farm to the cheese factory in refrigerated milk trucks. On arrival it is tested for quality, flavor, and odor.
- *Cooking:* The fresh milk is pasteurized to destroy any harmful bacteria. This milk is pumped into cooking vats. Special starter cultures (bacteria) are added to the warm milk and change a very small amount of the milk sugar into lactic acid. Coloring is added to produce a consistent cheese color. Finally, a substance to promote coagulation and firm the curd is added.
- *Cutting the curd:* After about 30 minutes, the vat of milk will set up. In this action a soft curd, consisting of casein and milk fat, is formed. Stainless steel knives cut the soft curd into 1/4-inch pieces. Then the temperature in the

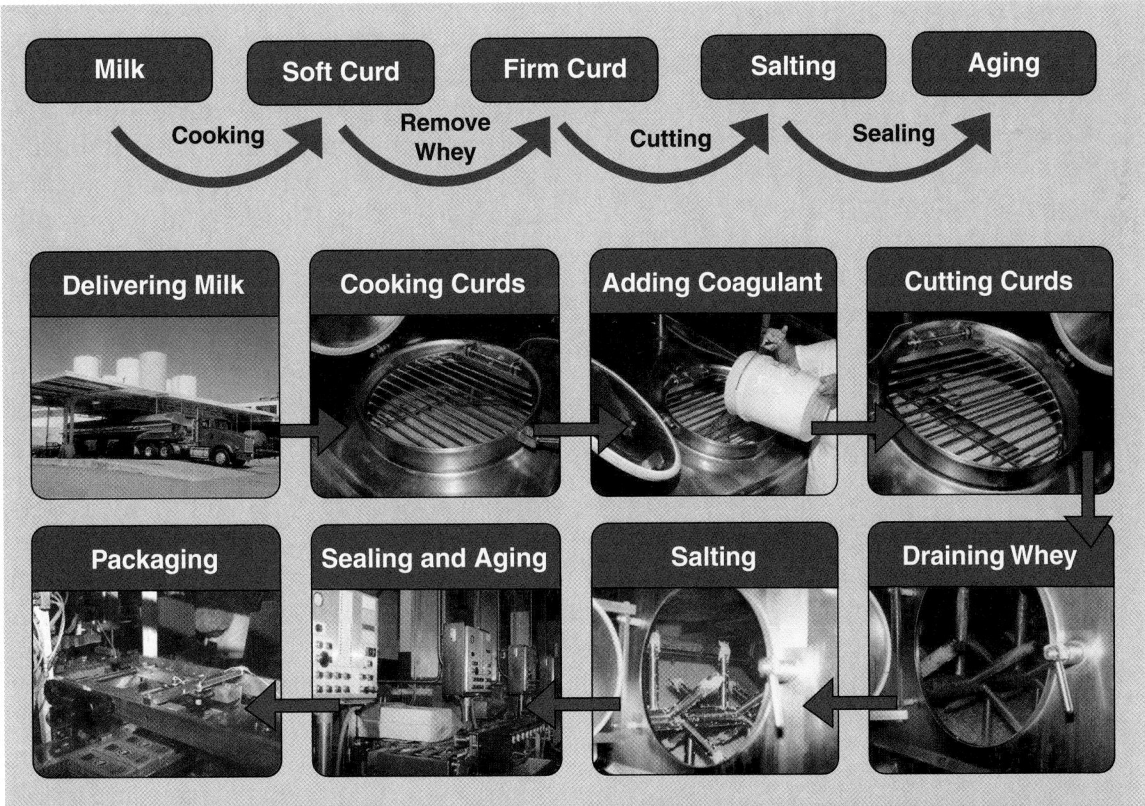

Figure 30-7. In the cheese-making process, milk curds are cooked, separated from the whey, cut, and sealed. (Tillamook County Creamery Association)

cooking vat is raised to about 100° F to drive out moisture and firm the curd. The liquid part in the vat is called whey.

- *Making cheddar:* The curds and whey are pumped to the cheddar-making machine. The whey is removed, and the curd is matted on a wide belt inside the machine. During the cheddaring process, a chemical change occurs. The curd particles adhere to each other, giving them a stringy consistency. When the proper acidity is reached in the curd, the cheddar mat is forced through the curd mill, which chops the large slabs into small, 3-inch-long bits.
- *Salting:* These curd chunks are passed through a salting chamber. In this machine, a thin layer of salt is applied to the surface of each curd. The salted curd is stirred, and it absorbs the salt. When the salt is completely absorbed, the curds are transferred to pressing towers, where a vacuum draws off the excess moisture in the cheese.
- *Sealing and aging:* After 30 minutes in the pressing towers, large blocks of cheese are cut from the base of the towers. These blocks are placed in plastic bags. A vacuum is drawn, and the package of cheese is sealed. The block is now contained in an airtight and moisture-proof bag. The sealed blocks are transported to a cooling room and held for about 24 hours at temperatures near 38°F. From there, they are palletized and placed in storage for aging and curing at about 42°F.
- *Packaging:* After aging for a set number of days, the cheese blocks are removed from their aging bags. Each batch of cheese is tested and graded for quality, **Figure 30-8.** The blocks are cut into specified sizes and packaged for various markets.

Figure 30-8. This inspector is checking cheese for unwanted odors. (Wisconsin Milk Marketing Board)

Food Preservation

Food preservation technologies are used to keep food from being unfit for human use. They reduce or eliminate the effects of bruising and insects. Preservation controls microorganisms such as bacteria, yeast, and molds. These techniques may destroy unwanted enzymes (proteins) present in raw foods. These enzymes cause chemical and physical changes that naturally occur after harvesting. They contribute to food spoilage.

Food preservation technologies also help eliminate moisture or temperature conditions that encourage the growth of microorganisms. These organisms can produce unwanted changes in the taste and appearance of the food. They can also cause food-borne illness.

Food processing can improve the nutritional value of foods. For example, high heat can destroy unwanted factors present in many foods. Prolonged boiling destroys the harmful lectins present in such foods as red kidney beans.

Some preservation techniques can reduce the quality of food. They can produce food that has lower nutritional value and poorer texture and flavor.

A number of different technologies are used for food preservation. These technologies include drying, curing and smoking, canning, aseptic packaging, refrigeration and freezing, controlled atmosphere storage, fermentation, pasteurization, irradiation, and preservatives.

Drying

Drying is the oldest method people have used to preserve food. Drying involves removing water from a food material. Removing water preserves the food because microorganisms that cause food spoilage need water to grow.

Drying includes three basic methods of food preservation. Sun drying allows foods to dry naturally in the sun. Hot-air drying exposes food to a blast of hot air. Freeze drying uses a vacuum chamber to draw water out of frozen foods. In freeze drying, water escapes from the food by a process in which ice changes from a solid directly to a vapor without first becoming a liquid, **Figure 30-9.**

Dried foods keep well because the moisture content is so low that spoilage organisms cannot grow. Drying is used on a limited range of foods, including milk, eggs, instant coffee, and fruits. It will never replace canning and freezing as the primary way to preserve foods. These methods retain the taste, appearance, and nutritive value of fresh food better than drying does.

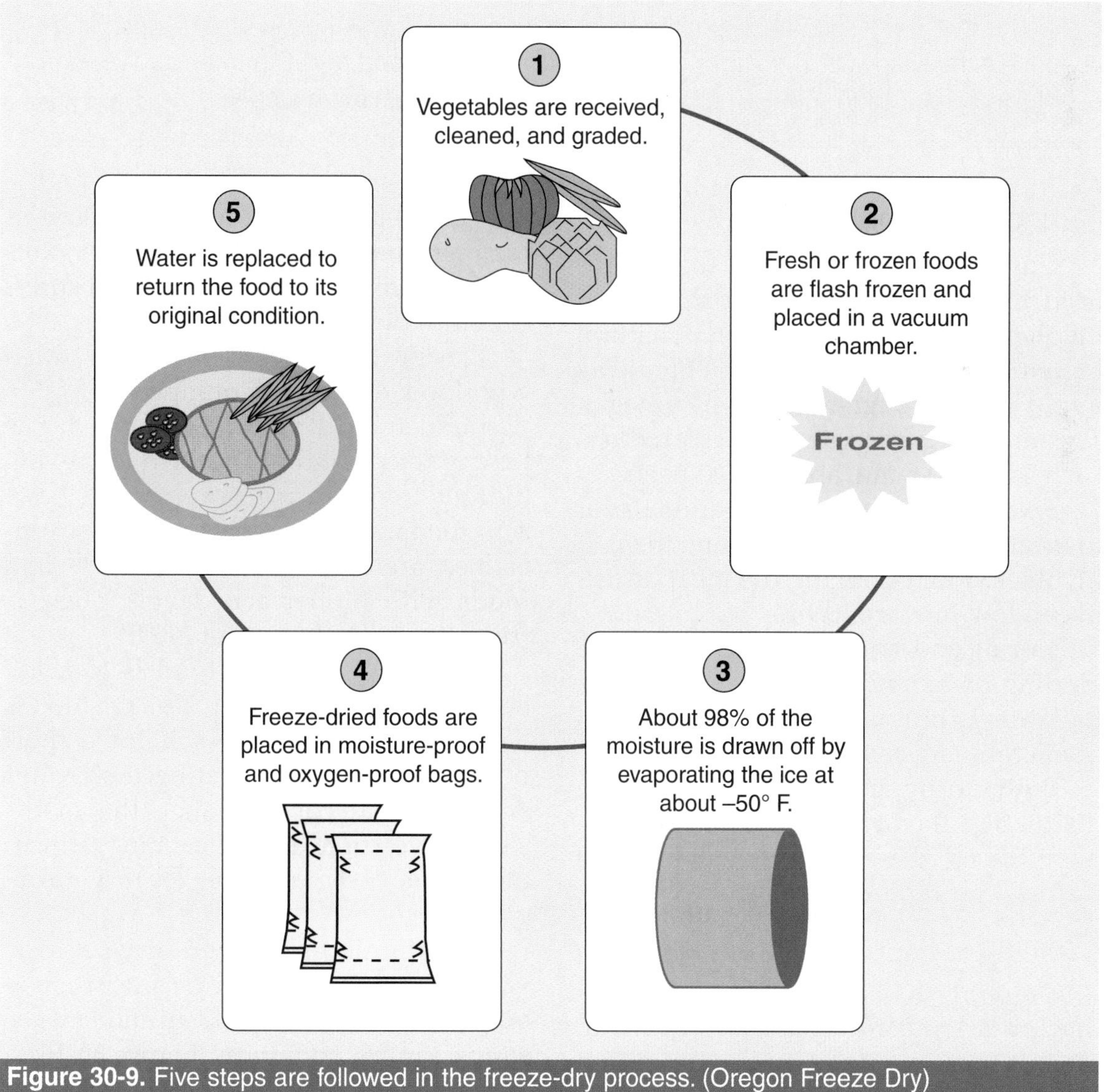

Figure 30-9. Five steps are followed in the freeze-dry process. (Oregon Freeze Dry)

Career Corner

Food Processing Workers

There is a number of different jobs in food processing. Two of these are food batch makers and food cooking machine operators. Food batch makers set up and operate equipment that mixes, blends, or cooks ingredients used in the manufacture of food products. Food cooking machine operators operate cooking equipment to prepare food products. These food-machine operators usually are trained on the job. They learn to operate equipment by watching and helping other workers. Employment opportunities for these workers are expected to increase more slowly than in other fields. Many food-processing workers are members of the United Food and Commercial Workers International Union and earn considerably more than minimum wage.

Curing and Smoking

Curing and smoking are techniques used to preserve meat and fish. ***Curing*** involves adding a combination of natural ingredients to the meat. These ingredients might be salt, sugar, spices, or vinegar. This process is used to produce products such as bacon, ham, and corned beef.

Smoking is a process that adds flavor to meat and fish while preserving them. It involves slowly cooking the meat or fish over a low-heat wood fire.

Curing and smoking preserve food by binding or removing water. This causes the water to be unavailable for the growth of microorganisms.

Both curing and smoking produce a distinctive flavor and color to food. In many cases, these techniques eliminate the need for refrigeration.

Canning

Canning is a preservation method that puts food into glass jars or metal cans. It preserves food by heating it in a vacuum-sealed container. This process removes oxygen from the container, kills microorganisms in the food, and destroys enzymes that could spoil the food.

During the process, the can is filled with food. Then air is pumped to form a vacuum, and the container is sealed. The food is heated and then cooled to prevent the food from becoming overcooked. Low-acid foods, such as meats and most vegetables, are heated to 240° to 265°F, and foods with higher acid levels, such as fruits, are heated to about 212°F.

In many cases, the food is *blanched* before canning. In this process the food is heated with steam or water for a short period of time. This heating inactivates enzymes that could change the food's color, flavor, or texture. Blanching reduces the volume of the vegetables by driving out gases.

Canning is used to preserve a wide variety of foods, including fruits, vegetables, jams and jellies, pickles, milk, meats, soups, sauces, and juices, **Figure 30-10.**

Canned foods can be stored without refrigeration for a long time. Because of the high temperatures used, canning affects the color, texture, flavor, and nutrient content of foods.

Aseptic Packaging

Aseptic packaging is commonly used for packaging milk and juice because they will keep for long periods of time without refrigeration. This process, like canning, uses heat to sterilize food. Unlike canning, however, the package and food are sterilized separately.

The containers are sterilized with hydrogen peroxide rather than with heat. This allows the use of lower-cost containers that would be destroyed by heat sterilization. The most commonly used containers are plastic bags and foil-lined cartons.

The food is sterilized more rapidly and at lower temperatures than in canning. This allows the food to have better flavor and to retain more nutrients.

Refrigeration and Freezing

Storage at low temperature slows many of the involved reactions that cause foods to spoil. The two low-temperature preservation techniques are refrigeration and freezing. ***Refrigeration*** maintains foods at temperatures from 32° to 40°F. Freezing keeps the foods at or below 32°F.

Refrigeration does not cause chemical or physical changes to food. Still, it preserves foods for a fairly short period of time. Foods that should be refrigerated include milk, eggs, meats, fish, and some fruits and vegetables.

Freezing is used on a wide range of foods, **Figure 30-11.** This process allows foods to be stored for longer periods because it greatly reduces enzyme activity and the growth of microbes. The ice crystals that form in the food disrupt the food's structure, however. When thawed, the food will have a softer texture. Also, food deterioration is rapid after thawing because the organisms in the food attack cells that were injured by ice crystals.

Figure 30-10. These carrots are being prepared for canning. (J. R. Simplot)

Controlled Atmosphere Storage

Fruits and vegetables can be stored in sealed environments where temperatures and humidity are controlled. In some cases, the composition of the air around the fruit is changed. Oxygen is often reduced while the level of carbon dioxide is increased. This change reduces the chance of food spoilage.

This controlled environment slows the reactions that lead to decomposition and decay. Controlled atmospheres can extend the storage life of fruits and vegetables by several months.

Fermentation

Fermentation uses microorganisms to break down complex organic compounds into simpler substances. Such compounds

Making and Freezing Commercial French Fries

Step 1: Getting potatoes—The potatoes come directly from the producers or from large warehouses. The potatoes are checked for solids content, grade, and sugar content.

Step 2: Peeling the Potatoes—A batch of potatoes are put into a large, hot, pressurized tank. After a set amount of time, the pressure is quickly released, and the potato skin is said to fly off. The potatoes are removed from the tank and sprayed with high-power water jets to remove any peel still clinging.

Step 3: Inspecting the Potatoes—The peeled potatoes pass an inspection line where sorters remove any defective potatoes.

Step 4: Cutting the Potatoes—The potatoes go through a pump that propels them at about 50 mph at some stationary blades, which chop the potatoes into strips.

Step 5: Inspecting the Strips—The small parts left from the outer edges of the potato are removed. The remaining strips are automatically inspected, and strips with black bits are removed. This is done at the rate of about 1,000 strips and chips each second!

Step 6: Blanching—The inspected strips are blanched on a moving conveyor chain that carries them through a large vat of hot water. This process removes excess sugars and gives the strips a consistent, uniform color.

Step 7: Drying—The blanched strips are partially dried as they are conveyed past blasts of hot air from both the top and bottom.

Step 8: Par-Frying—The strips are cooked for about a minute and a half in hot oil. The process is called a "par fry" or partial fry.

Step 9: The Deep Freeze—The strips enter blast freezing where the french fries travel down on the wire conveyor surrounded with air cooled to about –40° Fahrenheit.

Step 10: Packaging—After freezing, the product is bagged and/or boxed and shipped to customers.

Figure 30-11. The freezing of french fries allows them to be available longer. (J. R. Simplot)

as alcohol and acids are produced in fermentation. These compounds act as preservatives that reduce further microbial growth

In some cases, fermentation will spoil the product. In other cases, it is desirable. In these cases, microorganisms are added to foods.

For example, in the manufacture of yogurt and cheese, a sugar found in milk (lactose) is converted by bacteria into lactic acid. Fermentation is also used to produce alcoholic beverages, cheese, yeast bread, soy sauce, and cucumber pickles.

Pasteurization

Pasteurization uses heat to kill harmful microorganisms. For example, milk is usually heated to 145°F for 30 minutes, **Figure 30-12.** Pasteurization is commonly used for milk, fruit juices, beer, and wine.

In newer processes this technique is called Ultra-High Temperature (UHT) pasteurization. It uses higher temperatures and shorter heat times for foods in sterile packaging. The foods are heated to 280°F for 2 to 4 seconds. This rapid sterilization allows the food to have better flavor and retain more nutrients.

Irradiation

Irradiation uses gamma rays or X rays to kill most molds and bacteria that may be in the food. It is also used to delay the ripening of fruits and the sprouting of vegetables. This allows the produce to be stored for longer periods of time.

In this process, the food passes through a chamber where it is exposed to high-energy rays. Irradiation involves little heating and does not change the taste, texture, or nutritive value of food.

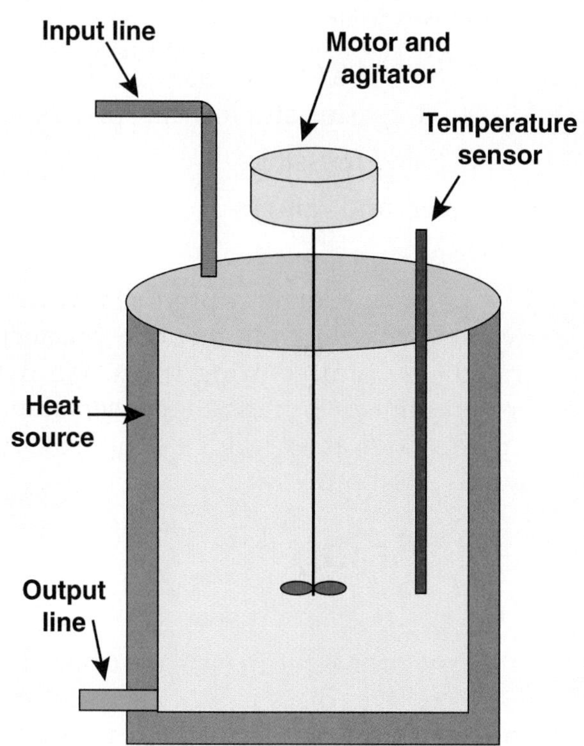

Figure 30-12. Shown here is a diagram of the operation of a batch pasteurizer. (Wisconsin Milk Marketing Board)

Preservatives

Preservatives are chemicals that are added to food in small amounts. These preservatives act in two ways. First, they delay the spoilage of the food. Second, they ensure that the food retains its quality. The first method uses sugar (jams and jellies), vinegar (pickles and meats), and salt (hams and bacon). The second method uses acids, sulfur dioxide, and other agents to slow the growth of microorganisms in food.

Secondary Food Processing

Secondary food-processing technology converts the ingredients produced in primary food-processing activities into edible products. It involves combining and processing ingredients and food products to their change properties.

There are literally thousands of end products of secondary food-processing activities. The shelves of a typical supermarket are evidence of this.

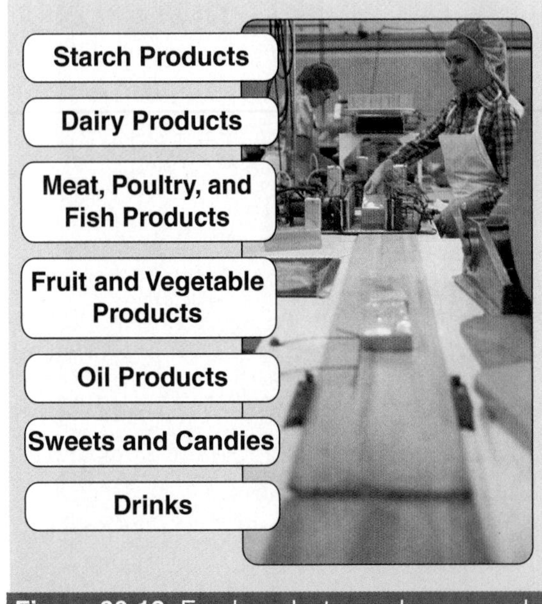

Figure 30-13. Food products can be grouped into seven major categories.

These products can be grouped in various ways. One way is as follows, **Figure 30-13:**

- Starch products
  - Bread
  - Cakes and cookies
  - Crackers
  - Pasta
- Dairy products
  - Milk, buttermilk, and cream
  - Butter
  - Cheese, cottage cheese, and yogurt
- Meat, poultry, and fish products
  - Fresh meat, fish, and poultry
  - Processed products (hot dogs and lunch meats, for example)
- Fruit and vegetable products
  - Fruit preserves and fillings
  - Vegetables dishes
- Oil products
  - Cooking oils
  - Margarine and low-fat spreads
  - Salad dressings
- Sweets and candies
- Drinks

The creation of these products involves two steps: developing the food products and then manufacturing them. We will first examine each of these steps separately and then look at a specific example of secondary food processing.

Developing Food Products

Many products can be made using similar basic ingredients. For example, flour and water are the basic ingredients for bread, biscuits, cookies, and pasta. Other ingredients are added to create the uniqueness of each product.

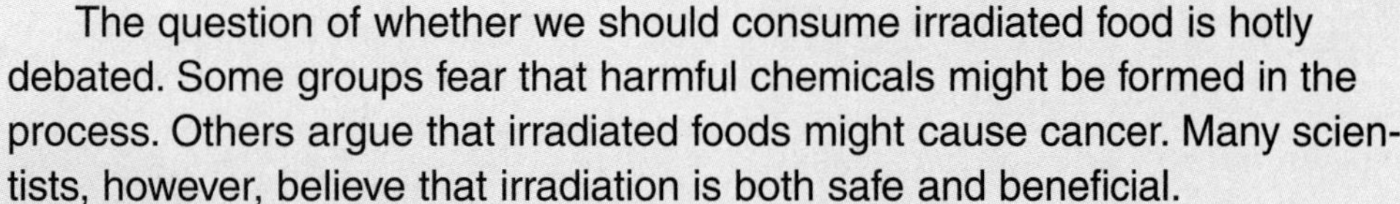

The question of whether we should consume irradiated food is hotly debated. Some groups fear that harmful chemicals might be formed in the process. Others argue that irradiated foods might cause cancer. Many scientists, however, believe that irradiation is both safe and beneficial.

To be able to make informed choices regarding this subject, we should know what irradiation of food involves. Irradiation itself is a form of radiant energy, which, as we learned in earlier chapters, is energy in the form of electromagnetic waves. The electromagnetic waves used in irradiation are primarily gamma rays, ultraviolet rays, and X rays. Scientists have found that when these rays penetrate food, chemical bonds are broken and the chemical composition of the food changes.

Food can be affected by the change in chemical bonds in many ways. For example, when a potato is exposed to a low level of these rays, the change in chemical bonds causes the sprouting to stop. At higher levels of exposure, the changes in chemical bonds cause such changes as the elimination of insects and the destruction of harmful bacteria.

The irradiation of food is sometimes chosen over the cooking of food (thermal energy) because even though chemical changes occur in both processes, fewer bonds are broken in irradiation. Therefore, the food is much fresher. Irradiation is also sometimes preferred to the addition of chemicals to destroy pests or extend storage life because of the possible hazards involved with certain chemicals.

Thus, with irradiation we can keep food longer and in better condition. Note, however, that we still use other forms of food preservation technologies with particular foods. For example, we use salt to preserve bacon, and we heat (pasteurization) to keep milk fresh. Can you think of some reasons for using these methods as opposed to irradiation?

Each product that is created from a series of ingredients is the result of a product development process. Companies use this process to develop an array of new or modified food products. The process of product development involves a series of steps that moves the product from the original idea to the consumer. These steps, some of which are shown in **Figure 30-14,** are:

1. New food product idea identification. A company identifies consumer trends and eating patterns. It studies new products on the market. It investigates advancements in food preparation techniques. It uses this information to brainstorm possible new products and line extensions (products that build on existing products).
2. Idea development. The company develops a number of new recipes. It specifies the ingredients and establishes their cost.

3. Small-scale testing. The company makes several versions of the product, usually using slightly different ingredients or processes. The products may be prototypes in the company's test kitchens. Company employees or small focus groups are asked to test the products and provide comments.
4. Sensory evaluation. The company asks trained personnel to comment on the appearance, taste, smell, and texture of the products.
5. Product modification. The company alters the composition and recipe of the product to incorporate the results of the initial test and sensory evaluations.
6. Pilot-plant production. The company scales up the product to pilot-plant production. This means the item is produced using small versions of the equipment that could be used in full-scale manufacture.
7. Consumer and sensory evaluation. The company tests the output of the pilot production to determine its potential of being a profitable product.
8. Product specification. The company develops the final product specifications. It develops the final ingredients and methods of production. These specifications will be used in producing each batch of the product.
9. Pilot production. The company readies the production plant for the product. It conducts production tests to determine the effectiveness of the production processes.
10. Continuous or large-scale product production. The company releases the product for full-scale production. The actual manufacture is often done in major processes such as measuring, mixing and cooking. The company carefully controls each of these processes to maintain product quality, promote food purity and safety, and reduce waste.

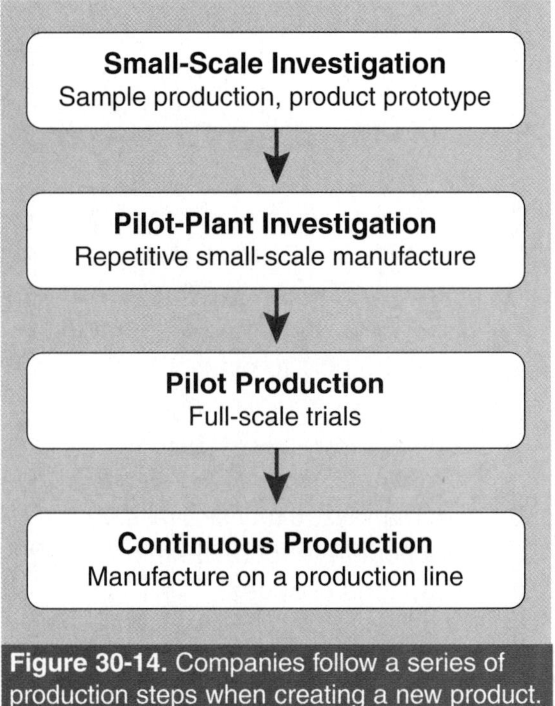

Figure 30-14. Companies follow a series of production steps when creating a new product.

11. Product packaging. The company places the product in appropriate packages. It places the individual packages in shipping containers ready for movement to the wholesaler or retailer.
12. Product promotion. The company announces the availability of the product through advertising and product promotion.

Not all companies develop products using this procedure. For example, some companies move products directly from the "test kitchen" stage to trials on factory production equipment. The decision on how to proceed depends on competitive pressure, economic factors, the size of the production facility, and the type of product being developed.

Manufacturing Food Products

Making a consumer food product involves a number of manufacturing processes, **Figure 30-15.** The procedure

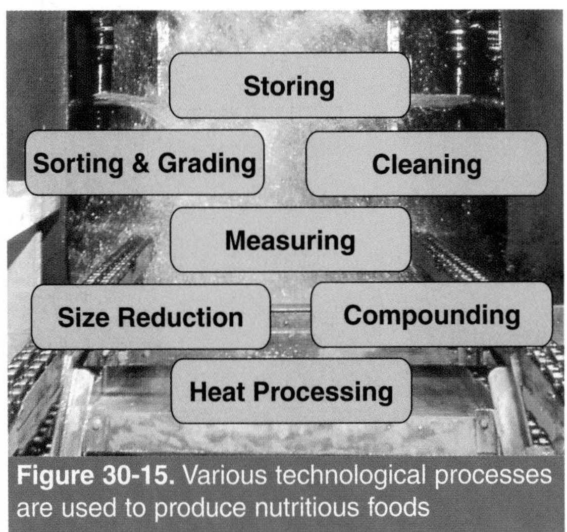

Figure 30-15. Various technological processes are used to produce nutritious foods

that most foods go through includes some of the following:

- Storing: Containing and protecting the condition of the raw materials
- Cleaning: Removing foreign matter from ingredients that will be used in the product
- Sorting and grading: Assessing the quality of the ingredients to be used in making the product
- Measuring: Weighing and measuring the wet and dry ingredients that are to be used in the product
- Size reduction: Trimming, slicing, and crushing the basic ingredients
- Compounding: Mixing and combining the ingredients for the product
- Heat processing: Cooking and cooling the mixture to form the product

These technological processes produce a consumer product that is nutritional and easy to use. Initially, food-processing technology produced a food unit such as bread, pasta, crackers, or stewed tomatoes. These products were combined at home to prepare dishes and complete meals. Now, however, people are demanding food that is easier and quicker to prepare. Today, food-processing technology provides both single unit–type foods and finished dishes. These finished dishes may be a meal entrée, such as lasagna, or a complete meal, such as that seen in a frozen dinner.

An Example of Secondary Food Processing: Pasta Making

The goal of pasta production is to convert flour into an edible product. Pasta production includes *blending* the ingredients to form a dough, *kneading* and mixing the dough, *forming* the final product, and then *drying* it, **Figure 30-16.**

Blending

The basic ingredients of pasta are durum wheat semolina and water. These ingredients are measured and added to a batch mixer. Additional ingredients such as eggs, spinach, tomato powder, and flavorings may be added at this time. They are mixed under vacuum to distribute moisture evenly throughout the flour. The semolina starts to absorb water and forms a crumb structure.

Kneading and Mixing

The lumps of semolina are kneaded (pressed together) to form a basic dough. This action forces the crumb structures together, which fuses the particles. The result is consistent dough.

Forming

An extruder is used to form the pasta shape. This machine uses the screw principle to push dough through a shaped die. These molds have round or oval holes through them to produce rods for products such as spaghetti. A steel rod can be placed in the center of each hole to form a tube-shaped product such as macaroni. The dough is scooped up by grooves in the screw. The rotating screw pushes the

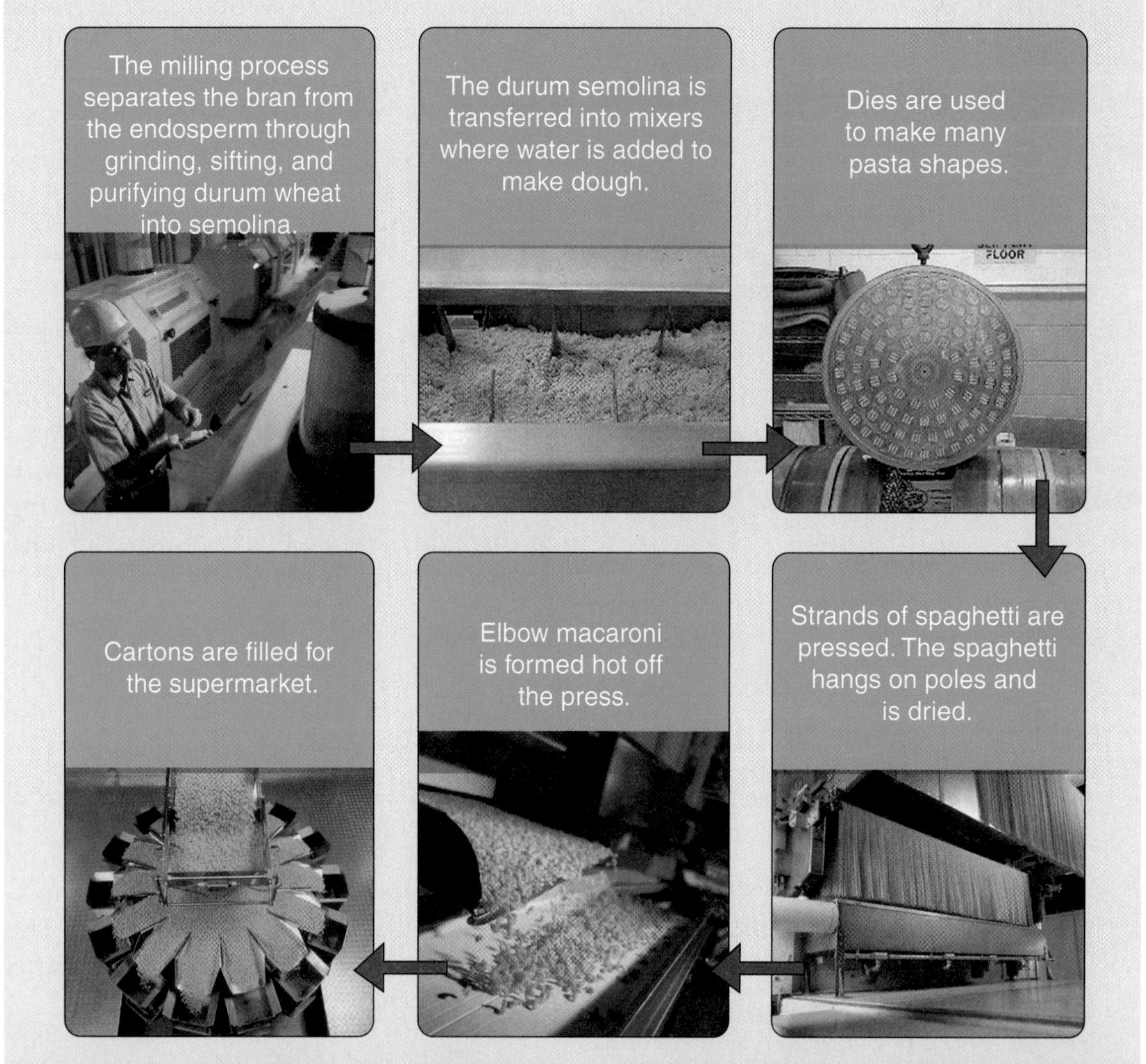

Figure 30-16. Making pasta involves several secondary food-processing techniques. (Dakota Growers Pasta Company)

dough toward the die opening. As the material passes through the die, it is shaped and formed. Cutters cut the developing tube of product to length.

Other pasta products, such as noodles, are formed in flat sheets. The shapes are stamped out of the sheets using shaped cutting dies.

Drying

The cut products pass through long tunnel dryers to remove excess moisture. Inside the dryers, very hot, moist air dries and cures the product.

Summary

Food-processing technology involves converting agricultural products into foods that have specific texture, appearance, and nutritional properties. It involves two basic types of processes: primary food processes and secondary food processes. Primary food processing involves two steps: material conversion and food preservation. Secondary food processing involves the development of the food products and their manufacture.

Key Terms

aseptic packaging
chemical processing
canning
curing
drying
fermentation
freezing
irradiation
mechanical processing
milling
pasteurization
preservatives
primary food-processing technology
refrigeration
secondary food-processing technology
smoking
thermal processing

Test Your Knowledge

Write your answers on a separate piece of paper. Please do not write in this book.

1. Name one way in which food sustains life.
2. Define the term *food-processing technology.*
3. Food processes that make basic food ingredients are called _______ processes.
4. Food processes that make finished food products are called _______ processes.
5. What are the major steps in milling flour?
6. The two types of roasting used for coffee are air roasting and _______ roasting.
7. Name one step in the chemical conversion process of making cheddar cheese.

For Questions 8–20, match the food preservation technology on the right to the correct description on the left. (Note: Some letters will be used more than once.)

Description	Food Preservation Technology
8. _____Using X rays to kill mold and bacteria	a. Drying
9. _____Heating a food in a vacuum-sealed container	b. Curing and smoking
10. _____Removing water by placing a food in the sun	c. Canning
11. _____Storing food in sealed areas with controlled environments	d. Aseptic packaging
12. _____Using chemicals to preserve food	e. Refrigeration and freezing
13. _____Treating meat with salt and adding flavor	f. Controlled atmosphere storage
14. _____Storing food at low temperatures	g. Fermentation
15. _____Using heat to kill microorganisms in milk and juices	h. Pasteurization
16. _____Sterilizing a package and then placing sterilized food in it	i. Irradiation
17. _____Using microorganisms to break down the food's structure	j. Preservatives
18. _____Using blanching to drive out gases	
19. _____Removing water by exposing food to hot air	
20. _____Sterilizing the food containers with hydrogen peroxide	

21. *True or false?* Brainstorming is one of the first activities in developing a food product.
22. The step of altering the composition of a food product to incorporate the results of the initial test is called _____.
23. Producing a scaled-up food product using small versions of the equipment is called _____.
24. Listing the ingredients of a food and the methods used to produce it is called _____.
25. List the major manufacturing processes in making a food product.

Applying Your Knowledge

1. Create a form like the one below. Take it to your neighborhood supermarket and find a specific product as an example of each type of preservation. Write the product name (for example, "Birdseye Frozen Baby Peas") in the space provided.

Preservation Process	Product Example
Drying	
Curing/smoking	
Canning	
Aseptic packaging	
Refrigeration	
Freezing	
Fermentation	
Pasteurization	

2. Bake a batch of cookies as an example of secondary food processing. Make a step-by-step list as you convert flour and other ingredients into an edible finished product. Share your cookies with the class.

Chapter 31
Medical and Health Technologies

Learning Objectives

After studying this chapter, you will be able to:

- Describe the roles of science and technology in wellness and illness.
- Name the four major factors involved in wellness.
- Explain how technology is applied to exercise and sports.
- Define medicine.
- Identify kinds of health care professionals.
- Describe how technological devices are used in diagnosing illnesses and physical conditions.
- Explain how diagnostic imaging systems work.
- Describe how technological devices are used to treat illnesses and physical conditions.
- Explain how a new drug is developed.
- Identify work done by biomedical engineers.
- Describe how technology is used in surgical procedures.

Throughout history, people have been concerned about living longer and better. They have sought ways to cure illnesses, repair damage to their bodies, and improve their health. Each of these challenges has led to the development of new technologies. People have created artifacts to care for, improve, and protect their health.

The search for healthy lives can be viewed from two perspectives, **Figure 31-1.** The first is wellness, and the second is illness. ***Wellness*** is a state of physical well-being. It is being considered healthy. Wellness is generally achieved through personal effort. It involves proper diet and regular exercise.

Illness can be described as a state of poor health. It is a result of a person being in an unhealthy condition. A disease or sickness may cause illness. In the broadest sense, it can be extended to include injuries caused by accidents.

Both science and technology play a role in wellness and illness. Science provides knowledge about the human body and health. It describes the natural processes that the body uses to maintain

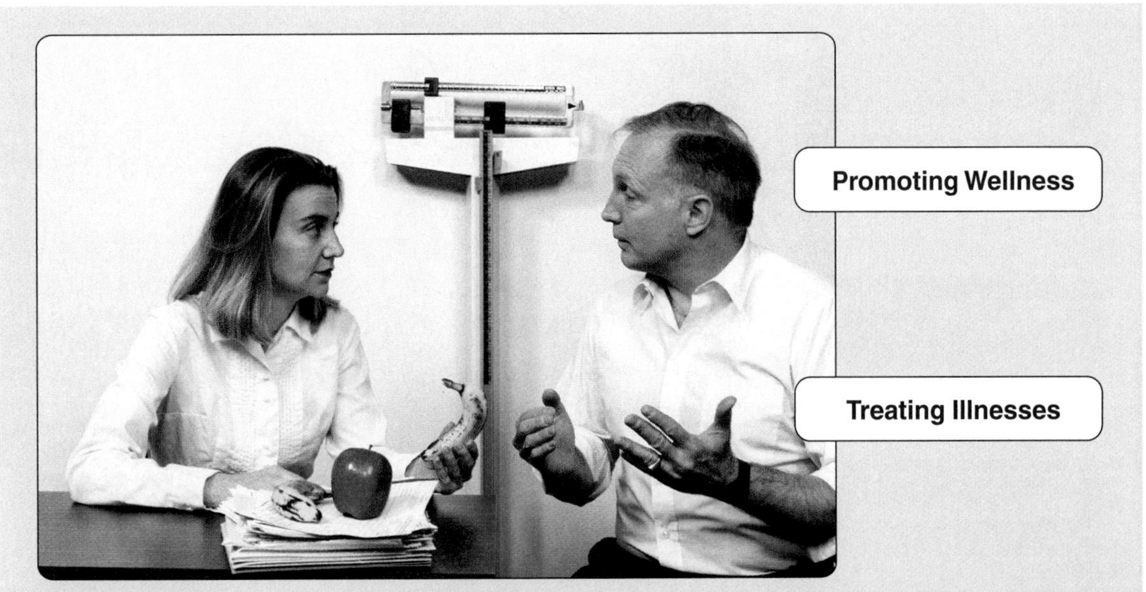

Figure 31-1. People are living longer because of wellness programs and modern ways of treating illnesses. (U.S. Department of Agriculture)

itself. Technology provides the tools and equipment needed to achieve wellness and treat illness. It helps people maintain and restore the body's processes and functions.

Technology and Wellness

Wellness involves actions that keep the body healthy. It contributes to physical fitness, which is a combination of good health and physical development. The objective of physical fitness is to maximize a person's health, strength, and endurance.

Wellness can be considered preventive medicine and treatment. It helps prevent people from becoming ill. Wellness often focuses on what people can do for themselves to maintain their well-being.

Wellness considers at least four major factors. These factors are nutrition and diet, environment, stress management, and physical fitness, **Figure 31-2.**

Wellness means that people should be concerned with what they eat. Proper nutrition is important to physical fitness and wellness. The energy a person can expend depends on nutrition. If the diet is inadequate, the fitness level will drop. Therefore, people should be conscious of the quantity and value of the food they consume, **Figure 31-3.** As we saw in the last chapter, proper food selection and processing involve many technological actions.

Wellness also means that individuals should be aware of the environment's effects on their body. For example, they should try to improve the quality of the air they breathe. They should control their exposure to direct sunlight. Technology has been applied to air and water purifiers. It has been used to develop sunscreen lotions, sun-filtering clothing, and other effective products.

The approach to wellness suggests that people should be aware and control the emotional stress they encounter. Moreover, they should keep their bodies fit through activity and exercise. Technology has also been applied to stress management and physical fitness through two areas that will be discussed further. These areas are exercise and sports.

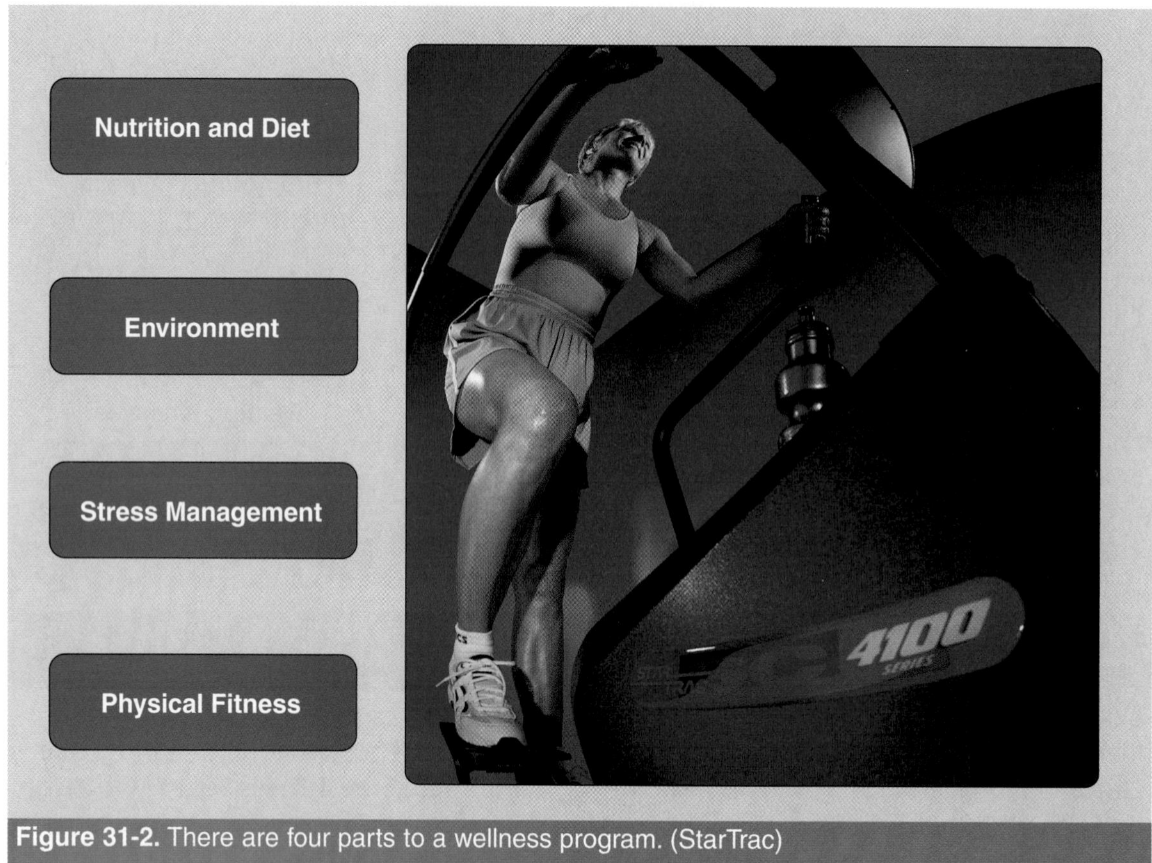

Figure 31-2. There are four parts to a wellness program. (StarTrac)

Figure 31-3. Wellness programs encourage people to select and eat proper foods.

Technology and Exercise

Exercise has been described as exerting the body or the mind, **Figure 31-4.** We can have physical or mental exercise. It is done for training or improving health. Two major types of exercise exist: anaerobic and aerobic.

Anaerobic exercise involves heavy work by a limited number of muscles. This type of exercise is maintained for short intervals of time. Examples of anaerobic exercise are weight lifting and sprinting. These activities are maintained only for short intervals. They increase strength and muscle mass but have limited benefit to cardiovascular health.

Aerobic exercise uses oxygen to keep large muscle groups moving continuously. The exercise must be maintained for at least twenty minutes. Aerobic exercise uses several major muscle groups throughout the body. This causes greater demands on the cardiovascular and respiratory systems to supply oxygen to the working muscles. Aerobic exercise includes walking, jogging, and swimming. This type of exercise is recommended for reducing the risk of heart disease. It also increases endurance.

Exercise can be done without special equipment. People can walk or jog to improve their health. Even these activities, however, often require technology. Special shoes and clothing have been designed and produced to help in these activities. Also, many medical professionals and fitness experts suggest that people use exercise equipment to improve their health and well-being. This equipment includes treadmills, stationary bikes, stairclimbers, rowers, and home gyms, **Figure 31-5.**

Treadmills are moving belts that allow people to walk or jog in place. They provide an aerobic fitness workout.

Stationary bikes are similar to a bicycle except that they do not move. They allow people to obtain the benefits of bicycling without leaving home. In addition, on most of these bikes the handlebars move and provide resistance. This feature provides an upper-body workout.

Stairclimbers allow people to obtain the benefits of climbing without having to have stairs available. These machines provide lower-body workouts.

Rowing machines simulate an actual rowboat with oars. The machine allows people to use their arms and legs without having a boat in the water.

Home gyms are multistation exercise machines that allow people to work on many different muscle groups. These machines can provide a complete workout program.

Technology and Sports

Sports are another way to promote wellness and physical fitness. Sports are games or contests that involve skill, physical strength, and endurance. Contests involving physical abilities can be traced back to prehistoric times, **Figure 31-6.** Many different types of sports have developed throughout history. Their value has been reinforced by medical proof of the benefits of physical exercise.

Figure 31-4. Exercise prepares people for the demands of everyday life. (U.S. Department of Agriculture)

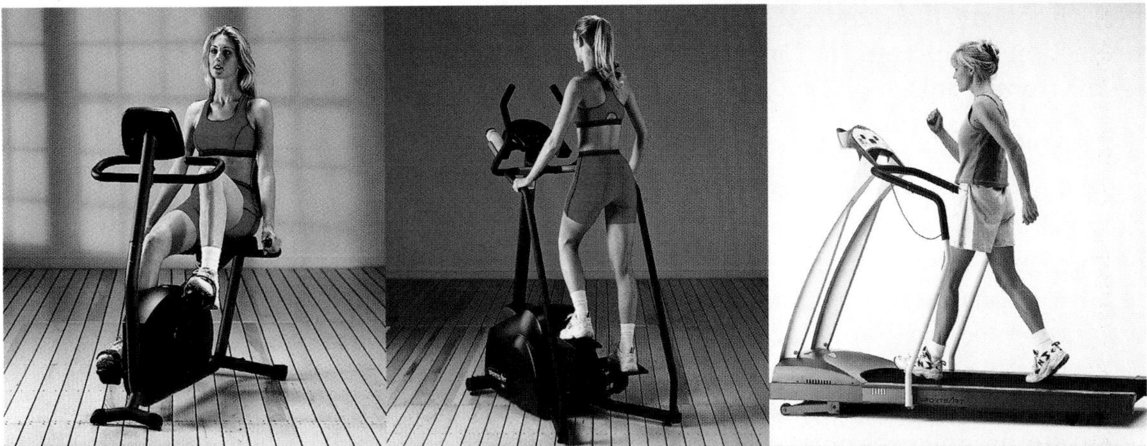

Figure 31-5. Among the many types of exercise equipment are stationary bicycles (left), stairclimbers (center), and treadmills (right). (SportArt)

Figure 31-6. This amphitheater in historic Pompeii was used for sporting events and other activities in ancient times.

This discussion of sports and wellness is focusing on using sports for physical fitness. Sports in this context deals with ordinary people using sports as a way to maintain health and well-being. We are not exploring sports as an economic activity here.

All sports require one or more types of technological products. These include playing fields or venues, game equipment, and personal protection. Each game, sport, or contest uses specific technologies, however. For example, football's playing venue is different from tennis's. Likewise, the game equipment for football (the ball and goal posts, for example) is different from tennis (the ball, racket, and net, for example). Similarly, the personal clothing and protection equipment for football are different from tennis.

Technology and Playing Venues

Each sport requires playing courts, fields, or buildings. These venues are part of our built environment. They are the result of construction technology, which was presented in Chapter 17.

Typically a playing venue has a playing surface, goals, and constraints, **Figure 31-7.** For example, golf uses grass-covered areas as a playing surface, holes for goals, and sand and water traps as constraints.

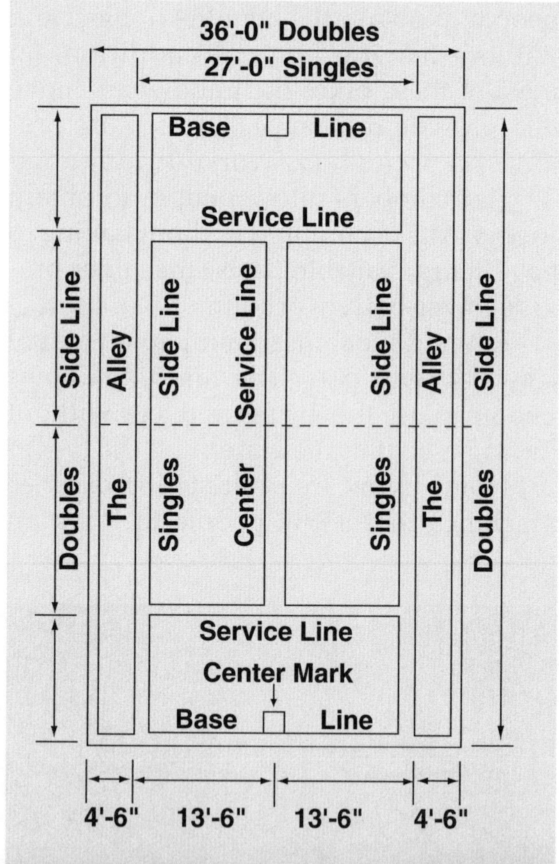

Figure 31-7. Shown here is a drawing for a tennis court.

Basketball uses a wood, concrete, or asphalt playing surface. It has a rim and backboard as a goal and painted lines as constraints. Skiing uses slopes covered with snow.

Technology is often used to improve the natural playing surfaces. For example, special grasses have been developed for golf courses. Fertilizers have been developed and manufactured to encourage the grass to grow. Special lawn grooming and moving equipment has been developed to maintain the courses. Likewise, snow-grooming equipment prepares and renews the surfaces of ski runs. Snow-making equipment has been developed to supplement natural snow for skiing, **Figure 31-8.** This equipment makes snow by breaking water into small particles with compressed air. The water is cooled as it moves through cold air. It forms small particles of ice and is distributed as snow on a surface.

Technology and Game Equipment

As we noted earlier, each sport has its own playing equipment. Hockey has pucks and sticks. Baseball has bats and balls. Tennis has rackets and balls. These manufactured products are designed and produced to specifications. These specifications are developed by groups who establish rules for playing the sport.

Technology and Personal Protection

Players in many sports wear special clothing and protective gear, **Figure 31-9.** The clothing may be made from fabrics that wick perspiration away from the body or shed rain. Special shoes may be developed to provide foot support and absorb the shock caused by running on hard surfaces.

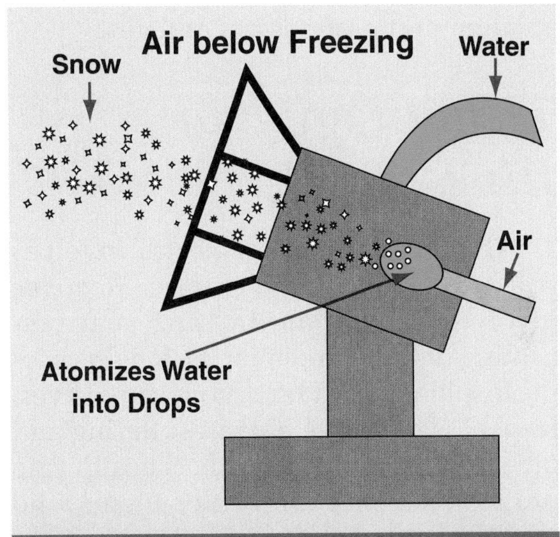

Figure 31-8. This diagram shows how snow-making equipment works.

Figure 31-9. The shoes on the left and the batting helmets on the right are examples of special equipment designed for sports.

Some sports can lead to bodily injury or muscle damage. Participants in these sports wear protective gear. Baseball players wear batting helmets to protect the head from wild pitches. Football players wear protective devices to protect the head and neck, the body, and the hands and feet.

The game and protective equipment are technological products. They were designed for a specific purpose and have a customer base. They are appropriate for amateur and professional athletes.

Technology and Illness

The second focus of health involves that area often called medicine. ***Medicine*** involves diagnosing, treating, and preventing diseases and injuries. It helps people live longer and lead more active lives. Its goal is to reduce human suffering and physical disability. It involves health care professionals working with patients who are ill or injured. It also includes searching for new drugs, treatments, and technology.

Disease is any change that interferes with the appearance, structure, or function of the body. Treating disease and injury requires a number of different health care professionals, **Figure 31-10.** These professions include:

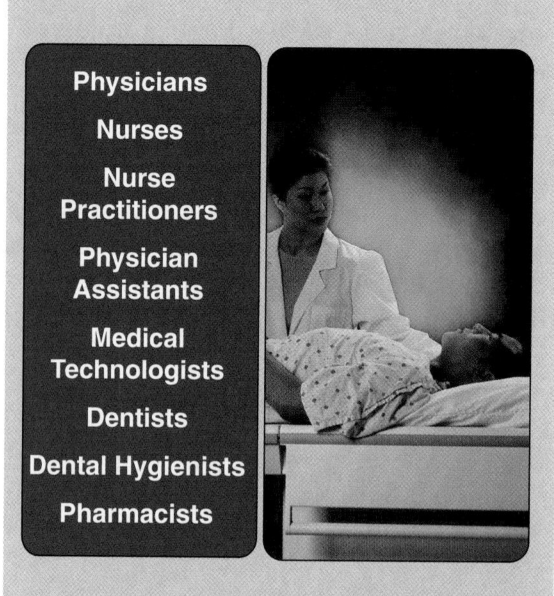

Figure 31-10. Health care professionals work to cure illnesses and to help people stay healthy.

- *Physicians:* Individuals who diagnose diseases and injuries. They administer appropriate treatment and advise patients on ways to stay healthy.
- *Nurses:* Individuals who help physicians diagnose and treat illnesses and injuries. They assist physicians during examinations, treatment, and surgery. They observe and record symptoms that patients exhibit. Nurses administer medications and provide care in hospitals and nursing homes.
- *Nurse practitioners:* Individuals who perform some of the basic duties that physicians once provided. They diagnose and treat common illnesses and prescribe medication.
- *Physician assistants:* Individuals who deliver basic health services under the supervision of a physician. They examine patients, order diagnostic tests and X-ray films, and prescribe drugs or other treatments.
- *Medical technologists:* Individuals who gather and analyze specimens to assist physicians in diagnosis and treatment.
- *Dentists:* Individuals who diagnose, treat, and help prevent diseases of the teeth and gums.
- *Dental hygienists:* Individuals who assist dentists in surgery. They also clean teeth and advise patients on proper techniques to prevent tooth and gum disease.
- *Pharmacists:* Individuals who dispense prescription drugs and advise people on their use.

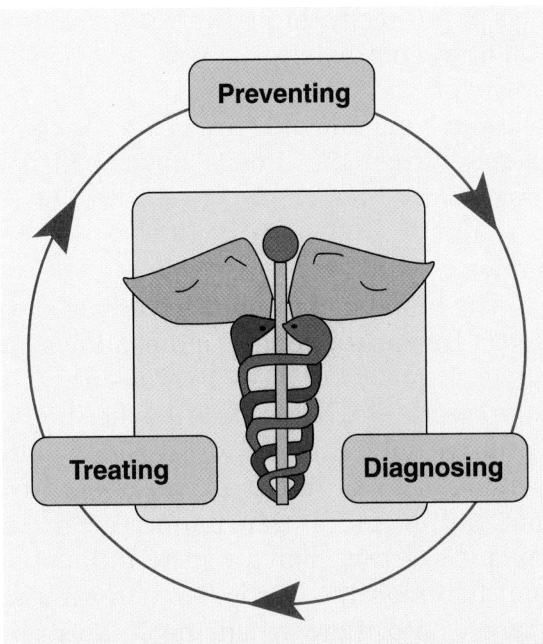

Figure 31-11. Health care professionals have three major goals.

This team of health care professionals depends on technology in their work. They use equipment and techniques to make their work more effective.

Goals of Medicine

Many people seek medical care because they are ill or injured. Health care professionals respond in three major ways, **Figure 31-11.**

First, the medical personnel diagnose the illness or condition. ***Diagnosis*** is performed by conducting interviews, physical examinations, and medical tests. The diagnostic process tries to determine the nature or cause of the condition.

Medical personnel then treat the illness or condition. ***Treatment*** involves applying medical procedures to cure diseases, heal injuries, or ease symptoms.

Prevention is the third common action taken by health care professionals. One method of prevention has already been discussed. This is through wellness programs that promote exercise and proper diet.

The other way is through immunization. This is a process of exposing the body to small amounts of a bacterium, protein, or virus to cause antibodies to form. These antibodies are cells or proteins that circulate in the blood and attack foreign bacteria, viruses, or fungi. These foreign substances can cause diseases.

Technology in Medicine

Many medical actions involve tools and equipment. Physicians and dentists use equipment to diagnose and treat illnesses and conditions. They use technology to extend the potential to deal with medical problems. We should note that the use of technology throughout medicine is different from medical technology. This term defines the narrow field of gathering and analyzing of specimens to assist physicians in diagnosing and treating illnesses.

Technology and Diagnosis

In the past, physicians depended on people to describe their symptoms. From this description, they planned treatment procedures. Physicians found, however, that descriptions were not always accurate and that many were hard to interpret. To deal with these problems, physicians and others have developed diagnostic equipment. Today, many different types of diagnostic devices exist. For this discussion, three major types will be examined here:

- Routine diagnostic equipment
- Noninvasive diagnostic equipment
- Invasive diagnostic equipment

Routine diagnostic equipment is used to gather general information about the patient. This equipment includes scales to determine the patient's weight and thermometers to determine body temperature. It also includes devices to such things as measure the oxygen in the blood, listen to heart rhythm (stethoscopes), and measure

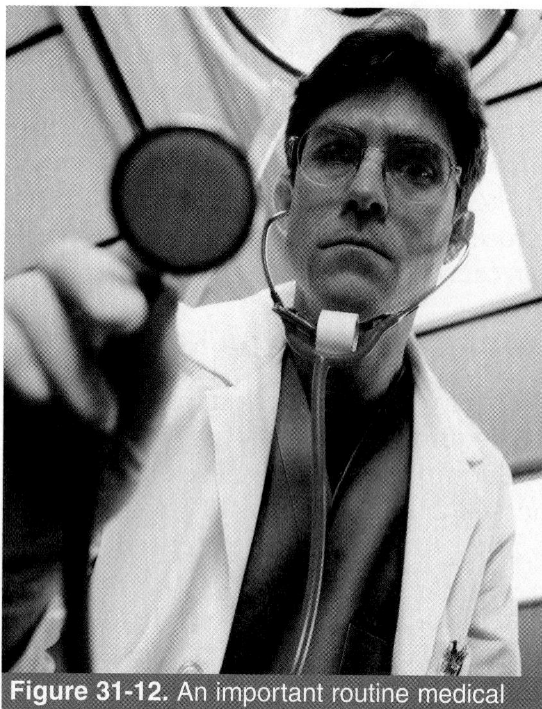

Figure 31-12. An important routine medical diagnostic tool is the stethoscope.

blood pressure, **Figure 31-12.** These items provide a baseline of general information.

Noninvasive diagnostic equipment gathers information about the patient _without entering the body. A typical example of this type of diagnostics is called ***radiology.*** It uses electromagnetic radiation (waves) and ultrasonics (high frequency sounds) to diagnose diseases and injuries, **Figure 31-13.**

Diagnostic radiology uses special equipment called body scanners or body imaging equipment. They produce images (pictures) of the body without entering it.

One of the most common diagnostic imaging machines is the ***X-ray machine.*** An X-ray machine is essentially a camera. It uses X rays instead of visible light, however, to expose the film. X rays are electromagnetic waves that are so short they can pass through solid materials such as paper and human tissue. Denser materials, however, such as metals and human bones, absorb some or all of the waves. Thus, if you put a piece of film under your hand and then pass X rays through your hand, the skin and tissue will let most of the X rays pass directly through. The film behind it will be completely exposed. The bones, however, will absorb most of the X rays. The film behind it will be not exposed completely. When the film is developed, an image of the bones in the hand will appear. Any fractures or joint deformity will be shown.

The first X-ray machines were used to detect fractures in bones and shadows on lungs. Later, substances that absorbed X rays were introduced into the body. Various cavities in the body are filled with a material to aid the X-ray process. This material is more transparent or more opaque to X rays than the surrounding tissue. It brings the particular organ more sharply into view when the X rays are taken. For example, a patient will swallow barium sulfate so that gastric ulcers can be located. In other cases, dyes are injected so that the heart, kidneys, or gallbladder can be X rayed. This allows people to produce radiographs or X-ray photographs of the desired area.

Look at **Figure 31-14.** It shows a diagram of an X-ray machine. It has an X-ray tube with a positive electrode (anode) and a negative electrode (cathode). The cathode is heated and gives off electrons. They travel

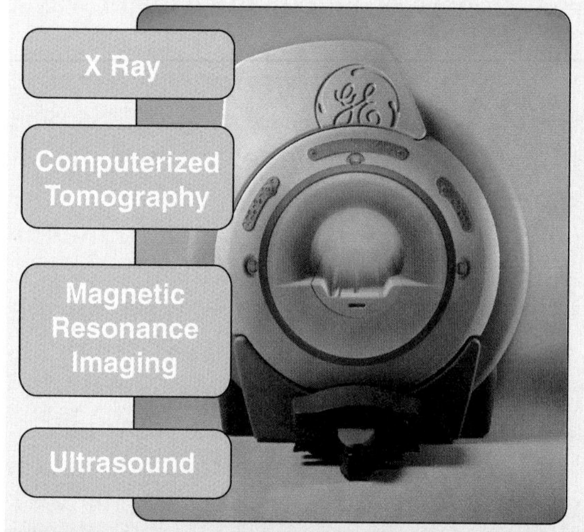

Figure 31-13. The four types of radiology are X ray, computerized tomography, magnetic resonance imaging, and ultrasound.

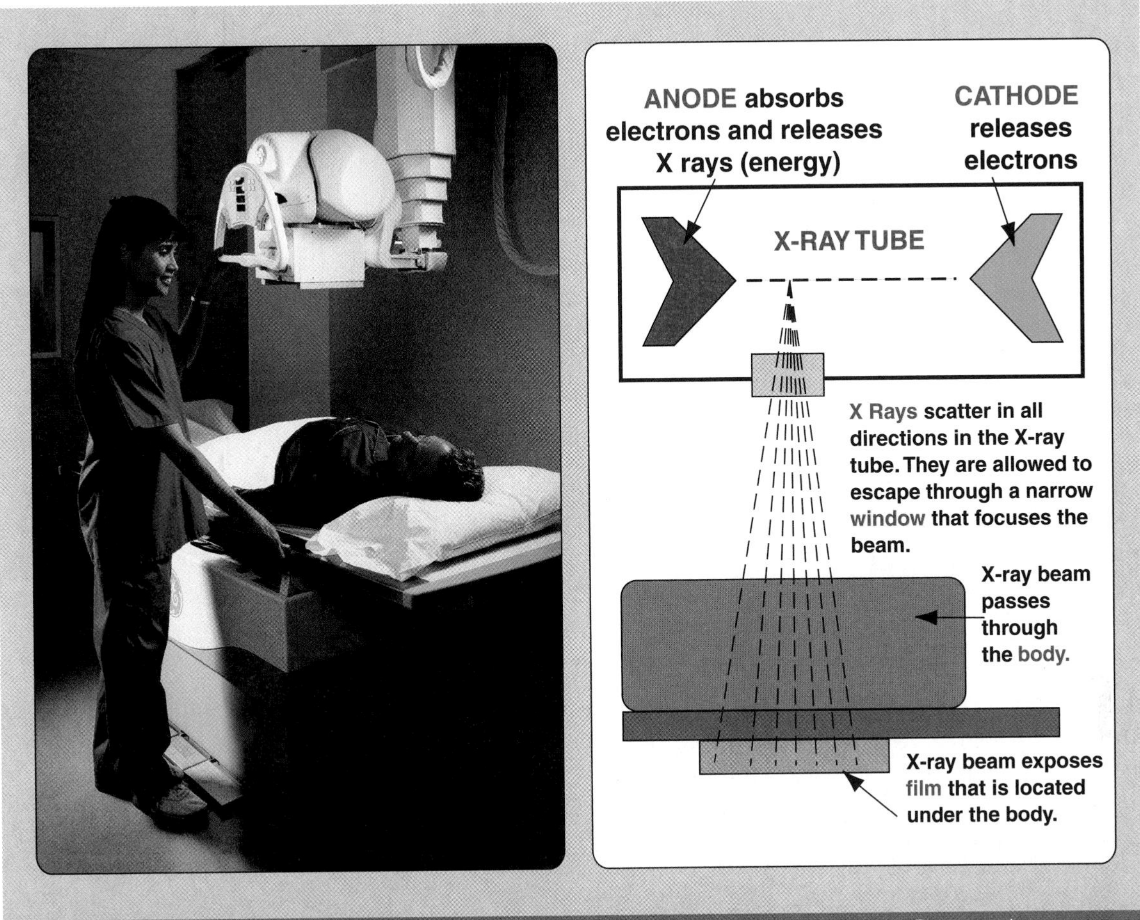

Figure 31-14. This diagram shows how an X-ray machine works. (GE Medical Systems)

toward the anode where they hit and release energy (X rays). The X rays are concentrated into a beam that leaves the X-ray tube. The beam is directed at the desired location on the body. There it passes through the body and exposes light-sensitive film. When the film is developed an image of the section of the body is revealed. Some newer X-ray machines do not use film but instead produce digital images that can be viewed on computer systems.

The major disadvantage of X rays is that the image is two-dimensional. It is a flat image taken of a three-dimensional object (the body). Therefore, depth is not shown. To deal with this shortcoming, people have developed *computerized tomography* (CT) scanners, **Figure 31-15.** A CT scanner will produce images of any part of the body without using dyes. The scanner rotates around a patient's body and sends a thin

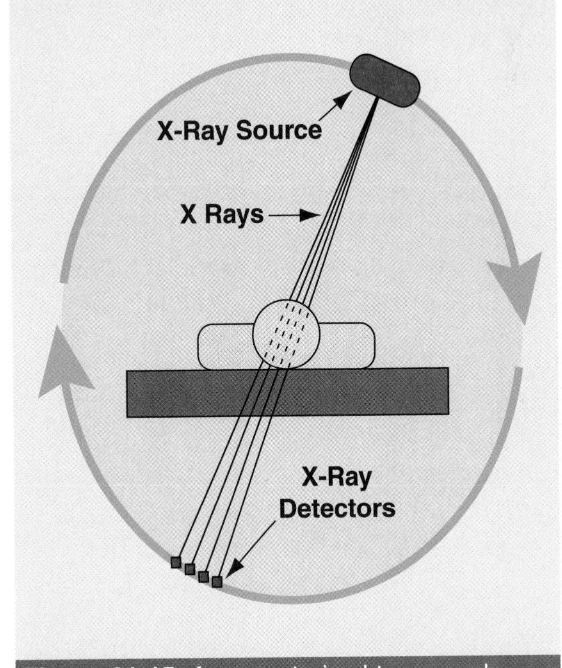

Figure 31-15. A computerized tomography scanner rotates around a person's body.

Technology Explained

Dialysis Machines

dialysis machine: A device that can do some of the functions of the human kidneys. It filters waste and fluid from the blood.

For any number of reasons, some people's kidneys stop functioning. They stop removing urea and certain salts from the blood. This condition is fatal, unless the person receives treatment. One treatment is called *hemodialysis.* It is the process of removing waste and excess fluid from the blood. An artificial kidney machine does this.

The word *dialysis* describes the movement of microscopic (very small) particles from one side of a semipermeable membrane to the other side of the membrane. Hemodialysis is a special type of dialysis. It involves the blood *(hemo). Hemodialysis* literally means "cleaning the blood."

In hemodialysis, the blood is pumped out of the patient's body to the artificial kidney machine, **Figure A.** In this dialysis machine, blood runs through tubes made of a semiporous *membrane.* Surrounding the tubes is a sterile liquid called the *dialysate solution.* This liquid is made up of water, sugars, and other components.

As the blood circulates, the red and white cells and other important components are too large to fit through the pores in the membranes. Impurities, such as urea and salt, however, pass through the membrane, **Figure B.** The dialysate solution, which is discarded, carries them away. Tubes connected to the kidney machine return the cleaned blood to the bloodstream.

Dialysis is a treatment for people in the late stages of chronic kidney failure. Trained professionals generally perform the procedure. Dialysis normally takes three to five hours to complete. Typical patients receive three treatments a week. Dialysis allows these people to maintain many of their normal activities after the treatments.

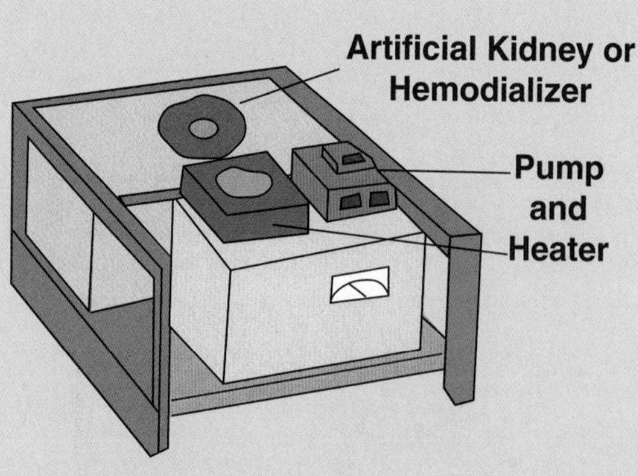

Figure A. These are the tools used for hemodialysis.

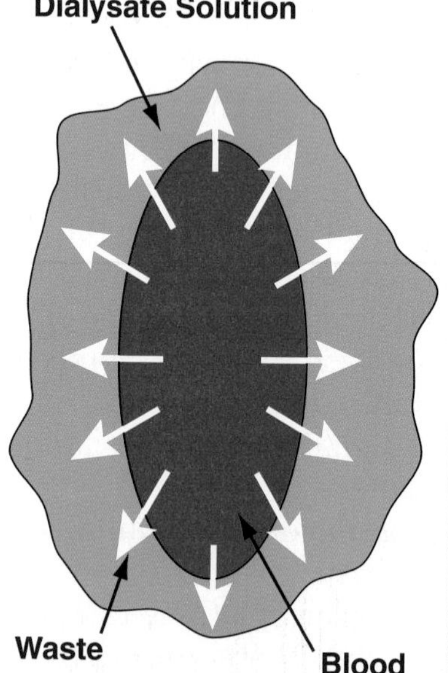

Figure B. Wastes pass from the blood, through the membrane of the tubing, into the dialysate solution.

Connections to Technology: Science

Aerodynamics

Many people use a bicycle simply to get somewhere or to exercise but do not realize the role that science plays in many areas of cycling. For example, the scientific study of the principles of aerodynamics has greatly affected such aspects of cycling as the design of the bicycle and the clothing a rider might wear for more effective cycling.

Generally speaking, aerodynamics is the study of the forces that act on an object as it moves through the air or other gaseous fluid. One of the major forces studied in aerodynamics is that of drag. *Drag* is the term used to describe the aerodynamic force that resists the forward movement of an object. When cycling, we encounter two kinds of drag: frictional drag and form drag. Both forces will slow a cyclist.

Scientists counter these forces in two ways. Frictional drag can be lessened by making the moving object smoother. Form drag can be lessened by streamlining the object. To counter drag, then, designers have created bicycles with such features as aerobars. These handlebars lower the rider's torso and help keep the elbows together. Drag is decreased because the torso and elbows are not blocking as much airflow.

Designers have also created different kinds of clothing to help with drag. For example, many professional riders wear such items as skinsuits, which are designed to reduce friction.

Bicycles also have other features that deal with the principles of aerodynamics. Can you think of two of these?

X-ray beam at many different points. Crystals opposite the beam pick up and record the absorption rates of the bone and tissue. A computer processes the data into a cross-sectional image of the part of the body being scanned. This digital image can be viewed immediately or stored for later use.

X rays can also be dangerous to use. They can cause damage to body parts. To deal with the hazards associated with X rays, people have developed new imaging techniques. One of these is ***magnetic resonance imaging*** (MRI), **Figure 31-16.** It can produce computer-developed cross-sectional images of any part of the body very quickly. The images are developed using magnetic waves rather than X rays.

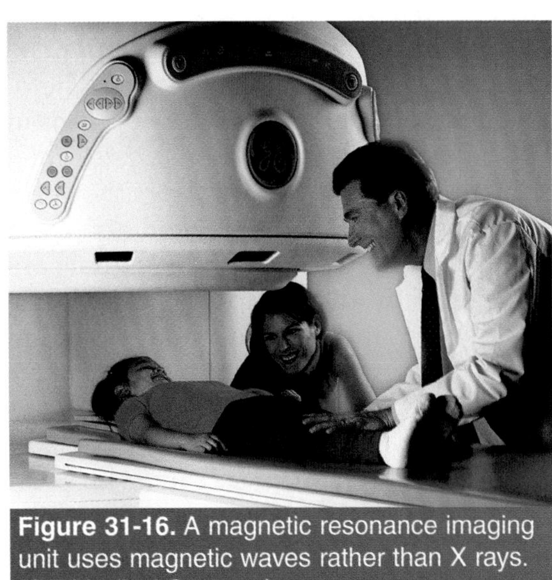

Figure 31-16. A magnetic resonance imaging unit uses magnetic waves rather than X rays. (GE Medical Systems)

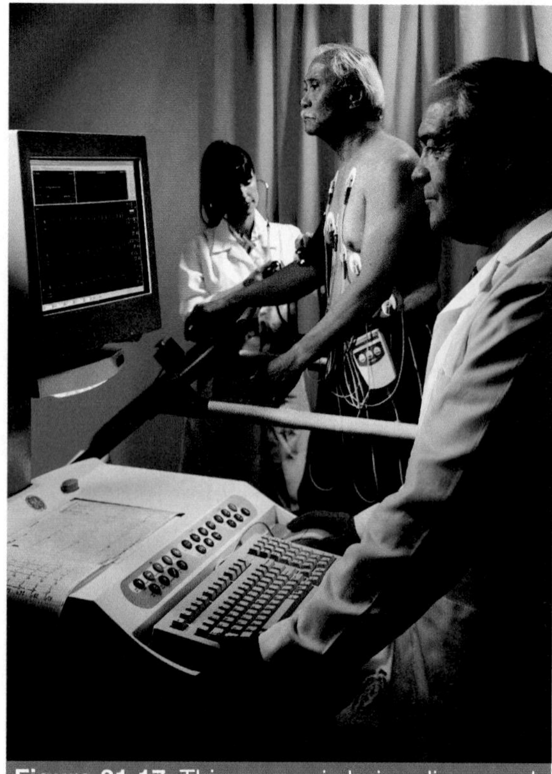

Figure 31-17. This person is being diagnosed with the use of an electrocardiograph machine. (GE Medical Systems)

An MRI unit is a large tube surrounded by a circular magnet. The patient lies on his or her back on a bed that is moved into the magnet. The body is positioned so that the part to be scanned is in the exact center of the magnetic field. At this point the scan can begin.

The MRI unit exposes the patient to radio waves in a strong magnetic field. The magnetic field lines up the protons in the tissues. A beam of radio waves then spins these protons. As they spin they produce signals that are picked up by a receiver in the scanner. These signals are processed by the computer to produce very sharp and detailed images of the body.

Another imaging technique is called ***ultrasound.*** It uses high-frequency sound waves and their echoes to develop an image of the body. The ultrasound machine subjects the body to high-frequency sound pulses using a probe. The sound waves travel into the body. There they hit a boundary between tissues. This boundary could be between soft tissue and bone or between fluid and soft tissue. Some of the sound waves are reflected back to the probe. Others travel further until they reach another boundary and get reflected. The probe picks up the reflected sound waves. They are processed by a computer to produce a still (photograph) or moving (television) image.

In a typical ultrasound procedure, millions of pulses are sent each second. The probe is moved along the surface of the body or angled to obtain different images.

Other technological devices are used in diagnosis besides imaging devices. A very important device is the ***electrocardiograph*** (EKG) machine, **Figure 31-17.** It produces a visual record of the heart's electrical activity. As the heart works, it sends off very small electrical signals that can be detected on the skin. The EKG machine uses electrodes attached to the skin to capture the signals. The machine amplifies the signals and it produces a graph of their values. This graph can be read by health care professionals to determine how the heart is functioning.

Another important diagnostic device is the ***endoscope,*** which allows a physician to look inside the body. An endoscope is a narrow flexible tube containing a number of fiber-optic fibers smaller in diameter than a human hair. The tube can be threaded through a natural opening, such as the throat, or through a small incision. Light sent through the fibers shines on an interior part of the body. This light is reflected back through the fibers to form a series of dots. One dot is produced by each fiber in the tube. The dots form a picture of an internal organ or other part of the body.

These examples are just a few of the many devices that have been designed and built to help diagnose illnesses and physical conditions. They show the dramatic

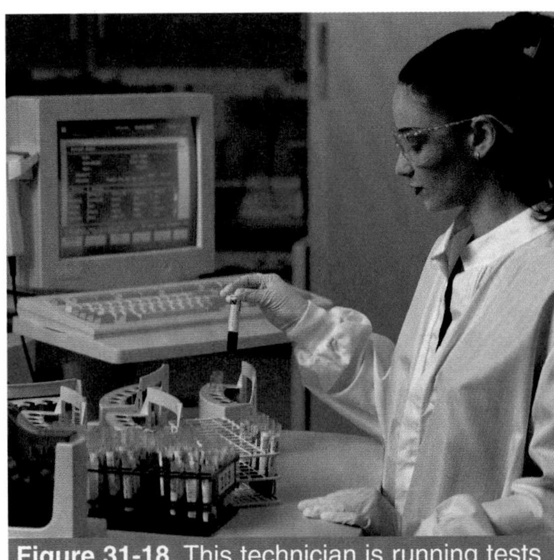

Figure 31-18. This technician is running tests of blood samples.

use of technology to help reduce human suffering.

Invasive diagnostic equipment is used when drawing and testing a blood sample. A blood test determines the chemical composition of a sample of blood. The sample is tested in a laboratory using a number of different technological procedures, **Figure 31-18.** The tests can detect the presence of specific chemicals associated with a disease. They can also detect an imbalance in the chemical composition of the blood. These data provide health care professionals with information needed to treat illnesses or physical conditions.

Invasive diagnostic equipment is also involved when taking tissue samples (biopsy) for laboratory examination. Many medical conditions, including cases of cancer, are diagnosed by removing a sample of tissue. A pathologist examines this tissue. This medical professional is a person who is trained to perform medical diagnoses by examining body tissues and fluids removed from the body.

Technology in Treatment

Treatment of illnesses and physical conditions can require drugs, specialized equipment, or both. Both of these approaches are the result of technology. They are the products of design and production actions.

Treatment with Drugs and Vaccines

Humans have always experimented with substances to treat pain and illness and to restore health. The development of new treatments and prevention techniques begins with the unmet medical needs of people. One common treatment method is drugs, **Figure 31-19.**

A ***drug*** is a substance used to prevent, diagnose, or treat a disease. It can also be used to prolong the life of patients with incurable conditions. Throughout history, drugs have improved the quality of life for people. For example, a vaccine, a special type of drug, is a substance administered to stimulate the immune system to produce antibodies against a disease. Vaccines have helped eliminate such diseases as smallpox and polio (poliomyelitis).

Drugs can be classified in many ways. These classifications include:

- Method of dispensing: Prescription and over the counter (proprietary) drugs. Drugs that are prescribed by physicians and dispensed by pharmacists are known as ***ethical drugs. Proprietary drugs*** are those considered safe for

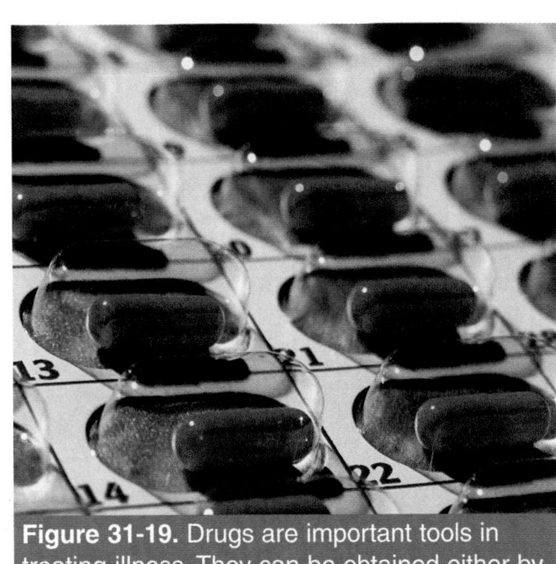

Figure 31-19. Drugs are important tools in treating illness. They can be obtained either by prescription or over the counter.

unsupervised use by consumers. They are available to everyone through retail outlets.

- Form: Pills, capsules, liquids, gases, **Figure 31-20.**
- Method of administering: Injection (shots), mouth, inhaling, and absorption (patches).
- Illness or condition treated: Cancer drugs, measles vaccines, and high blood pressure drugs, for example.

Today most drugs are the products of chemical laboratories. **Figure 31-21.** They are called synthetic (or human-made) drugs. They are developed because they can be controlled better than natural drugs. Not all synthetic drugs are totally new. Some are developed by altering the structure of existing substances. These new drugs are called analogs. This means they have a structure similar to another compound but have a slightly different composition. These new drugs may be more effective than the original drugs. Also, they may cause fewer side effects.

A number of new drugs have been developed by using gene splicing or recombinant DNA. This approach joins the DNA of a selected human cell to the DNA of a second organism, such as harmless bacterium. The new organism can produce the disease-fighting substance. This new substance is extracted from the bacterium and processed into a drug.

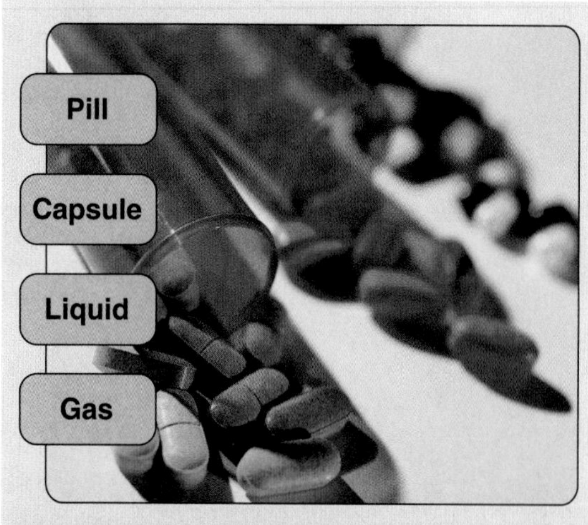

Figure 31-20. Drugs come in many different forms.

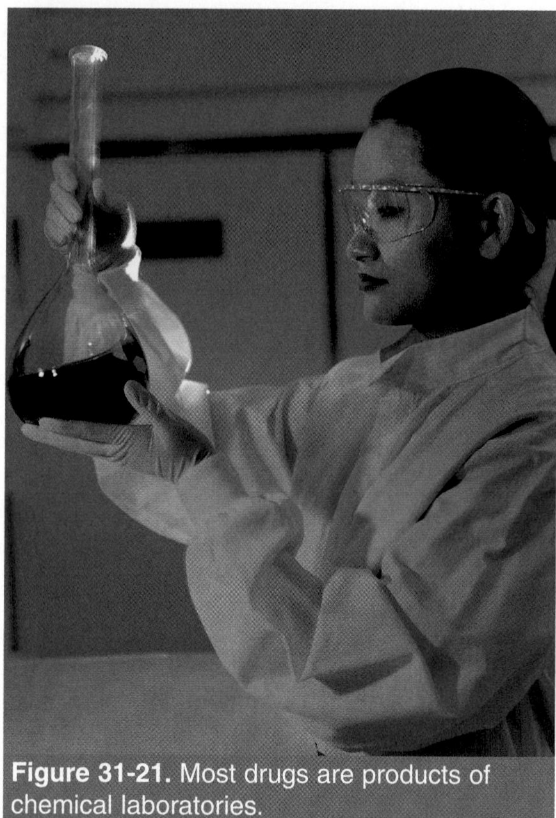

Figure 31-21. Most drugs are products of chemical laboratories.

The first drug produced using gene splicing was hormone insulin. It was created in 1982 by inserting a human insulin gene into *E. coli* bacteria.

The development of any new drug requires great amounts of time and money. It can take ten or more years and several millions of dollars.

The drug development process generally starts with a need to treat a disease or physical condition. Researchers start with an existing chemical substance that might have medical value. They may work with thousands of different substances before they find one that can serve as a drug.

Once a new substance that may have medical value is discovered, an extensive testing program starts, **Figure 31-22.** First, the drug is tested on small animals, such as rats and mice. If the tests show promising results, additional tests are conducted on larger animals, such as dogs and monkeys.

The tests are evaluated in terms of treating the disease and physical condition. Also, it must have a low level of toxicity (capability to poison a person). Drugs that are judged to be effective in animal tests are ready for the next level of testing. At this point, a request is made to the Food and Drug Administration (FDA) to conduct clinical tests.

If the FDA approves the request, the drug can be tested on humans. These tests are generally conducted in three phases. Each of these phases can take many months to complete.

During the first phase, the drug is given to a small number of healthy individuals. These tests are designed to determine the drug's effect on people.

If the drug passes this test, it moves into the next phase. Here it is given to a small number of people who have the disease or physical condition that the drug will treat. These individuals are divided into two groups. The first group will be given the drug. The other group is given an inert substance, such as sugar, that is used in place of an active drug. This inactive compound is called a placebo. During the test, neither the group nor the researchers know who is receiving the drug or who is receiving the placebo. This research technique is called a double-blind study. It keeps participants from influencing the results of the study.

Figure 31-22. Before a drug is made available to treat illnesses, it must go through animal tests and three stages of clinical tests.

Drugs that pass the second phase move into a final phase of testing. Here the drug is tested with a much larger group of people. The goal of these tests is to determine specific doses, side effects, interactions with other drugs, and other information. The data from these tests will be used in drug labeling.

When the third phase is complete, the results of the tests are submitted to the FDA for approval. The agency must decide whether the drug is effective and safe. Also, it must weigh its benefits against any risks that may be present. If the FDA determines that the drug meets its criteria, it approves the drug for use.

Treatment with Medical Equipment

Biomedical engineers develop a great deal of diagnostic and treatment equipment, **Figure 31-23.** These professionals combine engineering with medicine to improve health care. They define and solve problems in biology and medicine.

The work done by biomedical engineers includes:

- Designing life-support equipment, such as cardiac pacemakers, defibrillators, or artificial kidneys
- Designing artificial body parts, such as hearts, blood vessels, joints, arms, and legs
- Designing computer systems to monitor patients while in surgery and intensive care
- Designing sensors for the blood's chemistry
- Designing instruments and devices for therapeutic uses, such as a laser system for eye surgery or a catheter to clear out a blocked blood vessel
- Designing medical imaging and treatment systems

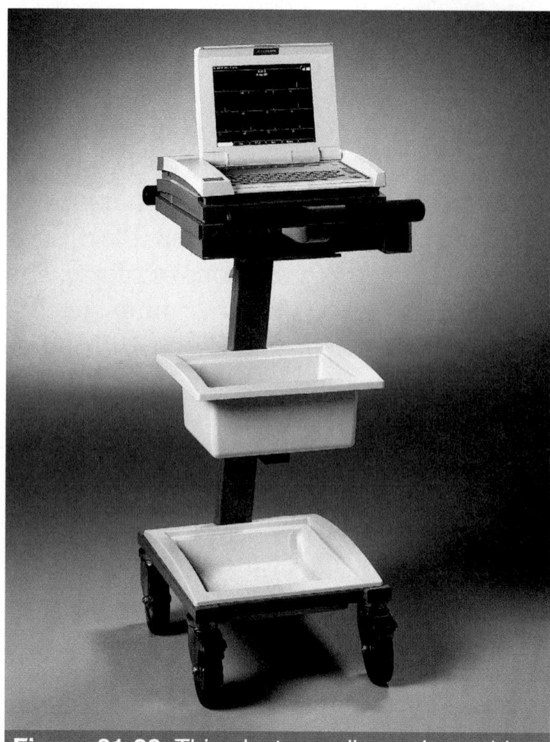

Figure 31-23. This electrocardiograph machine is the result of many people's work, including biomedical engineers. (GE Medical Systems)

Health care professionals use these and other technological systems to treat injuries and diseases. In this section, we will examine three major treatments: radiation therapy, surgery, and emergency medicine. Obviously, these are only three among many treatment technologies.

Many types of cancers can be treated using ***radiation therapy.*** This treatment is called therapeutic (treatment of disease or disorders) radiology. This treatment uses high-energy radiation to treat the cancer cells. The technique works because the radiation destroys the cancer cells' ability to reproduce. Normal cells can recover from the effects of radiation better than cancer cells.

Sometimes radiation therapy is only part of a patient's treatment. Patients can be treated with radiation therapy and chemotherapy (chemicals or drugs). This treatment may be followed by surgery.

Another use of radiology is called ***intervention radiology.*** This technique uses images produced by radiology for nonsurgical treatment of ailments. These images allow the physician to guide a catheter (hollow, flexible tubes), balloon, and other tiny instruments through blood vessels and organs. An example of this approach is balloon angioplasty, which uses a balloon to open blocked arteries.

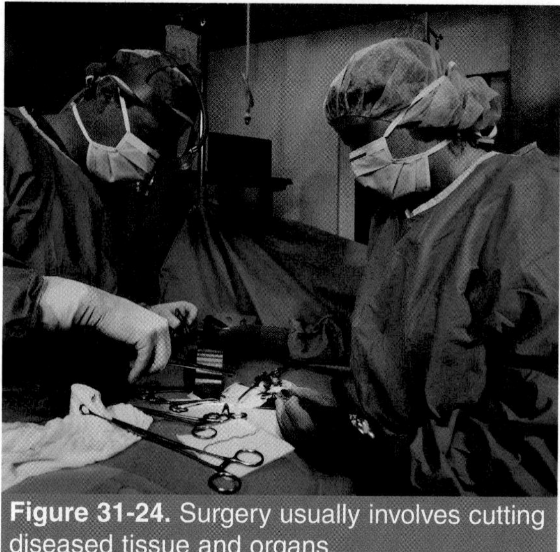

Figure 31-24. Surgery usually involves cutting diseased tissue and organs.

Surgery is a common way to treat a disease, **Figure 31-24.** Surgery treats diseases and injuries with procedures called operations. Surgery can be used to remove diseased organs, repair broken bones, and stop bleeding. Most surgery involves manually removing (cutting) diseased tissue and organs.

Newer technologies, however, are also being used for many types of surgery. High-frequency sound waves (ultrasound) can be used to break up kidney stones. Lasers use a beam of light to vaporize or destroy tissue. An endoscope can be used with special devices to operate on a particular area of the body.

In transplant surgery, organs removed from one person can be implanted into another person. Also, devices such as pacemakers can be implanted.

A practice closely related to surgery is prosthesis. A ***prosthesis*** is an artificial body part such as an artificial heart or arm. These devices are developed through

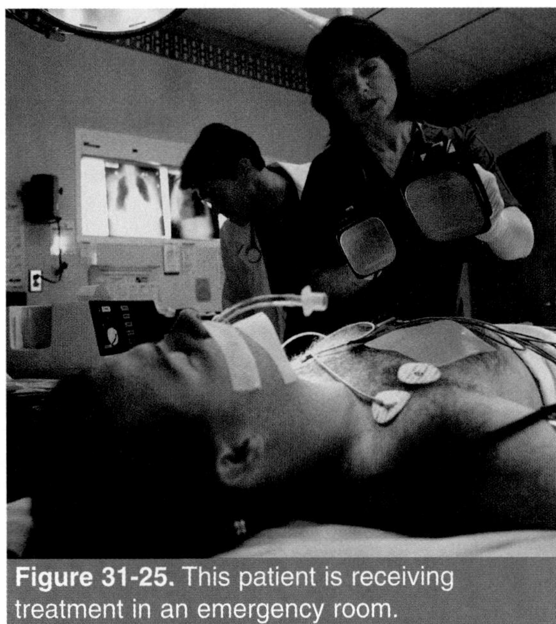
Figure 31-25. This patient is receiving treatment in an emergency room.

biomechanical engineering. This branch of engineering applies mechanical engineering principles and materials to surgery and prosthetics.

One of the oldest types of prosthesis is a denture. Dentures are removable artificial teeth that replace full-mouth (upper or lower) teeth. A similar treatment for a few missing teeth is a tooth-supported bridge. These artificial teeth fill a space between existing teeth. Another type of artificial tooth is the dental implant. Dental implants are replacement teeth set into the gums. They replace one or more missing teeth.

Emergency medicine deals with unexpected illnesses and injuries, **Figure 31-25.** In this area, for example, health care professionals deal with people who are injured in automobile accidents and on jobs. They treat people who become ill very quickly or suffer heart attacks and strokes. An emergency room contains many technological devices that allow health care professionals to diagnose and treat patients. Frequently their job is to determine what is wrong with the person and then stabilize his or her condition. Then more traditional treatments, such as surgery or radiation treatment, for example, can be administered.

Career Corner
Biomedical Engineers

Biomedical engineers combine biology and medicine with engineering to develop devices and procedures that solve medical and health-related problems. They develop devices used in various medical procedures (such as the computers used to analyze blood and the laser systems used in corrective eye surgery), artificial organs, and imaging systems (such as magnetic resonance, ultrasound, and X-ray systems). Most biomedical engineers have a background in one of the basic engineering specialties, such as mechanical or electronics, and specialized biomedical training. The specialties within biomedicine include biomaterials, biomechanics, medical imaging, rehabilitation engineering, and orthopedic engineering. Unlike many other engineering specialties, a graduate degree is recommended or required for many entry-level jobs.

Summary

Technology in medicine involves both wellness and illness. Wellness programs are designed to help people maintain their health. Illness treatment involves diagnosing, treating, and preventing diseases and injuries. Diagnosis determines what is wrong with the person. Treatment attempts to restore the person's health. Prevention deals with promoting wellness programs and immunizations. In wellness programs and illness treatment, technology plays a vital role. Both technological equipment and technological knowledge aid health care professionals in maintaining and restoring people's health.

Key Terms

aerobic exercise
anaerobic exercise
computerized tomography
dental hygienists
dentists
diagnosis
disease
drug
electrocardiograph
emergency medicine
endoscope
ethical drugs
illness
intervention radiology
invasive diagnostic equipment
magnetic resonance imaging
medical technologists
medicine
noninvasive diagnostic equipment
nurse practitioners
nurses
pharmacists
physician assistants
physicians
prevention
proprietary drugs
prosthesis
radiation therapy
radiology
routine diagnostic equipment
surgery
treatment
ultrasound
wellness
X-ray machine

Test Your Knowledge

Write your answers on a separate piece of paper. Please do not write in this book.

1. Name one difference between the roles of science and technology in regard to wellness and illness.
2. List the four major factors involved in wellness.
3. Name one technological artifact used in exercise designed to improve a person's well-being.
4. Name one technological artifact used in sports.
5. Medicine involves diagnosing, treating, and _____ diseases and injuries.
6. *True or false?* Nurse practitioners can prescribe medication.
7. Give one example of routine diagnostic equipment used by a health care professional.

8. Name one reason a physician might want a patient to have a computerized tomography scan instead of an X-ray procedure.
9. Briefly describe one step involved in ultrasound imaging.
10. What is an endoscope?
11. *True or false?* An electrocardiograph machine is an example of invasive diagnostic equipment.
12. What is a proprietary drug?
13. *True or false?* The second clinical phase of an FDA-approved drug involves a double-blind study.
14. Name one piece of equipment a biomedical engineer might design.
15. Give one example of a newer technological artifact used in surgical procedures.

Applying Your Knowledge

1. Make a list of all the foods (meals *and* snacks) you eat during a three-day period. Obtain a brochure or other information on healthy eating (your school nurse or a local drugstore should be able to supply the information). Compare your eating with the dietary guideline in the brochure. Are you eating a healthy diet, or are there some areas where you could improve?
2. For each of the sports listed, fill in the playing surface, goals, and constraints.

Sport	Playing surface	Goals	Constraints
Auto racing			
Archery			
Baseball			
Billiards			
Downhill skiing			
Hockey			
Swimming			
Soccer			

This activity develops the skills used in TSA's Medical Technology event.

Medical Technology

Activity Overview

In this activity, you will research a contemporary problem, issue, or technology related to medical technology, prepare a report, and create a display. You will prepare an oral presentation incorporating a PowerPoint presentation.

Your report must be contained in a three-ring binder and consist of the following items:

- Cover page
- Definition of problem, issue, or technology
- Report on topic (4–10 pages)
- Printout of PowerPoint slides (three slides per page)
- List of sources and references

Materials

- Three-ring binder
- Materials appropriate for a tabletop display (will vary greatly)
- Computer with PowerPoint (or similar) presentation software

Background Information

Selection. Before selecting the theme for your project, use brainstorming techniques to develop a list of possible themes. Some contemporary topics include:

- US healthcare system
- Minimally invasive surgery
- Home health testing
- Exercise

- Immunizations
- Diagnostic equipment
- Prescription medications
- Prevention programs and techniques

Research. Use a variety of sources to research your theme. Do not rely solely on information you find on the Internet. Use books and periodicals available at your local library. Research the historical developments of the topic. Did an individual develop the technology, or was the technology developed by a corporation? What were some previous technologies that allowed this to become a reality? How was the technology received by the public, and was the response expected?

PowerPoint presentation. When developing your PowerPoint presentation, consider the following design guidelines:

- Develop a general slide design and use it for all your slides.
- Keep the design simple—do not use more than two type fonts. Select type fonts that are easy to read. Be sure that the type size is large enough to be seen from the rear of the room in which you will be presenting.
- Include a title on each slide.
- Do not attempt to squeeze an abundance of information on a single slide—create multiple slides instead.

Guidelines

- Research should focus on any cultural, social, economic, or political impacts. Both opportunities and risks should be addressed.
- The display can be no larger than 18″ deep × 3′ wide × 3′ high.
- If a source of electricity is desired, only dry cells or photovoltaic cells can be used.
- The oral presentation can be up to 10 minutes in length.

Evaluation Criteria

Your project will be evaluated using the following criteria:

- Content and accuracy of report
- Attractiveness and creativeness of display
- Communication skills and presentation design of oral presentation

Section 9 Activities

Activity 9A

Design Problem

Background

Biomedical engineers develop a great deal of diagnostic and treatment equipment. They also develop artificial parts for human bodies and devices that allow people to do routine tasks.

Situation

You work with a volunteer agency that helps people with arthritis. Many of the clients of this agency have trouble bending over. They often complain that when they drop something, it is very difficult to reach it and pick it up.

Challenge

Design a device that will allow a person to pick up an object on the floor without bending over. The device should be easy to use and inexpensive to make.

Activity 9B

Fabrication Problem

Background

Drying is one of the oldest food preservation methods. Originally, it was done by salting the food and then drying it the sun, in open rooms, or on stoves. In 1795, the first dehydrator was developed to dry (dehydrate) fruits and vegetables. Today, dried foods have become a multimillion-dollar industry.

Foods can be dehydrated using the sun, a conventional oven, or an electric dehydrator.

- *Solar drying* is a type of sun drying in which solar energy is used to heat a specially designed unit that has adequate ventilation for removing moist air. This type of unit can develop temperatures that are up to 30° warmer than the outside temperature. The higher temperatures reduce the drying time.
- *Oven drying* is the easiest way to experiment with dehydration. It requires little initial investment, but it produces foods that are darker, more brittle, and less flavorful than foods dried by a dehydrator.
- *Electric dehydrators* use trays to hold the food that is being processed. The units have a heat source and ventilation system. They produce a product that is better than those produced by other drying methods.

Drying Process

Drying times in conventional ovens or dehydrators vary considerably depending on the amount of food dried, its moisture content, and room temperature and humidity. Some foods require several hours and others may take more than a day. It is important to control air temperature and circulation during the drying process.

Temperatures between 120°F and 140°F are recommended for drying fruits and vegetables. If the temperature is too low or the humidity is too high, the food will dry too slowly to prevent microbial growth. If the temperature is too high initially, a hard shell may develop on the outside, trapping moisture on the inside. If the temperature is too high at the end of the drying period, the food may scorch.

Challenge

You are a researcher for a food processing company. You want to develop a dried fruit snack that hikers can take with them on outings. You are to select a mixture of several fruits. Then, you are to experiment with drying them to develop a trail mix snack.

Materials and Equipment

- Electric dehydrator
- One pound of fruit

Procedure

1. Carefully read the operating instructions for the electric dehydrator.
2. Select fresh, good quality fruits and vegetables.
3. Trim away inedible and damaged portions.
4. Cut into halves, strips, or slices that will dry readily. Strips should be 1/8" to 1/4" thick.
5. If you are drying vegetables, they need to be blanched. Heat them in water to a temperature high enough to neutralize the natural enzymes.
6. Treat most fruits by dipping them in antioxidant such as citric/ascorbic acid mixtures.
7. Preheat the dehydrator to 125°F (52°C).
8. Place a single layer of food on the tray.
9. Stack the trays in the dehydrator and close the unit.
10. Gradually increase the temperature to 140°F (60°C). It takes 4 to 12 hours to dry fruits or vegetables in a dehydrator.
11. Examine the food often and turn trays frequently. Turn larger pieces, such as apricot halves, halfway through the drying time. Move the pieces from near the sides of the trays to the center.
12. After the drying process is complete, condition the food. This is done to equalize (evenly distribute) moisture left in the food after drying. To condition a food:
    - Allow the food to cool on trays.
    - Pour the food into a large, nonporous container until it is about 2/3 full.
    - Cover container and place in a warm and dry place. Stir contents at least once a day for 10 to 14 days.

Dried fruits can be eaten as snacks or soaked for one to two hours to re-hydrate them. Most vegetables are refreshed with water before use. This can be done by (1) soaking in water for one to two hours, (2) adding two cups of boiling water for each cup of food, or (3) adding dried vegetables directly to soups or stews.

Section 10

Managing a Technological Enterprise

Tomorrow's Technology Today
Warmup Jacket

If you've ever taken part in sports such as track or basketball, you probably have worn a warmup jacket over your uniform. "Warmup jacket," of course, doesn't mean that the jacket actually warms your body—you generate the heat, and the jacket merely traps it to keep you warm.

But now, there is a jacket that actually *does* provide the heat to warm up the person wearing it. Developed by a Massachusetts company, the jacket both generates and holds in the heat. This makes it ideal for people who must be outdoors in very cold conditions, such as military personnel, mountain climbers, or construction workers.

The principle involved is similar to the electric blanket you might have on your bed in the winter: an electric current causes wires in the fabric to give off heat. However, the electric blanket uses fairly large wires and has to be plugged into a wall outlet to operate. That wouldn't be very practical for use in a jacket, of course. The new jacket uses very thin wires that are heated by two small batteries, so the wearer's movements aren't restricted.

The breakthrough that made the new jacket possible was the development of stainless steel microfibers—electrical conductors thinner than a human hair. The microfibers can be woven right into the cloth of the jacket, and are as soft and flexible as the fabric itself. They can even make a trip through the washing machine without harm. Power is supplied by two rechargeable batteries that, together, weigh only slightly more than one-half pound.

Heating wires are concentrated in the chest area. Research has shown that heating the chest is the most effective method, since it warms the body's core, where heart and lungs are located. The blood circulates through the core, carrying heat to all parts of the body.

Batteries can presently provide five hours of heating at the "medium" setting of 108°F, or two-and-one-half hours at the "high" setting of 114°F. Researchers hope that improvements in battery technology will eventually make much longer periods possible.

To test the effectiveness of the system, a dramatic demonstration was conducted by a company that makes clothing and equipment for mountain climbers. The company testers outfitted a climber with a jacket, then placed him in a deep crevasse on a glacier and buried him in snow. He was told to wait until the extreme cold caused his body to start shaking violently, then switch on the jacket. Within minutes, the jacket warmed his body enough to stop the shaking.

The jackets are being evaluated by the armed services, and are being marketed to consumers on a limited basis by one company. Another company is testing use of the technology in a space-age version of the electric blanket. Researchers at the company that patented the technology are extending the idea to other articles of clothing, such as gloves.

One future application being discussed might be the coolest of all: finding a way to use the technology to provide air conditioning as well as heating.

Chapter 32
Organizing a Technological Enterprise

Learning Objectives

After studying this chapter, you will be able to:

- Explain how entrepreneurship and intrapreneurship differ.
- Define management.
- Identify the four functions of management.
- Name the three main forms of business ownership.
- List problems that can occur in the partnership form of business ownership.
- Explain why a corporation is said to be "like a person" in a legal sense.
- Define the role of a board of directors in a corporation.
- Describe the differences between equity financing and debt financing.

As you studied this book, you learned a great deal about technology. You have learned that technology consists of human-made devices designed to control and modify the natural environment. Technology is a system with goals, inputs, processes, and outputs. You also learned that technology occurs in seven broad areas: communication, construction, manufacturing, agriculture, energy and power, medicine, and transportation.

Technology and the Entrepreneur

Have you ever thought about where technology comes from? Technology is a product of the human mind, developed by people to serve people. At one time, most technology was developed, produced, and used by one person. However, in the modern world, *organization* is necessary—people use very complex systems to develop and produce technology, **Figure 32-1.** At the base of

Figure 32-1. Complex systems and a large organization are needed to produce a product such as this airplane. (United Airlines)

Figure 32-2. Entrepreneurs have the vision to recognize consumer wants and to devise ways to meet them with new products. (Goodyear Tire and Rubber Co.)

many of these systems are ***entrepreneurs***, **Figure 32-2.** These are people with very special talents, who look beyond present practices and products. They see new ways to meet human needs and wants. By focusing on what the customers value, entrepreneurs develop systems and products to meet desires and expectations. They might change the entire way something is being done.

A good example of ***entrepreneurship*** is the McDonald's restaurant chain. The first McDonald's was a small hamburger stand in southern California. The original owners had developed some innovative ways to make and sell their product, but didn't look beyond their local market. An outsider, Ray Kroc, saw greater possibilities. Under his leadership, the "fast-food" business was born. Kroc and his managers carefully studied the various jobs and developed special management techniques. They standardized the product, created effective training programs, and developed the chain into a worldwide organization. This is some of the work of the entrepreneur, to improve the use of the resources and create new products or markets.

The dictionary defines an entrepreneur as "any person taking the financial risks of starting a small business." But, this definition leaves out the aspect of entrepreneurial spirit, the spirit of innovation. People starting another beauty shop, delicatessen, or bakery are taking financial risks and may become successful business operators, but they are not *innovators.* They do not deal with change as an opportunity to produce a new product or service.

There also can be entrepreneurship within an existing small or large company. Entrepreneurship involves an attitude and an approach. It consists of searching for opportunities for change and responding to them. Large companies often encourage entrepreneurship within their organization. In fact, a new term has evolved to describe this action. The term is ***intrapreneurship***, which is "the application of entrepreneurial spirit and action within an existing company structure."

Technology and Management

Technology is *purposeful;* it is developed to meet a problem or opportunity. Identifying and responding to the need for change is only one part of developing technology, however. The production and use of technology must be managed. Therefore, *technology is a product of managed human activity,* **Figure 32-3.** Actually, all of your

Figure 32-3. The actions of these workers must be managed to produce boats efficiently.

Figure 32-4. Planning is an important part of managing a business. These managers are using a model of a boiler system as they plan a factory expansion. (American Electric Power)

actions are managed by yourself or by other people. This is not bad. ***Management*** is simply the act of planning, directing, and evaluating any activity, **Figure 32-4.** It can be as simple as managing personal expenditures, or as complicated as managing an industrial complex. Management involves ***authority*** (the right to direct actions) and ***responsibility*** (accountability for actions). Managers have the responsibility to make decisions that ensure the business is successful. Their authority may include hiring personnel, purchasing materials, developing products, setting pay rates, etc. Likewise, managers have the responsibility of protecting the rights of a company's owners, workers, and customers. This might include securing product patents, investing company funds wisely, providing a safe work environment, and producing a quality product.

Functions of Management

To carry out their duties, company managers perform four important functions, **Figure 32-5.** These are:

- ***Planning.*** This function includes setting long-term and short-term goals for the company or parts of it. It also involves selecting a course of action to meet those goals. Planning activities often result in an *action plan,* or plan of work. Such a plan lists what needs to be done, who will do it, and when it is to be done.
- ***Organizing.*** This involves structuring the company or workforce to address company goals. Typical activities include developing organizational charts, establishing chains of command, and determining the company's operating

Figure 32-5. Managers plan, organize, actuate, and control company activities. This manager is performing the control function by gathering information.

procedures. Organizing makes sure that people, materials, and equipment are in place to meet the action plan.

- *Actuating.* Initiating the work related to the action plan is known as *actuating*. This function could include training employees, issuing work orders, providing a motivational work environment, or solving production problems. Actuating causes plans to take form. For example, products and structures are built or services are provided.
- *Controlling.* Comparing results against the plan is the *controlling* function. Control actions ensure that resources are used properly and outputs meet stated standards. Typical terms applied to this area are inventory control, production control, quality control, and process control.

Authority and Responsibility

Managerial functions are carried out through an organizational structure, **Figure 32-6.** This structure typically begins with the ***owners*** of the business. The owners have ultimate control, or *final authority*, over company activities. It is their company. They can hire or fire management personnel, set policies, or close the business.

However, in most larger companies, the owners are not the managers. They frequently have other jobs or interests. Also, they often do not have the skills needed to manage a large, complex business. Such owners delegate responsibility to full-time managers. In many companies, the top

Figure 32-6. Managers have specific levels of authority and responsibility within a company.

manager is the ***president*** or the ***chief executive officer (CEO).*** In very large companies, these titles are usually held by two different people. In smaller companies, the two titles may be held by the same person. In either case, this top level of management is responsible for the entire company's operation. People at this level have *day-to-day* control of the company.

Few people can manage a company by themselves. Therefore, the top managers employ other managers to assist them. The number of managers and levels of management will vary with size and type of company. Larger companies generally have ***vice presidents*** that report to the president or CEO. Each vice president is responsible for some segment of the company. The segment might be a functional area such as marketing, finance, production, or engineering. In other cases, the vice president has a regional responsibility, such as foreign sales or West Coast operations.

Most vice presidents have a *scope of responsibilities* that includes a number of managers reporting to them. Regional sales managers may report to a vice president for sales. Plant managers often report to vice presidents in charge of production. This level of management is often called ***middle management.*** It is below ***top management*** (president and vice presidents), but above operating management.

The lowest levels of management directly oversee specific operations in the company. Managers at this level may be supervisors on the production floor, district sales managers, or human resources directors. They are the managers who are closest to the people who produce the company's products and services. They are often called ***supervisors*** or ***operating management.***

Risks and Rewards

As was noted earlier, most technology is developed and produced through the planning of industrial companies, **Figure 32-7.** The company designs, engineers, and produces the products, structures, transportation services, and communication media you depend on daily.

Figure 32-7. Most technology is developed and produced by industrial companies. Operation of any company, large or small, involves both risks and rewards. (The Coastal Corp.)

Everyone involved with a company has some basic elements in common. These elements are ***risks*** and ***rewards.*** Owners risk their money to finance the company. Banks also accept a level of risk. Banks and other lending institutions make loans to finance company growth. Employees risk missing other employment opportunities by working for the company. Consumers risk their money when they buy a product.

In return, the risk takers expect a reward. The owners want their investment to grow. They also expect periodic financial returns for the use of their money. Banks expect interest to be paid on their loans. Employees expect job promotions, pay raises, safe working conditions, and job security. Consumers expect performance and value in return for the money they spend on products and services.

Forming a Company

Companies are organized and operated under the laws and *mores* (accepted traditions and practices) of our society. There are several important features involved in the formation of a company. These features include:

- Selecting a type of ownership
- Establishing the enterprise
- Securing financing

Selecting a Type of Ownership

Business enterprises can be divided into two different sectors: public and private. ***Public enterprises*** are those enterprises that are controlled by the government or by a special form of corporation. They are generally those enterprises that meet two criteria:

- First, they are operated for the general welfare of the society.
- Second, they cannot or should not make a profit.

As an example, consider a police department, which is run like a business. Police are commissioned to protect people and their property. They have managers (police chiefs and captains) and workers (patrol officers, traffic officers, narcotics officers, etc.). If the police department had to show a profit and attract private investment, however, some aspects of law enforcement might get cut back. This could limit the department's market to the people who could pay for the service. The police would thus solve the crimes that showed the most profit.

Another example of public ownership is road construction. We pay for it through taxes. However, each segment of road does not have to show a profit. If it did, we would not have many of our rural roads.

Private enterprises are owned by individuals or groups of people. This ownership can be through a direct means of investment or through an indirect means, such as a pension or investment fund. Owners of private enterprises invest their money, take their risks, and hope to reap a profit. The owners, within legal limits, are free to select business activities, produce the products and services they choose, and divide the profits as they see fit.

Private enterprises can be either *publicly held* or *privately held.* A publicly held enterprise is one in which the public can purchase a portion of ownership in the form of shares of stock. Privately held enterprises are owned by individuals or a group of people, but do not offer stock for sale to members of the public.

There are three main types of private business ownership, **Figure 32-8.** The first type is a ***proprietorship.***

A proprietorship is a business with a single owner. The owner has complete control of the company. He or she sets goals, manages activities, and has the right to all business profits. A proprietorship is a fairly easy type of ownership to form.

Proprietors may have difficulty raising money, especially to start a new business.

The company's finances are typically limited to the owner's personal wealth or borrowing ability. However, banks may be hesitant to loan a large quantity of money to unproven businesses or individuals.

An additional problem of this type of business is limited knowledge. Many individuals do not have the skills and know-how needed to run all aspects of the business. This causes inefficiency in operating the company. Finally, the proprietor is responsible for all the debts the business incurs. He or she cannot separate business income and liabilities from their personal finances. This is a major disadvantage. It is called ***unlimited liability.***

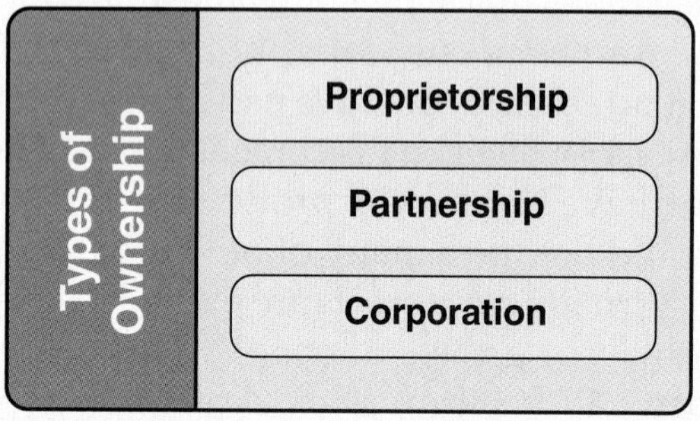

Proprietorships are the most common type of ownership in the United States. However, proprietorships are generally small retail, service, and farming businesses. Thus, the dollar impact of this form of ownership on the economy is considerably less than the impact made by large corporations.

A second form of private ownership is the ***partnership.*** Partnerships are businesses owned and operated by two or more people, **Figure 32-9.** Partnerships thus have more sources of money to finance the company. Also, interests and abilities of the partners may complement each other. One partner may be strong in production, while another may have sales skills.

Having more than one owner active in a business could cause confusion, however. Employees might receive conflicting

Types of Ownership
Proprietorship
Partnership
Corporation

Figure 32-8. A private enterprise may be owned by one person (a proprietorship), a small number of people (a partnership), or many investors (a corporation).

Figure 32-9. Many small businesses, such as farms, are partnerships.

directions. The partners could disagree when making important business decisions. Also, a partnership, like a proprietorship, has unlimited liability. This is a particularly touchy problem, since one partner can commit the entire partnership to financial risk.

The third form of ownership is the ***corporation.*** A corporation is a business in which investors have purchased partial ownership in the form of shares of stock. Investors may be individuals, other companies, or groups (such as a pension plan or investment club). Legally, the corporation is like a "person" — it can own property, sue or be sued, enter into contracts, and contribute to worthy causes.

Generally, the investor-owners of a corporation do not manage the business. They employ professional managers for this task. The owners invest their money and expect to receive a ***dividend*** (periodic payment from the company's profits) in return.

Since a corporation is a legal person, the company (not its owners) is responsible for the debts. This feature is called ***limited liability.*** It means that, if the company fails, an owner's loss is limited to the amount of money he or she had invested.

Establishing the Enterprise

Once the type of ownership is selected, the company must be established. There are few legal requirements for proprietorships and partnerships. In many cases, all that is needed is a license from the city or other local government where the business will operate. However, corporations are a different story. They are often large businesses that can have a serious financial impact on people and communities. For this reason, their formation is placed under state control. Each state establishes its own rules for forming a corporation. Corporations can then provide many products and services, **Figure 32-10.**

Filing Articles of Incorporation

A corporation, like a person, must be born. This process is begun by filing ***articles of incorporation*** with one of the states. The articles of incorporation serve as an application for a *corporate charter* (a "birth certificate" for the corporation). The state usually asks for the company name, type of business the company plans to enter, the location of the company offices, and the type and value of any stock that will be issued.

Receiving a Corporate Charter

The articles of incorporation are filed with the appropriate state office. State officials then review the articles. They try to determine if the business will operate legal activities and provide customers with appropriate products or services. If they believe that the business meets all state laws, a ***corporate charter*** (operating permit) is issued. This allows the company to conduct the specified business in the state. All the other states will recognize the corporate charter, and allow the company to conduct business within their borders.

Figure 32-10. Continuous manufacturing has brought people many new products, including these DVDs.

Approving Bylaws

An incorporated business must have a set of ***bylaws.*** These are the general rules under which the company will operate. A set of bylaws includes the information contained in the charter: name of the company, purpose of the business, and location of the corporate offices. In addition, the bylaws will list:

- Corporate officers
- Duties, term of office, and method of selecting corporate officers
- Number of directors, as well their duties and terms of office
- Date, location, and frequency of the board of directors' meetings
- Date and location of the annual stockholders' meeting
- Types of proposals that can be presented at the annual stockholders' meeting
- Procedure for changing the bylaws

Electing a Board of Directors

The charter and bylaws allow the company to operate. However, the stockholders want their investment to be wisely managed. This requires oversight and supervision. Many companies have hundreds or thousands of stockholders. Few can be, or wish to be, involved in managing the company. Therefore, a ***board of directors*** is elected to represent the interests of the stockholders. The directors are responsible for forming company policy and providing overall direction for the company.

A typical board of directors includes two groups of people. One group is made up of the top managers of the company. These individuals are known as ***inside directors.*** Other directors are not involved in the day-to-day operation of the company. They are selected to provide a different view of the company's operation. Since these people are outside the managerial structure, they are called ***outside directors.***

Directors are elected using a voting system similar to our political system. The main difference is that companies do not use a one-person/one-vote rule. Instead, they use a one-share/one-vote procedure. Each ***share of stock*** (equal portion of the total company) has a vote assigned to it. Each stockholder has as many votes as the shares of stock they own. Therefore, those who own a larger portion of the company and accept a larger risk have a greater say in forming company policy.

Financing Company Operations	
Equity Financing	Debt Financing
Sell shares of ownership (stock) in the company to investors	Borrow money from financial institutions or sell bonds to investors

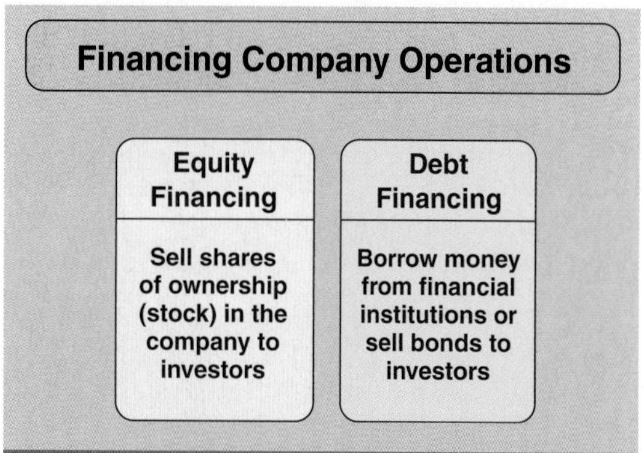

Figure 32-11. Two major methods of financing are available to build and expand a company.

Financing the Company

So far, you have learned how a company is formed. However, there is more to starting a company than completing paperwork. The company needs money to operate. There are two basic methods of raising operating funds. These are equity financing and debt financing, **Figure 32-11.**

Equity Financing

Equity financing involves selling portions of ownership in the company. This is an important way in which corporations are financed. The company is authorized

Career Corner

Top Executives

Top executives set specific goals and objectives for a company and direct the overall operations of businesses. They formulate policies and strategies so these objectives are met. There is a range of titles for top executives, including chief executive officer (CEO), chief operating officer (COO), chair of the board, and president.

Executives typically have spacious offices and large support staffs. They might travel to national and international sites to monitor operations and meet with customers and staff members. Many top executives have a bachelor's degree or an advanced degree. A top executive must have excellent personal skills and an analytical mind. Executives must be able to communicate clearly, provide leadership, and make difficult decisions.

by its charter to sell a specific number of shares of stock. Investors may buy these and, as a result, become an owner of part of the company. Owners receive certain rights with the shares they own. These include:

- The right to attend and to vote at the annual stockholders' meeting
- The right to sell their stock to another individual
- The right to receive the same dividend (portion of the company's profits) per share as other stockholders
- The right to a portion of the company's assets (property, money, etc.) if the company is liquidated

Debt Financing

Debt financing involves borrowing money from a financial institution or private investors. Banks and insurance companies loan corporations money to finance new buildings, equipment, and the company's daily operations. The banks and insurance companies charge interest for the use of their money. They charge their best (safest) customers a lower interest rate. This rate is called the ***prime interest rate*** or ***prime.*** Other borrowers pay a rate higher than the prime rate. This rate is often quoted in terms of the prime, such as "prime plus two percent."

Corporations can also sell debt securities called ***bonds.*** They are often sold in fairly large denominations, such as $5,000. These bonds are usually long-term securities, meaning that they will be in force for 10 to 20 years. The company pays quarterly or yearly interest on the face value (original value) of the bonds. At maturity (end of the bond's term), the company pays back the original investment.

Summary

Technology is developed by people for people. At the very foundation of this action are entrepreneurs. They are people that can see possibilities and are willing to take risks. Often, the results of an entrepreneur's work is a company. A company is a business enterprise that is organized and managed to produce goods and services. The company is operated by a group of people that are called managers. They plan, organize, actuate, and control company activities. The managers set goals, structure systems, direct operations, and measure results. Without management, technological activities would become less efficient. With proper management, technology promises to serve our needs.

Key Terms

actuating
articles of incorporation
authority
board of directors
bonds
bylaws
chief executive officer (CEO)
controlling
corporate charter
corporation
debt financing
dividend
entrepreneurs
entrepreneurship
equity financing
inside directors
intrapreneurship
limited liability
management
middle management
operating management
organizing
outside directors
owners
partnership
planning
president
prime
prime interest rate
private enterprises
proprietorship
public enterprises
responsibility
rewards
risks
share of stock
supervisors
top management
unlimited liability
vice presidents

Test Your Knowledge

Write your answers on a separate piece of paper. Please do not write in this book.

1. Entrepreneurship within an existing company is referred to as _____.
2. Management involves _____ and responsibility.
   a. credibility
   b. authority
   c. discipline
   d. personality

3. Describe the management function called "actuating."
4. When would an educational television station be considered a public enterprise?
5. List the three forms of business ownership.
6. Which form of business ownership provides the owners with limited liability?
7. Place the steps in forming a company, listed on the right, in their proper order:

____ First step	a. Prepare bylaws
____ Second step	b. Receive corporate charter
____ Third step	c. Elect corporate directors
____ Fourth step	d. File articles of incorporation

8. *True or false?* Like a person, a corporation can own property, enter into contracts, and sue or be sued.
9. A corporation's board of directors is responsible for:
   a. forming company policy
   b. representing the interests of the stockholders
   c. providing overall direction for the company
   d. all of the above
10. Equity financing involves the sale of _____; debt financing involves the sale of _____.

Applying Your Knowledge

1. Assume your class is going to produce and sell popcorn at school basketball games. Develop an organization chart for the enterprise. Indicate who has the most authority and the chain of command from that person to all other members of the enterprise.
2. Call or visit the office of a local stockbroker. Find out if you can obtain brochures or other simple explanations of stocks and bonds and how they are traded. Use the information as the basis for a written report. You may want to invite the person to speak to your class.

Chapter 33
Operating Technological Enterprises

Learning Objectives

After studying this chapter, you will be able to:

- Define an economic enterprise.
- Diagram the relationships among the five main managed areas of activity within a technological enterprise.
- Explain the difference between basic research and applied research.
- Name the four common systems of manufacturing.
- Define the term "quality control."
- Describe the channels of distribution used to move products from producer to the consumer.
- List the major types of programs that are included in industrial relations.
- Describe the different types of employee training.
- Explain how expenses, income, and profit are related.
- Identify the steps in the industry-consumer product cycle.

Societal Institutions

Over time, humans have developed a complex society to meet their wants and needs. Within this society are a series of institutions. Five basic institutions are listed in **Figure 33-1.** They are:

- *Family:* An institution that provides the foundation for social and economic actions; the basic unit within society
- *Religion:* An institution that develops and communicates values and beliefs about life and appropriate ways of living
- *Education:* An institution that communicates information, ideas, and skills from one person to another and from one generation to another
- *Political/legal:* An institution that establishes and enforces society's rules of behavior and conduct
- *Economic:* An institution that designs, produces, and delivers the basic goods and services required by the society

All of these institutions use technology because they are concerned with efficient and appropriate action. They apply

Figure 33-1. These five basic institutions are important to our society.

resources to meet human wants and needs. People in each institution use technical means to make their jobs more efficient.

However, almost all technology *originates* in the economic institution. For example, teachers work in the educational institution. They may use computers to make their teaching more efficient. But the computer is not a product of the educational institution. It is a product of the economic institution. In the same way, politicians are part of the political/legal institution. They may use television and printed material to help win an election. Again, television, printing presses, and all associated communication devices are not developed in the political system. They are outputs of the economic institution.

Figure 33-2. There are four major types of economic enterprises, but only the industrial enterprises actually develop and produce technology.

Economic Enterprises

Technology is directly associated with economic enterprises. ***Economic enterprises*** are organizations that engage in *business efforts directed toward making a profit*. This ***economic activity*** includes all trade in goods and services paid for with money, **Figure 33-2.**

In a free enterprise economic system, production and prices are determined by the principles of supply and demand. The government has little impact on markets. Businesses are free to produce any products. As consumer demands change over time, some businesses are created, some close, and some modify their production to meet current demands. In this system, business plays the important roles of owning production equipment and processes, determining product supply, and reacting to market demands.

Not all economic activity *develops* technology. Commercial trade includes the wholesale and retail merchants who form the link between producers and consumers. These traders do not change the products they distribute. They simply make them easily available for people to buy.

Banks, insurance companies, and stock brokerage firms provide financial services to people. They protect our wealth, buy and sell stocks and bonds, or insure our lives and possessions. Again, these companies use technology, but do not develop it.

Similarly, any number of service businesses repair products and structures. They service and maintain technological devices. They extend the useful life of such devices, but do not develop them.

Industry

Almost all design, development, and production of technology takes place within the type of economic enterprise called ***industry***, **Figure 33-3**. The term "industry" can have several meanings. One definition groups together all businesses that make similar products. Thus, we read about the steel industry or the electronics industry.

However, in this book, a more restricted definition is used: industry is *the area of economic activity that uses resources and systems to produce products, structures, and services with intent to make a profit.*

Areas of Industrial Activity

In each industry, a number of actions take place. These actions are designed to capture, develop, produce, and market creative ideas. These technological activities form the link from the inventor/innovator to the customer. There are thousands of individual actions that cause a new product or service to take shape. However, they can be gathered into five different areas of managed activity, **Figure 33-4.** These are:

- ***Research and development:*** The activities that might result in new or improved products and processes
- ***Production:*** The activities that develop methods for producing products or

Industry...
- Part of the economic system.
- Uses resources.
- Produces products, structures, and services.
- Intends to make a profit.

Figure 33-3. The design and production of goods such as these consumer products were carried out by industrial enterprises.

services and the activities that produce the desired outputs

- *Marketing:* The activities that encourage the flow of goods and services from the producer to the consumer
- *Industrial relations:* The activities that develop an efficient work force and the activities that maintain positive relations with the workers and the public
- *Financial affairs:* The activities that obtain, account for, and disburse funds

The first three activity areas are product- or service-centered. They directly contribute to the design, production, and delivery of the planned outputs. The other two areas are support areas. Financial affairs provides monetary support, while industrial relations contributes human or personnel support.

Research and Development

Research and development can be viewed as the "idea mill" of the enterprise. In this area, employees work with the true raw material of technology, human ideas. They convert what the mind envisions into physical products and services. These actions can be divided into three steps: research, development, and engineering.

Research

Research is the process of *scientifically seeking and discovering knowledge,* **Figure 33-5.** Research explores the universe systematically and with purpose. Research also determines, to a large extent, what technology we

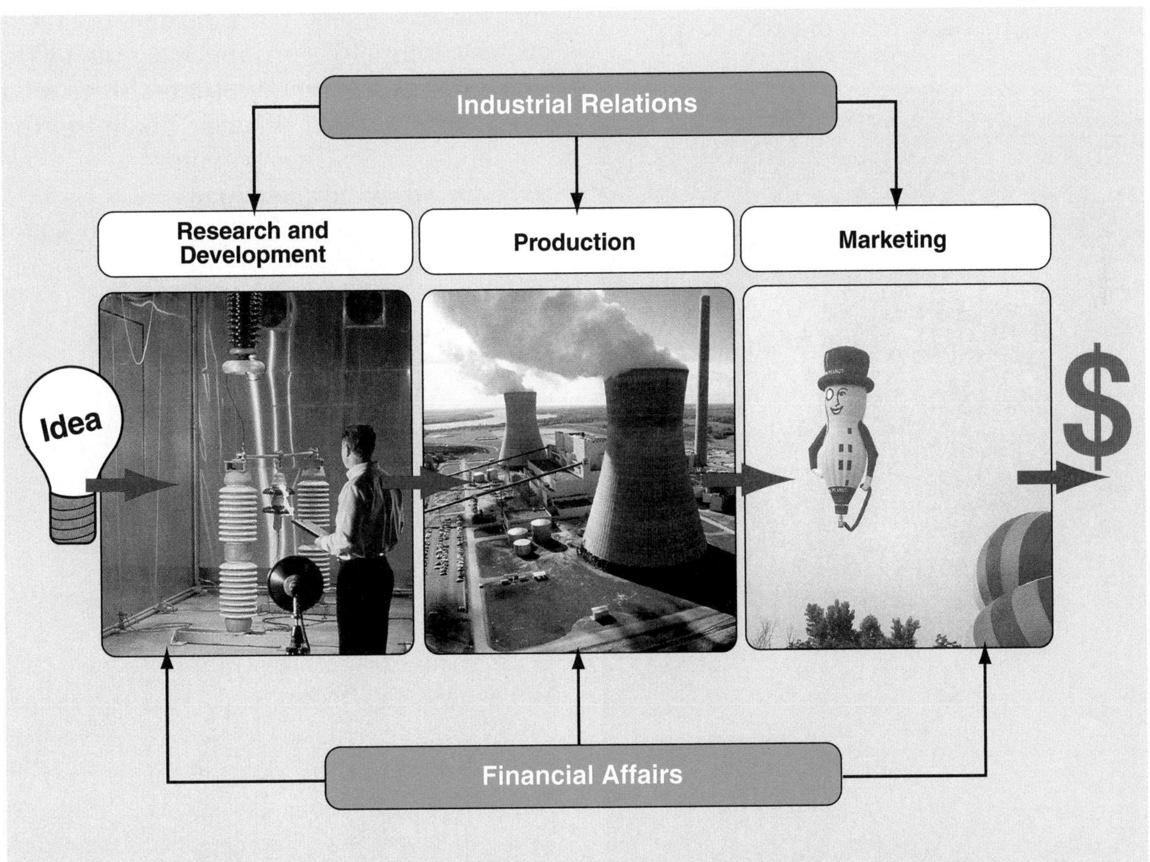

Figure 33-4. There are five major areas of managed activities in industry that change ideas into products, structures, or services. Notice how each area relates to the others.

will have in the future. It determines the type of human-built world we will live in.

There are two types of research. The first is ***basic research.*** Basic research seeks knowledge for its own sake. We conduct basic research to enlarge the scope and depth of human understanding. People working in basic research are not concerned about creating new products. Their focus is on generating knowledge.

The second type of research is ***applied research.*** This activity seeks to reach a commercial goal by selecting, applying, and adapting knowledge gathered during basic research. The focus of applied research is on *tangible* results such as products, structures, and technological systems.

Basic research and applied research complement one another. The former finds knowledge, while the latter finds a use for it. For example, basic research may develop knowledge about the reaction of different materials to high temperatures. Applied research might then determine which material is appropriate for the reentry shield of a spacecraft.

Figure 33-5. Research seeks to discover knowledge. Computers have proven a powerful aid for researchers. (AT&T)

Development

Development uses knowledge gained from research to derive specific answers to problems. Development converts knowledge into a physical form. The inputs for development are two-dimensional information such as sketches, drawings, or reports. The outputs are models of three-dimensional artifacts such as products or structures.

Development takes place in two areas, **Figure 33-6.** The first is ***product or structure development.*** This area of development applies knowledge to design new or improved products, structures, and services. Development may result in a totally new product or structure, or improve on one that already exists. For example, the bicycle was originally a product from the 1800s. The ten-speed bicycle was later developed from the standard bicycle. Likewise, the

Figure 33-6. The two types of development, product and process, were used in designing this aircraft. (Boeing)

laser printer uses many processes originally developed for photocopiers.

The second type of development is ***process development.*** This activity devises new or improved ways of completing tasks in manufacturing, construction, communication, agriculture, energy and power, medicine, or transportation. Process development may result in something totally new, such as fiber-optic communication. It may also improve on existing processes, such as inert gas welding, which was derived from standard arc welding.

Engineering

Developed products and structures must be built, and developed processes must be implemented. To do this, people need information. Product and structure engineering are responsible for this activity. ***Engineering*** develops the specifications for products, structures, processes, and services. This is done through two basic activities — design interpretation and engineering testing.

Design interpretation conveys the information needed to produce the product or structure. This includes three main types of documents: engineering and architectural drawings, bills of materials, and specification sheets.

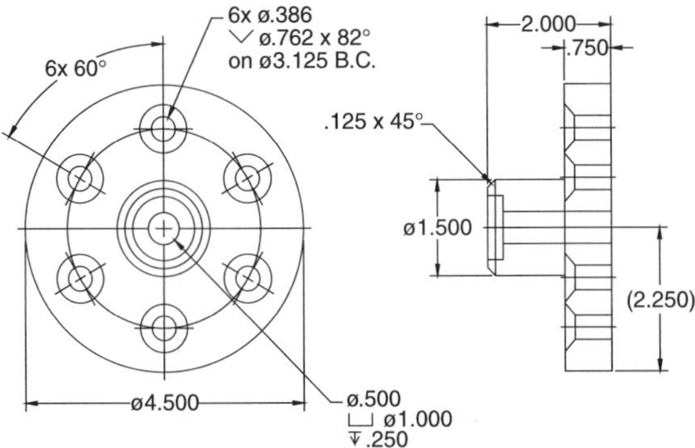

Figure 33-7. Engineering drawings provide detailed information on products or parts to be manufactured.

Engineering drawings convey the characteristics of manufactured products, **Figure 33-7.** A set of engineering drawings includes:

- Detail drawings that convey the size, shape, and surface finish of individual parts
- Assembly drawings that show how parts go together to produce assemblies and finished products
- Systems drawings that show the relationship of components in mechanical, fluidic (hydraulic and pneumatic), and electrical/electronic systems

Architectural drawings are used to specify characteristics of buildings and other structures, **Figure 33-8.** They include floor, plumbing, and electrical plans for a structure. The drawings also include elevations showing interior and exterior walls.

Figure 33-8. These engineers are reviewing architectural drawings for a oil refinery under construction. (Exxon Corp.)

However, drawings do not convey all the information needed to build products and structures. The people who implement designs need to know the quantities, types, and sizes of the materials and hardware needed. This information is included on a ***bill of materials.***

Finally, data about material characteristics are contained on ***specification sheets.*** You might look back in Chapter 12 for a more extensive presentation of these three types of documents.

Production

Ideas for products and structures are developed and specified by Research and Development. *Production,* then, must manufacture or construct the physical item. There are a number of different systems used to produce products and structures. The four common manufacturing systems are:

- ***Custom manufacturing.*** This system involves producing a limited quantity of a product to a customer's specifications. Generally, the product is produced only once. This system requires highly skilled workers and has a low production rate. This makes it an expensive system to operate. Examples of custom manufactured products are tailor-made clothing, some items of furniture, and the space shuttle.
- ***Intermittent (job-lot) manufacturing.*** With intermittent manufacturing, a group of products is manufactured to the company's or a customer's specification. The parts move through the manufacturing sequence in a single batch. All parts are processed at each workstation before the batch moves to the next station, **Figure 33-9.** Often, repeat orders for the product are expected. This manufacturing activity is relatively inexpensive, but considerable setup time is required between batches of new products.

Figure 33-9. These kitchen cabinets are being produced with an intermittent, or job-lot, manufacturing system. (American Woodmark)

- ***Continuous manufacturing.*** In continuous manufacturing, a production line manufactures or assembles products continuously. The materials flow down a manufacturing line that is specifically designed to produce that product, **Figure 33-10.** The parts flow from station to station in a steady stream. This type of manufacturing handles a high volume and has relatively low production costs. But, continuous manufacturing lines are fairly inflexible and can be used for very few different products. Many are dedicated to a single product.
- ***Flexible manufacturing.*** Flexible manufacturing is a computer-based manufacturing system that combines the

Figure 33-10. The continuous manufacturing process brought about rapid expansion in automobile production. This is the end of a continuous automobile body line. (DaimlerChrysler)

advantages of intermittent manufacturing with the advantages of continuous manufacturing. Thus, it makes possible short runs with low unit production cost. Machine setup and adjustment is computer-controlled. This permits quick and relatively inexpensive product changeovers.

Similar production systems are used in the other technologies such as construction. Housing can be built to an owner's specifications (custom-built), **Figure 33-11.** Other dwellings are built in tracts with a common plan used to make a large number of buildings.

An analogy could be made in transportation. Driving an automobile is much like custom manufacturing. It is flexible but relatively expensive. Rapid transit buses and trains are more like continuous manufacturing. They move people on set lines at lower cost.

Figure 33-11. Some homeowners choose to have their houses custom-built.

The actual production of products and structures can be divided into three important tasks, planning to produce, producing, and maintaining quality of the product or structure, **Figure 33-12.**

Technology Explained

Wind Tunnels

wind tunnel: A device used to test the aerodynamics of vehicles and structures under controlled conditions.

Wind resistance affects vehicles as they move people and cargo. This resistance has an impact on the operating efficiencies of the vehicles. In designing these vehicles, engineers often subject models to tests to maximize the efficiencies of the designs. One important test instrument in this quest is the wind tunnel.

Originally, this device was used solely to test airfoils during aircraft design activities. The uses of wind tunnels, however, have been expanded. With current concern for fuel efficiency, wind tunnels are now used extensively to ensure that vehicles offer the least amount of wind resistance possible. Wind tunnels are also used to test the wind patterns over and around buildings and various other structures, **Figure A.**

The wind tunnel is designed to pass high-speed air over a full-size or scale model of a vehicle or structure. An important design requirement is that a smooth and uniform flow of air be produced in the tunnel. To accomplish this, a number of wind tunnel designs have been produced. One of these is shown below, **Figure B.** Diagramed is a closed loop wind tunnel.

Most wind tunnels have fans or turbines developing airflow in large ducts. The diameter of the ducts increases as the air travels away from the fans. This reduces the airspeed, as well as frictional losses. The tunnels have mitered corners. This addition reduces the wind loss as the airflow changes directions. Also, in high-speed tunnel models, the air passes through cooling tubes to remove heat it gains while passing through the fans or turbines.

The air the fans produce has a swirling pattern. This airflow creates unreliable test results. Therefore, the airflow passes through an air-smoothing unit. This unit is a series of tubes removing the swirls and directing the air in a straight line.

As the air reaches the test chamber, the diameter of the tunnel shrinks rapidly. This causes the speed of the airflow to increase. The amount of the decrease in the tunnel's diameter controls the final airspeed in the test chamber.

Models in the chamber are carefully tested. Anemometers test the airspeed. Smoke might be introduced to visually observe the flow patterns of the air, as the air passes structures or vehicles. The vehicle itself might be attached to instruments to measure the lift and drag it develops as air passes.

Figure A. A wind tunnel is frequently used to test airflow patterns in industrial areas. (Colorado State University)

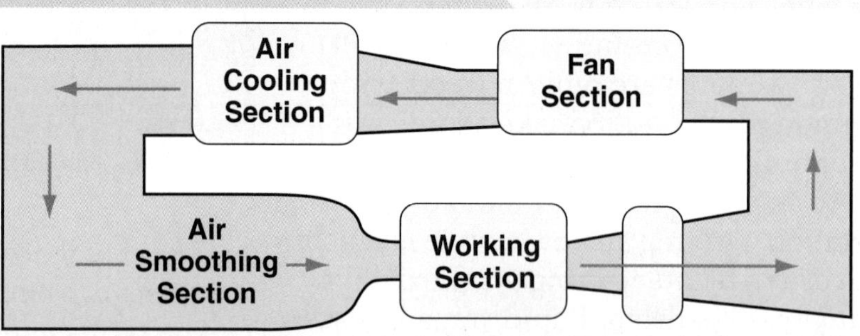

Figure B. This diagram shows the path the airflow follows in a common wind tunnel design.

Planning

Planning determines the sequence of operations needed to complete a particular task. Planning is the backbone of most production systems. It determines the needs for human, machine, and material resources. Planning also assigns people and tasks to various workstations.

Closely associated with planning and scheduling is production engineering. Production engineers design and install the system that is used to build the product or structure. They are concerned with the physical arrangement of the machines and workstations needed to produce the product.

Figure 33-12. These are the three tasks of producing. Shown are the planning and producing stages in the production of communication messages. (Hewlett-Packard)

Producing

Producing is the actual fabrication of the product. In manufacturing, it involves changing the form of materials to add to their worth or value. These activities include locating and securing material resources, producing standard stock, and manufacturing the products.

Product ***manufacturing processes*** are used to change the size, shape, combination, and composition of materials. These processes include using casting and molding, forming, and separating processes to size or shape materials. Conditioning processes change the internal properties of the material. Assembling processes put products together, and finishing processes protect or beautify their surfaces.

Construction processes are used to produce buildings and heavy engineering structures. Typical construction processes include preparing the building site, setting foundations, erecting superstructures, enclosing and finishing structures, and installing utility systems.

Communication processes are used to produce graphic and electronic media. Generally, communication messages are designed, prepared for production, produced, and delivered. This is done through the processes of encoding, storing, transmitting, receiving, and decoding operations.

Transportation processes are used to move people and cargo. They are used in land, water, air, and space systems. Typical transportation production processes include loading, moving, and unloading vehicles.

Maintaining Quality

Throughout these processes, a standard of perfection is maintained. Customers want products, structures, and systems to meet their needs and desires. For this to happen, a process called ***quality control*** is used. Quality control includes all systems and programs that *ensure the outputs of* *technological systems meet engineering standards and customer expectations.*

Often, people think "quality" means smooth, shiny, and exact size. This is not always true. A smooth, shiny road makes a poor driving surface. Cars would have difficulty controlling and braking on such a surface. Likewise, holding the length of a nail to a tolerance of ± .001" is inappropriate. The cost of manufacturing to that tolerance is too high for the product. The important quality consideration in a nail is holding power, not exact length. Quality can be measured only when a person knows how the product or part is to be used. The product's function dictates quality standards.

Inspection

An important part of a quality control program is ***inspection.*** The inspection process compares materials and products with set quality standards. There are three phases to an inspection program, **Figure 33-13.** The first phase inspects materials and purchased parts as they enter production operations. The second phase inspects work during production. The final phase inspects the end product or structure.

Inspection can be done on every product or on a representative sample of the products. Expensive, complex, or critical components and products are subjected to 100 percent inspection. This means that every part is inspected at least once. Products such as aircraft components and some medical devices are examples of outputs that receive 100 percent inspection.

Less expensive and less critical parts receive random inspections. A sample of the product is selected that represents a typical production run. The sample size and the frequency of inspections is determined using statistics (mathematically based predictions). Inspection of representative samples is part of a program called *statistical quality control.*

The selected sample is inspected. If it passes, the entire run is accepted. If the sample fails to meet the quality standards, the entire run is rejected. Rejected production lots can be dealt with in various ways: the run can be sorted to remove rejects (parts that fail to meet standards), the whole run can be discarded, or the run can be reworked.

Random inspection is used whenever it is cost-effective. Often, the cost of 100 percent inspection outweighs its value.

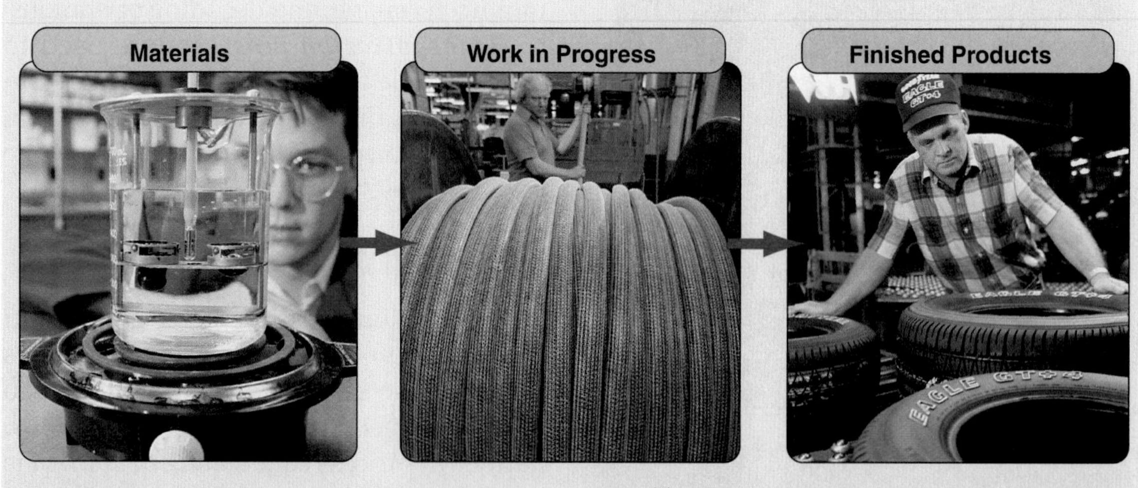

Figure 33-13. Quality control inspects materials entering the plant, work-in-progress, and, finally, finished products. (Goodyear Tire and Rubber Co.)

For example, it would be expensive to use 100 percent inspection on roofing nails. Also, the occasional defect that slips past random inspection can be discarded by the user without endangering the product or customers.

Price and Value

In addition to quality, customers expect value. Price and value are two different things. ***Price*** is what someone must pay to buy or use the product or service. Initial prices are established by businesses and reflect market conditions. ***Value*** is determined by the customer. Value is a measure of the functional worth that the customer sees in the product. The customer expects the product or structure to deliver service and satisfaction equal to or greater than its cost. Answering the question, "Was the product worth what I paid for it?" can establish its value.

Cost-cutting Systems

A number of new production systems have been developed to reduce product cost and, in turn, increase product value, **Figure 33-14.** One of these systems is ***computer-aided design (CAD),*** which reduces product design and engineering costs. ***Just-in-time (JIT)*** inventory control systems schedule materials to arrive at manufacturing when they are needed. This reduces warehousing costs. Flexible manufacturing, which was discussed earlier, reduces machine setup time.

Similar systems are used in the other technologies. Computer scheduling is used in construction to ensure that human and material resources are effectively used. Computer ticketing reduces transportation costs. Computer systems make layout and preparation of color illustrations for printed products more economical.

Marketing

Products, structures, or services are of little value to companies unless they can sell them to customers. The products and structures must be exchanged for money. This is the challenge for marketing personnel. Marketing efforts *promote, sell, and distribute products, structures, and services.* Specifically, marketing involves four important activities:

Figure 33-14. Computer systems, like this railroad vehicle tracking system, reduce costs and increase value. (Norfolk-Southern)

- ***Market research*** gathers information about the product's market, **Figure 33-15.** This could include data about who will buy the product and where these people are located, in addition to their age, gender, and marital status. Also, market research can measure the effectiveness of advertising campaigns, sales channels, or other marketing activities.
- ***Advertising*** includes the print and electronic messages that promote a company or its products. Advertising can also present ideas to promote safety or public health. Advertising is designed to cause people to take action (buy a certain product) or think differently (buckle your seat belt while riding in an automobile).

  Closely related to advertising is ***packaging,*** **Figure 33-16.** This activity deals with designing, producing, and filling containers. The packages are designed to promote the product through colorful or interesting designs. Packaging also protects the product during shipment and display. Finally, the packages must include information that helps the customer select and use the product wisely.
- ***Sales*** is the activity that involves the physical exchange of products for money. It includes sales planning, which develops selling methods and selects and trains sales personnel for their efforts. Sales, also, includes the act of selling. This involves approaching customers, presenting the product, and closing the sale. This series of steps is all part of sales operations. The end result of sales operations is an order from the customer and income for the company.

Figure 33-15. These people are participating in a quality audit program. This program examines customers' reactions to new recipes. (National Livestock and Meat Board)

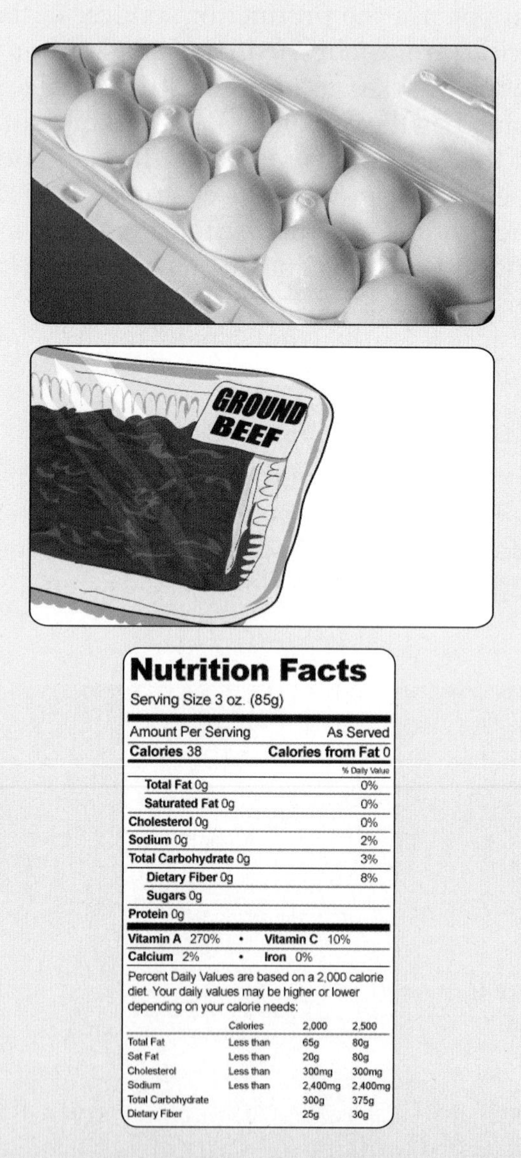

Figure 33-16. Packages are designed and produced to perform multiple functions. A package promotes the product, protects the product, and contains information to inform the customer about nutritional content and other topics.

- ***Distribution*** is physically moving the product from the producer to the consumer. This consumer may be another company or a retail customer. Consumer products follow at least three common channels, **Figure 33-17.** The product may move from the producer directly to the consumer. This channel is called direct sales. Sales of homes, encyclopedias, cosmetics, vacuum cleaners, and transportation services often use this channel.

In another channel, the producer sells the products or service directly to a retailer. The retailer then makes the items available to the customer. Franchised businesses, such as new automobile dealers or some restaurants, are an example of this distribution channel. This channel allows the producer to regulate the number of sales outlets and the quality of service that those outlets provide.

A third channel has the producer selling products to wholesalers. These businesses buy and take possession of the products. They, in turn, sell their commodities to retailers. The retailers then sell the commodities to the customers. In this channel, producers have little control over the retailers who are selling their products.

As described earlier, research and development develops the product, structure, or service. Production produces the item, while marketing promotes and sells it. But, these three activities cannot stand alone. They require money and people to make everything work. These requirements are the responsibility of two managed support areas — industrial relations and financial affairs.

Industrial Relations

You have learned that humans are the foundation for all technology. They create it, develop it, produce it, and use it. Thus, people are fundamental to all company operations. They work for the company, buy the company's products, and pass laws that regulate company operations.

Companies, therefore, are very concerned about their relationships with people. They nurture positive relationships with people by engaging in industrial relations (often called "human resources") activities. These activities can be grouped under three main programs. These programs are, as shown in **Figure 33-18:**

- ***Employee relations***: Programs that recruit, select, develop, and reward the company's employees
- ***Labor relations:*** Programs that deal with the employees' labor unions
- ***Public relations:*** Programs that communicate the company's policies and practices to governmental officials, community leaders, and the general population

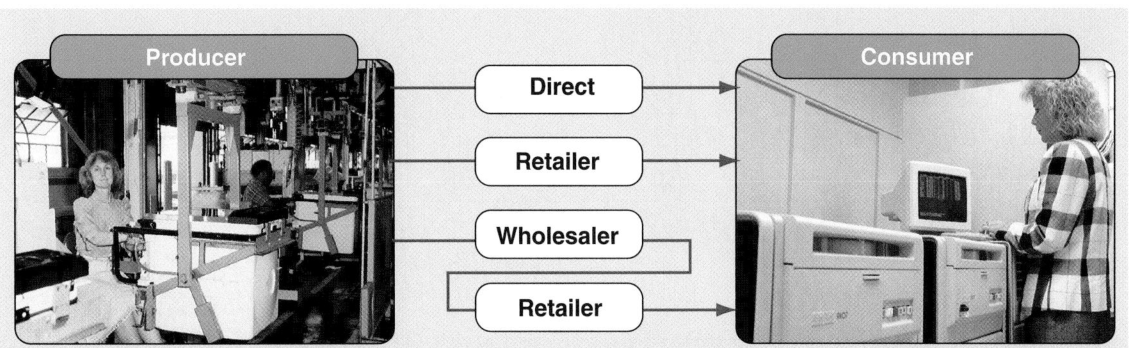

Figure 33-17. There are three different paths that move products from producers to consumers. (General Electric Co.)

Employee Relations

All companies have employees. These employees did not magically appear, fully qualified to work. They are the result of managed employee relations activities. These activities select, train, and reward the people that produce products and manage operations.

Selecting Employees

The first action in the employee relations process is called ***employment.*** This task involves determining the company's need for qualified workers. Then, applicants are acquired through a searching process called ***recruiting.*** Recruiting can be done through newspaper advertisements, school recruiting visits, or employment agencies. Other applicants may come to the company seeking employment on their own.

Next, job applicants go through a ***screening*** process. Screening allows qualified people to be selected from the applicant pool. Generally, this selection process starts with an application form to gather personal and work experience data. The promising applicants are interviewed to gather additional information, **Figure 33-19.** Some jobs require special abilities and knowledge. In these cases, applicants may be given a test to find those who qualify.

Training Employees

Successful applicants gain employment. However, few new employees are ready to begin work without training. Most need some basic training, and some may need special instruction. Basic information about the company and its rules and policies is provided to all workers. This is called induction training or orientation.

Special job skills can be provided through one of three programs. Simple skills are generally taught through ***on-the-job training.*** In this method, the new worker is trained at his or her workstation by an experienced worker or manager. More specialized skills may be developed in ***classroom training*** sessions. Classroom training involves qualified instructors providing information and demonstrating practices that each employee must learn.

Highly skilled workers are developed through ***apprenticeship training.*** Apprentices receive a combination of on-the-job and classroom training over an extended period of time. Apprenticeships usually last from two to four years, depending upon the skills to be learned.

In all three types of training, workplace safety is stressed. New employees are informed about company safety rules, shown how to work safely, and in many

Employee Relations
Labor Relations
Public Relations

Figure 33-18. Industrial relations develops and administers people-centered programs. (Harris Corp.)

Figure 33-19. This job applicant is being interviewed to gain information not gathered on the application form. (Inland Steel Co.)

Figure 33-20. These managers are receiving classroom training to improve communication skills.

cases, tested on safe work practices. Safety training is vital to providing a safe workplace. Federal and state agencies provide companies with safety rules and regulations, and are responsible for enforcing them. A principal source of these regulations is OSHA (Occupational Safety and Health Administration).

Executives and professional employees also receive training. This training may be called executive development, sales training, or managerial training. Since the work done by these employees does not involve making the products, most training sessions are given in classroom settings, **Figure 33-20.** In many companies, the total training program is called human resource development (HRD).

Rewarding Employees

People want to be recognized and rewarded for their work. Companies do this in two ways. First, they pay employees a wage, salary, or commission as a direct reward for work accomplished. A ***wage*** is a set rate that is paid for each hour worked. A ***salary*** is payment based on a longer period of time, such as a week, month, or year. Wage earners are often called hourly workers. Hourly workers are usually the production workers who build products, erect buildings, print products, or provide transportation services.

Salaried employees are usually technical and managerial workers. They develop products, engineer facilities, maintain financial records, and direct the work of other people. These employees often have more formal education than hourly workers. Salaried employees are held accountable more in terms of the amount of work that they do than the hours that they work.

Connections to Technology: **Mathematics**

Calculating Bids

The Rodriguez family want to add a 10′ x 10′ screened-in porch to their house. They ask Mr. Murphy, the sole proprietor of the A-Able Construction Company, to provide a bid for the work. Mr. Murphy decides he will need to charge enough to cover his costs and earn a 20 percent profit to finance new ventures. He calculates his bid as follows:

1. Rent equipment to clear the land and pour the foundation and footings	$ 400.00
2. Purchase materials (cement, lumber, screening, shingles, felt paper, fasteners, door)	$2900.00
3. Hire two laborers for approximately one week (8 hours/day) @ $25.00/hour each	$2000.00
Subtotal	$5300.00
4. Twenty percent profit ($5300.00 × 20 percent [.20])	$1060.00
Total	$6360.00

What would Mr. Murphy bid for the job if he found out that equipment rental would be $550 instead of $400 and if he decided he needed to make a 25 percent profit?

Some sales people are paid in a different way. Instead of wages or a salary, they receive a ***commission*** for each sale they make. The commission is usually a percentage of the total dollar value of the goods sold.

The second type of reward is called ***benefits.*** These are the insurance plans, vacations, holidays, and other programs provided by the company. These items cost the company money and, therefore, are a part of the total pay package for an employee.

Some companies also make use of special rewards called *bonuses* or *incentives.* These are typically awarded employees for performance that exceeds what is expected, or for suggestions that lead to improvements in efficiency, productivity, or workplace safety.

Labor Relations

In many larger companies, labor unions represent the employees. These companies require a labor relations program. This program works on two levels.

Figure 33-21. Community activities, such as entering a float in a parade, are part of a company's public relations program. The float is intended to help create a positive image for the company.

First, ***labor agreements,*** called "contracts," are negotiated between the company and the union. Labor agreements establish pay rates, hours, and working conditions for all employees covered by the contract. The agreements cover a specific period of time, generally ranging from one to three years.

During the contract period, disputes often arise over its interpretation. These disputes are called ***grievances.*** This is the second level of work for labor relations. Labor relations officials work with union representatives to settle the grievances.

Public Relations

Companies hire people, pay taxes, and have direct impacts on communities. Company managers form policies and have practices that they feel benefit the company. These practices often are subject to government regulation and may be affected by community pressures.

A company's public relations program is designed to gain acceptance for company operations and policies. The program informs governmental officials about the need for, and impact of, laws and regulations. Public relations also communicates with the community leaders so that local actions do not hamper the company's legitimate interests. Finally, public relations communicates with the general public. This communication presents the company as a positive force in the community, **Figure 33-21.** This image improves the company's ability to sell its products and to hire qualified workers.

Financial Affairs

Just as a company needs people, it also needs money. Companies must buy materials and equipment, pay wages and salaries, and rent or buy buildings. Taxes and insurance premiums must be paid. These actions can be shown in a simple flowchart, **Figure 33-22.** If you start at the

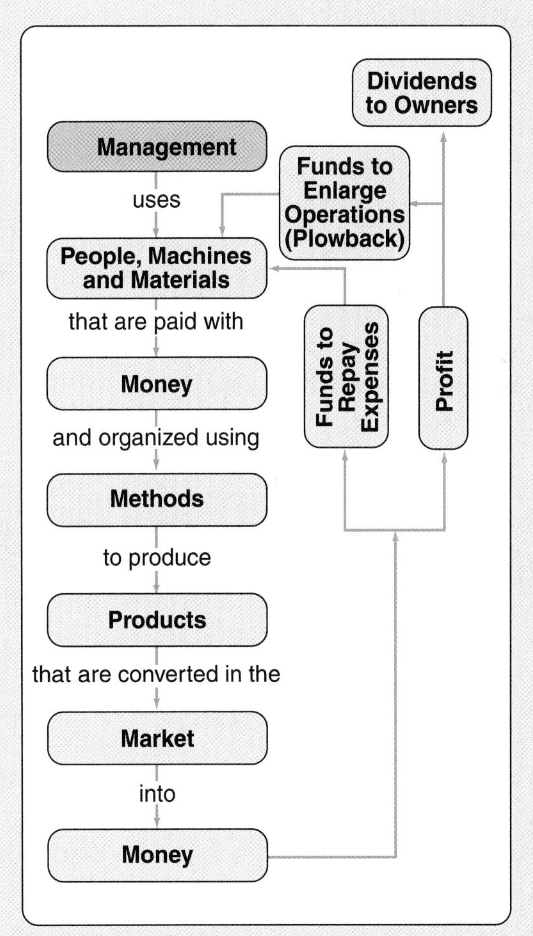

Figure 33-22. A simple flowchart that shows how money cycles within company activities.

Career Corner

Technical Illustrators

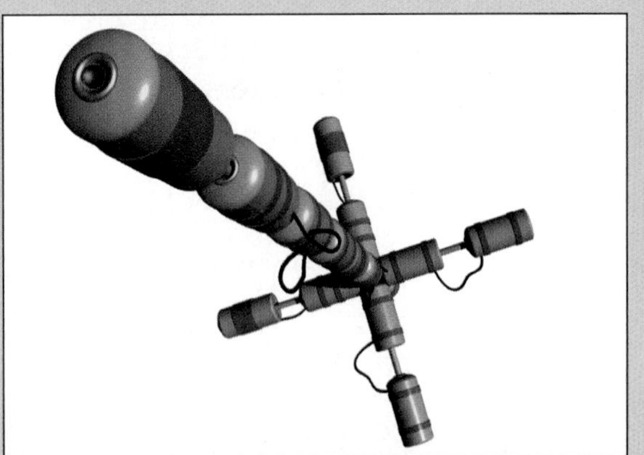

Technical illustrators create art to communicate ideas, thoughts, or feelings. They typically create pictures for publications, such as books, magazines, sales brochures, and advertisements, and for commercial products, such as textiles, wrapping paper, stationery, greeting cards, and calendars. Increasingly, illustrators prepare work directly on computers in a digital format.

Industrial enterprises, advertising agencies, publishing companies, and design firms employ illustrators. Many companies expect an applicant to have at least an associate's degree in technical illustration or a related field. Knowledge of computer graphics and skill in using visual display software are important.

upper left corner and read down you will see that:

Management employs people to use machines to change the form of materials, which is paid for by money and produces products, which are converted in the marketplace into money.

You will note that the word money appears twice. Money pays for resources: people's time, materials, and machines. This money is called ***expenses.*** Money is also the end result of sales: ***income.*** The difference between income and expenses is profit or loss. The goal of a company is to have more income than expenses. This makes a company profitable.

Profits must also be managed. ***Profits*** serve two important purposes. First, they can become ***retained earnings.*** These are profits that are held by the company and used to enlarge its operations. Profits are an important source of money for financing new products, plant expansions, and mergers.

Another portion of the profits is paid out by many companies as ***dividends.*** These are quarterly or annual payments to the stockholders. Dividends are the reward for investing in the company and sharing the risks of owning a business.

Managing the use of this money is the responsibility of financial affairs employees. They raise money, pay for insurance, collect from customers, and pay taxes. They also keep records of the financial transactions of the company. This area is called ***accounting.*** Each financial action is recorded as either an income or expense item.

Finally, financial affairs purchases the materials, machines, and other items needed to operate the company, **Figure 33-23.** Purchasing officers seek the "best" items in relation to company needs. The term "best" takes into account price, quality, and delivery date.

Figure 33-23. One responsibility of financial affairs personnel is purchasing materials needed by the company. (Brush Wellman)

Industry-Consumer Product Cycle

We have been looking at company activities as a linear action. These activities were presented as starting with research and development. Production was the next step and, finally, marketing. This path is correct if you look at a single-model or version of one product or structure. However, our system is much more complex. The economy is actually described as *dynamic*, or always changing, **Figure 33-24.** Products are developed, produced, and sold. Consumers select, use, maintain, and discard the products. They communicate their satisfaction or dissatisfaction with current products. This may cause companies to redesign existing products or, in some cases, to develop new ones. These new products are sold. They, in turn, are selected, used, maintained, and discarded. This cycle continues with a constant array of new products being developed and obsolete ones disappearing.

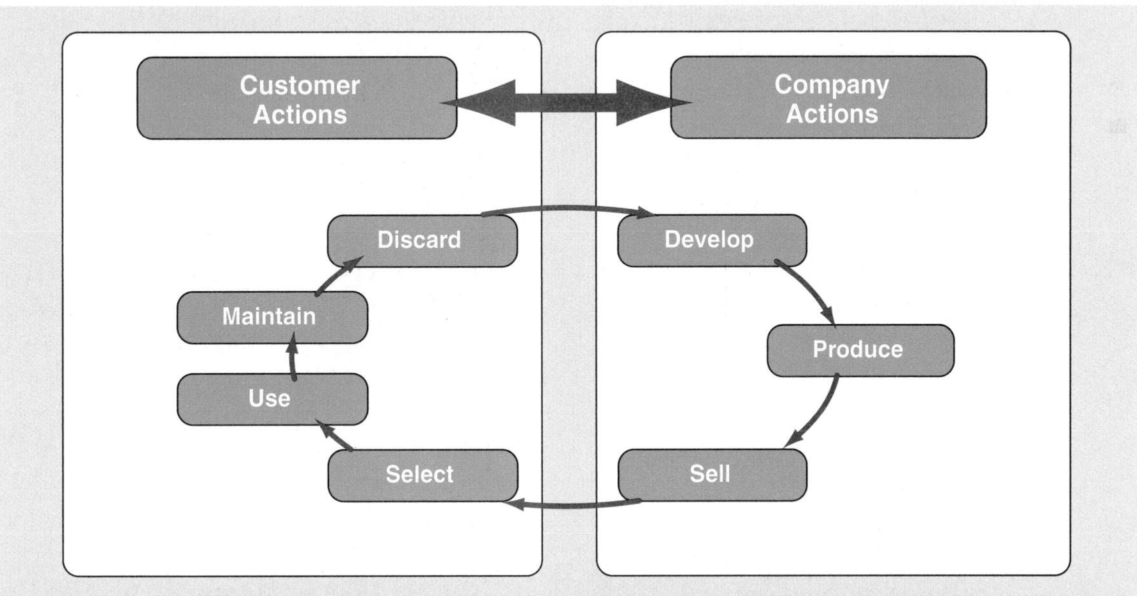

Figure 33-24. Customers' reactions to products cause companies to continually design, produce, and sell new items.

Summary

Most technology is developed and produced by industrial enterprises. These enterprises create a steady flow of products, structures, and services through five main areas of activity. Research and development develops and specifies the item; production produces it; marketing promotes, sells, and distributes it. These product- and structure-centered functions are supported by industrial relations and financial affairs activities. Industrial relations recruits and develops the workforce. It also deals with labor unions and promotes the company's image and policies. Financial affairs maintains financial records, pays the company's bills, and purchases material resources.

Key Terms

accounting
advertising
applied research
apprenticeship training
architectural drawings
basic research
benefit
bill of materials
classroom training
commission
communication processes
computer-aided design (CAD)
construction processes
continuous manufacturing
custom manufacturing
distribution
dividends
economic
economic activity
economic enterprises
education
employee relations
employment
engineering
engineering drawings
expenses
family
financial affairs
flexible manufacturing
grievance
income
industrial relations
industry
inspection
intermittent manufacturing
just-in-time (JIT)
labor agreements
labor relations
manufacturing processes
market research
marketing
on-the-job training
packaging
political/legal
price
process development
product development
production
profits
public relations
quality control
recruiting
religion
research and development
retained earnings
salary
sales
screening
specification sheets
structure development
transportation processes
value
wage

Test Your Knowledge

Write your answers on a separate piece of paper. Please do not write in this book.

1. In terms of technology, why is industry different from the other types of economic enterprises?

For Questions 2–8, match the area of activity on the right to the correct responsibility on the left. (Note: Some letters will be used more than once.)

Responsibility

2. _____Accounts for income and expenditures
3. _____Makes products or structures
4. _____Specifies characteristics of products
5. _____Promotes products
6. _____Responsible for quality control
7. _____Includes public relations
8. _____Purchases materials

Area of Activity

a. Research and development
b. Production
c. Marketing
d. Industrial relations
e. Financial affairs

9. Research that is focused on developing products is called _____ research.
10. List the four types of manufacturing systems.
11. The three phases of an inspection program are _____, work in progress, and _____.
12. *True or false?* Franchised businesses typically purchase products and necessary services through a wholesaler.
13. *True or false?* Many skilled workers are trained through apprenticeship programs.
14. To show a profit, a company must have more _____ than it does expenses.
15. List the four steps on the consumer side of the industry-consumer product cycle.

Applying Your Knowledge

1. Select a simple product, such as a kite. Apply the principles of research and development, production, and marketing to design, produce, and advertise it.
2. Set up a production line for chocolate suckers, cookies, or another food product. Describe how you will (1) plan for the product, (2) produce the product, and (3) maintain quality.

Chapter 34
Using and Assessing Technology

Learning Objectives

After studying this chapter, you will be able to:

- Describe how using a technological product is different from using a technological service.
- Explain how people can effectively select an appropriate product or technological service.
- Indicate where people can obtain information about operating products.
- Explain how products should be properly discarded.
- Describe ways societies assess technological advancements.

People working in various industries design and make technological products and systems. This group of people is a part of a larger community or nation. They have one type of job, and other people in the community have different jobs. However, one job that all people share is using these technological outputs, **Figure 34-1.** Each of us uses the products of technological effort every day.

Some people design and make...

... the product we all use.

Figure 34-1. A few people design and make the products that we all use. (Hewlett-Packard)

Using Technology

People are constantly generating needs and wants. They may want to travel to a vacation spot. They may want to cook food faster. They may want to have better health or higher energy levels. These and thousands of other needs and wants can be met with technological devices and systems.

These technological outputs can be grouped into two categories: technological products and technological services, **Figure 34-2.** ***Technological products*** are the artifacts that people build. They may be a manufactured product, a constructed structure, or a communication media (book, CD recording, etc.). ***Technological services*** are outputs that we use but do not own. These include transportation and communication services such as airline travel or television programming.

Figure 34-2. Technological outputs can be products or services. (Top) These gondolas are providing a service by delivering goods. (Bottom) This church is a constructed product.

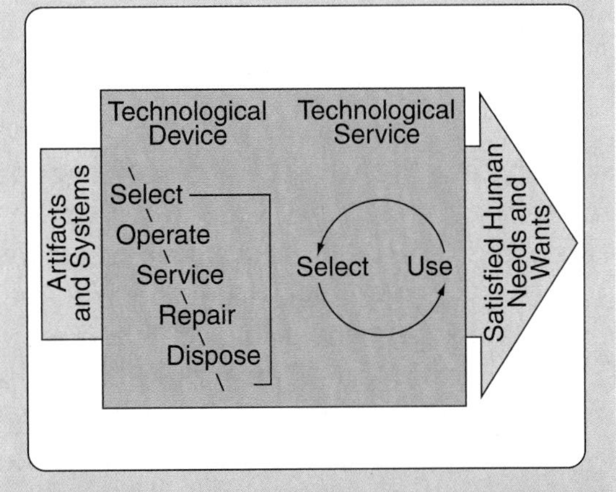

Figure 34-3. Five steps should be followed when using technological products and services.

Using products and systems is important to our lives. It can make life easier, more enjoyable, and more productive. However, it can also create frustration and negative impacts on people or the environment. This tension between good and bad, positive and negative means that people should use technology correctly.

But, how do people know they are using technology properly. It requires that they understand and apply the steps in proper technological use, **Figure 34-3.** For technological products these steps are:

- Selecting appropriate devices and systems.
- Operating devices and systems properly.
- Maintaining devices and systems.
- Repairing broken parts and systems.
- Disposing of worn out or obsolete devices and systems.

When we use technological services we do not use all these steps. We can only select and use the service. We do not operate the airline or the television network. We do not have a product to service or repair; nor do we have a product to discard.

Selecting Appropriate Devices and Systems

Look around you. You are surrounded with technological products. Someone selected each of these. They were not the only products that could have served the purpose. Consider electric lighting. People can choose from incandescent, halogen, or fluorescent lights. They could be of different lighting levels (brightness) such as 60 watt, 75 watt, or 100 watt. They could select ceiling, floor, or table lights.

Some people use trial and error to make product selections. They buy a product and try it. If it doesn't work well, they replace it with another product with different features. However, this is an inefficient and expensive approach to selecting an appropriate product.

A better selection method is to deal with the challenge logically, **Figure 34-4.** First, the person should determine the exact *need* that is being addressed. Then the need should be described in terms of the product's operation, price, quality, and similar features. From these descriptions, a list of alternative products that meet the need can be made. The features of each product should be identified. A list of advantages and disadvantages (pros and cons) of each product should be developed from these features. For example, purchase price, operational features, and maintenance requirements for each product could be identified. The ease of use could be explored. Its safety could be examined. Likewise, the appearance and styling could be considered. From this list of features and pros/cons, the most appropriate product could be selected. It might not be the cheapest or best operating product. But it should be the one that most closely meets the overall need.

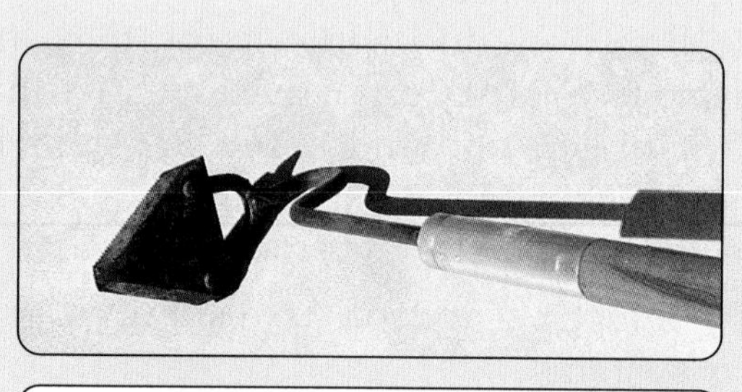

- Define need
- Describe need (establish criteria)
- List advantages and disadvantages
- Compare products
- Select best product

Figure 34-4. Product selection follows a logical sequence.

Operating Products and Systems

Once a product has been selected and purchased, the new owner must learn how to operate the device. Some products need little or no training to be able to use it, **Figure 34-5.** The owner already knows all he or she needs to know to operate the product. Few of us need to learn how to use a new pen or pencil.

Other products are replacements for an older model. Again, the owner knows a lot about the product's operation. He or she needs only to review the new product's operation and new features. For example, maybe you can effectively use a version of a word processing program. If you bought a new version, you would only need to learn how to use its new features.

Some products are new to the owner. He or she has never used it or one like it. Then, the person has a lot to learn. This would include learning how to:

- Unpack and set-up the product.
- Adjust the product for different operations.
- Correctly and safely operate the product.
- Care and maintain the product.
- Obtain service and repair.

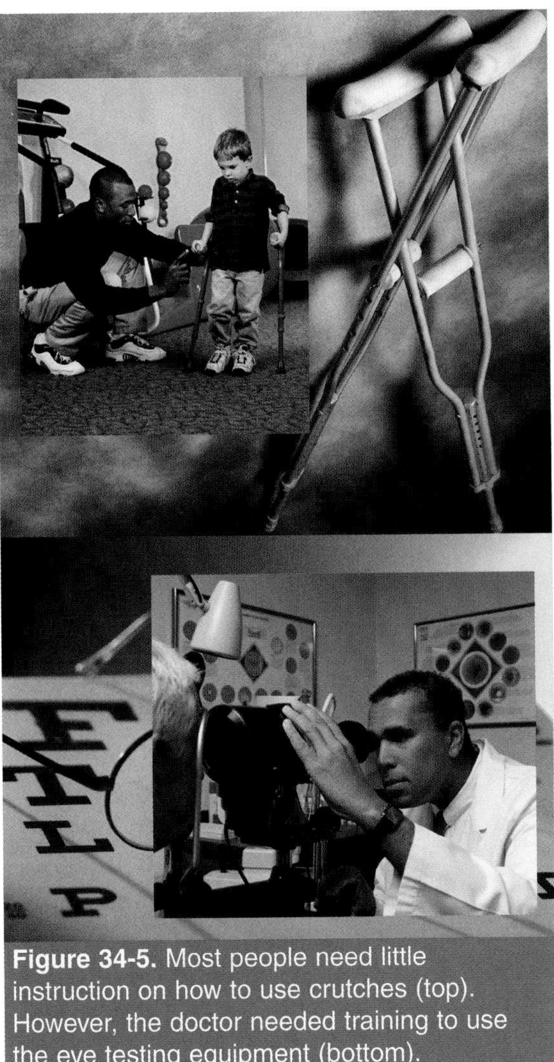

Figure 34-5. Most people need little instruction on how to use crutches (top). However, the doctor needed training to use the eye testing equipment (bottom).

There are several ways to learn this information. One way is to carefully read the owner's manual, **Figure 34-6.** The new owner should read and study the information to obtain information about the five elements listed earlier. Only after developing this information should the device be used.

Another way to obtain this information is through training. This could be from a person skilled in using the device. For example, a salesperson in a computer store could provide some basic information about setting up and using a simple computer system. The new owner could also seek more formal instruction. He or she could attend a seminar or class that teaches about using the device.

Servicing and Repairing Devices and Systems

Many products need ***maintenance*** during their life. This can include both ***service*** and ***repair.*** Servicing is doing routine tasks that keep the product operating. This includes cleaning the product, oiling moving parts, making simple operating adjustments, replacing filters, and other similar tasks, **Figure 34-7.** Most ***owners' manuals*** contain information needed to complete periodic servicing.

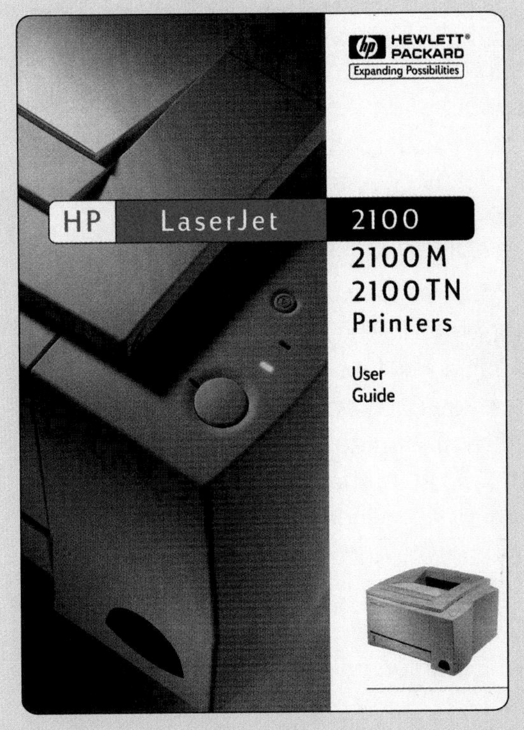

How to ...

- ... unpack and set up the product
- ... adjust the product for different operation
- ... correctly and safely operate the product
- ... care and maintain the product
- ... obtain service and repair

Figure 34.6. An owners' manual contains valuable information. (Hewlett Packard)

Technology Explained

Earth-Sheltered Building

earth-sheltered building: A structure built into the earth to take advantage of the insulation value and thermal properties of soil.

Energy efficient buildings are important to many people. A number of different building techniques can be used to make buildings more efficient. One way is to use *earth-sheltered construction.* This method of construction partially or completely covers the dwelling with soil. The purpose is to use the soil's thermal qualities to keep the house at an even temperature. The temperature of soil a few feet below the surface does not vary much throughout the year.

Using the earth to shelter a dwelling is not a new idea. Native Americans in Arizona used earth-sheltered buildings in 1000 A.D. This was *400 years* before Columbus discovered America. Using protective cliff faces to shelter communities was common throughout the southwestern United States. Today, a number of these sites have been uncovered and restored, **Figure A.**

Figure A. This cliff dwelling in the Walnut Creek National Monument in Arizona uses earth sheltering.

Earth-sheltered construction requires careful planning. Special problems exist, such as moisture, loads from soil, and orientation for solar heat gain. Buildings can be partially earth-sheltered or totally covered, **Figure B.** In the northern hemisphere, the building is normally sited so it faces south. This is so the house can absorb solar heat through large windows on the south face. The wall of windows helps make the front rooms light and airy. These are the daytime living rooms. Bathrooms and bedrooms can be placed toward the back of the structure. These rooms are normally used at night, when artificial light is needed.

This type of construction can be used for more than housing. It is also appropriate for theaters, shopping malls, convention centers, and warehouses. The function of these structures does not require windows or exterior views.

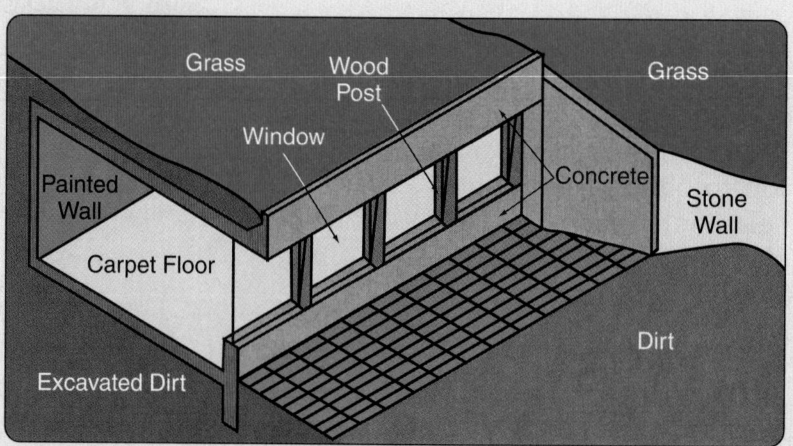

Figure B. This drawing shows the design for an earth-sheltered house that has the roof covered with earth. This house would be built using reinforced concrete.

Figure 34-7. Servicing involves simple tasks to keep equipment working, such as changing the abrasive on this machine.

Repairing devices involves replacing worn or broken parts, **Figure 34-8.** It also includes making major operating adjustments. The owner can do this type of work; however, most often, trained mechanics and technicians do it. For example, repairing an automobile transmission, a television set, or furnace is beyond most people's ability or interest.

Figure 34-8. Repairing involves replacing broken and worn parts. This worker is replacing timbers in a mine.

Disposing of Devices and Systems

Few products last forever. Most products have a useful life after which they may wear out or become obsolete. When this happens, the owner should properly dispose of the product or material, **Figure 34-9.** This could involve donating the product to a group that has use for it. For example, many charities and schools have use for computer systems that are not state-of-the-art. Many businesses that replace operating systems with newer, faster models donate the older systems to such groups.

Figure 34-9. These bins allow people to recycle waste paper, bottles, and cans.

Worn or obsolete products could be sold or given to a recycling operation. This group or company dismantles the product to retrieve good parts or materials. For example, automobile "wrecking yards" recycle good parts by removing and selling them to people who are repairing similar models. They also sell scrap metals to steel mills and aluminum processors.

Some products cannot be reused or recycled. The product may not have any useful parts. Recycling the materials may be impossible or not cost effective. For example, old dinnerware has little value.

Figure 34-10. Transportation services are selected, used, and evaluated.

Most ceramic materials cannot be recycled. Likewise, broken concrete slabs have little use or value. In this case, the materials should be sent to a landfill.

Selecting and Using Technological Services

Selecting and using a technological service requires fewer actions by the customer, **Figure 34-10.** Suppose you plan to travel across the country. You still need to establish your need (goal). This could be using the least amount of time to travel from point A to point B. Or it could be seeing as many historical sights as possible.

Once the need is defined, the possible ways to make the trip should be established. You could look at a road map and decide if you could make the trip in a car. You could explore flight, bus, and train schedules, routes, and prices. Armed with this information you could choose a method of travel.

After the trip is completed, you could evaluate the trip. Was it pleasant? Did you get where you were going on time? In short, you could decide if the service met your needs. If not, the next time you travel you could select a different mode of travel. Or you could choose a different company to take you to your destination.

You could use the same "select-use-evaluate" model for any technological service. It could be applied to a television program. You may want an entertainment program that is appropriate to all your family to watch at 7:00 p.m. on Tuesday. You could search the program listings, read the summaries, and select a program. You could then watch it. After you see the program, you could decide if it was entertaining and appropriate for your family. If not, then you would select a different program the next Tuesday evening.

Assessing Technology

Technological systems were not all developed in the past few years. They have been with people for several centuries. They date before computer systems and space travel. One of the most complex technological systems is a city. It arose during the Middle Ages and gathered people in one spot. Cities provided the workers for the industrial revolution. This launched the Industrial Revolution and a new way of working. This was called the division of labor. The task of making something was divided into a large number of jobs. Each job was assigned to a single worker. The model replaced the skilled worker who made a product from start to finish.

Figure 34-11. Early factories polluted the air.

At the same time agricultural productivity increased and medical science was developed. Most people saw technological advancements as a positive force. Life became better as the years passed. People were making more money. Many could work fewer hours and have more time for family activities. New and improved products were appearing almost daily. Homes were becoming larger and more comfortable. Better health care was increasing life expectancies.

However, all was not good in the country. World War I and the Great Depression caused some people to question technology's effects on people. This continued until the 1950s. A growing number of critics began to suggest that many products of technology had harmful aspects. The era of smog, caused by automobile exhausts, arrived in this country, **Figure 34-11.** Industrial wastes were polluting lakes, rivers, and groundwater. Pesticides, such as DDT, were entering the food chain.

Several plans to deal with technological impacts have been developed. Individuals evaluate devices and systems according to their personal needs. However, this is a personal view. A different type of evaluation is called technology assessment. It involves groups of people evaluating the impacts of technology on people, society, and the environment. It places the task of controlling technology on the government and the courts. These agencies provide a way of assessing the effects of technological innovations on human life. They conduct studies and review those done by other groups. In particular, they evaluate the economic, ethical, social, health, environmental, and political effects of various technological products and systems.

One model of ***technological assessment*** suggests what groups need to do, **Figure 34-12.** First the underlying goal for the technology needs to be identified. This should clearly describe why the technology was developed or used. It should list the desired outcomes for any new or improved technology. For example, the goal may be a fuel-efficient way to transport people quickly and safely from city-center hotels to the local airport.

Second, a way to measure the success in meeting the goal needs to be developed. This requires that criteria clearly describe the proposed outcomes. Using the airport transport example, several criteria could be developed. The system should be convenient to the major hotels in the downtown area. It should be fast and clean. It should have a schedule that meshes the major flights at the airport. It should produce at least 50 percent less emission than auto travel would

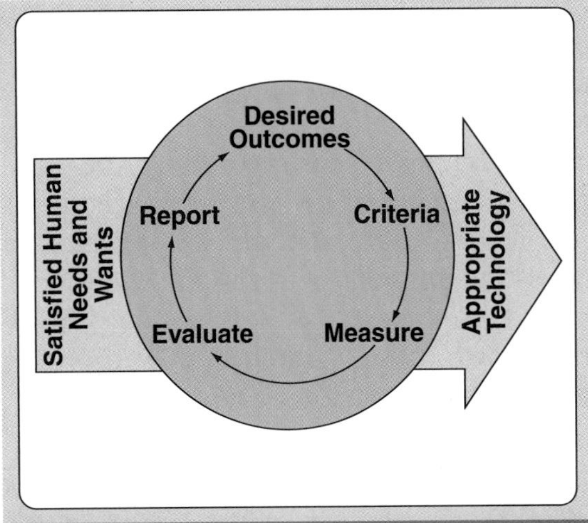

Figure 34-12. Technological assessment can be conducted in five steps.

Career Corner

Landscape Architects

Landscape architects design outside areas, such as residential yards, public parks, college campuses, shopping centers, and golf courses. They help plan the locations of buildings, roads, and walkways and determine the arrangement of plant materials, such as lawns, flowers, shrubs, and trees. These architects also work with environmental scientists and foresters to conserve or restore natural resources.

They create detailed plans for sites, which include soil slopes, vegetation, walkways, and landscaping details, such as fountains and statues. Often, these plans are prepared and presented using computer-aided design (CAD) systems and video simulation. Entry into the field requires a bachelor's or master's degree in landscape architecture. A successful landscape architect has creative vision, artistic talent, and good oral and written communication skills.

produce. It should be cost effective. It should run on public right-of-ways.

Third, the assessment needs to monitor and measure the impact of the technological innovation. A method to gather data needs to be developed. The system should gather information about each criterion of success. Again, using the transport example, interviews with riders could tell people about the convenience, cleanliness, and speed of the service. Monitoring equipment could measure vehicle emissions. Financial records contain data about cost effectiveness. The system's schedules and airline flight schedules contain data about the interface between the two modes of travel.

Fourth, people must review the data and draw conclusions. They need to evaluate the results of the measuring and monitoring action. They need to measure the results against the criteria.

Finally, the assessment group must prepare a report. It should present the problem and the criteria used to explore it. The report should summarize the findings. It should present the positive and negative aspects of the technology being assessed. And, most importantly, it should recommend action. For example, the report may recommend that the technology be abandoned, altered, and maintained.

The findings of technological assessment can be applied to political and legal actions. Local, state, and national laws and regulations can be developed. Lawsuits can be filed when the actions are found to be against existing laws and regulations. People can support or boycott the product or service.

Summary

People select and purchase technological products and services almost daily. They use products and systems to meet their needs and wants. With the application of technology, each person is responsible for using it correctly and wisely.

Likewise, communities, states, and the nation should be concerned about the impacts of technological use on people, the society, and the environment. These impacts can be measured through technological assessment.

Key Terms

maintenance
owners' manual
repair
service
technological assessment
technological product
technological service

Test Your Knowledge

Write your answers on a separate piece of paper. Please do not write in this book.

1. What are the common steps in using a technological product?
2. The first step in selecting an appropriate product is to determine the exact _____.
3. List the three common steps in using a technological service.
4. Evaluating the impact of technology on people, society, and the environment is called technology _____.
5. What are the five steps in assessing a technological product or system?

Applying Your Knowledge

Use the assessment model to deal with each of the following cases:

1. People of all walks of life are becoming more technology literate. People of all ages use telephone banking, ATM, the Internet, voice mail, e-mail, and fax machines, for example. The Internet is becoming a vehicle for information access and shopping. Is this good or bad?
2. Solar energy technology has failed to make the impact many people expected. It has been for many years, like the related research-and-development area of fuel cells, a "promising technology." What are the barriers to solar technology?
3. Hybrid cars use a combination of a gasoline engine and an electric motor/generator for power. Many people believe we should shift quickly to the new vehicle, which has high fuel efficiency and low emissions. What government and industry action is needed to make this happen?

Section 10 Activities

Activity 10A

Forming the Company

Background

Companies produce many technological devices and products. These enterprises are the products of human actions. They are formed and structured to efficiently use resources to produce artifacts meeting our wants and needs.

Situation

Students have mentioned they need a way to be informed about important sporting and social events at school. They also want a way to publicize happenings they feel are important. From these comments, you have concluded that an inexpensive personalized calendar would meet their needs and earn you a profit. This type of calendar can be produced with limited finances by using new computer software. See **Figure 10A-1.**

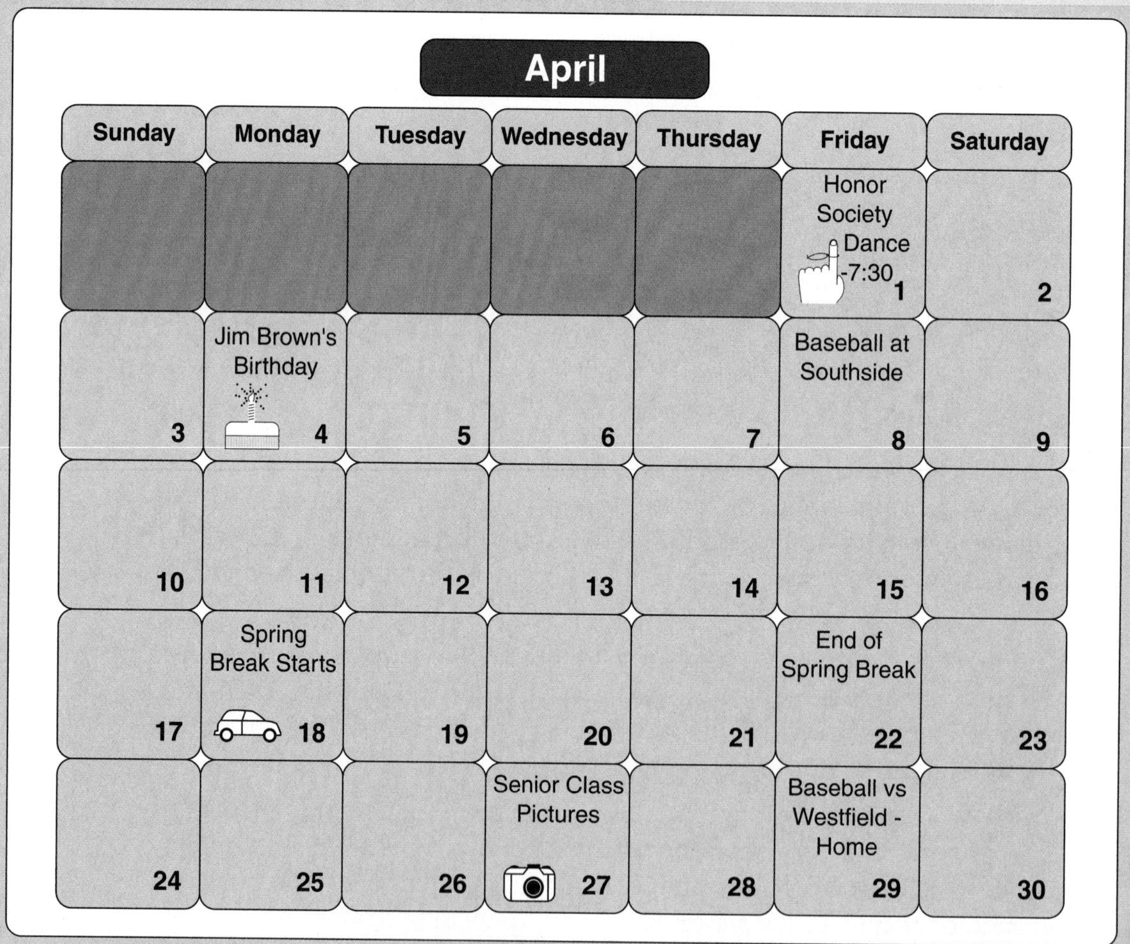

Figure 10A-1

Challenge

Organize a company to produce a nine-month calendar. The calendar should span the school year and have selected days personalized. Consider the tasks to be completed, and consider the managerial structure needed to complete them. Be sure to recognize that there are production and marketing tasks. Also, there are two distinct phases of the company's operations, which may require two different organizations. One operation can finance the company, design the calendar, and sell calendar entries. The other operation can maintain financial records, produce the calendars, and sell the finished products. See **Figure 10A-2.**

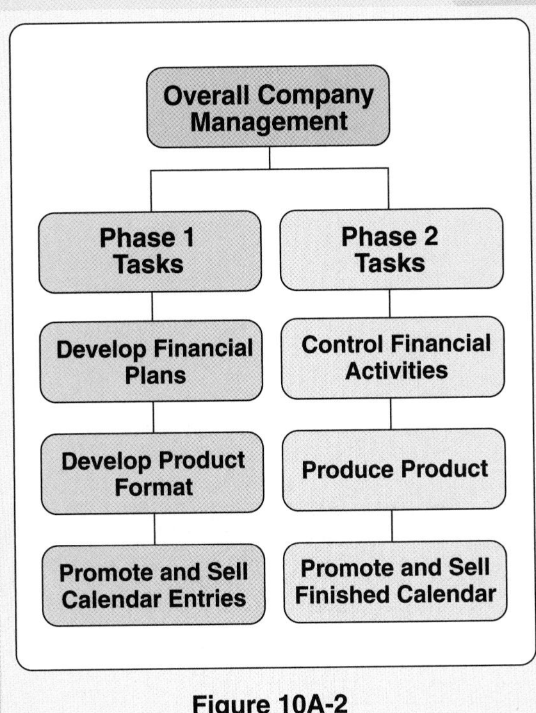

Figure 10A-2

Optional

- Write a one-page job description for each job in the organization chart.
- Develop a set of goals for each major department in the company.

Activity 10B

Operating the Company

Background

Companies are series of independent tasks that have been integrated into functioning enterprises. Each task must be planned for and carried out with efficiency.

Challenge

Identify, schedule, and complete the several tasks required to produce and market a personalized calendar for your school.

Procedure

Design and Development Department

1. Obtain software that can be used to produce a personalized calendar.
2. Follow the instructions to produce a calendar for a single month. This will acquaint the department members with the operation of the software.
3. Establish the layout for the calendar.
4. Produce a common layout sheet for the marketing group to use in selling calendar entries.

Calendar Entry Marketing Department

1. Determine the selling price for a calendar entry.
2. Develop a calendar entry order form.
3. Make posters to promote the sale of calendar entries.
4. Sell calendar entries.

Production Department

1. Receive calendar entry forms from the marketing department.

2. Enter data on calendar layouts.
3. Print a proof of the calendar.
4. Submit the proof to marketing for approval.
5. Correct calendar entries.
6. Print the master calendar.
7. Reproduce the calendar.

Calendar Marketing Department

1. Produce and distribute advertisements for the sale of calendars.
2. Select and train calendar salespeople.
3. Sell calendars.
4. Maintain sales records.

Finance Department

1. Set budgets for company operations.
2. Sell stock and maintain stockholder records.
3. Purchase materials and supplies.
4. Maintain all financial records.

Executive Committee (President and Vice Presidents)

1. Set deadlines for important activities.
2. Monitor progress in completing tasks.
3. Set budgets.
4. Establish selling prices for calendar entries and finished calendars.
5. After the calendar copies have been sold, close the company and liquidate the assets.

Managing a technological enterprise requires a great deal of planning and organization.

Section 11

Technological Systems in Modern Society

Tomorrow's Technology Today
The Eden Project

You may have gone to your neighborhood greenhouse and seen hundreds of species of plants, but you probably have never seen anything like the Eden Project. The Eden Project, in essence, is a gigantic garden that would cover 30 football fields! This enormous, multi-domed greenhouse is unlike anything before, and its creators have plans to continue to improve this incredible environmental center for its visitors.

Located in Cornwall, England, this unusual structure contains plants from around the world. The plants are housed in three biomes, each of which recreates a distinct climate found on our globe. The most extraordinary section is the Humid Tropics Biome, which consists of several domes representing the environment of a tropical rain forest. This greenhouse, which is big enough to house the Tower of London, contains hundreds of trees and other plants from rain forests in Africa, Asia, South America, and Australia, such as rubber, cocoa, vanilla orchids, and bamboo.

The second biome is another multi-domed greenhouse, the Warm Temperate Biome. This section has plants from temperate rain forests in California, Southern Africa, and the Mediterranean. Here, you can find grapevines, olives, orange groves, and hundreds of colorful flowers.

The final biome is the Roofless Biome, which is an open area that has beautiful and important plants from the Cornwall area and other areas with similar climates, such as Chile, the Himalayas, Asia, and Australia.

Besides viewing both familiar and exotic plants from around the globe in one place, visitors to the Eden Project can also learn about plants that have been important to human history and see how interrelated human life and plant life really are. In addition, exotic butterfly species have been introduced to the enclosed biomes. Depending on how successful this butterfly project proves to be, other animal species may be introduced to the greenhouses in the future.

The Eden Project does not only consist of the plant life biomes. Nature trails wind through the 30 acres of land, and informational kiosks and artistic pieces appear throughout the biomes. The Eden Project also frequently brings entertainment, special events, and educational workshops to the greenhouses for visitors. The Project's main goal is to educate people about the natural world, especially about using natural resources efficiently so they will continue to be available in the future.

There currently are plans to add another biome to this structure—a Semi-Arid Biome. This biome was part of the original plans for the project, and it is expected to be the final living element of the Project. The Eden team members are also working on building a conference center for seminars and meetings, and they are exploring the possibility of creating student housing near the site.

This global garden is unique in both its appearance and its purpose. It succeeds in bringing together science, art, technology, and communication to make a difference in the world. While the Eden Project is the only one of its kind, you can expect to see similar projects around the world in the coming years.

Chapter 35
Technology: A Societal View

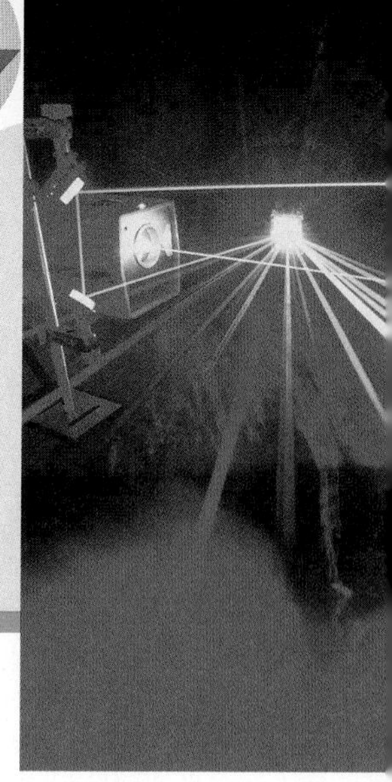

Learning Objectives

After studying this chapter, you will be able to:

- List some destructive natural forces that affect human lives.
- Give some examples of how technology is used to help to control natural forces.
- Describe the global impact of automobile use in North America.
- Distinguish between divergent thinking and convergent thinking as futuring tools.
- Describe the four major types of futures that are considered in futures research.
- List the three forces that are causing a condition called an environmental crisis.
- Trace the movement of the center of technical knowledge and power from east to west through human history.
- Identify several specific actions companies must take to remain competitive in a world economy.

You have been learning about technology for 34 chapters in this book. You've been introduced to many concepts and a great deal of information. Key to all this learning is one principle: *Technology is developed by people for people to use,* **Figure 35-1.** You've learned that technology is the product of human knowledge and ability. It is designed to help people modify and control the natural world. Also, technology is the sum total of all human-built systems and products. *Technology is the human-built world,* **Figure 35-2.**

Technology is neutral. By itself, technology does not affect people or the environment. How people *use* technology determines if it is good or bad, helpful or harmful. How we use technology can *help* or *harm* the world around us.

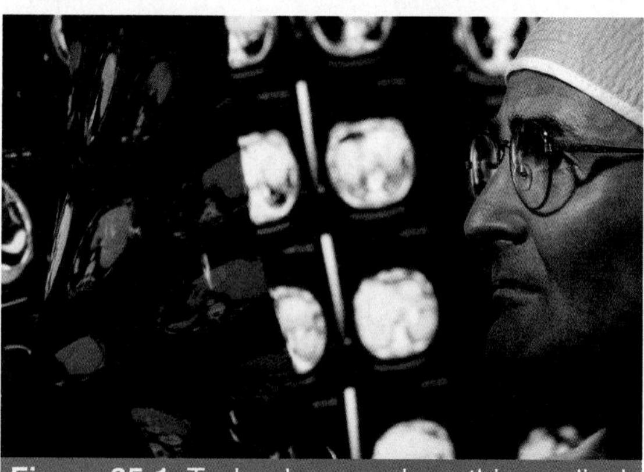

Figure 35-1. Technology, such as this medical imaging system, is developed by people for people to use.

Technology Controls and Harnesses Natural Forces

Natural forces affect, and sometimes disrupt, human life. Hurricanes wreck ships and destroy coastal settlements. Tornadoes can level entire sections of towns and cities. Floods wash away homes and carry away vital topsoil, reducing the productivity of croplands. Fires burn buildings, crops, and forests. Earthquakes shake structures until they collapse. In all these natural events, people can be killed and their possessions destroyed.

One of the earliest uses of technology was to harness natural forces. People started to design a human-built world that

Figure 35-2. The application of technology has greatly changed our world and our way of life. (DaimlerChrysler, GE Plastics)

Figure 35-3. People use technology to harness and control the forces of nature. Fire can destroy a forest (left) or can be harnessed for useful tasks, such as baking pastries (right).

could reduce natural destructive forces. Dams hold back floodwater and produce energy in hydroelectric plants. Fire is tamed to heat our homes and process industrial materials, **Figure 35-3.** The inner heat of the earth is captured by geothermal power plants. Wind is used to generate electricity. Solar energy provides heat and electricity for businesses and homes.

However, controlling and using natural forces is not enough. Humans are starting to understand that having control over the earth's future carries with it serious responsibilities. Many people realize that humans must protect the environment and the plants and animals that live with us on this planet. We must live in harmony with the natural world.

Technology Has a Global Impact

Responsible use of technology requires that we realize we live in a world very different from that of our grandparents. The technology that we develop and use today has impacts beyond our homes, cities, states, and countries. We must look at the impacts of our decisions on the entire world.

For example, the attachment that our society has to the automobile as its primary transportation vehicle has ***global impacts.*** The citizens of North America represent a small percentage of the world's population. However, they use a large percentage of the world's petroleum. Also, automobiles discharge great quantities of pollutants into the atmosphere.

Likewise, many North Americans are concerned about high birth rates in Third World countries. The impact of each new North American child on the world's resources, however, is many times greater than a child from the Third World. We must be concerned about population *and* resource use.

Technology and the Future

Since technology is a product of human activity, humans can control it. To do this, we must have an idea of the kind of future we want. This can be developed by using a research technique called ***futuring*** or ***futures research.*** This process helps people select the best of many possible courses of action, **Figure 35-4.** Futuring emphasizes five distinct features:

Figure 35-4. People can use futuring to determine the type of future they want.

- **Alternate avenues.** The futurist looks for many possible answers rather than *the* answer.
- **Different futures.** Traditional planning sees the future as a refinement of the present. A futurist looks for an entirely new future.
- **Rational decision-making.** Traditional planners rely heavily on statistical projections. They use mathematical formulas to help them predict events and impacts. A futurist uses logical thinking and considers the consequences when making decisions.
- **Designing the future.** Futurists are not concerned with improving present or past practices. They do not see the future as a variation of the present. They focus on predicting a possible future that can be created.
- **Interrelationships.** Traditional planners use linear models that suggest one step leads to the next step. Futurists see alternatives, cross-impacts, and leaps forward.

Using a futures approach, exciting new technologies can be developed. The futurist must have a dual view. One view must be of a present challenge or problem. The other view must be of the world of the future, where his or her grandchildren will live. Looking at both of these views requires a combination of short-term and long-term goals. The most important tool to develop these futures is the human mind. It is the only tool capable of reasoning and making value judgments.

Futuring will require two types of thinking. The first is ***divergent thinking.*** This type of thinking lets the mind soar. People are encouraged to explore all possible and, in many cases, impossible solutions. They look for interrelationships and connections. Out of this activity emerges a number of futures.

One possible future, however, must be selected as the one that will be "created." This requires ***convergent thinking.*** The final solution must receive focus and attention. Its positive and negative impacts must be carefully analyzed.

This analysis must be done with four types of futures in mind. The first is the ***social future.*** It suggests the type of relationships people want with each other. The second is the ***technological future.*** This looks at the type of human-built world we desire. The third is the ***biological future.*** It deals with the type of plant and animal life that we want, **Figure 35-5.** The fourth is the ***human-psyche future.*** This future deals with the mental condition of people. It stresses the spirit rather than

Figure 35-5. The biological future may hold new developments in the use of plant and animal products, especially those from the sea. New technologies impact the environment, people, and the society as a whole. (FMC Corp)

Technology Explained

Mechanization

mechanization: Replacing human labor with machines operated or controlled by people or other machines.

Increasing productivity is key to a rising standard of living. For increased productivity, more goods and services must be produced, while the consumption of materials or labor stays the same. One method of increasing productivity is mechanizing work. Mechanization is moving tasks from human operation and control to mechanical control. This action reduces the human labor component in each job.

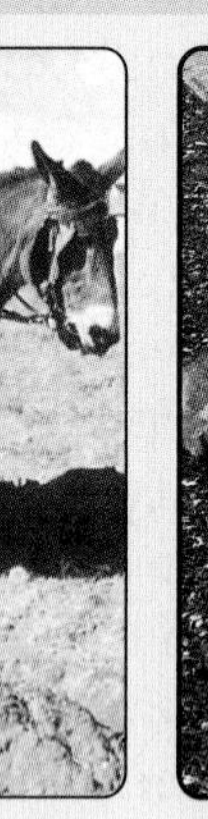

Task	Animal Power Mid 1800s	Mechanized Mid 1900s
Plowing	477 Hours 24 Minutes	39 Hours 30 Minutes
Discing	57 Hours 36 Minutes	23 Hours 10 Minutes
Harrowing	36 Hours 15 Minutes	12 Hours 15 Minutes
Cultivating	477 Hours 24 Minutes	17 Hours 3 Minutes
Planting	43 Hours 42 Minutes	done while cultivating
Reaping	64 Hours 32 Minutes	not needed
Shocking	512 Hours 10 Minutes	not needed
Threshing/Combining	480 Hours 30 Minutes	22 Hours 24 Minutes
Total Time	**2,149 Hours 33 Minutes**	**114 Hours 22 Minutes**

Figure A. These are the approximate times devoted to the important tasks in planting and harvesting 80 acres of grain.

Early mechanization was done with simple procedures. Recent mechanization often uses complex machinery. Agriculture provides an example of this situation, **Figure A.** An animal drawn plow replaced the digging stick. Later, the horse drawn moldboard plow replaced the animal drawn plow. Currently, we have large tractors that pull complex tilling machines, but even these machines are being replaced by no-till or minimum till farming practices, which reduce the amount of work it takes to prepare the soil for planting. In the industrial setting, robots, automatic machines, and other computer-controlled machines are reducing the amount of human labor needed for each job. These mechanized changes increase the level of education and skill required for remaining jobs.

the mind, attitude instead of physical condition. It is concerned about how people will feel about life and themselves.

These futures are listed separately, but they are actually interconnected. For example, new technologies directly impact how people relate to one another. Television dramatically changed family life, recreation, and the entertainment industry. Technology also changes the natural environment. Acid rain caused by automobile emissions and coal-burning electric plants has destroyed forests in Canada and the eastern United States. Likewise, technology has changed how we view ourselves. Some people feel threatened by technology, while other people feel empowered by it.

Technology: Challenges and Promises

It is impossible to explore how each new technology has impacted our lives and how it will impact the future. However, we can explore some examples to provide a foundation for personal study. Later, you can use these examples as you evaluate other technologies.

Three issues that are widely discussed are energy use, environmental protection, and global economic competition. Let's look at these issues in terms of their challenges and promises.

Energy Use

The world as we know it would come to a grinding halt without energy. Almost everything in the human-built environment depends on energy. Therefore, the supply and use of energy resources is very important.

Every person uses energy resources. Many of these are exhaustible, or ***nonrenewable resources.*** When we burn them all, there will be no more. The supply is said to be ***finite***—there is a limited quantity of the resource available. One of these resources is petroleum. It is still the fuel that powers most transportation vehicles. Other exhaustible energy resources are coal and natural gas.

The shrinking supply of these resources is a major concern, particularly in the case of petroleum. We are challenged to reduce our dependence on these resources. One alternative is to shift to ***renewable resources.*** These are the resources that have a life cycle. They are the products of farming, forestry, and fishing. However, these resources are in limited supply at any one time. For example, in many less-developed countries, wood is the primary fuel for cooking. As population grows, people must roam a greater distance to find the firewood they need. Thus, large regions are being stripped bare of trees. Also, using wood for fuel eliminates it as a source of building material.

Likewise, corn can be used to make ethanol. Ethanol can be used as a fuel. However, if we shift large quantities of corn from food production to fuel production, world hunger could be worsened.

Shifting from exhaustible resources to renewable resources may be a partial solution. Another solution is make greater use of ***inexhaustible resources,*** **Figure 35-6.** The most common of these are solar, wind, and water energy. We can generate electricity using any or all of these three energy sources. However, this will require a large expenditure of money and human energy. It will also cover large tracts of land with solar and wind generators. Additionally, it will take time to develop the technology to fully use these resources. For example, solar-powered automobiles are now an interesting experiment, **Figure 35-7.** Practical vehicles of this type are years away.

An immediate solution to our energy problems is to *use energy more efficiently.* This will require that people think about

Figure 35-6. The future will require a shift from exhaustible energy sources, such as petroleum, to renewable resources such as wood, and inexhaustible sources like solar power.

and change their lifestyles. Members of society will have to ask themselves a number of very difficult questions: Should people drive to work alone in a personal car? Should people heat their homes to 75°F (24°C) in the winter and cool them to 65°F (18°C) in the summer? Should we make buildings more costly by using more insulation and installing double- and triple-glazed windows? Should people be strongly encouraged, through taxes or fees, to use public transportation instead of their cars? Should gas-guzzling cars be removed from the market?

Environmental Protection

Open space, clean air, land for a home or farm, and safe drinking water were once viewed as a birthright for people in the U.S. and Canada. We simply expected them to be available. Today, we know better. We have discovered that unwise use of technology can threaten our quality of life.

This understanding was slow to come. These issues have been brought into sharper focus in recent years. People have participated in what can be called an "environmental revolution." It has resulted in three basic principles:

Figure 35-7. Solar-powered cars like this one may be a practical transportation vehicle in the future. (AC-Rochester)

Figure 35-8. A healthy environment, with unpolluted air and water, is essential for a good quality of life.

- The natural environment has a direct effect on the safety and health of people, **Figure 35-8.**
- The long-term survival of any civilization is based on wisely managing natural resources.
- A healthy natural environment is essential for human life.

Protecting the environment involves studying the relationship between human population and the use of technology. This relationship directly impacts a number of environmental conditions. These include the climate, the supply of food, water, energy, and material resources.

Many scientists say that we are creating an environmental crisis. This means that we must take action or the environment will be permanently damaged. Three important forces that are contributing to this crisis are overpopulation, resource depletion, and pollution.

Overpopulation

From the dawn of civilization to the end of the 1700s, world population grew to about one billion people. During the 1800s, the population increased at higher rates. It reached a total of about 1.7 billion people by 1900. World population grew even more rapidly during 1900s. The U.S. Census Bureau reported that the world's population reached three billion about 1960, four billion around 1975, five billion about 1987, and six billion by 1999. It predicts that the population will reach seven billion in 2013 and eight billion in 2028.

The world's population is not evenly distributed. The number of people per square mile of cultivated land varies greatly. In Europe, this density is three times that of North America. In south Asia it is four times, and in China is eight times that of North America.

This is a "good news/bad news" situation. The good news is that people are living longer. Technology has given us more food, better health care, better disease control, and better sanitation. The bad news is that this is true only in developed countries. "Third world" countries are experiencing most of the population growth. Ninety-six percent of world population increase now occurs in the developing regions of Africa, Asia, and Latin America. The economies of these countries cannot support rapidly growing populations. This leads to tremendous hardships. Many people in third-world countries go to sleep hungry each night. Millions of people starve to death each year and still more die of disease.

As we continue to control diseases, use diplomacy instead of war to solve international conflicts, and improve health care, the population will grow even faster. The increased population will contribute to the other two problems: pollution and resource depletion. Concerted effort will be needed to limit growth of the world's population.

Resource Depletion

Each living person places demands on the resources of the planet. Many material resources have a finite supply. Once used,

Figure 35-9. Farming is one source of renewable material resources and food.

they are gone *forever.* These resources include metal ores, petroleum, natural gas, sulfur, and gypsum.

As with energy resources, we can shift from using nonrenewable resources to employing renewable resources. These are primarily the food and fiber produced by farming, forestry, and fishing, **Figure 35-9.** However, the fertility and availability of land and the productivity of the oceans limit their supply. If we take too much from the water and land, they will be damaged.

An example of "taking too much" is the intensive forestry practices used in many parts of the world. The logs, limbs, and bark of the tree are taken away to be processed into products; nothing is left behind. This may seem wise, but it is not. The forest floor is left without the limbs and the bark. These parts of the tree would normally break down and provide nutrients for other plants and trees. As a result, the soil becomes less fertile. Coupled with the acid rain that is falling on many forests, this produces trees that are not as healthy as before. The production of wood fiber is reduced.

Likewise, intensive farming practices deplete the soil. This makes heavy applications of commercial fertilizers necessary. Also, using prime farmland for housing forces us to cultivate less productive areas. We also strain precious water resources to irrigate crops growing in the desert. We draw the water from rivers and wells. This, in turn, reduces the amount of water downstream and depletes aquifers. (***Aquifers*** are underground water-bearing layers of rock, sand, or gravel.)

A problem related to these issues is land use. Throughout the country, conflicts are raging between environmentalists (people who are concerned about leaving the environment unspoiled) and commercial interests, **Figure 35-10.** An example of one controversy is the spotted owl habitat in the forests of Oregon and Washington. Commercial interests want to cut virgin timber and manage the forest for maximum fiber production. Environmentalists want most of these forests left untouched as a habitat for the owls.

Debate continues over how much wilderness is enough. Some people want large tracts of range and forest land to

Figure 35-10. There is a continuing debate over land use in the United States. On one side are those who want land kept in a wilderness state. On the other side are people who want more land for lumber or agricultural production.

remain untouched. They want only hiking trails in these areas. Few people dispute the need to save some unspoiled areas. The debate is over the size and number of these areas. The difficulty with these debates is that there are no clear-cut answers. There are only opinions of what is right or wrong, good or bad.

As with energy, short-term solutions to resource depletion lie in better use of our material, land, and water resources. We must use our land, but not abuse it. Some of it should be set aside as nature preserves. Other areas should be considered multi-use land. These would combine recreational and commercial uses. Hunters, campers, ranchers, farmers, and loggers would all use the land. Finally, some tracts of land would be devoted to commercial, residential, and transportation uses.

We must reconsider how we use the earth's resources. We must learn to use all materials more efficiently. This will mean buying fewer items that we really don't need. Buying products that will last becomes very important; so will maintaining and repairing them instead of discarding them. Finally, when a product can no longer be used, the materials must be recycled.

Pollution

People living in some major cities rarely see clear skies. Their vision is often reduced by haze and their health is affected by smog. Water in many parts of the world is unsafe to drink. In some areas, the land is contaminated with hazardous waste. People cannot live on this land or travel over it. All of these things are called ***pollution,*** **Figure 35-11.** Pollution is most often a product of human activity.

Pollution sometimes has been brought to our attention in dramatic ways. In 1969, the Cuyahoga River in northern Ohio literally caught fire. The river, which runs through the city of Cleveland, had become severely polluted with debris and flammable liquids. The incident made many people more aware of the pollution issue. It led to passage of new laws to protect the waterways, and efforts to clean up streams, rivers, and lakes.

Pollution affects more than the air we breathe, the land we walk upon, and the water we drink. Many scientists think that it is changing the climate of our planet. In the spring of 1983, scientists observed a brown layer of air pollution over the Arctic region. Later, other scientists discovered a

Figure 35-11. Pollution can affect the land, water, and air that we depend on for life.

hole in the ozone layer over the Antarctic. This hole has grown larger over the years. Since that time a controversy has raged. Some scientists say that the hole will let more ultraviolet light into the atmosphere. They believe these rays, combined with increased levels of carbon dioxide and other gases, will cause the earth to retain more heat. This will cause global warming with higher land and water temperatures. Scientists are concerned that a warmer climate will melt the polar ice caps and cause the oceans to rise. In turn, large coastal areas, including some major cities, would be flooded. Scientists call this problem the ***greenhouse effect.***

Other scientists feel that global warming is part of a natural cycle. They say that warming and cooling has occurred a number of times over history. No one is sure what will happen. However, it is an area of great concern. Changes have been called for in our technological actions. Certain chemicals in air conditioners, refrigerators, foam containers, and aerosol sprays are blamed for part of the problem. These chemicals are being removed from the market as replacements are developed.

A plan is needed to protect the atmosphere and the ozone layer. Failure to act could cause a significant change in the climate of our world. The greenhouse effect could turn lush farmland into desert, cause the extinction of some species of plants and animals, and create widespread human suffering.

Global Economic Competition

A third issue directly related to people and technology deals with the distribution of wealth and industrial power. Some historians suggest that technical knowledge and power has moved steadily around the world from east to west. At one time, China was a global power and the source of many innovations. This center of power moved across what is now India into the area around the Mediterranean. That area was the dominant economic center about 2000 years ago. Later, northern Europe became the economic leader with the Industrial Revolution of the 1800s. This was followed by the dominance of the United States in the 1900s. Now we are seeing the area called the Pacific Rim become more important. Japan, Korea, China and other Far East countries are challenging the industries of North America and Europe. This represents nearly a full cycle of industrial development around the globe.

Figure 35-12. People often work together in teams to design and produce better products. (Texaco)

Today, we understand economics better than at any time in history. This allows countries on the back side of the economic wave to resist losing economic power. They can take a number of actions, including:

- **Changing management styles.** Companies are redefining the role of workers and managers. They are creating teams that design and produce products, **Figure 35-12.** These teams include designers, engineers, production workers, quality control specialists, marketing people, and managers. Managers are not seen as bosses. Managers help others do their work better and with greater ease.
- **Increasing their use of computers.** Competitive companies are assigning routine work to computers. This means there is less manual labor and more computer-aided or computer-controlled work, **Figure 35-13.** We now hear about an "alphabet" of computer actions:

Figure 35-13. In many business and industrial operations, manual methods have been replaced by computer applications.

- CAD (computer-aided design) systems make drawings easy to produce, correct, and store.
- CIM (computer-integrated manufacturing) ties many manufacturing actions to the computer. Machine control, quality control, parts movement, and an array of other operations are monitored by computer systems.
- Just-in-time (JIT) inventory control. This computer system monitors material orders so that supplies arrive at the plant just before they are needed. Also, finished products are made only when they are needed.
- Many purchasing and warehousing operations are also computer-controlled. These systems include warehousing systems in which robot vehicles store and retrieve parts.
- **Producing world-class products.** Successful companies now make products that meet the needs of customers around the world. This means that the products must function well, be fairly priced, and deliver excellent value to the customer. No longer can a local or national area provide a safe, protected market for a company. Political forces are causing the world to become one large market for all countries.
- **Using flexible, automated manufacturing systems.** Traditional manufacturing is based on long production runs. Semiskilled workers on a manufacturing line made a single product. This type of production is being replaced by automated production lines called flexible manufacturing systems, **Figure 35-14.** Such computer-controlled systems can produce a number of different products with simple tooling changes. Flexible manufacturing is cost-effective for small quantities of products.

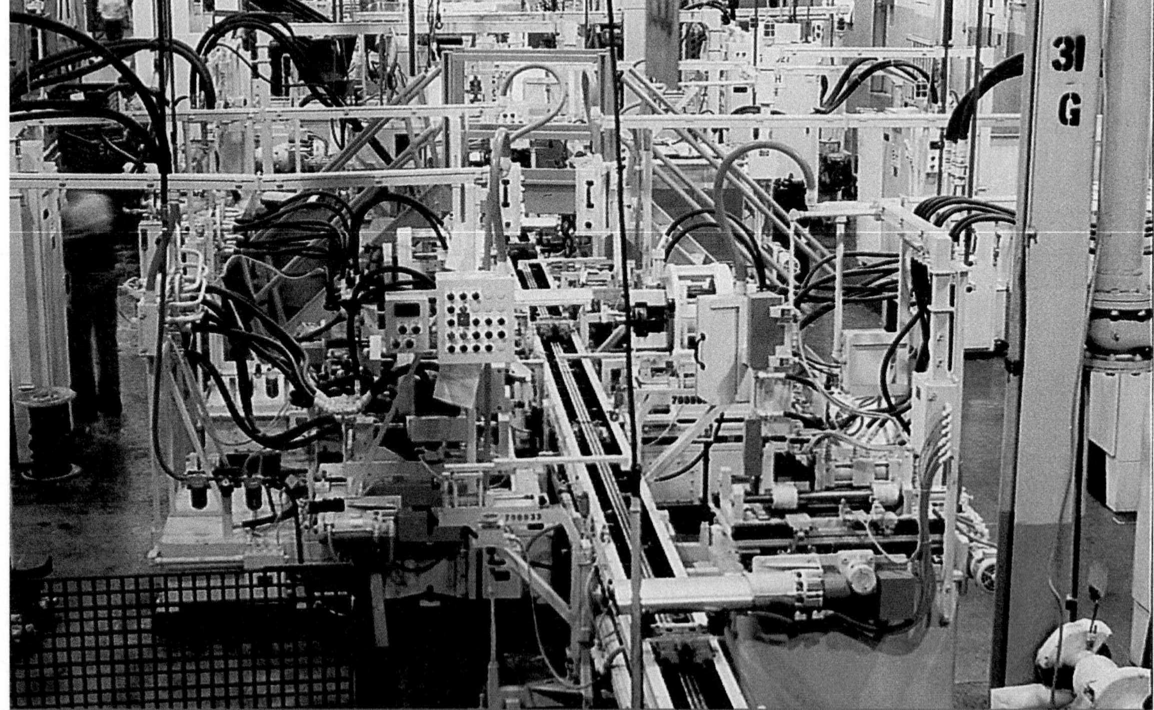

Figure 35-14. Today, products are often produced on flexible, automated production lines. Computer control allows fast, simple changeover to produce different types of products. (Arvin Industries)

Summary

Technology and human life cannot be separated. We use technology, depend on technology, and feel we must have technology. It lets us travel easily, communicate quickly, and live in comfort. Technology has also caused us concern. Poor application of technology pollutes the air, water, and land around us. It can threaten our very lives.

The challenge facing all people is to determine the type of future we want. Then, we must develop the technology that will promote that future. Likewise, we will have to change our personal habits and preferences. Some technologies that we now depend upon may need to be modified or abandoned. Only then can future generations enjoy a good quality of life.

Key Terms

aquifers
biological future
convergent thinking
divergent thinking
finite
futures research
futuring
global impacts
greenhouse effect
human-psyche future
inexhaustible resources
nonrenewable resources
pollution
renewable resources
social future
technological future

Test Your Knowledge

Write your answers on a separate piece of paper. Please do not write in this book.

1. *True or false?* Technology, by itself, does not affect people or the environment.
2. Name at least three destructive natural forces that impact humans.
3. Hydroelectric dams and "wind farms" are examples of harnessing ________ forces.
4. North America uses a large percentage of the world's petroleum production. What product of technology accounts for a major portion of that consumption?
5. Describe the two types of thinking used in the futuring process.
6. Which of the following is *NOT* one of the types of future considered in futuring research?
   a. Social
   b. Economic
   c. Technological
   d. Biological
   e. Human-psyche
7. Some people feel threatened by _______; others feel empowered by it.

8. What are the three forces contributing to the current environmental crisis?
9. Northern Europe became the center of technical knowledge and power during the _______ Revolution of the 1800s.
10. Companies are adopting new, computer-based technologies, such as CAD, CIM, and JIT to remain competitive in the global economy. What do those three abbreviations stand for?

Applying Your Knowledge

1. Choose an environmental problem and gather information about it. Summarize your knowledge on a form like the one shown.
2. Prepare a drawing or a photomontage that shows an "ideal" future that you would like to live in.

Problem:
Factors that cause or contribute to the problem:
Possible solution #1:
Problems that solution #1 might cause:
Possible solution #2:
Problems that solution #2 might cause:

Career Corner

Public Relations

Public relations involves managing the public image of a company or an individual. The public relations personnel deal with all aspects of communicating an organization's message to the general public and specialized segments of the population. The field of public relations is growing, as it broadens into the area of international communications.

People entering this field must have effective writing skills and a background in media production. Many public relations personnel are university graduates with majors in journalism, mass communication, or business. Careers in public relations include publicity managers, press agents, lobbyists, corporate public affairs specialists, and public opinion researchers.

Chapter 36
Technology: A Personal View

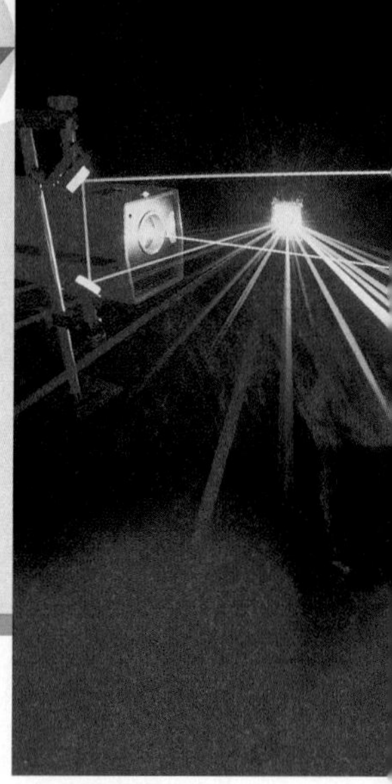

Learning Objectives

After studying this chapter, you will be able to:

- Describe how jobs have evolved from the colonial period, through the Industrial Revolution, to the information age.
- Discuss the typical levels of technology-based jobs and the types of education required for each.
- List the factors that should be considered in selecting a job.
- Identify the three factors affecting job satisfaction.
- Describe the ways that individuals exercise control over technology.
- List some technological activities introduced as fiction that have become actual processes.

Your personal life is highly dependent on the technology people have developed. In the brief span of 50 years, life has changed dramatically for citizens of developed countries. We live in different housing, travel on different systems, have different products to purchase, and communicate in ways far different from in the past. To help investigate the changes affecting everyday life in the twenty-first century, let us look at five areas:

- Technology and lifestyle
- Technology and employment
- Technology and individual control
- Technology and major concerns
- Technology and new horizons

Technology and Lifestyle

Each person lives in a specific way, or has a lifestyle. A ***lifestyle*** is what a person does with business and family life—their work, social, and recreational activities. See **Figure 36-1.**

Colonial Life and Technology

The lifestyle during America's colonial period was a harsh contrast to that of today. Housing was simple and modest. Most products were designed and produced to meet basic human needs. There

Figure 36-1. What kind of lifestyle do you think this father and son enjoy?

were few decorative items available, and those items were mostly owned by the wealthy. Transportation systems included horses, animal-drawn wagons, and simple boats. The communication systems available were crude.

Most people lived on farms. A few people practiced basic crafts. These people were the carpenters, blacksmiths, and other tradespeople needed to produce basic products the community required. The men and boys did most of the field work on the farm and practiced the trades. Women and girls tended gardens and did household work, **Figure 36-2.**

Everyone had to work long hours, six days a week, to raise small amounts of food. The seventh day was set aside by most colonial families for church activities. People did not take vacations, and they celebrated few holidays.

Figure 36-2. This woman is demonstrating the old craft of candlemaking, a common task in colonial life.

With the exception of slaves, however, most people were their own bosses. They owned their farms and stores, or they practiced their crafts as independent workers. Each person was an owner, manager, and worker, rolled into one.

The Industrial Revolution

The Industrial Revolution of the mid-1800s changed this lifestyle. A series of events took place. First, advanced technology was developed for the farm. This technology included the moldboard plow, the reaper, and the steam tractor. These and other devices made farmers more efficient. More land could be farmed, and more food could be produced, by fewer people, to sustain the society. Second, the movement to the land west of the Appalachians allowed for larger and more efficient farms. During this period, the percentage of the workforce engaged in farming began dropping rapidly. At the start of the Industrial Revolution, more than 90 percent of the workforce was engaged in farming. Today, fewer than 3 percent of workers are employed on farms. Third, a large number of people from Europe immigrated to the United States during this time. These immigrants, plus the farmers who were no longer needed to till the soil, provided a vast labor supply.

This labor supply was a basic resource for the factory system that was then being developed. The local tradespeople working in their shops could no longer meet the demand for goods. Centralized manufacturing operations were replacing the functions of the tradespeople. Such operations included several important factors, **Figure 36-3:**

- *Professional management,* which established procedures, employed resources, and supervised work.
- *Division of labor,* which assigned portions of the total job to individual workers. Each worker did only part of a total job, allowing them to quickly develop the specialized skill needed to do the assigned task.
- *Continuous manufacturing techniques,* which increased production speed. Raw materials generally entered a production line at one end, and finished products left the line at the other end.
- *Material handling devices,* which were

Figure 36-3. Many current manufacturing plants are based on the principles of the Industrial Revolution. (DaimlerChrysler Corp.)

used to move the products from workstation to workstation. Workers remained at their stations, and the products moved to them.

- *Interchangeable parts,* which allowed production of large quantities of uniform products.

Low wages and poor working conditions in the factories caused widespread worker unrest. Labor unions were formed to give the workers a voice in determining working conditions and pay rates. Bloody battles erupted between the workers and management. The government usually supported managerial positions, resisting attempts of the unions to deal with the issues.

These conflicts were finally settled with changes in governmental attitudes, new laws, and different management stances. This led to a strong industrial period for the country. The period was characterized by broad employment opportunities. The workers enjoyed a high standard of living. The 40-hour, five-day workweek with a number of holidays and paid vacation time became fairly common.

Figure 36-4. This robot is placing cartons on a pallet. This task was formerly done by semi-skilled workers. (Cincinnati Milicron)

The Information Age

The development of the computer changed the Industrial Revolution. During the industrial period, the company that could efficiently process the greatest amount of material was the most successful. This required major investment in large continuous manufacturing plants. These were characterized by the huge automobile and steel manufacturing operations. They employed thousands of people and used millions of tons of materials.

The computer allowed development of a new type of manufacturing, *flexible manufacturing*. This type of manufacturing can quickly and inexpensively respond to change. People with few skills are replaced with computer-controlled machines, **Figure 36-4.** The workers who remain have more training and motivation to work. They accept change and responsibility more readily than the workers of the Industrial Revolution.

Also, management has changed greatly. Management is less distant from the workers. The entire workforce is seen as a team, with each person having an area of responsibility, **Figure 36-5.** Managers may be responsible for setting goals and controlling money. Workers are responsible for producing products. Everyone is responsible, however, for work procedures and product quality.

These changes have given people a new lifestyle. Those who are able to change and adjust to the demands of the new age can live very differently. They are better informed, work more with their brains than their muscles, travel greater distances, and have more control over their work, **Figure 36-6.**

Figure 36-5. In today's industry, managers and workers work as a team. They cooperate to reach company goals. (Air Products Co.)

Figure 36-6. Workers in modern industry more often work with their brains than with their muscles.

Technology and Employment

Lifestyle and employment are closely connected. Most people need the money they earn through working to afford the type of life they want. Some general requirements for most jobs in the future can be identified. Fundamental to almost all jobs will be a high school education. Specialized technical training beyond that level quite often will be needed, **Figure 36-7.** This may be technical training in career centers and community colleges or a university education. In addition, workers in the information age must be willing to do the following:

- Pursue additional education and training throughout their work lives
- Accept job and career changes several times during their work lives
- Work in teams, and place team goals above personal ambitions
- Exercise leadership, and accept responsibility for their work

Figure 36-7. Specialized training beyond high school is required for many technical jobs. (General Electric Co.)

Types of Technical Jobs

There will be a wide variety of jobs requiring technical knowledge. They include five levels, as shown in **Figure 36-8.** The first is the technically trained ***production worker.*** These workers include the people who process materials and make products in manufacturing companies, erect structures, and operate transportation vehicles. ***Technicians*** work closely with production workers, but they do more specialized jobs. Typically, technicians set up and repair equipment, service machinery, conduct product tests and laboratory experiments, work in dental and medical laboratories, and do quality control testing. ***Technologists*** are highly trained technical employees. They form the bridge between

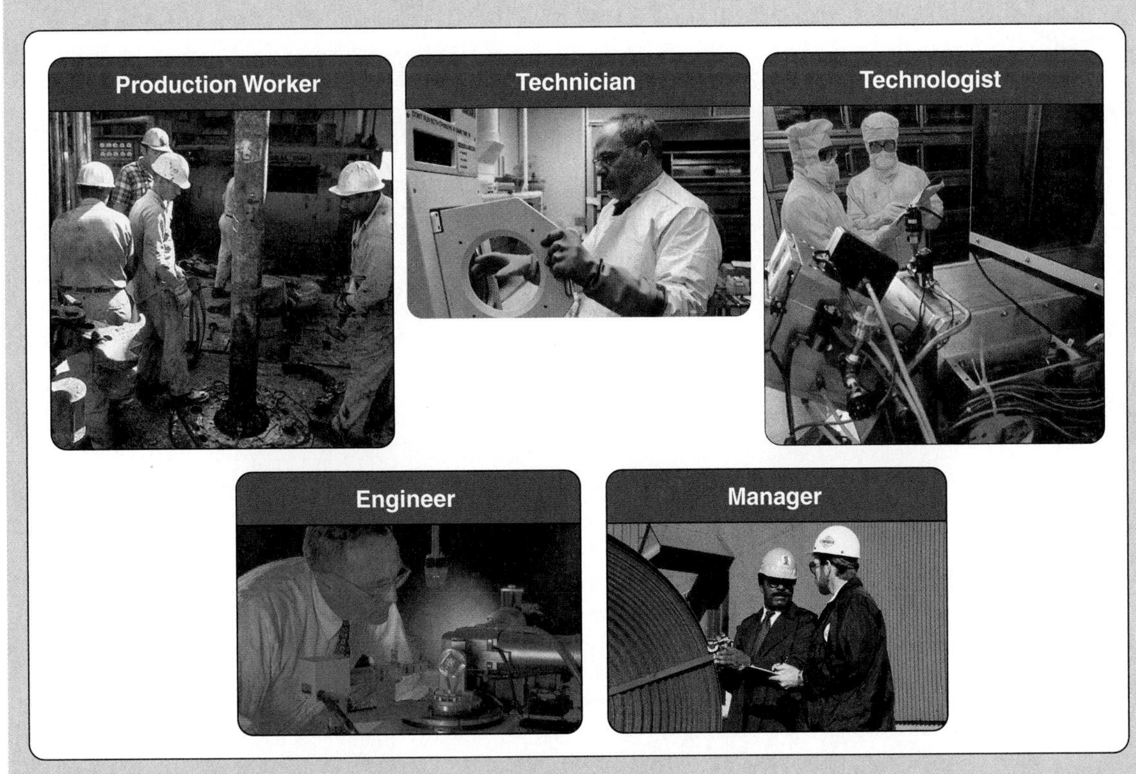

Figure 36-8. People with technology backgrounds or an aptitude for technology can become production workers, technicians, technologists, engineers, or managers. Many people will work in more than one of these areas during their careers. (American Petroleum Institute, Goodyear Tire and Rubber Co., AT&T, Inland Steel Co.)

the engineers who design systems and the workers and technicians who must implement them. Many companies have a need for people with advanced technological knowledge. ***Engineers*** design products and structures, conduct research, and develop production processes and systems. Technically trained ***managers*** can set goals, plot courses of action, and motivate people to work together. They are people-oriented leaders who also have technical knowledge.

Selecting a Job

People should not just take the first job they can find. Whenever possible, a job should match your interests and abilities. When deciding on employment opportunities, you should consider at least three factors: lifestyle, job requirements, and job satisfaction.

Lifestyle

Each of us wants to live comfortably. A person's job has a direct impact on both life at work and life away from it. Some people like to travel and meet new people. An industrial sales job could provide these aspects. A factory job that pays well and has good holiday and vacation benefits can also meet this need.

Job Requirements

A variety of job requirements affect each employee. In selecting employment, a person should consider job requirements from three important points. The first requirement considers your freedom to organize the tasks assigned to the job and the level of accountability that goes with the job. The second consideration is the balance of working with data, machines, and people, **Figure 36-9.** The last factor is the level of education a job requires.

Job Satisfaction

Job satisfaction is a description of how happy a worker is with his job. Three factors strongly affecting job satisfaction are values, recognition, and pay. A job should match the values of the person doing it. Some jobs allow for more visible recognition than others. The money you receive in exchange for your work is also important.

Job Skills and Employment

Each person seeking a job has a set of job skills. These are the activities a person does well. Each job has a set of skill requirements. For example, an engineer must be able to use mathematics, science,

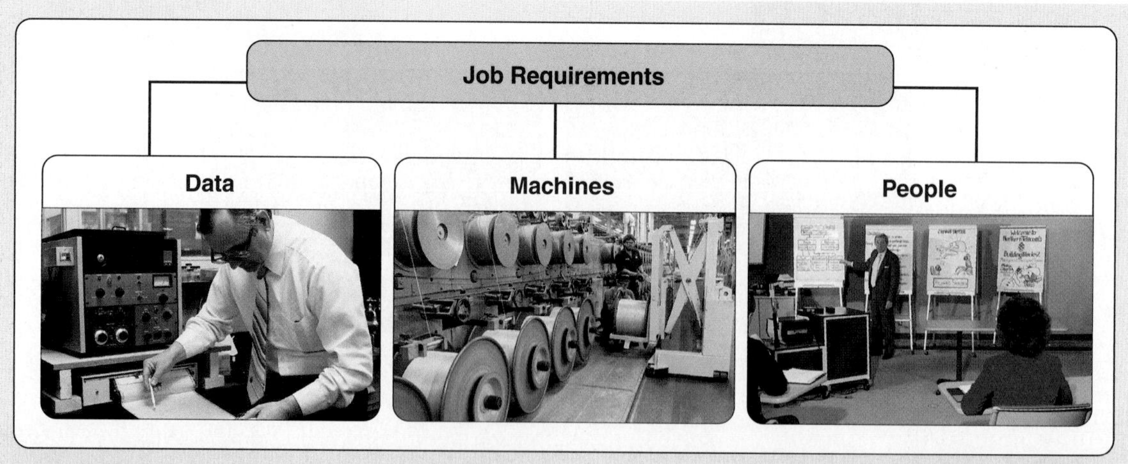

Figure 36-9. In industry or business, each job will deal, in varying degrees, with machines, people, and data. Which areas do you prefer? (Goodyear Tire and Rubber Co.)

and technology to solve problems. An accountant possesses mathematics and accounting skills. A taxi driver needs to know the city's streets, major hotels, and tourist sites and how to drive a car. The challenge is for people to find jobs that match their own sets of skills.

In addition to specific job skills, all applicants need general job skills. General job skills are skills that can be used in many different jobs. They are skills that can be taken from one job to another. These skills are transferable skills. General job skills should be developed during grade school and high school. Most specific job skills are gained through additional education and training activities.

General job skills can be divided into six major groups: language and communication skills, thinking skills, information skills, socioethical skills, people skills, and personal skills. ***Language and communication skills*** are the abilities to read, write, and speak the language. These skills include the following:

- **Reading.** Reading skills involve the ability to locate text information and then identify and comprehend relevant details.
- **Writing.** Written communication skills include the ability to write clearly and for a variety of audiences and purposes.
- **Speaking.** Oral communication skills involve the ability to explain and present ideas in clear, concise ways to a variety of audiences.

Thinking skills involve the ability to use mental processes to address problems and issues. These skills include the following:

- **Problem solving.** Problem-solving skills include the ability to recognize a problem, create possible solutions, implement a selected solution, and evaluate the solution.
- **Creative thinking.** Creative thinking skills involve the ability to use imagination to develop and combine ideas or information in new ways.
- **Decision making.** Decision-making skills include the ability to identify a goal and alternate courses of action, evaluate the advantages and disadvantages of each alternative, and then select the best action.
- **Visualizing.** Visualization skills involve the ability to envision a three-dimensional object presented by an engineering drawing or a schematic drawing.

Information skills are the abilities to locate, select, and use information. These skills include the following:

- **Information literacy.** Information literacy skills include the ability to locate, evaluate, use, and cite information properly.
- **Information technology.** Information technology skills involve the ability to use computer systems to acquire, organize, and present information. See **Figure 36-10.**

Socioethical skills involve understanding the implications of actions on people, society, and the environment. These skills include the following:

Figure 36-10. Information technology skills are required for most jobs.

- **Ethics.** Ethical skills include the ability to see the implications of actions on people, society, and the environment and then to act in legally and morally responsible ways.
- **Sociability.** Social skills involve the ability to show interest in, understanding of, and respect for the feelings and actions of others.

People skills are the abilities needed to work with people in a cooperative way. These skills include the following:

- **Leadership.** Leadership skills include the ability to encourage and persuade people to accept and act on a common goal.
- **Teamwork.** Team skills involve the ability to work cooperatively with others to complete a project. See **Figure 36-11**.
- **Diversity.** Diversity skills include the ability to work with people from different ethnic, social, or educational backgrounds.

Personal skills involve the ability to grow and manage personal actions on a job. These skills include the following:

- **Self-management.** Self-management skills involve the ability to manage time, adapt to change, be aware of personal responsibilities, and maintain physical and mental health.
- **Self-learning.** Self-learning skills include the ability to recognize the need for new knowledge and seek ways to meet the demand.
- **Goal.** Goal skills involve the ability to set goals and work toward meeting them.

Figure 36-11. Teamwork is an essential general job skill.

In addition to general and specific job skills, employers look for proper work attitudes. Employers look for cooperation, dependability, a good work ethic, and respect. This means the employee is expected to do the following:

- Cooperate with supervisors, other employees, and customers.
- Arrive at work on time and complete tasks in a timely manner.
- Put in an honest day's work for a day's pay.
- Show respect for others, the company, and himself.

Today's workplace emphasizes *equality*—that is, the idea that all employees are to be treated alike. *Harassment* (an offensive and unwelcome action against another person) and *discrimination* (treating someone differently due to a personal characteristic such as age, sex, or race) are not tolerated. These negative behaviors often result in termination of employment.

Technology and Individual Control

Technology holds great promise and hidden dangers. People will make the difference, **Figure 36-12.** We often say *they* should control this, or *they* should do that, or *they* should stop doing something else. The harsh reality is that "they" will never accomplish anything. Only when someone says "I," instead of "they," does anything meaningful get done.

Figure 36-12. This quiet fishing lake was once a strip mine. After the coal was removed, people worked together to restore the area. (American Electric Power Co.)

The future lies in the hands of people who believe they can make a difference. Examples such as Thomas Edison, the Wright brothers, and George Washington Carver might come to mind. These individuals did not wait for a group to do something. They pursued their own visions of what was important and needed by society. This type of action requires people to understand technology. People must also comprehend the political and economic systems directing technology's development and implementation.

Individuals control technology in a number of ways. The first is through the role of the ***consumer.*** This involves the following:

- Selecting proper products, structures, and services
- Using products, structures, and services properly
- Maintaining and servicing the products and structures they own
- Properly disposing worn or obsolete items

Consumer action causes appropriate technology to be developed, **Figure 36-13,** and helps inappropriate technology disappear. When a product or service does not sell, its production stops.

An individual also has ***political power.*** All companies must operate under governmental regulations. These regulations include the Occupational Safety and Health Act, the Pure Food and Drug Act, the Environmental Protection Act, and the Clean Water Act. Few laws are passed solely because an elected official thinks they are important. Most of them come from people asking their elected representatives to deal with a problem or concern.

Finally, an individual can make a difference as an activist. ***Activists*** use public opinion to shape practices and societal values. Some people see this as a bad thing. They are disturbed by the actions and words of people whose views are different from their own. Actually, an activist is simply a person practicing his or her constitutional rights to freedom of speech and assembly. As a citizen, you have the ability to meet with others to discuss and promote your point of view on a subject.

Technology and Major Concerns

Technology concerns many people. People are worried about pollution, energy use, and unemployment. Technology is designed, produced, and used, however, by people. These issues can only be resolved through human action. A few areas of

Figure 36-13. The types of products and structures you demand as a consumer make a difference. Solar heating, for example, is becoming increasingly popular. This solar-heated home will save energy without harm to the environment. (U.S. Department of Energy)

concern are at the forefront of our attention. These concerns include the following:

- **Nuclear power and waste disposal.** Is nuclear power generation an appropriate activity? How do we design and build safe and reliable nuclear power plants? Can the radioactive waste nuclear power plants produce be disposed of safely?
- **Technological unemployment.** Should technology that causes unemployment be applied? Does a company have a responsibility to provide training and other benefits to workers who become unemployed through the adoption of new technologies? Should foreign products that cause technological unemployment be barred from domestic markets? What are the individual worker's responsibilities in seeking training to deal with changing job requirements?
- **Genetic engineering.** Is it right to change the genetic structure of living organisms? Who will decide what genetic engineering activities are appropriate? Is it acceptable to alter the genetic structure of humans? How are religious and technological conflicts going to be dealt with?
- **Energy use.** How can we reduce our dependence on petroleum? Can society's reliance on the private automobile be changed, **Figure 36-14?** What alternate energy sources and converters should be developed? How can we control the environmental damage burning fossil fuels causes? Should mass transit be financed with tax money? Should we place high taxes on gasoline?
- **Land use.** What are the rights of landowners? Should the desires of the majority overrule individual

landowner rights? What responsibilities do governmental officials have for public lands? How do you balance environmental protection issues with economic issues, **Figure 36-15?**

- **Pollution.** Should products polluting the environment be banned from manufacture and use, and should strict pollution controls apply equally to individuals and companies? What type of evidence is needed before a product can be banned? Should there be a pollution tax on fuels that damage the environment? How do you handle the economic and social impact of banning products? Should we limit the use of wood as a fuel, and should the solid waste (garbage) from one state be allowed to enter another state?

This book does not attempt to provide answers to these or other technological problems. Remember, the *right answer*

Figure 36-14. Greater use of mass transit will be needed in the future to minimize environmental damage and relieve traffic congestion in urban areas.

Figure 36-15. Economic considerations must be balanced with concern for the environment as population continues to grow.

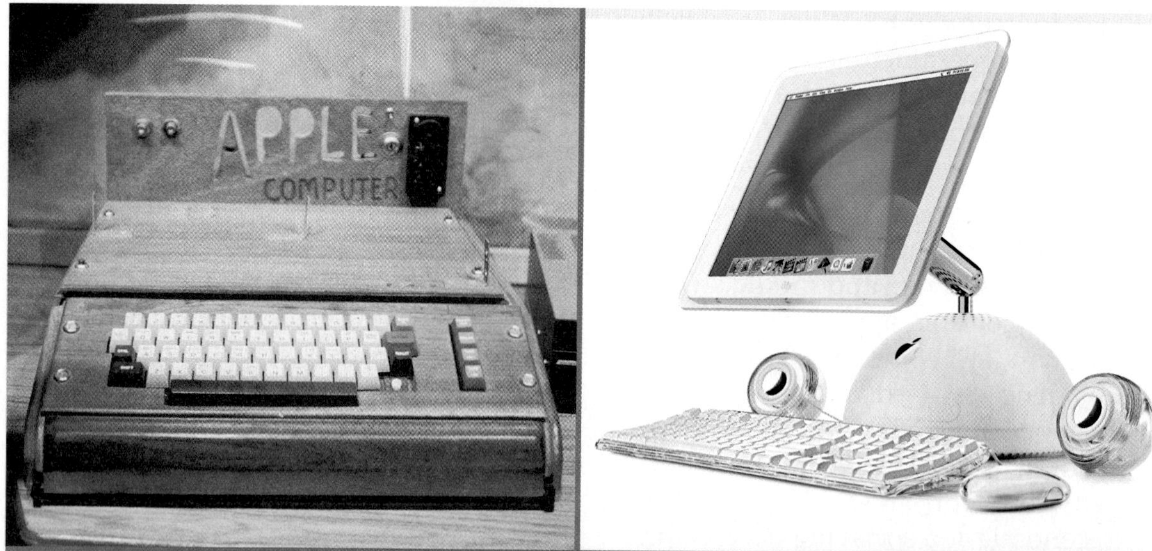

Figure 36-16. In only about 30 years, the Apple computer moved from crude prototype to a polished, powerful microcomputer. (Apple Computer)

according to one person or group is often rejected by other people and other groups. The best we can hope for is *an answer* most people will support.

Technology and New Horizons

Technology continues to be developed at a very rapid pace, **Figure 36-16.** Many ideas that seem impossible now will be commonplace in the near future. This is not new. In the 1800s, Jules Verne wrote a fictional story called *20,000 Leagues Under the Sea.* It dealt with the then-impossible feat of traveling under the ocean in a submarine. In the 1940s and 1950s, people read a fictional comic strip in which the characters used the then-ridiculous means of rocket ships to travel in space. What is fiction today might become an everyday part of life for your children and grandchildren.

Many futuristic ideas have been proposed. One is mining the resources of outer space and our oceans. Will we have colonies on the moon or other planets that extract and process precious mineral resources? Our oceans are the last of the vast resource beds on earth. Will they be mined, or will concerns regarding pollution keep them off-limits?

Figure 36-17. In the future, we may build complex manufacturing systems in space. (NASA)

Manufacturing in space, **Figure 36-17,** has been proposed and tried on a small scale. Experiments have shown that protein crystals can be grown in the ***microgravity*** (very low gravity) of space. Earth-grown crystals are often small and flawed. Those

Career Corner
Technology Education Teachers

A teacher must have the ability to effectively organize and deliver material to students. Communication skills are necessary to facilitate comprehension and understanding, whether the subject is technology, mathematics, science, reading, or English. Effective teachers are good communicators. Technology education teachers must possess all the skills of a good teacher and know and understand how people have developed, produced, and used tools, materials, and machines to control and improve their lives. These teachers are developed in technology teacher education programs in a number of universities and are certified to teach by the state in which they work.

grown in space can be larger and more complex. Some experiments have dealt with growing zeolite crystals. These crystals might be used in portable kidney dialysis machines and in cleanup efforts with radioactive waste. Earth's atmosphere and gravitational pull adversely affect some manufacturing processes. Will these processes be moved into space, and will we have space stations where manufacturing is routinely done?

A third future technological activity is commercial space travel. To date, space travel is government-financed. Most of it is restricted to military and scientific missions or communication satellite launching. A few private citizens have paid to have a ride into space. Will we routinely be traveling into space and back some day, and will future generations take vacation trips to the moon or nearby planets?

These are only a few examples of possible technological advancements. Society has moved from the horse-drawn wagon to space travel in a single lifetime. What could be next?

Summary

Technology is as old as humanity. Technology helps us lead better lives by enhancing our abilities. Technology provides better jobs and helps us evolve new opportunities. Technology also threatens us with pollution, a potential loss of personal worth, and possibly even nuclear disaster. How we design, produce, and use technology makes the difference. The choices are yours to make. You now know more about technology. Using technological knowledge, you can select products wisely, pursue appropriate employment, and better understand societal issues. You can be a better consumer, worker, and citizen.

Key Terms

activist
consumer
engineer
information skills
language and communication skills
lifestyle
manager
microgravity
people skills
personal skills
political power
production worker
socioethical skills
technician
technologist
thinking skills

Test Your Knowledge

Write your answers on a separate piece of paper. Please do not write in this book.

1. Before the Industrial Revolution of the mid-1800s, more than ________ percent of the workforce was engaged in farming.
2. What two factors made a large labor supply available for factories during the Industrial Revolution?
3. *True or false?* Most jobs in the future will require specialized technical training beyond high school.
4. What are the five categories of technical jobs described in this chapter?
5. When considering a job, you should look at whether it places emphasis on data, machines, or ________ .
6. List three ways that an individual can exercise control over the type of technology that is developed.
7. List some examples of technology that were once found only in fiction but are now part of everyday life.

Applying Your Knowledge

1. Select a technological idea that is now considered fiction. Write a description of how it will be used as an everyday item or process in the future.
2. Read a book about a person or persons who developed new technology. Write a report on their attitude toward change and criticism as well as their inventions or innovations.

Section 11 Activities

Activity 11A

Design Problem

Background

Technology, by itself, is neither good nor bad. How a technological object is used can create benefits or drawbacks. Each person should strive to use technology wisely to make the future better, protect the environment, and help people live in harmony with nature.

Challenge

Identify an important issue to people today that technology impacts. List the positive and negative factors related to the issue. Develop a plan to maximize the positive features and reduce the negative impacts.

Activity 11B

Production Problem

Background

Humans can control the use of technology. Many people, however, do not know about the impacts technology has on individuals and society.

Challenge

You are the communications director of a citizens group. Your group is concerned about public policy issues. It has determined that people in your community are not participating in the local recycling program. Design a 60-second public service commercial for television. The commercial needs to explain the importance of recycling to your community. Identify the materials that can be recycled and the benefits of recycling.

Equipment

- Storyboard forms
- Pencils and felt-tip pens
- Video camera, recorder, and monitor

Procedure

Designing the Commercial

1. Select the theme of the commercial.
2. List the major points that will be emphasized in the commercial.
3. Develop a storyboard for the commercial. Use photocopies of **Figure 11B-1.**
4. Write a script for the actors.
5. Develop a shot chart for the director and cameraperson to follow.

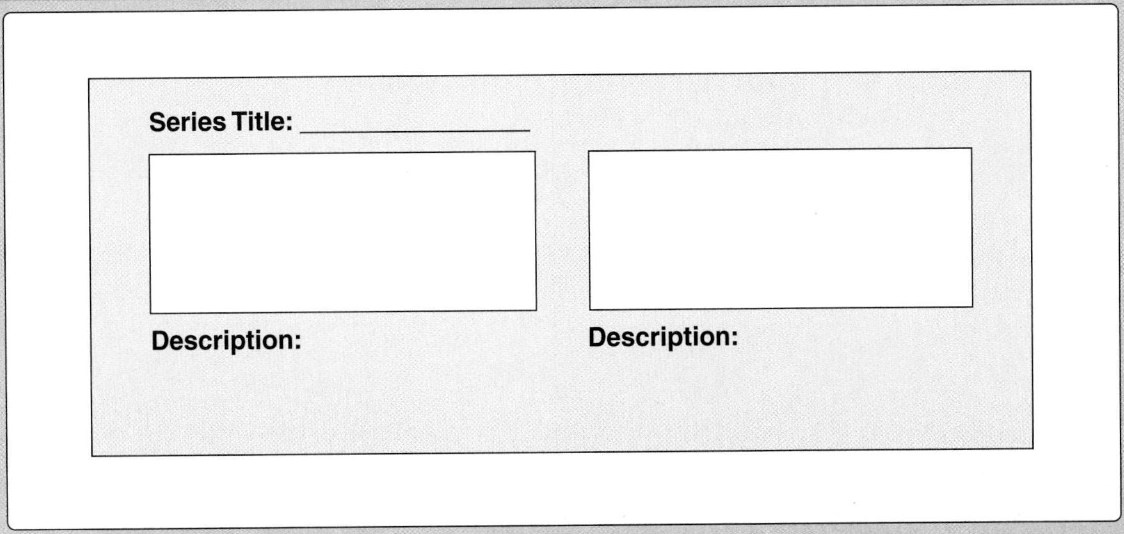

Series Title: ____________

Description:	**Description:**

Figure 11B-1

Producing the Commercial

1. Recruit and select actors for the commercial.
2. Present the script to the actors and have them rehearse their parts.
3. Walk through the commercial with the director. Have the actors and camera operators block (plan) their movements.
4. Record the commercial.
5. Edit in any titles you need.
6. Present the commercial to an audience. Ask for their reactions.

Technology impacts our personal lives in many ways. While it often provides benefits for society, the negative effects it may have on the environment must always be considered.

Technical Terms

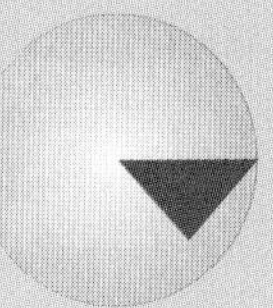

A

absorb: to take in.

accounting: the area in which financial affairs employees keep records of the financial transactions of the company.

active collectors: solar collectors that use pumps to circulate the water that collects, stores, and distributes the heat they convert from solar energy.

activist: an individual who uses public opinion to shape practices and societal values.

actuating: initiating the work related to an action plan.

adhesive bonding: a bonding technique that uses substances with high adhesive forces to hold parts together.

adjusting devices: devices that modify a system to produce better outputs.

adjustment: the step in the repair process in which misaligned parts are adjusted.

advertising: an activity that includes the print and electronic messages that promote a company or its products.

aerobic exercise: exercise that uses oxygen to keep large muscle groups moving continuously.

agricultural and related biotechnologies: the technologies used in growing food and producing natural fibers.

agriculture: people using materials, information, and machines to produce food and natural fibers.

agriculture technology: a technology that uses machines and equipment to help plant, grow, and harvest crops and to raise livestock.

air transportation: systems that use airplanes and helicopters to lift passengers and cargo into the air so that they can be moved from place to place.

airfoil: an object designed to produce some directional motion when in movement relative to the air.

altering: extending the useful life of a product.

alternating current: the flow of electrons in both directions along a conductor, reversing at regular intervals.

alumina: aluminum oxide. It is the input to the second phase of making aluminum.

amplitude: the height of a wave. It measures the strength of the wave.

amplitude modulation: a system that merges a message onto a carrier wave by changing the strength of the carrier signal.

anaerobic digestion: a controlled decaying process that takes place without oxygen.

anaerobic exercise: exercise that involves heavy work by a limited number of muscles.

analytical systems: systems that mathematically or scientifically make comparisons.

animal husbandry: the type of agriculture that involves breeding, feeding, and training animals.

annealing: a process used to soften and remove internal stress in a part.

antibiotics: substances derived from microorganisms that are able to kill other microorganisms.

apogee: the farthest distance away from the earth.

appearance: a factor to consider before buying a product.

applied research: a type of research that seeks to reach a commercial goal by selecting, using, and adapting knowledge gathered during basic research.

apprenticeship training: a program through which highly skilled workers are developed.

aquaculture: growing and harvesting water organisms in controlled conditions.

aquifers: underground water-bearing layers of rock, sand, or gravel.

arbors: spindles or shafts used to hold table saw blades and milling cutters.

arch bridges: bridges that use curved members to support the deck.

architectural drawings: documents used to specify characteristics of buildings and other structures.

area: the size of the surface an object covers.

armature: the inner magnet in an electric motor. It is an electromagnet that can rotate.

articles of incorporation: an application for a corporate charter.
artifacts: objects made by humans.
artificial intelligence (AI): intelligence exhibited by a manufactured device or system.
aseptic packaging: a process that uses heat to separately sterilize a food and its package.
assembling: the manufacturing process in which materials and parts are brought together to make a finished product.
assembling process: a process that connects parts together to make assemblies and products.
assembly drawings: engineering drawings that show how parts fit together.
audience assessment: actions that study audience needs and wants. It is the first step in the process of communication design.
audio: sound.
authority: the right to direct actions.
automatic control systems: systems that can monitor, compare, and adjust a system without human interference.
automatic transmissions: transmissions that use valves to change hydraulic pressure so the transmission shifts its input and output ratios.
avionics: the instruments that help pilots monitor and properly control aircraft.

B

backbones: fiber-optic communication lines built by large communications companies to connect regions in their systems.
balance: the design principle of having the information on both sides of a centerline appear equal in visual weight.
baler: a machine used to gather, compact, and contain hay.
ballast: extra weight added to a locomotive to give it better traction.
band saw: a saw that uses a blade made of a continuous strip or band of metal. These bands usually have teeth on one edge.
barbed-wire fences: fences consisting of wood or steel posts with strands of barbed wire attached to them.
basic research: a type of research that seeks knowledge for its own sake.
bauxite: aluminum ore.
beam bridges: bridges that use concrete or steel beams to support the deck.
benefit: a type of reward provided by a company.
berries: plants grown in many parts of the country and cultivated for their edible parts.
bill of materials: a document containing information regarding the materials and hardware needed to complete a project.
billets: long, square pieces of steel.
biochemical processes: processes that use chemical reactions caused by fungi, enzymes, or other microorganisms.
biofuels: organic material that can be burned or converted into methane.
biogas: a mixture of methane and carbon dioxide produced by the bacterial decomposition of organic wastes and used as a fuel.
biological future: a type of future that deals with the sorts of plant and animal life we want.
biomass: a type of resource having a living origin.
biomass resource: vegetable and animal waste generated through biological actions. It is a source of renewable energy being actively considered as an alternate energy supply.
biotechnology: 1. practices that improve propagation, growing, and harvesting activities. **2.** using biological agents in processes to produce goods or services.
blast furnace: a furnace commonly used in iron smelting.
blimp: a lighter-than-air nonrigid airship.
blooms: short, rectangular pieces of steel used to produce structural shapes and rails.
blow molding: a process using mated dies to make bottles and jars from plastic and glass.
blowout: a dangerous and wasteful occurrence in which oil surges out of a well.
board of directors: a group of people elected to represent the interests of stockholders.
bonding: an assembling process that uses cohesive or adhesive forces to hold parts together.
bonding agent: a material used to achieve bonding.
bonds: debt securities sold by corporations in large denominations.

bow thrusters: maneuvering propellers mounted at right angles to the keels on large ships.

brainstorming: seeking creative solutions to an identified problem.

broadcast frequencies: frequencies above audible sound.

broadcast systems: systems that send radio waves through the air carrying the signal from the sender to the receiver.

Bronze Age: the stage in human history that took place after the Stone Age. During this time, farming developed, villages and towns started to appear, and copper and copper-based metal tools replaced stone tools.

browser: a software program that acts as an interface between a user and the World Wide Web.

bucker: a worker who removes the limbs and tops of trees.

buildings: structures erected to protect people, materials, and equipment from the outside environment.

buoyancy: the upward force exerted on an object immersed in a fluid.

buttress dam: a dam that uses its structure to hold back water.

bylaws: general rules under which an incorporated company will operate.

C

cable-wire fences: fences that consist of 3/8-inch steel wire cables stretched from one anchor post to another.

cam: a pear-shaped disk with an off-center pivot point, used to change rotating motion into reciprocating motion.

canning: a preservation method that puts food into glass jars or metal cans.

cant: the square center section of a log.

cantilever bridges: bridges that use trusses extending out like arms.

capacity: the amount of a substance an object can hold.

carrier frequency: a frequency used by radio broadcast systems that radiates from the transmitter.

casting: a process through which on-camera performers are employed.

casting and molding: a method of shaping parts or products by pouring liquid material into a shaped cavity.

casting and molding processes: processes in which a liquid material is poured into a cavity in a mold, where it solidifies into the proper shape and size.

catalyst: a chemical agent used to cause a reaction.

ceiling: the inside surface at the top of a room.

ceiling joists: beams, resting on the outside walls and some interior walls, that support the weight of the ceiling.

center lines: lines that pass through the center of holes.

central processing unit (CPU): the heart of a computer.

ceramic minerals: fine-grained minerals that are formable when wet and become hard when dried or fired.

channel: a carrier.

charge-coupled device (CCD): an image-capture sensor used in most digital cameras.

charging: the first step in the process of making steel in a basic oxygen furnace. When being charged, the furnace tilts to one side to receive pig iron, scrap steel, and flux.

charts: graphic models that show relationships between people, actions, or operations.

chemical action: a change (for example, changing a material from a liquid to a solid) caused by adding chemicals.

chemical conditioning: a type of conditioning process that uses chemical actions to change the properties of a material.

chemical energy: energy stored within a chemical substance.

chemical machining: a nontraditional machining process using chemical reactions to remove material from a workpiece.

chemical processes: processes that break down or build up materials by changing their chemical compositions.

chemical processing: using chemicals to change the form of materials. The chemicals change the structure of the material's molecules.

chief executive officer (CEO): the top manager in a company.

chip removal: a separating process using a tool to cut away excess material in the form of small pieces, or chips.

chop saw: a circular saw used to cut narrow strips of material to length.
chucks: attachments used to hold and rotate drills and router bits.
circular saws: saws that use a blade in the shape of a disk with teeth arranged around the edge.
civil engineering structures: structures such as bridges, airports, or highways that are designed by civil engineers.
civilized conditions: societies in which humans make tools, grow crops, engineer materials, and develop transportation systems.
classification: the division of a problem into major segments.
classroom training: a program in which specialized skills may be developed.
clear cutting: a logging method in which all trees, regardless of species or size, are removed from a plot of land.
closed-loop control: a type of control system that uses feedback.
coal: a combustible solid composed mostly of carbon.
cold bonding: a joining technique that uses extreme pressure to squeeze two parts to create a bond.
combine: a machine that can be used to harvest a wide range of grains and other seed crops.
commercial aviation: industry that makes money by transporting people and cargo in airplanes.
commercial buildings: buildings used for business and government purposes.
commercial ships: large ships used for transporting people and cargo for a profit.
commercial structures: buildings used to conduct business.
commercial transportation: enterprises that move people and goods for money.
commission: a salesperson's pay, consisting of a percentage of the total dollar value of goods sold by that person.
communication and information technologies: the technologies used in processing data into information and communicating ideas and information.
communication processes: processes used to produce graphic and electronic media.
communication technology: a system that uses technical means to transmit information or data from one place to another or from one person to another.
composed: designed.
composite materials: combinations of natural and synthetic materials that are used to create items with other desirable properties.
compounding: the act of mixing the parts of a fluid for casting.
comprehensive layout: a layout for a final design.
computer: an information-processing machine that has changed the way we handle information. It can store information, and its programs can be changed.
computer models: models produced by computers that help designers to test the effects of color on a product's appearance and acceptance, to test strengths of materials and structures, and to observe the product during normal operation.
computer-aided design (CAD): a computer-based system used to create, modify, and communicate a plan or product design. It reduces product design and engineering costs.
computer-controlled machining: a process that adds automatic control, provided by a computer program, to basic machines.
computerized tomography: a type of scanning that produces cross-sectional images of any part of the body without using dyes.
conceptual models: models that show a general view of the components and their relationships.
conditioning: an action altering and improving the internal structure of materials. This action will change the properties of the material.
conditioning processes: those in which heat, chemicals, or mechanical forces are used to change the internal structure of a material.
conduction: the movement of heat through a solid material or between two solid materials that touch each other.
conductor: a metal that conducts an electrical current.
constraints: the limits on a design.
construction: the processs of using manufactured goods and industrial materials to build structures on a site.

construction processes: the type of activities that produces structures using technological actions.

construction technologies: the technologies used in building structures for housing, business, transportation, and energy transmission.

consumer: a person who financially supports a technological system by spending money on products or services.

consumer products: outputs developed for end users in the product cycle.

contact printing: a simple procedure used to produce prints in which a negative is placed directly on top of a piece of light-sensitive photographic paper.

container ships: ships that carry quantities of goods sealed in large, box-like metal containers.

continuous manufacturing: a manufacturing system in which a production line assembles products continuously.

contrast: the design principle used to emphasize portions of a message.

control: 1. the feedback loop that causes management and production activities to change through evaluation, feedback, and corrective action. 2. the system that makes changes in speed and direction of a vehicle possible.

control system: a vehicular system that controls the speed and direction of the vehicle.

controlling: the function of comparing system outputs to the goals.

convection: the transfer of heat between or within fluids.

convergent thinking: a type of thinking that seeks to narrow and focus ideas until the most feasible solution is found.

conversion: a process that changes hydrocarbon molecules into different sizes, both smaller and larger.

conversion/processing: the final step in agricultural practices that changes a food product into a foodstuff for human use.

converted surface finish: a finish on the surface of a metal that has been chemically changed to protect the metal.

copy: text.

core: the center layer of plywood.

corporate charter: an operating permit.

corporate participation: the spirit of cooperation in developing and operating a technological system.

corporation: a legal entity people form to own an operation. It is a business in which investors have purchased partial ownership in the form of shares of stock.

cracking: the process of breaking heavier hydrocarbons into smaller ones.

crank: a pivot pin near the outside edge of a wheel or disk that changes reciprocating motion into rotating motion.

creative personnel: in the theater, film, and television fields, a group of people who develop scripts and design scenery.

creativity: the ability to see a need or a way of making life easier and design systems and products to meet the need or desire.

criteria: the features a product or system must have in order to meet the expectations of the customer.

crop production: the type of agriculture that grows large quantities of plants for food or other uses.

crossbands: the layers between the faces in plywood.

cultivator: a machine used to control weeds. It is a series of hoe-shaped blades that are pulled through the ground.

curing: a technique used to preserve meat that involves adding a combination of natural ingredients to the meat.

current technologies: the range of techniques used to produce most products and services today.

custom manufacturing: a manufacturing system that involves producing a limited quantity of a product to a customer's specifications.

cutting motion: an action that causes material to be removed from a workpiece.

cutting tool: a tool used in cutting actions.

cycle: a complete set of motions needed to produce a surge of power.

cylindrical grinders: grinders that use the lathe principle to machine a material. A workpiece is held in a chuck or between centers and rotated. A grinding wheel is rotated in the opposite direction.

D

data: the raw, unorganized facts and figures collected by people and machines.

debt financing: raising money by borrowing money from a financial institution or private investors.

decode: to change coded information back into a recognizable form.

decoding: putting a meaning to a message. Decoding is understanding the message so proper action can be taken.

dedicated access: a system in which a subscriber's computer is directly connected to the Internet at all times.

degrees of freedom: the limited number of ways or directions a vehicle can move.

delayed outputs: various chemicals that have accumulated over decades and are harmful to the environment.

dental hygienists: individuals who clean teeth and assist dentists.

dentists: individuals who diagnose, treat, and help prevent diseases of the teeth and gums.

depth of field: the range of distances in which a camera will capture objects in focus.

descriptive methods: methods designers use to record observations of present conditions.

design: the stage in which a message is developed.

desirable outputs: the outputs from agricultural and related biotechnology, communication and information, construction, energy and power, manufacturing, medicine, and transportation technologies that benefit people.

desktop publishing: a computer system that produces type and line illustration layouts for printed messages.

detail drawings: drawings that communicate the designer's specifications and contain all the information needed to manufacture a particular part.

detailed sketches: sketches that communicate the information needed to build a model of a product or structure.

developing: the first step in developing film in which chemicals are used to alter light-sensitive crystals in an emulsion.

development: the work technologists do by building products and structures to make lives better.

diagnosis: **1.** the area of medicine that involves using knowledge, technological devices, and other means to determine the causes of abnormal body conditions. It is performed by conducting interviews, physical examinations, and medical tests. **2.** the step in the repair process in which the cause of a problem is determined.

diagrams: graphic models that show the relationships between components in a system.

dial-up access: a system in which computers are connected to the Internet through a modem.

diaphragm: the aperture control of a camera. It regulates the amount of light that can enter at any given moment.

dies: forming tools made of hardened steel.

dimension lines: the lines between extension lines that have arrows pointing to the extension lines, indicating the range of a dimension.

dipping: a technique in which stock is run through a vat of molten metal.

direct active solar system: a type of active collector system that does not have a heat exchanger.

direct current: the flow of electrons in one direction through a conductor.

direct gain solar system: a system that allows radiant energy to enter a home through windows, heating inside surfaces.

direct waves: waves used by television and FM stations.

direction: the path that a light wave travels.

direction control: a type of control system that makes a vehicle change its direction.

direct-reading measurement tools: measurement tools an operator manipulates and reads.

dirigibles: rigid airships with metal frames covered with skins of fabric.

disease: any change that interferes with the appearance, structure, or function of the body.

distance: the separation between two points. It is also called *length.*

distance multiplier: a simple machine that increases the amount of movement applied to the work at hand.

distribution: physically moving a product from a producer to a consumer.

divergent thinking: broad thinking that lets the mind soar and seeks to create as many different solutions as possible.

dividend: a periodic payment to stockholders from a company's profits.

domain name: the address of a site on the Internet.

domestic transportation: transportation that takes place within the geographic boundaries of one country.

drawing machines: machines that pull materials through die openings to form them.

drift mining: a type of underground mining used when a coal vein extends to the surface of the earth. Miners dig a horizontal shaft directly into the vein.

drilling: the process of obtaining materials by pumping them through holes drilled into the earth.

drilling machines: separating machines that produce or enlarge holes using a rotating cutter for the cutting motion.

drip irrigation: a type of irrigation that delivers water slowly to the base of plants.

drug: a substance used to prevent, diagnose, or treat a disease.

dry-cargo ships: ships used to haul both crated and bulk cargo.

drying: **1.** a common thermal conditioning process that removes excess moisture from materials. **2.** a method used to preserve food that involves removing water from the material.

drywall: gypsum wallboard. It is used as an interior wall covering.

duplex systems: basic systems used in mobile communication that use two channels.

dynamic process: a process that is constantly changing or causing change.

E

earth orbit travel: a type of space travel represented by communication satellites and space shuttles.

economic: an institution that designs, produces, and delivers the basic goods and services a society requires.

economic activity: an activity including all trade in goods and services paid for with money.

economic enterprises: institutions that design, produce, and deliver the basic goods and services required by a society.

edger saw: a machine that has a number of blades on a shaft.

education: an institution that communicates information, ideas, and skills from one person to another and from one generation to another.

edutainment: creating a situation in which people want to gain information.

elastic range: the range between a material at rest and the material's yield point.

electric fences: fences that use electrical charges to contain animals in a field.

electrical and electronic controllers: devices that control other devices to adjust operation of machines.

electrical discharge machining: a nontraditional process that uses electrical sparks to make a cavity in a piece of metal.

electrical energy: energy associated with electrons moving along a conductor.

electrical or electronic sensors: a type of monitoring device that can be used to determine the frequency of or changes in electric current or electromagnetic waves.

electrocardiograph: a type of machine that produces a visual record of the heart's electrical activity.

electrochemical processes: processes that break down or build up materials by changing their chemical compositions.

electromechanical controllers: adjusting devices that use electromagnetic coils and forces to move control linkages and operate switches to adjust machines or other devices.

electronic commerce: a type of commerce involving selling products and services over the Internet.

electronic mail: a computer-based communication tool that allows messages to move over the Internet.

electronic publishing: complex systems that can function as typesetting and layout systems. They produce and combine text and illustrations into one layout.

electrostatic printing: a process for printing that uses a machine with a special drum. A strong light is reflected off an image and onto the drum, creating an electrically charged likeness of the image on the drum's surface. It is also called *copying* or *photocopying.*

emergency medicine: an area that deals with unexpected illnesses and injuries.

emerging technologies: new technologies that are not widely employed today, but might be commonly used in a later period of time.

emotions: feelings that can be communicated.

employee relations: programs that recruit, select, develop, and reward a company's employees.

employment: the task that involves determining a company's need for qualified workers. It is the first action in the employee relations process.

enamel: a varnish that has color pigment added.

encoding: the step in the communication process that involves changing a message into a format that can be transmitted.

endoscope: a narrow, flexible tube containing a number of fiber-optic fibers that allows a physician to look inside the body.

energy: the ability to do work.

energy and power technologies: the technologies used in converting and applying energy to power devices and systems.

energy conversion: the changing of one form of energy into another.

energy-processing converters: converters that process energy in various ways.

engineer: a person who conducts research and applies scientific and technological knowledge to the design and development of products, structures, and systems.

engineering: an activity that develops the specifications for products, structures, processes, and services.

engineering drawings: documents that communicate the basic information needed to construct a manufactured product or structure.

entertain: the goal of communication to amuse people as they participate in or observe events and performances.

entrepreneurs: people with very special talents who look beyond present practices and products and create businesses.

entrepreneurship: searching for opportunities for change and responding to them.

equity financing: raising money by selling a portion of ownership in a company.

ergonomics: the science of designing products and structures around the people who use them. It is also called *human factors analysis.*

ethical drugs: drugs prescribed by physicians and dispensed by pharmacists.

ethical information: information that describes the values people have regarding devices and systems.

evaporation: a way to extract minerals from the oceans using solar energy.

exhaustible: capable of being entirely used up.

exhaustible energy resources: materials that cannot be replaced.

exhaustible materials: materials that, once depleted, cannot be replaced by human action or nature.

exosphere: the last layer of space above the earth that blends directly into outer space.

expendable molds: molds that are destroyed to remove the cast item.

expenses: money that pays for resources.

experimental methods: methods designers use to compare different conditions.

exploded views: drawings that show the parts that make up a product, as if it were taken apart.

extension lines: lines that indicate the points from which measurements are taken.

external combustion engines: engines powered by steam.

external storage devices: devices on which data can be stored outside the basic computer circuitry.

extrusion: a process in which material is pushed through a hole in a die.

F

faces: the outside layers of plywood.

family: an institution that provides a foundation for social and economic actions.

fares: the costs of tickets.

fascia: a type of board used to finish the ends of rafters and an overhang.

fasteners: devices used to hold parts together.

feed motion: an action that brings new material in contact with a cutting tool and allows the cutting action to be continuous.

feedback: the process of using information about the output of a system to regulate the inputs to the system.

feller: a logger who uses a chain saw to cut down appropriate trees.

fermentation: a technology that uses microorganisms to break down complex organic compounds into simpler substances.

fertilizer: a liquid, powder, or pellet that contains important chemicals that will encourage and support plant growth.

fiberglass: strands of glass used as the matrix for composite materials and insulation.

fiber-optic cables: strands of glass used to transmit voice, television, and computer data at high speeds.

field magnet: the stationary outer magnet used in electric motors.

filmstrips: a series of transparencies designed to be viewed one at a time.

finances: the money and credit necessary for the economic system to operate.

financial affairs: the activities that obtain, account for, and disburse the money and physical resources and maintain the financial records needed to manage a system.

finishing: the process in which the surface of parts and products are coated or modified to protect them or make them more appealing to consumers.

finishing processes: secondary processes that protect products and enhance their appearance.

finite: having a limited quantity.

firing: a thermal conditioning process used for ceramic products.

first-class levers: the class of levers in which the fulcrum is between the load and the effort.

fission: the process of splitting atoms to release vast quantities of energy.

fixed-wing aircraft: passenger and cargo aircraft. See *rotary-wing aircraft.*

fixing: the step in developing film that removes unexposed silver halide crystals remaining in the emulsion.

flame cutting: cutting material to size and shape using burning gases.

flexible manufacturing: a computer-based manufacturing system that combines the advantages of intermittent manufacturing with the advantages of continuous manufacturing.

flexography: an adaptation of letterpress. It uses a plastic or rubber image carrier.

float glass: glass that is changed into sheets for windows and similar products by floating the molten glass on a bed of molten tin.

flood irrigation: systems that use large quantities of water advancing across fields.

floor joists: beams that carry the weight of the floor.

flow bonding: a method of joining materials that uses a metal alloy as a bonding agent.

flow coating: a process that passes a product under a flowing stream of finishing material.

fluid mining: a mining method in which hot water is pumped down a well into a mineral deposit. The water dissolves the mineral and is forced up a second well.

fluidic controllers: adjusting devices that use fluids to adjust machines or other devices.

foot-pounds: a measurement of the amount of energy needed to move an object from one location to another.

forage crops: plants grown for animal feed.

force multiplier: a simple machine that increases the force applied to the work at hand.

forced-air heating: a type of system in which furnaces heat air as a conduction medium.

forming: squeezing or stretching materials into a desired shape. It also includes bending, shaping, stamping, and crushing.

forming processes: those in which force applied by a die or roll is used to reshape materials.

Forstner bits: two-lipped woodcutters that produce flat-bottomed round holes.

fossil fuels: exhaustible resources that are mixtures of carbon and hydrogen.

foundation: the base of a structure.

fractional distillation: a process in which petroleum is pumped through tubes and heated until it becomes a series of hot liquids and vapors.

fractionating tower: a tower that is used to separate the different liquids and vapors in petroleum.

fracture point: the point at which a material cannot withstand any more force.

freezing: a low-temperature preservation technique that keeps foods at or below 32°F.

frequency: the number of cycles that pass some point in one second.

frequency division multiplexing: multiplexing that uses a separate frequency to transmit each message.

frequency modulation: a system that encodes a message on a carrier wave by changing its frequency.

fruits: plants grown in many areas of the country and cultivated for their edible parts.

f-stop numbers: numbers that identify the sizes of the openings in a camera's diaphragm.

fuel converters: devices that convert fuel into energy.

fulcrum: the support on a lever on which the lever arm rests and turns.

function: a factor to consider before buying a product.

furrow irrigation: a system that uses small ditches created between rows of plants.

fuselage: the body of an aircraft that contains the flight crew, passenger, and cargo units.

fusion: the process of combining two atoms into a new, larger atom to release large amounts of energy.

fusion bonding: a bonding technique that uses heat or solvents to melt the edges of a joint.

futures research: another term for *futuring.*

futuring: a research technique that helps people select the best of many possible courses of action. It is also called *futures research.*

G

galvanized steel: zinc-coated steel used for automobile parts and containers.

gases: materials that easily disperse and expand to fill any space.

gasification: a process in which a material is heated in the absence of oxygen.

gear and rack: a rotating gear that meshes with a bar that has gear teeth along its length. It changes rotating motion into linear motion.

gears: wheels with teeth on their circumference that change the direction of a rotating force.

gems: stones that are cut, polished, and prized for their beauty and hardness.

gene splicing: the process of producing an organism with a new set of traits.

general aviation: travel for pleasure or business in an aircraft owned by a person or business.

genetic engineering: a process that produces new pest-resistant and chemical-tolerant crops that help combat diseases.

genetic materials: organic materials that have life cycles and can be regenerated. They are obtained during the normal life cycles of plants or animals.

geometry dimensions: dimensions that indicate the shapes of features and the angles at which surfaces meet.

geosynchronous orbit: a type of orbit in which a satellite travels the same speed the earth is turning.

germination: birth.

glass: a material produced using thermal processes by solidifying molten silica in an amorphous state.

global impacts: effects the actions of a small percentage of the world's population have on the world as a whole.

goals: the reasons or purposes for a system.

grain: a widely grown crop that has large edible seeds. It is a member of the grass family.

grain drill: a seed planter that is pulled behind a tractor.

graphic communications: communications processes in which messages are visual and have two dimensions.

graphic models: models used to explore ideas for components and systems.

graphs: graphic models that allow designers to organize and plot data. They display numerical information that can be used to design products and assess testing results.

gravity dam: a dam in which the lake side is vertical, while the other side slopes outward.

gravure: a type of process that prints finely detailed items.

green chain: the area of a sawmill where lumber is sorted and stacked prior to drying.

greenhouse effect: the problem in which ultraviolet rays, combined with increased levels of carbon dioxide and other gases, will cause the earth to retain more heat.

green-sand casting: a casting process that uses expendable molds and sand held together with a binder.

grievance: a dispute that arises over a contract's interpretation.

grinding machines: machines that use bonded abrasives to cut material.

growth: a major step in agricultural practices that involves providing feed and water for animals or cultivating and watering crops.

guidance: the system that gathers and displays information so a vehicle can be kept on course.

guidance system: a vehicular system that provides information.

H

hammer: a device that delivers force to complete a forming action. Hammers drop or drive a ram down with a quick action.

hand tools: simple, handheld artifacts that require human muscle power, air, or electric power to make them work.

hardening: a process used to increase the hardness of a material.

hard-wired system: a system that sends its signals through a physical channel.

hardwood lumber: a type of lumber produced from trees that lose their leaves at the end of each growing season.

harmony: the design principle achieved by blending the parts of a design to create a pleasing message.

harvesting: a step in agricultural practices that involves removing edible parts of plants from trees and stocks and butchering animals to produce meat and other products for consumption. It is the process of gathering genetic materials from the earth or bodies of water at the proper stage of their life cycles.

head rig: a very large band saw that cuts narrow slabs from a log.

headers: parts of a building framework that carry the weight from the roof and ceiling across the door and window openings. They are held up by trimmer studs.

heat pump: a unit used in climate control that works as a cooling and heating system by capturing heat in the atmosphere.

heat treating: the thermal conditioning processes used on metals.

heavy engineering construction: erecting structures that are not buildings.

heavy engineering structures: structures, such as bridges, highways, and airports, that help our economy function effectively.

hertz: the unit of measurement for the number of cycles per second.

hidden lines: light, dotted lines used to show details that are hidden in one or more of the views in a drawing.

high technology (high tech): new technologies that are not in wide use today, but may become common in time. They are also called *emerging technologies.*

historical information: information about devices and systems that were developed to solve problems similar to the current problem to be solved.

historical methods: methods designers use to gather information from existing records.

horsepower: a measurement used to describe the power output of mechanical systems.

hot water heating: an indirect climate control system that uses water to carry heat.

hovercraft: a special type of boat suspended on a cushion of air.

hub-and-spoke system: a route pattern airlines use.

human information: information that will affect the acceptance and use of a device or system.

humanities: a type of knowledge that describes the relationships between and among groups of people.

human-psyche future: a type of future that deals with the mental condition of people.

human-to-human communication: a type of communication used to inform, persuade, and entertain other people.

human-to-machine communication: a type of communication system that starts, changes, or ends a machine's operations.

hydraulic systems: systems that use liquids as the transfer medium.

hydroelectric generating plant: a water-powered plant that uses a dam to develop a water reservoir.

hydrofoil: a type of special boat that has a normal hull and a set of underwater wings.

hydroponics: the growing of plants in nutrient solutions without soil.

hyperlinks: connections between one web page and other pages and web sites through URLs.

hypertext markup language: a type of coding that tells a receiving computer how a page should look.

I

idea: a mental image of what a person thinks something should be.

ideation: a process in which designers create many possible answers by letting their minds create solutions.

illness: a state of poor health.

illustration preparation: an activity required for useful communication. It includes sizing and converting line art and photographs.

illustrations: pictures and symbols that add interest and clarity to printed communication.

image carrier: a printing block.

immediate outputs: products or services that have been designed and produced for immediate use.

inclined plane: an application of the principle that it is easier to move up a slope than a vertical surface.

income: money that is the end result of sales or employment.

indirect active solar system: a system that has a series of collectors that absorb solar energy.

indirect gain solar system: a system that uses a black concrete wall with glass panels in front of it.

indirect-reading measurement tools: systems that bring sensors and computers together to automate measurement.

induction: the process in which magnetic lines of force increase and decrease in strength, causing electrons to flow in an adjacent wire.

industrial buildings: buildings that house the machines that make products.

industrial materials: materials that are the inputs to secondary manufacturing activities.

industrial products: items used by companies in conducting their businesses.

industrial relations: the activities that develop and manage programs, ensuring an efficient work force and positive relations among the company, its workers, and the public.

Industrial Revolution: the historical period from 1750 to 1850, when tremendous changes in technology occurred. Many machines and devices were invented, including the steam engine and the telegraph.

industrial structure: a building housing machines that make products, or used to store raw materials or finished products.

industry: an economic enterprise that uses resources and systems to produce products, structures, and services with intent to make a profit.

inexhaustible: unable to be entirely used up or consumed.

inexhaustible energy resource: a part of the solar weather system that exists on earth.

inexhaustible resources: resources that are incapable of being used up.

inform: the goal of communication to provide information about people, events, or relationships.

information: facts and figures, called *data,* that have been sorted and arranged for human use. It is vital to taking an active part in society.

Information Age: occurring after the Industrial Revolution, this time period places most importance on information processing and cooperative working relations between production workers and managers.

information processing: gathering, storing, manipulating, and retrieving information that can be found in books and photographs and on tape and film.

information skills: the abilities to locate, select, and use information.

infotainment: providing information in an entertaining way.

ink-jet printing: a printing process in which a computer generates a printed message made up of tiny ink dots.

inland waterway transportation: transportation on rivers, on lakes, and along coastal waterways.

inorganic materials: materials that do not come from living organisms.

input devices: devices that allow an operator to enter data into a computer's operating system.

input unit: a device used to enter data into a system.

inputs: materials that flow into the system and are consumed or processed by the system.

inside directors: top managers of a company who serve on the company's board of directors.

insolation: the solar energy available in a specific location at any given time.

inspection: the part of a quality control program that compares materials and products with set quality standards.

intended outputs: products or services designed and produced with a specific goal in mind.

interference: anything that impairs the accurate communication of a message.

intermittent manufacturing: a manufacturing system in which a group of products is manufactured to a company's or customer's specification.

intermodal shipping: cargo traveling on two or more modes of transport before reaching its destination.

internal combustion engines: common power sources in land vehicles in which fuel is burned inside the engine to convert energy from one form to another.

international transportation: a level of transportation that moves passengers and cargo between nations.

Internet: interconnected networks of computers that share information.

Internet access: the way a computer is connected to the Internet.

Internet protocol address: the identifying number assigned to each computer connected to the Internet.

Internet service provider: a company that forms a network into which home computers are connected.

interstate commerce: business dealings that extend across state lines.

intervention radiology: a technique that uses images produced by radiology for nonsurgical treatment of ailments.

intrapreneurship: the application of entrepreneurial spirit and action within an existing company structure.

invasive diagnostic equipment: a type of diagnostic device used when drawing and testing a blood sample.

investment casting: a casting process that uses expendable molds made from plaster.

ionosphere: the region lying just above the mesosphere. It is also called the *thermosphere.*

Iron Age: the historical period beginning around 1200 B.C. During this period, iron and steel became the primary materials for tools. Trade, transportation, and communication all improved, and civilization expanded.

irradiation: a technology that uses gamma rays or X rays to kill most molds and bacteria that may be in food.

irrigation: artificial watering to maintain plant growth.

isolated solar system: a system that uses solar collectors separate from the house.

isometric sketches: sketches in which the angles formed by the lines in the upper-right corner are equal.

iterative: repetitious.

J

jet engines: engines that power businesses and commercial aircraft.

joints: places where parts meet.

joules: newtons per meter.

judgmental systems: systems that use human opinions and values to enter into the control process.

just-in-time (JIT): a type of inventory control system that schedules materials to arrive at manufacturing sites when needed.

K

kilohertz: a basic unit of measurement in telecommunication. It equals 1000 cycles per second.

kilowatt hour: the work that 1000 watts will complete in one hour.

kinetic energy: energy involved in moving something.

knowledge: information learned and applied to a task.

L

labor agreements: contracts negotiated between a company and a union to establish pay rates, hours, and working conditions for all employees covered by the contracts.

labor relations: programs that deal with employees' labor unions.

lacquer: a solvent-based, synthetic coating that dries through solvent evaporation.

land transportation systems: transportation systems that move people and goods on the surface of the earth from place to place.

landscaping: the activities that help prevent erosion and improve the appearance of a site.

language and communication skills: the abilities to read, write, and speak the language.

laser machining: a nontraditional process that uses the intense light generated by a laser to cut material.

lathes: machines that produce a cutting motion by rotating the workpiece.

layout: the stage in which a message is put together. It is the physical act of designing a message.

legal information: information about the laws and regulations that will control the installation and operation of a device or system.

lehr: an annealing oven.

length: the separation between two points. It is also called *distance.*

lens: the part of a camera that focuses light on the film.

letterpress: a type of relief printing that uses metal plates or type as the image carrier.

lever: a simple machine that multiplies the force applied to it. It changes the direction of a linear force.

lever arm: a rod or bar on a lever that rests and turns on the fulcrum.

lifestyle: what a person does with business and family life.

lift: the force that holds or lifts a craft in the air. It is critical for all flight.

light meter: the part of a camera that measures the amount of light available for a photo.

lighter-than-air vehicles: air vehicles that use either a light gas or hot air to produce lift.

limited liability: the feature of a corporation that limits an owner's loss, if the company fails, to the amount of money he or she had invested.

linear motion: a cutting and feed motion in which the cutter or work moves in one direction along a straight line.

links: connections between one web page and other pages and web sites through URLs.

liquefaction: a process in which a biofuel is heated at moderate temperatures under high pressure.

liquid fuel rockets: rockets with two tanks. One contains the fuel, and the other contains oxygen.

liquidification: a process in which a material is heated in the absence of oxygen.

liquids: visible, fluid materials that will not normally hold their size and shape.

lithographic printing: a method of printing that uses a flat-surface image carrier. It is also called *offset lithography.*

local area network: a system used in a single building or site to connect several personal computers or workstations to a central server.

location dimensions: dimensions that indicate the position of features on an object.

locomotive: in rail systems, the vehicle in which the power and operator units are placed.

logging: the activity of cutting down, trimming, and hauling off timber.

lumber: wood used to make frameworks in residences or other types of structures.

lumber core plywood: a type of core used for plywood that is made from pieces of solid lumber that have been glued to form a sheet.

M

machine tools: machines used to make other machines.

machines: artifacts that amplify the speed, amount, or direction of a force. They transmit or change the application of power, force, or motion.

machine-to-human communication: a type of communication system used to display machine operating conditions.

machine-to-machine communication: the type of communication computer-controlled operations use.

machining: a separating process based on the motion of a tool against a workpiece.

magnetic (electromagnetic) sensors: monitoring devices that can be used to determine whether changes are occurring in the amount of current flowing in a circuit.

magnetic resonance imaging: an imaging technique that can produce computer-developed cross-sectional images of any part of the body very quickly.

maintenance: a type of program that strives to keep products or structures in good condition and in good working order.

management: the act of planning, directing, and evaluating any activity.

management processes: the actions people use to ensure that production processes operate efficiently and appropriately. These processes are also designed and used to guide and direct the design, development, production, and marketing of the technological device, service, structure, or system.

manager: a person who organizes and directs the work of others in a business, by setting goals, structuring tasks to be completed, assigning work, and monitoring results.

manned space flight: a space flight that carries human beings into space and returns them safely to the earth.

manual control systems: systems that require humans to adjust the process.

manufactured home: a special type of building mostly built in a factory.

manufacturing: changing raw materials into useful products for public use.

manufacturing processes: processes used to change the size, shape, combination, and composition of materials.

manufacturing technologies: the technologies used in converting materials into industrial and consumer products.

maritime shipping: water transportation on oceans and large inland lakes.

market research: an activity that gathers information about a product's market.

marketing: the stage of promoting, selling, and delivering a product, structure, or service.

mass: the quantity of matter present in an object.

mated dies: dies that have the desired shape machined into one or both halves of the die set.

material processing: changing the form of materials using tools and machines.

materials: natural, synthetic, or composite substances from which artifacts are made.

mathematical models: models that show relationships in terms of formulas.

maturity: the period of time in the life cycles of plants and animals when growth slows down as they reach older age.

measurement: the practice of comparing the qualities of an object to a standard.

mechanical: in graphics, a sheet that looks exactly like the finished message.

mechanical conditioning: a type of conditioning process that uses mechanical forces to change the internal structure of a material.

mechanical controllers: adjusting devices that use cams, levers, and other types of linkages to adjust machines or other devices.

mechanical energy: energy produced by motion of technological devices.

mechanical fastening: an assembling process that uses mechanical forces to hold parts together.

mechanical processes: processes that use mechanical forces to change the form of natural resources.

mechanical processing: a way food can be processed using machines to change the form of the food product physically.

mechanical sensors: monitoring devices that can be used to determine position of components, force applied, or movement of parts.

mechanical transmissions: transmissions that have a clutch between the engine and the transmission.

mechanics: skilled workers in service operations.

medical technologies: the technologies used in maintaining health and curing illnesses.

medical technologists: individuals who gather and analyze specimens to assist physicians in diagnosis and treatment.

medicine: an area that involves diagnosing, treating, and preventing diseases and injuries.

megahertz: a basic unit of measurement in telecommunication. It equals 1 million cycles per second.

memory: a major part of a computer where data and operating instructions are stored.

memory unit: the section of the computer that holds information and instructions.

merchant ships: cargo-carrying ships.

mesosphere: the layer that extends from 22 to 50 miles (35 to 80 km) above the earth.

metric system: a measurement standard based on a unit of length called a *meter.*

microgravity: very low gravity.

micrometer: a measuring device used to establish precise diameters.

Middle Ages: the historical period beginning around 400 A.D. It is known for its various upheavals, as tribes continually fought each other for territory, but technology still progressed. One of the major inventions of this time was printing from movable type, developed by Johann Gutenberg in 1445.

middle management: the level of management below the president and vice presidents of a company, but above operating management.

military ships: vessels owned by a government that provide for the defense of the country.

milling: grinding or processing.

milling machines: separating machines that use a rotating cutter for the cutting motion.

minerals: any substances with a specific chemical composition that occur naturally.

mining: the process of obtaining materials from the earth through shafts or pits.

mock-up: an appearance model designed to show people how a product or structure will look.

modeling: the process of simulating expected conditions to test design ideas.

molding: a process that gives size and shape to pieces of material.

motion pictures: a series of transparencies, shot over a span of time, that create the illusion of motion.

mud: a mixture of water, clay, and chemicals.

multiple-point tool: a cutting device on which a series of single-point tools are arranged.

multiplex systems: basic systems used in mobile communication that use multiple frequencies to accommodate different types of units.

multiplexing: a system used to increase the capacity of a waveguide.

multiview method: a drawing method that places one or more views of the object in one drawing.

N

natural gas: a combustible gas that occurs in porous rock.

natural material: a material that occurs naturally on earth.

negative film: film that produces a reverse image of the photographed scene.

network access points: points that allow a customer of one company to connect with a customer of another company.

networks: connections through which computers can communicate with other computers.

noise: unwanted sounds or signals that become mixed in with desired information.

nonfood crops: plants grown on farms (such as cotton) that are not for human consumption.

noninvasive diagnostic equipment: a type of diagnostic device that gathers information about a patient without entering the body.

nonmetallic minerals: substances that do not have metallic qualities.

nonrenewable resources: exhaustible energy resources.

nontraditional machining: processes that use electrical, sound, chemical, and light energy to size and shape materials.

nuclear energy: energy produced by the fusion of atoms.

nurse practitioners: individuals who perform some of the basic duties that physicians once provided.

nurses: individuals who help physicians diagnose and treat illnesses and injuries.

nuts: crops grown in selected parts of the world for their hard-shelled seeds.

O

object lines: the solid, dark lines in a drawing that outline an object and its major details.

oblique sketches: pictorial sketches that show the front view as if a person were looking directly at it.

obsolete technologies: technologies that can no longer efficiently meet human needs for products and services.

ocean mechanical energy conversion systems: systems that use the mechanical energy in the oceans to generate power.

ocean thermal energy conversion systems: systems that use the differences in temperature between the various depths of the ocean to generate power.

offset lithography: a method of printing that uses a flat-surface image carrier.

one-view drawings: drawings used to show the layout of flat, sheet metal parts.

on-the-job training: a program through which simple skills are taught at a workstation by an experienced worker.

open dies: simple dies consisting of two flat die halves.

open-loop control: a type of control system in which output information is not used to adjust the process.

open-pit mining: a type of mining used when a coal vein is not very deep underground.

operating management: the managers closest to the people producing a company's products and services.

optical sensors: a type of monitoring device that can be used to determine the level of light or changes in the intensity of light.

ores: minerals that have a metal chemically combined with other elements.

organic materials: materials that come from living organisms.

organizing: a function that involves dividing tasks into major segments and structuring a work force so goals can be met and resources can be assigned to complete each task.

orthographic assembly drawings: assembly drawings that use a single view to show the mating of parts.

orthographic projection: the projection of a single view of an object onto a drawing surface in which the lines of projection are perpendicular to the drawing surface.

outboard motor: a type of power source attached to the stern of a boat.

output devices: devices on which data can be viewed.

output unit: a device used to display and record the results of the processing unit's actions.

outputs: the results, good and bad, of the operation of any system.

outside directors: people outside of the managerial structure who are selected to serve on a company's board of directors. They provide a different view of the company's operation.

owners: individuals who have final authority over company activities.

owners' manual: manuals that contain information needed to complete periodic servicing of a device.

ozone layer: the upper part of the stratosphere.

P

packaging: an activity that deals with designing, producing, and filling containers.

page: a text file someone creates to share information or ideas.

pagination systems: complex and expensive computer systems that allow the operator to merge text and illustrations very accurately.

paints: coatings that dry through polymerization.

panchromatic: a type of black-and-white film that reacts to all colors of visible light and records them as shades of gray.

particleboard core plywood: plywood that has a core made of particleboard.

partnership: a form of private ownership in which businesses are owned and operated by two or more people.

passenger ships: vessels that carry people.

passive collectors: solar collectors that directly collect, store, and distribute the heat they convert from solar energy.

pasteup: a sheet that looks exactly like the finished message.

pasteurization: a technology that uses heat to kill harmful microorganisms.

pathways: the structures along which vehicles travel.

people skills: the abilities needed to work with people in a cooperative way.

performers: a group of people who deliver a message.

perigee: the point at which an orbit comes closest to earth.

permanent molds: molds that withstand repeated use.

personal skills: the abilities to grow and manage personal actions on a job.

personal transportation: travel using a vehicle owned by one person.

perspective sketches: sketches that show an object as the human eye or a camera would see it.

persuade: the goal of communication to convince people to act in a certain way.

petroleum: an oily, flammable, nonuniform mixture of a large number of different solid and liquid hydrocarbons.

pharmacists: individuals who dispense prescription drugs and advise people on their use.

photograph: a picture made on light-sensitive material using a camera.

photographic communication: the process of using photographs to communicate an idea or information.

photography: the act of producing a photograph.

photovoltaic cell: a converter that generates an electrical current when struck by light.

physical models: three-dimensional representations of reality.

physician assistants: individuals who deliver basic health services under the supervision of a physician.

physicians: individuals who diagnose diseases and injuries.

pickling: dipping a material in a solvent to remove unwanted materials. Also called *chemical cleaning*.

pictorial assembly drawings: assembly drawings that show an assembly using oblique, isometric, or perspective views.

pig iron: the basic input for steelmaking that results from thermal and chemical actions.

pile foundations: types of foundations used on wet, marshy, or sandy soils.

pivot sprinkler: a type of sprinkler system that uses one long line attached at one end to a water source.

place utility: a value provided by the movement of people and cargo from one place to another.

planing machines: machine tools that produce flat surfaces. They move the workpiece back and forth under the tool to generate the cutting motion.

planning: the process of setting goals and developing courses of action for a company or parts of the company to reach the goals.

plastic range: the range in which a material can be stretched, compressed, or bent.

plating: an electrolytic process.

pleasure craft: vessels owned by private citizens for recreation.

plow: a piece of tilling equipment that breaks, raises, and turns soil.

pneumatic systems: systems that use air as the transfer medium.

point of interest: the place to which your eye is drawn.

point of presence: a connection point that lets local users access a company's network.

political/legal: institutions that establish and enforce society's rules of behavior and conduct.

political power: the ability of a person or organization to gain the attention and cooperation of elected officials.

pollution: a product of human activity that diminishes air or water quality.

polymerization: a conversion process that causes small hydrocarbon molecules to join together.

positive transparencies: slides.

potable water: water safe for drinking.

potential energy: stored energy.

potential field: an area that has never produced oil or gas.

power: the rate at which work is done.

power generation system: a system that uses an engine as an energy converter.

power transmission system: a system that controls and directs the power of an engine to do work.

precision measurement: a type of measurement used when exact size is critical to the function of a device.

preservatives: chemicals added to food in small amounts to delay spoilage and ensure the food retains its quality.

president: the top manager in a company.

press fit: a fit in which friction between parts will cause them to remain together.

presses: devices in which force is delivered to complete a forming action. They slowly close die halves by lowering a ram to produce a squeezing action.

pressure bonding: a bonding technique that applies heat and pressure to a bond area.

prevention: the area of medicine that involves using knowledge, technological devices, and other means to help people maintain healthy bodies.

preventive maintenance: maintenance designed to prevent breakdowns.

price: what someone must pay to buy or use a product or service.

primary food-processing technology: a type of process that produces the basic ingredients for food.

primary processing: the step in which material resources are converted into industrial materials.

prime: the low interest rate banks and insurance companies charge their safest customers.

prime interest rate: another term for *prime.*

prime mover: a device that changes a natural source of energy into mechanical power.

primitive conditions: conditions determined by nature. Primitive societies were dependent on nature to provide food, shelter, and clothing.

printed graphic communication: a mass communication system that uses technology to communicate through a printed medium.

printing: a communication system that places the images of written words on a material.

private enterprises: enterprises owned by individuals or groups of people.

problem-solving/design process: the procedure used to develop technology that will attempt to satisfy people's technological needs and wants.

process development: a type of development that devises new or improved ways of completing tasks in manufacturing, construction, communication, or transportation.

processes: the steps needed to complete a series of identifiable tasks within a system.

processing unit: the part of the computer, also called a *central processing unit (CPU)* or *microprocessor,* that manipulates the data.

producer: an individual who oversees every element of a show.

product development: an area of development that applies knowledge to design new or improved products, structures, and services.

production: the stage of developing and operating systems for producing a product, structure, or service.

production personnel: In the film and television industries, a group of people who record the sights and sounds of a performance.

production processes: the actions completed to perform the function of the technological system.

production workers: people who process materials, build structures, operate transportation vehicles, service products, or produce and deliver communication products.

profit: the amount of money left over after all the expenses of a business have been paid.

program: the instructions a computer uses to process data and produce output.

programmable logic controller (PLC): a device that uses a microprocessor to control machines or processes.

projection printing: a basic technique used to produce prints by shining light through a negative onto a piece of photographic paper.

propagation: a step in agricultural practices that allows a biological organism to reproduce.

proportion: the design principle that deals with the relative sizes of the parts of a design.

proprietary drugs: drugs that are considered safe for unsupervised use by consumers.

proprietorship: a type of business ownership in which the business has a single owner.

propulsion: the system in a vehicle that generates motion through energy conversion and transmission.

propulsion system: a vehicular system that produces a force to propel the vehicle from its starting point to its destination.

prosthesis: an artificial body part developed through biomechanical engineering.

prototype: a working model of a system, assembly, or product that is built to test the operation, maintenance, and safety of the item.

proven reserves: producing oil or gas fields.

public enterprises: enterprises controlled by the government or a special form of corporation.

public relations: programs that communicate a company's policies and practices to governmental officials, community leaders, and the general population.

pulleys: grooved wheels attached to an axle that can be used to change the direction of a force, multiply force, or multiply distance.

pulleys and V-belt: a technique in which two pulleys with a V-belt stretched between them change the speed or power of a motion.

pyrolysis: a thermochemical process in which materials are heated in the absence of oxygen.

Q

quality control: a process that includes all the systems and programs that ensure the outputs of technological systems will meet engineering standards and customer expectations.

R

radial saw: a circular saw that moves a rotating blade across the workpiece.

radiant energy: energy in the form of electromagnetic waves.

radiation: heat transfer by using electromagnetic waves.

radiation therapy: a treatment that uses high-energy radiation to treat cancer cells.

radio waves: a series of frequencies within the electromagnetic spectrum extending from around 30 hertz to 300 gigahertz.

radiology: methods that use electromagnetic waves and ultrasonics to diagnose diseases and injuries.

rafters: angled boards that extend from the top plate of exterior walls to the ridge of the roof.

rail fences: fences used as border fences around farm buildings or homes.

random-access memory: memory that can be read or changed by both the computer and the user.

read-only memory: memory that can be read by the computer, but cannot be changed.

receiver: the end of the communication channel that gathers and decodes a message.

receiving: recognizing and accepting information.

reciprocating motion: a back-and-forth movement.

recruiting: a searching process through which job applicants are acquired.

recycling: a choice of disposal method in which the materials in a product or structure are reclaimed.

refined sketch: a sketch that merges ideas from two or more rough sketches.

refining: the second step in the process of making steel in a basic oxygen furnace.

reflect: to prevent passage of and cause to change direction.

reflected waves: waves used in a broadcast system in which a signal is bounced off the ionosphere.

refrigeration: a preservation technique that maintains foods at temperatures from 32°F to 40°F.

regulation: an order issued by an executive authority.

reinforced concrete: a material used in the frameworks of industrial and commercial buildings.

relief printing: a printing process that uses an image on a raised surface.

religion: an institution that develops and communicates values and beliefs about life and appropriate ways of living.

Renaissance: the historical period that began in the early 1300s in Italy and lasted until 1600. This period is known for new ideas in art, literature, history, and political science, but technological developments, such as the calculator and the telescope, also occurred.

renewable: capable of being used up, but replaceable with the normal life cycle of the energy source.

renewable energy resource: a biological material that can be grown and harvested.

renewable resources: resources that have a life cycle.

repair: the process of putting a broken, damaged, or defective product or building back into good working order.

replacement: the step in the repair process in which worn or broken parts are replaced.

resaw: a machine that uses a group of evenly spaced circular or scroll saw-type blades to cut many boards at once.

research: the work scientists do by gathering information to try to explain why something exists or happens in a certain way.

research and development: the stage of designing, developing, and specifying the characteristics of a product, structure, or service. These activities might result in new or improved products and processes.

residential buildings: structures in which people live.

residential structure: a place where people live.

responsibility: accountability for actions.

retained earnings: profits held by a company and used to enlarge its operations.

retrieving: a process that allows information to be brought back.

rewards: something received for some service or attainment.

rhythm: the design principle dealing with the flow of a communication.

risks: possibilities of loss.

robot: a mechanical device that can perform tasks automatically or with varying degrees of direct human control.

rolling machines: machines that use two rolls rotating in opposing directions to form a material.

rotary motion: a cutting and feed motion in which the work or the tool rotates.

rotary-wing aircraft: an aircraft that develops lift by spinning an airfoil. See *fixed-wing aircraft.*

rotating motion: a motion that uses round cutters or spins a workpiece around an axis.

rough sketches: drawings that show only basic ideas of the size, shape, and appearance of a product. They integrate and refine ideas generated in the thumbnails.

route: the path a vehicle follows.

routers: specialized computers that determine how to send information from one computer to another.

routine diagnostic equipment: a type of device used to gather general information about a patient.

rudder: a large flat plate at the stern of a ship, used to guide the vessel.

rule: a strip of metal, wood, or plastic with measuring marks on its face. It is the most common linear measurement device.

S

salary: payment based on work over a period of time, such as a month or year.

sales: the activity that involves the physical exchange of products for money.

sawing machines: machines that use blades with teeth to cut materials to desired sizes and shapes.

schedules: lists of departure and arrival times for trips.

science: knowledge of the natural world.

scientific information: information about natural laws and principles that must be considered in developing a solution.

scientists: people who generally develop a basic knowledge of physics, materials science, geology, or chemistry to help create products and processes.

scope: panorama.

screen printing: a printing process that uses stencils with openings the shape of the message.

screening: a type of process that allows qualified people to be selected from an applicant pool.

screw: an inclined plane wrapped around a shaft.

script: a document that identifies characters, develops a situation, and communicates a story.

scroll saws: saws with a blade that is a strip of metal with teeth on one edge.

search engines: special sites on the Internet that operate on the principle of key words.

secondary food-processing technology: a type of process used to make finished food products.

secondary manufacturing process: the action used to change industrial materials into products.

secondary processes: manufacturing processes that change industrial materials into industrial equipment and consumer products.

secondary processing: a type of activity that changes industrial materials into products.

second-class levers: levers in which the load is between the effort and the fulcrum.

seed-tree cutting: a logging method in which all trees, regardless of species, are removed from a large area except three or four per acre. These trees are used to reseed the area.

seismographic study: an accurate way to explore for petroleum and natural gas using shock waves.

selective cutting: a logging method in which mature trees of a desired species are selected and cut from a plot of land.

separating: using tools to shear or machine away unwanted material.

separating process: a process in which tools or machines are used to remove excess material to make an object of the correct size and shape.

separation: a process used in petroleum refineries that breaks petroleum into major hydrocarbon groups.

server: a special computer used to store programs and data for a network.

service: routine tasks that keep a product operating.

servicing: maintenance, repair, and reconditioning. It is the scheduled adjustment, lubrication, or cleaning required to keep a product or structure operating properly.

shaft mining: a type of underground mining that requires a vertical shaft to reach a coal deposit.

shaped die: a die used to form plastic objects.

shaping machine: a metalworking machine tool that produces flat surfaces. They move a single-point tool back and forth over the workpiece to produce the cutting motion.

share of stock: a portion of the ownership of a company.

shearing: a separating process that uses opposing edges of blades, knives, or dies to fracture unwanted material away from a workpiece.

sheathing: covering.

shell molding: a casting process that uses expendable molds and a sand and resin mixture.

shutter: the device that opens and closes to permit or prevent light from entering a camera.

silk screening: a printing process that uses stencils mounted on silk fabric. It is also called *screen printing*.

sill: wood piece attached to the top of a foundation.

simplex system: a basic system used in mobile communication that uses the same channel for both base and mobile transmissions.

simulation: the process of imitating expected conditions to test design ideas.

single-point tool: a simple cutting device with a cutting edge on the end or along the edge of a rod, bar, or strip.

site preparation: the step in a construction project that involves removing existing buildings, structures, brush, and trees that will interfere with locating the new structure.

size dimensions: dimensions that indicate the size and major features of an object.

skelps: strips of steel used to form pipes.

slab foundation: a type of foundation used for structures erected on soft soils.

slabs: wide, flat pieces of steel.

slide: a single transparency designed to be viewed independently.

slope mining: a type of underground mining used when a coal vein is not too deep under the ground. Miners dig a sloping shaft to reach the coal.

smelting: the thermal process using heat to extract metals from their ores.

smith forging: a common process using open dies.

smoking: a process that adds flavor to meat and fish, while preserving them.

social future: a type of future that suggests the type of relationships people want with each other.

socioethical skills: skills involving the understanding of the implications of actions on people, society, and the environment.

soffit: the underside of a building's roof edge.

softwood lumber: a type of lumber produced from needle-bearing trees. It is used for construction, for shipping containers and crates, and for railroad ties.

solar converter: an inexhaustible energy converter that uses the constant energy source of the sun.

solar weather system: the natural cycle that starts with solar energy.

sole plate: the strip at the bottom of a framed wall.

sole proprietorship: a business or operation owned by one person.

solid fuel rockets: rockets that use a powder or spongelike mixture of fuel and oxidizer.

solid model: a complex computer model that takes into account both the surface and the interior substance of an object.

solids: materials that hold their size and shape and can support loads.

space transportation system: a mode of transportation that uses manned and unmanned flights to explore the universe.

spade bits: flat cutters on the end of a shaft, used to drill holes.

specification sheets: documents that communicate the properties a material must possess for a specific application.

speed control: a type of control system that makes a vehicle go faster through acceleration or slower by braking or coasting.

spraying: a process that uses air to carry fine particles of finishing materials to the surface of a product.

spread foundations: types of foundations used on rock and in hard soils.

sprinkler irrigation: an irrigation system that produces artificial rain to water crops.

squares: angle-measuring devices that have blades at right angles to the heads.

stamping: a process in which both forming and cutting are done.

standard measurement: a type of measurement in which the exact size of a part is not critical to the function of a product.

standard views: drawings that show a product in one piece, as it would be after it is assembled.

steel: an alloy of iron and carbon used for frameworks in industrial and commercial buildings.

Stone Age: the earliest period in history. First, simple stone tools were used to cut and pound vegetables and cut meat from animal carcasses; later, pointed stone hunting tools were developed. Humans also learned how to harness fire for heating, cooking, and protection.

stop bath: an acidic solution that neutralizes developers.

stopping: a step in developing film that stops the chemical action of a developer.

storing: processes that allow information to be retained for later use.

stratosphere: the region above the troposphere.

stroke: the movement of a piston from one end of a cylinder to another.

structural system: a vehicular system that helps people arrive safely and in comfort and protects cargo.

structure: the system that provides spaces for devices in vehicles.

structure development: an area of development that applies knowledge to design new or improved structures.

studs: uprights nailed to a sole plate.

subfloor: a base, usually made from plywood or particleboard, that goes under tile, carpeting, or other flooring materials.

submersible: a type of special boat that can travel on the surface of or under water.

substrate: a material on which a printing is applied.

superstructure: the framework of a building or tower constructed on a foundation. It also includes the pipes for pipelines, surfaces for roads and airport runways, and tracks for railroads.

supervisors: the managers closest to the people producing a company's products and services.

support staff: nonmanagerial workers who carry out such tasks as keeping financial records, maintaining sales documents, and developing personnel systems.

support system: the external operations and facilities that maintain transportation systems.

surface grinders: grinding machines that work on the metal planer principle.

surface mining: a type of coal mining used when the coal vein is not very deep underground.

surface model: a three-dimensional computer model that is a wire frame with a sheet.

surgery: a way to treat diseases and injuries with operations.

suspension: the subsystem that maintains a vehicle on a pathway.

suspension bridges: bridges that use cables to carry loads.

suspension system: a vehicular system that produces proper support for the weight of the vehicle and cargo.

swather: a machine that cuts and windrows hay in one pass over a field.

synergism: the concept that solutions proposed by one or more individuals in a group will often cause other members of the group to think of more ideas. The number of ideas generated by the group is more than the number they could develop if everyone worked alone.

synthetic materials: human-made materials.

system: a group of parts working together in a predictable way, designed to achieve a goal.

systems drawings: drawings used to show how parts in a system relate to each other and work together.

T

table saw: a circular saw that uses a linear feed of the material. The workpiece is pushed into the rotating blade to generate the cut.

tail assembly: a structure that provides steering capability for an aircraft.

tankers: large vessels used to move liquids across oceans.

tapping: the final step in the process of making steel in a basic oxygen furnace.

technical data sheets: documents that communicate the specifications for manufactured products.

technical graphic communication: a type of system that prepares and reproduces engineering drawings and technical illustrations.

technicians: skilled workers in laboratories and product-testing facilities. They work closely with production workers, but do more specialized jobs.

technological assessment: an evaluation that involves groups of people evaluating the impacts of technology on people, society, and the environment.

technological future: a type of future that looks at the type of human-built world we desire.

technological knowledge: knowledge of the human-built world used to design, produce, and use tools and materials.

technological opportunity: a "nonproblem" condition that can be improved with technology.

technological problem: a problem that can affect individuals and groups of people. These problems can be solved with devices or systems.

technological product: an artifact built by people.

technological service: outputs we use, but do not own.

technologically literate: having the understanding and ability to direct new technology.

technologist: a highly trained technical employee.

telecommunication: a communication process that depends on electromagnetic waves to carry a message over a distance.

temperature: the measurement of how hot or cold a material is.

tempering: a process used to relieve internal stress in a part.

terminals: the structures where transportation activities begin and end. They house passenger and cargo storage and loading facilities.

testing: the step in the repair process in which a repaired product is checked to ensure it works properly.

thermal conditioning: conditioning processes using heat.

thermal energy: heat energy.

thermal processes: the types of processes that use heat to melt and reform natural resources.

thermal processing: food-processing method that uses heat as the primary energy to convert the food.

thermal sensors: monitoring devices that can be used to determine changes in temperature.

thermochemical conversion: a basic process that produces a chemical reaction by applying heat.

thinking skills: skills involving the ability to use mental processes to address problems and issues.

third-class levers: levers in which the effort is placed between the load and the fulcrum.

three-view drawings: multiview drawings used to show the size and shape of rectangular and complex parts.

thumbnail sketches: sketches that allow a graphic designer to experiment with various arrangements of copy and illustrations.

tillage: the process of breaking and pulverizing soil to condition a seedbed.

timber cruising: a process in which foresters measure the diameter and height of trees to find stands of trees that can be economically harvested.

time: the measurement of how long an event lasts. It is a key resource in developing and operating technological systems.

time division multiplexing: multiplexing that divides time into very brief segments.

tin plate: tin-coated steel used to make food cans.

tolerance: a number indicating the amount of deviation that will be allowed in a dimension.

tools: artifacts humans use to expand their capabilities.

top management: presidents and vice presidents.

top plate: a double ribbon of 2 x 4s.

tractor: **1.** the part of a long-distance truck where the power and operator units are placed. **2.** a power source that provides power to pull all types of farm equipment.

trailers: cargo units attached to a tractor.

transducers: technological devices that change energy of one form into energy of another form.

transmit: to send a message from one person or place to another.

transmitter: a sender.

transmitting: **1.** to send a coded message from a sender to a receiver. **2.** to pass through a material.

transportation: all activities that relocate humans or their possessions.

transportation processes: processes used to move people and cargo.

transportation technology: the technology used in moving people and cargo from one place to another.

treating: a petroleum-refining process that adds or removes chemicals to change the properties of petroleum products.

treatment: the area of medicine that involves using knowledge and technological devices and applying medical procedures to fight diseases, heal injuries, or ease symptoms.

trim saw: a machine that has a series of spaced blades.

troposphere: the lowest region of space above the earth in which spacecraft and satellites operate.

truss: a triangle-shaped structure that includes both the rafter and ceiling joist in one unit.

truss bridges: bridges that use small parts arranged in triangles to support the deck.

turbofan engine: an engine used in most commercial aircraft in use today. It operates at lower speeds than a turbojet engine.

turbojet engine: a type of jet engine that was developed during World War II.

turboprop engine: a variation of the jet engine that operates efficiently at low speeds.

turning machine: a separating machine that rotates a workpiece against a single-point tool to produce a cutting motion.

twist drills: shafts of steel with points on the ends to produce chips.

two-view drawings: multiview drawings used to show the size and shape of cylindrical parts.

typesetting: an activity that produces the words of a message.

U

ultrahigh frequency (UHF): a radio frequency between superhigh frequency and very high frequency (VHF).

ultrasound: an imaging technique that uses high-frequency sound waves and their echoes to develop an image of the body.

underground mining: a type of mining that requires shafts in the earth to reach the coal deposits.

undesirable outputs: outputs that are not wanted.

uniform resource locator: a code that is a method of naming documents or sites on the World Wide Web.

unintended outputs: outputs that were not considered when the system was designed.

unlimited liability: a disadvantage in which a proprietor cannot separate business income and liabilities from his personal finances.

unmanned space flights: space flights that use rockets to place payloads into orbit.

U.S. customary: a type of system used as a measurement standard in the United States today.

utilities: the systems of a structure that provide water, electricity, heat, cooling, and communications.

V

value: a measure of the functional worth a customer sees in a product. It is a factor to consider before buying a product.

variety: the design principle that makes a message unique and interesting.

varnish: a clear finish made from a mixture of oil, resin, solvent, and a drying agent.

vegetables: an important farm crop. They have edible leaves, stems, roots, and seeds that provide important vitamins and minerals for the daily diet.

vehicle: a technological artifact designed to carry people and cargo on a pathway.

vehicular systems: the onboard technical systems that make a vehicle work.

veneer: a thin sheet of wood that is sliced, sawed, or peeled from a log.

veneer core plywood: the most common type of core used for plywood.

very high frequency (VHF): a radio frequency between ultrahigh frequency (UHF) and high frequency.

vice presidents: managers who report to the president or CEO of a company.

video: visual.

volume: the amount of space an object occupies or encloses.

W

wage: a set rate paid for each hour worked.

wastewater: the water from sinks, showers, tubs, toilets, and washing machines. It is drained away by part of the plumbing system.

water cycle: a cycle powered partially by solar energy.

water transportation systems: systems that use water to support a vehicle.

waterwheel: a series of paddles that extend into flowing water, which produces a rotating mechanical motion.

watt: one joule of work per second.

wavelength: the distance from the beginning to the end of one wave cycle.

wedge: a device used to split and separate materials and to grip parts.

weight: the force of the earth's pull on a mass.

wellness: a state of physical well-being.

"what if" scenarios: outrageous proposals that could lead to solutions after their good and bad points have been investigated.

wheel and axle: a shaft attached to a disk.

wide area network: a type of computer network that covers large geographic areas.

windmill: a wind-driven wheel that produces a rotating mechanical motion.

windrows: bands of hay.

wing: the part of an aircraft that separates the air into two streams, providing lift.

wire frame model: three-dimensional computer model developed by connecting all the edges of an object.

work: applying a force that moves a mass a distance in the direction of the applied force.

World Wide Web: a computer-based network of information resources.

woven-wire fences: fences that use posts with a special wire product attached to them.

X

x-ray machine: a diagnostic imaging machine that essentially is a camera.

Y

yarding: the process of gathering logs in a central location.

yield point: the point at which a material will not return to its original shape after being stretched.

Index